SELECTED SOLUTIONS MANUAL

MARY BETH KRAMER • *University of Delaware*
KATHLEEN THRUSH SHAGINAW • *Particular Solutions, Inc.*

Principles of Chemistry
A MOLECULAR APPROACH

NIVALDO J. TRO

Prentice Hall
New York Boston San Francisco
London Toronto Sydney Tokyo Singapore Madrid
Mexico City Munich Paris Cape Town Hong Kong Montreal

Project Editor: Jennifer Hart
Acquisitions Editor: Dan Kaveney
Editor in Chief, Chemistry and Geosciences: Nicole Folchetti
Marketing Manager: Erin Gardner
Managing Editor, Chemistry and Geosciences: Gina M. Cheselka
Project Manager, Science: Ed Thomas
Operations Specialist: Amanda A. Smith
Supplement Cover Manager: Paul Gourhan
Supplement Cover Designer: Tina Krivoshein
Cover Art: Quade Paul

Pearson Prentice Hall
Pearson Education, Inc.
Upper Saddle River, NJ 07458

Printed in the United States of America

10 9 8 7 6 5 4 3 2 1

ISBN-13: 978-0-321-58638-4
ISBN-10: 0-321-58638-7

Prentice Hall
is an imprint of

PEARSON

www.pearsonhighered.com

Table of Contents

Chapter 1
Matter, Measurement, and Problem Solving

1. a) This statement is a theory because it attempts to explain why. It is not possible to observe individual atoms.

   b) This statement is an observation.

   c) This statement is a law, because it summarizes many observations and can explain future behavior.

   d) This statement is an observation.

3. a) If we divide the mass of the oxygen by the mass of the carbon the result is always 4/3.

   b) If we divide the mass of the oxygen by the mass of the hydrogen the result is always 16.

   c) These observations suggest that the masses of elements in molecules are ratios of whole numbers (4 and 3; and 16 and 1, respectively).

   d) Atoms combine in small whole number ratios and not as random weight ratios.

5. a) Sweat is a homogeneous mixture of water, sodium chloride, and other components.

   b) Carbon dioxide is a pure substance that is a compound (two or more elements bonded together).

   c) Aluminum is a pure substance that is an element (element #13 in the periodic table).

   d) Vegetable soup is a heterogeneous mixture of broth, chunks of vegetables, and extracts from the vegetables.

7.

substance	pure or mixture	Type (element or compound)
aluminum	pure	element
apple juice	mixture	neither – homogeneous mixture
hydrogen peroxide	pure	compound
chicken soup	mixture	neither – heterogeneous mixture

9. a) pure substance that is a compound (one type of molecule that contains two different elements)

   b) heterogeneous mixture (two different molecules that are segregated into regions)

   c) homogeneous mixture (two different molecules that are randomly mixed)

   d) pure substance is an element (individual atoms of one type)

11. a) physical property (color can be observed without making or breaking chemical bonds)

    b) chemical property (must observe by making or breaking chemical bonds)

    c) physical property (the phase can be observed without making or breaking chemical bonds)

    d) physical property (density can be observed without making or breaking chemical bonds)

    e) physical property (mixing does not involve making or breaking chemical bonds, so this can be observed without making or breaking chemical bonds)

13. a) chemical property (burning involves breaking and making bonds, so bonds must be broken and made to observe this property)

b) physical property (sublimation is a phase change and so can be observed without making or breaking chemical bonds)

c) physical property (odor can be observed without making or breaking chemical bonds)

d) chemical property (burning involves breaking and making bonds, so bonds must be broken and made to observe this property)

15. a) chemical change (new compounds are formed as methane and oxygen react to form carbon dioxide and water)

b) physical change (vaporization is a phase change and does not involve the making or breaking of chemical bonds)

c) chemical change (new compounds are formed as propane and oxygen react to form carbon dioxide and water)

d) chemical change (new compounds are formed as the metal in the frame is converted to oxides)

17. a) physical change (vaporization is a phase change and does not involve the making or breaking of chemical bonds)

b) chemical change (new compounds are formed)

c) physical change (vaporization is a phase change and does not involve the making or breaking of chemical bonds)

19. a) To convert from °F to °C, first find the equation that relates these two quantities. $°C = \frac{°F - 32}{1.8}$ Now substitute °F into the equation and compute the answer. Note: The number of digits reported in this answer follow significant figure conventions, covered in section 1.6. $°C = \frac{°F - 32}{1.8} = \frac{0.}{1.8} = 0.\ °C$

b) To convert from K to °F, first find the equations that relate these two quantities.

$K = °C + 273.15$ and $°C = \frac{°F - 32}{1.8}$ Since these equations do not directly express K in terms of °F, you must combine the equations and then solve the equation for °F. Substituting for °C:

$K = \frac{°F - 32}{1.8} + 273.15$ rearrange $K - 273.15 = \frac{°F - 32}{1.8}$ rearrange $1.8\,(K - 273.15) = (°F - 32)$

finally $°F = 1.8\,(K - 273.15) + 32$ Now substitute K into the equation and compute the answer.

$°F = 1.8\,(77\ K - 273.15) + 32 = 1.8\,(-196\ K) + 32 = -353 + 32 = -321\ °F$

c) To convert from °F to °C, first find the equation that relates these two quantities. $°C = \frac{°F - 32}{1.8}$ Now substitute °F into the equation and compute the answer. $°C = \frac{-109°F - 32°F}{1.8} = \frac{-141}{1.8} = -78.3\ °C$

d) To convert from °F to K, first find the equations that relate these two quantities.

$K = °C + 273.15$ and $°C = \frac{°F - 32}{1.8}$ Since these equations do not directly express K in terms of °F, you must combine the equations and then solve the equation for K. Substituting for °C

$K = \frac{°F - 32}{1.8} + 273.15$

Now substitute °F into the equation and compute the answer.

$K = \frac{(98.6 - 32)}{1.8} + 273.15 = \frac{66.6}{1.8} + 273.15 = 37.0 + 273.15 = 310.2\ K$

21. To convert from °F to °C, first find the equation that relates these two quantities. $°C = \frac{°F - 32}{1.8}$ Now substitute °F into the equation and compute the answer. Note: The number of digits reported in this answer follow significant figure conventions, covered in section 1.6. $°C = \frac{-80.°F - 32°F}{1.8} = \frac{-112}{1.8} = -62.2\ °C$

Begin by finding the equation that relates the quantity that is given (°C) and the quantity you are trying to find (K). $K = °C + 273.15$ Since this equation gives the temperature in K directly, simply substitute in the correct value for the temperature in °C and compute the answer. $K = -62.2\ °C + 273.15 = 210.9\ K$.

23. Use Table 1.2 to determine the appropriate prefix multiplier and substitute the meaning into the expressions.

a) 10^{-9} implies "nano" so 1.2×10^{-9} m = 1.2 nanometers = 1.2 nm

b) 10^{-15} implies "femto" so 22×10^{-15} s = 22 femtoseconds = 22 fs

c) 10^{9} implies "giga" so 1.5×10^{9} g = 1.5 gigagrams = 1.5 Gg

d) 10^{6} implies "mega" so 3.5×10^{6} L = 3.5 megaliters = 3.5 ML

25. b) **Given:** 515 km **Find:** dm

Conceptual Plan: **km → m → dm**

$$\frac{1000\ m}{1 km} \quad \frac{10\ dm}{1 m}$$

Solution: $515\ \cancel{km} \times \frac{1000\ \cancel{m}}{1\ \cancel{km}} \times \frac{10\ dm}{1\ \cancel{m}} = 5.15 \times 10^{6}\ dm$

Check: The units (dm) are correct. The magnitude of the answer (10^6) makes physical sense because a decimeter is a much smaller unit than a kilometer.

Given: 515 km **Find:** cm

Conceptual Plan: **km → m → cm**

$$\frac{1000\ m}{1 km} \quad \frac{100\ cm}{1 m}$$

Solution: $515\ \cancel{km} \times \frac{1000\ \cancel{m}}{1\ \cancel{km}} \times \frac{100\ cm}{1\ \cancel{m}} = 5.15 \times 10^{7}\ cm$

Check: The units (cm) are correct. The magnitude of the answer (10^7) makes physical sense because a centimeter is a much smaller unit than a kilometer and a decimeter.

c) **Given:** 122.355 s **Find:** ms

Conceptual Plan: **s → ms**

$$\frac{1000\ ms}{1 s}$$

Solution: $122.355\ \cancel{s} \times \frac{1000\ ms}{1\ \cancel{s}} = 1.22355 \times 10^{5}\ ms$

Check: The units (ms) are correct. The magnitude of the answer (10^5) makes physical sense because a millisecond is a much smaller unit than a second.

Given: 122.355 s **Find:** ks

Conceptual Plan: **s → ks**

$$\frac{1\ ks}{1000\ s}$$

Solution: $122.355\ \cancel{s} \times \frac{1\ ks}{1000\ \cancel{s}} = 1.22355 \times 10^{-1}\ ks = 0.122355\ ks$

Check: The units (ks) are correct. The magnitude of the answer (10^{-1}) makes physical sense because a kilosecond is a much larger unit than a second.

d) **Given:** 3.345 kJ **Find:** J
Conceptual Plan: **kJ → J**

$$\frac{1000\ \text{J}}{1\text{kJ}}$$

Solution: $3.345\ \text{kJ} \times \frac{1000\ \text{J}}{1\text{kJ}} = 3.345 \times 10^3\ \text{J}$

Check: The units (J) are correct. The magnitude of the answer (10^3) makes physical sense because a joule is a much smaller unit than a kilojoule.
Given: 3.345×10^3 J (from above) **Find:** mJ
Conceptual Plan: **J → mJ**

$$\frac{1000\ \text{mJ}}{1\ \text{J}}$$

Solution: $3.345 \times 10^3\ \text{J} \times \frac{1000\ \text{mJ}}{1\text{J}} = 3.345 \times 10^6\ \text{mJ}$

Check: The units (mJ) are correct. The magnitude of the answer (10^6) makes physical sense because a millijoule is a much smaller unit than a joule.

27. **Given:** 1 m square 1 m^2 **Find:** cm^2
Conceptual Plan: $\mathbf{1\ m^2} \rightarrow \mathbf{cm^2}$

$$\frac{100\ \text{cm}}{1\ \text{m}}$$

Notice that for squared units, the conversion factors must be squared.

Solution: $1\ \text{m}^2 \times \frac{(100\ \text{cm})^2}{(1\ \text{m})^2} = 1 \times 10^4\ \text{cm}^2$

Check: The units of the answer are correct and the magnitude makes sense. The unit centimeter is smaller than a meter, so the value in square centimeters should be larger than in square meters.

29. **Given:** $m = 2.49$ g, $V = 0.349\ \text{cm}^3$ **Find:** d in g/cm^3 and compare to pure copper.
Conceptual Plan: $\boldsymbol{m, V} \rightarrow \boldsymbol{d}$

$$d = m/V$$

Compare to the published value. d (pure copper) = 8.96 g/cm^3 (This value is in Table 1.4.)

Solution: $d = \frac{2.49\ \text{g}}{0.349\ \text{cm}^3} = 7.13\ \frac{\text{g}}{\text{cm}^3}$

The density of the penny is much smaller than the density of pure copper (7.13 g/cm^3 < 8.96 g/cm^3) so the penny is not pure copper.
Check: The units (g/cm^3) are correct. The magnitude of the answer seems correct. Many coins are layers of metals, so it is not surprising that the penny is not pure copper.

31. **Given:** $m = 4.10 \times 10^3$ g, $V = 3.25$ L **Find:** d in g/cm^3
Conceptual Plan: $\boldsymbol{d, V} \rightarrow \boldsymbol{m}$ **then** $\mathbf{L \rightarrow cm^3}$

$$d = m/V \qquad \frac{1000\ \text{cm}^3}{1\ \text{L}}$$

Solution: $d = \frac{4.10 \times 10^3\ \text{g}}{3.25\ \text{L}} \times \frac{1\ \text{L}}{1000\ \text{cm}^3} = 1.26\ \frac{\text{g}}{\text{cm}^3}$

Check: The units (g/cm^3) are correct. The magnitude of the answer seems correct.

33. a) **Given:** $d = 1.11$ g/cm^3, $V = 417$ mL **Find:** m
Conceptual Plan: $\boldsymbol{d, V} \rightarrow \boldsymbol{m}$ **then** $\mathbf{cm^3 \rightarrow mL}$

$$d = m/V \qquad \frac{1\ \text{mL}}{1\ \text{cm}^3}$$

Solution: $d = m/V$ Rearrange by multiplying both sides of equation by V. $m = d \times V$

$$m = 1.11\ \frac{\text{g}}{\text{cm}^3} \times \frac{1\ \text{cm}^3}{1\ \text{mL}} \times 417\ \text{mL} = 4.63 \times 10^2\ \text{g}$$

Check: The units (g) are correct. The magnitude of the answer seems correct considering the value of the density is about 1 g/cm^3.

b) **Given:** $d = 1.11$ g/cm^3, $m = 4.1$ kg **Find:** V in L
Conceptual Plan: $d, V \rightarrow m$ **then kg → g and cm^3 → L**

$$d = m / V \qquad \frac{1000 \text{ g}}{1 \text{ kg}} \qquad \frac{1 \text{ L}}{1000 \text{ cm}^3}$$

Solution: $d = m/V$ Rearrange by multiplying both sides of equation by V and dividing both sides of the equation by d.

$$V = \frac{m}{d} = \frac{4.1 \text{ kg}}{1.11 \frac{\text{g}}{\text{cm}^3}} \times \frac{1000 \text{ g}}{1 \text{ kg}} = 3.7 \times 10^3 \text{ cm}^3 \times \frac{1 \text{ L}}{1000 \text{ cm}^3} = 3.7 \text{ L}$$

Check: The units (L) are correct. The magnitude of the answer seems correct considering the value of the density is about 1 g/cm^3.

35. In order to obtain the readings, look to see where the bottom of the meniscus lies. Estimate the distance between two markings on the device.
   a) 73.5 mL – the meniscus appears to be about half way between the 73 mL and the 74 mL marks.

   b) 88.2 °C – the mercury is between the 84 °C mark and the 85 °C mark, but it is closer to the lower number.

   c) 645 mL – the meniscus appears to be just above the 640 mL mark.

37. Remember that
   1. Interior zeroes (zeroes between two numbers) are significant.
   2. Leading zeroes (zeroes to the left of the first non-zero number) are not significant. They only serve to locate the decimal point.
   3. Trailing zeroes (zeroes at the end of a number) are categorized as follows:
      - trailing zeroes after a decimal point are always significant.
      - trailing zeroes before an implied decimal point are ambiguous and should be avoided by using scientific notation or by inserting a decimal point at the end of the number.

   a) 1,050,501 km
   b) 0.0020 m
   c) 0.000000000000002 s
   d) 0.001090 cm

39. Remember all of the rules from section 1.7.
   a) Three significant figures. The 3, 1, and the 2 are significant (rule 1). The leading zeroes only mark the decimal place and are therefore not significant (rule 3).

   b) Ambiguous. The 3, 1, and the 2 are significant (rule 1). The trailing zeroes occur before an implied decimal point and are therefore ambiguous (rule 4). Without more information, we would assume 3 significant figures. It is better to write this as 3.12×10^5 to indicate three significant figures or as 3.12000×10^5 to indicate six (rule 4).

   c) Three significant figures. The 3, 1, and the 2 are significant (rule 1).

   d) Five significant figures. The 1's, 3, 2, and 7 are significant (rule 1).

   e) Ambiguous. The 2 is significant (rule 1). The trailing zeroes occur before an implied decimal point and are therefore ambiguous (rule 4). Without more information, we would assume one significant figure. It is better to write this as 2×10^3 to indicate one significant figure or as 2.000×10^3 to indicate four (rule 4).

41. a) This is not exact because π is an irrational number. 3.14 only shows three of the infinite number of significant figures that π has.

b) This is an exact conversion, because it comes from a definition of the units, and so has an unlimited number of significant figures.

c) This is a measured number and so it is not an exact number. There are 2 significant figures.

d) This is an exact conversion, because it comes from a definition of the units, and so has an unlimited number of significant figures.

43. a) 156.9 – the 8 is rounded up since the next digit is a 5.

b) 156.8 – the last two digits are dropped since 4 is less than 5.

c) 156.8 – the last two digits are dropped since 4 is less than 5.

d) 156.9 – the 8 is rounded up since the next digit is a 9 which is greater than 5.

45. a) $9.15 \div 4.970 = 1.84$ – Three significant figures are allowed to reflect the three significant figures in the least precisely known quantity (9.15).

b) $1.54 \times 0.03060 \times 0.69 = 0.033$ – Two significant figures are allowed to reflect the two significant figures in the least precisely known quantity (0.69). The intermediate answer (0.03251556) is rounded up since the first non-significant digit is a 5.

c) $27.5 \times 1.82 \div 100.04 = 0.500$ – Three significant figures are allowed to reflect the three significant figures in the least precisely known quantity (27.5 and 1.82). The intermediate answer (0.50029988) is truncated since the first non-significant digit is a 2, which is less than 5.

d) $(2.290 \times 10^6) \div (6.7 \times 10^4) = 34$ – Two significant figures are allowed to reflect the two significant figures in the least precisely known quantity (6.7×10^4). The intermediate answer (34.17910448) is truncated since the first non-significant digit is a 1, which is less than 5.

47. a)

$$\begin{array}{r} 43.7 \\ -\ 2.341 \\ \hline 41.359 \end{array} = 41.4$$

Round the intermediate answer to one decimal place to reflect the quantity with the fewest decimal places (43.7). Round the last digit up since the first non-significant digit is 5.

b)

$$\begin{array}{r} 17.6 \\ +\ 2.838 \\ +\ 2.3 \\ +\ 110.77 \\ \hline 133.508 \end{array} = 133.5$$

Round the intermediate answer to one decimal place to reflect the quantity with the fewest decimal places (2.3). Truncate non-significant digits since the first non-significant digit is 0.

c)

$$\begin{array}{r} 19.6 \\ +\ 58.33 \\ -\ 4.974 \\ \hline 72.956 \end{array} = 73.0$$

Round the intermediate answer to one decimal place to reflect the quantity with the fewest decimal places (19.6). Round the last digit up since the first non-significant digit is 5.

d)

$$\begin{array}{r} 5.99 \\ -\ 5.572 \\ \hline 0.418 \end{array} = 0.42$$

Round the intermediate answer to two decimal places to reflect the quantity with the fewest decimal places (5.99). Round the last digit up since the first non-significant digit is 8.

49. Perform operations in parentheses first. Keep track of significant figures in each step, by noting which is the last significant digit in an intermediate result.

a) (24.6681 x 2.38) + 332.58 =

$$\begin{array}{r} 58.\underline{7}10078 \\ +\ \ 332.58 \\ \hline 391.290078 \end{array} = 391.3$$

The first intermediate answer has one significant digit to the right of the decimal, because it is allowed three significant figures (reflecting the quantity with the fewest significant figures (2.38)). Underline the most significant digit in this answer. Round the next intermediate answer to one decimal place to reflect the quantity with the fewest decimal places (58.7). Round the last digit up since the first non-significant digit is 9.

b) $$\frac{(85.3 - 21.489)}{0.0059} = \frac{63.\underline{8}11}{0.0059} = 1.\underline{0}81542 \times 10^4 = 1.1 \times 10^4$$

The first intermediate answer has one significant digit to the right of the decimal, to reflect the quantity with the fewest decimal places (85.3). Underline the most significant digit in this answer. Round the next intermediate answer to two significant figures to reflect the quantity with the fewest significant figures (0.0059). Round the last digit up since the first non-significant digit is 8.

c) (512 ÷ 986.7) + 5.44 =

$$\begin{array}{r} 0.51\underline{8}9014 \\ +\ \ \ \ 5.44 \\ \hline 5.9589014 \end{array} = 5.96$$

The first intermediate answer has three significant figures and three significant digits to the right of the decimal, reflecting the quantity with the fewest significant figures (512). Underline the most significant digit in this answer. Round the next intermediate answer to two decimal places to reflect the quantity with the fewest decimal places (5.44). Round the last digit up since the first non-significant digit is 8.

d) [(28.7 x 10^5) ÷ 48.533] + 144.99 =

$$\begin{array}{r} 59\underline{1}35.01 \\ +\ \ \ 144.99 \\ \hline 59280.01 \end{array} = 59300 = 5.93 \times 10^4$$

The first intermediate answer has three significant figures, reflecting the quantity with the fewest significant figures (28.7 x 10^5). Underline the most significant digit in this answer. Since the number is so large this means that when the addition is performed, the most significant digit is the 100's place. Round the next intermediate answer to the 100's places and put in scientific notation to remove any ambiguity. Note that the last digit is rounded up since the first non-significant digit is 8.

51. a) **Given:** 154 cm **Find:** in

Conceptual Plan: **cm → in**

$$\frac{1\ \text{in}}{2.54\ \text{cm}}$$

Solution: $$154\ \cancel{\text{cm}} \times \frac{1\ \text{in}}{2.54\ \cancel{\text{cm}}} = 60.\underline{6}2992\ \text{in} = 60.6\ \text{in}$$

Check: The units (in) are correct. The magnitude of the answer (60.6) makes physical sense because an inch is a larger unit than a cm. Three significant figures are allowed because 154 cm has three significant figures.

b) **Given:** 3.14 kg **Find:** g

Conceptual Plan: **kg → g**

$$\frac{1000\ \text{g}}{1\ \text{kg}}$$

Solution: $$3.14\ \cancel{\text{kg}} \times \frac{1000\ \text{g}}{1\ \cancel{\text{kg}}} = 3.14 \times 10^3\ \text{g}$$

Check: The units (g) are correct. The magnitude of the answer (10^3) makes physical sense because a kg is a much larger unit than a gram. Three significant figures are allowed because 3.14 kg has three significant figures.

c) **Given:** 3.5 L **Find:** qt
Conceptual Plan: **L → qt**

$$\frac{1.057\ \text{qt}}{1\ \text{L}}$$

Solution: $3.5\ \cancel{\text{L}} \times \frac{1.057\ \text{qt}}{1\ \cancel{\text{L}}} = 3.6995\ \text{qt} = 3.7\ \text{qt}$

Check: The units (qt) are correct. The magnitude of the answer (3.7) makes physical sense because a L is a smaller unit than a qt. Two significant figures are allowed because 3.5 L has two significant figures. Round the last digit up because the first non-significant digit is a 9.

d) **Given:** 109 mm **Find:** in
Conceptual Plan: **mm → m → in**

$$\frac{1\ \text{m}}{1000\ \text{mm}} \quad \frac{39.37\ \text{in}}{1\ \text{m}}$$

Solution: $109\ \cancel{\text{mm}} \times \frac{1\ \cancel{\text{m}}}{1000\ \cancel{\text{mm}}} \times \frac{39.37\ \text{in}}{1\ \cancel{\text{m}}} = 4.2\underline{9}133\ \text{in} = 4.29\ \text{in}$

Check: The units (in) are correct. The magnitude of the answer (4) makes physical sense because a mm is a much smaller unit than an in. Three significant figures are allowed because 109 mm has three significant figures.

53. **Given:** 10.0 km **Find:** minutes **Other:** running pace = 7.5 miles per hour
Conceptual Plan: **km → mi → hr → min**

$$\frac{0.6214\ \text{mi}}{1\ \text{km}} \quad \frac{1\text{hr}}{7.5\ \text{mi}} \quad \frac{60\ \text{min}}{1\ \text{hr}}$$

Solution: $10.0\ \cancel{\text{km}} \times \frac{0.6214\ \cancel{\text{mi}}}{1\ \cancel{\text{km}}} \times \frac{1\ \cancel{\text{hr}}}{7.5\ \cancel{\text{mi}}} \times \frac{60\ \text{min}}{1\ \cancel{\text{hr}}} = 4\underline{9}.712\ \text{min} = 50.\ \text{min} = 5.0 \times 10^1\ \text{min}$

Check: The units (min) are correct. The magnitude of the answer (50) makes physical sense because she is running almost 7.5 miles (which would take her 60 min = 1 hr). Two significant figures are allowed because of the limitation of 7.5 mi/hr (two significant figures). Round the last digit up because the first non-significant digit is a 7.

55. **Given:** 14 km/L **Find:** miles per gallon
Conceptual Plan: $\frac{\textbf{km}}{\textbf{L}} \rightarrow \frac{\textbf{mi}}{\textbf{L}} \rightarrow \frac{\textbf{mi}}{\textbf{gal}}$

$$\frac{0.6214\ \text{mi}}{1\ \text{km}} \quad \frac{3.785\ \text{L}}{1\ \text{gallon}}$$

Solution: $\frac{14\ \cancel{\text{km}}}{1\ \cancel{\text{L}}} \times \frac{0.6214\ \text{mi}}{1\ \cancel{\text{km}}} \times \frac{3.785\ \cancel{\text{L}}}{1\ \text{gallon}} = 3\underline{2}.927986\ \frac{\text{miles}}{\text{gallon}} = 33\ \frac{\text{miles}}{\text{gallon}}$

Check: The units (mi/gal) are correct. The magnitude of the answer (33) makes physical sense because the dominating factor is that a L is much smaller than a gallon, so the answer should go up. Two significant figures are allowed because of the limitation of 14 km/L (two significant figures). Round the last digit up because the first non-significant digit is a 9.

57. a) **Given:** 195 m^2 **Find:** km^2
Conceptual Plan: $\textbf{m}^2 \rightarrow \textbf{km}^2$

$$\frac{(1\ \text{km})^2}{(1000\ \text{m})^2}$$

Notice that for squared units, the conversion factors must be squared.

Solution: $195\ \cancel{\text{m}^2} \times \frac{(1\ \text{km})^2}{(1000\ \cancel{\text{m}})^2} = 1.95 \times 10^{-4}\ \text{km}^2$

Check: The units (km^2) are correct. The magnitude of the answer (10^{-4}) makes physical sense because a kilometer is a much larger unit than a meter.

b) **Given:** 195 m^2 **Find:** dm^2
Conceptual Plan: $m^2 \rightarrow dm^2$

$$\frac{(10\ dm)^2}{(1\ m)^2}$$

Notice that for squared units, the conversion factors must be squared.

Solution: $195\ m^2 \times \frac{(10\ dm)^2}{(1\ m)^2} = 1.95 \times 10^4\ dm^2$

Check: The units (dm^2) are correct. The magnitude of the answer (10^4) makes physical sense because a decimeter is a much smaller unit than a meter.

c) **Given:** 195 m^2 **Find:** cm^2
Conceptual Plan: $m^2 \rightarrow cm^2$

$$\frac{(100\ cm)^2}{(1\ m)^2}$$

Notice that for squared units, the conversion factors must be squared.

Solution: $195\ m^2 \times \frac{(100\ cm)^2}{(1\ m)^2} = = 1.95 \times 10^6\ cm^2$

Check: The units (cm^2) are correct. The magnitude of the answer (10^6) makes physical sense because a centimeter is a much smaller unit than a meter.

59. **Given:** 435 acres **Find:** square miles **Other:** 1 acre = 43,560 ft^2, 1 mile = 5280 ft
Conceptual Plan: acres $\rightarrow ft^2 \rightarrow mi^2$

$$\frac{43560\ ft^2}{1\ acre} \qquad \frac{(1\ mi)^2}{(5280\ ft)^2}$$

Notice that for squared units, the conversion factors must be squared.

Solution: $435\ acres \times \frac{43560\ ft^2}{1\ acre} \times \frac{(1\ mi)^2}{(5280\ ft)^2} = 0.67\underline{9}6875\ mi^2 = 0.680\ mi^2$

Check: The units (mi^2) are correct. The magnitude of the answer (0.7) makes physical sense because an acre is much smaller than a mi^2, so the answer should go down several orders of magnitude. Three significant figures are allowed because of the limitation of 435 acres (three significant figures). Round the last digit up because the first non-significant digit is a 7.

61. **Given:** 14 lbs **Find:** mL **Other:** 80 mg/0.80 mL and 15 mg/kg body
Conceptual Plan: lb $\rightarrow$ kg body $\rightarrow$ mg $\rightarrow$ mL

$$\frac{1\ kg\ body}{2.205\ lb} \qquad \frac{15\ mg}{1\ kg\ body} \qquad \frac{0.80\ mL}{80\ mg}$$

Solution: $14\ lb \times \frac{1\ kg\ body}{2.205\ lb} \times \frac{15\ mg}{1\ kg\ body} \times \frac{0.80\ mL}{80\ mg} = 0.9\underline{5}23809524\ mL = 0.95\ mL$

Check: The units (cm^3) are correct. The magnitude of the answer (1 mL) makes physical sense because it is reasonable amount of liquid to give to a baby. Two significant figures are allowed because of the statement in the problem. Truncate the last digit because the first non-significant digit is a 2.

63. **Given:** solar year **Find:** seconds **Other:** 60 seconds/minute; 60 minutes/ hour; 24 hours/solar day; and 365.24 solar days/solar year

Conceptual Plan: yr $\rightarrow$ day $\rightarrow$ hr $\rightarrow$ min $\rightarrow$ sec

$$\frac{365.24\ day}{1\ solar\ yr} \qquad \frac{24\ hr}{1 day} \qquad \frac{60\ min}{1\ hr} \qquad \frac{60\ sec}{1\ min}$$

Solution:

$$1\ solar\ yr \times \frac{365.24\ day}{1\ solar\ yr} \times \frac{24\ hr}{1 day} \times \frac{60\ min}{1\ hr} \times \frac{60\ sec}{1\ min} = 3.155\underline{6}736 \times 10^7\ sec = 3.1557 \times 10^7\ sec$$

Check: The units (seconds) are correct. The magnitude of the answer (10^7) makes physical sense because each conversion factor increases the value of the answer – a second is many orders of magnitude smaller than a year. Five significant figures are allowed because all conversion factors are assumed to be exact, except for the 365.24 days/ solar year (five significant figures). Round up the last digit because the first non-significant digit is a 7.

65. a) Extensive – the volume of a material depends on how much there is present.

b) Intensive – the boiling point of a material is independent of how much material you have, so these values can be published in reference tables.

c) Intensive – the temperature of a material depends on how much there is present.

d) Intensive – the electrical; conductivity of a material is independent of how much material you have, so these values can be published in reference tables.

e) Extensive - the energy contained in material depends on how much there is present. Many times energy is expressed in terms of Joules/mole, which then turns this quantity into an intensive property.

67. **Given:** 130 °X = 212 °F and 10 °X = 32 °F **Find:** temperature where °X = °F.
Conceptual Plan: Use data to derive an equation relating °X and °F. Then set °F = °X = z and solve for z.
Solution: Assume a linear relationship between the two temperatures (y = mx + b).
Let y = °F and let x = °X.
The slope of the line (m) is the relative change in the two temperature scales:

$$m = \frac{\Delta\ °F}{\Delta\ °X} = \frac{212\ °F - 32\ °F}{130\ °X - 10\ °X} = \frac{180\ °F}{120\ °X} = 1.5$$

Solve for intercept (b) by plugging one set of temperatures into the equation:
$y = 1.5\,x + b \rightarrow 32 = (1.5)(10) + b \rightarrow 32 = 15 + b \rightarrow b = 17 \rightarrow °F = (1.5)\ °X + 17$
Set °F = °X = z and solve for z.
$z = 1.5\,z + 17 \rightarrow -17 = 1.5\,z - z \rightarrow -17 = 0.5\,z \rightarrow z = -34 \rightarrow -34°F = -34\ °X$
Check: The units (°F and °X) are correct. Plugging the result back into the equation confirms that the calculations were done correctly. The magnitude of the answer seems correct, since it is known that the result is not between 32°F and 212 °F. The numbers are getting closer together as the temperature is dropped.

69. a) $1.76 \times 10^{-3}/8.0 \times 10^2 = 2.2 \times 10^{-6}$ Two significant figures are allowed to reflect the quantity with the fewest significant figures (8.0×10^2).

b) Write all figures so that the decimal points can be aligned:

	0.0187	
+	0.0002	All quantities are known to four places to the right of the decimal place,
–	0.0030	so the answer should be reported to four places to the right of the
	0.0159	decimal place or three significant figures.

c) $[(136000)(0.000322)/0.082](129.2) = 6.899910244 \times 10^4 = 6.9 \times 10^4$ Round the intermediate answer to two significant figures to reflect the quantity with the fewest significant figures (0.082). Round up the last digit since the first non-significant digit is 9.

71. a) **Given:** cylinder dimensions: length = 22 cm, radius = 3.8 cm, *d*(gold) = 19.3 g/cm³ and *d*(sand) = 3.00 g/cm³
Find: m(gold) and m(sand)
Conceptual Plan: $l, r \rightarrow V$ **then** $d, V \rightarrow m$

$V = l\pi r^2 \qquad d = m/V$

Solution: $V(\text{gold}) = V(\text{sand}) = (22\ \text{cm})(\pi)(3.8\ \text{cm})^2 = 9\underline{9}8.0212\ \text{cm}^3$ $d = m/V$ Rearrange by multiplying both sides of equation by *V*. $\rightarrow m = d \times V$

$$m(\text{gold}) = \left(19.3\ \frac{\text{g}}{\text{cm}^3}\right) \times (9\underline{9}8.0212\ \text{cm}^3) = 1.\underline{9}26181 \times 10^4\ \text{g} = 1.9 \times 10^4\ \text{g}$$

Check: The units (g) are correct. The magnitude of the answer seems correct considering the value of the density is ~20 g/cm^3. Two significant figures are allowed to reflect the significant figures in 22 cm and 3.8 cm. Truncate the non-significant digits because the first non-significant digit is a 2.

$$m(\text{sand}) = \left(3.00 \frac{g}{\cancel{cm^3}}\right) \times (998.0212\ \cancel{cm^3}) = 2.99206 \times 10^3\ g = 3.0 \times 10^3\ g$$

Check: The units (g) are correct. The magnitude of the answer seems correct considering the value of the density is 3 g/cm^3. This number is much lower than the gold mass. Two significant figures are allowed to reflect the significant figures in 22 cm and 3.8 cm. Round the last digit up because the first non-significant digit is a 9.

b) Comparing the two values 1.9×10^4 g versus 3.0×10^3 g shows a difference in weight of almost a factor of 10. This difference should be enough to trip the alarm and alert the authorities to the presence of the thief.

73. **Given:** 3.5 lb of titanium **Find:** volume in in^3 **Other:** density of titanium is 4.5 g/cm^3

Conceptual Plan: **lb → g** **then** ***m, d* → *V*** **then cm^3 → in^3**

$$\frac{453.6\ g}{1\ lb} \qquad d = m/V \qquad \frac{(1\ in)^3}{(2.54\ cm)^3}$$

Solution: $3.5\ \cancel{lb} \times \frac{453.6\ g}{1\ \cancel{lb}} = 1.5876 \times 10^3\ g$

$d = m/V$ Rearrange by multiplying both sides of the equation by V and dividing both sides of the equation by d. $\rightarrow V = \frac{m}{d} = \frac{1.5876 \times 10^3\ \cancel{g}}{4.5 \frac{\cancel{g}}{cm^3}} = 3.528 \times 10^2\ cm^3 = 3.5 \times 10^2\ \cancel{cm^3} \times \frac{(1\ in)^3}{(2.54\ \cancel{cm})^3} = 22\ in^3$

Check: The units (in^3) are correct. The magnitude of the answer seems correct considering many grams we have. Two significant figures are allowed to reflect the significant figures in 3.5 lb. Truncate the non-significant digits because the first non-significant digit is a 2.

75. **Given:** cylinder dimensions: length = 2.16 in, radius = 0.22 in, m= 41 g **Find:** density (g /cm^3)

Conceptual Plan: **in → cm** **then** ***l, r* → *V*** **then *m, V* → *d***

$$\frac{2.54\ cm}{1\ in} \qquad V = l\pi r^2 \qquad d = m/V$$

Solution: $2.16\ \cancel{in} \times \frac{2.54\ cm}{1\ \cancel{in}} = 5.4864\ cm = l$ $\qquad 0.22\ \cancel{in} \times \frac{2.54\ cm}{1\ \cancel{in}} = 0.5588\ cm = r$

$V = l\pi r^2 = (5.4864\ cm)(\pi)(0.5588\ cm)^2 = 5.3820798\ cm^3$

$d = \frac{m}{V} = \frac{41\ g}{5.3820798\ cm^3} = 7.6178729 \frac{g}{cm^3} = 7.6 \frac{g}{cm^3}$

Check: The units (g/cm^3) are correct. The magnitude of the answer seems correct considering the value of the density of iron (a major component in steel) is 7.86 g/cm^3. Two significant figures are allowed to reflect the significant figures in 0.22 in and 41 g. Truncate the non-significant digits because the first non-significant digit is a 2.

77. **Given:** 185 cubic yards (yd^3) of H_2O **Find:** mass of the H_2O (pounds) **Other:** $d(H_2O) = 1.00$ g/cm^3 at 0°C

Conceptual Plan: **yd^3 → m^3 → cm^3 → g → lb**

$$\frac{(1\ m)^3}{(1.094\ yd)^3} \quad \frac{(100\ cm)^3}{(1m)^3} \quad \frac{1.00\ g}{1.00\ cm^3} \quad \frac{1\ lb}{453.59\ g}$$

Solution:

$$185\ \cancel{yd^3} \times \frac{(1\ \cancel{m})^3}{(1.094\ \cancel{yd})^3} \times \frac{(100\ \cancel{cm})^3}{(1\ \cancel{m})^3} \times \frac{1.00\ \cancel{g}}{1.00\ \cancel{cm^3}} \times \frac{1\ lb}{453.59\ \cancel{g}} = 3.114987377 \times 10^5\ lbs = 3.11 \times 10^5\ lbs$$

Check: The units (lb) are correct. The magnitude of the answer (10^5) makes physical sense because a pool is not a small object. Three significant figures are allowed because the conversion factor with the least precision is the density (1.00 g/cm^3 - 3 significant figures) and the initial size has three significant figures. Truncate after the last digit because the first non-significant digit is a 4.

79. **Given:** 15 liters of gasoline **Find:** kilometers **Other:** 52 mi/gal in the city
Conceptual Plan: **L → gal → mi → km**

$$\frac{1\text{ gallon}}{3.785\text{ L}} \quad \frac{52\text{ mi}}{1.0\text{ gallon}} \quad \frac{1\text{ km}}{0.6214\text{ mi}}$$

Solution: $15\text{ L} \times \frac{1\text{ gallon}}{3.785\text{ L}} \times \frac{52\text{ mi}}{1.0\text{ gallon}} \times \frac{1\text{ km}}{0.6214\text{ mi}} = 3.\underline{3}16327941 \times 10^2\text{ km} = 3.3 \times 10^2\text{ km}$

Check: The units (km) are correct. The magnitude of the answer (10^2) makes physical sense because the dominating conversion factor is the mileage, which increases the answer. Two significant figures are allowed because the conversion factor with the least precision is 52 mi/gallon (2 significant figures) and the initial volume (15 L) has 2 significant figures. Truncate the last digit because the first non-significant digit is a 1. It is best to put the answer in scientific notation so that it is unambiguous how many significant figures are expressed.

81. **Given:** radius of nucleus of the hydrogen atom = 1.0×10^{-13} cm; radius of the hydrogen atom = 52.9 pm.
Find: percent of volume occupied by nucleus (%)
Conceptual Plan: **cm → m then pm → m then r → V then V_{atom}, $V_{nucleus}$ → % $V_{nucleus}$**

$$\frac{1\text{ m}}{100\text{ cm}} \quad \frac{1\text{ m}}{10^{12}\text{ pm}} \quad V = (4/3)\pi r^3 \quad \%V_{nucleus} = \frac{V_{nucleus}}{V_{atom}} \times 100\%$$

Solution: $1.0 \times 10^{-13}\text{ cm} \times \frac{1\text{ m}}{100\text{ cm}} = 1.0 \times 10^{-15}\text{ m}$ and $52.9\text{ pm} \times \frac{1\text{ m}}{10^{12}\text{ pm}} = 5.29 \times 10^{-11}\text{ m}$

$V = (4/3)\pi r^3$ Substitute into %V equation.

$$\%V_{nucleus} = \frac{V_{nucleus}}{V_{atom}} \times 100\% \rightarrow \%V_{nucleus} = \frac{(4/3)\,\pi r_{nucleus}^3}{(4/3)\,\pi r_{atom}^3} \times 100\%$$ Simplify equation.

$$\%V_{nucleus} = \frac{r_{nucleus}^3}{r_{atom}^3} \times 100\%$$ Substitute numbers and calculate result.

$$\%V_{nucleus} = \frac{(1.0 \times 10^{-15}\text{ m})^3}{(5.29 \times 10^{-11}\text{ m})^3} \times 100\% = (1.\underline{8}90359168 \times 10^{-5})^3 \times 100\% = 6.\underline{7}55118685 \times 10^{-13}\ \%$$

$= 6.8 \times 10^{-13}$ %

Check: The units (%) are correct. The magnitude of the answer seems correct (10^{-13} %), since a proton is so small. Two significant figures are allowed to reflect the significant figures in 1.0×10^{-13} cm. Round up the last digits because the first non-significant digit is a 5.

83. **Given:** mass of black hole (BH) = 1×10^3 suns; radius of black hole = one-half the radius of our moon.
Find: density (g/cm^3) **Other:** radius of our sun = 7.0×10^5 km; average density of our sun = 1.4×10^3 kg/m^3; diameter of the moon = 2.16×10^3 miles
Conceptual Plan: $d_{BH} = m_{BH}/V_{BH}$
Calculate m_{BH}: r_{sun} → V_{sun} km$^3_{sun}$ → m$^3_{sun}$ V_{sun}, d_{sun} → m_{sun} m_{sun} → m_{BH} kg → g

$$V = (4/3)\pi r^3 \quad \frac{(1000\text{ m})^3}{(1\text{ km})^3} \quad d_{sun} = \frac{m_{sun}}{V_{sun}} \quad m_{BH} = (1 \times 10^3) \times m_{sun} \quad \frac{1000\text{ g}}{1\text{ kg}}$$

Calculate V_{BH}: d_{moon} → r_{moon} → r_{BH} mi → km → m → cm r → V

$$r_{moon} = \tfrac{1}{2}\,d_{moon} \quad r_{BH} = \tfrac{1}{2}\,r_{moon} \quad \frac{1\text{ km}}{0.6214\text{ mi}} \quad \frac{1000\text{ m}}{1\text{ km}} \quad \frac{100\text{ cm}}{1\text{ m}} \quad V = (4/3)\pi r^3$$

Substitute into $d_{BH} = m_{BH}/V_{BH}$
Solution: Calculate m_{BH} $V_{sun} = (4/3)\,\pi\, r_{sun}^3 = (4/3)\,\pi\,(7.0 \times 10^5\text{ km})^3 = 1.\underline{4}3675504 \times 10^{18}\text{ km}^3$

$$1.\underline{4}3675504 \times 10^{18}\text{ km}^3 \times \frac{(1000\text{ m})^3}{(1\text{ km})^3} = 1.\underline{4}3675504 \times 10^{27}\text{ m}^3$$

$d_{sun} = m_{sun}/V_{sun}$ Solve for m by multiplying both sides of the equation by V_{sun}. $m_{sun} = V_{sun} \times d_{sun}$
$m_{sun} = (1.\underline{4}3675504 \times 10^{27}\text{ m}^3)(1.4 \times 10^3\text{ kg/m}^3) = 2.\underline{0}11457056 \times 10^{30}$ kg
$m_{BH} = (1 \times 10^3) \times m_{sun} = (1 \times 10^3) \times (2.\underline{0}11457056 \times 10^{30}\text{ kg}) = 2.\underline{0}11457056 \times 10^{33}$ kg

$$2.\underline{0}11457056 \times 10^{33}\text{ kg} \times \frac{1000\text{ g}}{1\text{ kg}} = 2.\underline{0}11457056 \times 10^{36}\text{ g}$$

Calculate V_{BH} $r_{moon} = \tfrac{1}{2}\,d_{moon} = \tfrac{1}{2}\,(2.16 \times 10^3\text{ miles}) = 1.08 \times 10^3$ miles

$r_{BH} = ½\, r_{moon} = ½\,(1.08 \times 10^3 \text{ miles}) = 540. \text{ miles}$

$$540. \text{ miles} \times \frac{1 \text{ km}}{0.6214 \text{ mi}} \times \frac{1000 \text{ m}}{1 \text{ km}} \times \frac{100 \text{ cm}}{1 \text{ m}} = 8.6\underline{9}00547 \times 10^7 \text{ cm}$$

$$V = (4/3)\,\pi r^3 = (4/3)\,\pi\,(8.6\underline{9}00547 \times 10^7 \text{ cm})^3 = 2.7\underline{4}888228 \times 10^{24} \text{ cm}^3$$

Substitute into $$d_{BH} = \frac{m_{BH}}{V_{BH}} = \frac{2.0\underline{1}1457056 \times 10^{36} \text{ g}}{2.7\underline{4}888228 \times 10^{24} \text{ cm}^3} = 7.\underline{3}1737339 \times 10^{11} \frac{\text{g}}{\text{cm}^3} = 7.3 \times 10^{11} \frac{\text{g}}{\text{cm}^3}$$

Check: The units (g/cm^3) are correct. The magnitude of the answer seems correct (10^{12}), since we expect extremely high numbers for black holes. Two significant figures are allowed to reflect the significant figures in the radius of our sun (7.0 x 10^5 km) and the average density of the sun (1.4 x 10^3 kg/m^3). Truncate the non-significant digits because the first non-significant digit is a 4.

85. **Given:** cubic nanocontainers with an edge length = 25 nanometers
Find: a) volume of one nanocontainer; b) grams of oxygen could be contained by each nanocontainer; c) grams of oxygen inhaled per hour; d) minimum number of nanocontainers per hour; and e) minimum volume of nanocontainers.
Other: (pressurized oxygen) = 85 g/L; 0.28 g of oxygen per liter; average human inhales about 0.50 L of air per breath and takes about 20 breaths per minute; and adult total blood volume = ~ 5 L.
Conceptual Plan:

a) nm → m → cm then $l \rightarrow V$ then cm^3 → L

$$\frac{1 \text{ m}}{10^9 \text{ nm}} \quad \frac{100 \text{ cm}}{1 \text{ m}} \quad V = l^3 \quad \frac{1 \text{ L}}{1000 \text{ cm}^3}$$

b) L → g pressurized oxygen

$$\frac{85 \text{ g oxygen}}{1 \text{ L nanocontainers}}$$

c) hr → min → breaths → L_{air} → g_{O2}

$$\frac{60 \text{ min}}{1 \text{ hr}} \quad \frac{20 \text{ breath}}{1 \text{ min}} \quad \frac{0.50 \text{ L}_{air}}{1 \text{ breath}} \quad \frac{0.28 \text{ g}_{CO}}{1 \text{ L}_{air}}$$

d) grams oxygen → number nanocontainers

$$\frac{1 \text{ nanocontainer}}{\text{part (b) grams of oxygen}}$$

e) number nanocontainers → volume nanocontainers

$$\frac{\text{part (a) volume}}{\text{of 1 nanocontainer}}$$

Solution:

a) $$25 \text{ nm} \times \frac{1 \text{ m}}{10^9 \text{ nm}} \times \frac{100 \text{ cm}}{1 \text{ m}} = 2.5 \times 10^{-6} \text{ cm}$$

$$V = l^3 = (2.5 \times 10^{-6} \text{ cm})^3 = 1.\underline{5}625 \times 10^{-17} \text{ cm}^3 \times \frac{1 \text{ L}}{1000 \text{ cm}^3} = 1.\underline{5}625 \times 10^{-20} \text{ L} = 1.6 \times 10^{-20} \text{ L}$$

b) $$1.\underline{5}625 \times 10^{-20} \text{ L} \times \frac{85 \text{ g oxygen}}{1 \text{ L nanocontainers}} = 1.\underline{3}28125 \times 10^{-18} \frac{\text{g pressurized } O_2}{\text{nanocontainer}}$$

$$= 1.3 \times 10^{-18} \frac{\text{g pressurized } O_2}{\text{nanocontainer}}$$

c) $$1 \text{ hr} \times \frac{60 \text{ min}}{1 \text{ hr}} \times \frac{20 \text{ breath}}{1 \text{ min}} \times \frac{0.50 \text{ L}_{air}}{1 \text{ breath}} \times \frac{0.28 \text{ g}O_2}{1 \text{ L}_{air}} = 1.\underline{6}8 \times 10^2 \text{ g oxygen} = 1.7 \times 10^2 \text{ g oxygen}$$

d) $$1.\underline{6}8 \times 10^2 \text{ g oxygen} \times \frac{1 \text{ nanocontainer}}{1.3 \times 10^{-18} \text{ g of oxygen}} = 1.\underline{2}92307692 \times 10^{20} \text{ nanocontainers}$$

$$= 1.3 \times 10^{20} \text{ nanocontainers}$$

e) $$1.\underline{2}92307692 \times 10^{20} \text{ nanocontainers} \times \frac{1.5625 \times 10^{-20} \text{ L}}{\text{nanocontainer}} = 2.\underline{0}19230769 \text{ L} = 2.0 \text{ L}$$

This volume is much too large to be feasible, since the volume of blood in the average human is 5 L.

Check:

a) The units (L) are correct. The magnitude of the answer (10^{-20}) makes physical sense because these are very, very tiny containers. Two significant figures are allowed, reflecting the significant figures in the starting dimension (25 nm – 2 significant figures). Round up the last digit because the first non-significant digit is a 6.

b) The units (g) are correct. The magnitude of the answer (10^{-18}) makes physical sense because these are very, very tiny containers and very few molecules can fit inside. Two significant figures are allowed, reflecting the significant figures in the starting dimension (25 nm) and the given concentration (85 g/L) – 2 significant figures in each. Truncate the non-significant digits because the first non-significant digit is a 2.

c) The units (g oxygen) are correct. The magnitude of the answer (10^{2}) makes physical sense because of the conversion factors involved and the fact that air is not very dense. Two significant figures are allowed, because it is stated in the problem. Round up the last digit because the first non-significant digit is an 8.

d) The units (nanocontainers) are correct. The magnitude of the answer (10^{20}) makes physical sense because these are very, very tiny containers and we need a macroscopic quantity of oxygen in these containers. Two significant figures are allowed, reflecting the significant figures in both of the quantities in the calculation – 2 significant figures. Round up the last digit because the first non- significant digit is a 9.

e) The units (L) are correct. The magnitude of the answer (2) makes physical sense because, the magnitudes of the numbers in this step. Two significant figures are allowed reflecting the significant figures in both of the quantities in the calculation – 2 significant figures. Truncate the non-significant digits because the first non-significant digit is a 1.

87. c) is the best representation. When solid carbon dioxide (dry ice) sublimes, it changes phase from a solid to a gas. Phase changes are physical changes, so no molecular bonds are broken. This diagram shows molecules with one carbon atom and two oxygen atoms bonded together in every molecule. The other diagrams have no carbon dioxide molecules.

89. In order to determine which number is large, the units need to be compared. There is a factor of 1000 between grams and kg, in the numerator. There is a factor of $(100)^3$ or 1,000,000 between cm^3 and m^3. This second factor more than compensates for the first factor. Thus Substance A with a density of 1.7 g/cm^3 is denser than Substance B with a density of 1.7 kg/m^3.

91. Remember that: An observation is the information collected when studying phenomena. A law is a concise statement that summarizes observed behaviors and observations and predicts future observations. A theory attempts to explain why the observed behavior is happening.

a) This statement is most like a law, because it summarizes many observations and can explain future behavior – many places and many days.

b) This statement is a theory because it attempts to explain why – gravitational forces.

c) This statement is most like an observation, because it is information collected in order to understand tidal behavior.

d) This statement is most like a law, because it summarizes many observations and can explain future behavior – many places and many days and months.

Chapter 2
Atoms and Elements

1. **Given:** 1.50 g hydrogen; 12.0 g oxygen **Find:** grams water vapor
Conceptual Plan: total mass reactants = total mass products
Solution: Mass of reactants = 1.50 g hydrogen + 12.0 g oxygen = 13.5 grams
Mass of products = mass of reactants = 13.5 grams water vapor.
Check: According to the Law of Conservation of Mass, matter is not created or destroyed in a chemical reaction, so, since water vapor is the only product the masses of hydrogen and oxygen must combine to form the mass of water vapor.

3. **Given:** sample 1: 38.9 g carbon, 448 g chlorine; sample 2: 14.8 g carbon, 134 g chlorine
Find: consistent with definite proportions.
Conceptual Plan: determine mass ratio of sample 1 and 2 and compare.

$$\frac{\text{mass of chlorine}}{\text{mass of carbon}}$$

Solution: Sample 1: $\frac{448\text{ g chorine}}{38.9\text{ g carbon}} = 11.5$ Sample 2: $\frac{134\text{ g chlorine}}{14.8\text{ g carbon}} = 9.05$

Results are not consistent with the law of definite proportions because the ratio of chlorine to carbon is not the same.
Check: According to the Law of Definite Proportions, the mass ratio of one element to another is the same for all samples of the compound.

5. **Given:** mass ratio sodium to fluorine = 1.21:1; sample = 28.8 g sodium **Find:** g fluorine
Conceptual Plan: g sodium → g fluorine

$$\frac{\text{mass of fluorine}}{\text{mass of sodium}}$$

Solution: $28.8\text{ g }\cancel{\text{sodium}} \times \frac{1\text{ g fluorine}}{1.21\text{ g }\cancel{\text{sodium}}} = 23.8\text{ g fluorine}$

Check: The units of the answer, g fluorine are correct. The magnitude of the answer is reasonable since it is less than the grams of sodium.

7. **Given:** 1 gram osmium: sample 1 = 0.168 g oxygen; sample 2 = 0.3369 g oxygen
Find: consistent with multiple proportions.
Conceptual Plan: determine mass ratio of oxygen

$$\frac{\text{mass of oxygen sample 2}}{\text{mass of oxygen sample 1}}$$

Solution: $\frac{0.3369\text{ g oxygen}}{0.168\text{ g oxygen}} = 2.00$ Ratio is a small whole number. Results are consistent with multiple proportions
Check: According to the Law of Multiple Proportions, when two elements form two different compounds, the masses of element B that combine with 1 g of element A can be expressed as a ratio of small whole numbers.

9. **Given:** sulfur dioxide = 3.49 g oxygen and 3.50 g sulfur, sulfur trioxide = 6.75 g oxygen and 4.50 g sulfur.
Find: mass oxygen per g S for each compound and then determine the mass ratio of oxygen

$$\frac{\text{mass of oxygen in sulfur dioxide}}{\text{mass of sulfur in sulfur dioxide}} \quad \frac{\text{mass of oxygen in sulfur trioxide}}{\text{mass of sulfur in sulfur trioxide}} \quad \frac{\text{mass of oxyen in sulfur trioxide}}{\text{mass of oxyen in sulfur dioxide}}$$

Solution: sulfur dioxide $= \frac{3.49\text{ g oxygen}}{3.50\text{ g sulfur}} = \frac{0.997\text{ g oxygen}}{1\text{ g sulfur}}$ sulfur trioxide $= \frac{6.75\text{ g oxygen}}{4.50\text{ g sulfur}} = \frac{1.50\text{ g oxygen}}{1\text{ g sulfur}}$

$$\frac{1.50\text{ g oxygen in sulfur trioxide}}{0.997\text{ g oxygen in sulfur dioxide}} = \frac{1.50}{1} = \frac{3}{2}$$

Ratio is in small whole numbers and is consistent with multiple proportions.
Check: According to the Law of Multiple Proportions, when two elements form two different compounds, the masses of element B that combine with 1 g of element A can be expressed as a ratio of small whole numbers.

11. a) Sulfur and oxygen atoms have the same mass. INCONSISTENT with Dalton's atomic theory because only atoms of the same element have the same mass.

b) All cobalt atoms are identical. CONSISTENT with Dalton's atomic theory because all atoms of a given element have the same mass and other properties that distinguish them from atoms of other elements.

c) Potassium and chlorine atoms combine in a 1:1 ratio to form potassium chloride. CONSISTENT with Dalton's atomic theory because atoms combine in simple whole number ratios to form compounds.

d) Lead atoms can be converted into gold. INCONSISTENT with Dalton's atomic theory because atoms of one element cannot change into atoms of another element.

13. a) The volume of an atom is mostly empty space. CONSISTENT with Rutherford's nuclear theory because most of the volume of the atom is empty space, throughout which tiny, negatively charged electrons are dispersed.

b) The nucleus of an atom is small compared to the size of the atom. CONSISTENT with Rutherford's nuclear theory because most of the atom's mass and all of its positive charge are contained in a small core called the nucleus.

c) Neutral lithium atoms contain more neutrons than protons. INCONSISTENT with Rutherford's nuclear theory because it did not distinguish where the mass of the nucleus came from other than from the protons.

d) Neutral lithium atoms contain more protons than electrons. INCONSISTENT with Rutherford's nuclear theory because there are as many negatively charged particles outside the nucleus as there are positively charged particles within the nucleus.

15. **Given:** drop A = -6.9×10^{-19} C; drop B = -9.2×10^{-19} C; drop C = -11.5×10^{-19} C; drop D = -4.6×10^{-19} C
Find: The charge on a single electron
Conceptual Plan: determine the ratio of charge for each set of drops.

$$\frac{\text{charge on drop 1}}{\text{charge on drop 2}}$$

Solution: $\dfrac{-6.9 \times 10^{-19}\text{C drop A}}{-4.6 \times 10^{-19}\text{ C drop D}} = 1.5 \qquad \dfrac{-9.2 \times 10^{-19}\text{C drop B}}{-4.6 \times 10^{-19}\text{ C drop D}} = 2 \qquad \dfrac{-11.5 \times 10^{-19}\text{C drop C}}{-4.6 \times 10^{-19}\text{ C drop D}} = 2.5$

The ratios obtained are not whole numbers, but can be converted to whole numbers by multiplying by 2. Therefore, the charge on the electron has to be 1/2 the smallest value experimentally obtained. The charge on the electron = -2.3×10^{-19} C.
Check: The units of the answer, Coulombs, are correct. The magnitude of the answer is reasonable since all the values experimentally obtained are integer multiples of -2.3×10^{-19}.

17. **Given:** charge on body = $-15\ \mu$C. **Find:** number of electrons, mass of the electrons
Conceptual Plan: μC → C → number of electrons → mass of electrons

$$\frac{1\text{ C}}{10^6\ \mu\text{C}} \qquad \frac{1\text{ electron}}{-1.60 \times 10^{-19}\text{C}} \qquad \frac{9.10 \times 10^{-28}\text{ g}}{1\text{ electron}}$$

Solution: $-15\ \cancel{\mu\text{C}} \times \dfrac{1\ \cancel{\text{C}}}{10^6\ \cancel{\mu\text{C}}} \times \dfrac{1\text{ electron}}{-1.60 \times 10^{-19}\ \cancel{\text{C}}} = 9.\underline{3}75 \times 10^{13}\text{ electrons} = 9.4 \times 10^{13}\text{ electrons}$

$$9.\underline{3}75 \times 10^{13}\ \cancel{\text{electrons}} \times \frac{9.10 \times 10^{-28}\text{ g}}{1\ \cancel{\text{electron}}} = 8.5 \times 10^{-14}\text{ g}$$

Check: The units of the answers, number of electrons and grams, are correct. The magnitude of the answers are reasonable since the charge on an electron and the mass of an electron are very small.

19. a) True: protons and electrons have equal and opposite charges.

b) True: protons and electrons have opposite charge so they will attract each other.

c) True: the mass of the electron is much less than the mass of the neutron.

d) False: the mass of the proton and the mass of the neutron are about the same.

21. **Given:** mass of proton **Find:** number of electron in equal mass
Conceptual Plan: mass of protons → number of electrons

$$\frac{1 \text{ electron}}{0.00091 \times 10^{-27} \text{ kg}}$$

Solution: $1.67262 \times 10^{-27} \cancel{\text{kg}} \times \dfrac{1 \text{ electron}}{0.00091 \times 10^{-27} \cancel{\text{kg}}} = 1.8 \times 10^3$ electrons

Check: The units of the answer, electrons, are correct. The magnitude of the answer is reasonable since the mass of the electron is much less than the mass of the proton.

23. For each of the isotopes: determine Z (the number of protons) from the periodic table then determine A (protons + neutrons). Then, write the symbol in the form ${}^{A}_{Z}X$.

a) The sodium isotope with 12 neutrons: Z = 11; A = 11 + 12 = 23. ${}^{23}_{11}Na$

b) The oxygen isotope with 8 neutrons: Z = 8; A = 8 + 8 = 16. ${}^{16}_{8}O$

c) The aluminum isotope with 14 neutrons: Z = 13; A = 13 + 14 = 27 ${}^{27}_{13}Al$

d) The iodine isotope with 74 neutrons: Z = 53; A = 53 + 74 = 127 ${}^{127}_{53}I$

25. a) ${}^{14}_{7}N$: Z = 7 ; A = 14: protons = Z = 7; neutrons = A – Z = 14 – 7 = 7

b) ${}^{23}_{11}Na$: Z = 11: A = 23: protons = Z = 11; neutrons = A – Z = 23 – 11 = 12

c) ${}^{222}_{86}Rn$: Z = 86: A = 222: protons = Z = 86; neutrons = A – Z = 222 – 86 = 136

d) ${}^{208}_{82}Pb$: Z = 82: A = 208: protons = Z = 82; neutrons = A – Z = 208 – 82 = 126

27. Carbon – 14: A = 14, Z = 6: ${}^{14}_{6}C$ # protons = Z = 6 # neutrons = A – Z = 14 – 6 = 8

29. In a neutral atom the number of protons = the number of electrons = Z. For an ion, electrons are lost (cations) or gained (anions)

a) Ni^{2+}: Z = 28 = protons; Z – 2 = 26 = electrons

b) S^{2-}: Z = 16 = protons; Z + 2 = 18 = electrons

c) Br^{-}: Z = 35 = protons; Z + 1 = 36 = electrons

d) Cr^{3+}: Z = 24 = protons; Z – 3 = 21 = electrons

31. Main group metal atoms will lose electrons to form a cation with the same number of electrons as the nearest, previous noble gas.
Nonmetal atoms will gain electrons to form an anion with the same number of electrons as the nearest noble gas.

a) O^{2-} O is a nonmetal and has 8 electrons. It will gain electrons to form an anion. The nearest noble gas is neon with 10 electrons, so O will gain 2 electrons.

b) K^{+} K is a main group metal and has 19 electrons. It will lose electrons to form a cation. The nearest noble gas is argon with 18 electrons, so K will lose 1 electron.

c) Al^{3+} Al is a main group metal and has 13 electrons. It will lose electrons to form a cation. The nearest noble gas is neon with 10 electrons, so Al will lose 3 electrons.

d) Rb^+ Rb is a main group metal and has 37 electrons. It will lose electrons to form a cation. The nearest noble gas is krypton with 36 electrons, so Rb will lose 1 electron.

33. Main group metal atoms will lose electrons to form a cation with the same number of electrons as the nearest, previous noble gas.
Nonmetal atoms will gain electrons to form an anion with the same number of electrons as the nearest noble gas.

Symbol	Ion Formed	Number of Electrons in Ion	Number of Protons in Ion
Ca	Ca^{2+}	**18**	**20**
Be	Be^{2+}	2	**4**
Se	**Se^{2-}**	**36**	34
In	**In^{3+}**	**46**	49

35. a) Na — Sodium is a metal

b) Mg — Magnesium is a metal

c) Br — Bromine is a nonmetal

d) N — Nitrogen is a nonmetal

e) As — Arsenic is a metalloid

37. a) tellurium — Te is in group 6A and is a main group element

b) potassium — K is in group 1A and is a main group element

c) vanadium — V is in group 5B and is a transition element

d) manganese — Mn is in group 7B and is a transition element

39. a) sodium — Na is in group 1A and is an alkali metal

b) iodine — I is in group 7A and is a halogen

c) calcium — Ca is in group 2A and is an alkaline earth metal

d) barium — Ba is in group 2A and is an alkaline earth metal

e) krypton — Kr is in group 8A and is a noble gas

41. a) N and Ni would not be similar. Nitrogen is a nonmetal, nickel is a metal.

b) Mo and Sn would not be most similar. Although both are metals, molybdenum is a transition metal and tin is a main group metal.

c) Na and Mg would not be similar. Although both are main group metals, sodium is in group 1A and magnesium is in group 2A.

d) Cl and F would be most similar. Chlorine and fluorine are both in group 7A. Elements in the same group have similar chemical properties.

e) Si and P would not be most similar. Silicon is a metalloid and phosphorus is a nonmetal.

43. **Given:** Rb – 85; mass = 84.9118 amu; 72.15%: Rb – 87; mass = 86.9092 amu; 27.85 % **Find:** atomic mass Rb

Conceptual Plan: % abundance → fraction and then find atomic mass

$\frac{\%\text{ abundance}}{100}$ Atomic mass = $\sum_n$ (fraction of isotope n) x (mass of isotope n)

Solution: Fraction Rb - 85 = $\frac{72.15}{100}$ = 0.7215 Fraction Rb - 87 = $\frac{27.85}{100}$ = 0.2785

Atomic mass = $\sum_n$ (fraction of isotope n) x (mass of isotope n)

= 0.7215(84.9116 amu) + 0.2785(86.9092 amu) = 85.47 amu

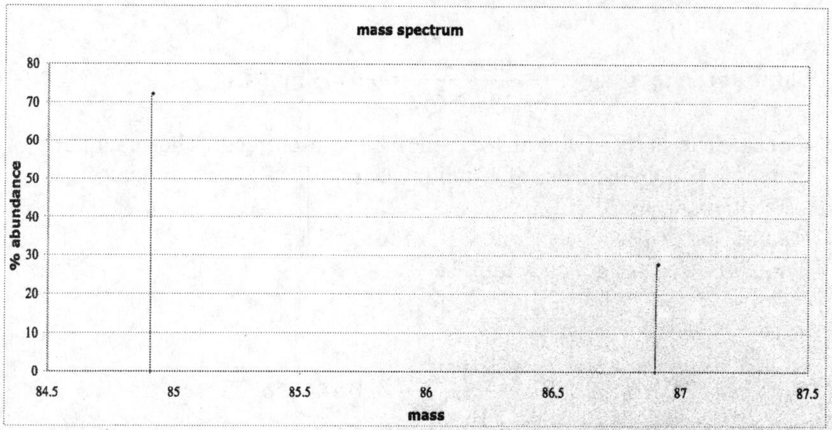

Check: Units of the answer, amu, are correct. The magnitude of the answer is reasonable because it lies between 84.9116 amu and 86.9092 amu and is closer to 84.9118, which has the higher % abundance. The mass spectrum is reasonable because it has two mass lines corresponding to the two isotopes and the line at 84.9116 is about 2.5 times larger than the line at 86.9092.

45. **Given:** Isotope – 1; mass = 120.9038 amu; 57.4%: Isotope – 2; mass = 122.9042 amu;
Find: atomic mass of the atom and identify the atom
Conceptual Plan:
% abundance isotope 2 → and then % abundance → fraction and then find atomic mass

100% – % abundance Isotope 1 $\frac{\%\text{ abundance}}{100}$ Atomic mass = $\sum_n$ (fraction of isotope n) x (mass of isotope n)

Solution: 100.0% - 57.4 % Isotope 1 = 42.6 % Isotope 2

Fraction Isotope 1 = $\frac{57.4}{100}$ = 0.574 Fraction Isotope 2 = $\frac{42.6}{100}$ = 0.426

Atomic mass = $\sum_n$ (fraction of isotope n) x (mass of isotope n) = 0.574(120.9038 amu) + 0.426(122.9042 amu) = 121.8 amu

From the periodic table Sb has a mass of 121.757 amu, so it is the closest mass and the element is antimony.

Check: The units of the answer, amu, are correct. The magnitude of the answer is reasonable because it lies between 120.9038 and 122.9042 and is slightly less than halfway between the two values because the lower value has a slightly greater abundance.

47. **Given:** 3.8 mol sulfur **Find:** atoms of sulfur
Conceptual Plan: mol S → atoms S

$\frac{6.022 \times 10^{23}\text{ atoms}}{\text{mol}}$

Solution: 3.8 $\cancel{\text{mol S}}$ x $\frac{6.022 \times 10^{23}\text{ atoms S}}{\cancel{\text{mol S}}}$ = 2.3 x 10^{24} atoms S

Check: The units of the answer, atoms S, are correct. The magnitude of the answer is reasonable since there is more than 1 mole of material present.

49. a) **Given:** 11.8 g Ar **Find:** mol Ar
Conceptual Plan: g Ar → mol Ar

$\frac{1\text{ mol Ar}}{39.948\text{ g Ar}}$

Solution: $11.8 \text{ g Ar} \times \dfrac{1 \text{ mol Ar}}{39.948 \text{ g Ar}} = 0.295 \text{ mol Ar}$

Check: The units of the answer, mol Ar, are correct. The magnitude of the answer is reasonable since there is less than the mass of 1 mol present.

b) **Given:** 3.55 g Zn **Find:** mol Zn
Conceptual Plan: g Zn → mol Zn

$\dfrac{1 \text{mol Zn}}{65.39 \text{g Zn}}$

Solution: $3.55 \text{ g Zn} \times \dfrac{1 \text{ mol Zn}}{65.39 \text{ g Zn}} = 0.0543 \text{ mol Zn}$

Check: The units of the answer, mol Zn, are correct. The magnitude of the answer is reasonable since there is less than the mass of 1 mol present.

c) **Given:** 26.1 g Ta **Find:** mol Ta
Conceptual Plan: g Ta → mol Ta

$\dfrac{1 \text{mol Ta}}{180.948 \text{g Ta}}$

Solution: $26.1 \text{ g Ta} \times \dfrac{1 \text{ mol Ta}}{180.948 \text{ g Ta}} = 0.144 \text{ mol Ta}$

Check: The units of the answer, mol Ta, are correct. The magnitude of the answer is reasonable since there is less than the mass of 1 mol present.

d) **Given:** 0.211 g Li **Find:** mol Li
Conceptual Plan: g Li → mol Li

$\dfrac{1 \text{mol Li}}{6.941 \text{g Li}}$

Solution: $0.211 \text{ g Li} \times \dfrac{1 \text{ mol Li}}{6.941 \text{ g Li}} = 0.0304 \text{ mol Li}$

Check: The units of the answer, mol Li, are correct. The magnitude of the answer is reasonable since there is less than the mass of 1 mol present.

51. **Given:** 3.78 g silver **Find:** atoms Ag
Conceptual Plan: g Ag → mol Ag → atoms Ag

$\dfrac{1 \text{ mol Ag}}{107.868 \text{g Ag}}$ $\dfrac{6.022 \times 10^{23} \text{ atoms}}{\text{mol}}$

Solution: $3.78 \text{ g Ag} \times \dfrac{1 \text{ mol Ag}}{107.868 \text{ g Ag}} \times \dfrac{6.022 \times 10^{23} \text{ atoms Ag}}{1 \text{ mol Ag}} = 2.11 \times 10^{22} \text{ atoms Ag}$

Check: The units of the answer, atoms Ag, are correct. The magnitude of the answer is reasonable since there is less than the mass of 1 mol of Ag present.

53. a) **Given:** 5.18 g P **Find:** atoms P
Conceptual Plan: g P → mol P → atoms P

$\dfrac{1 \text{ mol P}}{30.9738 \text{g P}}$ $\dfrac{6.022 \times 10^{23} \text{ atoms}}{\text{mol}}$

Solution: $5.18 \text{ g P} \times \dfrac{1 \text{ mol P}}{30.9738 \text{ g P}} \times \dfrac{6.022 \times 10^{23} \text{ atoms P}}{1 \text{ mol P}} = 1.01 \times 10^{23} \text{ atoms P}$

Check: The units of the answer, atoms P, are correct. The magnitude of the answer is reasonable since there is slightly less than the mass of 1 mol of P present.

b) **Given:** 2.26 g Hg **Find:** atoms Hg
Conceptual Plan: **g Hg → mol Hg → atoms Hg**

$$\frac{1 \text{ mol Hg}}{200.59 \text{ g Hg}} \quad \frac{6.022 \times 10^{23} \text{ atoms}}{\text{mol}}$$

Solution: $$2.26 \text{ g Hg} \times \frac{1 \text{ mol Hg}}{200.59 \text{ g Hg}} \times \frac{6.022 \times 10^{23} \text{ atoms Hg}}{1 \text{ mol Hg}} = 6.78 \times 10^{21} \text{ atoms Hg}$$

Check: The units of the answer, atoms Hg, are correct. The magnitude of the answer is reasonable since there is less than the mass of 1 mol of Hg present.

c) **Given:** 1.87 g Bi **Find:** atoms Bi
Conceptual Plan: **g Bi → mol Bi → atoms Bi**

$$\frac{1 \text{ mol Bi}}{208.98 \text{ g Bi}} \quad \frac{6.022 \times 10^{23} \text{ atoms}}{\text{mol}}$$

Solution: $$1.87 \text{ g Bi} \times \frac{1 \text{ mol Bi}}{208.98 \text{ g Bi}} \times \frac{6.022 \times 10^{23} \text{ atoms Bi}}{1 \text{ mol Bi}} = 5.39 \times 10^{21} \text{ atoms Bi}$$

Check: The units of the answer, atoms Bi, are correct. The magnitude of the answer is reasonable since there is less than the mass of 1 mol of Bi present.

d) **Given:** 0.082 g Sr **Find:** atoms Sr
Conceptual Plan: **g Sr → mol Sr → atoms Sr**

$$\frac{1 \text{ mol Sr}}{87.62 \text{ g Sr}} \quad \frac{6.022 \times 10^{23} \text{ atoms}}{\text{mol}}$$

Solution: $$0.082 \text{ g Sr} \times \frac{1 \text{ mol Sr}}{87.62 \text{ g Sr}} \times \frac{6.022 \times 10^{23} \text{ atoms Sr}}{1 \text{ mol Sr}} = 5.6 \times 10^{20} \text{ atoms Sr}$$

Check: The units of the answer, atoms Sr, are correct. The magnitude of the answer is reasonable since there is less than the mass of 1 mol of Sr present.

55. **Given:** 52 mg diamond (carbon) **Find:** atoms C
Conceptual Plan: **mg C → g C → mol C → atoms C**

$$\frac{1 \text{ g C}}{1000 \text{ mg C}} \quad \frac{1 \text{ mol C}}{12.011 \text{ g C}} \quad \frac{6.022 \times 10^{23} \text{ atoms}}{\text{mol}}$$

Solution: $$52 \text{ mg C} \times \frac{1 \text{ g C}}{1000 \text{ mg C}} \times \frac{1 \text{ mol C}}{12.011 \text{ g C}} \times \frac{6.022 \times 10^{23} \text{ atoms C}}{1 \text{ mol C}} = 2.6 \times 10^{21} \text{ atoms C}$$

Check: The units of the answer, atoms C, are correct. The magnitude of the answer is reasonable since there is less than the mass of 1 mol of C present.

57. **Given:** 1 atom platinum **Find:** g Pt
Conceptual Plan: **atoms Pt → mol Pt → g Pt**

$$\frac{1 \text{ mol}}{6.022 \times 10^{23} \text{ atoms}} \quad \frac{195.08 \text{ g Pt}}{1 \text{ mol Pt}}$$

Solution: $$1 \text{ atom Pt} \times \frac{1 \text{ mol Pt}}{6.022 \times 10^{23} \text{ atoms Pt}} \times \frac{195.08 \text{ g Pt}}{1 \text{ mol Pt}} = 3.239 \times 10^{-22} \text{ g Pt}$$

Check: The units of the answer, g Pt, are correct. The magnitude of the answer is reasonable since there is only 1 atom in the sample.

59. **Given:** 7.83 g HCN sample 1: 0.290 g H; 4.06 g N. 3.37 g HCN sample 2 **Find:** g C in sample 2
Conceptual Plan:
g HCN sample 1 → g C in HCN sample 1 → ratio g C to g HCN → g C in HCN sample 2

$$\text{g HCN - g H - g N} \qquad \frac{\text{g C}}{\text{g HCN}} \qquad \text{g HCN} \times \frac{\text{g C}}{\text{g HCN}}$$

Solution: $$7.83 \text{ g HCN} - 0.290 \text{ g H} - 4.06 \text{ g N} = 3.48 \text{ g C}$$

$$3.37 \text{ g HCN} \times \frac{3.48 \text{ g C}}{7.83 \text{ g HCN}} = 1.50 \text{ g C}$$

Check: The units of the answer, g C, are correct. The magnitude of the answer is reasonable since the sample size is about half the original sample size, the g C are about half the original g C.

61. **Given:** In CO mass ratio O:C = 1.33:1; In compound X, mass ratio O:C = 2:1. **Find:** formula of X
Conceptual Plan: determine the mass ratio of O:O in the 2 compounds.

Solution: For 1 gram of C: $\frac{2\text{ g O in compound X}}{1.33\text{ g O in CO}} = 1.5$

So, the ratio of O to C in compound X has to be 1.5:1 and the formula is C_2O_3.

Check: The answer is reasonable since it fulfills the criteria of multiple proportions and the mass ratio of O:C is 2:1.

63. **Given:** $^4He^{2+} = 4.00151$ amu; **Find:** charge to mass ratio C/kg
Conceptual Plan: determine total charge on $^4He^{2+}$ and then amu $^4He^{2+} \rightarrow$ g $^4He^{2+} \rightarrow$ kg $^4He^{2+}$

$$\frac{+1.60218 \times 10^{-19}\text{ C}}{\text{proton}} \qquad \frac{1\text{ g}}{1.66054 \times 10^{-24}\text{ amu}} \quad \frac{1\text{ kg}}{1000\text{ g}}$$

Solution:

$$\frac{2\text{ protons}}{1\text{ atom }^4He^{2+}} \times \frac{+1.60218 \times 10^{-19}\text{ C}}{\text{proton}} = \frac{3.20436 \times 10^{-19}\text{ C}}{\text{atom }^4He^{2+}}$$

$$\frac{4.00151\text{ amu}}{1\text{ atom }^4He^{2+}} \times \frac{1.66054 \times 10^{-24}\text{ g}}{1\text{ amu}} \times \frac{1\text{ kg}}{1000\text{ g}} = \frac{6.644\underline{6}6742 \times 10^{-27}\text{ kg}}{1\text{ atom }^4He^{2+}}$$

$$\frac{3.20436 \times 10^{-19}\text{ C}}{\text{atom }^4He^{2+}} \times \frac{1\text{ atom }^4He^{2+}}{6.644\underline{6}6742 \times 10^{-27}\text{ kg}} = 4.82245 \times 10^{7}\text{ C/kg}$$

Check: the units of the answer, C/kg, are correct. The magnitude of the answer is reasonable when compared to the charge to mass ratio of the electron.

65. $^{236}_{90}Th$ A – Z = number of neutrons. 236 – 90 = 146 neutrons. So, any nucleus with 146 neutrons is an isotone of $^{236}_{90}Th$.

Some would be: $^{238}_{92}U$; $^{239}_{93}Np$; $^{241}_{95}Am$; $^{237}_{91}Pa$; $^{235}_{89}Ac$; $^{244}_{98}Cf$ etc.

67.

Symbol	Z	A	Number protons	Number electrons	Number neutrons	Charge
O^{2-}	8	**16**	**8**	**10**	8	2 –
Ca^{2+}	20	**40**	**20**	**18**	20	**2+**
Mg^{2+}	**12**	25	**12**	**10**	13	2+
N^{3-}	**7**	14	**7**	10	**7**	**3 –**

69. **Given:** r(nucleus) = 2.7 fm; r(atom) = 70 pm (assume 2 significant figures)
Find: vol(nucleus); vol(atom), % vol(nucleus)
Conceptual Plan:

r(nucleus)(fm) → r(nucleus)(pm) → vol(nucleus) and then r(atom) → vol(atom) and then % vol

$$\frac{10^{-15}\text{ m}}{1\text{ fm}} \quad \frac{1\text{ pm}}{10^{-12}\text{ m}} \qquad V = \frac{4}{3}\pi r^3 \qquad V = \frac{4}{3}\pi r^3 \qquad \frac{\text{vol(nucleus)}}{\text{vol(atom)}} \times 100$$

Solution: $2.7\text{ fm} \times \frac{10^{-15}\text{ m}}{\text{fm}} \times \frac{1\text{ pm}}{10^{-12}\text{ m}} = 2.7 \times 10^{-3}\text{pm}$ $\qquad V_{nucleus} = \frac{4}{3}\pi\,(2.7 \times 10^{-3}\text{pm})^3 = 8.2 \times 10^{-8}\text{ pm}^3$

$V_{atom} = \frac{4}{3}\pi\,(70\text{ pm})^3 = 1.4 \times 10^{6}\text{ pm}^3$ $\qquad \frac{8.2 \times 10^{-8}\text{ pm}^3}{1.4 \times 10^{6}\text{ pm}^3} \times 100\% = 5.9 \times 10^{-12}\%$

Check: The units of the answer, % vol, are correct. The magnitude of the answer is reasonable because the nucleus only occupies a very small % of the vol of the atom.

71. **Given:** 6.022×10^{23} pennies **Find:** the amount in dollars; the dollars/person
Conceptual Plan: pennies → dollars → dollars/person

$$\frac{1\,\text{dollar}}{100\,\text{pennies}} \qquad 6.5\text{ billion people}$$

Solution:

$$6.022 \times 10^{23}\ \text{pennies} \times \frac{1\,\text{dollar}}{100\ \text{pennies}} = 6.022 \times 10^{21}\,\text{dollars} \qquad \frac{6.022 \times 10^{21}\ \text{dollars}}{6.5 \times 10^{9}\ \text{people}} = 9.3 \times 10^{11}\,\text{dollars / person},$$

They are billionaires

73. **Given:** O = 15.9994 amu when C = 12.011 amu **Find:** mass O when C = 12.00 amu
Conceptual Plan: determine ratio O:C for ^{12}C system then use the same ratio when C = 12.00

$$\frac{\text{mass O}}{\text{mass C}}$$

Solution: Based on ^{12}C = 12.00; O = 15.9994 and C = 12.011 so, $\frac{\text{mass O}}{\text{mass C}} = \frac{15.9994\,\text{amu}}{12.011\,\text{amu}} = \frac{1.33206\,\text{amu O}}{1\,\text{amu C}}$

Based on C = 12.00, the ratio has to be the same,

$$12.000\ \text{amu C} \times \frac{1.33206\ \text{amu O}}{1\ \text{amu C}} = 15.985\ \text{amu O}$$

Check: The units of the answer, amu O, are correct. The magnitude of the answer is reasonable because the value for the new mass basis is smaller then the original mass basis, therefore, the mass of O should be less.

75. **Given:** Cu sphere: r = 0.935 in; d = 8.96 g/cm^3 **Find:** number of Cu atoms
Conceptual Plan: r in inch → r in cm → vol sphere → g Cu → mol Cu → atoms Cu

$$\frac{2.54\,\text{cm}}{1\,\text{inch}} \qquad V = \frac{4}{3}\pi r^3 \qquad \frac{8.96\,\text{g}}{\text{cm}^3} \qquad \frac{1\,\text{mol Cu}}{63.546\,\text{g}} \qquad \frac{6.022 \times 10^{23}\ \text{atoms}}{\text{mol}}$$

Solution: $0.935\ \text{in} \times \frac{2.54\,\text{cm}}{\text{in}} = 2.3\underline{7}49\,\text{cm}$

$$\frac{4}{3}\pi(2.3\underline{7}49\ \text{cm})^3 \times \frac{8.96\ \text{g}}{\text{cm}^3} \times \frac{1\ \text{mol Cu}}{63.546\ \text{g}} \times \frac{6.022 \times 10^{23}\ \text{atoms Cu}}{1\ \text{mol Cu}} = 4.76 \times 10^{24}\ \text{atoms Cu}$$

Check: The units of the answer, atoms Cu, are correct. The magnitude of the answer is reasonable because there are about 8 mol Cu present.

77. **Given:** Li – 6 = 6.01512 amu; Li – 7 = 7.01601 amu; B = 6.941 amu
Find: % abundance Li – 6 and Li – 7
Conceptual Plan: Let x = fraction Li – 6 then 1 – x = fraction Li – 7 → abundances

$$\text{Atomic mass} = \sum_n (\text{fraction of isotope n}) \times (\text{mass of isotope n})$$

Solution:

$$\text{Atomic mass} = \sum_n (\text{fraction of isotope n}) \times (\text{mass of isotope n})$$

$$6.941 = (x)(6.01512\ \text{amu}) + (1 - x)(7.01601\ \text{amu})$$

$$0.07501 = 1.00089\ x$$

$$x = 0.07494 \qquad 1 - x = 0.92506$$

Li – 6 = 0.07494 x 100 = 7.494 % and Li – 7 = 0.92506 x 100 = 92.506 %

Check: The units of the answer, %, which gives the relative abundance of each isotope, are correct. The relative abundances are reasonable because Li has an atomic mass closer to the mass of Li – 7 than to Li – 6.

79. **Given:** sun: d = 1.4 g/cm^3;r = 7 x 10^8 m; 100 billion stars/galaxy; 10 billion galaxies/universe
Find: number of atoms in the universe
Conceptual Plan: r (star) in m → r (star) in cm → vol (star) → g H/star → mol H star → atoms H/star

$$\frac{100\,\text{cm}}{\text{m}} \qquad V = \frac{4}{3}\pi r^3 \qquad \frac{1.4\,\text{g H}}{\text{cm}^3} \qquad \frac{1\,\text{mol H}}{1.008\,\text{g}} \qquad \frac{6.022 \times 10^{23}\ \text{atoms}}{\text{mol}}$$

→ atoms H/galaxy → atoms H/universe

$\frac{100 \times 10^9 \text{ stars}}{\text{galaxy}}$ $\frac{10 \times 10^9 \text{ galaxies}}{\text{universe}}$

Solution: $7 \times 10^8 \not{m} \times \frac{100 \text{ cm}}{\not{m}} = 7 \times 10^{10}$ cm

$$\frac{4}{3}\pi \frac{(7 \times 10^{10} \not{cm})^3}{\not{star}} \times \frac{1.4 \not{g H}}{\not{cm}^3} \times \frac{1 \not{mol H}}{1.008 \not{g H}} \times \frac{6.022 \times 10^{23} \not{atoms H}}{\not{mol H}} \times \frac{100 \times 10^9 \not{stars}}{\not{galaxy}} \times \frac{10 \times 10^9 \not{galaxies}}{\text{universe}}$$

$= 1 \times 10^{78}$ atoms/universe

Check: The units of the answer, atoms/universe, are correct.

81. a) This is the Law of Definite Proportions: All samples of a given compound, regardless of their source or how they were prepared, have the same proportions of their constituent elements.

b) This is the Law of Conservation of Mass: In a chemical reaction, matter is neither created nor destroyed.

c) This is the Law of Multiple Proportions: When two elements form two different compounds, the masses of element B that combine with 1 g of element A can be expressed as a ratio of small whole numbers. In this example the ratio of O from hydrogen peroxide to O from water = 16:8 → 2:1, a small whole number ratio.

83. **Given:** a. Cr: 55.0 g; atomic mass = 52 g/mol b. Ti: 45.0 g; atomic mass = 48 g/mol and c. Zn: 60.0 g; atomic mass = 65 g/mol

Find: which has the greatest mol, and which has the greatest mass

Conceptual Plan: without calculation, compare grams of material to g/mol for each.

Solution: Cr would have the greatest mole amount of the elements. It is the only one whose mass is greater than the molar mass. Zn would be the greatest mass amount because it is the largest mass value.

Chapter 3
Molecules, Compounds, and Chemical Equations

1. The chemical formula gives you the kind of atom and the number of each atom in the compound.
   a) $Ca_3(PO_4)_2$ contains: 3 calcium atoms, 2 phosphorus atoms, and 8 oxygen atoms

   b) $SrCl_2$ contains: 1 strontium atom and 2 chlorine atoms

   c) KNO_3 contains: 1 potassium atom, 1 nitrogen atom, and 3 oxygen atoms

   d) $Mg(NO_2)_2$ contains: 1 magnesium atom, 2 nitrogen atoms, and 4 oxygen atoms

3. a) 1 blue = nitrogen, 3 white = hydrogen: NH_3

   b) 2 black = carbon, 6 white = hydrogen: C_2H_6

   c) 1 yellow – green = sulfur, 3 red = oxygen: SO_3

5. a) Neon is an element and it is not one of the elements that exist as diatomic molecules, therefore it is an atomic element.

   b) Fluorine is one of the elements that exist as diatomic molecules, therefore it is a molecular element.

   c) Potassium is not one of the elements that exist as diatomic molecules, therefore it is an atomic element.

   d) Nitrogen is one of the elements that exist as diatomic molecules, therefore it is a molecular element.

7. a) CO_2 is a compound composed of a nonmetal and a nonmetal, therefore it is a molecular compound.

   b) $NiCl_2$ is a compound composed of a metal and a nonmetal, therefore it is an ionic compound.

   c) NaI is a compound composed of a metal and a nonmetal, therefore it is an ionic compound.

   d) PCl_3 is a compound composed of a nonmetal and a nonmetal, therefore it is a molecular compound.

9. a) white – hydrogen: a molecule composed of two of the same element, therefore it is a molecular element.

   b) blue – nitrogen, white – hydrogen: a molecule composed of a nonmetal and a nonmetal, therefore it is a molecular compound.

   c) purple – sodium: a substance composed of all the same atoms, therefore it is an atomic element.

11. To write the formula for an ionic compound: 1) Write the symbol for the metal cation and its charge and the symbol for the nonmetal anion and its charge. 2) Adjust the subscript on each cation and anion to balance the overall charge. 3) Check that the sum of the charges of the cations equals the sum of the charges of the anions.
    a) magnesium and sulfur: Mg^{2+} S^{2-} MgS cations 2+; anions 2-

    b) barium and oxygen: Ba^{2+} O^{2-} BaO cations 2+; anions 2-

    c) strontium and bromine: Sr^{2+} Br^{-} $SrBr_2$ cation 2+; anions 2(1-) = 2-

    d) beryllium and chlorine: Be^{2+} Cl^{-} $BeCl_2$ cations 2+; anions 2(1-) = 2-

13. To write the formula for an ionic compound: 1) Write the symbol for the metal cation and its charge and the symbol for the polyatomic anion and its charge. 2) Adjust the subscript on each cation and anion to balance the overall charge. 3) Check that the sum of the charges of the cations equals the sum of the charges of the anions.
    Cation = barium: Ba^{2+}
    a) hydroxide: OH^{-} $Ba(OH)_2$ cation 2+, anion 2(1-) = 2-

b) chromate: CrO_4^{2-} $BaCrO_4$ cation 2+; anion 2-

c) phosphate: PO_4^{3-} $Ba_3(PO_4)_2$ cation 3(2+)=6+; anion 2(3-) = 6-

d) cyanide: CN^- $Ba(CN)_2$ cation 2+; anion 2(1-) = 2-

15. To name a binary ionic compound: name the metal cation followed by the base name of the anion + ide.
    a) Mg_3N_2: the cation is magnesium, the anion is from nitrogen which becomes nitride: magnesium nitride.

    b) KF: the cation is potassium, the anion is from fluorine which becomes fluoride: potassium fluoride.

    c) Na_2O: the cation is sodium, the anion is from oxygen which becomes oxide: sodium oxide.

    d) Li_2S: the cation is lithium, the anion is from sulfur which becomes sulfide: lithium sulfide.

17. To name an ionic compound with a metal cation that can have more than one charge: name the metal cation followed by parentheses with the charge in roman numerals followed by the base name of the anion + ide.
    a) $SnCl_4$: the charge on Sn must be 4+ for the compound to be charge neutral: the cation is tin(IV), the anion is from chlorine which becomes chloride; tin(IV) chloride.

    b) PbI_2: the charge on Pb must be 2+ for the compound to be charge neutral: the cation is lead(II), the anion is from iodine which becomes iodide; lead(II) iodide.

    c) Fe_2O_3: the charge on Fe must be 3+ for the compound to be charge neutral; the cation is iron(III), the anion is from oxygen which becomes oxide: iron(III) oxide.

    d) CuI_2: the charge on Cu must be 2+ for the compound to be charge neutral; the cation is copper(II), the anion is from iodine which becomes iodide: copper(II) iodide.

19. To name these compounds you must first decide if the metal cation is invariant or can have more than one charge. Then, name the metal cation followed by the base name of the anion + ide.
    a) SnO: Sn can have more than one charge. The charge on Sn must be 2+ for the compound to be charge neutral: the cation is tin(II), the anion is from oxygen which becomes oxide: tin(II) oxide.

    b) Cr_2S_3: Cr can have more than one charge. The charge on Cr must be 3+ for the compound to be charge neutral: the cation is chromium(III), the anion is from sulfur which becomes sulfide: chromium(III) sulfide.

    c) RbI: Rb is invariant. The cation is rubidium, the anion is from iodine which becomes iodide: rubidium iodide.

    d) $BaBr_2$: Ba is invariant. The cation is barium, the anion is from bromine which becomes bromide: barium bromide.

21. To name these compounds you must first decide if the metal cation is invariant or can have more than one charge. Then, name the metal cation followed by the name of the polyatomic anion.
    a) $CuNO_2$: Cu can have more than one charge. The charge on Cu must be 1+ for the compound to be charge neutral. The cation is copper(I), the anion is nitrite: copper(I) nitrite

    b) $Mg(C_2H_3O_2)_2$: Mg is invariant. The cation is magnesium, the anion is acetate: magnesium acetate.

    c) $Ba(NO_3)_2$: Ba is invariant. The cation is barium, the anion is nitrate: barium nitrate.

    d) $Pb(C_2H_3O_2)_2$: Pb can have more than one charge. The charge on Pb must be 2+ for the compound to be charge neutral. The cation is lead(II), the anion is acetate: lead(II) acetate.

    e) $KClO_3$: K is invariant. The cation is potassium, the anion is chlorate: potassium chlorate.

f) $PbSO_4$: Pb can have more than one charge. The charge on Pb must be 2+ for the compound to be charge neutral. The cation is lead(II), the anion is sulfate: lead(II) sulfate.

23. To write the formula for an ionic compound: 1) Write the symbol for the metal cation and its charge and the symbol for the nonmetal anion or polyatomic anion and its charge. 2) Adjust the subscript on each cation and anion to balance the overall charge. 3) Check that the sum of the charges of the cations equals the sum of the charges of the anions.

a) sodium hydrogen sulfite:	Na^+	HSO_3^-	$NaHSO_3$	cation 1+; anion 1-
b) lithium permanganate:	Li^+	MnO_4^-	$LiMnO_4$	cation 1+; anion 1-
c) silver nitrate:	Ag^+	NO_3^-	$AgNO_3$	cation 1+; anion 1-
d) potassium sulfate:	K^+	SO_4^{2-}	K_2SO_4	cation 2(1+) = 2+; anion 2-
e) rubidium hydrogen sulfate:	Rb^+	HSO_4^-	$RbHSO_4$	cation 1+; anion 1-
f) potassium hydrogen carbonate:	K^+	HCO_3^-	$KHCO_3$	cation 1+; anion 1-

25. Hydrates are named the same way as other ionic compounds with the addition of the term *prefix*hydrate, where the prefix is the number of water molecules associated with each formula unit.

a) $CoSO_4 \bullet 7H_2O$ — cobalt(II) sulfate heptahydrate

b) iridium(III) bromide tetrahydrate — $IrBr_3 \bullet 4H_2O$

c) $Mg(BrO_3)_2 \bullet 6H_2O$ — magnesium bromate hexahydrate

d) potassium carbonate dihydrate — $K_2CO_3 \bullet 2H_2O$

27. a) CO — The name of the compound is the name of the first element, *carbon*, following by the base name of the second element, *ox*, prefixed by *mono-* to indicate one and given the suffix – *ide*; carbon monoxide.

b) NI_3 — The name of the compound is the name of the first element, *nitrogen*, followed by the base name of the second element, *iod*, prefixed by *tri-* to indicate three and given the suffix – *ide*: nitrogen triiodide.

c) $SiCl_4$ — The name of the compound is the name of the first element, *silicon*, followed by the base name of the second element, *chlor*, prefixed by *tetra-* to indicate four and given the suffix – *ide*: silicon tetrachloride.

d) N_4Se_4 — The name of the compound is the name of the first element, *nitrogen*, prefixed by *tetra-* to indicate four followed by the base name of the second element, *selen*, prefixed by *tetra-* to indicate four and given the suffix – *ide*: tetranitrogen tetraselenide.

e) I_2O_5 — The name of the compound is the name of the first element, *iodine*, prefixed by di to indicate two followed by the base name of the second element, *ox*, prefixed by *penta-* to indicate five and given the suffix – *ide*: diiodine pentaoxide.

29. a) phosphorus trichloride: PCl_3

b) chlorine monoxide: ClO

c) disulfur tetrafluoride: S_2F_4

d) phosphorus pentafluoride: PF_5

e) diphosphorus pentasulfide: P_2S_5

31\. a) HI: the base name of I is *iod* so the name is hydroiodic acid

b) HNO_3: the oxyanion is *nitrate,* which ends in *–ate;* therefore, the name of the acid is nitric acid.

c) H_2CO_3: the oxyanion is *carbonate,* which ends in *–ate;* therefore, the name of the acid is carbonic acid.

d) $HC_2H_3O_2$: the oxyanion is *acetate,* which ends in *–ate;* therefore, the name of the acid is acetic acid.

33\. a) hydrofluoric acid: HF

b) hydrobromic acid: HBr

c) sulfurous acid: H_2SO_3

35\. To find the formula mass, we sum the atomic masses of each atom in the chemical formula.

a) NO_2 formula mass = 1 x (atomic mass N) + 2 x (atomic mass O)
= 1 x (14.01 amu) + 2 x (16.00 amu)
= 46.01 amu

b) C_4H_{10} formula mass = 4 x (atomic mass C) + 10 x (atomic mass H)
= 4 x (12.01 amu) + 10 x (1.008 amu)
= 58.12 amu

c) $C_6H_{12}O_6$ formula mass = 6 x (atomic mass C) + 12 x (atomic mass H) + 6 x (atomic mass O)
= 6 x (12.01 amu) + 12 x (1.008 amu) + 6 x (16.00 amu)
= 180.16 amu

d) $Cr(NO_3)_3$ formula mass = 1 x (atomic mass Cr) + 3 x (atomic mass N) + 9 x (atomic mass O)
= 1 x (52.00 amu) + 3 x (14.01 amu) + 9 x (16.00 amu)
= 238.0 amu

37\. a) **Given:** 6.5 g H_2O **Find:** number of molecules

Conceptual Plan: **g H_2O → mole H_2O → number H_2O molecules**

$$\frac{1 \text{ mol}}{18.02 \text{ g } H_2O} \qquad \frac{6.022 \times 10^{23} \ H_2O \text{ molecules}}{\text{mol } H_2O}$$

Solution: $$6.5 \ \cancel{\text{g } H_2O} \times \frac{1 \ \cancel{\text{mol } H_2O}}{18.02 \ \cancel{\text{g } H_2O}} \times \frac{6.022 \times 10^{23} \ H_2O \text{ molecules}}{\cancel{\text{mol } H_2O}} = 2.2 \times 10^{23} \ H_2O \text{ molecules}$$

Check: Units of the answer, H_2O molecules, are correct. The magnitude is appropriate because it is smaller than Avogadro's number as expected since we have less than 1 mole of H_2O.

b) **Given:** 389 g CBr_4 **Find:** number of molecules

Conceptual Plan: **g CBr_4 → mole CBr_4 → number CBr_4 molecules**

$$\frac{1 \text{ mol}}{331.6 \text{ g } CBr_4} \qquad \frac{6.022 \times 10^{23} \ CBr_4 \text{ molecules}}{\text{mol } CBr_4}$$

Solution: $$389 \ \cancel{\text{g } CBr_4} \times \frac{1 \ \cancel{\text{mol } CBr_4}}{331.6 \ \cancel{\text{g } CBr_4}} \times \frac{6.022 \times 10^{23} \ CBr_4 \text{ molecules}}{\cancel{\text{mol } CBr_4}} = 7.06 \times 10^{23} \ CBr_4 \text{ molecules}$$

Check: Units of the answer, CBr_4 molecules, are correct. The magnitude is appropriate because it is large than Avogadro's number as expected since we have more than 1 mole of CBr_4.

c) **Given:** 22.1 g O_2**Find:** number of molecules

Conceptual Plan: **g O_2 → mole O_2 → number O_2 molecules**

$$\frac{1 \text{ mol}}{32.00 \text{ g } O_2} \qquad \frac{6.022 \times 10^{23} \ O_2 \text{ molecules}}{\text{mol } O_2}$$

Solution: $22.1\ \text{g O}_2 \times \frac{1\ \text{mol O}_2}{32.00\ \text{g O}_2} \times \frac{6.022 \times 10^{23}\ \text{O}_2\ \text{molecules}}{\text{mol O}_2} = 4.16 \times 10^{23}\ \text{O}_2\ \text{molecules}$

Check: Units of the answer, O_2 molecules, are correct. The magnitude is appropriate because it is smaller than Avogadro's number as expected since we have less than 1 mole of O_2.

d) **Given:** 19.3 g C_8H_{10} **Find:** number of molecules

Conceptual Plan: **g C_8H_{10} → mole C_8H_{10} → number C_8H_{10} molecules**

$$\frac{1\ \text{mol}}{106.16\ \text{g C}_8\text{H}_{10}} \qquad \frac{6.022 \times 10^{23}\ \text{C}_8\text{H}_{10}\ \text{molecules}}{\text{mol C}_8\text{H}_{10}}$$

Solution:

$$19.3\ \text{g C}_8\text{H}_{10} \times \frac{1\ \text{mol C}_8\text{H}_{10}}{106.16\ \text{g C}_8\text{H}_{10}} \times \frac{6.022 \times 10^{23}\ \text{C}_8\text{H}_{10}\ \text{molecules}}{\text{mol C}_8\text{H}_{10}} = 1.09 \times 10^{23}\ \text{C}_8\text{H}_{10}\ \text{molecules}$$

Check: Units of the answer, C_8H_{10} molecules, are correct. The magnitude is appropriate because it is smaller than Avogadro's number as expected since we have less than 1 mole of C_8H_{10}.

39. **Given:** 1 H_2O molecule **Find:** mass in g

Conceptual Plan: **number H_2O molecules → mole H_2O → g H_2O**

$$\frac{1\ \text{mol H}_2\text{O}}{6.022 \times 10^{23}\ \text{H}_2\text{O molecules}} \qquad \frac{18.02\ \text{g H}_2\text{O}}{1\ \text{mol H}_2\text{O}}$$

Solution: $1\ \text{H}_2\text{O molecule} \times \frac{1\ \text{mol H}_2\text{O}}{6.022 \times 10^{23}\ \text{H}_2\text{O molecules}} \times \frac{18.02\ \text{g H}_2\text{O}}{1\ \text{mol H}_2\text{O}} = 2.992 \times 10^{-23}\ \text{g H}_2\text{O}$

Check: Units of the answer, grams H_2O, are correct. The magnitude is appropriate because there is much less than Avogadro's number of molecules so we have much less than 1 mole of H_2O.

41. **Given:** 1.8×10^{17} $C_{12}H_{22}O_{11}$ molecule **Find:** mass in mg

Conceptual Plan: **number $C_{12}H_{22}O_{11}$ molecules → mole $C_{12}H_{22}O_{11}$ → g $C_{12}H_{22}O_{11}$ → mg $C_{12}H_{22}O_{11}$**

$$\frac{1\ \text{mol C}_{12}\text{H}_{22}\text{O}_{11}}{6.022 \times 10^{23}\ \text{C}_{12}\text{H}_{22}\text{O}_{11}\ \text{molecules}} \qquad \frac{342.3\ \text{g C}_{12}\text{H}_{22}\text{O}_{11}}{1\ \text{mol C}_{12}\text{H}_{22}\text{O}_{11}} \qquad \frac{1 \times 10^3\ \text{mg C}_{12}\text{H}_{22}\text{O}_{11}}{1\ \text{g C}_{12}\text{H}_{22}\text{O}_{11}}$$

Solution:

$$1.8 \times 10^{17}\ \text{C}_{12}\text{H}_{22}\text{O}_{11}\ \text{molecules} \times \frac{1\ \text{mol C}_{12}\text{H}_{22}\text{O}_{11}}{6.022 \times 10^{23}\ \text{C}_{12}\text{H}_{22}\text{O}_{11}\ \text{molecules}} \times \frac{342.3\ \text{g C}_{12}\text{H}_{22}\text{O}_{11}}{1\ \text{mol C}_{12}\text{H}_{22}\text{O}_{11}} \times \frac{1 \times 10^3\ \text{mg C}_{12}\text{H}_{22}\text{O}_{11}}{1\ \text{g C}_{12}\text{H}_{22}\text{O}_{11}}$$

$= 0.10\ \text{mg C}_{12}\text{H}_{22}\text{O}_{11}$

Check: Units of the answer, milligrams $C_{12}H_{22}O_{11}$, are correct. The magnitude is appropriate because there is much less than Avogadro's number of molecules so we have much less than 1 mole of $C_{12}H_{22}O_{11}$.

43. a) **Given:** CH_4 **Find:** mass percent C

Conceptual Plan: $\text{mass } \%\text{C} = \frac{1 \times \text{molar mass C}}{\text{molar mass CH}_4} \times 100$

Solution:

$$1 \times \text{molar mass C} = 1(12.01\text{g/mol}) = 12.01\ \text{g C}$$

$$\text{molar mass CH}_4 = 1(12.01\ \text{g/mol}) + 4(1.008\ \text{g/mol}) = 16.04\ \text{g/mol}$$

$$\text{mass } \%\ \text{C} = \frac{1 \times \text{molar mass C}}{\text{molar mass CH}_4} \times 100\%$$

$$= \frac{12.01\ \text{g/mol}}{16.04\ \text{g/mol}} \times 100\%$$

$$= 74.87\ \%$$

Check: Units of the answer, %, are correct. The magnitude is reasonable because it is between 0 and 100% and carbon is the heaviest element.

b) **Given:** C_2H_6 **Find:** mass percent C

Conceptual Plan: $\text{mass \%C} = \frac{2 \text{ x molar mass C}}{\text{molar mass } C_2H_6} \text{ x } 100$

Solution:

$$2 \text{ x molar mass C} = 2(12.01 \text{g/mol}) = 24.02 \text{ g C}$$

$$\text{molar mass } C_2H_6 = 2(12.01 \text{ g/mol}) + 6(1.008 \text{ g/mol}) = 30.07 \text{ g/mol}$$

$$\text{mass \% C} = \frac{2 \text{ x molar mass C}}{\text{molar mass } C_2H_6} \text{ x } 100\%$$

$$= \frac{24.02 \cancel{\text{g/mol}}}{30.07 \cancel{\text{g/mol}}} \text{ x } 100\%$$

$$= 79.89 \%$$

Check: Units of the answer, %, are correct. The magnitude is reasonable because it is between 0 and 100% and carbon is the heaviest element.

c) **Given:** C_2H_2 **Find:** mass percent C

Conceptual Plan: $\text{mass \%C} = \frac{2 \text{ x molar mass C}}{\text{molar mass } C_2H_2} \text{ x } 100$

Solution:

$$2 \text{ x molar mass C} = 2(12.01 \text{g/mol}) = 24.02 \text{ g C}$$

$$\text{molar mass } C_2H_2 = 2(12.01 \text{ g/mol}) + 2(1.008 \text{ g/mol}) = 26.04 \text{ g/mol}$$

$$\text{mass \% C} = \frac{2 \text{ x molar mass C}}{\text{molar mass } C_2H_2} \text{ x } 100\%$$

$$= \frac{24.02 \cancel{\text{g/mol}}}{26.04 \cancel{\text{g/mol}}} \text{ x } 100\%$$

$$= 92.26 \%$$

Check: Units of the answer, %, are correct. The magnitude is reasonable because it is between 0 and 100% and carbon is the heaviest element.

d) **Given:** C_2H_5Cl **Find:** mass percent C

Conceptual Plan: $\text{mass \%C} = \frac{2 \text{ x molar mass C}}{\text{molar mass } C_2H_5Cl} \text{ x } 100$

Solution:

$$2 \text{ x molar mass C} = 2(12.01 \text{g/mol}) = 24.02 \text{ g C}$$

$$\text{molar mass } C_2H_5Cl = 2(12.01 \text{ g/mol}) + 5(1.008 \text{ g/mol}) + 1(35.45 \text{ g/mol}) = 64.51 \text{ g/mol}$$

$$\text{mass \% C} = \frac{2 \text{ x molar mass C}}{\text{molar mass } C_2H_5Cl} \text{ x } 100\%$$

$$= \frac{24.02 \cancel{\text{g/mol}}}{64.51 \cancel{\text{g/mol}}} \text{ x } 100\%$$

$$= 37.23 \%$$

Check: Units of the answer, %, are correct. The magnitude is reasonable because it is between 0 and 100% and chlorine is heavier than carbon.

45. **Given:** NH_3 **Find:** mass percent N

Conceptual Plan: $\text{mass \%N} = \frac{1 \text{ x molar mass N}}{\text{molar mass } NH_3} \text{ x } 100$

Solution:

$$1 \text{ x molar mass N} = 1(14.01 \text{g/mol}) = 14.01 \text{ g N}$$

$$\text{molar mass } NH_3 = 3(1.008 \text{ g/mol}) + (14.01 \text{ g/mol}) = 17.03 \text{ g/mol}$$

$$\text{mass \% N} = \frac{1 \text{ x molar mass N}}{\text{molar mass } NH_3} \text{ x } 100\%$$

$$= \frac{14.01 \text{ g/mol}}{17.03 \text{ g/mol}} \times 100\%$$

$= 82.27\ \%$

Check: Units of the answer, %, are correct. The magnitude is reasonable because it is between 0 and 100% and nitrogen is the heaviest.

Given: $CO(NH_2)_2$ **Find:** mass percent N

Conceptual Plan: $\text{mass } \%N = \frac{2 \text{ x molar mass N}}{\text{molar mass } CO(NH_2)_2} \text{ x } 100$

Solution:

$2 \text{ x molar mass N} = 1(14.01 \text{g/mol}) = 28.02 \text{ g N}$

$\text{molar mass } CO(NH_2)_2 = (12.01 \text{ g/mol}) + (16.00 \text{ g/mol}) + 2(14.01 \text{ g/mol}) + 4(1.008 \text{ g/mol}) = 60.06 \text{ g/mol}$

$$\text{mass } \% \text{ N} = \frac{2 \text{ x molar mass N}}{\text{molar mass } CO(NH_2)_2} \text{ x } 100\%$$

$$= \frac{28.02 \text{ g/mol}}{60.06 \text{ g/mol}} \text{ x } 100\%$$

$= 46.65\ \%$

Check: Units of the answer, %, are correct. The magnitude is reasonable because it is between 0 and 100% and there are two nitrogen and only one carbon and one oxygen.

Given: NH_4NO_3 **Find:** mass percent N

Conceptual Plan: $\text{mass } \%N = \frac{2 \text{ x molar mass N}}{\text{molar mass } NH_4NO_3} \text{ x } 100$

Solution:

$2 \text{ x molar mass N} = 2(14.01 \text{g/mol}) = 28.02 \text{ g N}$

$\text{molar mass } NH_4NO_3 = 2(14.01 \text{ g/mol}) + 4(1.008 \text{ g/mol}) + 3(16.00 \text{ g/mol}) = 80.05 \text{ g/mol}$

$$\text{mass } \% \text{ N} = \frac{2 \text{ x molar mass N}}{\text{molar mass } NH_4NO_3} \text{ x } 100\%$$

$$= \frac{28.02 \text{ g/mol}}{80.05 \text{ g/mol}} \text{ x } 100\%$$

$= 35.00\ \%$

Check: Units of the answer, %, are correct. The magnitude is reasonable because it is between 0 and 100% and the mass of nitrogen is less than the mass of oxygen and there are two nitrogen and three oxygen.

Given: $(NH_4)_2SO_4$ **Find:** mass percent N

Conceptual Plan: $\text{mass } \%N = \frac{2 \text{ x molar mass N}}{\text{molar mass } (NH_4)_2SO_4} \text{ x } 100$

Solution:

$2 \text{ x molar mass N} = 2(14.01 \text{g/mol}) = 28.02 \text{ g N}$

$\text{molar mass } (NH_4)_2SO_4 = 2(14.01 \text{ g/mol}) + 8(1.008 \text{ g/mol}) + (32.07 \text{ g/mol}) + 4(16.00 \text{ g/mol}) = 132.15 \text{ g/mol}$

$$\text{mass } \% \text{ N} = \frac{2 \text{ x molar mass N}}{\text{molar mass } (NH_4)_2SO_4} \text{ x } 100\%$$

$$= \frac{28.02 \text{ g/mol}}{132.15 \text{ g/mol}} \text{ x } 100\%$$

$= 21.20\ \%$

Check: Units of the answer, %, are correct. The magnitude is reasonable because it is between 0 and 100% and the mass of nitrogen is less than the mass of oxygen and sulfur.

The fertilizer with the highest nitrogen content is NH_3 because it has the highest %Nat 82.27% N.

47. **Given:** 55.5 g CuF_2: 37.42 % F **Find:** g F in CuF_2
Conceptual Plan: **g CuF_2 → g F**

$$\frac{37.42\text{ g F}}{100.0\text{ g CuF}_2}$$

Solution: $$55.5\text{ g CuF}_2 \times \frac{37.42\text{ g F}}{100.0\text{ g CuF}_2} = 20.77 = 20.8\text{ g F}$$

Check: Units of the answer, g F, are correct. The magnitude is reasonable because it is less than the original mass.

49. **Given:** 150 μg I; 76.45% I in KI **Find:** μg KI
Conceptual Plan: **μg I → g I → g KI → μg KI**

$$\frac{1\text{ g I}}{1\times10^6\ \mu\text{g I}} \quad \frac{100.0\text{ g KI}}{76.45\text{ g I}} \quad \frac{1\times10^6\ \mu\text{g KI}}{1\text{ g KI}}$$

Solution: $$150\ \mu\text{g I} \times \frac{1\text{ g I}}{1\text{x}10^6\ \mu\text{g I}} \times \frac{100.0\text{ g KI}}{76.45\text{ g I}} \times \frac{1\times10^6\ \mu\text{g KI}}{1\text{ g KI}} = 196\ \mu\text{g KI}$$

Check: Units of the answer, μg KI, are correct. The magnitude is reasonable because it is greater than the original mass.

51. a) red – oxygen, white – hydrogen: 2H:O H_2O
b) black – carbon, white – hydrogen: 4H:C CH_4
c) black – carbon, white – hydrogen, red – oxygen: 2C:O:6H CH_3CH_2OH or C_2H_6O

53. a) **Given:** 0.0885 mol C_4H_{10} **Find:** mol H atoms
Conceptual Plan: **mol C_4H_{10} → mole H atom**

$$\frac{10\text{ mol H}}{1\text{ mol C}_4\text{H}_{10}}$$

Solution: $$0.0885\text{ mol C}_4\text{H}_{10} \times \frac{10\text{ mol H}}{1\text{ mol C}_4\text{H}_{10}} = 0.885\text{ mol H atoms}$$

Check: Units of the answer, mol H atoms, are correct. The magnitude is reasonable because it is greater than the original mol C_4H_{10}.

b) **Given:** 1.3 mol CH_4 **Find:** mol H atoms
Conceptual Plan: **mol CH_4 → mole H atom**

$$\frac{4\text{ mol H}}{1\text{ mol CH}_4}$$

Solution: $$1.3\text{ mol CH}_4 \times \frac{4\text{ mol H}}{1\text{ mol CH}_4} = 5.2\text{ mol H atoms}$$

Check: Units of the answer, mol H atoms, are correct. The magnitude is reasonable because it is greater than the original mol CH_4.

c) **Given:** 2.4 mol C_6H_{12} **Find:** mol H atoms
Conceptual Plan: **mol C_6H_{12} → mole H atom**

$$\frac{12\text{ mol H}}{1\text{ mol C}_6\text{H}_{12}}$$

Solution: $$2.4\text{ mol C}_6\text{H}_{12} \times \frac{12\text{ mol H}}{1\text{ mol C}_6\text{H}_{12}} = 29\text{ mol H atoms}$$

Check: Units of the answer, mol H atoms, are correct. The magnitude is reasonable because it is greater than the original mol C_6H_{12}.

d) **Given:** 1.87 mol C_8H_{18} **Find:** mol H atoms
Conceptual Plan: **mol C_8H_{18} → mole H atom**

$$\frac{18\text{ mol H}}{1\text{ mol C}_8\text{H}_{18}}$$

Solution: $1.87\ \text{mol}\ C_8H_{18} \times \dfrac{18\ \text{mol H}}{1\ \text{mol}\ C_8H_{18}} = 33.7\ \text{mol H atoms}$

Check: Units of the answer, mol H atoms, are correct. The magnitude is reasonable because it is greater than the original mol C_8H_{18}.

55. a) **Given:** 8.5 g NaCl **Find:** g Na

Conceptual Plan: g NaCl → mole NaCl → mol Na → g Na

$$\frac{1\ \text{mol NaCl}}{58.44\ \text{g NaCl}} \quad \frac{1\ \text{mol Na}}{1\ \text{mol NaCl}} \quad \frac{22.99\ \text{g Na}}{1\ \text{mol Na}}$$

Solution: $8.5\ \text{g NaCl} \times \dfrac{1\ \text{mol NaCl}}{58.44\ \text{g NaCl}} \times \dfrac{1\ \text{mol Na}}{1\ \text{mol NaCl}} \times \dfrac{22.99\ \text{g Na}}{1\ \text{mol Na}} = 3.3\ \text{g Na}$

Check: Units of the answer, g Na, are correct. The magnitude is reasonable because it is less than the original g NaCl.

b) **Given:** 8.5 g Na_3PO_4 **Find:** g Na

Conceptual Plan: g Na_3PO_4 → mole Na_3PO_4 → mol Na → g Na

$$\frac{1\ \text{mol}\ Na_3PO_4}{163.94\ \text{g}\ Na_3PO_4} \quad \frac{3\ \text{mol Na}}{1\ \text{mol}\ Na_3PO_4} \quad \frac{22.99\ \text{g Na}}{1\ \text{mol Na}}$$

Solution: $8.5\ \text{g}\ Na_3PO_4 \times \dfrac{1\ \text{mol}\ Na_3PO_4}{163.94\ \text{g}\ Na_3PO_4} \times \dfrac{3\ \text{mol Na}}{1\ \text{mol}\ Na_3PO_4} \times \dfrac{22.99\ \text{g Na}}{1\ \text{mol Na}} = 3.6\ \text{g Na}$

Check: Units of the answer, g Na, are correct. The magnitude is reasonable because it is less than the original g Na_3PO_4.

c) **Given:** 8.5 g $NaC_7H_5O_2$ **Find:** g Na

Conceptual Plan: g $NaC_7H_5O_2$ → mole $NaC_7H_5O_2$ → mol Na → g Na

$$\frac{1\ \text{mol}\ NaC_7H_5O_2}{144.10\ \text{g}\ NaC_7H_5O_2} \quad \frac{1\ \text{mol Na}}{1\ \text{mol}\ NaC_7H_5O_2} \quad \frac{22.99\ \text{g Na}}{1\ \text{mol Na}}$$

Solution: $8.5\ \text{g}\ NaC_7H_5O_2 \times \dfrac{1\ \text{mol}\ NaC_7H_5O_2}{144.10\ \text{g}\ NaC_7H_5O_2} \times \dfrac{1\ \text{mol Na}}{1\ \text{mol}\ NaC_7H_5O_2} \times \dfrac{22.99\ \text{g Na}}{1\ \text{mol Na}} = 1.4\ \text{g Na}$

Check: Units of the answer, g Na, are correct. The magnitude is reasonable because it is less than the original g $NaC_7H_5O_2$.

d) **Given:** 8.5 g $Na_2C_6H_6O_7$ **Find:** g Na

Conceptual Plan: g $Na_2C_6H_6O_7$ → mole $Na_2C_6H_6O_7$ → mol Na → g Na

$$\frac{1\ \text{mol}\ Na_2C_6H_6O_7}{236.1\ \text{g}\ Na_2C_6H_6O_7} \quad \frac{2\ \text{mol Na}}{1\ \text{mol}\ Na_2C_6H_6O_7} \quad \frac{22.99\ \text{g Na}}{1\ \text{mol Na}}$$

Solution:

$$8.5\ \text{g}\ Na_2C_6H_6O_7 \times \frac{1\ \text{mol}\ Na_2C_6H_6O_7}{236.1\ \text{g}\ Na_2C_6H_6O_7} \times \frac{2\ \text{mol Na}}{1\ \text{mol}\ Na_2C_6H_6O_7} \times \frac{22.99\ \text{g Na}}{1\ \text{mol Na}} = 1.7\ \text{g}\ Na_2C_6H_6O_7$$

Check: Units of the answer, g Na, are correct. The magnitude is reasonable because it is less than the original g $Na_2C_6H_6O_7$.

57. a) **Given:** 1.651 g Ag; 0.1224 g O **Find:** empirical formula

Conceptual Plan:

convert mass to mol of each element → write pseudoformula → write empirical formula

$$\frac{1\ \text{mol Ag}}{107.9\ \text{g Ag}} \quad \frac{1\ \text{mol O}}{16.00\ \text{g O}} \qquad \text{divide by smallest number}$$

Solution: $1.651\ \text{g Ag} \times \dfrac{1\ \text{mol Ag}}{107.9\ \text{g Ag}} = 0.01530\ \text{mol Ag}$

$0.1224\ \text{g O} \times \dfrac{1\ \text{mol O}}{16.00\ \text{g O}} = 0.007650\ \text{mol O}$

$Ag_{0.01530}O_{0.007650}$

$Ag_{\frac{0.01530}{0.007650}}O_{\frac{0.007650}{0.007650}} \rightarrow Ag_2O$

The correct empirical formula is Ag_2O

b) **Given:** 0.672 g Co; 0.569 g As; 0.486 g O **Find:** empirical formula
Conceptual Plan:
convert mass to mol of each element → write pseudoformula → write empirical formula

$\frac{1 \text{ mol Co}}{58.93 \text{ g Co}} \quad \frac{1 \text{ mol As}}{74.92 \text{ g As}} \quad \frac{1 \text{ mol O}}{16.00 \text{ g O}}$ divide by smallest number

Solution: $0.672 \text{ g Co} \times \frac{1 \text{ mol Co}}{58.93 \text{ g Co}} = 0.0114 \text{ mol Co}$

$0.569 \text{ g As} \times \frac{1 \text{ mol As}}{74.92 \text{ g As}} = 0.00759 \text{ mol O}$

$0.486 \text{ g O} \times \frac{1 \text{ mol O}}{16.00 \text{ g O}} = 0.0304 \text{ mol O}$

$Co_{0.0114}As_{0.00759}O_{0.0304}$

$Co_{\frac{0.0114}{0.00759}}As_{\frac{0.00759}{0.00759}}O_{\frac{0.0304}{0.00759}} \rightarrow Co_{1.5}As_1O_4$

$Co_{1.5}As_1O_4 \times 2 \rightarrow Co_3As_2O_8$

The correct empirical formula is $Co_3As_2O_8$

c) **Given:** 1.443 g Se; 5.841 g Br **Find:** empirical formula
Conceptual Plan:
convert mass to mol of each element → write pseudoformula → write empirical formula

$\frac{1 \text{ mol Se}}{78.96 \text{ g Se}} \quad \frac{1 \text{ mol Br}}{79.90 \text{ g Br}}$ divide by smallest number

Solution: $1.443 \text{ g Se} \times \frac{1 \text{ mol Se}}{78.96 \text{ g Se}} = 0.01828 \text{ mol Se}$

$5.841 \text{ g Br} \times \frac{1 \text{ mol Br}}{79.90 \text{ g Br}} = 0.07310 \text{ mol Br}$

$Se_{0.01828}Br_{0.07310}$

$Se_{\frac{0.01828}{0.01828}}Br_{\frac{0.07310}{0.01828}} \rightarrow SeBr_4$

The correct empirical formula is $SeBr_4$

59. a) **Given:** In a 100 g sample: 74.03 g C, 8.70 g H, 17.27 g N **Find:** empirical formula
Conceptual Plan:
convert mass to mol of each element → write pseudoformula → write empirical formula

$\frac{1 \text{ mol C}}{12.01 \text{ g C}} \quad \frac{1 \text{ mol H}}{1.008 \text{ g H}} \quad \frac{1 \text{ mol N}}{14.01 \text{ g N}}$ divide by smallest number

Solution: $74.03 \text{ g C} \times \frac{1 \text{ mol C}}{12.01 \text{ g C}} = 6.164 \text{ mol C}$

$8.70 \text{ g H} \times \frac{1 \text{ mol H}}{1.008 \text{ g H}} = 8.63 \text{ mol H}$

$17.27 \text{ g N} \times \frac{1 \text{ mol N}}{14.01 \text{ g N}} = 1.233 \text{ mol N}$

$C_{6.164}H_{8.63}N_{1.233}$

$C_{\frac{6.164}{1.233}}H_{\frac{8.63}{1.233}}N_{\frac{1.233}{1.233}} \rightarrow C_5H_7N$

The correct empirical formula is C_5H_7N

b) **Given:** In a 100 g sample: 49.48 g C, 5.19 g H, 28.85 g N, 16.48 g O **Find:** empirical formula
Conceptual Plan:
convert mass to mol of each element → write pseudoformula → write empirical formula

$\frac{1 \text{ mol C}}{12.01 \text{ g C}} \quad \frac{1 \text{ mol H}}{1.008 \text{ g H}} \quad \frac{1 \text{ mol N}}{14.01 \text{ g N}} \quad \frac{1 \text{ mol O}}{16.00 \text{ g O}}$ divide by smallest number

Solution: $49.48\ \text{g C} \times \dfrac{1\ \text{mol C}}{12.01\ \text{g C}} = 4.120\ \text{mol C}$

$5.19\ \text{g H} \times \dfrac{1\ \text{mol H}}{1.008\ \text{g H}} = 5.15\ \text{mol H}$

$28.85\ \text{g N} \times \dfrac{1\ \text{mol N}}{14.01\ \text{g N}} = 2.059\ \text{mol N}$

$16.48\ \text{g O} \times \dfrac{1\ \text{mol O}}{16.00\ \text{g O}} = 1.030\ \text{mol O}$

$C_{4.120}H_{5.15}N_{2.059}O_{1.030}$

$C_{\frac{4.120}{1.030}}H_{\frac{5.15}{1.030}}N_{\frac{2.059}{1.030}}O_{\frac{1.030}{1.030}} \rightarrow C_4H_5N_2O$

The correct empirical formula is $C_4H_5N_2O$

61. **Given:** 0.77 mg N, 6.61 mg N_xCl_y **Find:** empirical formula
Conceptual Plan:
Find mg Cl → convert mg to g for each element → convert mass to mol of each element →

mg N_xCl_y - mg N $\quad \dfrac{1\ \text{g}}{1000\ \text{mg}} \quad \dfrac{1\ \text{mol N}}{14.01\ \text{g N}} \quad \dfrac{1\ \text{mol Cl}}{35.45\ \text{g Cl}}$

write pseudoformula → write empirical formula
divide by smallest number

Solution: 6.61 mg N_xCl_y – 0.77 mg N = 5.84 mg Cl

$0.77\ \text{mg N} \times \dfrac{1\ \text{g N}}{1000\ \text{mg N}} \times \dfrac{1\ \text{mol N}}{14.01\ \text{g N}} = 5.5 \times 10^{-5}\ \text{mol N}$

$5.84\ \text{mg Cl} \times \dfrac{1\ \text{g Cl}}{1000\ \text{mg Cl}} \times \dfrac{1\ \text{mol Cl}}{35.45\ \text{g Cl}} = 1.6 \times 10^{-4}\ \text{mol Cl}$

$N_{5.5 \times 10^{-5}}Cl_{1.6 \times 10^{-4}}$

$N_{\frac{5.5\times10^{-5}}{5.5\times10^{-5}}}Cl_{\frac{1.6\times10^{-4}}{5.5\times10^{-5}}} \rightarrow NCl_3$

The correct empirical formula is NCl_3

63. a) **Given:** empirical formula = C_6H_7N, molar mass = 186.24 g/mol **Find:** molecular formula

Conceptual Plan: molecular formula = empirical formula x n $\quad n = \dfrac{\text{molar mass}}{\text{empirical formula mass}}$

Solution: empirical formula mass = 6(12.01 g/mol) + 7(1.008 g/mol) + 1(14.01 g/mol) = 93.13 g/mol

$n = \dfrac{\text{molar mass}}{\text{formula molar mass}} = \dfrac{186.24\ \text{g/mol}}{93.13\ \text{g/mol}} = 1.998 = 2$

molecular formula $= C_6H_7N \times 2$
$= C_{12}H_{14}N_2$

b) **Given:** empirical formula = C_2HCl, molar mass = 181.44 g/mol **Find:** molecular formula

Conceptual Plan: molecular formula = empirical formula x n $\quad n = \dfrac{\text{molar mass}}{\text{empirical formula mass}}$

Solution: empirical formula mass = 2(12.01 g/mol) + 1(1.008 g/mol) + 1(35.45 g/mol) = 60.48 g/mol

$n = \dfrac{\text{molar mass}}{\text{formula molar mass}} = \dfrac{181.44\ \text{g/mol}}{60.48\ \text{g/mol}} = 3$

molecular formula $= C_2HCl \times 3$
$= C_6H_3Cl_3$

c) **Given:** empirical formula = $C_5H_{10}NS_2$, molar mass = 296.54 g/mol **Find:** molecular formula

Conceptual Plan: molecular formula = empirical formula x n $\quad n = \dfrac{\text{molar mass}}{\text{empirical formula mass}}$

Solution: empirical formula mass = 5(12.01 g/mol) + 10(1.008 g/mol) + 1(14.01 g/mol) + 2(32.07) = 148.28 g/mol

$$n = \frac{\text{molar mass}}{\text{formula molar mass}} = \frac{296.54 \text{ g/mol}}{148.28 \text{ g/mol}} = 2$$

molecular formula $= C_5H_{10}NS_2 \times 2$
$= C_{10}H_{20}N_2S_4$

65. **Given:** 33.01 g CO_2, 13.51 g H_2O **Find:** empirical formula
Conceptual Plan:
mass CO_2, H_2O → mol CO_2, H_2O → mol C, mol H → pseudoformula → empirical formula

$\frac{1 \text{ mol } CO_2}{44.01 \text{ g } CO_2}$ $\frac{1 \text{ mol } H_2O}{18.02 \text{ g } H_2O}$ $\frac{1 \text{ mol C}}{1 \text{ mol } CO_2}$ $\frac{2 \text{ mol H}}{1 \text{ mol } H_2O}$ divide by smallest number

Solution:

$$33.01 \text{ g } CO_2 \times \frac{1 \text{ mol } CO_2}{44.01 \text{ g } CO_2} = 0.7500 \text{ mol } CO_2$$

$$13.51 \text{ g } H_2O \times \frac{1 \text{ mol } H_2O}{18.02 \text{ g } H_2O} = 0.7497 \text{ mol } H_2O$$

$$0.7500 \text{ mol } CO_2 \times \frac{1 \text{ mol C}}{1 \text{ mol } CO_2} = 0.7500 \text{ mol C}$$

$$0.7497 \text{ mol } H_2O \times \frac{2 \text{ mol H}}{1 \text{ mol } H_2O} = 1.499 \text{ mol H}$$

$C_{0.7500}H_{1.499}$

$C_{\frac{0.7500}{0.7500}}H_{\frac{1.499}{0.7500}} \rightarrow CH_2$

The correct empirical formula is CH_2

67. **Given:** 4.30 g sample, 8.59 g CO_2, 3.52 g H_2O **Find:** empirical formula
Conceptual Plan:
mass CO_2, H_2O → mol CO_2, H_2O → mol C, mol H → mass C, mass H, mass O → mol O →

$\frac{1 \text{ mol } CO_2}{44.01 \text{ g } CO_2}$ $\frac{1 \text{ mol } H_2O}{18.02 \text{ g } H_2O}$ $\frac{1 \text{ mol C}}{1 \text{ mol } CO_2}$ $\frac{2 \text{ mol H}}{1 \text{ mol } H_2O}$ $\frac{12.01 \text{ g C}}{1 \text{ mol C}}$ $\frac{1.008 \text{ g H}}{1 \text{ mol H}}$ g sample - gC - g H $\frac{1 \text{ mol O}}{16.00 \text{ g O}}$

pseudoformula → empirical formula
divide by smallest number

Solution:

$$8.59 \text{ g } CO_2 \times \frac{1 \text{ mol } CO_2}{44.01 \text{ g } CO_2} = 0.195 \text{ mol } CO_2$$

$$3.52 \text{ g } H_2O \times \frac{1 \text{ mol } H_2O}{18.02 \text{ g } H_2O} = 0.195 \text{ mol } H_2O$$

$$0.195 \text{ mol } CO_2 \times \frac{1 \text{ mol C}}{1 \text{ mol } CO_2} = 0.195 \text{ mol C}$$

$$0.195 \text{ mol } H_2O \times \frac{2 \text{ mol H}}{1 \text{ mol } H_2O} = 0.390 \text{ mol H}$$

$$0.195 \text{ mol C} \times \frac{12.01 \text{ g C}}{1 \text{ mol C}} = 2.34 \text{ g C}$$

$$0.390 \text{ mol } H_2O \times \frac{1.008 \text{ g H}}{1 \text{ mol H}} = 0.393 \text{ g H}$$

4.30 g - 2.34 g - 0.393 g = 1.57 g O

$$1.57 \text{ g O} \times \frac{1 \text{ mol O}}{16.00 \text{ g O}} = 0.0979 \text{ mol O}$$

$C_{0.195}H_{0.390}O_{0.0979}$

$C_{\frac{0.195}{0.0979}}H_{\frac{0.390}{0.0979}}O_{\frac{0.0979}{0.0979}} \rightarrow C_2H_4O$

The correct empirical formula is C_2H_4O

69. **Conceptual Plan: write a skeletal reaction → balance atoms in more complex compounds → balance elements that occur as free elements → clear fractions**

Solution: Skeletal reaction: $SO_2(g) + O_2(g) + H_2O(l) \rightarrow H_2SO_4(aq)$

Balance O: $SO_2(g) + 1/2O_2(g) + H_2O(l) \rightarrow H_2SO_4(aq)$

Clear fraction: $2SO_2(g) + O_2(g) + 2H_2O(l) \rightarrow 2H_2SO_4(aq)$

Check:

left side	right side
2 S atoms	2 S atoms
8 O atoms	8 O atoms
4 H atoms	4 H atoms

71. **Conceptual Plan: write a skeletal reaction → balance atoms in more complex compounds → balance elements that occur as free elements → clear fractions**

Solution: Skeletal reaction: $Na(s) + H_2O(l) \rightarrow H_2(g) + NaOH(aq)$

Balance H: $Na(s) + H_2O(l) \rightarrow 1/2H_2(g) + NaOH(aq)$

Clear fraction: $2Na(s) + 2H_2O(l) \rightarrow H_2(g) + 2NaOH(aq)$

Check:

left side	right side
2 Na atoms	2 Na atoms
4 H atoms	4 H atoms
2 O atoms	2 O atoms

73. **Conceptual Plan: write a skeletal reaction → balance atoms in more complex compounds → balance elements that occur as free elements → clear fractions**

Solution: Skeletal reaction: $C_{12}H_{22}O_{11}(aq) + H_2O(l) \rightarrow C_2H_5OH(aq) + CO_2(g)$

Balance H: $C_{12}H_{22}O_{11}(aq) + H_2O(l) \rightarrow 4C_2H_5OH(aq) + CO_2(g)$

Balance C: $C_{12}H_{22}O_{11}(aq) + H_2O(l) \rightarrow 4C_2H_5OH(aq) + 4CO_2(g)$

Check:

left side	right side
12 C atoms	12 C atoms
24 H atoms	24 H atoms
12 O atoms	12 O atoms

75. a) **Conceptual Plan: write a skeletal reaction → balance atoms in more complex compounds → balance elements that occur as free elements → clear fractions**

Solution: Skeletal reaction: $PbS(s) + HBr(aq) \rightarrow PbBr_2(s) + H_2S(g)$

Balance Br: $PbS(s) + 2HBr(aq) \rightarrow PbBr_2(s) + H_2S(g)$

Check:

left side	right side
1 Pb atom	1 Pb atom
1 S atom	1 S atom
2 H atoms	2 H atoms
2 Br atoms	2 Br atoms

b) **Conceptual Plan: write a skeletal reaction → balance atoms in more complex compounds → balance elements that occur as free elements → clear fractions**

Solution: Skeletal reaction: $CO(g) + H_2(g) \rightarrow CH_4(g) + H_2O(l)$

Balance H: $CO(g) + 3H_2(g) \rightarrow CH_4(g) + H_2O(l)$

Check:

left side	right side
1 C atom	1 C atom
1 O atom	1 O atom
6 H atoms	6 H atoms

c) **Conceptual Plan: write a skeletal reaction → balance atoms in more complex compounds → balance elements that occur as free elements → clear fractions**

Solution: Skeletal reaction: $HCl(aq) + MnO_2(s) \rightarrow MnCl_2(aq) + H_2O(l) + Cl_2(g)$

Balance Cl: $4HCl(aq) + MnO_2(s) \rightarrow MnCl_2(aq) + H_2O(l) + Cl_2(g)$

Balance O: $4HCl(aq) + MnO_2(s) \rightarrow MnCl_2(aq) + 2H_2O(l) + Cl_2(g)$

Check:

left side	right side
4 H atoms	4 H atoms
4 Cl atoms	4 Cl atoms
1 Mn atom	1 Mn atom
2 O atoms	2 O atoms

d) **Conceptual Plan: write a skeletal reaction → balance atoms in more complex compounds → balance elements that occur as free elements → clear fractions**

Solution: Skeletal reaction: $C_5H_{12}(l) + O_2(g) \rightarrow CO_2(g) + H_2O(l)$
Balance C: $C_5H_{12}(l) + O_2(g) \rightarrow 5CO_2(g) + H_2O(l)$
Balance H: $C_5H_{12}(l) + O_2(g) \rightarrow 5CO_2(g) + 6H_2O(l)$
Balance O: $C_5H_{12}(l) + 8O_2(g) \rightarrow 5CO_2(g) + 6H_2O(l)$

Check:

left side	right side
5 C atoms	5 C atoms
12 H atoms	12 H atoms
16 O atoms	16 O atoms

77. a) **Conceptual Plan: balance atoms in more complex compounds → balance elements that occur as free elements → clear fractions**

Solution: Skeletal reaction: $CO_2(g) + CaSiO_3(s) + H_2O(l) \rightarrow SiO_2(s) + Ca(HCO_3)_2(aq)$
Balance C: $2CO_2(g) + CaSiO_3(s) + H_2O(l) \rightarrow SiO_2(s) + Ca(HCO_3)_2(aq)$

Check:

left side	right side
2 C atoms	2 C atoms
8 O atoms	8 O atoms
1 Ca atom	1 Ca atom
1 Si atom	1 Si atom
2 H atoms	2 H atoms

b) **Conceptual Plan: balance atoms in more complex compounds → balance elements that occur as free elements → clear fractions**

Solution: Skeletal reaction: $Co(NO_3)_3(aq) + (NH_4)_2S(aq) \rightarrow Co_2S_3(s) + NH_4NO_3(aq)$
Balance S: $Co(NO_3)_3(aq) + 3(NH_4)_2S(aq) \rightarrow Co_2S_3(s) + NH_4NO_3(aq)$
Balance Co: $2Co(NO_3)_3(aq) + 3(NH_4)_2S(aq) \rightarrow Co_2S_3(s) + NH_4NO_3(aq)$
Balance N: $2Co(NO_3)_3(aq) + 3(NH_4)_2S(aq) \rightarrow Co_2S_3(s) + 6NH_4NO_3(aq)$

Check:

left side	right side
2 Co atoms	2 Co atoms
12 N atoms	12 N atoms
18 O atoms	18 O atoms
24 H atoms	24 H atoms
3 S atoms	3 S atoms

c) **Conceptual Plan: balance atoms in more complex compounds → balance elements that occur as free elements → clear fractions**

Solution: Skeletal reaction: $Cu_2O(s) + C(s) \rightarrow Cu(s) + CO(g)$
Balance Cu: $Cu_2O(s) + C(s) \rightarrow 2Cu(s) + CO(g)$

Check:

left side	right side
2 Cu atoms	2 Cu atoms
1 O atom	1 O atom
1 C atom	1 C atom

d) **Conceptual Plan: balance atoms in more complex compounds → balance elements that occur as free elements → clear fractions**

Solution: Skeletal reaction: $H_2(g) + Cl_2(g) \rightarrow HCl(g)$
Balance Cl: $H_2(g) + Cl_2(g) \rightarrow 2HCl(g)$

Check:

left side	right side
2 H atom	2 H atom
2 Cl atom	2 Cl atom

79. a) composed of metal cation and polyatomic anion – inorganic compound
b) composed of carbon and hydrogen – organic compound
c) composed of carbon, hydrogen, and oxygen – organic compound
d) composed of metal cation and nonmetal anion – inorganic compound

81. **Given:** 145 mL C_2H_5OH, d = 0.789g/cm^3 **Find:** number of molecules

Conceptual Plan: cm^3 → mL: mL C_2H_5OH → g C_2H_5OH → mol C_2H_5OH → molecules C_2H_5OH

$$\frac{1\ cm^3}{1\ mL} \quad \frac{1\ mL\ C_2H_5OH}{0.789\ g\ C_2H_5OH} \quad \frac{1\ mol\ C_2H_5OH}{46.07\ g\ C_2H_5OH} \quad \frac{6.022 \times 10^{23}\ molecules\ C_2H_5OH}{1\ mol\ C_2H_5OH}$$

Solution:

$$145\ \cancel{mL\ C_2H_5OH} \times \frac{0.789\ \cancel{g\ C_2H_5OH}}{\cancel{cm^3}} \times \frac{1\ \cancel{cm^3}}{1\ \cancel{mL}} \times \frac{1\ mol\ C_2H_5OH}{46.07\ \cancel{g\ C_2H_5OH}} \times \frac{6.022 \times 10^{23}\ molecules\ C_2H_5OH}{1\ \cancel{mol\ C_2H_5OH}}$$

$= 1.50 \times 10^{24}$ molecules C_2H_5OH

Check: Units of answer, molecules C_2H_5OH, are correct. The magnitude is reasonable because we had more than 2 moles of C_2H_5OH and we have more than 2 times Avogadro's number of molecules.

83. a) To write the formula for an ionic compound: 1) Write the symbol for the metal cation and its charge and the symbol for the nonmetal anion or polyatomic anion and its charge. 2) Adjust the subscript on each cation and anion to balance the overall charge. 3) Check that the sum of the charges of the cations equals the sum of the charges of the anions.

potassium chromate: K^+ CrO_4^{2-}; K_2CrO_4 cation 2(1+) = 2+; anion 2-

Given: K_2CrO_4 **Find:** mass percent of each element

Conceptual Plan: %K, then %Cr, then %O

$$\text{mass \% K} = \frac{2 \times \text{molar mass K}}{\text{molar mass } K_2CrO_4} \times 100 \quad \text{mass \% Cr} = \frac{1 \times \text{molar mass Cr}}{\text{molar mass } K_2CrO_4} \times 100 \quad \text{mass \% O} = \frac{4 \times \text{molar mass O}}{\text{molar mass } K_2CrO_4} \times 100$$

molar mass of K = 39.10 g/mol, molar mass Cr = 52.00 g/mol, molar mass O = 16.00 g/mol

Solution: molar mass K_2CrO_4 = 2(39.10 g/mol) + 1(52.00 g/mol) + 4(16.00 g/mol) = 194.20 g/mol

2 x molar mass K = 2(39.10 g/mol) = 78.20 g K

$$\text{mass \% K} = \frac{2 \times \text{molar mass K}}{\text{molar mass } K_2CrO_4} \times 100\% = \frac{78.20\ \cancel{g/mol}}{194.20\ \cancel{g/mol}} \times 100\% = 40.27\ \%$$

1 x molar mass Cr = 1(52.00 g/mol) = 52.00 g Cr

$$\text{mass \% Cr} = \frac{1 \times \text{molar mass Cr}}{\text{molar mass } K_2CrO_4} \times 100\% = \frac{52.00\ \cancel{g/mol}}{194.20\ \cancel{g/mol}} \times 100\% = 26.78\ \%$$

4 x molar mass O = 4(16.00 g/mol) = 64.00 g O

$$\text{mass \% O} = \frac{4 \times \text{molar mass O}}{\text{molar mass } K_2CrO_4} \times 100\% = \frac{64.00\ \cancel{g/mol}}{194.20\ \cancel{g/mol}} \times 100\% = 32.96\ \%$$

Check: Units of the answer, %, are correct. The magnitude is reasonable because each is between 0 and 100% and the total is 100%.

b) To write the formula for an ionic compound: 1) Write the symbol for the metal cation and its charge and the symbol for the nonmetal anion or polyatomic anion and its charge. 2) Adjust the subscript on each cation and anion to balance the overall charge. 3) Check that the sum of the charges of the cations equals the sum of the charges of the anions.

Lead(II)phosphate: Pb^{2+} PO_4^{3-}; $Pb_3(PO_4)_2$ cation 3(2+) = 6+; anion 2(3-) = 6-

Given: $Pb_3(PO_4)_2$ **Find:** mass percent of each element

Conceptual Plan: %Pb, then % P, then %O

$$\text{mass \% Pb} = \frac{3 \times \text{molar mass Pb}}{\text{molar mass } Pb_3(PO_4)_2} \times 100 \quad \text{mass \% P} = \frac{2 \times \text{molar mass P}}{\text{molar mass } Pb_3(PO_4)_2} \times 100 \quad \text{mass \% O} = \frac{8 \times \text{molar mass O}}{\text{molar mass } Pb_3(PO_4)_2} \times 100$$

Solution: molar mass $Pb_3(PO_4)_2$ = 3(207.2 g/mol) + 2(30.97 g/mol) + 8(16.00 g/mol) = 811.5 g/mol

3 x molar mass Pb = 3(207.2 g/mol) = 621.6 g Pb

$$\text{mass \% Pb} = \frac{3 \times \text{molar mass Pb}}{\text{molar mass } Pb_3(PO_4)_2} \times 100\% = \frac{621.6\ \cancel{g/mol}}{811.5\ \cancel{g/mol}} \times 100\% = 76.60\ \%$$

2 x molar mass P = 2(30.97 g/mol) = 61.94 g P

$$\text{mass \% P} = \frac{2 \text{ x molar mass P}}{\text{molar mass } Pb_3(PO_4)_2} \text{ x } 100\%$$

$$= \frac{61.94 \text{ g/mol}}{811.5 \text{ g/mol}} \text{ x } 100\%$$

= 7.632 %

4 x molar mass O = 8(16.00 g/mol) = 128.0 g O

$$\text{mass \% O} = \frac{8 \text{ x molar mass O}}{\text{molar mass } Pb_3(PO_4)_2} \text{ x } 100\%$$

$$= \frac{128.0 \text{ g/mol}}{811.5 \text{ g/mol}} \text{ x } 100\%$$

= 15.77 %

Check: Units of the answer, %, are correct. The magnitude is reasonable because each is between 0 and 100% and the total is 100%.

c) sulfurous acid: H_2SO_3

Given: H_2SO_3 **Find:** mass percent of each element

Conceptual Plan: %H, then %S, then %O

$$\text{mass \% H} = \frac{2 \text{ x molar mass H}}{\text{molar mass } HSO_3} \text{ x } 100 \qquad \text{mass \% S} = \frac{1 \text{ x molar mass S}}{\text{molar mass } HSO_3} \text{ x } 100 \qquad \text{mass \% O} = \frac{3 \text{ x molar mass O}}{\text{molar mass } HSO_3} \text{ x } 100$$

Solution: molar mass H_2SO_3 = 2(1.008 g/mol) + 1(32.07 g/mol) + 3(16.00 g/mol) = 82.0$\underline{8}$6 g/mol

2 x molar mass H = 1(1.008 g/mol) = 2.016 g H 1 x molar mass S = 1(32.07 g/mol) = 32.06 g S

$$\text{mass \% H} = \frac{2 \text{ x molar mass H}}{\text{molar mass } H_2SO_3} \text{ x } 100\%$$

$$= \frac{2.016 \text{ g/mol}}{82.0\underline{8}6 \text{ g/mol}} \text{ x } 100\%$$

= 2.456 %

$$\text{mass \% S} = \frac{1 \text{ x molar mass S}}{\text{molar mass } H_2SO_3} \text{ x } 100\%$$

$$= \frac{32.07 \text{ g/mol}}{82.0\underline{8}6 \text{ g/mol}} \text{ x } 100\%$$

= 39.07 %

3 x molar mass O = 3(16.00 g/mol) = 48.00 g O

$$\text{mass \% O} = \frac{3 \text{ x molar mass O}}{\text{molar mass } H_2SO_3} \text{ x } 100\%$$

$$= \frac{48.00 \text{ g/mol}}{82.0\underline{8}6 \text{ g/mol}} \text{ x } 100\%$$

= 58.48 %

Check: Units of the answer, %, are correct. The magnitude is reasonable because each is between 0 and 100% and the total is 100%.

d) To write the formula for an ionic compound: 1) Write the symbol for the metal cation and its charge and the symbol for the nonmetal anion or polyatomic anion and its charge. 2) Adjust the subscript on each cation and anion to balance the overall charge. 3) Check that the sum of the charges of the cations equals the sum of the charges of the anions.

cobalt(II)bromide: Co^{2+} Br^- ; $CoBr_2$ cation 2+ = 2+; anion 2(1-) = 2-

Given: $CoBr_2$ **Find:** mass percent of each element

Conceptual Plan: %Co, then %Br

$$\text{mass \% Co} = \frac{1 \text{ x molar mass Co}}{\text{molar mass } CoBr_2} \text{ x } 100 \qquad \text{mass \% Br} = \frac{2 \text{ x molar mass Br}}{\text{molar mass } CoBr_2} \text{ x } 100$$

Solution: molar mass $CoBr_2$ = (58.93 g/mol) + 2(79.90 g/mol) = 218.7 g/mol

2 x molar mass Co = 1(58.93 g/mol) = 58.93 g Co 1 x molar mass Br = 2(79.90 g/mol) = 159.8 g Br

$$\text{mass \% Co} = \frac{1 \text{ x molar mass Co}}{\text{molar mass } CoBr_2} \text{ x } 100\%$$

$$= \frac{58.93 \text{ g/mol}}{218.7 \text{ g/mol}} \text{ x } 100\%$$

= 26.94 %

$$\text{mass \% Br} = \frac{2 \text{ x molar mass Br}}{\text{molar mass } CoBr_2} \text{ x } 100\%$$

$$= \frac{159.8 \text{ g/mol}}{218.7 \text{ g/mol}} \text{ x } 100\%$$

= 73.07 %

Check: Units of the answer, %, are correct. The magnitude is reasonable because each is between 0 and 100% and the total is 100%.

85. **Given:** 25 g CF_2Cl_2/mo. **Find:** g Cl /yr.
Conceptual Plan: g CF_2Cl_2/mo → g Cl/mo → g Cl/yr

$$\frac{70.09\text{ g Cl}}{120.91\text{ g }CF_2Cl_2} \qquad \frac{12\text{ mo.}}{1\text{ yr}}$$

Solution: $$\frac{25\text{ g }CF_2Cl_2}{\text{mo.}} \times \frac{70.90\text{ g Cl}}{120.91\text{ g }CF_2Cl_2} \times \frac{12\text{ mo.}}{1\text{ yr}} = 1.8 \times 10^2\text{ g Cl/yr}$$

Check: Units of answer, g Cl, is correct. Magnitude is reasonable because it is less than the total CF_2Cl_2 /yr.

87. **Given:** MCl_3, 65.57% Cl **Find:** identify M
Conceptual Plan: g Cl → mol Cl → mol M → atomic mass M

$$\frac{1\text{ mol Cl}}{35.45\text{ g Cl}} \qquad \frac{1\text{ mol M}}{3\text{ mol Cl}} \qquad \frac{\text{g M}}{\text{mol M}}$$

Solution: in 100 g sample: 65.57 g Cl, 34.43 g M

$$65.57\text{ g Cl} \times \frac{1\text{ mol Cl}}{35.45\text{ g Cl}} \times \frac{1\text{ mol M}}{3\text{ mol Cl}} = 0.6165\text{ mol M} \qquad \frac{34.43\text{ g M}}{0.6165\text{ mol M}} = 55.84\text{ g/mol M}$$

molar mass of 55.84 = Fe
The identity of M = Fe

89. **Given:** In a 100 g sample: 79.37 g C, 8.88 g H, 11.75 g O, molar mass = 272.37g/mol
Find: molecular formula
Conceptual Plan:
convert mass to mol of each element → pseudoformula → empirical formula → molecular formula

$$\frac{1\text{ mol C}}{12.01\text{ g C}} \quad \frac{1\text{ mol H}}{1.008\text{ g H}} \quad \frac{1\text{ mol O}}{16.00\text{ g O}} \qquad \text{divide by smallest number} \qquad \text{empirical formula x n}$$

Solution: $$79.37\text{ g C} \times \frac{1\text{ mol C}}{12.01\text{ g C}} = 6.609\text{ mol C}$$

$$8.88\text{ g H} \times \frac{1\text{ mol H}}{1.008\text{ g H}} = 8.81\text{ mol H}$$

$$11.75\text{ g O} \times \frac{1\text{ mol O}}{16.00\text{ g O}} = 0.7344\text{ mol O}$$

$$C_{6.609}H_{8.81}O_{0.7344}$$

$$C_{\frac{6.609}{0.7344}}H_{\frac{8.81}{0.7344}}O_{\frac{0.7344}{0.7344}} \rightarrow C_9H_{12}O$$

The correct empirical formula is $C_9H_{12}O$

empirical formula mass = 9(12.01 g/mol) + 12(1.008 g/mol) + 1(16.00 g/mol) = 136.19 g/mol

$$n = \frac{\text{molar mass}}{\text{formula molar mass}} = \frac{272.37\text{ g/mol}}{136.19\text{ g/mol}} = 2$$

molecular formula = $C_9H_{12}O \times 2 = C_{18}H_{24}O_2$

91. **Given:** 13.42 g sample, 39.61 g CO_2, 9.01 g H_2O, molar mass = 268.34 g/mol
Find: molecular formula
Conceptual Plan:
mass CO_2, H_2O → mol CO_2, H_2O → mol C, mol H → mass C, mass H, mass O → mol O →

$$\frac{1\text{ mol }CO_2}{44.01\text{ g }CO_2} \quad \frac{1\text{ mol }H_2O}{18.02\text{ g }H_2O} \qquad \frac{1\text{ mol C}}{1\text{ mol }CO_2} \quad \frac{2\text{ mol H}}{1\text{ mol }H_2O} \quad \frac{12.01\text{ g C}}{1\text{ mol C}} \quad \frac{1.008\text{ g H}}{1\text{ mol H}} \quad \text{g sample - gC - g H} \quad \frac{1\text{ mol O}}{16.00\text{ g O}}$$

pseudoformula → empirical formula → molecular formula
divide by smallest number   empirical formula x n

$$39.61\text{ g }CO_2 \times \frac{1\text{ mol }CO_2}{44.01\text{ g }CO_2} = 0.9000\text{ mol }CO_2$$

$$9.01\text{ g }H_2O \times \frac{1\text{ mol }H_2O}{18.02\text{ g }H_2O} = 0.5000\text{ mol }H_2O$$

$$0.9000 \text{ mol } CO_2 \times \frac{1 \text{ mol C}}{1 \text{ mol } CO_2} = 0.9000 \text{ mol C}$$

$$0.5000 \text{ mol } H_2O \times \frac{2 \text{ mol H}}{1 \text{ mol } H_2O} = 1.000 \text{ mol H}$$

$$0.9000 \text{ mol C} \times \frac{12.01 \text{ g C}}{1 \text{ mol C}} = 10.81 \text{ g C}$$

$$1.000 \text{ mol } H_2O \times \frac{1.008 \text{ g H}}{1 \text{ mol H}} = 1.008 \text{ g H}$$

13.42 g - 10.81 g - 1.008 g = 1.60 g O

$$1.60 \text{ g O} \times \frac{1 \text{ mol O}}{16.00 \text{ g O}} = 0.100 \text{ mol O}$$

$$C_{0.9000}H_{1.000}O_{0.100}$$

$$C_{\frac{0.9000}{0.100}}H_{\frac{1.000}{0.100}}O_{\frac{0.100}{0.100}} \rightarrow C_9H_{10}O$$

The correct empirical formula is $C_9H_{10}O$

empirical formula mass = 9(12.01 g/mol) + 10(1.008 g/mol) + 1(16.00 g/mol) = 134.2 g/mol

$$n = \frac{\text{molar mass}}{\text{formula molar mass}} = \frac{268.34 \text{ g/mol}}{134.2 \text{ g/mol}} = 2$$

molecular formula $= C_9H_{10}O \times 2 = C_{18}H_{20}O_2$

93. **Given:** 4.93 g $MgSO_4\bullet xH_2O$, 2.41 g $MgSO_4$ **Find:** value of x

Conceptual Plan: g $MgSO_4$ → mol $MgSO_4$ $\frac{1 \text{ mol } MgSO_4}{120.38 \text{ g } MgSO_4}$ **g H_2O → mol H_2O** $\frac{1 \text{ mol } H_2O}{18.02 \text{ g } H_2O}$ **Determine mole ratio** $\frac{\text{mol } HO_2}{\text{mol } MgSO_4}$

Solution:

$$2.41 \text{ g } MgSO_4 \times \frac{1 \text{ mol } MgSO_4}{120.38 \text{ g } MgSO_4} = 0.0200 \text{ mol } MgSO_4$$

Determine g H_2O: 4.93 g $MgSO_4\bullet xH_2O$ - 2.41 g $MgSO_4$ = 2.52 g H_2O

$$2.52 \text{ g } H_2O \times \frac{1 \text{ mol } H_2O}{18.02 \text{ g } H_2O} = 0.140 \text{ mol } H_2O$$

$$\frac{0.140 \text{ mol } H_2O}{0.0200 \text{ mol } MgSO_4} = 7$$

x = 7

95. **Given:** molar mass = 177 g/mol, g C = 8(g H) **Find:** molecular formula

Conceptual Plan: C_xH_yBrO

Solution: in 1 mol compound, let x = mol C and y = mol H, assume mol Br = 1, mol O = 1

177 g/mol = x(12.01 g/mol) + y(1.008 g/mol) + 1(79.90 g/mol) + 1(16.00 g/mol)

x(12.01 g/mol) = 8 {y(1.008 g/mol)}

177 g/mol = 8y(1.008 g/mol) + y(1.008 g/mol) + 79.90 g/mol + 16.00 g/mol

81 = 9y(1.008)

y = 9 = mol H

x(12.01) = 9(1.008)

x = 6 = mol C

molecular formula = C_6H_9BrO

Check: molar mass = 6(12.01 g/mol) + 9(1.008 g/mol) + 1(79.90 g/mol) + 1(16.00 g/mol) = 177.0 g/mol

97. **Given:** 23.5 mg $C_{17}H_{22}ClNO_4$ **Find:** total number of atoms

Conceptual Plan: mg compound → g compound → mol compound → mol atoms → number of atoms

$$\frac{1 \text{ g}}{1000 \text{ mg}} \quad \frac{1 \text{ mol}}{339.8 \text{ g}} \quad \frac{45 \text{ mol atoms}}{1 \text{ mol compound}} \quad \frac{6.022 \times 10^{23} \text{ atoms}}{1 \text{ mol atoms}}$$

Solution: $23.5 \text{ mg} \times \frac{1 \text{ g}}{1000 \text{ mg}} \times \frac{1 \text{ mol cpd}}{339.8 \text{ g}} \times \frac{45 \text{ mol atoms}}{1 \text{ mol cpd}} \times \frac{6.022 \times 10^{23} \text{ atoms}}{\text{mol}} = 1.87 \times 10^{21} \text{ atoms}$

Check: The units of the answer, number of atoms, is correct. The magnitude of the answer is reasonable since the molecule is so complex.

99. **Given:** MCl_3, 2.395 g sample, 3.606×10^{-2} mol Cl **Find:** atomic mass M
Conceptual Plan: mol Cl → g Cl → g X

$\frac{35.45 \text{ g Cl}}{1 \text{ mol Cl}}$ g sample - g Cl

mol Cl → mol M → atomic mass M

$\frac{1 \text{ mol M}}{3 \text{ mol Cl}}$ $\frac{\text{g M}}{\text{mol M}}$

Solution:

$$3.606 \times 10^{-2} \text{ mol Cl} \times \frac{35.45 \text{ g}}{1 \text{ mol Cl}} = 1.278 \text{ g Cl}$$

$$2.395 \text{ g} - 1.278 \text{ g} = 1.117 \text{ g M}$$

$$3.606 \times 10^{-2} \text{ mol Cl} \times \frac{1 \text{ mol M}}{3 \text{ mol Cl}} = 1.202 \times 10^{-2} \text{ mol M}$$

$$\frac{1.117 \text{ g M}}{0.01202 \text{ mol M}} = 92.93 \text{ g/mol M}$$

molar mass of M = 92.93 g/mol

101. **Given:** Sample 1:1.00 g X, 0.472 g Z, X_2Z_3; Sample 2: 1.00 g X, 0.630 g Z; Sample 3: 1.00 g X, 0.789 g Z
Find: empirical formula for samples 2 and 3
Conceptual Plan: moles X remains constant, determine relative moles of Z for the 3 samples.
Solution: Let X = atomic mass X, Z = atomic mass Z

$$n_X = \frac{1.00 \text{ g X}}{X} \qquad n_Z = \frac{0.472 \text{ g Z}}{Z}$$

for sample 1: $\frac{n_X}{n_Z} = \frac{2}{3}$

for sample 2: $\frac{0.630 \text{ g}}{0.472 \text{ g}} = 1.33$, so, mol = $1.33n_Z$

mol ratio: $\frac{n_X}{1.33n_Z} = \frac{2}{(1.33)3} = \frac{2}{4} = \frac{1}{2}$

Empirical formula sample 2: XZ_2

for sample 3: $\frac{0.789 \text{ g}}{0.472 \text{ g}} = 1.67$, so, mol = $1.67n_Z$

mol ratio: $\frac{n_X}{1.67n_Z} = \frac{2}{(1.67)3} = \frac{2}{5}$

Empirical formula sample 3: X_2Z_5

103. **Given:** 50.0 g S, 100 g Cl_2, 150 g mixture S_2Cl_2 and SCl_2 **Find:** % S_2Cl_2
Conceptual Plan: total mol S = 2(mol S_2Cl_2) + mol SCl_2; mol S_2Cl_2 → g S_2Cl_2 → %S_2Cl_2

$\frac{134.9 \text{ g}}{1 \text{ mol } S_2Cl_2}$ $\frac{\text{g } S_2Cl_2}{150 \text{ g sample}} \times 100$

then, S_2Cl_2 = 134.9 g/mol, SCl_2 = 103.0 g/mol, let x = mol S_2Cl_2, y = mol SCl_2
x(134.9) = g S in S_2Cl_2, y(103.0) = g S in SCl_2

Solution:

$$\text{mol S} = 50.0 \text{ g S} \times \frac{1 \text{ mol S}}{32.1 \text{ g S}} = 1.56 \text{ mol S}$$

2x = mol S in S_2Cl_2, y = mol S in SCl_2

$2x + y = 1.56$

$x(134.9) + y(103.0) = 150.0$

$134.9x + 103.0(1.56 - 2x) = 150.0$

$71.1x = 10.44$

$x = 0.147$

$y = 1.27$

$$0.147 \text{ mol } S_2Cl_2 \times \frac{134.9 \text{ g } S_2Cl_2}{1 \text{ mol } S_2Cl_2} = 19.8 \text{ g } S_2Cl_2$$

$$\frac{19.8 \text{ g } S_2Cl_2}{150 \text{ g sample}} \times 100 = 13.2\ \%S_2Cl_2$$

Check: Units of answer, % S_2Cl_2, are correct. Magnitude is reasonable since it is between 0 and 100% and mol of S_2Cl_2 are less than mol SCl_2.

105. **Given:** coal = 2.55%S, H_2SO_4,1.0 metric ton coal **Find:** metric ton H_2SO_4 produced

Conceptual Plan: metric ton coal → kg coal → kg S → kg H_2SO_4 → metric ton H_2SO_4

$$\frac{1000 \text{ kg}}{\text{metric ton}} \quad \frac{2.55 \text{ kg S}}{100 \text{ kg coal}} \quad \frac{98.09 \text{ kg } H_2SO_4}{32.07 \text{ kg S}} \quad \frac{\text{metric ton}}{1000 \text{ kg}}$$

Solution:

$$1.0 \text{ metric ton coal} \times \frac{1000 \text{ kg coal}}{1 \text{ metric ton coal}} \times \frac{2.55 \text{ kg S}}{100 \text{ kg coal}} \times \frac{98.09 \text{ kg } H_2SO_4}{32.07 \text{ kg S}} \times \frac{1 \text{ metric ton } H_2SO_4}{1000 \text{ kg } H_2SO_4}$$

$= 0.078$ metric ton H_2SO_4

Check: Units of the answer, metric ton H_2SO_4, are correct. Magnitude is reasonable since it is more than 2.55% of a metric ton and the mass of H_2SO_4 is greater than the mass of S.

107. The sphere in the molecular models represents the electron cloud of the atom. On this scale, the nucleus would be too small to see.

109. The statement is incorrect because a chemical formula is based on the ratio of atoms combined, not the ratio of grams combined. The statement should read: "The chemical formula for ammonia (NH_3) indicates that ammonia contains three hydrogen atoms to each nitrogen atom."

111. H_2SO_4: Atomic mass S is approximately twice atomic mass O, both are much greater than atomic mass H. The order of % mass: % O > % S > % H

Chapter 4
Chemical Quantities and Aqueous Reactions

1. **Given:** 4.9 moles C_6H_{14} **Find:** balanced reaction, moles O_2 required
Conceptual Plan: balance the reaction then mol $C_6H_{14} \rightarrow$ mol O_2

$2\ C_6H_{14}(g) + 19\ O_2(g) \rightarrow 12\ CO_2(g) + 14\ H_2O(g)$ $\quad \dfrac{19 \text{ mol } O_2}{2 \text{ mol } C_6H_{14}}$

Solution: $4.9 \text{ mol } C_6H_{14} \times \dfrac{19 \text{ mol } O_2}{2 \text{ mol } C_6H_{14}} = 47 \text{ mol } O_2$

Check: The units, mol O_2, are correct. The magnitude is reasonable because much more O_2 is needed than C_6H_{14}.

3. a) **Given:** 1.3 mol N_2O_5 **Find:** mol NO_2
Conceptual Plan: mol $N_2O_5 \rightarrow$ mol NO_2

$\dfrac{4\ NO_2}{2\ N_2O_5}$

Solution: $1.3 \text{ mol } N_2O_5 \times \dfrac{4 \text{ mol } NO_2}{2 \text{ mol } N_2O_5} = 2.6 \text{ mol } NO_2$

Check: The units of the answer, mol NO_2, are correct. The magnitude is reasonable since it is greater than mol N_2O_5.

b) **Given:** 5.8 mol N_2O_5 **Find:** mol NO_2
Conceptual Plan: mol $N_2O_5 \rightarrow$ mol NO_2

$\dfrac{4\ NO_2}{2\ N_2O_5}$

Solution: $5.8 \text{ mol } N_2O_5 \times \dfrac{4 \text{ mol } NO_2}{2 \text{ mol } N_2O_5} = 11.6 \text{ mol } NO_2 = 12 \text{ mol } NO_2$

Check: The units of the answer, mol NO_2, are correct. The magnitude is reasonable since it is greater than mol N_2O_5.

c) **Given:** 10.5 g N_2O_5 **Find:** mol NO_2
Conceptual Plan: g $N_2O_5 \rightarrow$ mol $N_2O_5 \rightarrow$ molNO$_2$

$\dfrac{1 \text{ mol } N_2O_5}{108.02 \text{ g } N_2O_5}$ $\quad \dfrac{4\ NO_2}{2\ N_2O_5}$

Solution: $10.5 \text{ g } N_2O_5 \times \dfrac{1 \text{ mol } N_2O_5}{108.02 \text{ g } N_2O_5} \times \dfrac{4 \text{ mol } NO_2}{2 \text{ mol } N_2O_5} = 0.194 \text{ mol } NO_2$

Check: The units of the answer, mol NO_2, are correct. The magnitude is reasonable since 10 g is about 0.1 mol N_2O_5 and the answer is greater than mol N_2O_5.

d) **Given:** 1.55 kg N_2O_5 **Find:** mol NO_2

Conceptual Plan: kg $N_2O_5 \rightarrow$ g $N_2O_5 \rightarrow$ mol $N_2O_5 \rightarrow$ molNO$_2$

$\dfrac{1000 \text{ g } N_2O_5}{\text{kg } N_2O_5}$ $\quad \dfrac{1 \text{ mol } N_2O_5}{108.02 \text{ g } N_2O_5}$ $\quad \dfrac{4\ NO_2}{2\ N_2O_5}$

Solution: $1.55 \text{ kg } N_2O_5 \times \dfrac{1000 \text{ g } N_2O_5}{\text{kg } N_2O_5} \times \dfrac{1 \text{ mol } N_2O_5}{108.02 \text{ g } N_2O_5} \times \dfrac{4 \text{ mol } NO_2}{2 \text{ mol } N_2O_5} = 28.7 \text{ mol } NO_2$

Check: The units of the answer, mol NO_2, are correct. The magnitude is reasonable since 1.5 kg is about 14 mol N_2O_5 and the answer is greater than mol N_2O_5.

5. **Given:** 3 mol SiO_2 **Find:** mol C, mol SiC, mol CO

Conceptual Plan: mol SiO_2 → mol C → mol SiC → mol CO

$$\frac{3\text{ C}}{SiO_2} \quad \frac{SiC}{SiO_2} \quad \frac{2\text{ CO}}{SiO_2}$$

Solution:

$$3\ \cancel{\text{mol } SiO_2} \times \frac{3\text{ mol C}}{\cancel{\text{mol } SiO_2}} = 9\text{ mol C} \qquad 3\ \cancel{\text{mol } SiO_2} \times \frac{\text{mol SiC}}{\cancel{\text{mol } SiO_2}} = 3\text{ mol SiC}$$

$$3\ \cancel{\text{mol } SiO_2} \times \frac{2\text{ mol CO}}{\cancel{\text{mol } SiO_2}} = 6\text{ mol CO}$$

Given: 6 mol C **Find:** mol SiO_2, mol SiC, mol CO

Conceptual Plan: mol C → mol SiO_2 → mol SiC → mol CO

$$\frac{SiO_2}{3\text{ C}} \quad \frac{SiC}{3\text{ C}} \quad \frac{2\text{ CO}}{3\text{ C}}$$

Solution:

$$6\ \cancel{\text{mol C}} \times \frac{\text{mol } SiO_2}{3\ \cancel{\text{mol C}}} = 2\text{ mol } SiO_2 \qquad 6\ \cancel{\text{mol C}} \times \frac{\text{mol SiC}}{3\ \cancel{\text{mol C}}} = 2\text{ mol SiC}$$

$$6\ \cancel{\text{mol C}} \times \frac{2\text{ mol CO}}{3\ \cancel{\text{mol C}}} = 4\text{ mol CO}$$

Given: 10 mol CO **Find:** mol SiO_2, mol C, mol SiC

Conceptual Plan: mol CO → mol SiO_2 → mol C → mol SiC

$$\frac{SiO_2}{2\text{ CO}} \quad \frac{3\text{ C}}{2\text{ CO}} \quad \frac{SiC}{2\text{ CO}}$$

Solution:

$$10\ \cancel{\text{mol CO}} \times \frac{\text{mol } SiO_2}{2\ \cancel{\text{mol CO}}} = 5.0\text{ mol } SiO_2 \qquad 10\ \cancel{\text{mol C}} \times \frac{3\text{ mol C}}{2\ \cancel{\text{mol CO}}} = 15\text{ mol C}$$

$$10\ \cancel{\text{mol CO}} \times \frac{\text{mol SiC}}{2\ \cancel{\text{mol CO}}} = 5.0\text{ mol SiC}$$

Given: 2.8 mol SiO_2 **Find:** mol C, mol SiC, mol CO

Conceptual Plan: mol SiO_2 → mol C → mol SiC → mol CO

$$\frac{3\text{ C}}{SiO_2} \quad \frac{SiC}{SiO_2} \quad \frac{2\text{ CO}}{SiO_2}$$

Solution:

$$2.8\ \cancel{\text{mol } SiO_2} \times \frac{3\text{ mol C}}{\cancel{\text{mol } SiO_2}} = 8.4\text{ mol C} \qquad 2.8\ \cancel{\text{mol } SiO_2} \times \frac{\text{mol SiC}}{\cancel{\text{mol } SiO_2}} = 2.8\text{ mol SiC}$$

$$2.8\ \cancel{\text{mol } SiO_2} \times \frac{2\text{ mol CO}}{\cancel{\text{mol } SiO_2}} = 5.6\text{ mol CO}$$

Given: 1.55 mol C **Find:** mol SiO_2, mol SiC, mol CO

Conceptual Plan: mol C → mol SiO_2 → mol SiC → mol CO

$$\frac{SiO_2}{3\text{ C}} \quad \frac{SiC}{3\text{ C}} \quad \frac{2\text{ CO}}{3\text{ C}}$$

Solution:

$$1.55\ \cancel{\text{mol C}} \times \frac{3\text{ mol } SiO_2}{3\ \cancel{\text{mol C}}} = 0.517\text{ mol } SiO_2 \qquad 1.55\ \cancel{\text{mol C}} \times \frac{\text{mol SiC}}{3\ \cancel{\text{mol C}}} = 0.517\text{ mol SiC}$$

$$1.55\ \cancel{\text{mol C}} \times \frac{2\text{ mol CO}}{3\ \cancel{\text{mol C}}} = 1.03\text{ mol CO}$$

SiO_2	C	SiC	CO
3	9	3	6
2	**6**	2	4
5.0	15	5.0	**10**
2.8	8.4	2.8	5.6
0.517	**1.55**	0.517	1.03

7. **Given:** 3.2 g Fe **Find:** g HBr; g H_2
Conceptual Plan: **g Fe → mol Fe → mol HBr → g HBr**

$$\frac{\text{mol Fe}}{55.8\text{ g Fe}} \quad \frac{2\text{ mol HBr}}{\text{mol Fe}} \quad \frac{80.9\text{ g HBr}}{\text{mol HBr}}$$

g Fe → mol Fe → mol H_2 → g H_2

$$\frac{\text{mol Fe}}{55.8\text{ g Fe}} \quad \frac{1\text{ mol H}_2}{\text{mol Fe}} \quad \frac{2.02\text{ g H}_2}{\text{mol H}_2}$$

Solution:
$$3.2\text{ g Fe} \times \frac{1\text{ mol Fe}}{55.8\text{ g Fe}} \times \frac{2\text{ mol HBr}}{1\text{ mol Fe}} \times \frac{80.9\text{ g HBr}}{1\text{ mol HBr}} = 9.3\text{ g HBr}$$

$$3.2\text{ g Fe} \times \frac{1\text{ mol Fe}}{55.8\text{ g Fe}} \times \frac{1\text{ mol H}_2}{1\text{ mol Fe}} \times \frac{2.02\text{ g H}_2}{1\text{ mol H}_2} = 0.12\text{ g H}_2$$

Check: Units of answers, g HBr, g H_2, are correct. The magnitude of the answers is reasonable because molar mass HBr is greater than Fe and molar mass H_2 is much less than Fe.

9. a) **Given:** 2.5 g Ba **Find:** g $BaCl_2$
Conceptual Plan: **g Ba → mol Ba → mol $BaCl_2$ → g $BaCl_2$**

$$\frac{\text{mol Ba}}{137.33\text{ g Ba}} \quad \frac{1\text{ mol BaCl}_2}{1\text{ mol Ba}} \quad \frac{208.23\text{ g BaCl}_2}{1\text{ mol BaCl}_2}$$

Solution:
$$2.5\text{ g Ba} \times \frac{1\text{ mol Ba}}{137.33\text{ g Ba}} \times \frac{1\text{ mol BaCl}_2}{1\text{ mol Ba}} \times \frac{208.23\text{ g BaCl}_2}{1\text{ mol BaCl}_2} = 3.8\text{ g BaCl}_2$$

Check: Units of answer, g $BaCl_2$, are correct. The magnitude of the answer is reasonable because it is larger than grams Ba.

b) **Given:** 2.5 g CaO **Find:** g $CaCO_3$
Conceptual Plan: **g CaO → mol CaO → mol $CaCO_3$ → g $CaCO_3$**

$$\frac{\text{mol CaO}}{56.08\text{ g CaO}} \quad \frac{\text{mol CaCO}_3}{1\text{ mol CaO}} \quad \frac{100.09\text{ g CaCO}_3}{\text{mol CaCO}_3}$$

Solution:
$$2.5\text{ g CaO} \times \frac{1\text{ mol CaO}}{56.08\text{ g CaO}} \times \frac{1\text{ mol CaCO}_3}{1\text{ mol CaO}} \times \frac{100.09\text{ g CaCO}_3}{1\text{ mol CaCO}_3} = 4.5\text{ g CaCO}_3$$

Check: Units of answer, g $CaCO_3$, are correct. The magnitude of the answer is reasonable because it is larger than grams CaO.

c) **Given:** 2.5 g Mg **Find:** g MgO
Conceptual Plan: **g Mg → mol Mg → mol MgO → g MgO**

$$\frac{\text{mol Mg}}{24.30\text{ g Mg}} \quad \frac{\text{mol MgO}}{\text{mol Mg}} \quad \frac{40.30\text{ g MgO}}{\text{mol MgO}}$$

Solution:
$$2.5\text{ g Mg} \times \frac{1\text{ mol Mg}}{24.30\text{ g Mg}} \times \frac{1\text{ mol MgO}}{1\text{ mol Mg}} \times \frac{40.30\text{ g MgO}}{1\text{ mol MgO}} = 4.1\text{ g MgO}$$

Check: Units of answer, g MgO, are correct. The magnitude of the answer is reasonable because it is larger than grams Mg.

d) **Given:** 2.5 g Al **Find:** g Al_2O_3
Conceptual Plan: **g Al → mol Al → mol Al_2O_3 → g Al_2O_3**

$$\frac{\text{mol Al}}{26.98\text{ g Al}} \quad \frac{2\text{ mol }Al_2O_3}{4\text{ mol Al}} \quad \frac{101.96\text{ g }Al_2O_3}{\text{mol }Al_2O_3}$$

Solution: $2.5\text{ g Al} \times \frac{1\text{ mol Al}}{26.98\text{ g Al}} \times \frac{1\text{ mol }Al_2O_3}{1\text{ mol Al}} \times \frac{101.96\text{ g }Al_2O_3}{1\text{ mol }Al_2O_3} = 4.7\text{ g }Al_2O_3$

Check: Units of answer, g Al_2O_3, are correct. The magnitude of the answer is reasonable because it is larger than grams Al.

11. a) **Given:** 4.85 g NaOH **Find:** g HCl
Conceptual Plan: g NaOH → mol NaOH → mol HCl → g HCl

$$\frac{\text{mol NaOH}}{40.01\text{ g NaOH}} \quad \frac{1\text{ mol HCl}}{1\text{ mol NaOH}} \quad \frac{36.46\text{ g HCl}}{1\text{ mol HCl}}$$

Solution: $4.85\text{ g NaOH} \times \frac{1\text{ mol NaOH}}{40.01\text{ g NaOH}} \times \frac{1\text{ mol HCl}}{1\text{ mol NaOH}} \times \frac{36.46\text{ g HCl}}{1\text{ mol HCl}} = 4\ 42\text{ g HCl}$

Check: Units of answer, g HCl, are correct. The magnitude of the answer is reasonable since it is less than g NaOH.

b) **Given:** 4.85 g $Ca(OH)_2$ **Find:** g HNO_3
Conceptual Plan: g $Ca(OH)_2$ → mol $Ca(OH)_2$ → mol HNO_3 → g HNO_3

$$\frac{\text{mol }Ca(OH)_2}{74.10\text{ g }Ca(OH)_2} \quad \frac{2\text{ mol }HNO_3}{1\text{ mol }Ca(OH)_2} \quad \frac{63.02\text{ g }HNO_3}{1\text{ mol }HNO_3}$$

Solution: $4.85\text{ g }Ca(OH)_2 \times \frac{1\text{ mol }Ca(OH)_2}{74.10\text{ g }Ca(OH)_2} \times \frac{2\text{ mol }HNO_3}{1\text{ mol }Ca(OH)_2} \times \frac{63.02\text{ g }HNO_3}{1\text{ mol }HNO_3} = 8.25\text{ g }HNO_3$

Check: Units of answer, g HNO_3, are correct. The magnitude of the answer is reasonable since it is more than g $Ca(OH)_2$.

c) **Given:** 4.85 g KOH **Find:** g H_2SO_4
Conceptual Plan: g KOH → mol KOH → mol H_2SO_4 → g H_2SO_4

$$\frac{\text{mol KOH}}{56.11\text{ g KOH}} \quad \frac{1\text{ mol }H_2SO_4}{2\text{ mol KOH}} \quad \frac{98.09\text{ g }H_2SO_4}{1\text{ mol }H_2SO_4}$$

Solution: $4.85\text{ g NaOH} \times \frac{1\text{ mol KOH}}{56.11\text{ g KOH}} \times \frac{1\text{ mol }H_2SO_4}{2\text{ mol KOH}} \times \frac{98.09\text{ g }H_2SO_4}{1\text{ mol }H_2SO_4} = 4\ 24\text{ g }H_2SO_4$

Check: Units of answer, g H_2SO_4, are correct. The magnitude of the answer is reasonable since it is less than g KOH.

13. a) **Given:** 2 mol Na; 2 mol Br_2 **Find:** Limiting reactant
Conceptual Plan: mol Na → mol NaBr

$$\frac{2\text{ mol NaBr}}{2\text{ mol Na}}$$

→ smallest mol amount determines limiting reactant

mol Br_2 → mol NaBr

$$\frac{2\text{ mol NaBr}}{1\text{ mol }Br_2}$$

Solution:

$$2\text{ mol Na} \times \frac{2\text{ mol NaBr}}{2\text{ mol Na}} = 2\text{ mol NaBr}$$

$$2\text{ mol }Br_2 \times \frac{2\text{ mol NaBr}}{1\text{ mol }Br_2} = 4\text{ mol NaBr}$$

Na is limiting reactant

Check: Answer is reasonable since Na produced smallest amount of product.

b) **Given:** 1.8 mol Na; 1.4 mol Br_2 **Find:** Limiting reactant

Conceptual Plan: mol Na → mol NaBr

$$\frac{2 \text{ mol NaBr}}{2 \text{ mol Na}}$$

→ smallest mol amount determines limiting reactant

mol Br_2 → mol NaBr

$$\frac{2 \text{ mol NaBr}}{1 \text{ mol Br}_2}$$

Solution: $1.8 \cancel{\text{mol Na}} \times \frac{2 \text{ mol NaBr}}{2 \cancel{\text{mol Na}}} = 1.8 \text{ mol NaBr}$

$1.4 \cancel{\text{mol Br}_2} \times \frac{2 \text{ mol NaBr}}{1 \cancel{\text{mol Br}_2}} = 2.8 \text{ mol NaBr}$

Na is limiting reactant

Check: Answer is reasonable since Na produced smallest amount of product.

c) **Given:** 2.5 mol Na; 1 mol Br_2 **Find:** Limiting reactant

Conceptual Plan: mol Na → mol NaBr

$$\frac{2 \text{ mol NaBr}}{2 \text{ mol Na}}$$

→ smallest mol amount determines limiting reactant

mol Br_2 → mol NaBr

$$\frac{2 \text{ mol NaBr}}{1 \text{ mol Br}_2}$$

Solution:

$2.5 \cancel{\text{mol Na}} \times \frac{2 \text{ mol NaBr}}{2 \cancel{\text{mol Na}}} = 2.5 \text{ mol NaBr}$

$1 \cancel{\text{mol Br}_2} \times \frac{2 \text{ mol NaBr}}{1 \cancel{\text{mol Br}_2}} = 2 \text{ mol NaBr}$

Br_2 is limiting reactant

Check: Answer is reasonable since Br_2 produced smallest amount of product.

d) **Given:** 12.6 mol Na; 6.9 mol Br_2 **Find:** Limiting reactant

Conceptual Plan: mol Na → mol NaBr

$$\frac{2 \text{ mol NaBr}}{2 \text{ mol Na}}$$

→ smallest mol amount determines limiting reactant

mol Br_2 → mol NaBr

$$\frac{2 \text{ mol NaBr}}{1 \text{ mol Br}_2}$$

Solution:

$12.6 \cancel{\text{mol Na}} \times \frac{2 \text{ mol NaBr}}{2 \cancel{\text{mol Na}}} = 12.6 \text{ mol NaBr}$

$6.9 \cancel{\text{mol Br}_2} \times \frac{2 \text{ mol NaBr}}{1 \cancel{\text{mol Br}_2}} = 13.8 \text{ mol NaBr}$

Na is limiting reactant

Check: Answer is reasonable since Na produced smallest amount of product.

15. The greatest number of Cl_2 molecules will be formed from reaction mixture b and would be 3 molecules Cl_2.

a) **Given:** 7 molecules HCl, 1 molecule O_2 **Find:** Theoretical yield Cl_2

Conceptual Plan: molecule HCl → molecules Cl_2

$$\frac{2 \text{ molecule Cl}_2}{4 \text{ molecule HCl}}$$

→ smallest molecule amount determines limiting reactant

molecules O_2 → molecules Cl_2

$$\frac{2\text{ molecules } Cl_2}{1\text{ molecules } O_2}$$

Solution:

$$7\ \cancel{\text{molecules HCl}} \times \frac{2\text{ molecules } Cl_2}{4\ \cancel{\text{molecules HCl}}} = 3\text{ molecules } Cl_2$$

$$1\ \cancel{\text{molecules } O_2} \times \frac{2\text{ molecules } Cl_2}{1\ \cancel{\text{molecules } O_2}} = 2\text{ molecules } Cl_2$$

Theoretical Yield = 2 molecules Cl_2

b) **Given:** 6 molecules HCl, molecule O_2 **Find:** Theoretical yield Cl_2
Conceptual Plan: molecule HCl → molecules Cl_2

$$\frac{2\text{ molecule } Cl_2}{4\text{ molecule HCl}}$$

→ smallest molecule amount determines limiting reactant

molecules O_2 → molecules Cl_2

$$\frac{2\text{ molecules } Cl_2}{1\text{ molecules } O_2}$$

Solution:

$$6\ \cancel{\text{molecules HCl}} \times \frac{2\text{ molecules } Cl_2}{4\ \cancel{\text{molecules HCl}}} = 3\text{ molecules } Cl_2$$

$$3\ \cancel{\text{molecules } O_2} \times \frac{2\text{ molecules } Cl_2}{1\ \cancel{\text{molecules } O_2}} = 6\text{ molecules } Cl_2$$

Theoretical Yield = 3 molecules Cl_2

c) **Given:** 4 molecules HCl, 5 molecule O_2 **Find:** Theoretical yield Cl_2
Conceptual Plan: molecule HCl → molecules Cl_2

$$\frac{2\text{ molecule } Cl_2}{4\text{ molecule HCl}}$$

→ smallest molecule amount determines limiting reactant

molecules O_2 → molecules Cl_2

$$\frac{2\text{ molecules } Cl_2}{1\text{ molecules } O_2}$$

Solution:

$$4\ \cancel{\text{molecules HCl}} \times \frac{2\text{ molecules } Cl_2}{4\ \cancel{\text{molecules HCl}}} = 2\text{ molecules } Cl_2$$

$$5\ \cancel{\text{molecules } O_2} \times \frac{2\text{ molecules } Cl_2}{1\ \cancel{\text{molecules } O_2}} = 10\text{ molecules } Cl_2$$

Theoretical Yield = 2 molecules Cl_2
Check: The units of the answer, molecules Cl_2, is correct. The answer is reasonable based on the limiting reactant in each mixture.

17. a) **Given:** 4 mol Ti, 4 mol Cl_2 **Find:** Theoretical yield $TiCl_4$
Conceptual Plan: mol Ti → mol $TiCl_4$

$$\frac{1\text{ mol } TiCl_4}{1\text{ mol Ti}}$$

→ smallest mol amount determines limiting reactant

mol Cl_2 → mol $TiCl_4$

$$\frac{1\text{ mol } TiCl_4}{2\text{ mol } Cl_2}$$

Solution: $4\ \cancel{\text{mol Ti}} \times \dfrac{1\text{ mol } TiCl_4}{1\ \cancel{\text{mol Ti}}} = 4\text{ mol } TiCl_4$

$$4\ \text{mol Cl}_2 \times \frac{1\ \text{mol TiCl}_4}{2\ \text{mol Cl}_2} = 2\ \text{mol TiCl}_4$$

Theoretical Yield = 2 mol $TiCl_4$

Check: Units of the answer, mol $TiCl_4$, are correct. Answer is reasonable since Cl_2 produced smallest amount of product and is the limiting reactant.

b) **Given:** 7 mol Ti, 17 mol Cl_2 **Find:** Theoretical yield $TiCl_4$

Conceptual Plan: mol Ti → mol $TiCl_4$

$$\frac{1\ \text{mol TiCl}_4}{1\ \text{mol Ti}}$$ → **smallest mol amount determines limiting reactant**

mol Cl_2 → mol $TiCl_4$

$$\frac{1\ \text{mol TiCl}_4}{2\ \text{mol Cl}_2}$$

Solution:

$$7\ \text{mol Ti} \times \frac{1\ \text{mol TiCl}_4}{1\ \text{mol Ti}} = 7\ \text{mol TiCl}_4$$

$$17\ \text{mol Cl}_2 \times \frac{1\ \text{mol TiCl}_4}{2\ \text{mol Cl}_2} = 8.5\ \text{mol TiCl}_4$$

Theoretical Yield = 7 mol $TiCl_4$

Check: Units of the answer, mol $TiCl_4$, are correct. Answer is reasonable since Ti produced smallest amount of product and is the limiting reactant.

c) **Given:** 12.4 mol Ti, 18.8 mol Cl_2 **Find:** Theoretical yield $TiCl_4$

Conceptual Plan: mol Ti → mol $TiCl_4$

$$\frac{1\ \text{mol TiCl}_4}{1\ \text{mol Ti}}$$ → **smallest mol amount determines limiting reactant**

mol Cl_2 → mol $TiCl_4$

$$\frac{1\ \text{mol TiCl}_4}{2\ \text{mol Cl}_2}$$

Solution:

$$12.4\ \text{mol Ti} \times \frac{1\ \text{mol TiCl}_4}{1\ \text{mol Ti}} = 12.4\ \text{mol TiCl}_4$$

$$18.8\ \text{mol Cl}_2 \times \frac{1\ \text{mol TiCl}_4}{2\ \text{mol Cl}_2} = 9.4\ \text{mol TiCl}_4$$

Theoretical Yield = 9.4 mol $TiCl_4$

Check: Units of the answer, mol $TiCl_4$, are correct. Answer is reasonable since Cl_2 produced smallest amount of product and is the limiting reactant.

19. a) **Given:** 2.0 g Al, 2.0 g Cl_2 **Find:** Theoretical yield in g $AlCl_3$

Conceptual Plan: g Al → mol Al → mol $AlCl_3$

$$\frac{1\ \text{mol Al}}{26.98\ \text{g Al}} \quad \frac{2\ \text{mol AlCl}_3}{2\ \text{mol Al}}$$ → **smallest mol amount determines limiting reactant**

g Cl_2 → mol Cl_2 → mol $AlCl_3$

$$\frac{1\ \text{mol Cl}_2}{70.91\ \text{g Cl}_2} \quad \frac{2\ \text{mol AlCl}_3}{3\ \text{mol Cl}_2}$$

then: mol $AlCl_3$ → g $AlCl_3$

$$\frac{133.34\ \text{g AlCl}_3}{\text{mol AlCl}_3}$$

Solution:

$$2.0\ \text{g Al} \times \frac{1\ \text{mol Al}}{26.98\ \text{g Al}} \times \frac{2\ \text{mol AlCl}_3}{2\ \text{mol Al}} = 0.074\ \text{mol AlCl}_3$$

$$2.0\ \text{g Cl}_2 \times \frac{1\ \text{mol Cl}_2}{70.90\ \text{g Cl}_2} \times \frac{2\ \text{mol AlCl}_3}{3\ \text{mol Cl}_2} = 0.01\underline{8}8\ \text{mol AlCl}_3$$

$$0.01\underline{8}8\ \text{mol AlCl}_3 \times \frac{133.34\ \text{g AlCl}_3}{\text{mol AlCl}_3} = 2.5\ \text{g AlCl}_3$$

Check: Units of the answer, g $AlCl_3$, are correct. Answer is reasonable since Cl_2 produced smallest amount of product and is the limiting reactant.

b) **Given:** 7.5 g Al, 24.8 g Cl_2 **Find:** Theoretical yield in g $AlCl_3$

Conceptual Plan: g Al → mol Al → mol $AlCl_3$

$$\frac{1\ \text{mol Al}}{26.98\ \text{g Al}} \quad \frac{2\ \text{mol AlCl}_3}{2\ \text{mol Al}}$$ **→ smallest mol amount determines limiting reactant**

g Cl_2 → mol Cl_2 → mol $AlCl_3$

$$\frac{1\ \text{mol Cl}_2}{70.91\ \text{g Cl}_2} \quad \frac{2\ \text{mol AlCl}_3}{3\ \text{mol Cl}_2}$$

then: mol $AlCl_3$ → g $AlCl_3$

$$\frac{133.34\ \text{g AlCl}_3}{\text{mol AlCl}_3}$$

Solution:

$$7.5\ \text{g Al} \times \frac{1\ \text{mol Al}}{26.98\ \text{g Al}} \times \frac{2\ \text{mol AlCl}_3}{2\ \text{mol Al}} = 0.2780\ \text{mol AlCl}_3$$

$$24.8\ \text{g Cl}_2 \times \frac{1\ \text{mol Cl}_2}{70.90\ \text{g Cl}_2} \times \frac{2\ \text{mol AlCl}_3}{3\ \text{mol Cl}_2} = 0.23\underline{3}2\ \text{mol AlCl}_3$$

$$0.23\underline{3}2\ \text{mol AlCl}_3 \times \frac{133.34\ \text{g AlCl}_3}{\text{mol AlCl}_3} = 31.1\ \text{g AlCl}_3$$

Check: Units of the answer, g $AlCl_3$, are correct. Answer is reasonable since Cl_2 produced smallest amount of product and is the limiting reactant.

c) **Given:** 0.235 g Al, 1.15 g Cl_2 **Find:** Theoretical yield in g $AlCl_3$

Conceptual Plan: g Al → mol Al → mol $AlCl_3$

$$\frac{1\ \text{mol Al}}{26.98\ \text{g Al}} \quad \frac{2\ \text{mol AlCl}_3}{2\ \text{mol Al}}$$ **→ smallest mol amount determines limiting reactant**

g Cl_2 → mol Cl_2 → mol $AlCl_3$

$$\frac{1\ \text{mol Cl}_2}{70.91\ \text{g Cl}_2} \quad \frac{2\ \text{mol AlCl}_3}{3\ \text{mol Cl}_2}$$

then: mol $AlCl_3$ → g $AlCl_3$

$$\frac{133.34\ \text{g AlCl}_3}{\text{mol AlCl}_3}$$

Solution:

$$0.235\ \text{g Al} \times \frac{1\ \text{mol Al}}{26.98\ \text{g Al}} \times \frac{2\ \text{mol AlCl}_3}{2\ \text{mol Al}} = 0.0087\underline{1}0\ \text{mol AlCl}_3$$

$$1.15\ \text{g Cl}_2 \times \frac{1\ \text{mol Cl}_2}{70.90\ \text{g Cl}_2} \times \frac{2\ \text{mol AlCl}_3}{3\ \text{mol Cl}_2} = 0.01081\ \text{mol AlCl}_3$$

$$0.0087\underline{1}0\ \text{mol AlCl}_3 \times \frac{133.34\ \text{g AlCl}_3}{\text{mol AlCl}_3} = 1.16\ \text{g AlCl}_3$$

Check: Units of the answer, g $AlCl_3$, are correct. Answer is reasonable since Al produced smallest amount of product and is the limiting reactant.

21. **Given:** 28.5 g KCl; 25.7 g Pb^{2+}; 29.4 g $PbCl_2$ **Find:** limiting reactant, theoretical yield $PbCl_2$, % yield
Conceptual Plan: **g KCl → mol KCl → mol $PbCl_2$**

$$\frac{1 \text{ mol KCl}}{74.55 \text{ g KCl}} \quad \frac{1 \text{ mol PbCl}_2}{2 \text{ mol KCl}}$$

→ smallest mol amount determines limiting reactant

g Pb^{2+} → mol Pb^{2+} → mol $PbCl_2$

$$\frac{1 \text{ mol Pb}^{2+}}{207.2 \text{ g Pb}^{2+}} \quad \frac{1 \text{ mol PbCl}_2}{1 \text{ mol Pb}^{2+}}$$

then: mol $PbCl_2$ → g $PbCl_2$

$$\frac{278.1 \text{ g PbCl}_2}{\text{mol PbCl}_2}$$

then: determine % yield

$$\frac{\text{actual yield g PbCl}_2}{\text{theoretical yield g PbCl}_2} \times 100$$

Solution:

$$28.5 \text{ g KCl} \times \frac{1 \text{ mol KCl}}{74.55 \text{ g KCl}} \times \frac{1 \text{ mol PbCl}_2}{2 \text{ mol KCl}} = 0.19\underline{1}1 \text{ mol PbCl}_2$$

$$25.7 \text{ g Pb}^{2+} \times \frac{1 \text{ mol Pb}^{2+}}{207.2 \text{ g Pb}^{2+}} \times \frac{1 \text{ mol PbCl}_2}{1 \text{ mol Pb}^{2+}} = 0.1240 \text{ mol PbCl}_2$$

$$0.12\underline{4}0 \text{ mol PbCl}_2 \times \frac{278.1 \text{ g PbCl}_2}{1 \text{ mol PbCl}_2} = 34.\underline{5} \text{ g PbCl}_2$$

$$\frac{29.4 \text{ g PbCl}_2}{34.\underline{5} \text{ g PbCl}_2} \times 100\% = 85.2\%$$

Check: The theoretical yield has the correct units, g $PbCl_2$, and has a reasonable magnitude compared to the mass of Pb^{2+}, the limiting reactant. The % yield is reasonable, under 100%.

23. **Given:** 136.4 kg NH_3; 211.4 kg CO_2; 168.4 kg CH_4N_2O **Find:** limiting reactant, theoretical yield CH_4N_2O, % yield

Conceptual Plan: kg NH_3 → g NH_3 → mol NH_3 → mol CH_4N_2O

$$\frac{1000 \text{ g}}{1 \text{ kg}} \quad \frac{1 \text{ mol NH}_3}{17.03 \text{ g NH}_3} \quad \frac{1 \text{ mol CH}_4\text{N}_2\text{O}}{2 \text{ mol NH}_3}$$

→ smallest amount determines limiting reactant

kg CO_2 → g CO_2 → mol CO_2 → mol CH_4N_2O

$$\frac{1000 \text{ g}}{1 \text{ kg}} \quad \frac{1 \text{ mol CO}_2}{44.01 \text{ g CO}_2} \quad \frac{1 \text{ mol CH}_4\text{N}_2\text{O}}{1 \text{ mol CO}_2}$$

then: mol CH_4N_2O → g CH_4N_2O → kg CH_4N_2O

$$\frac{60.06 \text{ g CH}_4\text{N}_2\text{O}}{1 \text{ mol CH}_4\text{N}_2\text{O}} \quad \frac{1 \text{ kg}}{1000 \text{ g}}$$

then: determine % yield

$$\frac{\text{actual yield kg CH}_4\text{N}_2\text{O}}{\text{theoretical yield kg CH}_4\text{N}_2\text{O}} \times 100$$

Solution:

$$136.4 \text{ kg NH}_3 \times \frac{1000 \text{ g}}{\text{kg}} \times \frac{1 \text{ mol NH}_3}{17.03 \text{ g NH}_3} \times \frac{1 \text{ mol CH}_4\text{N}_2\text{O}}{2 \text{ mol NH}_3} = 400\underline{4}.7 \text{ mol CH}_4\text{N}_2\text{O}$$

$$211.4 \text{ kg CO}_2 \times \frac{1000 \text{ g}}{\text{kg}} \times \frac{1 \text{ mol CO}_2}{44.01 \text{ g CO}_2} \times \frac{1 \text{ mol CH}_4\text{N}_2\text{O}}{1 \text{ mol CO}_2} = 480\underline{3}.2 \text{ mol CH}_4\text{N}_2\text{O}$$

$$400\underline{4}.7 \text{ mol CH}_4\text{N}_2\text{O} \times \frac{60.06 \text{ g CH}_4\text{N}_2\text{O}}{1 \text{ mol CH}_4\text{N}_2\text{O}} \times \frac{\text{kg}}{1000 \text{ g}} = 240.\underline{5}2 \text{ kg CH}_4\text{N}_2\text{O}$$

$$\frac{168.4 \text{ kg CH}_4\text{N}_2\text{O}}{240.\underline{5}2 \text{ kg CH}_4\text{N}_2\text{O}} \times 100\% = 70.01\%$$

Check: The theoretical yield has the correct units, kg CH_4N_2O, and has a reasonable magnitude compared to the mass of NH_3, the limiting reactant. The % yield is reasonable, under 100%.

25. a) **Given:** 4.3 mol LiCl; 2.8 L solution **Find:** Molarity LiCl
Conceptual Plan: mol LiCl, L solution → Molarity

$$\text{molarity (M)} = \frac{\text{amount of solute (in moles)}}{\text{volume of solution (in L)}}$$

Solution: $\frac{4.3 \text{ mol LiCl}}{2.8 \text{ L solution}} = 1.5 \text{ M}$

Check: The units of the answer, M, are correct. The magnitude of the answer is reasonable. Concentrations are usually between 0 M and 18 M.

b) **Given:** 22.6 g $C_6H_{12}O_6$; 1.08 L solution **Find:** Molarity $C_6H_{12}O_6$
Conceptual Plan: g $C_6H_{12}O_6$ → mol $C_6H_{12}O_6$, L solution → Molarity

$$\frac{\text{mol } C_6H_{12}O_6}{180.16 \text{ g } C_6H_{12}O_6} \qquad \text{molarity (M)} = \frac{\text{amount of solute (in moles)}}{\text{volume of solution (in L)}}$$

Solution: $22.6 \text{ g } C_6H_{12}O_6 \times \frac{1 \text{ mol } C_6H_{12}O_6}{180.16 \text{ g } C_6H_{12}O_6} = 0.12\underline{5}4 \text{ mol } C_6H_{12}O_6$

$$\frac{0.12\underline{5}4 \text{ mol } C_6H_{12}O_6}{1.08 \text{ L solution}} = 0.116 \text{ M}$$

Check: The units of the answer, M, are correct. The magnitude of the answer is reasonable. Concentrations are usually between 0 M and 18 M.

c) **Given:** 45.5 mg NaCl; 154.4 mL solution **Find:** Molarity NaCl
Conceptual Plan: mg NaCl → g NaCl → mol NaCl, and mL solution → L solution then Molarity

$$\frac{\text{g NaCl}}{1000 \text{ mg NaCl}} \quad \frac{\text{mol NaCl}}{58.45 \text{ g NaCl}} \qquad \frac{\text{L solution}}{1000 \text{ mL solution}} \quad \text{molarity (M)} = \frac{\text{amount of solute (in moles)}}{\text{volume of solution (in L)}}$$

Solution: $45.5 \text{ mg NaCl} \times \frac{1 \text{ g}}{1000 \text{ mg}} \times \frac{1 \text{ mol NaCl}}{58.45 \text{ g NaCl}} = 7.7\underline{8}4 \times 10^{-4} \text{ mol NaCl}$

$$154.4 \text{ mL solution} \times \frac{1 \text{ L}}{1000 \text{ mL}} = 0.1544 \text{ L}$$

$$\frac{7.7\underline{8}4 \times 10^{-4} \text{ mol NaCl}}{0.1544 \text{ L}} = 0.00504 \text{ M NaCl}$$

Check: The units of the answer, M, are correct. The magnitude of the answer is reasonable. Concentrations are usually between 0 M and 18 M.

27. a) **Given:** 0.556 L; 2.3 M KCl **Find:** mol KCl
Conceptual Plan: volume solution x M = mol

volume solution (L) x M = mol

Solution: $0.556 \text{ L solution} \times \frac{2.3 \text{ mol KCl}}{\text{L solution}} = 1.3 \text{ mol KCl}$

Check: Units of answer, mol KCl, are correct. The magnitude is reasonable since it is less than 1 L solution.

b) **Given:** 1.8 L; 0.85 M KCl **Find:** mol KCl
Conceptual Plan: volume solution x M = mol

volume solution (L) x M = mol

Solution: $1.8 \text{ L solution} \times \frac{0.85 \text{ mol KCl}}{\text{L solution}} = 1.5 \text{ mol KCl}$

Check: Units of answer, mol KCl, are correct. The magnitude is reasonable since it is less than 2 L solution.

c) **Given:** 114 mL; 1.85 M KCl **Find:** mol KCl
Conceptual Plan: mL solution → L solution, then volume solution x M = mol

$$\frac{1 \text{ L}}{1000 \text{ mL}} \qquad \text{volume solution (L) x M = mol}$$

Solution: $114 \text{ mL solution} \times \frac{1 \text{ L}}{1000 \text{ mL}} \times \frac{1.85 \text{ mol KCl}}{\text{L solution}} = 0.211 \text{ mol KCl}$

Check: Units of answer, mol KCl, are correct. The magnitude is reasonable since it is less than 1 L solution.

29. **Given:** 400.0 mL; 1.1 M $NaNO_3$ **Find:** g $NaNO_3$

Conceptual Plan: mL solution → L solution, then volume solution x M = mol $NaNO_3$

$\frac{\text{L solution}}{\text{1000 mL solution}}$ volume solution (L) x M = mol

then mol $NaNO_3$ → g $NaNO_3$

$\frac{85.01 \text{ g } NaNO_3}{\text{mol } NaNO_3}$

Solution: $400.0 \text{ mL solution} \times \frac{1 \text{ L}}{1000 \text{ mL}} \times \frac{1.1 \text{ mol } NaNO_3}{\text{L solution}} \times \frac{85.01 \text{ g}}{\text{mol } NaNO_3} = 37 \text{ g } NaNO_3$

Check: Units of answer, g $NaNO_3$, are correct. The magnitude is reasonable for the concentration and volume of solution.

31. **Given:** $V_1 = 123$ mL;$M_1 = 1.1$ M;$V_2 = 500.0$ mL **Find:** M_2

Conceptual Plan: mL → L then V_1, M_1, V_2 → M_2

$\frac{1 \text{ L}}{1000 \text{ mL}}$ $V_1M_1 = V_2M_2$

Solution: $123 \text{ mL} \times \frac{1 \text{ L}}{1000 \text{ mL}} = 0.123 \text{ L}$ $500.0 \text{ mL} \times \frac{1 \text{ L}}{1000 \text{ mL}} = 0.5000 \text{ L}$

$$M_2 = \frac{V_1M_1}{V_2} = \frac{(0.123 \text{ L})(1.1 \text{ M})}{(0.5000 \text{ L})} = 0.27 \text{ M}$$

Check: Units of the answer, M, are correct. The magnitude of the answer is reasonable since it is less than the original concentration.

33. **Given:** $V_1 = 50$ mL;$M_1 = 12$ M; $M_2 = 0.100$ M **Find:** V_2

Conceptual Plan: mL → L then V_1, M_1, M_2 → V_2

$\frac{1 \text{ L}}{1000 \text{ mL}}$ $V_1M_1 = V_2M_2$

Solution: $50 \text{ mL} \times \frac{1 \text{ L}}{1000 \text{ mL}} = 0.050 \text{ L}$

$$V_2 = \frac{V_1M_1}{M_2} = \frac{(0.050 \text{ L})(12 \text{ M})}{(0.100 \text{ M})} = 6.0 \text{ L}$$

Check: Units of the answer, L, are correct. The magnitude of the answer is reasonable since the new concentration is much less than the original, the volume must be larger.

35. **Given:** 95.4 mL, 0.102 M $CuCl_2$; 0.175 M Na_3PO_4 **Find:** volume Na_3PO_4

Conceptual Plan: mL $CuCl_2$→ L $CuCl_2$ → mol $CuCl_2$ → mol Na_3PO_4 → L Na_3PO_4 → mL Na_3PO_4

$\frac{1 \text{ L}}{1000 \text{ mL}}$ $\frac{0.102 \text{ mol } CuCl_2}{\text{L}}$ $\frac{2 \text{ mol } Na_3PO_4}{3 \text{ mol } CuCl_2}$ $\frac{1 \text{ L}}{0.175 \text{ mol } Na_3PO_4}$ $\frac{1000 \text{ mL}}{\text{L}}$

Solution:

$$95.4 \text{ mL } CuCl_2 \times \frac{1 \text{ L}}{1000 \text{ mL}} \times \frac{0.102 \text{ mol } CuCl_2}{1 \text{ L}} \times \frac{2 \text{ mol } Na_3PO_4}{3 \text{ mol } CuCl_2} \times \frac{1 \text{ L}}{0.175 \text{ mol } Na_3PO_4} \times \frac{1000 \text{ mL}}{1 \text{ L}}$$

$= 37.1 \text{ mL } Na_3PO_4$

Check: Units of answer, mL Na_3PO_4, are correct. The magnitude of the answer is reasonable since the concentration of Na_3PO_4 is greater.

37. **Given:** 25.0 g H_2; 6.0 M H_2SO_4 **Find:** volume H_2SO_4

Conceptual Plan: g H_2 → mol H_2 → mol H_2SO_4 → L H_2SO_4

$$\frac{2.016\ \text{g}\ H_2}{1\ \text{mol}\ H_2} \qquad \frac{3\ \text{mol}\ H_2SO_4}{3\ \text{mol}\ H_2} \qquad \frac{1\ \text{L}}{6.0\ \text{mol}\ H_2SO_4}$$

Solution: $25.0\ \text{g}\ H_2 \times \frac{1\ \text{mol}\ H_2}{2.016\ \text{g}\ H_2} \times \frac{3\ \text{mol}\ H_2SO_4}{3\ \text{mol}\ H_2} \times \frac{1\ \text{L}}{6.0\ \text{mol}\ H_2SO_4} = 2.1\ \text{L}\ H_2SO_4$

Check: The units, L H_2SO_4, are correct. The magnitude is reasonable since there are approximately 12 mol H_2 and the mole ratio is 1:1.

39. a) CsCl is an ionic compound. An aqueous solution is an electrolyte solution, so it conducts electricity.

b) CH_3OH is a molecular compound that does not dissociate. An aqueous solution is a nonelectrolyte solution, so it does not conduct electricity.

c) $Ca(NO_3)_2$ is an ionic compound. An aqueous solution is an electrolyte solution, so it conducts electricity.

d) $C_6H_{12}O_6$ is a molecular compound that does not dissociate. An aqueous solution is a nonelectrolyte solution, so it does not conduct electricity.

41. a) $AgNO_3$ is soluble. Compounds containing NO_3^- are always soluble with no exceptions. The ions in solution are $Ag^+(aq)$ and $NO_3^-(aq)$.

b) $Pb(C_2H_3O_2)_2$ is soluble. Compounds containing $C_2H_3O_2^-$ are always soluble with no exceptions. The ions in solution are $Pb^{2+}(aq)$ and $C_2H_3O_2^-(aq)$.

c) KNO_3 is soluble. Compounds containing K^+ are always soluble with no exceptions. The ions in solution are $K^+(aq)$ and $NO_3^-(aq)$.

d) $(NH_4)_2S$ is soluble. Compounds containing NH_4^+ are always soluble with no exceptions. The ions in solution are $NH_4^+(aq)$ and $S^{2-}(aq)$.

43. a) $LiI(aq) + BaS(aq) \rightarrow$ Possible products: Li_2S and BaI_2. Li_2S is soluble. Compounds containing S^{2-} are normally insoluble but Li^+ is an exception. BaI_2 is soluble. Compounds containing I^- are normally soluble and Ba^{2+} is not an exception. $LiI(aq) + BaS(aq) \rightarrow$ No Reaction

b) $KCl(aq) + CaS(aq) \rightarrow$ Possible products: K_2S and $CaCl_2$. K_2S is soluble. Compounds containing S^{2-} are normally insoluble but K^+ is an exception. $CaCl_2$ is soluble. Compounds containing Cl^- are normally soluble and Ca^{2+} is not an exception. $KCl(aq) + CaS(aq) \rightarrow$ No Reaction

c) $CrBr_2(aq) + Na_2CO_3(aq) \rightarrow$ Possible products: $CrCO_3$ and NaBr. $CrCO_3$ is insoluble. Compounds containing CO_3^{2-} are normally insoluble and Cr^{2+} is not an exception. NaBr is soluble. Compounds containing Br^- are normally soluble and Na^+ is not an exception.
$CrBr_2(aq) + Na_2CO_3(aq) \rightarrow CrCO_3(s) + 2\ NaBr(aq)$

d) $NaOH(aq) + FeCl_3(aq) \rightarrow$ Possible products NaCl and $Fe(OH)_3$. NaCl is soluble. Compounds containing Na^+ are normally soluble, no exceptions. $Fe(OH)_3$ is insoluble. Compounds containing OH^- are normally insoluble and Fe^{3+} is not an exception.
$3\ NaOH(aq) + FeCl_3(aq) \rightarrow 3\ NaCl(aq) + Fe(OH)_3(s)$

45. a) $K_2CO_3(aq) + Pb(NO_3)_2(aq) \rightarrow$ Possible products: KNO_3 and $PbCO_3$. KNO_3 is soluble. Compounds containing K^+ are always soluble, no exceptions. $PbCO_3$ is insoluble. Compounds containing CO_3^{2-} are normally insoluble and Pb^{2+} is not an exception.
$K_2CO_3(aq) + Pb(NO_3)_2(aq) \rightarrow 2\ KNO_3(aq) + PbCO_3(s)$

b) $Li_2SO_4(aq) + Pb(C_2H_3O_2)_2(aq) \rightarrow$ Possible products: $LiC_2H_3O_2$ and $PbSO_4$. $LiC_2H_3O_2$ is soluble. Compounds containing Li^+ are always soluble, no exceptions. $PbSO_4$ is insoluble. Compounds containing $SO_4^{\,2-}$ are normally soluble but, Pb^{2+} is an exception.
$Li_2SO_4(aq) + Pb(C_2H_3O_2)_2(aq) \rightarrow 2\ LiC_2H_3O_2(aq) + PbSO_4(s)$

c) $Cu(NO_3)_2(aq) + MgS(s) \rightarrow$ Possible products: CuS and $Mg(NO_3)_2$. CuS is insoluble. Compounds containing S^{2-} are normally insoluble and Cu^{2+} is not an exception. $Mg(NO_3)_2$ is soluble. Compounds containing $NO_3^{\,-}$ are always soluble, no exceptions. $Cu(NO_3)_2(aq) + MgS(s) \rightarrow CuS(s) + Mg(NO_3)_2(aq)$

d) $Sr(NO_3)_2(aq) + KI(aq) \rightarrow$ Possible products: SrI_2 and KNO_3. SrI_2 is soluble. Compounds containing I^- are normally soluble and Sr^{2+} is not an exception. KNO_3 is soluble. Compounds containing K^+ are always soluble, no exceptions. $Sr(NO_3)_2(aq) + KI(aq) \rightarrow$ No Reaction

47. a) $H^+(aq) +$ ~~Cl~~$^-(aq) +$ ~~Li~~$^+(aq) + OH^-(aq) \rightarrow H_2O(l) +$ ~~Li~~$^+(aq) +$ ~~Cl~~$^-(aq)$
$H^+(aq) + OH^-(aq) \rightarrow H_2O(l)$

b) ~~Mg~~$^{2+}(aq) + S^{2-}(aq) + Cu^{2+}(aq) + 2$ ~~Cl~~$^-(aq) \rightarrow CuS(s) +$ ~~Mg~~$^{2+}(aq) + 2$ ~~Cl~~$^-(aq)$
$Cu^{2+}(aq) + S^{2-}(aq) \rightarrow CuS(s)$

c) ~~Na~~$^+(aq) + OH^-(aq) + H^+(aq) +$ ~~NO_3~~$^-(aq) \rightarrow H_2O(l) +$ ~~Na~~$^+(aq) +$ ~~NO_3~~$^-(aq)$
$H^+(aq) + OH^-(aq) + \rightarrow H_2O(l)$

d) 6 ~~Na~~$^+(aq) + 2\ PO_4^{\,3-}(aq) + 3\ Ni^{2+}(aq) + 6$ ~~Cl~~$^-(aq) \rightarrow Ni_3(PO_4)_2(s) + 6$ ~~Na~~$^+(aq) + 6$ ~~Cl~~$^-(aq)$
$3\ Ni^{2+}(aq) + 2\ PO_4^{\,3-}(aq) \rightarrow Ni_3(PO_4)_2(s)$

49. $Hg_2^{\,2+}(aq) +$ ~~$2NO_3$~~$^-(aq) +$ ~~$2Na$~~$^+(aq) + 2\ Cl^-(aq) \rightarrow Hg_2Cl_2(s) +$ ~~$2Na$~~$^+(aq) +$ ~~$2NO_3$~~$^-(aq)$
$Hg_2^{\,2+}(aq) + 2\ Cl^-(aq) \rightarrow Hg_2Cl_2(s)$

51. Skeletal reaction: $HBr(aq) + KOH(aq) \rightarrow H_2O(l) + KBr(aq)$
acid base water salt
Net ionic equation: $H^+(aq) + OH^-(aq) \rightarrow H_2O(l)$

53. a) Skeletal reaction: $H_2SO_4(aq) + Ca(OH)_2(aq) \rightarrow H_2O(l) + CaSO_4(s)$
acid base water salt
Balanced reaction: $H_2SO_4(aq) + Ca(OH)_2(aq) \rightarrow 2\ H_2O(l) + CaSO_4(s)$

b) Skeletal reaction: $HClO_4(aq) + KOH(aq) \rightarrow H_2O(l) + KClO_4(aq)$
acid base water salt
Balanced reaction: $HClO_4(aq) + KOH(aq) \rightarrow H_2O(l) + KClO_4(aq)$

c) Skeletal reaction: $H_2SO_4(aq) + NaOH(aq) \rightarrow H_2O(l) + Na_2SO_4(aq)$
acid base water salt
Balanced reaction: $H_2SO_4(aq) + 2\ NaOH(aq) \rightarrow 2\ H_2O(l) + Na_2SO_4(aq)$

55. a) Skeletal reaction: $HBr(aq) + NiS(S) \rightarrow NiBr_2(aq) + H_2S(g)$
gas
Balanced reaction: $2\ HBr(aq) + NiS(s) \rightarrow NiBr_2(aq) + H_2S(g)$

b) Skeletal reaction: $NH_4I(aq) + NaOH(aq) \rightarrow NH_4OH(aq) + NaI(aq) \rightarrow H_2O(l) + NH_3(g) + NaI(aq)$
decomposes gas
Balanced reaction: $NH_4I(aq) + NaOH(aq) \rightarrow H_2O(l) + NH_3(g) + NaI(aq)$

c) Skeletal reaction: $HBr(aq) + Na_2S(aq) \rightarrow NaBr(aq) + H_2S(g)$
gas
Balanced reaction: $2\ HBr(aq) + Na_2S(aq) \rightarrow 2\ NaBr(aq) + H_2S(g)$

d) Skeletal reaction:
$HClO_4(aq) + Li_2CO_3(aq) \rightarrow H_2CO_3(aq) + LiClO_4(aq) \rightarrow H_2O(l) + CO_2(g) + LiClO_4(aq)$
decomposes gas
Balanced reaction: $2\ HClO_4(aq) + Li_2CO_3(aq) \rightarrow H_2O(l) + CO_2(g) + 2\ LiClO_4(aq)$

57. a) Ag. The oxidation state of Ag = 0. The oxidation state of an atom in a free element is 0.

b) Ag^+. The oxidation state of Ag^+ = +1. The oxidation state of a monatomic ion is equal to its charge.

c) CaF_2. The oxidation state of Ca = +2, the oxidation state of F = – 1. The oxidation state of a Group 2A metal always has an oxidation state of +2, the oxidation of F is – 1 since the sum of the oxidation states in a neutral formula unit = 0.

d) H_2S. The oxidation state of H = + 1, the oxidation state of S = – 2. The oxidation state of H when listed first is +1, the oxidation state of S is – 2 since S is in group 6A and the sum of the oxidation states in a neutral molecular unit = 0.

e) $CO_3{}^{2-}$. The oxidation state of C = +4, the oxidation state of O = – 2. The oxidation state of O is normally – 2, the oxidation state of C is deduced from the formula since the sum of the oxidation states must equal the charge on the ion. (C ox state) + 4(O ox state) = – 2: (C ox state) + 2(– 2) = – 2, so C ox state = + 4.

f) $CrO_4{}^{2-}$. The oxidation state of Cr = +6, the oxidation state of O = – 2. The oxidation state of O is normally – 2, the oxidation state of Cr is deduced from the formula since the sum of the oxidation states must equal the charge on the ion. (Cr ox state) + 4(O ox state) = – 2; (Cr ox state) + 2(– 2) = – 2, so Cr ox state = + 6.

59. a) CrO. The oxidation state of Cr = +2, the oxidation state of O = – 2. The oxidation state of O is normally – 2, the oxidation state of Cr is deduced from the formula since the sum of the oxidation states must = 0.
(Cr ox state) + (O ox state) = 0; (Cr ox state) + (– 2) = 0, so Cr = +2.

b) CrO_3. The oxidation state of Cr = +6, the oxidation state of O = – 2. The oxidation state of O is normally – 2, the oxidation state of Cr is deduced from the formula since the sum of the oxidation states must = 0.
(Cr ox state) + 3(O ox state) = 0; (Cr ox state) +3 (– 2) = 0, so Cr = +6.

c) Cr_2O_3. The oxidation state of Cr = +3, the oxidation state of O = – 2. The oxidation state of O is normally – 2, the oxidation state of Cr is deduced from the formula since the sum of the oxidation states must = 0.
2(Cr ox state) +3 (O ox state) = 0; 2(Cr ox state) + 3(– 2) = 0, so Cr = +3.

61. a) $4\ Li(s) + O_2(g) \rightarrow 2\ Li_2O(s)$
oxidation states; 0 0 +1 –2
This is a redox reaction since Li increases in oxidation number (oxidation) and O decreases in number (reduction). O_2 is the oxidizing agent, Li is the reducing agent.

b) $Mg(s) + Fe^{2+}(aq) \rightarrow Mg^{2+}(aq) + Fe(s)$
oxidation states; 0 +2 +2 0
This is a redox reaction since Mg increases in oxidation number (oxidation) and Fe decreases in number (reduction). Fe^{2+} is the oxidizing agent, Mg is the reducing agent.

c) $Pb(NO_3)_2(aq) + Na_2SO_4(aq) \rightarrow PbSO_4(s) + 2\ NaNO_3(aq)$
oxidation states; +2 +5 - 2 +1 +6 - 2 +2 +6 -2 +1 +5 -2
This is a not a redox reaction since none of the atoms undergoes a change in oxidation number.

d) $HBr(aq) + KOH(aq) \rightarrow H_2O(l) + KBr(aq)$
oxidation states; +1 - 1 +1 – 2 +1 +1 -2 +1 -2
This is a not a redox reaction since none of the atoms undergoes a change in oxidation number.

63. a) Skeletal reaction: $S(s) + O_2(g) \rightarrow SO_2(g)$
Balanced reaction: $S(s) + O_2(g) \rightarrow SO_2(g)$

b) Skeletal reaction: $C_3H_6(g) + O_2(g) \rightarrow CO_2(g) + H_2O(g)$
Balance C: $C_3H_6(g) + O_2(g) \rightarrow 3CO_2(g) + H_2O(g)$
Balance H: $C_3H_6(g) + O_2(g) \rightarrow 3CO_2(g) + 3H_2O(g)$
Balance O: $C_3H_6(g) + 9/2\ O_2(g) \rightarrow 3CO_2(g) + 3H_2O(g)$
Clear fraction: $2C_3H_6(g) + 9O_2(g) \rightarrow 6CO_2(g) + 6H_2O(g)$

c) Skeletal reaction: $Ca(s) + O_2(g) \rightarrow CaO$
Balance O: $Ca(s) + O_2(g) \rightarrow 2CaO$
Balance Ca: $2Ca(s) + O_2(g) \rightarrow 2CaO$

d) Skeletal reaction: $C_5H_{12}S(l) + O_2(g) \rightarrow CO_2(g) + H_2O(g) + SO_2(g)$
Balance C: $C_5H_{12}S(l) + O_2(g) \rightarrow 5CO_2(g) + H_2O(g) + SO_2(g)$
Balance H: $C_5H_{12}S(l) + O_2(g) \rightarrow 5CO_2(g) + 6H_2O(g) + SO_2(g)$
Balance S: $C_5H_{12}S(l) + O_2(g) \rightarrow 5CO_2(g) + 6H_2O(g) + SO_2(g)$
Balance O: $C_5H_{12}S(l) + 9O_2(g) \rightarrow 5CO_2(g) + 6H_2O(g) + SO_2(g)$

65. **Given:** In 100 g solution, 20.0 g $C_2H_6O_2$; density of solution = 1.03 g/mL **Find:** M of solution
Conceptual Plan: **g $C_2H_6O_2 \rightarrow$ mol $C_2H_6O_2$ and g solution $\rightarrow$ mL solution $\rightarrow$ L solution**

$$\frac{1\ \text{mol}\ C_2H_6O_2}{62.06\ \text{g}\ C_2H_6O_2} \qquad \frac{1.00\ \text{mL}}{1.03\ \text{g}} \qquad \frac{1\ \text{L}}{1000\ \text{mL}}$$

then M $C_2H_6O_2$

$$M = \frac{\text{mol}\ C_2H_6O_2}{\text{L solution}}$$

Solution:

$$20.0\ \text{g}\ C_2H_6O_2 \times \frac{1\ \text{mol}\ C_2H_6O_2}{62.06\ \text{g}\ C_2H_6O_2} = 0.32\underline{2}2\ \text{mol}\ C_2H_6O_2$$

$$100.0\ \text{g solution} \times \frac{1.00\ \text{mL solution}}{1.03\ \text{g solution}} \times \frac{1\ \text{L}}{1000\ \text{mL}} = 0.097\underline{0}8\ \text{L}$$

$$M = \frac{0.32\underline{2}2\ \text{mol}\ C_2H_6O_2}{0.097\underline{0}8\ \text{L}} = 3.32\ \text{M}$$

Check: The units of the answer, M $C_2H_6O_2$, are correct. The magnitude of the answer is reasonable since the concentration of solutions is usually between 0 and 18 M.

67. **Given:** 2.5 g $NaHCO_3$ **Find:** g HCl
Conceptual Plan: g $NaHCO_3 \rightarrow$ mol $NaHCO_3 \rightarrow$ mol HCl $\rightarrow$ g HCl

$$\frac{1\ \text{mol}\ NaHCO_3}{84.02\ \text{g}\ NaHCO_3} \qquad \frac{1\ \text{mol HCl}}{1\ \text{mol}\ NaHCO_3} \qquad \frac{36.46\ \text{g HCl}}{1\ \text{mol HCl}}$$

Solution: $HCl(aq) + NaHCO_3(aq) \rightarrow H_2O(l) + CO_2(g) + NaCl(aq)$

$$2.5\ \text{g}\ NaHCO_3 \times \frac{1\ \text{mol}\ NaHCO_3}{84.02\ \text{g}\ NaHCO_3} \times \frac{1\ \text{mol HCl}}{1\ \text{mol}\ NaHCO_3} \times \frac{36.46\ \text{g HCl}}{1\ \text{mol HCl}} = 1.1\ \text{g HCl}$$

Check: The units of the answer, g HCl, are correct. The magnitude of the answer is reasonable since the molar mass of HCl is less than the molar mass of $NaHCO_3$.

69. **Given:** 1.0 kg C_8H_{18} **Find:** kg CO_2
Conceptual Plan: kg $C_8H_{18} \rightarrow$ g $C_8H_{18} \rightarrow$ mol $C_8H_{18} \rightarrow$ mol $CO_2 \rightarrow$ g $CO_2 \rightarrow$ kg CO_2

$$\frac{1000\ \text{g}}{\text{kg}} \qquad \frac{1\ \text{mol}\ C_8H_{18}}{114.22\ \text{g}\ C_8H_{18}} \qquad \frac{16\ \text{mol}\ CO_2}{2\ \text{mol}\ C_8H_{18}} \qquad \frac{44.01\ \text{g}\ CO_2}{1\ \text{mol}\ CO_2} \qquad \frac{\text{kg}}{1000\ \text{g}}$$

Solution: $2\ C_8H_{18}(g) + 25\ O_2(g) \rightarrow 16\ CO_2(g) + 18\ H_2O(g)$

$$1.0\ \text{kg}\ C_8H_{18} \times \frac{1000\ \text{g}}{\text{kg}} \times \frac{1\ \text{mol}\ C_8H_{18}}{114.22\ \text{g}\ C_8H_{18}} \times \frac{16\ \text{mol}\ CO_2}{2\ \text{mol}\ C_8H_{18}} \times \frac{44.01\ \text{g}\ CO_2}{1\ \text{mol}\ CO_2} \times \frac{\text{kg}}{1000\ \text{g}} = 3.1\ \text{kg}\ CO_2$$

Check: The units of the answer, kg CO_2, are correct. The magnitude of the answer is reasonable since the ratio of CO_2 to C_8H_{18} is 8:1.

71. **Given:** 3.00 mL $C_4H_6O_3$, d = 1.08 g/mL; 1.25 g $C_7H_6O_3$; 1.22 g $C_9H_8O_4$ **Find:** limiting reactant, theoretical yield $C_9H_8O_4$ and % yield $C_9H_8O_4$

Conceptual Plan: **mL $C_4H_6O_3$ → g $C_4H_6O_3$ → mol $C_4H_6O_3$ → mol $C_9H_8O_4$**

$$\frac{1.08\ g\ C_4H_6O_3}{1.00\ mL\ C_4H_6O_3} \quad \frac{1\ mol\ C_4H_6O_3}{102.09\ g\ C_4H_6O_3} \quad \frac{1\ mol\ C_9H_8O_4}{1\ mol\ C_4H_6O_3}$$

→ smallest amount determines limiting reactant

g $C_7H_6O_3$ → mol $C_7H_6O_3$ → mol $C_9H_8O_4$

$$\frac{1\ mol\ C_7H_6O_3}{138.12\ g\ C_7H_6O_3} \quad \frac{1\ mol\ C_9H_8O_4}{1\ mol\ C_7H_6O_3}$$

then: mol $C_9H_8O_4$ → g $C_9H_8O_4$ **then: determine % yield**

$$\frac{180.1\ g\ C_9H_8O_4}{mol\ C_9H_8O_4} \qquad \frac{\text{actual yield g } C_9H_8O_4}{\text{theoretical yield g } C_9H_8O_4} \times 100$$

Solution:

$$3.00\ \cancel{mL\ C_4H_6O_3} \times \frac{1.08\ \cancel{g\ C_4H_6O_3}}{\cancel{mL\ C_4H_6O_3}} \times \frac{1\ \cancel{mol\ C_4H_6O_3}}{102.09\ \cancel{g\ C_4H_6O_3}} \times \frac{1\ mol\ C_9H_8O_4}{1\ \cancel{mol\ C_4H_6O_3}} = 0.031\underline{7}4\ mol\ C_9H_8O_4$$

$$1.25\ \cancel{g\ C_7H_6O_3} \times \frac{1\ \cancel{mol\ C_7H_6O_3}}{138.12\ \cancel{g\ C_7H_6O_3}} \times \frac{1\ mol\ C_9H_8O_4}{1\ \cancel{mol\ C_7H_6O_3}} = 0.0090\underline{5}0\ mol\ C_9H_8O_4$$

$$0.0090\underline{5}0\ \cancel{mol\ C_9H_8O_4} \times \frac{180.1\ g\ C_9H_8O_4}{1\ \cancel{mol\ C_9H_8O_4}} = 1.6\underline{3}0\ g\ C_9H_8O_4$$

$$\frac{1.22\ \cancel{g\ C_9H_8O_4}}{1.6\underline{3}0\ \cancel{g\ C_9H_8O_4}} \times 100 = 74.8\%$$

Check: The theoretical yield has the correct units, g $C_9H_8O_4$, and has a reasonable magnitude compared to the mass of $C_7H_6O_3$, the limiting reactant. The % yield is reasonable, under 100%.

73. **Given:** (a) 11 molecules H_2, 2 molecules O_2; (b) 8 molecules H_2, 4 molecules O_2; (c) 4 molecules H_2, 5 molecules O_2; (d) 3 molecules H_2, 6 molecules O_2 **Find:** loudest explosion based on equation

Conceptual Plan: loudest explosion will occur in the balloon with the mol ratio closest to the balanced equation and that contains the most H_2

Solution: $2H_2(g) + O_2(g) \rightarrow H_2O(l)$

Balloon (a) has enough O_2 to react with 4 molecules H_2; balloon (b) has enough O_2 to react with 8 molecules H_2; balloon (c) has enough O_2 to react with 10 molecules H_2; and balloon (d) has enough O_2 for 3 molecules of H_2 to react. Therefore, balloon (b) will have the loudest explosion because it has the most H_2 that will react.

Check: Answer seems correct since it has the most H_2 with enough O_2 in the balloon to completely react.

75. a) Skeletal reaction: $HCl(aq) + Hg_2(NO_3)_2(aq) \rightarrow Hg_2Cl_2(s) + HNO_3(aq)$
    Balance Cl: $2HCl(aq) + Hg_2(NO_3)_2(aq) \rightarrow Hg_2Cl_2(s) + 2HNO_3(aq)$

b) Skeletal reaction: $KHSO_3(aq) + HNO_3(aq) \rightarrow H_2O(l) + SO_2(g) + KNO_3(aq)$
   Balanced reaction: $KHSO_3(aq) + HNO_3(aq) \rightarrow H_2O(l) + SO_2(g) + KNO_3(aq)$

c) Skeletal reaction: $NH_4Cl(aq) + Pb(NO_3)_2(aq) \rightarrow PbCl_2(s) + NH_4NO_3(aq)$
   Balance Cl: $2NH_4Cl(aq) + Pb(NO_3)_2(aq) \rightarrow PbCl_2(s) + NH_4NO_3(aq)$
   Balance N: $2NH_4Cl(aq) + Pb(NO_3)_2(aq) \rightarrow PbCl_2(s) + 2NH_4NO_3(aq)$

d) Skeletal reaction: $NH_4Cl(aq) + Ca(OH)_2(aq) \rightarrow NH_3(g) + H_2O(l) + CaCl_2(aq)$
   Balance Cl: $2NH_4Cl(aq) + Ca(OH)_2(aq) \rightarrow NH_3(g) + H_2O(l) + CaCl_2(aq)$
   Balance N: $2NH_4Cl(aq) + Ca(OH)_2(aq) \rightarrow 2NH_3(g) + H_2O(l) + CaCl_2(aq)$
   Balance H: $2NH_4Cl(aq) + Ca(OH)_2(aq) \rightarrow 2NH_3(g) + 2H_2O(l) + CaCl_2(aq)$

77. **Given:** 1.5 L solution, 0.050 M $CaCl_2$, 0.085 M $Mg(NO_3)_2$ **Find:** g Na_3PO_4

Conceptual Plan: V,M $CaCl_2$ → mol $CaCl_2$ and V,M $Mg(NO_3)_2$ → mol $Mg(NO_3)_2$

V x M = mol   V x M = mol

then (mol $CaCl_2$ + mol $Mg(NO_3)_2$) → Na_3PO_4 → g Na_3PO_4

$$\frac{2 \text{ mol } Na_3PO_4}{3 \text{ mol } (CaCl_2 + Mg(NO_3)_2)} \quad \frac{163.97 \text{ g } Na_3PO_4}{1 \text{ mol } Na_3PO_4}$$

Solution: $3CaCl_2(aq) + 2Na_3PO_4(aq) \rightarrow Ca_3(PO_4)_2(s) + 6\ NaCl(aq)$

$3\ Mg(NO_3)_2(aq) + 2Na_3PO_4(aq) \rightarrow Mg_3(PO_4)_2(s) + 6\ NaCl(aq)$

$1.5 \text{ L x } 0.050 \text{ M } CaCl_2 = 0.07\underline{5} \text{ mol } CaCl_2$

$1.5 \text{ L x } 0.085 \text{ M } Mg(NO_3)_2 = 0.1\underline{2}75 \text{ mol } Mg(NO_3)_2$

$$0.2\underline{0}25 \text{ mol } \cancel{CaCl_2 \text{ and } Mg(NO_3)_2} \times \frac{2 \text{ mol } \cancel{Na_3PO_4}}{3 \text{ mol } \cancel{CaCl_2 \text{ and } Mg(NO_3)_2}} \times \frac{163.97 \text{ g mol } Na_3PO_4}{\cancel{\text{mol } Na_3PO_4}} = 22 \text{ g mol } Na_3PO_4$$

Check: The units of the answer, g Na_3PO_4, are correct. The magnitude of the answer is reasonable since it is needed to remove both the Ca and Mg ions.

79. **Given:** 1.0 L, 0.10 M OH^- **Find:** g Ba
Conceptual Plan: VM → mol OH^- → mol $Ba(OH)_2$ → mol BaO → mol Ba → g Ba

$$V \times M = \text{mol} \quad \frac{1 \text{ mol } Ba(OH)_2}{2 \text{ mol } OH^-} \quad \frac{1 \text{ mol BaO}}{1 \text{ mol } Ba(OH)_2} \quad \frac{1 \text{ mol Ba}}{1 \text{ mol BaO}} \quad \frac{137.3 \text{ g Ba}}{1 \text{ mol Ba}}$$

Solution: $BaO(s) + H_2O(l) \rightarrow Ba(OH)_2(aq)$

$$1.0 \cancel{\text{L}} \times \frac{0.10 \cancel{\text{mol } OH^-}}{\cancel{\text{L}}} \times \frac{1 \cancel{\text{mol } Ba(OH)_2}}{2 \cancel{\text{mol } OH^-}} \times \frac{1 \cancel{\text{mol BaO}}}{1 \cancel{\text{mol } Ba(OH)_2}} \times \frac{1 \cancel{\text{mol Ba}}}{1 \cancel{\text{mol BaO}}} \times \frac{137.3 \text{ g Ba}}{1 \cancel{\text{mol Ba}}} = 6.9 \text{ g Ba}$$

Check: The units of the answer, g Ba, are correct. The magnitude is reasonable since the molar mass of Ba is large and there are 2 moles hydroxide per mole Ba.

81. **Given:** 30.0% $NaNO_3$, \$9.00/ 100 lb; 20.0 % $(NH_4)_2SO_4$, \$8.10/ 100 lb **Find:** cost / lb N
Conceptual Plan: **mass fertilizer → mass $NaNO_3$ → mass N → cost/lb N**

$$\frac{30.0 \text{ lb } NaNO_3}{100 \text{ lb fertilizer}} \quad \frac{16.48 \text{ lb N}}{100 \text{ lb } NaNO_3} \quad \frac{\$9.00}{100 \text{ lb fertilizer}}$$

and: mass fertilizer → mass $(NH_4)_2SO_4$ → mass N → cost/lb N

$$\frac{20.0 \text{ lb } (NH_4)_2SO_4}{100 \text{ lb fertilizer}} \quad \frac{21.2 \text{ lb N}}{100 \text{ lb } (NH_4)_2SO_4} \quad \frac{\$8.10}{100 \text{ lb fertilizer}}$$

Solution:

$$100 \cancel{\text{ lb fertilizer}} \times \frac{30.0 \cancel{\text{ lb } NaNO_3}}{100 \cancel{\text{ lb fertilizer}}} \times \frac{16.48 \text{ lb N}}{100 \cancel{\text{ lb } NaNO_3}} = 4.9\underline{4}4 \text{ lb N}$$

$$\frac{\$9.00}{100 \cancel{\text{ lb fertilizer}}} \times \frac{100 \cancel{\text{ lb fertilizer}}}{4.9\underline{4}4 \text{ lb N}} = \$1.82/ \text{ lb N}$$

$$100 \cancel{\text{ lb fertilizer}} \times \frac{20.0 \cancel{\text{ lb } (NH_4)_2SO_4}}{100 \cancel{\text{ lb fertilizer}}} \times \frac{21.2 \text{ lb N}}{100 \cancel{\text{ lb } (NH_4)_2SO_4}} = 4.2\underline{4}0 \text{ lb N}$$

$$\frac{\$8.10}{100 \cancel{\text{ lb fertilizer}}} \times \frac{100 \cancel{\text{ lb fertilizer}}}{4.2\underline{4} \text{ lb N}} = \$1.91/ \text{ lb N}$$

The more economical fertilizer is the $NaNO_3$ because it costs less/ lb N.

Check: The units of the cost, \$/lb N, are correct. The answer is reasonable because you compare the cost/lb N directly.

83. **Given:** 24.5 g Au, 24.5 g BrF_3, 24.5 g KF **Find:** g $KAuF_4$
Conceptual Plan: **g Au → mol Au → mol $KAuF_4$**

$$\frac{1 \text{ mol Au}}{196.97 \text{ g Au}} \quad \frac{2 \text{ mol } KAuF_4}{2 \text{ mol Au}}$$

g BrF_3 → mol BrF_3 → mol $KAuF_4$ **→ smallest mol amount determines limiting reactant**

$$\frac{1 \text{ mol } BrF_3}{136.9 \text{ g } BrF_3} \quad \frac{2 \text{ mol } KAuF_4}{2 \text{ mol } BrF_3}$$

g KF → mol KF → mol $KAuF_4$

$$\frac{1 \text{ mol KF}}{58.10 \text{ g KF}} \quad \frac{2 \text{ mol } KAuF_4}{2 \text{ mol KF}}$$

then: mol $KAuF_4$ → g $KAuF_4$

$$\frac{312.07 \text{ g } KAuF_4}{\text{mol } KAuF_4}$$

$$2\,Au(s) + 2BrF_3(l) + 2KF(s) \rightarrow Br_2(l) + 2KAuF_4(s)$$

oxidation states; 0 +3 – 1 +1 – 1 0 +1 +3 – 1

This is a redox reaction since Au increases in oxidation number (oxidation) and Br decreases in number (reduction). BrF_3 is the oxidizing agent, Au is the reducing agent.

Solution:

$$24.5 \text{ g Au} \times \frac{1 \text{ mol Au}}{196.97 \text{ g Au}} \times \frac{2 \text{ mol } KAuF_4}{2 \text{ mol Au}} = 0.12\underline{4}4 \text{ mol } KAuF_4$$

$$24.5 \text{ g } BrF_3 \times \frac{1 \text{ mol } BrF_3}{136.90 \text{ g } BrF_3} \times \frac{2 \text{ mol } KAuF_4}{2 \text{ mol } BrF_3} = 0.17\underline{9}0 \text{ mol KAuF}$$

$$24.5 \text{ g KF} \times \frac{1 \text{ mol KF}}{58.10 \text{ g KF}} \times \frac{2 \text{ mol } KAuF_4}{2 \text{ mol KF}} = 0.42\underline{1}7 \text{ mol } KAuF_4$$

$$0.12\underline{4}4 \text{ mol } KAuF_4 \times \frac{312.07 \text{ g } KAuF_4}{1 \text{ mol } KAuF_4} = 38.8 \text{ g } KAuF_4$$

Check: Units of the answer, g $KAuF_4$, are correct. The magnitude of the answer is reasonable compared to the mass of the limiting reactant Au.

85. **Given:** solution may contain Ag^+, Ca^{2+}, and Cu^{2+} **Find:** determine which ions are present.

Conceptual Plan: test the solution sequentially with NaCl, Na_2SO_4, and Na_2CO_3 and see if precipitates form.

Solution: original solution + NaCl yields no reaction: Ag^+ not present: since chlorides are normally soluble but Ag^+ is an exception.

Original solution with Na_2SO_4 yields a precipitate and solution 2. The precipitate is $CaSO_4$, so Ca^{2+} is present. Sulfates are normally soluble but Ca^{2+} is an exception.

Solution 2 with Na_2CO_3 yields a precipitate. The precipitate is $CuCO_3$, so Cu^{2+} is present. All carbonates are insoluble.

NET IONIC EQUATIONS:

$$Ca^{2+}(aq) + SO_4^{\,2-}(aq) \rightarrow CaSO_4(s)$$
$$Cu^{2+}(aq) + CO_3^{\,2-}(aq) \rightarrow CuCO_3(s)$$

Check: the answer is reasonable since two different precipitates formed and all the Ca^{2+} was removed before the carbonate was added.

87 **Given:** 15.2 billion L lake water, 1.8×10^{-5} M H_2SO_4, 8.7×10^{-6} M HNO_3 **Find:** kg $CaCO_3$ needed to neutralize

Conceptual Plan: Vol lake → mol H_2SO_4 → mol H^+ and vol lake → mol HNO_3 → mol H^+

vol x M = mol $\quad \frac{2 \text{ mol } H^+}{1 \text{ mol } H_2SO_4} \quad$ vol x M = mol $\quad \frac{1 \text{ mol } H^+}{1 \text{ mol } HNO_3}$

Then: total mol H^+ → mol $CO_3^{\,2-}$ → mol $CaCO_3$ → g $CaCO_3$ → kg $CaCO_3$

$$\frac{1 \text{ mol } CO_3^{2-}}{2 \text{ mol } H^+} \quad \frac{1 \text{ mol } CaCO_3}{1 \text{ mol } CO_3^{2-}} \quad \frac{100.09 \text{ g } CaCO_3}{1 \text{ mol } CaCO_3} \quad \frac{\text{kg}}{1000 \text{ g}}$$

Solution: $2H^+(aq) + CO_3^{\,2-}(aq) \rightarrow H_2O(l) + CO_2(g)$

$$15.2 \times 10^9 \text{ L} \times \frac{1.8 \times 10^{-5} \text{ mol } H_2SO_4}{\text{L soln}} \times \frac{2 \text{ mol } H^+}{\text{mol } H_2SO_4} = 5\underline{4}7200 \text{ mol } H^+$$

$$15.2 \times 10^9 \text{ L} \times \frac{8.7 \times 10^{-6} \text{ mol } HNO_3}{\text{L soln}} \times \frac{1 \text{ mol } H^+}{\text{mol } HNO_3} = 1\underline{3}2240 \text{ mol } H^+$$

$$679440 \text{ mol H}^+ \times \frac{1 \text{ mol CO}_3^{2-}}{2 \text{ mol H}^+} \times \frac{1 \text{ mol CaCO}_3}{\text{mol CO}_3^{2-}} \times \frac{100.09 \text{ g CaCO}_3}{1 \text{ mol CaCO}_3} \times \frac{\text{kg}}{1000 \text{ g}} = 3.4 \times 10^4 \text{ kg CaCO}_3$$

Check: Units of the answer, kg $CaCO_3$, are correct. The magnitude of the answer is reasonable based on the size of the lake.

89. **Given:** 45 μg Pb/dL blood, Vol = 5.0 L, 1 mol succimer($C_4H_6O_4S_2$) = 1 mol Pb **Find:** mass $C_4H_6O_4S_2$ in mg

Conceptual Plan: Volume blood L → Volume blood dL → μg Pb → g Pb → mol Pb →

$$\frac{10 \text{ dL}}{\text{L}} \quad \frac{45 \text{ μg}}{\text{dL}} \quad \frac{10^6 \text{ μg}}{\text{g}} \quad \frac{\text{mol Pb}}{207.2 \text{ g Pb}} \quad \frac{1 \text{ mol succimer}}{1 \text{ mol Pb}}$$

mol succimer → g succimer → mg succimer

$$\frac{182.23 \text{ g succimer}}{1 \text{ mol succimer}} \quad \frac{1000 \text{ mg succimer}}{1 \text{ g succimer}}$$

Solution:

$$5.0 \text{ L blood} \times \frac{10 \text{ dL}}{\text{L}} \times \frac{45 \text{ μg}}{\text{dL}} \times \frac{1 \text{ g}}{10^6 \text{ μg}} \times \frac{1 \text{ mol Pb}}{207.2 \text{ g}} \times \frac{1 \text{ mol succimer}}{1 \text{ mol Pb}} \times \frac{182.23 \text{ g succimer}}{1 \text{ mol succimer}} \times \frac{1000 \text{ mg}}{\text{g}}$$

= 2.0 mg succimer

Check: The units of the answer, mg succimer, are correct. The magnitude is reasonable for the volume of blood and the concentration.

91. **Given:** 250 g sample, 67.2 mol % Al **Find:** Theoretical yield in g of Mn

Conceptual Plan: mol % Al → g Al and mol % MnO_2 → g MnO_2, then mass % Al

$$\frac{26.98 \text{ g Al}}{\text{mol Al}} \quad \frac{86.94 \text{ g MnO}_2}{\text{mol MnO}_2} \quad \frac{\text{g Al}}{\text{total g}} \times 100$$

then: sample → g Al → mol Al → mol Mn

$$\frac{38.86 \text{ g Al}}{100 \text{ g sample}} \quad \frac{\text{mol Al}}{26.98 \text{ g Al}} \quad \frac{3 \text{ mol Mn}}{4 \text{ mol Al}}$$

→ smallest mol amount determines limiting reactant

sample → g MnO_2 → mol MnO_2 → mol Mn

$$\frac{61.14 \text{ g MnO}_2}{100 \text{ g sample}} \quad \frac{\text{mol MnO}_2}{86.94 \text{ g MnO}_2} \quad \frac{1 \text{ mol Mn}}{\text{mol MnO}_2}$$

then mol Mn → g Mn

$$\frac{54.94 \text{ g Mn}}{\text{mol Mn}}$$

Solution: $4Al(s) + 3\,MnO_2(s) \rightarrow 3Mn + 2Al_2O_3(s)$

Assume 1 mole: $0.672 \text{ mol Al} \times \frac{26.98 \text{ g Al}}{\text{mol Al}} = 18.13 \text{ g Al}$

$$0.328 \text{ mol MnO}_2 \times \frac{86.94 \text{ g MnO}_2}{\text{mol MnO}_2} = 28.52 \text{ g MnO}_2$$

$$\frac{18.13 \text{ g Al}}{(18.13 \text{ g Al} + 28\ 52 \text{ g MnO}_2)} \times 100 = 38.86 \text{ \% Al}$$

So: 61.14 % MnO_2

$$250 \text{ g sample} \times \frac{38.86 \text{ g Al}}{100 \text{ g sample}} \times \frac{\text{mol Al}}{26.98 \text{ g Al}} \times \frac{3 \text{ mol Mn}}{4 \text{ mol Al}} = 2.7\underline{0}1 \text{ mol Mn}$$

$$250 \text{ g sample} \times \frac{61.14 \text{ g MnO}_2}{100 \text{ g sample}} \times \frac{\text{mol MnO}_2}{86.94 \text{ g MnO}_2} \times \frac{1 \text{ mol Mn}}{1 \text{ mol MnO}_2} = 1.7\underline{5}8 \text{ mol Mn}$$

$$1.7\underline{5}8 \text{ mol Mn} \times \frac{54.94 \text{ g Mn}}{1 \text{ mol Mn}} = 96.6 \text{ g Mn}$$

Check: the units of the answer, g Mn, are correct. The magnitude of the answer is reasonable based on the amount of the limiting reactant, MnO_2.

93. The correct answer is d. The molar mass of K and O_2 are comparable. Since the stoichiometry has a ratio of 4 mol K to 1 mol O_2, K will be the limiting reactant when mass of K is less than 4 times the mass of O_2.

95. **Given:** 1 M solution contains 8 particles **Find:** amount of solute or solvent needed to obtain new concentration

Conceptual Plan: determine amount of solute particles in each new solution, then determine if solute (if the number is greater) or solvent (if the number is less) needs to be added to obtain the new concentration.

Solution: Solution (a) contains 12 particles solute. Concentration is greater than the original, so solute needs to be added. $12 \text{ particles} \times \frac{1 \text{ mol}}{8 \text{ particles}} = 1.5 \text{ mol}$ $(1.5 \text{ mol} - 1.0 \text{ mol}) = 0.5 \text{ mol solute added.}$

$$0.5 \text{ mol solute} \times \frac{8 \text{ particles}}{1 \text{ mol solute}} = 4 \text{ solute particles added}$$

Solution (a) is obtained by adding 4 particles solute to 1 L of original solution

Solution (b) contains 4 particles. Concentration is less than the original so solvent needs to be added.

$4 \text{ particles} \times \frac{1 \text{ mol}}{8 \text{ particles}} = 0.5 \text{ mol solute}$ So, 1 L solution contains 0.5 mol = 0.5M

$(1 \text{ M})(1 \text{ L}) = (0.5 \text{ M})(x)$ $x = 2 \text{ L}$

Solution (b) is obtained by diluting 1 L of the original solution to 2 L

Solution (c) contains 6 particles. Concentration is less than the original so solvent needs to be added.

$6 \text{ particles} \times \frac{1 \text{ mol}}{8 \text{ particles}} = 0.75 \text{ mol solute}$ So, 1 L solution contains 0.75 mol = 0.75M

$(1 \text{ M})(1 \text{ L}) = (0.75 \text{ M})(x)$ $x = 2 \text{ L}$

Solution (c) is obtained by diluting 1 L of the original solution to 1.3 L

Chapter 5
Gases

1. a) **Given:** 24.9 in Hg **Find:** atm
Conceptual Plan: **in Hg → atm**

$$\frac{1\,\text{atm}}{29.92\,\text{in Hg}}$$

Solution: $24.9\,\text{in Hg} \times \frac{1\,\text{atm}}{29.92\,\text{in Hg}} = 0.832\,\text{atm}$

Check: The units (atm) are correct. The magnitude of the answer (<1) makes physical sense because we started with less than 29.92 in Hg.

b) **Given:** 24.9 in Hg **Find:** mmHg
Conceptual Plan: **Use answer from part a) then convert atm → mmHg**

$$\frac{760\,\text{mm Hg}}{1\,\text{atm}}$$

Solution: $0.832\,\text{atm} \times \frac{760\,\text{mmHg}}{1\,\text{atm}} = 632\,\text{mmHg}$

Check: The units (mmHg) are correct. The magnitude of the answer (< 760 mmHg) makes physical sense because we started with less than 1 atm.

c) **Given:** 24.9 in Hg **Find:** psi
Conceptual Plan: **Use answer from part a) then convert atm → psi**

$$\frac{14.7\,\text{psi}}{1\,\text{atm}}$$

Solution: $0.832\,\text{atm} \times \frac{14.7\,\text{psi}}{1\,\text{atm}} = 12.2\,\text{psi}$

Check: The units (psi) are correct. The magnitude of the answer (< 14.7 psi) makes physical sense because we started with less than 1 atm.

d) **Given:** 24.9 in Hg **Find:** Pa
Conceptual Plan: **Use answer from part a) then convert atm → Pa**

$$\frac{101{,}325\,\text{Pa}}{1\,\text{atm}}$$

Solution: $0.832\,\text{atm} \times \frac{101{,}325\,\text{Pa}}{1\,\text{atm}} = 8.43\text{x}10^4\,\text{Pa}$

Check: The units (mmHg) are correct. The magnitude of the answer (< 760 mmHg) makes physical sense because we started with less than 1 atm.

3. a) **Given:** 31.85 in Hg **Find:** mmHg
Conceptual Plan: **in Hg → mmHg**

$$\frac{25.4\,\text{mmHg}}{1\,\text{in Hg}}$$

Solution: $31.85\,\text{in Hg} \times \frac{25.4\,\text{mmHg}}{1\,\text{in Hg}} = 809.0\,\text{mmHg}$

Check: The units (mmHg) are correct. The magnitude of the answer (809) makes physical sense because inches are larger than mm.

b) **Given:** 31.85 in Hg **Find:** atm
Conceptual Plan: **Use answer from part a) then convert mmHg → atm**

$$\frac{1\,\text{atm}}{760\,\text{mmHg}}$$

Solution: $809.0\,\text{mmHg} \times \frac{1\,\text{atm}}{760\,\text{mmHg}} = 1.064\,\text{atm}$

Check: The units (atm) are correct. The magnitude of the answer (>1) makes physical sense because we started with more than 760 mmHg.

c) **Given:** 31.85 in Hg **Find:** torr

Conceptual Plan: **Use answer from part a) then convert** **mmHg → torr**

$$\frac{1\,\text{torr}}{1\,\text{mmHg}}$$

Solution: $809.0\ \text{mmHg} \times \frac{1\ \text{torr}}{1\ \text{mmHg}} = 809.0\ \text{torr}$

Check: The units (torr) are correct. The magnitude of the answer (809) makes physical sense because both units are of the same size.

d) **Given:** 31.85 in Hg **Find:** kPa
Conceptual Plan: **Use answer from part b) then convert** **atm → Pa → kPa**

$$\frac{101{,}325\,\text{Pa}}{1\,\text{atm}} \qquad \frac{1\,\text{kPa}}{1000\,\text{Pa}}$$

Solution: $1.064\ \text{atm} \times \frac{101{,}325\ \text{Pa}}{1\ \text{atm}} \times \frac{1\ \text{kPa}}{1000\ \text{Pa}} = 107.8\ \text{kPa}$

Check: The units (kPa) are correct. The magnitude of the answer (108) makes physical sense because we started with more than 1 atm and there are ~101 kPa in an atm.

5. **Given:** $V_1 = 2.8$ L, $P_1 = 755$ mmHg, and $V_2 = 3.7$ L **Find:** P_2
Conceptual Plan: $V_1, P_1, V_2 \rightarrow P_2$

$$P_1 V_1 = P_2 V_2$$

Solution:

$P_1 V_1 = P_2 V_2$ Rearrange to solve for P_2. $P_2 = P_1 \frac{V_1}{V_2} = 755\ \text{mmHg} \times \frac{2.8\ \text{L}}{3.7\ \text{L}} = 570\ \text{mmHg} = 5.7 \times 10^2\ \text{mmHg}$

Check: The units (mmHg) are correct. The magnitude of the answer (570 mmHg) makes physical sense because Boyle's Law indicates that as the volume increases, the pressure decreases.

7. **Given:** $V_1 = 48.3$ mL, $T_1 = 22$ °C, and $T_2 = 87$ °C **Find:** V_2
Conceptual Plan: **°C → K then** $V_1, T_1, T_2 \rightarrow V_2$

$$\text{K} = {^\circ}\text{C} + 273.15 \qquad \frac{V_1}{T_1} = \frac{V_2}{T_2}$$

Solution: $T_1 = 22\ {^\circ}\text{C} + 273.15 = 295$ K and $T_2 = 87\ {^\circ}\text{C} + 273.15 = 360.$ K

$\frac{V_1}{T_1} = \frac{V_2}{T_2}$ Rearrange to solve for V_2. $V_2 = V_1 \frac{T_2}{T_1} = 48.3\ \text{mL} \times \frac{360\ \text{K}}{295\ \text{K}} = 58.9\ \text{mL}$

Check: The units (mL) are correct. The magnitude of the answer (59 mL) makes physical sense because Charles' Law indicates that as the volume increases, the temperature increases.

9. **Given:** $V_1 = 2.76$ L, $n_1 = 0.128$ mol, and $\Delta n = 0.073$ mol **Find:** V_2
Conceptual Plan: $n_1 \rightarrow n_2$ **then** $V_1, n_1, n_2 \rightarrow V_2$

$$n_1 + \Delta n = n_2 \qquad \frac{V_1}{n_1} = \frac{V_2}{n_2}$$

Solution: $n_2 = 0.128\ \text{mol} + 0.073\ \text{mol} = 0.201\ \text{mol}$

$\frac{V_1}{n_1} = \frac{V_2}{n_2}$ Rearrange to solve for V_2. $V_2 = V_1 \frac{n_2}{n_1} = 2.76\ \text{L} \times \frac{0.201\ \text{mol}}{0.128\ \text{mol}} = 4.33\ \text{L}$

Check: The units (L) are correct. The magnitude of the answer (4.33 L) makes physical sense because Avogadro's Law indicates that as the number of moles increases, the volume increases.

11. **Given:** $n = 0.118$ mol, $P = 0.97$ atm, and $T = 305$ K **Find:** V
Conceptual Plan: $n, P, T \rightarrow V$

$$PV = nRT$$

Solution: $PV = nRT$ Rearrange to solve for V. $V = \frac{nRT}{P} = \frac{0.118\ \text{mol} \times 0.08206\ \frac{\text{L} \cdot \text{atm}}{\text{mol} \cdot \text{K}} \times 305\ \text{K}}{0.97\ \text{atm}} = 3.0\ \text{L}$

Check: The units (L) are correct. The magnitude of the answer (3 L) makes sense because, as you will see in the next section, one mole of an ideal gas under standard conditions (273 K and 1 atm) occupies 22.4 L. Although these are not standard conditions, they are close enough for a ballpark check of the answer. Since this gas sample contains 0.118 moles, a volume of 3 L is reasonable.

13. **Given:** $V = 28.5$ L, $P = 1.8$ atm, and $T = 298$ K **Find:** n

Conceptual Plan: $V, P, T \rightarrow n$

$PV = nRT$

Solution: $PV = nRT$ Rearrange to solve for n. $n = \frac{PV}{RT} = \frac{1.8\ \text{atm} \times 28.5\ \text{L}}{0.08206\ \frac{\text{L}\cdot\text{atm}}{\text{mol}\cdot\text{K}} \times 298\ \text{K}} = 2.1\ \text{mol}$

Check: The units (mol) are correct. The magnitude of the answer (2 mol) makes sense because, as you will see in the next section, one mole of an ideal gas under standard conditions (273 K and 1 atm) occupies 22.4 L. Although these are not standard conditions, they are close enough for a ballpark check of the answer. Since this gas sample has a volume of 28.5 L, and a pressure of 1.8 atm, ~ 2 mol is reasonable.

15. **Given:** $P_1 = 36.0$ psi (gauge P), $V_1 = 11.8$ L, $T_1 = 12.0$ °C, $V_2 = 12.2$ L, and $T_2 = 65.0$ °C
Find: P_2 and compare to $P_{max} = 38.0$ psi (gauge P)
Conceptual Plan: **°C → K and gauge P → psi → atm then $P_1, V_1, T_1, V_2, T_2 \rightarrow P_2$**

K = °C + 273.15   psi = gauge P + 14.7 $\frac{1\ \text{atm}}{14.7\ \text{psi}}$ $\frac{P_1V_1}{T_1} = \frac{P_2V_2}{T_2}$

Solution: $T_1 = 12.0\ °\text{C} + 273.15 = 285.2$ K and $T_2 = 65.0\ °\text{C} + 273.15 = 338.2$ K

$P_1 = 36.0\ \text{psi (gauge P)} + 14.7 = 50.7\ \text{psi} \times \frac{1\ \text{atm}}{14.7\ \text{psi}} = 3.4\underline{4}898\ \text{atm}$

$P_{max} = 38.0\ \text{psi (gauge P)} + 14.7 = 52.7\ \text{psi} \times \frac{1\ \text{atm}}{14.7\ \text{psi}} = 3.59\ \text{atm}$

$\frac{P_1V_1}{T_1} = \frac{P_2V_2}{T_2}$ Rearrange to solve for P_2. $P_2 = P_1 \frac{V_1}{V_2}\frac{T_2}{T_1} = 3.4\underline{4}898\ \text{atm} \times \frac{11.8\ \text{L}}{12.2\ \text{L}} \times \frac{338.2\ \text{K}}{285.2\ \text{K}} = 3.96\ \text{atm}$

This exceeds the maximum tire rating of 3.59 atm or 38.0 psi (gauge P).
Check: The units (atm) are correct. The magnitude of the answer (3.95 atm) makes physical sense because the relative increase in T is greater than the relative increase in V, so P should increase.

17. **Given:** $m\ (CO_2) = 28.8$ g, $P = 742$ mmHg, and $T = 22$ °C **Find:** V
Conceptual Plan: **°C → K and mmHg → atm and g → mol then $n, P, T \rightarrow V$**

K = °C + 273.15 $\frac{1\ \text{atm}}{760\ \text{mm Hg}}$ $\frac{1\ \text{mol}}{44.01\ \text{g}}$ $PV = nRT$

Solution: $T_1 = 22\ °\text{C} + 273.15 = 295$ K, $P = 742\ \text{mmHg} \times \frac{1\ \text{atm}}{760\ \text{mmHg}} = 0.97\underline{6}316\ \text{atm}$,

$n = 28.8\ \text{g} \times \frac{1\ \text{mol}}{44.01\ \text{g}} = 0.65\underline{4}397\ \text{mol}$ $PV = nRT$ Rearrange to solve for V.

$V = \frac{nRT}{P} = \frac{0.65\underline{4}397\ \text{mol} \times 0.08206\ \frac{\text{L}\cdot\text{atm}}{\text{mol}\cdot\text{K}} \times 295\ \text{K}}{0.97\underline{6}316\ \text{atm}} = 16.2\ \text{L}$

Check: The units (L) are correct. The magnitude of the answer (16 L) makes sense because one mole of an ideal gas under standard conditions (273 K and 1 atm) occupies 22.4 L. Although these are not standard conditions, they are close enough for a ballpark check of the answer. Since this gas sample contains 0.65 moles, a volume of 16 L is reasonable.

19. **Given:** sample a = 5 gas particles, sample b = 10 gas particles, and sample c = 8 gas particles, with all temperatures and volumes the same **Find:** sample with largest P
Conceptual Plan: $n, V, T \rightarrow P$

$PV = nRT$

Solution: $PV = nRT$ Since V and T are constant, this means that $P \propto n$. The sample with the largest number of gas particles will have the highest P. $P_b > P_c > P_a$.

21. **Given:** $P_1 = 755$ mmHg, $T_1 = 25$ °C, and $T_2 = 1155$ °C **Find:** P_2
Conceptual Plan: **°C → K and mmHg → atm then $P_1, T_1, T_2 \rightarrow P_2$**

K = °C + 273.15 $\frac{1\ \text{atm}}{760\ \text{mmHg}}$ $\frac{P_1}{T_1} = \frac{P_2}{T_2}$

Solution: $T_1 = 25\ °\text{C} + 273.15 = 298$ K and $T_2 = 1155\ °\text{C} + 273.15 = 1428$ K

$P = 755\ \text{mmHg} \times \frac{1\ \text{atm}}{760\ \text{mmHg}} = 0.99\underline{3}421\ \text{atm}$ $\frac{P_1}{T_1} = \frac{P_2}{T_2}$ Rearrange to solve for P_2.

$$P_2 = P_1\frac{T_2}{T_1} = 0.99\underline{3}421\text{ atm} \times \frac{1428\text{ K}}{298\text{ K}} = 4.76\text{ atm}$$

Check: The units (atm) are correct. The magnitude of the answer (5 atm) makes physical sense because there is a significant increase in *T*, which will increase *P* significantly.

23. **Given:** STP and *m* (Ne) = 10.0 g **Find:** *V*

Conceptual Plan: **g → mol → *V***

$\frac{1\text{ mol}}{20.18\text{ g}}$ $\frac{22.414\text{ L}}{1\text{ mol}}$

Solution: $10.0\text{ g} \times \frac{1\text{ mol}}{20.18\text{ g}} \times \frac{22.414\text{ L}}{1\text{ mol}} = 11.1\text{ L}$

Check: The units (L) are correct. The magnitude of the answer (11 L) makes sense because one mole of an ideal gas under standard conditions (273 K and 1 atm) occupies 22.4 L and we have about 0.5 mol.

25. **Given:** H_2, *P* = 1655 psi, and *T* = 20.0 °C **Find:** *d*

Conceptual Plan: **°C → K and psi → atm then *P, T*, $\mathfrak{M}$ → *d***

$K = °C + 273.15$ $\frac{1\text{ atm}}{14.7\text{ psi}}$ $d = \frac{P\mathfrak{M}}{RT}$

Solution: $T = 20.0\text{ °C} + 273.15 = 293.2\text{ K}$ $P = 1655\text{ psi} \times \frac{1\text{ atm}}{14.7\text{ psi}} = 11\underline{2}.585\text{ atm}$

$$d = \frac{P\mathfrak{M}}{RT} = \frac{11\underline{2}.585\text{ atm} \times 2.016\frac{\text{g}}{\text{mol}}}{0.08206\frac{\text{L atm}}{\text{K mol}} \times 293.2\text{ K}} = 9.43\frac{\text{g}}{\text{L}}$$

Check: The units (g/L) are correct. The magnitude of the answer (9 g/L) makes physical sense because this is a high pressure, so the gas density will be on the high side.

27. **Given:** *V* = 248 mL, *m* = 0.433 g, *P* = 745 mmHg, and *T* = 28 °C **Find:** $\mathfrak{M}$

Conceptual Plan: **°C → K mmHg → atm mL → L then *V, m* → *d* then *d, P, T* → $\mathfrak{M}$**

$K = °C + 273.15$ $\frac{1\text{ atm}}{760\text{ mmHg}}$ $\frac{1\text{ L}}{1000\text{ mL}}$ $d = \frac{m}{V}$ $d = \frac{P\mathfrak{M}}{RT}$

Solution: $T = 28\text{ °C} + 273.15 = 301\text{ K}$ $P = 745\text{ mmHg} \times \frac{1\text{ atm}}{760\text{ mmHg}} = 0.98\underline{0}263\text{ atm}$

$V = 248\text{ mL} \times \frac{1\text{ L}}{1000\text{ mL}} = 0.248\text{ L}$ $d = \frac{m}{V} = \frac{0.433\text{ g}}{0.248\text{ L}} = 1.7\underline{4}597\text{ g/L}$ $d = \frac{P\mathfrak{M}}{RT}$ Rearrange to solve for $\mathfrak{M}$.

$$\mathfrak{M} = \frac{dRT}{P} = \frac{1.7\underline{4}597\frac{\text{g}}{\text{L}} \times 0.08206\frac{\text{L}\cdot\text{atm}}{\text{K}\cdot\text{mol}} \times 301\text{ K}}{0.98\underline{0}263\text{ atm}} = 44.0\text{ g/mol}$$

Check: The units (g/mol) are correct. The magnitude of the answer (44 g/mol) makes physical sense because this is a reasonable number for a molecular weight of a gas.

29. **Given:** *m* = 38.8 mg, *V* = 224 mL, *T* = 55 °C, and *P* = 886 torr **Find:** $\mathfrak{M}$

Conceptual Plan: **mg → g mL → L °C → K torr → atm then *V, m* → *d* then *d, P, T* → $\mathfrak{M}$**

$\frac{1\text{ g}}{1000\text{ mg}}$ $\frac{1\text{ L}}{1000\text{ mL}}$ $K = °C + 273.15$ $\frac{1\text{ atm}}{760\text{ torr}}$ $d = \frac{m}{V}$ $d = \frac{P\mathfrak{M}}{RT}$

Solution: $m = 38.8\text{ mg} \times \frac{1\text{ g}}{1000\text{ mg}} = 0.0388\text{ g}$ $V = 224\text{ mL} \times \frac{1\text{ L}}{1000\text{ mL}} = 0.224\text{ L}$ $T = 55\text{ °C} + 273.15 = 328\text{ K}$

$P = 886\text{ torr} \times \frac{1\text{ atm}}{760\text{ torr}} = 1.1\underline{6}5789\text{ atm}$ $d = \frac{m}{V} = \frac{0.0388\text{ g}}{0.224\text{ L}} = 0.17\underline{3}214\text{ g/L}$ $d = \frac{P\mathfrak{M}}{RT}$

Rearrange to solve for $\mathfrak{M}$. $$\mathfrak{M} = \frac{dRT}{P} = \frac{0.17\underline{3}214\frac{\text{g}}{\text{L}} \times 0.08206\frac{\text{L atm}}{\text{K mol}} \times 328\text{ K}}{1.1\underline{6}5789\text{ atm}} = 4.00\text{ g/mol}$$

Check: The units (g/mol) are correct. The magnitude of the answer (4 g/mol) makes physical sense because this is a reasonable number for a molecular weight of a gas, especially since the density is on the low side.

31. **Given:** $P_{N2} = 325$ torr, $P_{O2} = 124$ torr, $P_{He} = 209$ torr, *V* = 1.05 L, and *T* = 25.0 °C

Find: m_{N2}, m_{O2}, m_{He}

Conceptual Plan: °C → K and torr → atm and P, V, T → n then mol → g

$K = °C + 273.15$ $\quad \frac{1\ atm}{760\ torr}$ $\quad PV = nRT$ $\quad \mathfrak{M}$

and $P_{N2}, P_{O2}, P_{He} \rightarrow P_{Total}$

$P_{Total} = P_{N2} + P_{O2} + P_{He}$

Solution: $T_1 = 25.0\ °C + 273.15 = 298.2\ K$, $\quad PV = nRT$ Rearrange to solve for n.

$$n = \frac{PV}{RT} \quad P_{N2} = 325\ torr \times \frac{1\ atm}{760\ torr} = 0.42\underline{7}632\ atm \quad n_{N2} = \frac{0.42\underline{7}632\ atm \times 1.05\ L}{0.08206\ \frac{L \cdot atm}{mol \cdot K} \times 298.2\ K} = 0.018\underline{3}493\ mol$$

$$0.018\underline{3}493\ mol \times \frac{28.02\ mol}{1\ mol} = 0.514\ g\ N_2$$

$$P_{O2} = 124\ torr \times \frac{1\ atm}{760\ torr} = 0.16\underline{3}158\ atm \quad n_{O2} = \frac{0.16\underline{3}158\ atm \times 1.05\ L}{0.08206\ \frac{L \cdot atm}{mol \cdot K} \times 298.2\ K} = 0.0070\underline{0}097\ mol$$

$$0.0070\underline{0}097\ mol \times \frac{32.00\ mol}{1\ mol} = 0.224\ g\ O_2$$

$$P_{He} = 209\ torr \times \frac{1\ atm}{760\ torr} = 0.275\ atm \quad n_{He} = \frac{0.275\ atm \times 1.05\ L}{0.08206\ \frac{L \cdot atm}{mol \cdot K} \times 298.2\ K} = 0.011\underline{8}000\ mol$$

$$0.011\underline{8}000\ mol \times \frac{4.003\ mol}{1\ mol} = 0.0472\ g\ He \text{ and}$$

$P_{Total} = P_{N2} + P_{O2} + P_{He} = 0.428\ atm + 0.163\ atm + 0.275\ atm = 0.866\ atm$

Check: The units (g and atm) are correct. The magnitude of the answer (1 g) makes sense because gases are not very dense and these pressures are < 1 atm. Since all of the pressures are small, the total is < 1 atm.

33. **Given:** $m\ (CO_2) = 1.20\ g$, $V = 755\ mL$, $P_{N2} = 725\ mmHg$, and $T = 25.0\ °C$ **Find:** P_{Total}

Conceptual Plan: mL → L and °C → K and g → mol and $n, P, T \rightarrow V$ then atm → mmHg

$\frac{1\ L}{1000\ mL}$ $\quad K = °C + 273.15$ $\quad \frac{1\ mol}{44.01\ g}$ $\quad PV = nRT$ $\quad \frac{760\ mmHg}{1\ atm}$

finally $P_{CO2}, P_{N2} \rightarrow P_{Total}$

$P_{Total} = P_{CO2} + P_{N2}$

Solution: $V = 755\ mL \times \frac{1\ L}{1000\ mL} = 0.755\ L$ $\quad T = 25.0\ °C + 273.15 = 298.2\ K$,

$n = 1.20\ g \times \frac{1\ mol}{44.01\ g} = 0.027\underline{2}665\ mol$, $\quad PV = nRT$ Rearrange to solve for P.

$$P = \frac{nRT}{V} = \frac{0.027\underline{2}665\ mol \times 0.08206\ \frac{L \cdot atm}{mol \cdot K} \times 298.2\ K}{0.755\ L} = 0.88\underline{3}735\ atm$$

$$P_{CO2} = 0.88\underline{3}735\ atm \times \frac{760\ mmHg}{1\ atm} = 672\ mmHg$$

$P_{Total} = P_{CO2} + P_{N2} = 672\ mmHg + 725\ mmHg = 1397\ mmHg$

Check: The units (mmHg) are correct. The magnitude of the answer (1400 mmHg) makes sense because it must be greater than 725 mmHg.

35. **Given:** $m\ (N_2) = 1.25\ g$, $m\ (O_2) = 0.85\ g$, $V = 1.55\ L$, and $T = 18\ °C$ **Find:** $\chi_{N2}, \chi_{O2}, P_{N2}, P_{O2}$

Conceptual Plan: g → mol then $n_{N2}, n_{O2} \rightarrow \chi_{N2}$ and $n_{N2}, n_{O2} \rightarrow \chi_{O2}$ °C → K

$\mathfrak{M}$ $\quad \chi_{N2} = \frac{n_{N2}}{n_{N2} + n_{O2}}$ $\quad \chi_{O2} = \frac{n_{O2}}{n_{N2} + n_{O2}}$ $\quad K = °C + 273.15$

then $n, V, T \rightarrow P$

$PV = nRT$

Solution: $n_{N2} = 1.25\ g \times \frac{1\ mol}{28.02\ g} = 0.044\underline{6}110\ mol$, $n_{O2} = 0.85\ g \times \frac{1\ mol}{32.00\ g} = 0.02\underline{6}563\ mol$,

$$T = 18\ °C + 273.15 = 291\ K, \quad \chi_{N2} = \frac{n_{N2}}{n_{N2} + n_{O2}} = \frac{0.044\underline{6}110\ mol}{0.044\underline{6}110\ mol + 0.02\underline{6}563\ mol} = 0.62\underline{6}792 = 0.627,$$

$\chi_{O2} = \frac{n_{O2}}{n_{N2} + n_{O2}} = \frac{0.02\underline{6}563 \text{ mol}}{0.044\underline{6}110 \text{ mol} + 0.02\underline{6}563 \text{ mol}} = 0.37$ We can also calculate this as

$\chi_{O2} = 1 - \chi_{N2} = 1 - 0.6\underline{2}6792 = 0.37\underline{3}208 = 0.373$ $\quad PV = nRT$ Rearrange to solve for P. $\quad P = \frac{nRT}{V}$

$$P_{N2} = \frac{0.044\underline{6}11 \text{ mol} \times 0.08206 \frac{\text{L} \cdot \text{atm}}{\text{mol} \cdot \text{K}} \times 291 \text{ K}}{1.55 \text{ L}} = 0.687 \text{ atm}$$

$$P_{O2} = \frac{0.02\underline{6}563 \text{ mol} \times 0.08206 \frac{\text{L} \cdot \text{atm}}{\text{mol} \cdot \text{K}} \times 291 \text{ K}}{1.55 \text{ L}} = 0.409 \text{ atm}$$

Check: The units (none and atm) are correct. The magnitude of the answers makes sense because the mole fractions should total 1 and since the weight of N_2 is greater than O_2, its mole fraction is larger. The number of moles is <<1, so we expect the pressures to be <1 atm, given the V (1.55 L).

37. **Given:** $T = 30.0$ °C, $P_{\text{Total}} = 732$ mmHg, and $V = 722$ mL **Find:** P_{H2} and m_{H2}
Conceptual Plan: $T \rightarrow P_{H2O}$ (Table 5.3) then $P_{\text{Total}}, P_{H2O} \rightarrow P_{H2}$ ($P_{Total} = P_{H2O} + P_{H2}$) then mmHg → atm ($\frac{1 \text{ atm}}{760 \text{ mmHg}}$) and mL → L ($\frac{1 \text{ L}}{1000 \text{ mL}}$)
and °C → K ($K = °C + 273.15$) $\quad P, V, T \rightarrow n$ ($PV = nRT$) then mol → g ($\frac{2.016 \text{ g}}{1 \text{ mol}}$)

Solution: Table 5.3 states that $P_{H2O} = 31.86$ mmHg $\quad P_{Total} = P_{H2O} + P_{H2}$
Rearrange to solve for P_{H2}. $P_{H2} = P_{Total} - P_{H2O} = 732 \text{ mmHg} - 31.86 \text{ mmHg} = 700. \text{ mmHg}$

$P_{H2} = 700. \text{ mmHg} \times \frac{1 \text{ atm}}{760 \text{ mmHg}} = 0.92\underline{1}052 \text{ atm}$ $\quad V = 722 \text{ mL} \times \frac{1 \text{ L}}{1000 \text{ mL}} = 0.722 \text{ L}$,

$T = 30.0 \text{ °C} + 273.15 = 303.2 \text{ K}$, $PV = nRT$ Rearrange to solve for n. $n = \frac{PV}{RT}$

$$n_{H2} = \frac{0.92\underline{1}052 \text{ atm} \times 0.722 \text{ L}}{0.08206 \frac{\text{L} \cdot \text{atm}}{\text{mol} \cdot \text{K}} \times 303.2 \text{ K}} = 0.026\underline{7}277 \text{ mol} \quad \text{then} \quad 0.026\underline{7}277 \text{ mol} \times \frac{2.016 \text{ g}}{1 \text{ mol}} = 0.0539 \text{ g } H_2$$

Check: The units (g) are correct. The magnitude of the answer (<< 1 g) makes sense because gases are not very dense, hydrogen is light, the volume is small, and the pressure is ~1 atm.

39. **Given:** $T = 25$ °C, $P_{\text{Total}} = 748$ mmHg, and $V = 0.951$ L **Find:** P_{H2} and m_{H2}
Conceptual Plan: $T \rightarrow P_{H2O}$ (Table 5.3) then $P_{\text{Total}}, P_{H2O} \rightarrow P_{H2}$ ($P_{Total} = P_{H2O} + P_{H2}$) then mmHg → atm ($\frac{1 \text{ atm}}{760 \text{ mmHg}}$) and mL → L ($\frac{1 \text{ L}}{1000 \text{ mL}}$)
and °C → K ($K = °C + 273.15$) $\quad P, V, T \rightarrow n$ ($PV = nRT$) then mol → g ($\frac{2.016 \text{ g}}{1 \text{ mol}}$)

Solution: Table 5.3 states that $P_{H2O} = 23.78$ mmHg $\quad P_{Total} = P_{H2O} + P_{H2}$
Rearrange to solve for P_{H2}. $\quad P_{H2} = P_{Total} - P_{H2O} = 748 \text{ mmHg} - 23.78 \text{ mmHg} = 724 \text{ mmHg}$

$P_{H2} = 724 \text{ mmHg} \times \frac{1 \text{ atm}}{760 \text{ mmHg}} = 0.95\underline{2}632 \text{ atm}$ $\quad T = 25 \text{ °C} + 273.15 = 298 \text{ K}$, $\quad PV = nRT$

Rearrange to solve for n. $n_{H2} = \frac{PV}{RT} = \frac{0.95\underline{2}632 \text{ atm} \times 0.951 \text{ L}}{0.08206 \frac{\text{L} \cdot \text{atm}}{\text{mol} \cdot \text{K}} \times 298 \text{ K}} = 0.037\underline{0}474 \text{ mol}$

$0.037\underline{0}474 \text{ mol} \times \frac{2.016 \text{ g}}{1 \text{ mol}} = 0.0747 \text{ g } H_2$

Check: The units (g) are correct. The magnitude of the answer (<< 1 g) makes sense because gases are not very dense, hydrogen is light, the volume is small, and the pressure is ~1 atm.

41. **Given:** m (C) = 15.7 g, $P = 1.0$ atm, and $T = 355$ K **Find:** V
Conceptual Plan: g C → mol C ($\frac{1 \text{ mol}}{12.01 \text{ g C}}$) → mol H_2 ($\frac{1 \text{ mol } H_2}{1 \text{ mol C}}$) then n (mol H_2), P, T → V ($PV = nRT$)

Solution: $15.7 \text{ g C} \times \frac{1 \text{ mol C}}{12.01 \text{ g C}} \times \frac{1 \text{ mol } H_2}{1 \text{ mol C}} = 1.3\underline{0}724 \text{ mol } H_2$, $PV = nRT$ Rearrange to solve for V.

$$V = \frac{nRT}{P} = \frac{1.30724 \text{ mol} \times 0.08206 \frac{\text{L} \cdot \text{atm}}{\text{mol} \cdot \text{K}} \times 355 \text{ K}}{1.0 \text{ atm}} = 38 \text{ L}$$

Check: The units (L) are correct. The magnitude of the answer (38 L) makes sense because we have more than one mole of gas, and so we expect more than 22 L.

43. **Given:** $P = 748$ mmHg, $T = 86$ °C, and m (CH_3OH) = 25.8 g, and **Find:** V_{H2} and V_{CO}
Conceptual Plan: g CH_3OH → mol CH_3OH → mol H_2 and mmHg → atm and °C → K

$\frac{1 \text{ mol } CH_3OH}{32.04 \text{ g } CH_3OH}$ $\frac{2 \text{ mol } H_2}{1 \text{ mol } CH_3OH}$ $\frac{1 \text{ atm}}{760 \text{ mmHg}}$ $K = °C + 273.15$

then n (mol H_2), P, T → V and mol H_2 → mol CO then n (mol CO), P, T → V

$PV = nRT$ $\frac{1 \text{ mol CO}}{2 \text{ mol } H_2}$ $PV = nRT$

Solution: $25.8 \text{ g } CH_3OH \times \frac{1 \text{ mol } CH_3OH}{32.04 \text{ g } CH_3OH} \times \frac{2 \text{ mol } H_2}{1 \text{ mol } CH_3OH} = 1.61049 \text{ mol } H_2$,

$P_{H2} = 748 \text{ mmHg} \times \frac{1 \text{ atm}}{760 \text{ mmHg}} = 0.984211 \text{ atm}$, $T = 86 \text{ °C} + 273.15 = 359 \text{ K}$, $PV = nRT$

Rearrange to solve for V. $V = \frac{nRT}{P}$ $V_{H2} = \frac{1.61049 \text{ mol} \times 0.08206 \frac{\text{L} \cdot \text{atm}}{\text{mol} \cdot \text{K}} \times 359 \text{ K}}{0.984211 \text{ atm}} = 48.2 \text{ L } H_2$

$1.61049 \text{ mol } H_2 \times \frac{1 \text{ mol CO}}{2 \text{ mol } H_2} = 0.80525 \text{ mol CO}$, $V_{co} = \frac{0.80525 \text{ mol} \times 0.08206 \frac{\text{L} \cdot \text{atm}}{\text{mol} \cdot \text{K}} \times 359 \text{ K}}{0.984211 \text{ atm}} = 24.1 \text{ L CO}$

Check: The units (L) are correct. The magnitude of the answer (48 L and 24 L) makes sense because we have more than one mole of hydrogen gas and half that of CO and so we expect significantly more than 22 L for hydrogen and half that for CO.

45. **Given:** $V = 11.8$ L, and STP **Find:** m (NaN_3)
Conceptual Plan: V_{N2} → mol N_2 → mol NaN_3 → g NaN_3

$\frac{1 \text{ mol } N_2}{22.414 \text{ L } N_2}$ $\frac{2 \text{ mol } NaN_3}{3 \text{ mol } N_2}$ $\frac{65.01 \text{ g } NaN_3}{1 \text{ mol } NaN_3}$

Solution: $11.8 \text{ L } N_2 \times \frac{1 \text{ mol } N_2}{22.414 \text{ L } N_2} \times \frac{2 \text{ mol } NaN_3}{3 \text{ mol } N_2} \times \frac{65.01 \text{ g } NaN_3}{1 \text{ mol } NaN_3} = 22.8 \text{ g } NaN_3$

Check: The units (g) are correct. The magnitude of the answer (23 g) makes sense because, we have about a half a mole of nitrogen gas, which translates to even fewer moles of NaN_3 and so we expect significantly less than 65 g.

47. **Given:** $V_{CH4} = 25.5$ L, $P_{CH4} = 732$ torr, and $T = 25$ °C; mixed with $V_{H2O} = 22.8$ L, $P_{H2O} = 702$ torr, and $T = 125$ °C; forms $P_{H2} = 26.2$ L at STP **Find:** % Yield
Conceptual Plan: CH_4: torr → atm and °C → K and P, V, T → n_{CH4} → n_{H2}

$\frac{1 \text{ atm}}{760 \text{ torr}}$ $K = °C + 273.15$ $PV = nRT$ $\frac{3 \text{ mol } H_2}{1 \text{ mol } CH_4}$

H_2O: torr → atm and °C → K and P, V, T → n_{H2O} → n_{H2}

$\frac{1 \text{ atm}}{760 \text{ torr}}$ $K = °C + 273.15$ $PV = nRT$ $\frac{3 \text{ mol } H_2}{1 \text{ mol } H_2O}$

Select smaller n_{H2} as theoretical yield.
then L_{H2} → mol H_2 (actual yield) finally actual yield, theoretical yield → % Yield

$\frac{1 \text{ mol } H_2}{22.414 \text{ L } H_2}$ $\% \text{ Yield} = \frac{\text{actual yield}}{\text{theoretical yield}} \times 100 \%$

Solution: CH_4: $P_{CH4} = 732 \text{ torr} \times \frac{1 \text{ atm}}{760 \text{ torr}} = 0.963158 \text{ atm}$, $T = 25 \text{ °C} + 273.15 = 298 \text{ K}$, $PV = nRT$

Rearrange to solve for n. $n = \frac{PV}{RT}$ $n_{CH4} = \frac{0.963158 \text{ atm} \times 25.5 \text{ L}}{0.08206 \frac{\text{L} \cdot \text{atm}}{\text{mol} \cdot \text{K}} \times 298 \text{ K}} = 1.00436 \text{ mol } CH_4$

$1.00436 \text{ mol } CH_4 \times \frac{3 \text{ mol } H_2}{1 \text{ mol } CH_4} = 3.01308 \text{ mol } H_2$

H_2O: $P_{H2O} = 702 \text{ torr} \times \frac{1 \text{ atm}}{760 \text{ torr}} = 0.92\underline{3}684 \text{ atm}$, $T = 125\ °C + 273.15 = 398 \text{ K}$, $n = \frac{PV}{RT}$

$$n_{H2O} = \frac{0.92\underline{3}684 \text{ atm} \times 22.8 \text{ L}}{0.08206 \frac{\text{L} \cdot \text{atm}}{\text{mol} \cdot \text{K}} \times 398 \text{ K}} = 0.64\underline{4}828 \text{ mol } H_2O \qquad 0.64\underline{4}828 \text{ mol } H_2O \times \frac{3 \text{ mol } H_2}{1 \text{ mol } H_2O} = 1.9\underline{3}448 \text{ mol } H_2$$

Water is the limiting reagent since the moles of hydrogen generated is lower.

Theoretical yield = $1.9\underline{3}448$ mol H_2. $26.2 \text{ L } H_2 \times \frac{1 \text{ mol} H_2}{22.414 \text{ L } H_2} = 1.1\underline{6}891 \text{mol} H_2$ = actual yield

$$\% \ Yield = \frac{actual\ yield}{theoretical\ yield} \times 100\,\% = \frac{1.1\underline{6}891 \text{ mol } H_2}{1.9\underline{3}448 \text{ mol } H_2} \times 100\,\% = 60.4\,\%$$

Check: The units (%) are correct. The magnitude of the answer (60 %) makes sense because it is between 0 and 100 %.

49. a) Yes, since the average kinetic energy of a particle is proportional to the temperature in kelvins and the two gases are at the same temperature, they have the same average kinetic energy.

b) No, since the helium atoms are lighter, they must move faster to have the same kinetic energy as argon atoms.

c) No, since the Ar atoms are moving slower to compensate for their larger mass, they will exert the same pressure on the walls of the container.

d) Since He is lighter, it will have the faster rate of effusion.

51. **Given:** F_2, Cl_2, Br_2, and $T = 298$ K **Find:** u_{rms} KE_{avg} for each gas and relative rates of effusion
Conceptual Plan: $\mathfrak{M}$, T → u_{rms} → KE_{avg}

$$u_{rms} = \sqrt{\frac{3RT}{\mathfrak{M}}} \qquad KE_{avg} = \frac{1}{2} N_A m u_{rms}^2 = \frac{3}{2} RT$$

Solution:

$$F_2: \mathfrak{M} = \frac{38.00 \text{ g}}{1 \text{ mol}} \times \frac{1 \text{ kg}}{1000 \text{ g}} = 0.03800 \text{ kg/mol}, \quad u_{rms} = \sqrt{\frac{3RT}{\mathfrak{M}}} = \sqrt{\frac{3 \times 8.314 \frac{\text{J}}{\text{K} \times \text{mol}} \times 298 \text{ K}}{0.03800 \frac{\text{kg}}{\text{mol}}}} = 442 \text{ m/s}$$

$$Cl_2: \mathfrak{M} = \frac{70.90 \text{ g}}{1 \text{ mol}} \times \frac{1 \text{ kg}}{1000 \text{ g}} = 0.07090 \text{ kg/mol}, \quad u_{rms} = \sqrt{\frac{3RT}{\mathfrak{M}}} = \sqrt{\frac{3 \times 8.314 \frac{\text{J}}{\text{K} \times \text{mol}} \times 298 \text{ K}}{0.07090 \frac{\text{kg}}{\text{mol}}}} = 324 \text{ m/s}$$

$$Br_2: \mathfrak{M} = \frac{159.80 \text{ g}}{1 \text{ mol}} \times \frac{1 \text{ kg}}{1000 \text{ g}} = 0.15980 \text{ kg/mol}, \quad u_{rms} = \sqrt{\frac{3RT}{\mathfrak{M}}} = \sqrt{\frac{3 \times 8.314 \frac{\text{J}}{\text{K} \times \text{mol}} \times 298 \text{ K}}{0.15980 \frac{\text{kg}}{\text{mol}}}} = 216 \text{ m/s}$$

All molecules have the same kinetic energy:

$$KE_{avg} = \frac{3}{2} RT = \frac{3}{2} \times 8.314 \frac{\text{J}}{\text{K} \cdot \text{mol}} \times 298 \text{ K} = 3.72 \times 10^3 \text{ J/mol}$$

Since rate of effusion is proportional to $\sqrt{\frac{1}{\mathfrak{M}}}$, F_2 will have the fastest rate and Br_2 will have the slowest rate.

Check: The units (m/s) are correct. The magnitude of the answer (200 – 450 m/s) makes sense because it is consistent with what was seen in the text, and the heavier the molecule, the slower the molecule.

53. **Given:** $^{238}UF_6$ and $^{235}UF_6$ U-235 = 235.054 amu, U-238 = 238.051 amu
Find: ratio of effusion rates $^{238}UF_6$ / $^{235}UF_6$
Conceptual Plan: $\mathfrak{M}$ ($^{238}UF_6$), $\mathfrak{M}$ ($^{235}UF_6$) → Rate ($^{238}UF_6$) / Rate ($^{235}UF_6$)

$$\frac{Rate(^{238}UF_6)}{Rate(^{235}UF_6)} = \sqrt{\frac{\mathfrak{M}(^{235}UF_6)}{\mathfrak{M}(^{238}UF_6)}}$$

Solution: $^{238}UF_6$: $\mathfrak{M} = \frac{352.05\ \text{g}}{1\ \text{mol}} \times \frac{1\ \text{kg}}{1000\ \text{g}} = 0.35205\ \text{kg/mol}$,

$^{235}UF_6$: $\mathfrak{M} = \frac{349.05\ \text{g}}{1\ \text{mol}} \times \frac{1\ \text{kg}}{1000\ \text{g}} = 0.34905\ \text{kg/mol}$,

$$\frac{Rate(^{238}UF_6)}{Rate(^{235}UF_6)} = \sqrt{\frac{\mathfrak{M}(^{235}UF_6)}{\mathfrak{M}(^{238}UF_6)}} = \sqrt{\frac{0.34905\ \text{kg/mol}}{0.35205\ \text{kg/mol}}} = 0.99574$$

Check: The units (none) are correct. The magnitude of the answer (<1) makes sense because, the heavier molecule has the lower effusion rate since it moves slower.

55. **Given:** Ne and unknown gas; and Ne effusion in 76 s and unknown in 155 s
Find: identify unknown gas
Conceptual Plan: **$\mathfrak{M}$ (Ne), Rate (Ne), Rate (Unk) → $\mathfrak{M}$ (Kr)**

$$\frac{Rate(Ne)}{Rate(Unk)} = \sqrt{\frac{\mathfrak{M}(Unk)}{\mathfrak{M}(Ne)}}$$

Solution: Ne: $\mathfrak{M} = \frac{20.18\ \text{g}}{1\ \text{mol}} \times \frac{1\ \text{kg}}{1000\ \text{g}} = 0.02018\ \text{kg/mol}$, $\frac{Rate(Ne)}{Rate(Unk)} = \sqrt{\frac{\mathfrak{M}(Unk)}{\mathfrak{M}(Ne)}}$ Rearrange to solve for $\mathfrak{M}(Unk)$. $\mathfrak{M}(Unk) = \mathfrak{M}(Ne)\left(\frac{Rate(Ne)}{Rate(Unk)}\right)^2$ Since Rate α 1/(effusion time),

$$\mathfrak{M}(Unk) = \mathfrak{M}(Ne)\left(\frac{Time(Unk)}{Time(Ne)}\right)^2 = 0.02018\frac{\text{kg}}{\text{mol}} x \left(\frac{155\ \text{s}}{76\ \text{s}}\right)^2 = 0.084\frac{\text{kg}}{\text{mol}} \times \frac{1000\ \text{g}}{1\ \text{kg}} = 84\ \text{g/mol} \text{ or Kr.}$$

Check: The units (g/mol) are correct. The magnitude of the answer (>Ne) makes sense because, Ne effused faster and so must be lighter.

57. Gas A has the higher molar mass, since it has the slower average velocity. Gas B will have the higher effusion rate, since it has the higher velocity.

59. The postulate that the volume of the gas particles is small compared to the space between them breaks down at high pressure. At high pressures the number of molecules increases, so the volume of the gas particles becomes larger and since the spacing between the particles is smaller, the molecules themselves occupy a significant portion of the volume.

61. **Given:** Ne, $n = 1.000$ mol, $P = 500.0$ atm, and $T = 355.0$ K **Find:** V(ideal) and V(van der Waals)
Conceptual Plan: ***n, P, T* → *V*** **and** ***n, P, T* → *V***

$$PV = nRT \qquad \left(P + \frac{an^2}{V^2}\right)(V - nb) = nRT$$

Solution: $PV = nRT$ Rearrange to solve for V.

$$V = \frac{nRT}{P} = \frac{1.000\ \text{mol} \times 0.08206\ \frac{\text{L} \cdot \text{atm}}{\text{mol} \cdot \text{K}} \times 355.0\ \text{K}}{500.0\ \text{atm}} = 0.05826\ \text{L}$$

$\left(P + \frac{an^2}{V^2}\right)(V - nb) = nRT$ Rearrange to solve to: $V = \frac{nRT}{\left(P + \frac{an^2}{V^2}\right)} + nb$

Using a = 0.211 L^2 atm/mol^2 and b = 0.0171 L/mol from Table 5.5, and the V from the ideal gas law above, solve for V by successive approximations.

$$V = \frac{1.000\ \text{mol} \times 0.08206\ \frac{\text{L} \cdot \text{atm}}{\text{mol} \cdot \text{K}} \times 355.0\ \text{K}}{500.0\ \text{atm} + \frac{0.211\frac{\text{L}^2 \cdot \text{atm}}{\text{mol}^2} \times (1.000\ \text{mol})^2}{(0.05826\ \text{L})^2}} + \left(1.000\ \text{mol} \times 0.0171\frac{\text{L}}{\text{mol}}\right) = 0.068\underline{9}15\ \text{L}$$

Plug in this new value.

$$V = \frac{1.000\ \text{mol} \times 0.08206\ \frac{\text{L}\cdot\text{atm}}{\text{mol}\cdot\text{K}} \times 355.0\ \text{K}}{500.0\ \text{atm} + \frac{0.211\ \frac{\text{L}^2\cdot\text{atm}}{\text{mol}^2} \times (1.000\ \text{mol})^2}{(0.068915\ \text{L})^2}} + \left(1.000\ \text{mol} \times 0.0171\ \frac{\text{L}}{\text{mol}}\right) = 0.070\underline{6}09\ \text{L}$$

Plug in this new value.

$$V = \frac{1.000\ \text{mol} \times 0.08206\ \frac{\text{L}\cdot\text{atm}}{\text{mol}\cdot\text{K}} \times 355.0\ \text{K}}{500.0\ \text{atm} + \frac{0.211\ \frac{\text{L}^2\cdot\text{atm}}{\text{mol}^2} \times (1.000\ \text{mol})^2}{(0.070609\ \text{L})^2}} + \left(1.000\ \text{mol} \times 0.0171\ \frac{\text{L}}{\text{mol}}\right) = 0.070\underline{8}17\ \text{L}$$

Plug in this new value.

$$V = \frac{1.000\ \text{mol} \times 0.08206\ \frac{\text{L}\cdot\text{atm}}{\text{mol}\cdot\text{K}} \times 355.0\ \text{K}}{500.0\ \text{atm} + \frac{0.211\ \frac{\text{L}^2\cdot\text{atm}}{\text{mol}^2} \times (1.000\ \text{mol})^2}{(0.070817\ \text{L})^2}} + \left(1.000\ \text{mol} \times 0.0171\ \frac{\text{L}}{\text{mol}}\right) = 0.070\underline{8}42\ \text{L} = 0.0708\ \text{L}$$

The two values are different because we are at very high pressures. The pressure is corrected from 500.0 atm to 542.1 atm and the final volume correction is 0.0171 L.

Check: The units (L) are correct. The magnitude of the answer (~0.06 L) makes sense because we are at such a high pressure and have one mole of gas.

63. **Given:** m (penny) = 2.482 g, T = 25 °C, V = 0.899 L, and P_{Total} = 791 mmHg
Find: % Zn in penny
Conceptual Plan: $T \rightarrow P_{H2O}$ (Table 5.3) then $P_{\text{Total}}, P_{H2O} \rightarrow P_{H2}$ ($P_{Total} = P_{H2O} + P_{H2}$) then mmHg → atm ($\frac{1\ \text{atm}}{760\ \text{mmHg}}$) and °C → K (K = °C + 273.15)

and $P, V, T \rightarrow n_{H2} \rightarrow n_{Zn} \rightarrow g_{Zn} \rightarrow$ % Zn

$PV = nRT$; $\frac{1\ \text{mol Zn}}{1\ \text{mol H}_2}$; $\frac{65.39\ \text{g Zn}}{1\ \text{mol Zn}}$; $\%\ Zn = \frac{g_{Zn}}{g_{penny}} \times 100\ \%$

Solution: Table 5.3 states that P_{H2O} = 23.78 mmHg at 25 °C $P_{Total} = P_{H2O} + P_{H2}$ Rearrange to solve for P_{H2}.
$P_{H2} = P_{Total} - P_{H2O} = 791\ \text{mmHg} - 23.78\ \text{mmHg} = 767\ \text{mmHg}$

$$P_{H2} = 767\ \text{mmHg} \times \frac{1\ \text{atm}}{760\ \text{mmHg}} = 1.0\underline{0}95\ \text{atm} \quad \text{then} \quad T = 25\ °\text{C} + 273.15 = 298\ \text{K}, \quad PV = nRT$$

Rearrange to solve for n. $$n_{H2} = \frac{PV}{RT} = \frac{1.0\underline{0}95\ \text{atm} \times 0.899\ \text{L}}{0.08206\ \frac{\text{L}\cdot\text{atm}}{\text{mol}\cdot\text{K}} \times 298\ \text{K}} = 0.037\underline{1}123\ \text{mol}$$

$$0.037\underline{1}123\ \text{mol H}_2 \times \frac{1\ \text{mol Zn}}{1\ \text{mol H}_2} \times \frac{65.39\ \text{g Zn}}{1\ \text{mol Zn}} = 2.4\underline{2}677\ \text{g Zn}$$

$$\%\ Zn = \frac{g_{Zn}}{g_{penny}} \times 100\ \% = \frac{2.4\underline{2}677\ \text{g}}{2.482\ \text{g}} \times 100\ \% = 97.8\ \%\ \text{Zn}$$

Check: The units (% Zn) are correct. The magnitude of the answer (98 %) makes sense because it should be between 0 and 100 %. We expect about 1/22 a mole of gas, since our conditions are close to STP and we have ~ 1 L of gas.

65. **Given:** V = 255 mL, m (flask) = 143.187 g, m (flask + gas) = 143.289 g, P = 267 torr, and T = 25 °C
Find: $\mathfrak{M}$
Conceptual Plan: °C → K (K = °C + 273.15) torr → atm ($\frac{1\ \text{atm}}{760\ \text{torr}}$) mL → L ($\frac{1\ \text{L}}{1000\ \text{mL}}$) m (flask), m (flask + gas) → m (gas) (m (gas) = m (flask + gas) – m (flask))

then $V, m \rightarrow d$ ($d = \frac{m}{V}$) then $d, P, T, \rightarrow \mathfrak{M}$ ($d = \frac{P\mathfrak{M}}{RT}$)

Solution: $T = 25\ °\text{C} + 273.15 = 298\ \text{K}$, $\quad P = 267\ \text{torr} \times \frac{1\ \text{atm}}{760\ \text{torr}} = 0.35\underline{1}316\ \text{atm}$,

$V = 255\ \text{mL} \times \frac{1\ \text{L}}{1000\ \text{mL}} = 0.255\ \text{L}$,

$m\ (gas) = m\ (flask + gas) - m\ (flask) = 143.289\ \text{g} - 143.187\ \text{g} = 0.102\ \text{g}$, $d = \frac{m}{V} = \frac{0.102\ \text{g}}{0.255\ \text{L}} = 0.400\ \text{g/L}$,

$d = \frac{P\mathfrak{M}}{RT}$ Rearrange to solve for $\mathfrak{M}$.

$$\mathfrak{M} = \frac{dRT}{P} = \frac{0.400\ \frac{\text{g}}{\text{L}} \times 0.08206\ \frac{\text{L atm}}{\text{K mol}} \times 298\ \text{K}}{0.35\underline{1}316\ \text{atm}} = 27.8\ \text{g/mol}$$

Check: The units (g/mol) are correct. The magnitude of the answer (28 g/mol) makes physical sense because this is a reasonable number for a molecular weight of a gas.

67. **Given:** $V = 158$ mL, m (gas) = 0.275 g, $P = 556$ mmHg, $T = 25$ °C, gas = 82.66 % C and 17.34 % H.
Find: Molecular formula
Conceptual Plan: **°C → K** ($K = °C + 273.15$) **mmHg → atm** ($\frac{1\ \text{atm}}{760\ \text{mmHg}}$) **mL → L** ($\frac{1\ \text{L}}{1000\ \text{mL}}$) **then V, m → d** ($d = \frac{m}{V}$)

then d, P, T, → $\mathfrak{M}$ ($d = \frac{P\mathfrak{M}}{RT}$) **then % C, % H, $\mathfrak{M}$ → formula** ($\#C = \frac{\mathfrak{M}\ 0.8266\ g\ C}{12.01\ \frac{g\ C}{mol\ C}}$, $\#H = \frac{\mathfrak{M}\ 0.1734\ g\ H}{1.008\ \frac{g\ H}{mol\ H}}$)

Solution: $T = 25\ °\text{C} + 273.15 = 298\ \text{K}$, $P = 556\ \text{mmHg} \times \frac{1\ \text{atm}}{760\ \text{mmHg}} = 0.73\underline{1}579\ \text{atm}$,

$V = 158\ \text{mL} \times \frac{1\ \text{L}}{1000\ \text{mL}} = 0.158\ \text{L}$, $d = \frac{m}{V} = \frac{0.275\ \text{g}}{0.158\ \text{L}} = 1.7\underline{4}051\ \text{g/L}$, $d = \frac{P\mathfrak{M}}{RT}$ Rearrange to solve for $\mathfrak{M}$.

$$\mathfrak{M} = \frac{dRT}{P} = \frac{1.7\underline{4}051\ \frac{\text{g}}{\text{L}} \times 0.08206\ \frac{\text{L atm}}{\text{K mol}} \times 298\ \text{K}}{0.73\underline{1}579\ \text{atm}} = 58.2\ \text{g/mol},$$

$$\#C = \frac{\mathfrak{M} \times 0.8266\ g\ C}{12.01\ \frac{g\ C}{mol\ C}} = \frac{58.2\ \frac{\text{g HC}}{\text{mol HC}} \times \frac{0.8266\ \text{g C}}{1\ \text{g HC}}}{12.01\ \frac{\text{g C}}{\text{mol C}}} = 4.00\ \frac{\text{mol C}}{\text{mol HC}}$$

$$\#H = \frac{\mathfrak{M} \times 0.1734\ g\ H}{1.008\ \frac{g\ H}{mol\ H}} = \frac{58.2\ \frac{\text{g HC}}{\text{mol HC}} \times \frac{0.1734\ \text{g H}}{1\ \text{g HC}}}{1.008\ \frac{\text{g H}}{\text{mol H}}} = 10.0\ \frac{\text{mol H}}{\text{mol HC}}$$ Formula is C_4H_{10} or butane.

Check: The answer came up with integer number of C and H atoms in the formula and a molecular weight (58 g/mol) that is reasonable for a gas.

69. **Given:** m (NiO) = 24.78 g, $T = 40.0$ °C, and $P_{Total} = 745$ mmHg **Find:** V_{O2}
Conceptual Plan: **T → P_{H2O}** (Table 5.3) **then P_{Total}, P_{H2O} → P_{O2}** ($P_{Total} = P_{H2O} + P_{O2}$) **then mmHg → atm** ($\frac{1\ \text{atm}}{760\ \text{mmHg}}$) **and °C → K** ($K = °C + 273.15$)

and g_{NiO} → n_{NiO} ($\frac{1\ \text{mol NiO}}{74.69\ \text{g NiO}}$) **→ n_{O2}** ($\frac{1\ \text{mol}\ O_2}{2\ \text{mol NiO}}$) **then P, V, T → n_{O2}** ($PV = nRT$)

Solution: Table 5.3 states that $P_{H2O} = 55.40$ mmHg at 40°C $P_{Total} = P_{H2O} + P_{O2}$ Rearrange to solve for P_{O2}.
$P_{O2} = P_{Total} - P_{H2O} = 745\ \text{mmHg} - 55.40\ \text{mmHg} = 689.\underline{6}\ \text{mmHg}$

$P_{O_2} = 689.\underline{6}\ \text{mmHg} \times \frac{1\ \text{atm}}{760\ \text{mmHg}} = 0.90\underline{7}368\ \text{atm}$ $T = 40.0\ °\text{C} + 273.15 = 313.2\ \text{K}$,

$24.78\ \text{g NiO} \times \frac{1\ \text{mol NiO}}{74.69\ \text{mol NiO}} \times \frac{1\ \text{mol}\ O_2}{2\ \text{mol NiO}} = 0.165\underline{8}857\ \text{mol}\ O_2$ $PV = nRT$

Rearrange to solve for V. $$V_{O_2} = \frac{nRT}{P} = \frac{0.165\underline{8}857\ \text{mol} \times 0.08206\ \frac{\text{L} \cdot \text{atm}}{\text{mol} \cdot \text{K}} \times 313.2\ \text{K}}{0.90\underline{7}368\ \text{atm}} = 4.70\ \text{L}$$

Check: The units (L) are correct. The magnitude of the answer (5 L) makes sense because we have much less than 0.5 mole of NiO, so we get less than a mole of oxygen. Thus we expect a volume much less than 22 L.

71. **Given:** HCl, K_2S to H_2S, $V_{H2S} = 42.9$ mL, $P_{H2S} = 752$ mmHg, and $T = 25.8$ °C **Find:** $m(K_2S)$
Conceptual Plan: read description of reaction and convert words to equation then **°C → K** ($K = °C + 273.15$)
and mmHg → atm ($\frac{1\text{ atm}}{760\text{ mmHg}}$) **and mL → L** ($\frac{1\text{ L}}{1000\text{ mL}}$) **then** $P, V, T \rightarrow n_{H2S}$ ($PV = nRT$) $\rightarrow n_{K2S}$ ($\frac{1\text{ mol }K_2S}{1\text{ mol }H_2S}$) $\rightarrow g_{K2S}$ ($\frac{1\text{ mol }K_2S}{110.27\text{ g }K_2S}$)

Solution: $2\text{ HCl}(aq) + K_2S(s) \rightarrow H_2S(g) + 2\text{ KCl}(aq)$

$$T = 25.8\text{ °C} + 273.15 = 299.0\text{ K},\ P_{H2S} = 752\ \cancel{\text{mmHg}} \times \frac{1\text{ atm}}{760\ \cancel{\text{mmHg}}} = 0.98\underline{9}474\text{ atm},$$

$$V_{H2S} = 42.9\ \cancel{\text{mL}} \times \frac{1\text{ L}}{1000\ \cancel{\text{mL}}} = 0.0429\text{ L} \quad PV = nRT \quad \text{Rearrange to solve for } n_{H2S}.$$

$$n_{H2S} = \frac{PV}{RT} = \frac{0.98\underline{9}474\ \cancel{\text{atm}} \times 0.0429\ \cancel{\text{L}}}{0.08206\ \frac{\cancel{\text{L}}\cdot\cancel{\text{atm}}}{\text{mol}\cdot\cancel{\text{K}}} \times 299.0\ \cancel{\text{K}}} = 0.0017\underline{3}005\text{ mol}$$

$$0.0017\underline{3}005\ \cancel{\text{mol }H_2S} \times \frac{1\ \cancel{\text{mol }K_2S}}{1\ \cancel{\text{mol }H_2S}} \times \frac{110.27\text{ g }K_2S}{1\ \cancel{\text{mol }K_2S}} = 0.191\text{ g }K_2S$$

Check: The units (g) are correct. The magnitude of the answer (0.1 g) makes sense because we have such a small volume of gas generated.

73. **Given:** $T = 22$ °C, $P = 1.02$ atm, and $m = 11.83$ g **Find:** V_{Total}
Conceptual Plan: °C → K ($K = °C + 273.15$) **and** $g_{(NH4)2CO3} \rightarrow n_{(NH4)2CO3}$ ($\frac{1\text{ mol }(NH_4)_2CO_3}{96.09\text{ g }(NH_4)_2CO_3}$) $\rightarrow n_{Gas}$ ($\frac{(2+1+1=4)\text{ mol gas}}{1\text{ mol }NH_4CO_3}$) **then** $P, n, T \rightarrow V$ ($PV = nRT$)

Solution: $T = 22\text{ °C} + 273.15 = 295\text{ K}$,

$$11.83\ \cancel{\text{g }NH_4CO_3} \times \frac{1\ \cancel{\text{mol }(NH_4)_2CO_3}}{96.09\ \cancel{\text{g }(NH_4)_2CO_3}} \times \frac{4\text{ mol gas}}{1\ \cancel{\text{mol }(NH_4)_2CO_3}} = 0.49\underline{2}455\text{ mol gas}$$

$$PV = nRT \text{ Rearrange to solve for } V_{gas}.\ V_{gas} = \frac{nRT}{P} = \frac{0.49\underline{2}455\ \cancel{\text{mol gas}} \times 0.08206\ \frac{\text{L}\cdot\cancel{\text{atm}}}{\cancel{\text{mol}}\cdot\cancel{\text{K}}} \times 295\ \cancel{\text{K}}}{1.02\ \cancel{\text{atm}}} = 11.7\text{ L}$$

Check: The units (L) are correct. The magnitude of the answer (12 L) makes sense because we have about a half a mole of gas generated.

75. **Given:** He and air; $V = 855$ mL, $P = 125$ psi, $T = 25$ °C, $\mathfrak{M}(\text{air}) = 28.8$ g/mol **Find:** $\Delta = m(\text{air}) - m(\text{He})$
Conceptual Plan: mL → L ($\frac{1\text{ L}}{1000\text{ mL}}$) **and psi → atm** ($\frac{1\text{ atm}}{14.7\text{ psi}}$) **and °C → K** ($K = °C + 273.15$) **then** $P, T, \mathfrak{M} \rightarrow d$ ($d = \frac{P\mathfrak{M}}{RT}$)
then $d, V \rightarrow m$ ($d = \frac{m}{V}$) $m(\text{air}), m(\text{He}) \rightarrow \Delta$ ($\Delta = m(air) - m(He)$)

Solution: $V = 855\ \cancel{\text{mL}} \times \frac{1\text{ L}}{1000\ \cancel{\text{mL}}} = 0.855\text{ L},\ P = 125\ \cancel{\text{psi}} \times \frac{1\text{ atm}}{14.7\ \cancel{\text{psi}}} = 8.5\underline{0}340\text{ atm},\ T = 25\text{ °C} + 273.15 = 298\text{K},$

$$d_{air} = \frac{P\mathfrak{M}}{RT} = \frac{8.5\underline{0}340\ \cancel{\text{atm}} \times 28.8\ \frac{\text{g air}}{\cancel{\text{mol air}}}}{0.08206\ \frac{\text{L}\cdot\cancel{\text{atm}}}{\cancel{\text{K}}\cdot\cancel{\text{mol}}} \times 298\ \cancel{\text{K}}} = 10.\underline{0}147\ \frac{\text{g air}}{\text{L}},\ d = \frac{m}{V} \text{ Rearrange to solve for } m.\quad m = dV$$

$$m_{air} = 10.\underline{0}147\ \frac{\text{g air}}{\cancel{\text{L}}} \times 0.8554\ \cancel{\text{L}} = 8.5\underline{6}657\text{ g air},\ d_{He} = \frac{P\mathfrak{M}}{RT} = \frac{8.5\underline{0}340\ \cancel{\text{atm}} \times 4.03\ \frac{\text{g He}}{\cancel{\text{mol He}}}}{0.08206\ \frac{\text{L}\cdot\cancel{\text{atm}}}{\cancel{\text{K}}\cdot\cancel{\text{mol}}} \times 298\ \cancel{\text{K}}} = 1.4\underline{0}136\ \frac{\text{g He}}{\text{L}},$$

$$m_{air} = 1.4\underline{0}136\ \frac{\text{g He}}{\cancel{\text{L}}} \times 0.855\ \cancel{\text{L}} = 1.1\underline{9}816\text{ g He},\ \Delta = m(air) - m(He) = 8.5\underline{6}657\text{ g air} - 1.1\underline{9}816\text{ g He} = 7.37\text{ g}$$

Check: The units (g) are correct. We expect the difference to be less than the difference in the molecular weights since we have less than a mole of gas.

77. **Given:** $Flow_{NO}$ = 335 L/s, P_{NO} = 22.4 torr, T_{NO} = 955 K, P_{NH3} = 755 torr, and T_{NO} = 298 K, and NH_3 purity = 65.2 % **Find:** $Flow_{NH3}$
Conceptual Plan: **torr → atm** then P_{NO}, V_{NO}/s, T_{NO} → n_{NO}/s → n_{NH3}/s (pure)

$\frac{1\text{ atm}}{760\text{ torr}}$ $\quad PV = nRT \quad$ $\frac{4\text{ mol NH}_3}{4\text{ mol NO}}$

then n_{NH3}/s (pure) → n_{NH3}/s (impure) then n_{NH3}/s (impure), P_{NH3}, T_{NH3} → V_{NH3}/s

$\frac{100\text{ mol NH}_3\text{ impure}}{65.2\text{ mol NH}_3\text{ pure}}$ $\quad PV = nRT$

Solution: $P_{NO} = 22.4\text{ torr} \times \frac{1\text{ atm}}{760\text{ torr}} = 0.0294737\text{ atm}$, $P_{NH3} = 755\text{ torr} \times \frac{1\text{ atm}}{760\text{ torr}} = 0.993421\text{ atm}$

$PV = nRT$ Rearrange to solve for n_{NO}. Note that we can substitute V/s for V and get n/s as a result.

$$\frac{n_{NO}}{s} = \frac{PV}{RT} = \frac{0.0294737\text{ atm} \times 335\text{ L/s}}{0.08206\frac{\text{L}\cdot\text{atm}}{\text{mol}\cdot\text{K}} \times 955\text{ K}} = 0.125992\frac{\text{mol NO}}{\text{s}}$$

$$0.125992\frac{\text{mol NO}}{\text{s}} \times \frac{4\text{ mol NH}_3}{4\text{ mol NO}} \times \frac{100\text{ mol NH}_3\text{ impure}}{65.2\text{ mol NH}_3\text{ pure}} = 0.193240\frac{\text{mol NH}_3\text{ impure}}{\text{s}} \qquad PV = nRT$$

Rearrange to solve for V_{NH3}. Note that we can substitute n/s for n and get V/s as a result.

$$\frac{V_{NH3}}{s} = \frac{nRT}{P} = \frac{0.193240\frac{\text{mol NH}_3\text{ impure}}{\text{s}} \times 0.08206\frac{\text{L}\cdot\text{atm}}{\text{mol}\cdot\text{K}} \times 298\text{ K}}{0.993421\text{ atm}} = 4.76\frac{\text{L}}{\text{s}}\text{ impure NH}_3$$

Check: The units (L) are correct. The magnitude of the answer (5 L/s) makes sense because we expect it to be less than for the NO. The NO is at a very low concentration and a high temperature, when this converts to a much higher pressure and lower temperature this will go down significantly, even though the ammonia is impure. From a practical standpoint, you would like a low flow rate to make it economical.

79. **Given:** l = 30.0 cm, w = 20.0 cm, h = 15.0 cm, 14.7 psi **Find:** Force (lbs)
Conceptual Plan: ***l, w, h*** **→ Surface Area, SA (cm²) → Surface Area(in²) → Force**

$SA = 2(lh) + 2(wh) + 2(lw) \quad \frac{(1\text{ in})^2}{(2.54\text{ cm})^2} \quad \frac{14.7\text{ lbs}}{1\text{ in}^2}$

Solution:

$SA = 2(lh) + 2(wh) + 2(lw) = 2(30.0\text{ cm} \times 15.0\text{ cm}) + 2(20.0\text{ cm} \times 15.0\text{ cm}) + 2(30.0\text{ cm} \times 20.0\text{ cm}) = 2700\text{ cm}^3$

$2700\text{ cm}^2 \times \frac{(1\text{ in})^2}{(2.54\text{ cm})^2} = 418.50\text{ in}^2$, $418.50\text{ in}^2 \times \frac{14.7\text{ lbs}}{1\text{ in}^2} = 6150\text{ lbs}$ The can would be crushed.

Check: The units (lbs) are correct. The magnitude of the answer (6150 lbs) is not unreasonable since there is a large surface area.

81. **Given:** V_1 = 160.0 L, P_1 = 1855 psi, 3.5 L/balloon, P_2 = 1.0 atm = 14.7 psi and T = 298 K
Find: # balloons
Conceptual Plan: $V_1, P_1, P_2 \rightarrow V_2$ then **L → # balloons**

$P_1V_1 = P_2V_2 \quad \frac{1\text{ balloon}}{3.5\text{ L}}$

Solution: $P_1V_1 = P_2V_2$ Rearrange to solve for V_2. $V_2 = \frac{P_1}{P_2}V_1 = \frac{1855\text{ psi}}{14.7\text{ psi}} \times 160.0\text{ L} = 20190.5\text{ L}$,

$20190.5\text{ L} \times \frac{1\text{ balloon}}{3.5\text{ L}} = 5800\text{ balloons}$

Check: The units (balloons) are correct. The magnitude of the answer (5800) is reasonable since a store does not want to buy a new helium tank very often.

83. **Given:** r_1 = 2.5 cm, P_1 = 4.00 atm, T = 298 K, and P_2 = 1.00 atm **Find:** r_2
Conceptual Plan: $r_1 \rightarrow V_1 \quad V_1, P_1, P_2 \rightarrow V_2$ then $V_2 \rightarrow r_2$

$V = \frac{4}{3}\pi r^3 \quad P_1V_1 = P_2V_2 \quad V = \frac{4}{3}\pi r^3$

Solution: $V = \frac{4}{3}\pi r^3 = \frac{4}{3} \text{ x } \pi \text{ x } (2.5 \text{ cm})^3 = 6\underline{5}.450 \text{ cm}^3$ $P_1V_1 = P_2V_2$ Rearrange to solve for V_2.

$V_2 = \frac{P_1}{P_2} V_1 = \frac{4.00 \text{ atm}}{1.00 \text{ atm}} \text{ x } 6\underline{5}.450 \text{ cm}^3 = 2\underline{6}1.80 \text{ cm}^3$, $V = \frac{4}{3}\pi r^3$

Rearrange to solve for r. $r = \sqrt[3]{\frac{3V}{4\pi}} = \sqrt[3]{\frac{3 \text{x } 2\underline{6}1.80 \text{ cm}^3}{4 \text{ x } \pi}} = 4.0 \text{ cm}$

Check: The units (cm) are correct. The magnitude of the answer (4 cm) is reasonable since the bubble will expand as the pressure is decreased.

85. **Given:** 2.0 mol CO : 1.0 mol O_2, $V = 2.45$ L, $P_1 = 745$ torr, $P_2 = 552$ torr, and $T = 5552$ °C
Find: % reacted
Conceptual Plan: from $PV = nRT$ we know that $P \alpha n$, looking at the chemical reaction we see that 2 + 1 = 3 moles of gas gets converted to 2 moles of gas. If all the gas reacts, $P_2 = 2/3 P_1$.
Calculate $-\Delta P$ for 100 % reacted and for actual case. Then calculate % reacted.

$-\Delta P \text{ 100 \% } reacted = P_1 - \frac{2}{3}P_1$ $-\Delta P \text{ } actual = P_1 - P_2$ $\% \text{ } reacted = \frac{\Delta P \text{ } actual}{\Delta P \text{ 100\% } reacted} x \text{ 100 \%}$

Solution: $-\Delta P \text{ 100 \% } reacted = P_1 - \frac{2}{3}P_1 = 745 \text{ torr} - \frac{2}{3} 745 \text{ torr} = 24\underline{8}.333 \text{ torr}$,

$-\Delta P \text{ } actual = P_1 - P_2 = 745 \text{ torr} - 552 \text{ torr} = 193 \text{ torr}$,

$\% \text{ } reacted = \frac{\Delta P \text{ } actual}{\Delta P \text{100\% } reacted} \text{ x } 100 \% = \frac{193 \text{ torr}}{24\underline{8}.333 \text{ torr}} \text{ x } 100 \% = 77.7 \%$

Check: The units (%) are correct. The magnitude of the answer (78 %) makes sense because the pressure dropped most of the way to the pressure if all of the reactants had reacted. Note: There are many ways to solve this problem, including calculating the moles of reactants and products using $PV = nRT$.

87. **Given:** $P(\text{Total})_1 = 2.2$ atm = CO + O_2, $P(\text{Total})_2 = 1.9$ atm = CO + O_2 + CO_2, $V = 1.0$ L, $T = 1.0 \text{ x } 10^3$ K
Find: CO_2 made
Conceptual Plan: $P(\text{Total})_1 = 2.2$ atm = $P(\text{CO})_1 + P(O_2)_1$, $P(\text{Total})_2 = 1.9$ atm = $P(\text{CO})_2 + P(O_2)_2 + P(CO_2)_2$, Let x = amount of $P(O_2)$ reacted. From stoichiometry: $P(\text{CO})_2 = P(\text{CO})_1 - 2x$, $P(O_2)_2 = P(O_2)_1 - x$, $P(CO_2)_2 = 2x$. Thus $P(\text{Total})_2 = 1.9$ atm = $P(\text{CO})_1 - 2x + P(O_2)_1 - x + 2x = P(\text{Total})_1 - x$. Using the initial conditions: 1.9 atm = 2.2 atm − x. So x = 0.3 atm and since $2x = P(CO_2)_2 = 0.6$ atm then

P, V, T* → *n
$PV = nRT$

Solution: $PV = nRT$ Rearrange to solve for n. $n = \frac{PV}{RT} = \frac{0.6 \text{ atm x } 1.0 \text{ L}}{0.08206 \frac{\text{L} \cdot \text{atm}}{\text{mol} \cdot \text{K}} \text{ x } 1000 \text{ K}} = 0.007 \text{ mol}$

Check: The units (mol) are correct. The magnitude of the answer (0.007 mol) makes sense because we have such a small volume, at a very high temperature and such a small pressure. All of these lead us to expect a very small number of moles.

89. **Given:** $h_1 = 22.6$ m, $T_1 = 22$ °C, and $h_2 = 23.8$ m **Find:** T_2
Conceptual Plan: °C → K since $V_{cylinder} \alpha h$ we do not need to know r to use V_1, T_1, T_2 → V_2

K = °C + 273.15 $V = \pi r^2 h$ $\frac{V_1}{T_1} = \frac{V_2}{T_2}$

Solution: $T_1 = 22 \text{ °C} + 273.15 = 295\text{K}$, $\frac{V_1}{T_1} = \frac{V_2}{T_2}$ Rearrange to solve for T_2.

$T_2 = T_1 \text{ x } \frac{V_2}{V_1} = T_1 \text{ x } \frac{\pi r^2 l_2}{\pi r^2 l_1} = 295 \text{ K x } \frac{23.8 \text{ m}}{22.6 \text{ m}} = 311 \text{ K}$

Check: The units (K) are correct. We expect the temperature to increase since the volume increased.

91. **Given:** He, $V = 0.35$ L, $P_{max} = 88$ atm, and $T = 299$ K **Find:** m_{He}
Conceptual Plan: *P, V, T* → *n* then mol → g
$PV = nRT$ $\mathfrak{M}$

Solution: $PV = nRT$ Rearrange to solve for n.

$$n_{He} = \frac{PV}{RT} = \frac{88 \text{ atm} \times 0.35 \text{ L}}{0.08206 \frac{\text{L} \cdot \text{atm}}{\text{mol} \cdot \text{K}} \times 299 \text{ K}} = 1.2553 \text{ mol}, \quad 1.2553 \text{ mol} \times \frac{4.003 \text{ g}}{1 \text{ mol}} = 5.0 \text{ g He}$$

Check: The units (g) are correct. The magnitude of the answer (5 g) makes sense because the high pressure and the low volume cancel out (remember 22 L / mol at STP) and so we expect ~ 1 mol and so ~ 4 g.

93. **Given:** 15.0 mL HBr in 1.0 min; and 20.3 mL unknown hydrocarbon gas in 1.0 min
Find: formula of unknown gas
Conceptual Plan: Since these are gases under the same conditions $V \propto n$, V, time → Rate then

$$Rate = \frac{V}{time}$$

$\mathfrak{M}$ (HBr), Rate (HBr), Rate (Unk) → $\mathfrak{M}$ (Unk)

$$\frac{Rate(HBr)}{Rate(U)} = \sqrt{\frac{\mathfrak{M}(U)}{\mathfrak{M}(HBr)}}$$

Solution: $Rate(HBr) = \frac{V}{time} = \frac{15.0 \text{ mL}}{1.0 \text{ min}} = 15.0 \frac{\text{mL}}{\text{min}}$, $Rate(Unk) = \frac{V}{time} = \frac{20.3 \text{ mL}}{1.0 \text{ min}} = 20.3 \frac{\text{mL}}{\text{min}}$,

$\frac{Rate(HBr)}{Rate(Unk)} = \sqrt{\frac{\mathfrak{M}(Unk)}{\mathfrak{M}(HBr)}}$ Rearrange to solve for $\mathfrak{M}(Unk)$.

$$\mathfrak{M}(Unk) = \mathfrak{M}(HBr)\left(\frac{Rate(HBr)}{Rate(Unk)}\right)^2 = 80.91 \frac{\text{g}}{\text{mol}} \times \left(\frac{15.0 \frac{\text{mL}}{\text{min}}}{20.3 \frac{\text{mL}}{\text{min}}}\right)^2 = 44.2 \frac{\text{g}}{\text{mol}}$$ The formula is C_3H_8, propane.

Check: The units (g/mol) are correct. The magnitude of the answer (< HBr) makes sense because the unknown diffused faster and so must be lighter.

95. **Given:** CH_4: $V = 155$ mL at STP; O_2: $V = 885$ mL at STP; NO: $V = 55.5$ mL at STP; mixed in a flask: $V = 2.0$ L, $T = 275$ K, and 90.0 % of limiting reagent used. **Find:** P's of all components and P_{Total}.
Conceptual Plan: CH_4: mL → L → mol$_{CH4}$ → mol$_{CO2}$ and

$$\frac{1 \text{ L}}{1000 \text{ mL}} \quad \frac{1 \text{ mol}}{22.414 \text{ L}} \quad \frac{1 \text{ mol } CO_2}{5 \text{ mol NO}}$$

O_2: mL → L → mol$_{O2}$ → mol$_{CO2}$ and NO: mL → L → mol$_{NO}$ → mol$_{CO2}$

$$\frac{1 \text{ L}}{1000 \text{ mL}} \quad \frac{1 \text{ mol}}{22.414 \text{ L}} \quad \frac{1 \text{ mol } CO_2}{5 \text{ mol } O_2} \qquad \frac{1 \text{ L}}{1000 \text{ mL}} \quad \frac{1 \text{ mol}}{22.414 \text{ L}} \quad \frac{1 \text{ mol } CO_2}{5 \text{ mol NO}}$$

the smallest yield determines the limiting reagent then initial mol$_{NO}$ → reacted mol$_{NO}$ → final mol$_{NO}$
NO is the limiting reagent 90.0 % 0.100 x initial mol$_{no}$

reacted mol$_{NO}$ → reacted mol$_{CH4}$ then initial mol$_{CH4}$, reacted mol$_{CH4}$ → final mol$_{CH4}$ then

$\frac{1 \text{ mol } CH_4}{5 \text{ mol NO}}$ initial mol$_{CH4}$ – reacted mol$_{CH4}$ = final mol$_{CH4}$

final mol$_{CH4}$, V, T → final P_{CH4} and reacted mol$_{NO}$ → reacted mol$_{O2}$ then

$PV = nRT$ $\frac{5 \text{ mol } O_2}{5 \text{ mol NO}}$

initial mol$_{O2}$, reacted mol$_{O2}$ → final mol$_{O2}$ then final mol$_{O2}$, V, T → final P_{O2} and
initial mol$_{O2}$ – reacted mol$_{O2}$ = final mol$_{O2}$ $PV = nRT$

final mol$_{NO}$, V, T → final P_{NO} and theoretical mol$_{CO2}$ from NO → final mol$_{CO2}$
$PV = nRT$ 90.0 %

final mol$_{CO2}$, V, T → P_{CO2} then final mol$_{CO2}$ → mol$_{H2O}$ then mol$_{H2O}$, V, T → P_{H2O} and

$PV = nRT$ $\frac{1 \text{ mol } H_2O}{1 \text{ mol } CO_2}$ $PV = nRT$

final mol$_{CO2}$ → mol$_{NO2}$ then mol$_{NO2}$, V, T → P_{NO2} and final mol$_{CO2}$ → mol$_{OH}$ then

$\frac{1 \text{ mol } NO_2}{1 \text{ mol } CO_2}$ $PV = nRT$ $\frac{2 \text{ mol OH}}{1 \text{ mol } CO_2}$

mol$_{OH}$, V, T → P_{OH} finally P_{CH4}, P_{O2}, P_{NO}, P_{CO2}, P_{H2O}, P_{NO2}, P_{OH} → P_{Ttoal}

$PV = nRT$ $P_{Total} = \sum P$

Solution: CH_4: $155 \text{ mL} \times \frac{1 \text{ L}}{1000 \text{ mL}} \times \frac{1 \text{ mol } CH_4}{22.414 \text{ L}} \times \frac{1 \text{ mol } CO_2}{1 \text{ mol } CH_4} = 0.00691532 \text{ mol } CO_2$,

O_2: $885\ \text{mL} \times \dfrac{1\ \text{L}}{1000\ \text{mL}} \times \dfrac{1\ \text{mol}\ O_2}{22.414\ \text{L}} = 0.0394\underline{8}42\ \text{mol}\ O_2 \times \dfrac{1\ \text{mol}\ CO_2}{5\ \text{mol}\ O_2} = 0.0078\underline{9}685\ \text{mol}\ CO_2$

NO: $55.5\ \text{mL} \times \dfrac{1\ \text{L}}{1000\ \text{mL}} \times \dfrac{1\ \text{mol NO}}{22.414\ \text{L}} \times \dfrac{1\ \text{mol}\ CO_2}{5\ \text{mol NO}} = 0.000495\underline{2}26\ \text{mol}\ CO_2$.

$0.000495\underline{2}26\ \text{mol}\ CO_2$ is the smallest yield, so NO is the limiting reagent.

$55.5\ \text{mL} \times \dfrac{1\ \text{L}}{1000\ \text{mL}} \times \dfrac{1\ \text{mol NO}}{22.414\ \text{L}} = 0.0024\underline{7}613\ \text{mol NO}$

reacted mol NO = 0.900 x *mol NO* = 0.900 x 0.0024$\underline{7}$613 mol NO = 0.0022$\underline{2}$852 mol NO ,

unreacted mol NO = 0.100 x *mol NO* = 0.100 x 0.0024$\underline{7}$613 mol NO = 0.00024$\underline{7}$613 mol NO ,

$0.0022\underline{2}852\ \text{mol NO} \times \dfrac{1\ \text{mol}\ CH_4}{5\ \text{mol NO}} = 0.00044\underline{5}704\ \text{mol}\ CH_4$ reacted ,

$0.0069\underline{1}532\ \text{mol}\ CH_4 - 0.00044\underline{5}704\ \text{mol}\ CH_4\ \text{reacted} = 0.0064\underline{6}962\ \text{mol}\ CH_4$ then $PV = nRT$

Rearrange to solve for P. $P = \dfrac{nRT}{V} = \dfrac{0.0064\underline{6}962\ \text{mol} \times 0.08206\ \dfrac{\text{L}\cdot\text{atm}}{\text{mol}\cdot\text{K}} \times 275\ \text{K}}{2.0\ \text{L}} = 0.0730\ \text{atm}\ CH_4$

$0.0022\underline{2}852\ \text{mol NO} \times \dfrac{5\ \text{mol}\ O_2}{5\ \text{mol NO}} = 0.0022\underline{2}852\ \text{mol}\ O_2$ reacted ,

$0.039\underline{4}842\ \text{mol}\ O_2 - 0.0022\underline{2}852\ \text{mol}\ O_2\ \text{reacted} = 0.037\underline{2}557\ \text{mol}\ O_2$

$$P = \frac{nRT}{V} = \frac{0.037\underline{2}557\ \text{mol} \times 0.08206\ \frac{\text{L}\cdot\text{atm}}{\text{mol}\cdot\text{K}} \times 275\ \text{K}}{2.0\ \text{L}} = 0.420\ \text{atm}\ O_2$$

$$P = \frac{nRT}{V} = \frac{0.00024\underline{7}613\ \text{mol} \times 0.08206\ \frac{\text{L}\cdot\text{atm}}{\text{mol}\cdot\text{K}} \times 275\ \text{K}}{2.0\ \text{L}} = 0.00279\ \text{atm NO},$$

$0.0022\underline{2}852\ \text{mol NO} \times \dfrac{1\ \text{mol}\ CO_2}{5\ \text{mol NO}} = 0.00044\underline{5}704\ \text{mol}\ CO_2$

$$P = \frac{nRT}{V} = \frac{0.00044\underline{5}704\ \text{mol} \times 0.08206\ \frac{\text{L}\cdot\text{atm}}{\text{mol}\cdot\text{K}} \times 275\ \text{K}}{2.0\ \text{L}} = 0.00503\ \text{atm}\ CO_2$$

$0.0022\underline{2}852\ \text{mol NO} \times \dfrac{1\ \text{mol}\ H_2O}{5\ \text{mol NO}} = 0.00044\underline{5}704\ \text{mol}\ H_2O$

$$P = \frac{nRT}{V} = \frac{0.00044\underline{5}704\ \text{mol} \times 0.08206\ \frac{\text{L}\cdot\text{atm}}{\text{mol}\cdot\text{K}} \times 275\ \text{K}}{2.0\ \text{L}} = 0.00503\ \text{atm}\ H_2O$$

$0.0022\underline{2}852\ \text{mol NO} \times \dfrac{5\ \text{mol}\ NO_2}{5\ \text{mol NO}} = 0.0022\underline{2}852\ \text{mol}\ NO_2$

$$P = \frac{nRT}{V} = \frac{0.0022\underline{2}852\ \text{mol} \times 0.08206\ \frac{\text{L}\cdot\text{atm}}{\text{mol}\cdot\text{K}} \times 275\ \text{K}}{2.0\ \text{L}} = 0.0251\ \text{atm}\ NO_2$$

$0.0022\underline{2}852\ \text{mol NO} \times \dfrac{2\ \text{mol OH}}{5\ \text{mol NO}} = 0.00089\underline{1}408\ \text{mol OH}$

$$P = \frac{nRT}{V} = \frac{0.00089\underline{1}408\ \text{mol} \times 0.08206\ \frac{\text{L}\cdot\text{atm}}{\text{mol}\cdot\text{K}} \times 275\ \text{K}}{2.0\ \text{L}} = 0.0101\ \text{atm OH}$$

$P_{Total} = \sum P$

= 0.0730 atm + 0.420 atm +0.00279 atm + 0.00503 atm + 0.00503 atm + 0.0251 atm + 0.0101 atm

= 0.541 atm

Check: The units (atm) are correct. The magnitude of the answers is reasonable. The limiting reagent has the lowest pressure. The product pressures are in line with the ratios of the stoichiometric coefficients.

97. **Given:** $P_{CH4} + P_{C2H6} = 0.53$ atm, $P_{CO2} + P_{H2O} = 2.2$ atm **Find:** χ_{CH4}
Conceptual Plan: Write balanced reactions to determine change in moles of gas for CH_4 and C_2H_6.

$2\ CH_4\ (g) + 4\ O_2\ (g) \rightarrow 4\ H_2O\ (g) + 2\ CO_2\ (g)$ and $2\ C_2H_6\ (g) + 7\ O_2\ (g) \rightarrow 6\ H_2O\ (g) + 4\ CO_2\ (g)$ thus $\frac{6 \text{ mol gases}}{2 \text{ mol } CH_4}$ $\frac{10 \text{ mol gases}}{2 \text{ mol } C_2H_6}$

write expression for final pressure, substituting in data given → χ_{CH4}

$$\chi_{CH4} = \frac{n_{CH4}}{n_{CH4} + n_{C2H6}} \text{ and } \chi_{C2H6} = 1 - \chi_{CH4}$$

$$P_{CH4} = \chi_{CH4} P_{Total} \quad P_{C2H6} = \chi_{C2H6} P_{Total} \quad P_{Final} = \left(\chi_{CH4} P_{Total}\ x\ \frac{6 \text{ mol gases}}{2 \text{ mol } CH_4}\right) + \left((1 - \chi_{CH4}) P_{Total}\ x\ \frac{10 \text{ mol gases}}{2 \text{ mol } C_2H_6}\right)$$

Solution: $P_{Final} = \left(\chi_{CH4} \text{ x } 0.53 \text{ atm x } \frac{6 \text{ mol gases}}{2 \text{ mol } CH_4}\right) + \left((1 - \chi_{CH4}) \text{ x } 0.53 \text{ atm x } \frac{10 \text{ mol gases}}{2 \text{ mol } C_2H_6}\right) = 2.2 \text{ atm}$

Rearrange to solve for $\chi_{CH4} = 0.42$.
Check: The units (none) are correct. The magnitude of the answer (0.42) makes sense because if it were all methane the final pressure would have been 1.59 atm and if it were all ethane the final pressure would have been 2.65 atm. Since we are closer to the latter pressure, we expect the mole fraction of methane to be less than 0.5.

99. Since the passengers have more mass than the balloon, they have more momentum than the balloon. The passengers will continue to travel in their original direction longer. The car is slowing so the relative position of the passengers is to move forward and the balloons to move backwards. The opposite happens upon acceleration.

101. Since each gas will occupy 22.414 L / mole at STP and we have 2 moles of gas, we will have a volume of 44.828 L.

103. Br_2 would deviate the most from ideal behavior since it is the largest of the three.

Chapter 6
Thermochemistry

1. a) **Given:** 3.55×10^4 J **Find:** cal

Conceptual Plan: **J → cal**

$$\frac{1\text{ cal}}{4.184\text{ J}}$$

Solution: $$3.55 \times 10^4 \text{ J} \times \frac{1\text{ cal}}{4.184\text{ J}} = 8.48 \times 10^3 \text{ cal}$$

Check: The units (cal) are correct. The magnitude of the answer (8000) makes physical sense because a calorie is larger than a Joule, so the answer decreases.

b) **Given:** 1025 Cal **Find:** J

Conceptual Plan: **Cal → J**

$$\frac{4184\text{ J}}{1\text{ Cal}}$$

Solution: $$1025 \text{ Cal} \times \frac{4184\text{ J}}{1\text{ Cal}} = 4.289 \times 10^6 \text{ J}$$

Check: The units (J) are correct. The magnitude of the answer (10^6) makes physical sense because a Calorie is much larger than a Joule, so the answer increases.

c) **Given:** 355 kJ **Find:** cal

Conceptual Plan: **kJ → J → cal**

$$\frac{1000\text{ J}}{1\text{ kJ}} \quad \frac{1\text{ cal}}{4.184\text{ J}}$$

Solution: $$355 \text{ kJ} \times \frac{1000\text{ J}}{1\text{ kJ}} \times \frac{1\text{ cal}}{4.184\text{ J}} = 8.48 \times 10^4 \text{ cal}$$

Check: The units (cal) are correct. The magnitude of the answer (10^4) makes physical sense because a calorie is much smaller than a kJ, so the answer increases.

d) **Given:** 125 kWh **Find:** J

Conceptual Plan: **kWh → J**

$$\frac{3.60 \times 10^6\text{ J}}{1\text{kWh}}$$

Solution: $$125 \text{ kWh} \times \frac{3.60 \times 10^6\text{ J}}{1\text{ kWh}} = 4.50 \times 10^8 \text{ J}$$

Check: The units (J) are correct. The magnitude of the answer (10^8) makes physical sense because a kWh is much larger than a Joule, so the answer increases.

3. a) **Given:** 2155 Cal **Find:** J

Conceptual Plan: **Cal → J**

$$\frac{4184\text{ J}}{1\text{ Cal}}$$

Solution: $$2155 \text{ Cal} \times \frac{4184\text{ J}}{1\text{ Cal}} = 9.017 \times 10^6 \text{ J}$$

Check: The units (J) are correct. The magnitude of the answer (10^6) makes physical sense because a Calorie is much larger than a Joule, so the answer increases.

b) **Given:** 2155 Cal **Find:** kJ

Conceptual Plan: **Cal → J → kWh**

$$\frac{4184\text{ J}}{1\text{ Cal}} \quad \frac{1\text{ kJ}}{1000\text{ J}}$$

Solution: $$2155 \text{ Cal} \times \frac{4184\text{ J}}{1\text{ Cal}} \times \frac{1\text{ kJ}}{1000\text{ J}} = 9.017 \times 10^3 \text{ kJ}$$

Check: The units (kJ) are correct. The magnitude of the answer (10^3) makes physical sense because a Calorie is larger than a kJ, so the answer increases.

c) **Given:** 2155 Cal **Find:** kWh
Conceptual Plan: **Cal → J → kWh**

$\frac{4184 J}{1 Cal}$ $\frac{1 kWh}{3.60 \times 10^6 J}$

Solution: $2155 \text{ Cal} \times \frac{4184 \text{ J}}{1 \text{ Cal}} \times \frac{1 \text{ kWh}}{3.60 \times 10^6 \text{ J}} = 2.50 \text{ kWh}$

Check: The units (kWh) are correct. The magnitude of the answer (3) makes physical sense because a Calorie is much smaller than a kWh, so the answer decreases.

5. d) $\Delta E_{sys} = -\ \Delta E_{surr}$

7. a) The energy exchange is primarily heat since the skin (part of the surroundings) is cooled. There is a small expansion (work) since water is being converted from a liquid to a gas. The sign of ΔE_{sys} is positive since the surroundings cool.

b) The energy exchange is primarily work. The sign of ΔE_{sys} is negative since the system is expanding (doing work on the surroundings).

c) The energy exchange is primarily heat. The sign of ΔE_{sys} is positive since the system is being heated by the flame.

9. **Given:** 415 kJ heat released; 125 kJ work done on surroundings **Find:** ΔE_{sys}
Conceptual Plan: interpret language to determine the sign of the two terms then $q, w \rightarrow \Delta E_{sys}$

$\Delta E = q + w$

Solution: since heat is released from the system to the surroundings, $q = -$ 415 kJ; since the system is doing work on the surroundings, $w = -$ 125 kJ. $\Delta E = q + w = -$ 415 kJ $-$ 125 kJ $= -$ 540. kJ $= -5.40 \times 10^2$ kJ

Check: The units (kJ) are correct. The magnitude of the answer (–540) makes physical sense because both terms are negative.

11. **Given:** 655 J heat absorbed; 344 J work done on surroundings **Find:** ΔE_{sys}
Conceptual Plan: interpret language to determine the sign of the two terms then $q, w \rightarrow \Delta E_{sys}$

$\Delta E = q + w$

Solution: since heat is absorbed by the system, $q = +$ 655 J; since the system is doing work on the surroundings, $w = -$ 344 J. $\Delta E = q + w =$ 655 J $-$ 344 J $=$ 311 J

Check: The units (J) are correct. The magnitude of the answer (+300) makes physical sense because heat term dominates over the work term.

13. Cooler A had more ice after 3 hours because most of the ice in cooler B was melted in order to cool the soft drinks that started at room temperature. In cooler A the drinks were already cold and so the ice only needed to maintain this cool temperature.

15. **Given:** 1.50 L water, $T_i = 25.0$ °C, $T_f = 100.0$ °C, d = 1.0 g/mL **Find:** q
Conceptual Plan:
L → mL → g and pull C_s from Table 6.4 and T_i, T_f → ΔT then m, C_s, ΔT → q

$\frac{1000 \text{ mL}}{1 \text{ L}}$ $\frac{1.0 \text{ g}}{1.0 \text{ mL}}$ $4.18 \frac{\text{J}}{\text{g} \cdot °\text{C}}$ $\Delta T = T_f - T_i$ $q = m\, C_s\, \Delta T$

Solution: $1.50 \text{ L} \times \frac{1000 \text{ mL}}{1 \text{ L}} \times \frac{1.0 \text{ g}}{1.0 \text{ mL}} = 1\underline{5}00 \text{ g}$ and $\Delta T = T_f - T_i = 100.0\ °\text{C} - 25.0\ °\text{C} = 75.0\ °\text{C}$

then $q = m\, C_s\, \Delta T = 1\underline{5}00 \text{ g} \times 4.18 \frac{\text{J}}{\text{g} \cdot °\text{C}} \times 75.0\ °\text{C} = 4.7 \times 10^5 \text{ J}$

Check: The units (J) are correct. The magnitude of the answer (10^6) makes physical sense because there is such a large mass, a significant temperature change, and a high specific heat capacity material.

17. a) **Given:** 25 g gold, $T_i = 27.0$ °C, $q = 2.35$ kJ **Find:** T_f

Conceptual Plan:

kJ → J and pull C_s from Table 6.4 then $m, C_s, q \rightarrow \Delta T$ then $T_i, \Delta T \rightarrow T_f$

$\frac{1000\,\text{g}}{1\,\text{g}}$ $\quad 0.128 \frac{\text{J}}{\text{g}\cdot°\text{C}}$ $\quad q = m\,C_s\,\Delta T$ $\quad \Delta T = T_f - T_i$

Solution: $2.35\,\text{kJ} \times \frac{1000\text{ J}}{1\text{ kJ}} = 2350\text{ J}$ then $q = m\,C_s\,\Delta T$ Rearrange to solve for ΔT.

$$\Delta T = \frac{q}{m\,C_s} = \frac{2350\text{ J}}{25\text{ g} \times 0.128 \frac{\text{J}}{\text{g}\cdot°\text{C}}} = 734.375\text{ °C}$$ finally $\Delta T = T_f - T_i$ Rearrange to solve for T_f.

$$T_f = \Delta T + T_i = 734.375\text{ °C} + 27.0\text{ °C} = 760\text{ °C}$$

Check: The units (°C) are correct. The magnitude of the answer (760) makes physical sense because there is such a large heat absorbed, such a small mass, and specific heat capacity. The temperature change should be very large.

b) **Given:** 25 g silver, $T_i = 27.0$ °C, $q = 2.35$ kJ **Find:** T_f

Conceptual Plan:

kJ → J and pull C_s from Table 6.4 then $m, C_s, q \rightarrow \Delta T$ then $T_i, \Delta T \rightarrow T_f$

$\frac{1000\,\text{g}}{1\,\text{g}}$ $\quad 0.235 \frac{\text{J}}{\text{g}\cdot°\text{C}}$ $\quad q = m\,C_s\,\Delta T$ $\quad \Delta T = T_f - T_i$

Solution: $2.35\,\text{kJ} \times \frac{1000\text{ J}}{1\text{ kJ}} = 2350\text{ J}$ then $q = m\,C_s\,\Delta T$ Rearrange to solve for ΔT.

$$\Delta T = \frac{q}{m\,C_s} = \frac{2350\text{ J}}{25\text{ g} \times 0.235 \frac{\text{J}}{\text{g}\cdot°\text{C}}} = 400\text{ °C}$$ finally $\Delta T = T_f - T_i$ Rearrange to solve for T_f.

$$T_f = \Delta T + T_i = 400\text{ °C} + 27.0\text{ °C} = 430\text{ °C}$$

Check: The units (°C) are correct. The magnitude of the answer (430) makes physical sense because there is such a large amount of heat absorbed, such a small mass, and specific heat capacity. The temperature change should be very large. The temperature change should be less than that of the gold because the specific heat capacity is greater.

c) **Given:** 25 g aluminum, $T_i = 27.0$ °C, $q = 2.35$ kJ **Find:** T_f

Conceptual Plan:

kJ → J and pull C_s from Table 6.3 then $m, C_s, q \rightarrow \Delta T$ then $T_i, \Delta T \rightarrow T_f$

$\frac{1000\,\text{g}}{1\,\text{g}}$ $\quad 0.903 \frac{\text{J}}{\text{g}\cdot°\text{C}}$ $\quad q = m\,C_s\,\Delta T$ $\quad \Delta T = T_f - T_i$

Solution: $2.35\,\text{kJ} \times \frac{1000\text{ J}}{1\text{ kJ}} = 2350\text{ J}$ then $q = m\,C_s\,\Delta T$ Rearrange to solve for ΔT.

$$\Delta T = \frac{q}{m\,C_s} = \frac{2350\text{ J}}{25\text{ g} \times 0.903 \frac{\text{J}}{\text{g}\cdot°\text{C}}} = 104.10\text{ °C}$$ finally $\Delta T = T_f - T_i$ Rearrange to solve for T_f.

$$T_f = \Delta T + T_i = 104.10\text{ °C} + 27.0\text{ °C} = 130\text{ °C}$$

Check: The units (°C) are correct. The magnitude of the answer (130) makes physical sense because there is such a large heat absorbed, and such a small mass. The temperature change should be less than that of the silver because the specific heat capacity is greater.

d) **Given:** 25 g water, $T_i = 27.0$ °C, $q = 2.35$ kJ **Find:** T_f

Conceptual Plan:

kJ → J and pull C_s from Table 6.4 then $m, C_s, q \rightarrow \Delta T$ then $T_i, \Delta T \rightarrow T_f$

$\frac{1000\,\text{g}}{1\,\text{g}}$ $\quad 4.18 \frac{\text{J}}{\text{g}\cdot°\text{C}}$ $\quad q = m\,C_s\,\Delta T$ $\quad \Delta T = T_f - T_i$

Solution: $2.35 \text{ kJ} \times \frac{1000 \text{ J}}{1 \text{ kJ}} = 23\underline{5}0 \text{ J}$ then $q = m\, C_s\, \Delta T$ Rearrange to solve for ΔT.

$$\Delta T = \frac{q}{m\, C_s} = \frac{23\underline{5}0 \text{ J}}{25 \text{ g} \times 4.18 \frac{\text{J}}{\text{g} \cdot °\text{C}}} = 2\underline{2}.491 \text{ °C}$$ finally $\Delta T = T_f - T_i$ Rearrange to solve for T_f.

$$T_f = \Delta T + T_i = 2\underline{2}.491 \text{ °C} + 27.0 \text{ °C} = 49 \text{ °C}$$

Check: The units (°C) are correct. The magnitude of the answer (130) makes physical sense because there is such a large heat absorbed, and such a small mass. The temperature change should be less than that of the aluminum because the specific heat capacity is greater.

19. **Given:** $V_i = 0.0$ L, $V_f = 2.5$ L, $P = 1.1$ atm **Find:** w (J)
Conceptual Plan: $V_i, V_f \rightarrow \Delta V$ **then** $P, \Delta V \rightarrow w$ **(L atm)** $\rightarrow w$ **(J)**

$\Delta V = V_f - V_i$ $w = -P\,\Delta V$ $\frac{101.3 \text{ J}}{1 \text{ L} \cdot \text{atm}}$

Solution: $\Delta V = V_f - V_i = 2.5 \text{ L} - 0.0 \text{ L} = 2.5 \text{ L}$ then

$$w = -P\,\Delta V = -1.1 \text{ atm} \times 2.5 \text{ L} \times \frac{101.3 \text{ J}}{1 \text{ L} \cdot \text{atm}} = -280 \text{ J}$$

Check: The units (J) are correct. The magnitude of the answer (–280) makes physical sense because this is an expansion (negative work) and we have ~ atmospheric pressure and a small volume of expansion.

21. **Given:** $q = 565$ J absorbed, $V_i = 0.10$ L, $V_f = 0.85$ L, $P = 1.0$ atm **Find:** ΔE_{sys}
Conceptual Plan: $V_i, V_f \rightarrow \Delta V$ **and interpret language to determine the sign of the heat then**

$\Delta V = V_f - V_i$ $q = +565$ J

then $P, \Delta V \rightarrow w$ **(L atm)** $\rightarrow w$ **(J) finally** $q, w \rightarrow \Delta E_{sys}$

$w = -P\,\Delta V$ $\frac{101.3 \text{ J}}{1 \text{ L atm}}$ $\Delta E = q + w$

Solution: $\Delta V = V_f - V_i = 0.85 \text{ L} - 0.10 \text{ L} = 0.75 \text{ L}$ then

$$w = -P\,\Delta V = -1.0 \text{ atm} \times 0.75 \text{ L} \times \frac{101.3 \text{ J}}{1 \text{ L} \cdot \text{atm}} = -7\underline{5}.975 \text{ J}$$ $\Delta E = q + w = +565 \text{ J} - 7\underline{5}.975 \text{ J} = 489 \text{ J}$

Check: The units (J) are correct. The magnitude of the answer (500) makes physical sense because the heat absorbed dominated the small expansion work (negative work).

23. **Given:** 1 mol fuel, 3452 kJ heat produced; 11 kJ work done on surroundings **Find:** ΔE_{sys}, ΔH
Conceptual Plan:
interpret language to determine the sign of the two terms then $q \rightarrow \Delta H$ **and** $q, w \rightarrow \Delta E_{sys}$

$\Delta H = q_P$ $\Delta E = q + w$

Solution: since heat is produced by the system to the surroundings, $q = -$ 3452 kJ; since the system is doing work on the surroundings, $w = -$ 11 kJ. $\Delta H = q_P = -3452$ kJ and

$$\Delta E = q + w = -3452 \text{ kJ} - 11 \text{ kJ} = -3463 \text{ kJ}$$

Check: The units (kJ) are correct. The magnitude of the answer (–3500) makes physical sense because both terms are negative. We expect significant amounts of energy from fuels.

25. a) combustion is an exothermic process, ΔH is negative

b) evaporation requires an input of energy and so it is endothermic, ΔH is positive

c) condensation is the reverse of evaporation, so it is exothermic, ΔH is negative

27. **Given:** 177 mL acetone (C_3H_6O), $\Delta H°_{rxn} = -1790$ kJ; d = 0.788 g/mL **Find:** q
Conceptual Plan: **mL acetone** $\rightarrow$ **g acetone** $\rightarrow$ **mol acetone** $\rightarrow$ q

$\frac{0.788 \text{ g}}{1 \text{ mL}}$ $\frac{1 \text{ mol}}{58.08 \text{ g}}$ $\frac{-1790 \text{ kJ}}{1 \text{ mol}}$

Solution: $177 \text{ mL} \times \frac{0.788 \text{ g}}{1 \text{ mL}} \times \frac{1 \text{ mol}}{58.08 \text{ g}} \times \frac{-1790 \text{ kJ}}{1 \text{ mol}} = -4.30 \times 10^3 \text{ kJ}$ or 4.30×10^3 kJ released

Check: The units (kJ) are correct. The magnitude of the answer (-10^3) makes physical sense because the enthalpy change is negative and we have more than a mole of acetone. We expect more than 1790 kJ to be released.

29. **Given:** pork roast, $\Delta H°_{rxn} = -2217$ kJ; q used = 1.6×10^3 kJ, 10 % efficiency **Find:** $m(CO_2)$
Conceptual Plan: q **used →** q **generated → mol** CO_2 **→ g** CO_2

$$\frac{100 \text{ kJ generated}}{10 \text{ kJ used}} \quad \frac{3 \text{ mol}}{2217 \text{ kJ}} \quad \frac{44.01 \text{ g}}{1 \text{ mol}}$$

Solution: $1.6 \times 10^3 \text{ kJ} \times \frac{100 \text{ kJ generated}}{10 \text{ kJ used}} \times \frac{3 \text{ mol } CO_2}{2217 \text{ kJ}} \times \frac{44.01 \text{ g } CO_2}{1 \text{ mol } CO_2} = 950 \text{ g } CO_2$

Check: The units (g) are correct. The magnitude of the answer (~1000) makes physical sense because the process is not very efficient and a lot of energy is needed.

31. $\Delta H_{rxn} = q_P$ and $\Delta E_{rxn} = q_V = \Delta H - P\,\Delta V$. Since combustions always involve expansions, expansions do work and so have a negative value. Combustions are always exothermic and so have a negative value. This means that ΔE_{rxn} is more negative than $\Delta H°_{rxn}$ and so A (–25.9 kJ) is the constant volume process and B (–23.3 kJ) is the constant pressure process.

33. **Given:** 0.514 g biphenyl ($C_{12}H_{10}$), bomb calorimeter, $T_i = 25.8$ °C, $T_f = 29.4$ °C, $C_{cal} = 5.86$ kJ/°C
Find: ΔE_{rxn}
Conceptual Plan:
T_i, T_f **→** ΔT **then** $\Delta T, C_{cal}$ **→** q_{cal} **→** q_{rxn} **then g** $C_{12}H_{10}$ **→ mol** $C_{12}H_{10}$

$$\Delta T = T_f - T_i \quad q_{cal} = C_{cal}\,\Delta T \quad q_{cal} = -q_{rxn} \quad \frac{1 \text{ mol}}{154.20 \text{ g}}$$

then q_{rxn}, **mol** $C_{12}H_{10}$ **→** ΔE_{rxn}

$$\Delta E_{rxn} = \frac{q_V}{mol\ C_{12}H_{10}}$$

Solution: $\Delta T = T_f - T_i = 29.4 \text{ °C} - 25.8 \text{ °C} = 3.6 \text{ °C}$ then $q_{cal} = C_{cal}\,\Delta T = 5.86 \frac{\text{kJ}}{\text{°C}} \times 3.6 \text{ °C} = 2\underline{1}.096 \text{ kJ}$

then $q_{cal} = -q_{rxn} = -2\underline{1}.096$ kJ and $0.514 \text{ g } C_{12}H_{10} \times \frac{1 \text{ mol } C_{12}H_{10}}{154.20 \text{ g } C_{12}H_{10}} = 0.00333\underline{3}333 \text{ mol } C_{12}H_{10}$ then

$$\Delta E_{rxn} = \frac{q_V}{mol\ C_{12}H_{10}} = \frac{-2\underline{1}.096 \text{ kJ}}{0.00333\underline{3}333 \text{ mol } C_{12}H_{10}} = -6.3 \times 10^3 \text{ kJ/mol}$$

Check: The units (kJ/mol) are correct. The magnitude of the answer (–6000) makes physical sense because there is such a large heat generated from a very small amount of biphenyl.

35. **Given:** 0.103 g zinc, coffee-cup calorimeter, $T_i = 22.5$°C, $T_f = 23.7$ °C, 50.0 mL solution, d (solution) = 1.0 g/mL, $C_{sol'n} = 4.18$ kJ/g °C **Find:** ΔH_{rxn}
Conceptual Plan:
T_i, T_f **→** ΔT **and mL sol'n → g sol'n then** $\Delta T, C_{cal}$ **→** q_{cal} **→** q_{rxn} **then**

$$\Delta T = T_f - T_i \quad \frac{1.0 \text{ g}}{1.0 \text{ mL}} \quad q_{cal} = m\,C_{sol'n}\,\Delta T \quad q_{sol'n} = -q_{rxn}$$

g Zn → mol Zn then q_{rxn}, **mol Zn →** ΔH_{rxn}

$$\frac{1 \text{ mol}}{65.37 \text{ g}} \quad \Delta H_{rxn} = \frac{q_P}{mol\ Zn}$$

Solution: $\Delta T = T_f - T_i = 23.7 \text{ °C} - 22.5 \text{ °C} = 1.2 \text{ °C}$ and $50.0 \text{ mL} \times \frac{1.0 \text{ g}}{1.0 \text{ mL}} = 50.0 \text{ g}$ then

$q_{sol'n} = m\,C_{sol'n}\,\Delta T = 50.0 \text{ g} \times 4.18 \frac{\text{J}}{\text{g} \cdot \text{°C}} \times 1.2 \text{ °C} = 2\underline{5}0.8 \text{ J}$ then $q_{sol'n} = -q_{rxn} = -2\underline{5}0.8$ J and

$0.103 \text{ g Zn} \times \frac{1 \text{ mol Zn}}{65.37 \text{ g Zn}} = 0.0015\underline{7}565 \text{ mol Zn}$ then

$$\Delta H_{rxn} = \frac{q_P}{mol\ Zn} = \frac{-250.8\ J}{0.001575 65\ mol\ Zn} = -1.6 \times 10^5\ J/mol = -1.6 \times 10^2\ kJ/mol$$

Check: The units (kJ/mol) are correct. The magnitude of the answer (–160) makes physical sense because there is such a large heat generated from a very small amount of zinc.

37. a) Since A + B → 2 C has ΔH_1 then 2 C → A + B will have a $\Delta H_2 = -\Delta H_1$. When the reaction direction is reversed, it changes from exothermic to endothermic (or vice versa), so the sign of ΔH changes.

b) Since A + ½ B → C has ΔH_1 then 2 A + B → 2 C will have a $\Delta H_2 = 2\ \Delta H_1$. When the reaction amount doubles, the amount of heat (or ΔH) doubles.

c) Since A → B + 2 C has ΔH_1 then ½ A → ½ B + C will have a $\Delta H_1' = ½\ \Delta H_1$. When the reaction amount is cut in half, the amount of heat (or ΔH) is cut in half. Then ½ B + C → ½ A will have a $\Delta H_2 = -\Delta H_1' = -½\ \Delta H_1$. When the reaction direction is reversed, it changes from exothermic to endothermic (or vice versa), so the sign of ΔH changes.

39. Since the first reaction has Fe_2O_3 as a product and the reaction of interest has it as a product, we need to reverse the first reaction. When the reaction direction is reversed, ΔH changes.

$Fe_2O_3\ (s) \rightarrow 2\ Fe\ (s) + 3/2\ O_2\ (g)$ $\quad \Delta H = +824.2$ kJ

Since the second reaction has 1 mole CO as a reactant and the reaction of interest has 3 moles of CO as a reactant, we need to multiply the second reaction and the ΔH by 3.

$3\ [CO\ (g) + 1/2\ O_2\ (g) \rightarrow CO_2\ (g)]$ $\quad \Delta H = 3(-282.7\ kJ) = -848.1$ kJ

Hess' Law states the ΔH of the net reaction is the sum of the ΔH of the steps.
The rewritten reactions are:

$Fe_2O_3\ (s) \rightarrow 2\ Fe\ (s) + \cancel{3/2\ O_2\ (g)}$ $\quad \Delta H = +824.2$ kJ

$3\ CO\ (g) + \cancel{3/2\ O_2\ (g)} \rightarrow 3\ CO_2\ (g)$ $\quad \Delta H = -848.1$ kJ

$Fe_2O_3\ (s) + 3\ CO\ (g) \rightarrow 2\ Fe\ (s) + 3\ CO_2\ (g)$ $\quad \Delta H_{rxn} = -23.9$ kJ

41. Since the first reaction has C_5H_{12} as a reactant and the reaction of interest has it as a product, we need to reverse the first reaction. When the reaction direction is reversed, ΔH changes.

$5\ CO_2\ (g) + 6\ H_2O\ (g) \rightarrow C_5H_{12}\ (l) + 8\ O_2\ (g)$ $\quad \Delta H = +3505.8$ kJ

Since the second reaction has 1 mole C as a reactant and the reaction of interest has 5 moles of C as a reactant, we need to multiply the second reaction and the ΔH by 5.

$5\ [C\ (s) + O_2\ (g) \rightarrow CO_2\ (g)]$ $\quad \Delta H = 5(-393.5\ kJ) = -1967.5$ kJ

Since the third reaction has 2 moles H_2 as a reactant and the reaction of interest has 6 moles of H_2 as a reactant, we need to multiply the third reaction and the ΔH by 3.

$3[2\ H_2\ (g) + O_2\ (g) \rightarrow 2\ H_2O\ (g)]$ $\quad \Delta H = 3(-483.5\ kJ) = -1450.5$ kJ

Hess' Law states the ΔH of the net reaction is the sum of the ΔH of the steps.
The rewritten reactions are:

$\cancel{5\ CO_2\ (g)} + \cancel{6\ H_2O\ (g)} \rightarrow C_5H_{12}\ (l) + \cancel{8\ O_2\ (g)}$ $\quad \Delta H = +3505.8$ kJ

$5\ C\ (s) + \cancel{5\ O_2\ (g)} \rightarrow \cancel{5\ CO_2\ (g)}$ $\quad \Delta H = -1967.5$ kJ

$6\ H_2\ (g) + \cancel{3\ O_2\ (g)} \rightarrow \cancel{6\ H_2O\ (g)}$ $\quad \Delta H = -1450.5$ kJ

$5\ C\ (s) + 6\ H_2\ (g) \rightarrow C_5H_{12}\ (l)$ $\quad \Delta H_{rxn} = +87.8$ kJ

43. a) $½\ N_2\ (g) + 3/2\ H_2\ (g) \rightarrow NH_3\ (g)$ $\quad \Delta H^\circ_f = -45.9$ kJ/mol

b) $C\ (s) + O_2\ (g) \rightarrow CO_2\ (g)$ $\quad \Delta H^\circ_f = -393.5$ kJ/mol

c) $Fe\ (s) + 3/2\ O_2\ (g) \rightarrow Fe_2O_3\ (s)$ $\quad \Delta H^\circ_f = -824.2$ kJ/mol

d) $C\ (s) + 2\ H_2\ (g) \rightarrow CH_4\ (g)$ $\quad \Delta H^\circ_f = -74.6$ kJ/mol

45. **Given:** $N_2H_4\ (l) + N_2O_4\ (g) \rightarrow 2\ N_2O\ (g) + 2\ H_2O\ (g)$ **Find:** $\Delta H°_{rxn}$

Conceptual Plan: $\Delta H^0_{rxn} = \sum n_P \Delta H^0_f(products) - \sum n_R \Delta H^0_f(reactants)$

Solution:

Reactant/Product	ΔH^0_f (kJ/mol from Appendix IIB)
$N_2H_4\ (l)$	50.6
$N_2O_4\ (g)$	11.1
$N_2O\ (g)$	81.6
$H_2O\ (g)$	– 241.8

Be sure to pull data for the correct formula and phase.

$$\Delta H^0_{rxn} = \sum n_P \Delta H^0_f(products) - \sum n_R \Delta H^0_f(reactants)$$
$$= [2(\Delta H^0_f(N_2O\ (g))) + 2(\Delta H^0_f(H_2O\ (g)))] - [1(\Delta H^0_f(N_2H_4\ (l))) + 1(\Delta H^0_f(N_2O_4\ (g)))]$$
$$= [2(81.6\text{ kJ}) + 2(-241.8\text{ kJ})] - [1(50.6\text{ kJ}) + 1(11.1\text{ kJ})]$$
$$= [-320.4\text{ kJ}] - [61.7\text{ kJ}]$$
$$= -382.1\text{ kJ}$$

Check: The units (kJ) are correct. The answer is negative, which means that the reaction is exothermic. The answer is dominated by the negative heat of formation of water.

47. a) **Given:** $C_2H_4\ (g) + H_2\ (g) \rightarrow C_2H_6\ (g)$ **Find:** $\Delta H°_{rxn}$

Conceptual Plan: $\Delta H^0_{rxn} = \sum n_P \Delta H^0_f(products) - \sum n_R \Delta H^0_f(reactants)$

Solution:

Reactant/Product	ΔH^0_f (kJ/mol from Appendix IIB)
$C_2H_4\ (g)$	52.4
$H_2\ (g)$	0.0
$C_2H_6\ (g)$	– 84.68

Be sure to pull data for the correct formula and phase.

$$\Delta H^0_{rxn} = \sum n_P \Delta H^0_f(products) - \sum n_R \Delta H^0_f(reactants)$$
$$= [1(\Delta H^0_f(C_2H_6\ (g)))] - [1(\Delta H^0_f(C_2H_4\ (g))) + 1(\Delta H^0_f(H_2\ (g)))]$$
$$= [1(-84.68\text{ kJ})] - [1(52.4\text{ kJ}) + 1(0.0\text{ kJ})]$$
$$= [-84.68\text{ kJ}] - [52.4\text{ kJ}]$$
$$= -137.1\text{ kJ}$$

Check: The units (kJ) are correct. The answer is negative, which means that the reaction is exothermic. Both hydrocarbon terms are negative, so the final answer is negative.

b) **Given:** $CO\ (g) + H_2O\ (g) \rightarrow H_2\ (g) + CO_2\ (g)$ **Find:** $\Delta H°_{rxn}$

Conceptual Plan: $\Delta H^0_{rxn} = \sum n_P \Delta H^0_f(products) - \sum n_R \Delta H^0_f(reactants)$

Solution:

Reactant/Product	ΔH^0_f (kJ/mol from Appendix IIB)
$CO\ (g)$	–110.5
$H_2O\ (g)$	– 241.8
$H_2\ (g)$	0.0
$CO_2\ (g)$	– 393.5

Be sure to pull data for the correct formula and phase.

$$\Delta H^0_{rxn} = \sum n_P \Delta H^0_f(products) - \sum n_R \Delta H^0_f(reactants)$$
$$= [1(\Delta H^0_f(H_2\ (g))) + 1(\Delta H^0_f(CO_2\ (g)))] - [1(\Delta H^0_f(CO\ (g))) + 1(\Delta H^0_f(H_2O\ (g)))]$$
$$= [1(0.0\text{ kJ}) + 1(-393.5\text{ kJ})] - [1(-110.5\text{ kJ}) + 1(-241.8\text{ kJ})]$$
$$= [-393.5\text{ kJ}] - [-352.3\text{ kJ}]$$
$$= -41.2\text{ kJ}$$

Check: The units (kJ) are correct. The answer is negative, which means that the reaction is exothermic.

c) **Given:** $3\ NO_2\ (g) + H_2O\ (l) \rightarrow 2\ HNO_3\ (aq) + NO\ (g)$ **Find:** $\Delta H°_{rxn}$

Conceptual Plan: $\Delta H^0_{rxn} = \sum n_P \Delta H^0_f(products) - \sum n_R \Delta H^0_f(reactants)$

Solution:

Reactant/Product	ΔH^0_f (kJ/mol from Appendix IIB)
$NO_2\ (g)$	33.2
$H_2O\ (l)$	– 285.8
$HNO_3\ (aq)$	–207
$NO\ (g)$	91.3

Be sure to pull data for the correct formula and phase.

$$\Delta H^0_{rxn} = \sum n_P \Delta H^0_f(products) - \sum n_R \Delta H^0_f(reactants)$$
$$= [2(\Delta H^0_f(HNO_3\ (aq))) + 1(\Delta H^0_f(NO\ (g)))] - [3(\Delta H^0_f(NO_2\ (g))) + 1(\Delta H^0_f(H_2O\ (l)))]$$
$$= [2(-207\ kJ) + 1(91.3\ kJ)] - [3(33.2\ kJ) + 1(-285.8\ kJ)]$$
$$= [-32\underline{2}.7\ kJ] - [-186.2\ kJ]$$
$$= -137\ kJ$$

Check: The units (kJ) are correct. The answer is negative, which means that the reaction is exothermic.

d) **Given:** $Cr_2O_3\ (s) + 3\ CO\ (g) \rightarrow 2\ Cr\ (s) + 3\ CO_2\ (g)$ **Find:** $\Delta H°_{rxn}$

Conceptual Plan: $\Delta H^0_{rxn} = \sum n_P \Delta H^0_f(products) - \sum n_R \Delta H^0_f(reactants)$

Solution:

Reactant/Product	ΔH^0_f (kJ/mol from Appendix IIB)
$Cr_2O_3\ (s)$	– 1139.7
$CO\ (g)$	–110.5
$Cr\ (s)$	0.0
$CO_2\ (g)$	– 393.5

Be sure to pull data for the correct formula and phase.

$$\Delta H^0_{rxn} = \sum n_P \Delta H^0_f(products) - \sum n_R \Delta H^0_f(reactants)$$
$$= [2(\Delta H^0_f(Cr\ (s))) + 3(\Delta H^0_f(CO_2\ (g)))] - [1(\Delta H^0_f(Cr_2O_3\ (s))) + 3(\Delta H^0_f(CO\ (g)))]$$
$$= [2(0.0\ kJ) + 3(-393.5\ kJ)] - [1(-1139.7\ kJ) + 3(-110.5\ kJ)]$$
$$= [-1180.5\ kJ] - [-1471.2\ kJ]$$
$$= 290.7\ kJ$$

Check: The units (kJ) are correct. The answer is positive, which means that the reaction is endothermic.

49. **Given:** form glucose ($C_6H_{12}O_6$) and oxygen from sunlight, carbon dioxide and water **Find:** $\Delta H°_{rxn}$

Conceptual Plan: **write balanced reaction then** $\Delta H^0_{rxn} = \sum n_P \Delta H^0_f(products) - \sum n_R \Delta H^0_f(reactants)$

Solution: $6\ CO_2\ (g) + 6\ H_2O\ (l) \rightarrow C_6H_{12}O_6\ (s) + 6\ O_2\ (g)$

Reactant/Product	ΔH^0_f (kJ/mol from Appendix IIB)
$CO_2\ (g)$	– 393.5
$H_2O\ (l)$	– 285.8
$C_6H_{12}O\ (s)$	– 1273.3
$O_2\ (g)$	0.0

Be sure to pull data for the correct formula and phase.

$$\Delta H^0_{rxn} = \sum n_P \Delta H^0_f(products) - \sum n_R \Delta H^0_f(reactants)$$
$$= [1(\Delta H^0_f(C_6H_{12}O\ (s))) + 6(\Delta H^0_f(O_2\ (g)))] - [6(\Delta H^0_f(CO_2\ (g))) + 6(\Delta H^0_f(H_2O\ (l)))]$$
$$= [1(-1273.3\ kJ) + 6(0.0\ kJ)] - [6(-393.5\ kJ) + 6(-285.8\ kJ)]$$
$$= [-1273.3\ kJ] - [-4076\ kJ]$$
$$= +\ 2803\ kJ$$

Check: The units (kJ) are correct. The answer is positive, which means that the reaction is endothermic. The reaction requires the input of light energy, so we expect that this will be an endothermic reaction.

51. **Given:** $2\ CH_3NO_2\ (l) + 3/2\ O_2\ (g) \rightarrow 2\ CO_2\ (g) + 3\ H_2O\ (l) + N_2\ (g)$ and $\Delta H°_{rxn} = -\ 1418.4$ kJ/mol
Find: $\Delta H°_f\ (CH_3NO_2\ (l))$
Conceptual Plan: fill known values into $\Delta H^0_{rxn} = \sum n_P \Delta H^0_f(products) - \sum n_R \Delta H^0_f(reactants)$ **and rearrange to solve for $\Delta H°_f\ (CH_3NO_2\ (l))$**
Solution:

Reactant/Product	ΔH^0_f (kJ/mol from Appendix IIB)
$O_2\ (g)$	0.0
$CO_2\ (g)$	− 393.5
$H_2O\ (l)$	− 285.8
$N_2\ (g)$	0.0

Be sure to pull data for the correct formula and phase.

$$H^0_{rxn} = \sum n_P \Delta H^0_f(products) - \sum n_R \Delta H^0_f(reactants)$$
$$= [2(\Delta H^0_f(CO_2\ (g))) + 3(\Delta H^0_f(H_2O\ (l))) + 1(\Delta H^0_f(N_2\ (g)))] - [2(\Delta H^0_f(CH_3NO_2\ (l))) + 3/2(\Delta H^0_f(O_2\ (g)))]$$
$$-1418.4\ kJ = [2(-393.5\ kJ) + 3(-285.8\ kJ) + 1(0.0\ kJ)] - [2(\Delta H^0_f(CH_3NO_2\ (l)) + 3/2(0.0\ kJ)]$$
$$-1418.4\ kJ = [-1644.4\ kJ] - [2(\Delta H^0_f(CH_3NO_2\ (l))]$$
$$\Delta H^0_f(CH_3NO_2\ (l)) = -\ 113.0\ kJ/mol$$

Check: The units (kJ/mol) are correct. The answer is negative, but not as negative as water and carbon dioxide, which is consistent with an exothermic combustion reaction. Also, the answer agrees with the published value for the heat of formation (– 113 kJ/mol).

53. **Given:** billiard ball$_A$ = system: $m_A = 0.17$ kg, $v_{A1} = 4.5$ m/s slows to $v_{A2} = 3.8$ m/s and $v_{A3} = 0$; ball$_B$: $m_B = 0.17$ kg, $v_{B1} = 0$ and $v_{B2} = 3.8$ m/s and $KE = ½\ mv^2$ **Find:** w, q, ΔE_{sys}
Conceptual Plan:
m, v* → *KE **then** ***KE$_{A3}$, KE$_{A1}$* → Δ*E$_{sys}$*** **and** ***KE$_{A2}$, KE$_{A1}$* → *q*** **and** ***KE$_{B2}$, KE$_{B1}$* → *w*$_B$**
$KE = ½\ mv^2$ $\quad \Delta E_{sys} = KE_{A3} - KE_{A1}$ $\quad q = KE_{A2} - KE_{A1}$ $\quad w_B = KE_{B2} - KE_{B1}$
Δ*E$_{sys}$*, *q* → *w*$_A$ verify that *w*$_A$ = − *w*$_B$ so that no heat is transferred to ball$_B$
$\Delta E = q + w$

Solution: $KE = ½\ mv^2$ since m is in kg and v is in m/s, KE will be in kg·m^2/s^2 which is joule.

$$KE_{A1} = ½\ (0.17\ kg)\left(4.5\frac{m}{s}\right)^2 = 1.\underline{7}213\ \frac{kg \cdot m^2}{s^2} = 1.\underline{7}213\ J\ ,$$

$$KE_{A2} = ½\ (0.17\ kg)\left(3.8\frac{m}{s}\right)^2 = 1.\underline{2}274\ \frac{kg \cdot m^2}{s^2} = 1.\underline{2}274\ J\ ,$$

$$KE_{A3} = ½\ (0.17\ kg)\left(0\frac{m}{s}\right)^2 = 0\ \frac{kg \cdot m^2}{s^2} = 0\ J\ ,\ KE_{B1} = ½\ (0.17\ kg)\left(0\frac{m}{s}\right)^2 = 0\ \frac{kg \cdot m^2}{s^2} = 0\ J\ \text{and}$$

$$KE_{B2} = ½\ (0.17\ kg)\left(3.8\frac{m}{s}\right)^2 = 1.\underline{2}274\ \frac{kg \cdot m^2}{s^2} = 1.\underline{2}274\ J\ .$$

$$\Delta E_{sys} = KE_{A3} - KE_{A1} = 0\ J - 1.\underline{7}213\ J = -\ 1.\underline{7}213\ J = -\ 1.7\ J\ ,$$

$$q = KE_{A2} - KE_{A1} = 1.\underline{2}274\ J - 1.\underline{7}213\ J = -\ 0.\underline{4}939\ J = -\ 0.5\ J\ ,$$

$w_B = KE_{B2} - KE_{B1} = 1.2274\ J - 0\ J = 1.2274\ J$ and $w = \Delta E - q = -1.7213\ J - -0.4939\ J = -1.2274\ J = -1.2\ J$.
Since $w_A = -w_B$ so that no heat is transferred to ball$_B$.
Check: The units (J) are correct. Since the ball is initially moving and is stopped at the end, it has lost energy (negative ΔE_{sys}). As the ball slows due to friction, it is releasing heat (negative q). The kinetic energy is transferred to a second ball, so it does work (w negative).

55. **Given:** $H_2O\ (l) \rightarrow H_2O\ (g)$ $\Delta H^\circ_{rxn} = +44.01$ kJ/mol; $\Delta T_{body} = -0.50$ °C, $m_{body} = 95$ kg, $C_{body} = 4.0$ J/g °C
Find: m_{H2O}
Conceptual Plan:
kg → g then m_{body}, ΔT, C_{body} → q_{body} → q_{rxn} (J) → q_{rxn} (kJ) → mol H_2O → g H_2O

$\frac{1000\ g}{1\ kg}$ $\quad q_{body} = m_{body}\, C_{body}\, \Delta T_{body} \quad q_{rxn} = -q_{body} \quad \frac{1\ kJ}{1000\ J} \quad \frac{1\ mol}{44.01\ kJ} \quad \frac{18.01\ g}{1\ mol}$

Solution: $95\ kg \times \frac{1000\ g}{1\ kg} = 95000\ g$ then

$$q_{body} = m_{body}\, C_{body}\, \Delta T_{body} = 95000\ g \times 4.0\ \frac{J}{g \cdot °C} \times (-0.50\ °C) = -190000\ J \text{ then}$$

$$q_{rxn} = -q_{body} = 190000\ J \times \frac{1\ kJ}{1000\ J} \times \frac{1\ mol}{44.01\ kJ} \times \frac{18.01\ g}{1\ mol} = 78\ g\,H_2O$$

Check: The units (g) are correct. The magnitude of the answer (78) makes physical sense because a person can sweat this much on a hot day.

57. **Given:** $H_2O\ (s) \rightarrow H_2O\ (l)$ $\Delta H^\circ_f\ (H_2O\ (s)) = -291.8$ kJ/mol; 355 mL beverage $T_{Bevi} = 25.0$ °C, $T_{Bevf} =$ 0.0 °C, $C_{Bev} = 4.184$ J/g °C, $d_{Bev} = 1.0$ g/mL **Find:** ΔH°_{rxn} (ice melting) and m_{ice}
Conceptual Plan: $\Delta H^0_{rxn} = \sum n_P \Delta H^0_f(products) - \sum n_R \Delta H^0_f(reactants)$ **mL → g and T_i, T_f → ΔT then**

$\frac{1.0\ g}{1.0\ mL} \quad \Delta T = T_f - T_i$

m_{H2O}, ΔT_{H2O}, C_{H2O} → q_{H2O} → q_{rxn} (J) → q_{rxn} (kJ) → mol ice → g ice

$q_{Bev} = m_{Bev}\, C_{Bev}\, \Delta T_{Bev} \quad q_{rxn} = -q_{Bev} \quad \frac{1\ kJ}{1000\ J} \quad \frac{1\ mol}{\Delta H^0_{rxn}} \quad \frac{18.01\ g}{1\ mol}$

Solution:

Reactant/Product	ΔH°_f (kJ/mol from Appendix IIB)
$H_2O\ (s)$	– 291.8
$H_2O\ (l)$	– 285.8

Be sure to pull data for the correct formula and phase.

$$\begin{aligned}\Delta H^0_{rxn} &= \sum n_P \Delta H^0_f(products) - \sum n_R \Delta H^0_f(reactants)\\ &= [1(\Delta H^0_f(H_2O\ (l)))] - [1(\Delta H^0_f(H_2O\ (s)))]\\ &= [1(-285.8\ kJ)] - [1(-291.8\ kJ)]\\ &= +6.0\ kJ\end{aligned}$$

$355\ mL \times \frac{1.0\ g}{1.0\ mL} = 355\ g$ and $\Delta T = T_f - T_i = 0.0\ °C - 25.0\ °C = -25.0\ °C$ then

$$q_{Bev} = m_{Bev}\, C_{Bev}\, \Delta T_{Bev} = 355\ g \times 4.184\ \frac{J}{g \cdot °C} \times (-25.0\ °C) = -37133\ J \text{ then}$$

$$q_{rxn} = -q_{Bev} = -37133\ J \times \frac{1\ kJ}{1000\ J} \times \frac{1\ mol}{-6.0\ kJ} \times \frac{18.01\ g}{1\ mol} = 110\ g\ ice$$

Check: The units (kJ and g) are correct. The answer is positive, which means that the reaction is endothermic. We expect an endothermic reaction because we know that heat must be added to melt ice. The magnitude of the answer (110 g) makes physical sense because it is much smaller than the weight of the beverage and it would fit in a glass with the beverage.

59. **Given:** 25.5 g aluminum, $T_{Ali} = 65.4$ °C, 55.2 g water, $T_{H2Oi} = 22.2$ °C, **Find:** T_f
Conceptual Plan: **pull C_s values from table** **then** $m, C_s, T_i \rightarrow T_f$

Al: $0.903 \frac{J}{g \cdot °C}$ H₂O: $4.18 \frac{J}{g \cdot °C}$ $q = m\,C_s\left(T_f - T_i\right)$ then set $q_{Al} = -\,q_{H2O}$

Solution: $q = m\,C_s\left(T_f - T_i\right)$ substitute in values and set $q_{Al} = -\,q_{H2O}$.

$$q_{Al} = m_{Al}\,C_{Al}\left(T_f - T_{Ali}\right) = 25.5\,\text{g} \times 0.903\,\frac{J}{\text{g} \cdot °C} \times \left(T_f - 65.4\,°C\right) =$$

$$-\,q_{H2O} = -\,m_{H2O}\,C_{H2O}\left(T_f - T_{H2Oi}\right) = -\,55.2\,\text{g} \times 4.18\,\frac{J}{\text{g} \cdot °C} \times \left(T_f - 22.2°C\right)$$

Rearrange to solve for T_f.

$$23.0265\,\frac{J}{°C} \times \left(T_f - 65.4\,°C\right) = -\,230.736\,\frac{J}{°C} \times \left(T_f - 22.2°C\right) \rightarrow$$

$$23.0265\,\frac{J}{°C}\,T_f - 1505.93\,J = -\,230.736\,\frac{J}{°C}\,T_f + 5122.34\,J \rightarrow$$

$$-\,5122.34\,J - 1505.93\,J = -230.736\,\frac{J}{°C}\,T_f - 23.0265\,\frac{J}{°C}\,T_f \rightarrow 6628.27\,J = 253.7625\,\frac{J}{°C}\,T_f \rightarrow$$

$$T_f = \frac{6628.27\,J}{253.7625\,\frac{J}{°C}} = 26.1\,°C$$

Check: The units (°C) are correct. The magnitude of the answer (26) makes physical sense because the heat transfer is dominated by the water (larger mass and larger specific heat capacity). The final temperature should be closer to the initial temperature of water than of aluminum.

61. **Given:** palmitic acid ($C_{16}H_{32}O_2$) combustion $\Delta H°_f$ ($C_{16}H_{32}O_2$ (*s*)) = – 208 kJ/mol; sucrose ($C_{12}H_{22}O_{11}$) combustion $\Delta H°_f$ ($C_{12}H_{22}O_{11}$ (*s*)) = – 2226.1 kJ/mol **Find:** $\Delta H°_{rxn}$ in kJ/mol and Cal/g
Conceptual Plan: **write balanced reaction then** $\Delta H^0_{rxn} = \sum n_P \Delta H^0_f(products) - \sum n_R \Delta H^0_f(reactants)$ **then**
kJ/mol → J/mol → Cal/mol → Cal/g

$\frac{1000\,J}{1\,kJ}$ $\frac{1\,Cal}{4184\,J}$ PA: $\frac{1\,mol}{256.42\,g}$ S: $\frac{1\,mol}{342.30\,g}$

Solution: combustion is combination with oxygen to form carbon dioxide and water (*l*)
$C_{16}H_{32}O_2$ (*s*) + 23 O_2 (*g*) → 16 CO_2 (*g*) + 16 H_2O (*l*)

Reactant/Product	ΔH^0_f (kJ/mol from Appendix IIB)
$C_{16}H_{32}O_2$ (*s*)	– 208
O_2 (*g*)	0.0
CO_2 (*g*)	– 393.5
H_2O (*g*)	– 285.8

Be sure to pull data for the correct formula and phase.

$$\Delta H^0_{rxn} = \sum n_P \Delta H^0_f(products) - \sum n_R \Delta H^0_f(reactants)$$
$$= [16(\Delta H^0_f(CO_2\,(g))) + 16(\Delta H^0_f(H_2O\,(l)))] - [1(\Delta H^0_f(C_{16}H_{32}O_2\,(s))) + 23(\Delta H^0_f(O_2\,(g)))]$$
$$= [16(-393.5\,kJ) + 16(-\,285.8\,kJ)] - [1(-208\,kJ) + 23(0.0\,kJ)]$$
$$= [-\,10868.8\,kJ] - [-208\,kJ]$$
$$= -\,10{,}660.8\,kJ/mol = -10{,}661\,kJ/mol$$

$$-10{,}660.8\,\frac{kJ}{mol} \times \frac{1000\,J}{1\,kJ} \times \frac{1\,Cal}{4184\,J} \times \frac{1\,mol}{256.42\,g} = -\,9.9378\,Cal/g$$

$C_{12}H_{22}O_{11}$ (*s*) + 12 O_2 (*g*) → 12 CO_2 (*g*) + 11 H_2O (*l*)

Reactant/Product	ΔH_f^0 (kJ/mol from Appendix IIB)
$C_{12}H_{22}O_{11}$ (*s*)	– 2226.1
O_2 (*g*)	0.0
CO_2 (*g*)	– 393.5
H_2O (*g*)	– 285.8

Be sure to pull data for the correct formula and phase.

$$\Delta H^0_{rxn} = \sum n_P \Delta H^0_f(products) - \sum n_R \Delta H^0_f(reactants)$$

$$= [12(\Delta H^0_f(CO_2\,(g))) + 11(\Delta H^0_f(H_2O\,(l)))] - [1(\Delta H^0_f(C_{12}H_{22}O_{11}\,(s))) + 12(\Delta H^0_f(O_2\,(g)))]$$

$$= [12(-393.5\text{ kJ}) + 11(-285.8\text{ kJ})] - [1(-2226.1\text{ kJ}) + 12(0.0\text{ kJ})]$$

$$= [-7865.8\text{ kJ}] - [-2226.1\text{ kJ}]$$

$$= -5639.7\text{ kJ/mol}$$

$$-5639.7\frac{\text{kJ}}{\text{mol}} \times \frac{1000\text{ J}}{1\text{ kJ}} \times \frac{1\text{ Cal}}{4184\text{ J}} \times \frac{1\text{ mol}}{342.30\text{ g}} = -3.938\text{ Cal/g}$$

Check: The units (kJ/mol and Cal/g) are correct. The magnitudes of the answers are consistent with the food labels we see every day.

63. At constant *P* $\Delta H_{rxn} = q_P$ and at constant *V* $\Delta E_{rxn} = q_V = \Delta H_{rxn} - P\,\Delta V$. $PV = nRT$ at constant *P*, and a constant number of moles of gas, as we change the *T* the only variable that can change is *V*, so $P\Delta V = nR\Delta T$. Substituting into the equation for ΔE_{rxn} we get $\Delta E_{rxn} = \Delta H_{rxn} - nR\Delta T$ or $\Delta H_{rxn} = \Delta E_{rxn} + nR\Delta T$.

65. **Given:** 16 g peanut butter, bomb calorimeter, T_i = 22.2 °C, T_f = 25.4 °C, C_{cal} = 120.0 kJ/°C
Find: calories in peanut butter
Conceptual Plan:
$T_i, T_f \rightarrow \Delta T$ **then** $\Delta T, C_{cal} \rightarrow q_{cal} \rightarrow q_{rxn}\text{ (kJ)} \rightarrow q_{rxn}\text{ (kJ)} \rightarrow q_{rxn}\text{ (Cal)}$

$\Delta T = T_f - T_i$; $q_{cal} = C_{cal}\,\Delta T$; $q_{rxn} = -q_{cal}$; $\frac{1000\text{ J}}{1\text{ kJ}}$; $\frac{1\text{ Cal}}{4184\text{ J}}$

then q_{rxn} **(Cal)** $\rightarrow$ **Cal/g**
÷16 g peanut butter

Solution: $\Delta T = T_f - T_i = 25.4\text{ °C} - 22.2\text{ °C} = 3.2\text{ °C}$ then $q_{cal} = C_{cal}\,\Delta T = 120.0\frac{\text{kJ}}{\text{°C}} \times 3.2\text{ °C} = 384\text{ kJ}$

then $q_{rxn} = -q_{cal} = -384\text{ kJ} \times \frac{1000\text{ J}}{1\text{ kJ}} \times \frac{1\text{ Cal}}{4184\text{ J}} = 91.778\text{ Cal}$ then $\frac{91.778\text{ Cal}}{16\text{ g}} = 5.7\text{ Cal/g}$

Check: The units (Cal/g) are correct. The magnitude of the answer (6) makes physical sense because there is a significant percentage of fat and sugar in peanut butter. The answer is in line with the answers in #93.

67. **Given:** V_1 = 20.0 L at P_1 = 3.0 atm; P_2 = 1.5 atm let expand at constant *T*. **Find:** *w*, *q*, ΔE_{sys}
Conceptual Plan: $V_1, P_1, P_2 \rightarrow V_2$ **then** $V_1, V_2 \rightarrow \Delta V$ **then** $P, \Delta V \rightarrow w\text{ (L atm)} \rightarrow w\text{ (J)}$

$P_1 V_1 = P_2 V_2$; $\Delta V = V_2 - V_1$; $w = -P\,\Delta V$; $\frac{101.3\text{ J}}{1\text{ L atm}}$

for an ideal gas ΔE_{sys} α *T*, so since this is a constant temperature process $\Delta E_{sys} = 0$ finally
$\Delta E_{sys}, w \rightarrow q$
$\Delta E = q + w$

Solution: $P_1 V_1 = P_2 V_2$ Rearrange to solve for V_2. $V_2 = V_1\frac{P_1}{P_2} = (20.0\text{ L}) \times \frac{3.0\text{ atm}}{1.5\text{ atm}} = 40.\text{ L}$ and

$\Delta V = V_2 - V_1 = 40.\text{ L} - 20.0\text{ L} = 20.\text{ L}$ then

$w = -P\,\Delta V = -1.5\text{ atm} \times 20.\text{ L} \times \frac{101.3\text{ J}}{1\text{ L}\cdot\text{atm}} = -3039\text{ J} = -3.0 \times 10^3\text{ J}$ $\Delta E = q + w$

Rearrange to solve for *q*. $q = \Delta E_{sys} - w = +0\text{ J} - (-3039\text{ J}) = 3.0 \times 10^3\text{ J}$

Check: The units (J) are correct. Since there is no temperature change, we expect no energy change (ΔE_{sys} =

0). The piston expands and so does work (negative work) and so heat is absorbed (positive q).

69. **Given:** 655 kWh/yr, coal is 3.2 % S, remainder is C, S emitted as SO_2 (*g*) and gets converted to H_2SO_4 when reacting with water **Find:** m (H_2SO_4)/yr
Conceptual Plan: write balanced reaction then $\Delta H^\circ_{rxn} = \sum n_P \Delta H^\circ_f(products) - \sum n_R \Delta H^\circ_f(reactants)$
(since the form of sulfur is not given, assume all heat is from combustion of only carbon) then
kWh → J → kJ → mol (C) → g (C) → g (S) → mol (H_2SO_4) → mol (H_2SO_4) → g (H_2SO_4)

$$\frac{3.60 \times 10^6\ \text{J}}{1\ \text{kWh}} \quad \frac{1\ \text{kJ}}{1000\ \text{J}} \quad \frac{\text{mol C}}{\Delta H^0_f\ (CO_2\ (g))} \quad \frac{12.01\ \text{g}}{1\ \text{mol}} \quad \frac{3.2\ \text{g S}}{(100.0 - 3.2)\ \text{g C}} \quad \frac{1\ \text{mol}}{32.06\ \text{g}} \quad \frac{1\ \text{mol}\ H_2SO_4}{1\ \text{mol S}} \quad \frac{98.09\ \text{g}}{1\ \text{mol}}$$

Solution: C (*s*) + O_2 (*g*) → CO_2 (*g*) This reaction is the heat of formation of CO_2 (*g*) so
$\Delta H^0_{rxn} = \Delta H^0_f\ (CO_2\ (g)) = -393.5$ kJ/mol then

$$655\ \text{kWh} \times \frac{3.60 \times 10^6\ \text{J}}{1\ \text{kWh}} \times \frac{1\ \text{kJ}}{1000\ \text{J}} \times \frac{\text{mol C}}{393.5\ \text{kJ}} \times \frac{12.01\ \text{g C}}{1\ \text{mol C}} \times \frac{3.2\ \text{g S}}{(100.0 - 3.2)\ \text{g C}} \times \frac{1\ \text{mol S}}{32.06\ \text{g S}} \times$$
$$\times \frac{1\ \text{mol}\ H_2SO_4}{1\ \text{mol S}} \times \frac{98.09\ \text{g}\ H_2SO_4}{1\ \text{mol}\ H_2SO_4} = 7.3 \times 10^3\ \text{g}\ H_2SO_4$$

Check: The units (g) are correct. The magnitude (7300) is reasonable, considering this is just 1 home.

71. **Given:** methane combustion, 100 % efficiency, $\Delta T = 10.0$ °C, house = 30.0 m x 30.0 m x 3.0 m, Cs (air) = 30 J/K·mol, 1.00 mol air = 22.4 L **Find:** m (CH_4)
Conceptual Plan:
l, *w*, h → V(m^3) → V(cm^3) → V(L) → mol (air) then m, C_s, ΔT → q_{air} (J)

$$V = l\,w\,h \quad \frac{(100\ \text{cm})^3}{(1\ \text{m})^3} \quad \frac{1\ \text{L}}{1000\ \text{cm}^3} \quad \frac{1\ \text{mol air}}{22.4\ \text{L}} \qquad q = m\,C_S\,\Delta T$$

then q_{air} (J) → q_{rxn} (J) → q(kJ) then write balanced reaction for methane combustion

$$q_{rxn} = -\,q_{air} \qquad \frac{1\ \text{kJ}}{1000\ \text{J}}$$

then $\Delta H^0_{rxn} = \sum n_P \Delta H^0_f(products) - \sum n_R \Delta H^0_f(reactants)$ **then q(kJ) → mol (CH_4) → g (CH_4)**

$$\Delta H^0_{rxn} \qquad \frac{16.04\ \text{g}}{1\ \text{mol}}$$

Solution: $V = l\,w\,h = 30.0\ \text{m} \times 30.0\ \text{m} \times 3.0\ \text{m} = 2\underline{7}00\ \text{m}^3$ then

$$2\underline{7}00\ \text{m}^3 \times \frac{(100\ \text{cm})^3}{(1\ \text{m})^3} \times \frac{1\ \text{L}}{1000\ \text{cm}^3} \times \frac{1\ \text{mol air}}{22.4\ \text{L}} = 1.\underline{2}0536 \times 10^5\ \text{mol air} \quad \text{then}$$

$$q = m\,C_s\,\Delta T = 1.\underline{2}0536 \times 10^5\ \text{mol} \times 30\ \frac{\text{J}}{\text{mol}\cdot{}^\circ\text{C}} \times 10.0\ {}^\circ\text{C} = \underline{3}.6161 \times 10^7\ \text{J} \times \frac{1\ \text{kJ}}{1000\ \text{J}} = \underline{3}.6161 \times 10^4\ \text{J}$$

CH_4 (*g*) + 2 O_2 (*g*) → CO_2 (*g*) + 2 H_2O (*g*)

Reactant/Product	ΔH°_f (kJ/mol from Appendix IIB)
CH_4 (*g*)	− 74.6
O_2 (*g*)	0.0
CO_2 (*g*)	− 393.5
H_2O (*g*)	− 241.8

Be sure to pull data for the correct formula and phase.

$$\begin{aligned}\Delta H^0_{rxn} &= \sum n_P \Delta H^0_f(products) - \sum n_R \Delta H^0_f(reactants)\\ &= [1(\Delta H^0_f(CO_2\ (g))) + 2(\Delta H^0_f(H_2O\ (g)))] - [1(\Delta H^0_f(CH_4\ (g))) + 2(\Delta H^0_f(O_2\ (g)))]\\ &= [(-393.5\ \text{kJ}) + 2(-241.8\ \text{kJ})] - [1(-74.6\ \text{kJ}) + 2(0.0\ \text{kJ})]\\ &= [-877.1\ \text{kJ}] - [-74.6\ \text{kJ}]\\ &= -802.5\ \text{kJ}\end{aligned}$$

$q_{rxn} = -q_{air} = -3.6161 \times 10^4 \text{ kJ} \times \frac{1 \text{ mol } CH_4}{-802.5 \text{ kJ}} \times \frac{16.04 \text{ g } CH_4}{1 \text{ mol } CH_4} = 722.8 \text{ g } CH_4 = 700 \text{ g } CH_4$

Check: The units (g) are correct. The magnitude (700) is not surprising since the volume of a house is large.

73. **Given:** m (ice) = 9.0 g; coffee: $T_l = 90.0$ °C, m = 120.0 g, $C_s = C_{H2O}$, $\Delta H°_{fus} = 6.0$ kJ/mol
Find: T_f of coffee
Conceptual Plan: $q_{ice} = -q_{coffee}$ **so g (ice) → mol (ice) →** q_{fus}**(kJ) →** q_{fus} **(J) →** q_{coffee} **(J) then**

$\frac{1 \text{ mol}}{18.01 \text{g}}$ $\frac{6.0 \text{ kJ}}{1 \text{ mol}}$ $\frac{1000 \text{ J}}{1 \text{ kJ}}$ $q_{coffee} = -q_{ice}$

q, m, C_s **→** ΔT **then** $T_i, \Delta T$ **→** T_2 **now we have slightly cooled coffee in contact with 0.0 °C water**

$q = mC_S\Delta T$ $\Delta T = T_2 - T_i$

so $q_{ice} = -q_{coffee}$ **with** m, C_s, T_i **→** T_f

$q = m\, C_s\left(T_f - T_i\right)$ then set $q_{Al} = -q_{H2O}$

Solution: $9.0 \text{ g} \times \frac{1 \text{ mol}}{18.01 \text{ g}} \times \frac{6.0 \text{ kJ}}{1 \text{ mol}} \times \frac{1000 \text{ J}}{1 \text{ kJ}} = 2.9983 \times 10^3 \text{ J}$, $q_{coffee} = -q_{ice} = -2.9983 \times 10^3 \text{ J}$

$q = mC_S\Delta T$ Rearrange to solve for ΔT. $\Delta T = \frac{q}{mC_S} = \frac{-2.9983 \times 10^3 \text{ J}}{120.0 \text{ g} \times 4.18 \frac{\text{J}}{\text{g} \cdot °\text{C}}} = -5.9775$ °C then

$\Delta T = T_2 - T_i$. Rearrange to solve for T_2. $T_2 = \Delta T + T_i = -5.9775\ °\text{C} + 90.0\ °\text{C} = 84.0225\ °\text{C}$

$q = m\, C_s\left(T_f - T_i\right)$ substitute in values and set $q_{H2O} = -q_{coffee}$.

$q_{H2O} = m_{H2O}\, C_{H2O}\left(T_f - T_{H2Oi}\right) = 9.0 \text{ g} \times 4.18 \frac{\text{J}}{\text{g} \cdot °\text{C}} \times \left(T_f - 0.0\ °\text{C}\right) =$

$-q_{coffee} = -m_{coffee}\, C_{coffee}\left(T_f - T_{coffee2}\right) = -120.0 \text{ g} \times 4.18 \frac{\text{J}}{\text{g} \cdot °\text{C}} \times \left(T_f - 84.0225°\text{C}\right)$

Rearrange to solve for T_f.

$9.0 \text{ g } T_f = -120.0 \text{ g}\left(T_f - 84.0225°\text{C}\right) \rightarrow 9.0 \text{ g } T_f = -120.0 \text{ g } T_f + 10082.7 \text{ g} \rightarrow$

$-10082.7 \text{ g} = -129.0 \frac{\text{g}}{°\text{C}} T_f \rightarrow T_f = \frac{-10082.7 \text{ g}}{-129.0 \frac{\text{g}}{°\text{C}}} = 78.2\ °\text{C}$

Check: The units (°C) are correct. The temperature is closer to the original coffee temperature since the mass of coffee is so much larger than the ice mass.

75. $KE = ½\, mv^2$, for an ideal gas $v = u_{rms} = \sqrt{\frac{3RT}{\mathfrak{M}}}$ and so $KE_{avg} = \frac{1}{2}N_A m u_{rms}^2 = \frac{3}{2}RT$ then

$\Delta E_{sys} = KE_2 - KE_1 = \frac{3}{2}RT_2 - \frac{3}{2}RT_1 = \frac{3}{2}R\Delta T$. At constant V $\Delta E_{sys} = C_V\Delta T$ so $C_V = \frac{3}{2}R$.

At constant P, $\Delta E_{sys} = q + w = q_P - P\Delta V = \Delta H - P\Delta V$ but since $PV = nRT$, for one mole of an ideal gas at constant P $P\,\Delta V = R\Delta T$ so $\Delta E_{sys} = q + w = q_P - P\Delta V = \Delta H - P\Delta V = \Delta H - R\Delta T$ then

$\frac{3}{2}R\Delta T = \Delta H - R\Delta T$ or $\Delta H = \frac{5}{2}R\Delta T = C_P\Delta T$ so $C_P = \frac{5}{2}R$.

77. d) only one answer is possible. $\Delta E_{sys} = -\Delta E_{surr}$

79. a) At constant P, $\Delta E_{sys} = q + w = q_P + w = \Delta H + w$ so $\Delta E_{sys} - w = \Delta H$

81. The aluminum cylinder will be cooler after one hour because it has a lower heat capacity than water (less heat needs to be pulled out for every °C temperature change).

83. The internal energy of a chemical system is the sum of its kinetic energy and its potential energy. It is this potential energy that is the energy source in an exothermic chemical reaction. Under normal circumstances, chemical potential energy (or simply chemical energy) arises primarily from the electrostatic forces between the protons and electrons that compose the atoms and molecules within the system. In an exothermic reaction, some bonds break and new ones form, and the protons and electrons go from an arrangement of higher potential energy to one of lower potential energy. As they rearrange, their potential energy is converted into kinetic energy. Heat is emitted in the reaction and so it feels hot to the touch.

Chapter 7
The Quantum-Mechanical Model of the Atom

1. **Given:** distance to sun = 1.496×10^8 km **Find:** time for light to travel from sun to Earth
Conceptual Plan: distance km → distance m → time

$$\frac{1000\text{ m}}{\text{km}} \qquad \text{time} = \frac{\text{distance}}{3.00 \times 10^8 \text{ m/s}}$$

Solution: $1.496 \times 10^8 \text{ km} \times \frac{1000 \text{ m}}{\text{km}} \times \frac{\text{s}}{3.00 \times 10^8 \text{ m}} = 499 \text{ s}$

Check: The units of the answer, seconds, is correct. The magnitude of the answer is reasonable, since it corresponds to about 8 min.

3. i) by increasing wavelength the order is: d) ultraviolet < c) infrared < b) microwave < a) radio waves

ii) by increasing energy the order is: a) radio waves < b) microwaves < c) infrared < d) ultraviolet

5. a) **Given:** $\lambda = 632.8$ nm **Find:** frequency (ν)
Conceptual Plan: nm → m → ν

$$\frac{\text{m}}{10^9 \text{ nm}} \qquad \nu = \frac{c}{\lambda}$$

Solution: $632.8 \text{ nm} \times \frac{\text{m}}{10^9 \text{ nm}} = 6.328 \times 10^{-7} \text{ m}$ $\qquad \nu = \frac{3.00 \times 10^8 \text{ m}}{\text{s}} \times \frac{1}{6.328 \times 10^{-7} \text{ m}} = 4.74 \times 10^{14} \text{ s}^{-1}$

Check: The units of the answer, s^{-1}, are correct. The magnitude of the answer seems reasonable since wavelength and frequency are inversely proportional.

b) **Given:** $\lambda = 503$ nm **Find:** frequency (ν)
Conceptual Plan: nm → m → ν

$$\frac{\text{m}}{10^9 \text{ nm}} \qquad \nu = \frac{c}{\lambda}$$

Solution: $503 \text{ nm} \times \frac{\text{m}}{10^9 \text{ nm}} = 5.03 \times 10^{-7} \text{ m}$ $\qquad \nu = \frac{3.00 \times 10^8 \text{ m}}{\text{s}} \times \frac{1}{5.03 \times 10^{-7} \text{ m}} = 5.96 \times 10^{14} \text{ s}^{-1}$

Check: The units of the answer, s^{-1}, are correct. The magnitude of the answer seems reasonable since wavelength and frequency are inversely proportional.

c) **Given:** $\lambda = 0.052$ nm **Find:** frequency (ν)
Conceptual Plan: nm → m → ν

$$\frac{\text{m}}{10^9 \text{ nm}} \qquad \nu = \frac{c}{\lambda}$$

Solution: $0.052 \text{ nm} \times \frac{\text{m}}{10^9 \text{ nm}} = 5.2 \times 10^{-9} \text{ m}$ $\qquad \nu = \frac{3.00 \times 10^8 \text{ m}}{\text{s}} \times \frac{1}{5.2 \times 10^{-9} \text{ m}} = 5.8 \times 10^{18} \text{ s}^{-1}$

Check: The units of the answer, s^{-1}, are correct. The magnitude of the answer seems reasonable since wavelength and frequency are inversely proportional.

7. a) **Given:** frequency (ν) from 5 a. = $4.74 \times 10^{14} \text{ s}^{-1}$ **Find:** Energy
Conceptual Plan: ν → E

$$E = h\nu \quad h = 6.626 \times 10^{-34} \text{ J s}$$

Solution: $6.626 \times 10^{-34} \text{ J s} \times \frac{4.74 \times 10^{14}}{\text{s}} = 3.14 \times 10^{-19} \text{ J}$

Check: The units of the answer, J, are correct. The magnitude of the answer is reasonable since we are talking about the energy of one photon.

b) **Given:** frequency (ν) from 5 b. = $5.96 \times 10^{14}\ s^{-1}$ **Find:** Energy
Conceptual Plan: $\nu \rightarrow E$

$$E = h\nu \quad h = 6.626 \times 10^{-34}\ J\ s$$

Solution: $6.626 \times 10^{-34}\ J\ \cancel{s} \times \dfrac{5.96 \times 10^{14}}{\cancel{s}} = 3.95 \times 10^{-19}\ J$

Check: The units of the answer, J, are correct. The magnitude of the answer is reasonable since we are talking about the energy of one photon.

c) **Given:** frequency (ν) from 5 c. = $5.8 \times 10^{18}\ s^{-1}$ **Find:** Energy
Conceptual Plan: $\nu \rightarrow E$

$$E = h\nu \quad h = 6.626 \times 10^{-34}\ J\ s$$

Solution: $6.626 \times 10^{-34}\ J\ \cancel{s} \times \dfrac{5.8 \times 10^{18}}{\cancel{s}} = 3.8 \times 10^{-15}\ J$

Check: The units of the answer, J, are correct. The magnitude of the answer is reasonable since we are talking about the energy of one photon.

9. **Given:** $\lambda = 532$ nm and $E_{pulse} = 4.67$ mJ **Find:** number of photons
Conceptual Plan: nm → m → E_{photon} → number of photons

$$\frac{m}{10^9\ nm} \quad E = \frac{hc}{\lambda};\ h = 6.626 \times 10^{-34}\ J\ s \quad \frac{E_{pulse}}{E_{photon}}$$

Solution:

$$532\ \cancel{nm} \times \frac{m}{10^9\ \cancel{nm}} = 5.32 \times 10^{-7}\ m \qquad E = \frac{6.626 \times 10^{-34}\ J\ \cancel{s} \times \dfrac{3.00 \times 10^8\ \cancel{m}}{\cancel{s}}}{5.32 \times 10^{-7}\ \cancel{m}} = 3.7\underline{3}64 \times 10^{-19}\ J/photon$$

$$4.67\ \cancel{mJ} \times \frac{\cancel{J}}{1000\ \cancel{mJ}} \times \frac{1\ photon}{3.7\underline{3}64 \times 10^{-19}\ \cancel{J}} = 1.25 \times 10^{16}\ photons$$

Check: The units of the answer, number of photons, are correct. The magnitude of the answer is reasonable for the amount of energy involved.

11. a) **Given:** $\lambda = 1500$ nm **Find:** E for 1 mol photons
Conceptual Plan: nm → m → E_{photon} → $E(J)_{mol}$ → $E(kJ)_{mol}$

$$\frac{m}{10^9\ nm} \quad E = \frac{hc}{\lambda};\ h = 6.626 \times 10^{-34}\ J\ s \quad \frac{mol}{6.022 \times 10^{23}\ photons} \quad \frac{kJ}{1000\ J}$$

Solution:

$$1500\ \cancel{nm} \times \frac{m}{10^9\ \cancel{nm}} = 1.500 \times 10^{-6}\ m \qquad E = \frac{6.626 \times 10^{-34}\ J\ \cancel{s} \times \dfrac{3.00 \times 10^8\ \cancel{m}}{\cancel{s}}}{1.500 \times 10^{-6}\ \cancel{m}} = 1.3\underline{2}52 \times 10^{-19}\ J/photon$$

$$\frac{1.3\underline{2}52 \times 10^{-19}\ \cancel{J}}{\cancel{photon}} \times \frac{6.022 \times 10^{23}\ \cancel{photons}}{mol} \times \frac{kJ}{1000\ \cancel{J}} = 79.8\ kJ/mol$$

Check: The units of the answer, kJ/mol, are correct. The magnitude of the answer is reasonable for a wavelength in the infrared region.

b) **Given:** $\lambda = 500$ nm **Find:** E for 1 mol photons
Conceptual Plan: nm → m → E_{photon} → E_{mol} → $E(kJ)_{mol}$

$$\frac{m}{10^9\ nm} \quad E = \frac{hc}{\lambda};\ h = 6.626 \times 10^{-34}\ J\ s \quad \frac{mol}{6.022 \times 10^{23}\ photons} \quad \frac{kJ}{1000\ J}$$

Solution: $500\ \cancel{nm} \times \dfrac{m}{10^9\ \cancel{nm}} = 5.00 \times 10^{-7}\ m \qquad E = \dfrac{6.626 \times 10^{-34}\ J\ \cancel{s} \times \dfrac{3.00 \times 10^8\ \cancel{m}}{\cancel{s}}}{5.00 \times 10^{-7}\ \cancel{m}} = 3.9\underline{7}56 \times 10^{-19}\ J/photon$

$$\frac{3.9756 \times 10^{-19}\ \text{J}}{\text{photon}} \times \frac{6.022 \times 10^{23}\ \text{photons}}{\text{mol}} \times \frac{\text{kJ}}{1000\ \text{J}} = 239\ \text{kJ/mol}$$

Check: The units of the answer, kJ/mol, are correct. The magnitude of the answer is reasonable for a wavelength in the visible region.

c) **Given:** $\lambda = 150$ nm **Find:** E for 1 mol photons

Conceptual Plan: **nm → m → E_{photon} → E_{mol} → $E(kJ)_{mol}$**

$$\frac{\text{m}}{10^9\ \text{nm}} \quad E=\frac{hc}{\lambda};\ h=6.626 \times 10^{-34}\ \text{J s} \quad \frac{\text{mol}}{6.022 \times 10^{23}\ \text{photons}} \quad \frac{\text{kJ}}{1000\ \text{J}}$$

Solution:

$$1.50\ \text{nm} \times \frac{\text{m}}{10^9\ \text{nm}} = 1.50 \times 10^{-7}\ \text{m} \qquad E = \frac{6.626 \times 10^{-34}\ \text{J s} \times \dfrac{3.00 \times 10^8\ \text{m}}{\text{s}}}{1.50 \times 10^{-7}\ \text{m}} = 1.3252 \times 10^{-18}\ \text{J/photon}$$

$$\frac{1.3252 \times 10^{-18}\ \text{J}}{\text{photon}} \times \frac{6.022 \times 10^{23}\ \text{photons}}{\text{mol}} \times \frac{\text{kJ}}{1000\ \text{J}} = 798\ \text{kJ/mol}$$

Check: The units of the answer, kJ/mol, are correct. The magnitude of the answer is reasonable for a wavelength in the ultraviolet region. Note: the energy increases from the IR to the Vis to the UV as expected.

13. The interference pattern would be a series of light and dark lines.

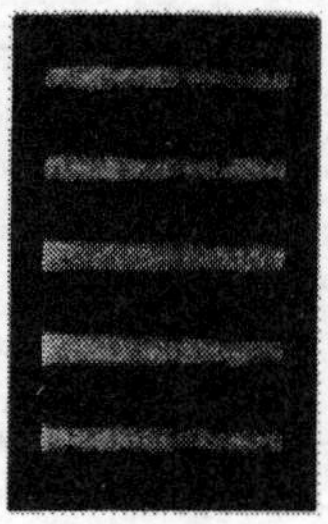

15. **Given:** v = 1.15 x 10^5 m/s **Find:** λ

Conceptual Plan: **v → λ**

$$\lambda = \frac{h}{mv}$$

Solution: $$\frac{6.626 \times 10^{-34}\ \dfrac{\text{kg} \cdot \text{m}^2}{\text{s}^2} \cdot \text{s}}{(9.11 \times 10^{-31}\ \text{kg})(1.15 \times 10^5\ \dfrac{\text{m}}{\text{s}})} = 6.32 \times 10^{-9}\ \text{m}$$

Check: The units of the answer, m, are correct. The magnitude of the answer is very small, as expected for the wavelength of an electron.

17. **Given:** m = 143 g; v = 95 mph **Find:** λ

Conceptual Plan: **m,v → λ**

$$\lambda = \frac{h}{mv}$$

Solution: $$\frac{6.626 \times 10^{-34}\ \dfrac{\text{kg} \cdot \text{m}^2}{\text{s}^2} \cdot \text{s}}{(143\ \text{g})\left(\dfrac{\text{kg}}{1000\ \text{g}}\right)\left(\dfrac{95\ \text{mi}}{\text{hr}}\right)\left(\dfrac{1.609\ \text{km}}{\text{mi}}\right)\left(\dfrac{1000\ \text{m}}{\text{km}}\right)\left(\dfrac{\text{hr}}{3600\ \text{s}}\right)} = 1.1 \times 10^{-34}\ \text{m}$$

The value of the wavelength, 1.1×10^{-34} m, is so small it will not have an effect on the trajectory of the baseball.

Check: The units of the answer, m, are correct. The magnitude of the answer is very small as would be expected for the de Broglie wavelength of a baseball.

19. Since the size of the orbital is determined by the *n* quantum, with the size increasing with increasing *n*, an electron in a 2*s* orbital is closer, on average, to the nucleus than an electron in a 3*s* orbital.

21. The value of *l* is an integer that lies between 0 and $n - 1$.
   a) When $n = 1$, *l* can only be: $1 = 0$.

   b) When $n = 2$, *l* can be: $l = 0$ or $l = 1$.

   c) When $n = 3$, *l* can be: $l = 0$, $l = 1$, or $l = 2$.

   d) When $n = 4$, *l* can be: $l = 0$, $l = 1$, $l = 2$, or $l = 3$.

23. Set c cannot occur together as a set of quantum numbers to specify an orbital. *l* must lie between 0 and $n - 1$, so for $n = 3$, 1 can only be as high as 2.

25. The 2*s* orbital would be the same shape as the 1*s* orbital but would be larger in size and the 3*p* orbitals would have the same shape as the 2*p* orbitals but would be larger in size. Also, the 2*s* and 3*p* orbitals would have more nodes.

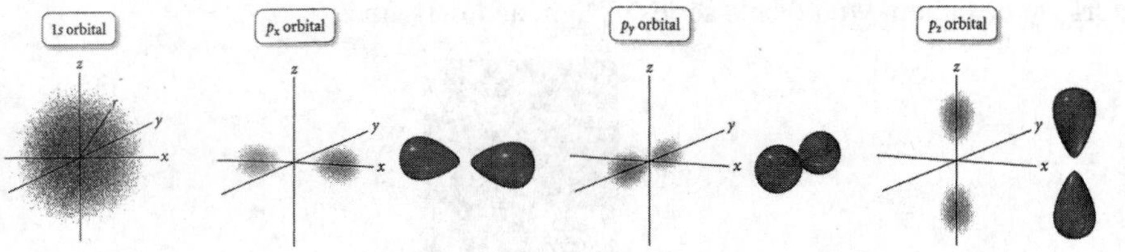

27. When the atom emits the photon of energy that was needed to raise the electron to the $n = 2$ level, the photon has the same energy as the energy absorbed to move the electron to the excited state. Therefore, the electron has to be in $n = 1$ (the ground state) following the emission of the photon.

29. According to the quantum-mechanical model, the higher the *n* level the higher the energy. So, the transition from 3*p* → 1*s* would be a greater energy difference than a transition from 2*p* → 1*s*. The lower energy transition would have the longer wavelength. Therefore, the 2*p* → 1*s* transition would produce a longer wavelength.

31. a) **Given:** $n = 2 \rightarrow n = 1$ **Find:** λ

   Conceptual Plan: $n = 1, n = 2 \rightarrow \Delta E_{atom} \rightarrow \Delta E_{photon} \rightarrow \lambda$

   $\Delta E_{atom} = E_1 - E_2$ $\quad \Delta E_{atom} \rightarrow -\Delta E_{photon}$ $\quad E = \dfrac{hc}{\lambda}$

   Solution:

   $\Delta E = E_1 - E_2$

   $$= -2.18 \times 10^{-18}\,\text{J}\left(\frac{1}{1^2}\right) - \left[-2.18 \times 10^{-18}\left(\frac{1}{2^2}\right)\right] = -2.18 \times 10^{-18}\,\text{J}\left[\left(\frac{1}{1^2}\right) - \left(\frac{1}{2^2}\right)\right] = -1.635 \times 10^{-18}\,\text{J}$$

   $$\Delta E_{photon} = -\Delta E_{atom} = 1.635 \times 10^{-18}\,\text{J} \quad \lambda = \frac{hc}{E} = \frac{(6.626 \times 10^{-34}\,\text{J} \cdot \text{s})(3.00 \times 10^{8}\,\text{m/s})}{1.635 \times 10^{-18}\,\text{J}} = 1.22 \times 10^{-7}\,\text{m}$$

   This transition would produce a wavelength in the UV region.

   Check: The units of the answer, m, are correct. The magnitude of the answer is reasonable since it is in the region of UV radiation.

b) **Given:** $n = 3 \rightarrow n = 1$ **Find:** λ

Conceptual Plan: $n = 1, n = 3 \rightarrow \Delta E_{atom} \rightarrow \Delta E_{photon} \rightarrow \lambda$

$\Delta E_{atom} = E_1 - E_3$ $\quad \Delta E_{atom} \rightarrow -\Delta E_{photon}$ $\quad E = \frac{hc}{\lambda}$

Solution:

$$\Delta E = E_1 - E_3$$

$$= -2.18 \times 10^{-18}\text{ J}\left(\frac{1}{1^2}\right) - \left[-2.18 \times 10^{-18}\left(\frac{1}{3^2}\right)\right] = -2.18 \times 10^{-18}\text{ J}\left[\left(\frac{1}{1^2}\right) - \left(\frac{1}{3^2}\right)\right] = -1.9\underline{3}8 \times 10^{-18}\text{ J}$$

$$\Delta E_{photon} = -\Delta E_{atom} = 1.9\underline{3}8 \times 10^{-18}\text{ J} \quad \lambda = \frac{hc}{E} = \frac{(6.626 \times 10^{-34}\text{ J} \cdot \text{s})(3.00 \times 10^8\text{ m/s})}{1.9\underline{3}8 \times 10^{-18}\text{ J}} = 1.03 \times 10^{-7}\text{ m}$$

This transition would produce a wavelength in the UV region.

Check: The units of the answer, m, are correct. The magnitude of the answer is reasonable since it is in the region of UV radiation.

c) **Given:** $n = 4 \rightarrow n = 2$ **Find:** λ

Conceptual Plan: $n = 2, n = 4 \rightarrow \Delta E_{atom} \rightarrow \Delta E_{photon} \rightarrow \lambda$

$\Delta E_{atom} = E_2 - E_4$ $\quad \Delta E_{atom} \rightarrow -\Delta E_{photon}$ $\quad E = \frac{hc}{\lambda}$

Solution:

$$\Delta E = E_2 - E_4$$

$$= -2.18 \times 10^{-18}\text{ J}\left(\frac{1}{2^2}\right) - \left[-2.18 \times 10^{-18}\left(\frac{1}{4^2}\right)\right] = -2.18 \times 10^{-18}\text{ J}\left[\left(\frac{1}{2^2}\right) - \left(\frac{1}{4^2}\right)\right] = -4.0\underline{8}7 \times 10^{-19}\text{ J}$$

$$\Delta E_{photon} = -\Delta E_{atom} = 4.0\underline{8}7 \times 10^{-19}\text{ J} \quad \lambda = \frac{hc}{E} = \frac{(6.626 \times 10^{-34}\text{ J} \cdot \text{s})(3.00 \times 10^8\text{ m/s})}{4.0\underline{8}7 \times 10^{-19}\text{ J}} = 4.86 \times 10^{-7}\text{ m}$$

This transition would produce a wavelength in the visible region.

Check: The units of the answer, m, are correct. The magnitude of the answer is reasonable since it is in the region of visible light.

d) **Given:** $n = 5 \rightarrow n = 2$ **Find:** λ

Conceptual Plan: $n = 2, n = 5 \rightarrow \Delta E_{atom} \rightarrow \Delta E_{photon} \rightarrow \lambda$

$\Delta E_{atom} = E_2 - E_5$ $\quad \Delta E_{atom} \rightarrow -\Delta E_{photon}$ $\quad E = \frac{hc}{\lambda}$

Solution:

$$\Delta E = E_2 - E_5$$

$$= -2.18 \times 10^{-18}\text{ J}\left(\frac{1}{2^2}\right) - \left[-2.18 \times 10^{-18}\left(\frac{1}{5^2}\right)\right] = -2.18 \times 10^{-18}\text{ J}\left[\left(\frac{1}{2^2}\right) - \left(\frac{1}{5^2}\right)\right] = -4.5\underline{7}8 \times 10^{-19}\text{ J}$$

$$\Delta E_{photon} = -\Delta E_{atom} = 4.5\underline{7}8 \times 10^{-19}\text{ J} \quad \lambda = \frac{hc}{E} = \frac{(6.626 \times 10^{-34}\text{ J} \cdot \text{s})(3.00 \times 10^8\text{ m/s})}{4.5\underline{7}8 \times 10^{-19}\text{ J}} = 4.34 \times 10^{-7}\text{ m}$$

This transition would produce a wavelength in the visible region.

Check: The units of the answer, m, are correct. The magnitude of the answer is reasonable since it is in the region of visible light.

33. **Given:** $n(\text{initial}) = 7$ $\lambda = 397$ nm **Find:** $n(\text{final})$

Conceptual Plan: $\lambda \rightarrow \Delta E_{photon} \rightarrow \Delta E_{atom} \rightarrow n = x, n = 7$

$E = \frac{hc}{\lambda}$ $\quad \Delta E_{photon} \rightarrow -\Delta E_{atom}$ $\quad \Delta E_{atom} = E_x - E_7$

Solution: $$E = \frac{hc}{\lambda} = \frac{(6.626 \times 10^{-34}\text{ J} \cdot \text{s})(3.00 \times 10^8\text{ m/s})}{(397\text{ nm})\left(\frac{\text{m}}{10^9\text{ nm}}\right)} = 5.0\underline{0}7 \times 10^{-19}\text{ J}$$

$\Delta E_{atom} = -\Delta E_{photon} = -5.0\underline{0}7 \times 10^{-19}$ J

$$\Delta E = E_x - E_7 \quad = -5.0\underline{0}7 \times 10^{-19} = -2.18 \times 10^{-18}\,J\left(\frac{1}{x^2}\right) - \left[-2.18 \times 10^{-18}\left(\frac{1}{7^2}\right)\right] = -2.18 \times 10^{-18}\,J\left[\left(\frac{1}{x^2}\right) - \left(\frac{1}{7^2}\right)\right]$$

$$0.2297 = \left(\frac{1}{x^2}\right) - \left(\frac{1}{7^2}\right) \qquad 0.25229 = \left(\frac{1}{x^2}\right) \qquad x^2 = 3.998 \qquad x = 2$$

Check: The answer is reasonable since it is an integer less than the initial value of 7.

35. **Given:** 348 kJ/mol **Find:** λ

Conceptual Plan: kJ/mol → kJ/molec → J/molec → λ

$$\frac{6.022 \times 10^{23}\ \text{C–C bonds}}{\text{mol C–C bonds}} \quad \frac{1000\,J}{kJ} \qquad E = \frac{hc}{\lambda}$$

Solution:
$$\frac{348\ \cancel{kJ}}{\cancel{\text{mol C-C bonds}}} \times \frac{\cancel{\text{mol C-C bonds}}}{6.022 \times 10^{23}\ \cancel{\text{C-C bonds}}} \times \frac{1000\,J}{\cancel{kJ}} = 5.7\underline{7}9 \times 10^{-19}\ J$$

$$\lambda = \frac{(6.626 \times 10^{-34}\ \cancel{J} \cdot \cancel{s})(3.00 \times 10^8\ m/\cancel{s})}{5.7\underline{7}9 \times 10^{-19}\ \cancel{J}} = 3.44 \times 10^{-7}\ m = 344\ nm$$

Check: The units of the answer, m or nm, is correct. The magnitude of the answer is reasonable since this wavelength is in the UV region.

37. **Given:** E_{pulse} = 5.0 watts; d = 5.5 mm; hole = 1.2 mm; λ = 532 nm **Find:** photons/s

Conceptual Plan: fraction of beam through hole → fraction of power and then E_{photon} → number photons/s

$$\frac{\text{area hole}}{\text{area beam}} \qquad \text{fraction x power} \qquad E = \frac{hc}{\lambda}\ \frac{\text{power / s}}{\text{E / photon}}$$

Solution:
$$A = \pi r^2 \qquad \frac{\pi(0.60\text{mm})^2}{\pi(2.75\text{mm})^2} = 0.0476 \qquad 0.0476 \times 5.0\ \text{watts} \times \frac{J/s}{\text{watt}} = 0.2\underline{3}8\ J/s$$

$$E_{photon} = \frac{(6.626 \times 10^{-34}\ J \cdot \cancel{s})(3.00 \times 10^8\ \cancel{m}/\cancel{s})}{(532\ \cancel{nm})\left(\frac{\cancel{m}}{10^9\ \cancel{nm}}\right)} = 3.7\underline{3}6 \times 10^{-19}\ J/\text{photon}$$

$$\frac{0.2\underline{3}8\ \cancel{J}/s}{3.7\underline{3}6 \times 10^{-19}\ \cancel{J}/\text{photon}} = 6.4 \times 10^{17}\ \text{photons/s}$$

Check: The units of the answer, number of photons/s, is correct. The magnitude of the answer is reasonable.

39. **Given:** KE = 506 eV **Find:** λ

Conceptual Plan: KE_{ev} → KE_J → v → λ

$$\frac{1.602 \times 10^{-19}\ J}{eV} \quad KE = 1/2\,mv^2 \quad \lambda = \frac{h}{mv}$$

Solution:
$$506\,eV\left(\frac{1.602 \times 10^{-19}\ J}{eV}\right)\left(\frac{\frac{kg \cdot m^2}{s^2}}{J}\right) = \frac{1}{2}\left(9.11 \times 10^{-31}\ kg\right) v^2$$

$$v^2 = \frac{506\ \cancel{eV}\left(\frac{1.602 \times 10^{-19}\ \cancel{J}}{\cancel{eV}}\right)\left(\frac{\frac{\cancel{kg} \cdot m^2}{s^2}}{\cancel{J}}\right)}{\frac{1}{2}\left(9.11 \times 10^{-31}\ \cancel{kg}\right)} = 1.7796 \times 10^{14}\ \frac{m^2}{s^2}$$

$v = 1.33 \times 10^7 \text{ m/s}$ $\quad \lambda = \dfrac{h}{mv} = \dfrac{6.626 \times 10^{-34} \frac{\cancel{kg} \cdot m^{\cancel{2}}}{\cancel{s}^{\cancel{2}}} \cdot \cancel{s}}{(9.11 \times 10^{-31} \cancel{kg})(1.33 \times 10^7 \cancel{m} / \cancel{s})} = 5.47 \times 10^{-11} \text{ m} = 0.0547 \text{ nm}$

Check: The units of the answer, m or nm, are correct. The magnitude of the answer is reasonable because a deBroglie wavelength is usually a very small number.

41. **Given:** $n = 1 \rightarrow n = \infty$ **Find:** E; λ
Conceptual Plan: $n = \infty, n = 1 \rightarrow \Delta E_{atom} \rightarrow \Delta E_{photon} \rightarrow \lambda$

$\Delta E_{atom} = E_\infty - E_1$ $\quad \Delta E_{atom} \rightarrow \Delta E_{photon}$ $\quad E = \dfrac{hc}{\lambda}$

Solution: $\Delta E = E_\infty - E_1 = 0 - \left[-2.18 \times 10^{-18} \left(\dfrac{1}{1^2}\right)\right] = +2.18 \times 10^{-18} \text{ J}$

$\Delta E_{photon} = -\Delta E_{atom} = +2.18 \times 10^{-18} \text{ J}$

$$\lambda = \frac{hc}{E} = \frac{(6.626 \times 10^{-34} \cancel{J} \cdot \cancel{s})(3.00 \times 10^8 \text{ m} / \cancel{s})}{2.18 \times 10^{-18} \cancel{J}} = 9.12 \times 10^{-8} \text{ m} = 91.2 \text{ nm}$$

Check: The units of the answers, J for E and m or nm for part 1, are correct. The magnitude of the answer is reasonable because it would require more energy to completely remove the electron than just moving it to a higher n level. This results in a shorter wavelength.

43. a) **Given:** $n = 1$ **Find:** number of orbitals if $l = 0 \rightarrow n$
Conceptual Plan: **value n → values l → values m_l → number of orbitals**
$l = 0 \rightarrow n$ $\quad m_l = -1 \rightarrow +1$ total m_l

Solution:

$n =$	1		
$l =$	0	1	
$m_l =$	0	- 1, 0, +1	

total 4 orbitals

Check: The total orbitals will be equal to the number of l sublevels^2

b) **Given:** $n = 2$ **Find:** number of orbitals if $l = 0 \rightarrow n$
Conceptual Plan: **value n → values l → values m_l → number of orbitals**
$l = 0 \rightarrow n$ $\quad m_l = -1 \rightarrow +1$ total m_l

Solution:

$n =$	2		
$l =$	0	1	2
$m_l =$	0	- 1, 0, +1	- 2,- 1,0,1,2

total 9 orbitals

Check: The total orbitals will be equal to the number of l sublevels^2

c) **Given:** $n = 3$ **Find:** number of orbitals if $l = 0 \rightarrow n$
Conceptual Plan: **value n → values l → values m_l → number of orbitals**
$l = 0 \rightarrow n$ $\quad m_l = -1 \rightarrow +1$ total m_l

Solution:

n =	3			
$l =$	0	1	2	3
$m_l =$	0	- 1, 0, +1	- 2,- 1,0,1,2	- 3,- 2,- 1,0,1,2,3

total 16 orbitals

Check: The total orbitals will be equal to the number of l sublevels^2

45. **Given:** λ = 1875 nm; 1282 nm; 1093 nm **Find:** equivalent transitions
Conceptual Plan: $\lambda \rightarrow E_{photon} \rightarrow E_{atom} \rightarrow n$

$E = \dfrac{hc}{\lambda}$ $\quad E_{photon} = -E_{atom}$ $\quad E = -2.18 \times 10^{-18} \text{ J}\left(\dfrac{1}{n_f^2} - \dfrac{1}{n_i^2}\right)$

Solution: Since the wavelength of the transitions are longer wavelengths than those obtained in the visual region, the electron must relax to a higher n level. Therefore, we can assume that the electron returns to the n = 3 level.

For $\lambda = 1875$ nm:

$$E=\frac{(6.626\times10^{-34}\,J\cdot \cancel{s})(3.00\times10^{8}\,\cancel{m}/\cancel{s})}{1875\,\cancel{nm}\left(\frac{\cancel{m}}{10^{9}\,\cancel{nm}}\right)}=1.060\times10^{-19}\,J \qquad 1.060\times10^{-19}\,J=-1.060\times10^{-19}\,J$$

$$-1.060\times10^{-19}\,J=-2.18\times10^{-18}\left(\frac{1}{3^2}-\frac{1}{n^2}\right);\ n=4$$

For $\lambda = 1282$ nm:

$$E=\frac{(6.626\times10^{-34}\,J\cdot \cancel{s})(3.00\times10^{8}\,\cancel{m}/\cancel{s})}{1282\,\cancel{nm}\left(\frac{\cancel{m}}{10^{9}\,\cancel{nm}}\right)}=1.551\times10^{-19}\,J \qquad 1.551\times10^{-19}\,J=-1.551\times10^{-19}\,J$$

$$-1.551\times10^{-19}=-2.18\times10^{-18}\left(\frac{1}{3^2}-\frac{1}{n^2}\right);\ n=5$$

For $\lambda = 1093$ nm:

$$E=\frac{(6.626\times10^{-34}\,J\cdot \cancel{s})(3.00\times10^{8}\,\cancel{m}/\cancel{s})}{1093\,\cancel{nm}\left(\frac{\cancel{m}}{10^{9}\,\cancel{nm}}\right)}=1.819\times10^{-19}\,J \qquad 1.819\times10^{-19}\,J=-1.819\times10^{-19}\,J$$

$$-1.819\times10^{-19}\,J=-2.18\times10^{-18}\left(\frac{1}{3^2}-\frac{1}{n^2}\right);\ n=6$$

Check: The values obtained are all integers, which is correct. The values of *n*; 4,5,6, are reasonable. The values of n increase as the wavelength decreases because the two n levels involved are further apart and more energy is released as the electron relaxes to the $n = 3$ level.

47. **Given:** $\Phi = 193$ kJ/mol **Find:** threshold frequency(ν)

Conceptual Plan: Φ kJ/ mol → Φ kJ/ atom → Φ J/ atom → ν

$$\frac{6.022\times10^{23}\,\text{atoms}}{\text{mol}} \qquad \frac{1000\,J}{kJ} \qquad \Phi = h\nu$$

Solution: $$\nu=\frac{\Phi}{h}=\frac{\left(\frac{193\,\cancel{kJ}}{\cancel{mol}}\right)\left(\frac{\cancel{mol}}{6.022\times10^{23}\,\text{atoms}}\right)\left(\frac{1000\,\cancel{J}}{\cancel{kJ}}\right)}{6.626\times10^{-34}\,\cancel{J}\cdot s}=4.84\times10^{14}\,s^{-1}$$

Check: The units of the answer, s^{-1}, are correct. The magnitude of the answer puts the frequency in the infrared range and is a reasonable answer.

49. **Given:** $\nu_{low} = 30 s^{-1}$ $\nu_{hi} = 1.5 \times 10^4\, s^{-1}$; speed = 344 m/s **Find:** $\lambda_{low} - \lambda_{hi}$

Conceptual Plan: ν_{low} → λ_{low} and $\nu_{hi} = \lambda_{hi}$ then $\lambda_{low} - \lambda_{hi}$

$$\lambda\nu = \text{speed}$$

Solution: $$\lambda=\frac{\text{speed}}{\nu} \qquad \lambda_{low}=\frac{344\,m/\cancel{s}}{30\,\cancel{s^{-1}}}=11\,m \qquad \lambda_{hi}=\frac{344\,m/\cancel{s}}{1.5\times10^{4}\,\cancel{s^{-1}}}=0.023\,m \qquad 11\,m-0.023\,m=11\,m$$

Check: The units of the answer, m, are correct. The magnitude is reasonable since the value is only determined by the low frequency value because of significant figures.

51. a. **Given:** $n = 1$, $n = 2$, $n = 3$, L = 155 pm **Find:** E_1, E_2, E_3

Conceptual Plan: *n* → E

$$E_n=\frac{n^2h^2}{8\,m\,L^2}$$

Solution:

$$E_1=\frac{1^2(6.626\times10^{-34}\,J\cdot s)^2}{8(9.11\times10^{-31}\,kg)(155\,\cancel{pm})^2\left(\frac{m}{10^{12}\,\cancel{pm}}\right)^2}=\frac{1(6.626\times10^{-34})^2\,J^2s^2}{8(9.11\times10^{-31}\,kg)(155\times10^{-12})^2\,m^2}$$

$$= \frac{1(6.626\times10^{-34})^2\left(\frac{\cancel{kg}\cdot \cancel{m^2}}{\cancel{s^2}}\right)J\,\cancel{s^2}}{8(9.11\times10^{-31}\,\cancel{kg})(155\times10^{-12})^2\,\cancel{m^2}} = 2.51\times10^{-18}\ J$$

$$E_2 = \frac{2^2(6.626\times10^{-34}\ J\cdot s)^2}{8(9.11\times10^{-31} kg)(155\ \cancel{pm})^2\left(\frac{m}{10^{12}\ \cancel{pm}}\right)^2} = \frac{4(6.626\times10^{-34})^2 J^2s^2}{8(9.11\times10^{-31}kg)(155\times10^{-12})^2 m^2}$$

$$= \frac{4(6.626\times10^{-34})^2\left(\frac{\cancel{kg}\cdot \cancel{m^2}}{\cancel{s^2}}\right)J\,\cancel{s^2}}{8(9.11\times10^{-31}\,\cancel{kg})(155\times10^{-12})^2\,\cancel{m^2}} = 1.00\times10^{-17}\ J$$

$$E_3 = \frac{3^2(6.626\times10^{-34}\ J\cdot s)^2}{8(9.11\times10^{-31} kg)(155\ \cancel{pm})^2\left(\frac{m}{10^{12}\ \cancel{pm}}\right)^2} = \frac{9(6.626\times10^{-34})^2 J^2s^2}{8(9.11\times10^{-31}kg)(155\times10^{-12})^2 m^2}$$

$$= \frac{9(6.626\times10^{-34})^2\left(\frac{\cancel{kg}\cdot \cancel{m^2}}{\cancel{s^2}}\right)J\,\cancel{s^2}}{8(9.11\times10^{-31}\,\cancel{kg})(155\times10^{-12})^2\,\cancel{m^2}} = 2.26\times10^{-17}\ J$$

Check: The units of the answers, J, are correct. The answers seem reasonable since the energy is increasing with increasing *n* level.

b. **Given:** $n = 1 \rightarrow n = 2$ and $n = 2 \rightarrow n = 3$ **Find:** λ

Conceptual Plan: $\boldsymbol{n = 1, n = 2 \rightarrow \Delta E_{atom} \rightarrow \Delta E_{photon} \rightarrow \lambda}$

$\Delta E_{atom} = E_2 - E_1$ $\Delta E_{atom} \rightarrow -\Delta E_{photon}$ $E = \frac{hc}{\lambda}$

Solution: Using the energies calculated in part a:

$E_2 - E_1 = (1.00 \times 10^{-17}\ J - 2.51 \times 10^{-18}\ J) = 7.49 \times 10^{-18}\ J$

$$\lambda = \frac{(6.626\times10^{-34}\ \cancel{J}\cdot\cancel{s})(3.00\times10^{8}\ m/\cancel{s})}{7.49\times10^{-18}\ \cancel{J}} = 2.65\times10^{-8}\ m = 26.5\ nm$$

$E_3 - E_2 = (2.26 \times 10^{-17}\ J - 1.00 \times 10^{-17}\ J) = 1.26 \times 10^{-17}\ J$

$$\lambda = \frac{(6.626\times10^{-34}\ \cancel{J}\cdot\cancel{s})(3.00\times10^{8}\ m/\cancel{s})}{1.26\times10^{-17}\ \cancel{J}} = 1.58\times10^{-8}\ m = 15.8\ nm$$

These wavelengths would lie in the UV region.

Check: The units of the answers, m, are correct. The magnitude of the answers is reasonable based on the energies obtained for the levels.

53. For the 1s orbital: In the Excel spreadsheet, call column A: r; and column B: ψ(1s). Make the values for r column A: 0 – 200. In column B, put the equation for the wavefunction written as: =(POWER(1/3.1415,1/2))*(1/POWER(53,3/2))*(EXP(-A2/53)). Go to make chart, choose xy scatter.

e.g., sample values

r	ψ(1s)
0	0.00146224
1	0.00143491
2	0.00140809
3	0.00138177
4	0.00135594
5	0.0013306
6	0.00130573

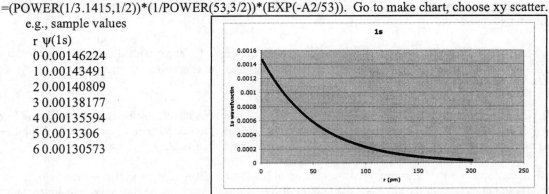

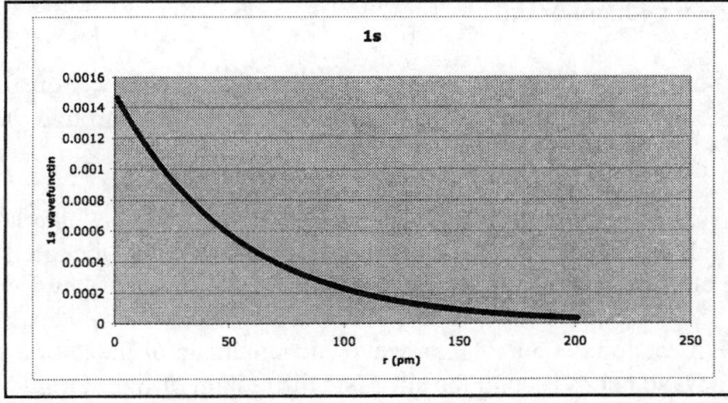

For the 2s orbital: In the same Excel spreadsheet, call column A: r; and column C: ψ(2s). Use the same values for r in column A: 0 – 200. In column C, put the equation for the wavefunction written as: =(POWER(1/((32)*(3.1415)),1/2))*(1/POWER(53,3/2))*(2-(A2/53))*(EXP(-A2/53)). Go to make chart, choose xy scatter.

e.g., sample values

r	ψ(2s)
0	0.000516979
1	0.00050253
2	0.000488441
3	0.000474702
4	0.000461307
5	0.000448247
6	0.000435513

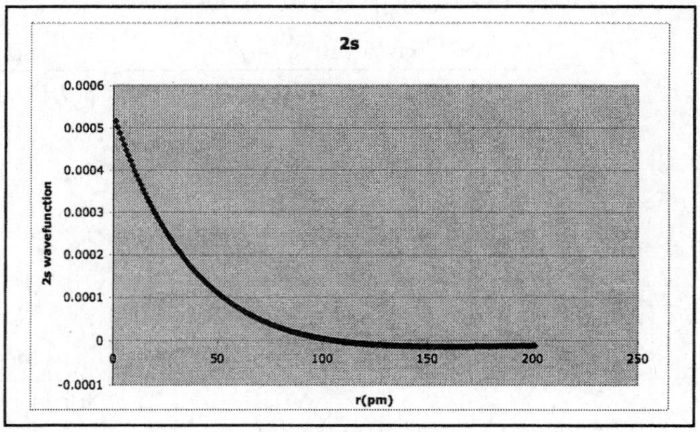

Note: The plot for the 2s orbital extends below the x axis. The x-intercept represents the radial node of the orbital.

55. **Given:** threshold frequency = $2.25 \times 10^{14} s^{-1}$; $\lambda = 5.00 \times 10^{-7}$ m **Find:** v of electron
Conceptual Plan: ν → Φ and then λ → E and then → KE → v

$\Phi = h\nu$ $\quad E = \frac{hc}{\lambda}$ $\quad KE = E - \Phi$ $\quad KE = 1/2\, mv^2$

Solution:

$$\Phi = (6.626 \times 10^{-34}\,J \cdot \cancel{s})(2.25 \times 10^{14}\,\cancel{s^{-1}}) = 1.4\underline{9}1 \times 10^{-19}\,J \qquad E = \frac{(6.626 \times 10^{-34}\,J \cdot \cancel{s})(3.00 \times 10^{8}\,\cancel{m}/\cancel{s})}{5.00 \times 10^{-7}\,\cancel{m}} = 3.9\underline{7}6 \times 10^{-19}\,J$$

$$KE = 3.9\underline{7}6 \times 10^{-19}\,J - 1.4\underline{9}1 \times 10^{-19}\,J = 2.485 \times 10^{-19}\,J \qquad v^2 = \frac{2.485 \times 10^{-19}\,\frac{\cancel{kg} \cdot m^2}{s^2}}{\frac{1}{2}(9.11 \times 10^{-31}\,\cancel{kg})} = 5.455 \times 10^{11}\,\frac{m^2}{s^2}$$

$v = 7.39 \times 10^5$ m/s

Check: The units of the answer, m/s, are correct. The magnitude of the answer is reasonable for the speed of an electron.

57. In the Bohr model of the atom, the electron travels in a circular orbit around the nucleus. It is a 2-dimensional model. The electron is constrained to move only from one orbit to another orbit. But, the electron is treated as a particle that behaves according to the laws of classical physics. The quantum-mechanical model of the atom is 3-dimensional. In this model, we treat the electron, an absolutely small particle, differently than we treat particles with classical physics. The electron is in an orbital, which gives us the probability of finding the electron within a volume of space.
Because the electron in the Bohr model is constrained to a circular orbit, it would theoretically be possible to know both the position and the velocity of the electron simultaneously. This contradicts the Heisenberg uncertainty principle which states that position and velocity are complementary terms which cannot both be known with precision.

59. a) Since the interference pattern is caused by single electrons interfering with themselves, the pattern remains the same even when the rate of the electrons passing through the slits is one electron per minute. It will simply take longer for the full pattern to develop.

b) When a light is placed behind the slits to determine which hole the electron passes through, the light flashes to indicate which hole the electron passed through, but the interference pattern is now absent. With the laser on, the electrons hit positions directly behind each slit, as if they were ordinary particles.

c) Diffraction occurs when a wave encounters an obstacle of a slit that is comparable in size to its wavelength. The wave bends around the slit. The diffraction of light through two slits separated by a distance comparable to

the wavelength of the light results in an interference pattern. Each slit acts as a new wave source, and the two new waves interfere with each other, which result in a pattern of bright and dark lines.

d) Since the mass of the bullets and their particle size are not absolutely small, the bullets will not produce an interference pattern when they pass through the slits. The de Broglie wavelength produced by the bullets will not be sufficiently large enough to interfere with the bullet trajectory and no interference pattern will be observed.

Chapter 8
Periodic Properties of the Elements

1. a) P Phosphorus has 15 electrons. Distribute two of these into the 1*s* orbital, two into the 2*s* orbital, six into the 2*p* orbital, two into the 3s orbital, and three into the 3*p* orbital. $1s^22s^22p^63s^23p^3$

   b) C Carbon has 6 electrons. Distribute two of these into the 1*s* orbital, two into the 2*s* orbital, and two into the 2*p* orbital. $1s^22s^22p^2$

   c) Na Sodium has 11 electrons. Distribute two of these into the 1*s* orbital, two into the 2*s* orbital, six into the 2*p* orbital, and one into the 3*s* orbital. $1s^22s^22p^63s^1$

   d) Ar Argon has 18 electrons. Distribute two of these into the 1*s* orbital, two into the 2*s* orbital, six into the 2*p* orbital, two into the 3*s* orbital, and six into the 3*p* orbital. $1s^22s^22p^63s^23p^6$

3. a) N Nitrogen has 7 electrons and has the electron configuration: $1s^22s^22p^3$. Draw a box for each orbital, putting the lowest energy orbital (1s) on the far left and proceeding to orbitals of higher energy to the right. Distribute the 7 electrons into the boxes representing the orbitals, allowing a maximum of two electrons per orbital and remembering Hund's rule. You can see from the diagram that nitrogen has 3 unpaired electrons.

1*s*	2*s*	2*p*
↓↑	↓↑	↑ | ↑ | ↑

   b) F Fluorine has 9 electrons and has the electron configuration: $1s^22s^22p^5$. Draw a box for each orbital, putting the lowest energy orbital (1s) on the far left and proceeding to orbitals of higher energy to the right. Distribute the 9 electrons into the boxes representing the orbitals, allowing a maximum of two electrons per orbital and remembering Hund's rule. You can see from the diagram that fluorine has 1 unpaired electron.

1*s*	2*s*	2*p*
↓↑	↓↑	↓↑ | ↓↑ | ↑

   c) Mg Magnesium has 12 electrons and has the electron configuration: $1s^22s^22p^63s^2$. Draw a box for each orbital, putting the lowest energy orbital (1s) on the far left and proceeding to orbitals of higher energy to the right. Distribute the 12 electrons into the boxes representing the orbitals, allowing a maximum of two electrons per orbital and remembering Hund's rule. You can see from the diagram that magnesium has no unpaired electrons.

1*s*	2*s*	2*p*	3*s*
↓↑	↓↑	↓↑ | ↓↑ | ↓↑	↓↑

   d) Al Aluminum has 13 electrons and has the electron configuration: $1s^22s^22p^63s^23p^1$. Draw a box for each orbital, putting the lowest energy orbital (1s) on the far left and proceeding to orbitals of higher energy to the right. Distribute the 13 electrons into the boxes representing the orbitals, allowing a maximum of two electrons per orbital and remembering Hund's rule. You can see from the diagram that aluminum has 1 unpaired electron.

1*s*	2*s*	2*p*	3*s*	3*p*
↓↑	↓↑	↓↑ | ↓↑ | ↓↑	↓↑	↑ | |

5. a) P The atomic number of P is 15. The noble gas that precedes P in the periodic table is neon, so the inner electron configuration is [Ne]. Obtain the outer electron configuration by tracing the elements between Ne and P and assigning electrons to the appropriate orbitals. Begin with [Ne]. Because P is in row 3, add two 3*s* electrons. Next add three 3*p* electrons as you trace across the *p* block to P which is in the third column of the p block.

   P $[Ne]3s^23p^3$

b) Ge The atomic number of Ge is 32. The noble gas that precedes Ge in the periodic table is argon, so the inner electron configuration is [Ar]. Obtain the outer electron configuration by tracing the elements between Ar and Ge and assigning electrons to the appropriate orbitals. Begin with [Ar]. Because Ge is in row 4, add two 4*s* electrons. Next, add ten 3*d* electrons as you trace across the *d* block. Finally add two 4*p* electrons as you trace across the *p* block to Ge which is in the second column of the *p* block.

Ge $[Ar]4s^2 3d^{10} 4p^2$

c) Zr The atomic number of Zr is 40. The noble gas that precedes Zr in the periodic table is krypton, so the inner electron configuration is [Kr]. Obtain the outer electron configuration by tracing the elements between Kr and Zr and assigning electrons to the appropriate orbitals. Begin with [Kr]. Because Zr is in row 5, add two 5*s* electrons. Next, add two 4*d* electrons as you trace across the *d* block to Zr which is in the second column.

Zr $[Kr]5s^2 4d^2$

d) I The atomic number of I is 53. The noble gas that precedes I in the periodic table is krypton, so the inner electron configuration is [Kr]. Obtain the outer electron configuration by tracing the elements between Kr and I and assigning electrons to the appropriate orbitals. Begin with [Kr]. Because I is in row 5, add two 5*s* electrons. Next, add ten 4*d* electrons as you trace across the d block. Finally add five 5*p* electrons as you trace across the *p* block to I which is in the fifth column of the *p* block.

I $[Kr]5s^2 4d^{10} 5p^5$

7. a) Li is in period 2, and the first column in the *s* block so Li has one 2*s* electron.

b) Cu is in period 4, and the ninth column in the *d* block (n – 1) so Cu should have nine 3*d* electrons, however, it is one of our exceptions, so it has ten 3*d* electrons.

c) Br is in period 4, and the fifth column of the *p* block, so Br has five 4*p* electrons.

d) Zr is in period 5, and the second column of the *d* block (n – 1), so Zr has two 4*d* electrons.

9. a) In period 4, an element with five valence electrons could be V or As.

b) In period 4, an element with four 4*p* electrons would be in the fourth column of the *p* block and is Se.

c) In period 4, an element with three 3*d* electrons would be in the third column of the *d* block (n – 1) and is V.

d) In period 4, an element with a complete outer shell would be in the sixth column of the *p* block and is Kr.

11. a) Ba is in column 2A, so it has two valence electrons.

b) Cs is in column 1A, so it has one valence electron.

c) Ni is in column 8 of the *d* block, so it has 10 valence electrons (8 from the *d* block and 2 from the *s* block).

d) S is in column 6A, so it has six valence electrons.

13. a) The outer electron configuration ns^2 would belong to a reactive metal in the alkaline earth family.

b) The outer electron configuration ns^2np^6 would belong to an unreactive nonmetal in the noble gas family.

c) The outer electron configuration ns^2np^5 would belong to a reactive nonmetal in the halogen family.

d) The outer electron configuration ns^2np^2 would belong to an element in the carbon family. If $n = 2$, the element is a nonmetal, if $n = 3$ or 4, the element is a metalloid, and if $n = 5$ or 6, the element is a metal.

15. The valence electrons in nitrogen would experience a greater effective nuclear charge. Be has 4 protons and N has 7 protons. Both atoms have 2 core electrons that predominately contribute to the shielding while the

valence electrons will contribute a slight shielding effect. So, Be has an effective nuclear charge of slightly more than 2+ and N has an effective nuclear charge of slightly more than 5+.

17. a) K(19) $[Ar]4s^1$ $Z_{eff} = Z - \text{core electrons} = 19 - 18 = 1+$

b) Ca(20) $[Ar]4s^2$ $Z_{eff} = Z - \text{core electrons} = 20 - 18 = 2+$

c) O(8) $[He]2s^22p^4$ $Z_{eff} = Z - \text{core electrons} = 8 - 2 = 6+$

d) C(6) $[He]2s^22p^2$ $Z_{eff} = Z - \text{core electrons} = 6 - 2 = 4+$

19. a) Al or In In atoms are larger than Al atoms because, as you trace the path between Al and In on the periodic table, you move down a column. Atomic size increases as you move down a column because the outermost electrons occupy orbitals with a higher principal quantum number that are therefore larger, resulting in a larger atom.

b) Si or N Si atoms are larger than N atoms because, as you trace the path between N and Si on the periodic table, you move down a column (atomic size increases) and then to the left across a period (atomic size increases). These effects add together for an overall increase.

c) P or Pb Pb atoms are larger than P atoms because, as you trace the path between P and Pb on the periodic table, you move down a column (atomic size increases) and then to the left across a period (atomic size increases). These effects add together for an overall increase.

d) C or F C atoms are larger than F atoms because, as you trace the path between C and F on the periodic table, you move to the right within the same period. As you move to the right across a period, the effective nuclear charge experienced by the outermost electrons increase, which results in a smaller size.

21. Ca, Rb, S, Si, Ge, F F is above and to the right of the other elements, so we start with F as the smallest atom. As you trace a path from F to S you move to the left (size increases) and down (size increases), then you move to Si to the left (size increases) then down to Ge (size increases) then to the left to Ca (size increases) then to the left and down to Rb (size increases). So, in order of increasing atomic radii: F < S < Si < Ge < Ca < Rb

23. a) O^{2-} Begin by writing the electron configuration of the neutral atom.
O $1s^22s^22p^4$
Since this ion has a 2 – charge, add two electrons to write the electron configuration of the ion.
O^{2-} $1s^22s^22p^6$ This is isoelectronic with Ar

b) Br^- Begin by writing the electron configuration of the neutral atom.
Br $[Ar]4s^23d^{10}4p^5$
Since this ion has a 1 – charge, add one electron to write the electron configuration of the ion.
Br^- $[Ar]4s^23d^{10}4p^6$ This is isoelectronic with Kr

c) Sr^{2+} Begin by writing the electron configuration of the neutral atom.
Sr $[Kr]5s^2$
Since this ion has a 2+ charge, remove two electrons to write the electron configuration of the ion.
Sr^{2+} [Kr]

d) Co^{3+} Begin by writing the electron configuration of the neutral atom.
Co $[Ar]4s^23d^7$
Since this ion has a 3+ charge, remove three electrons to write the electron configuration of the ion. Since it is a transition metal, remove the electrons from the 4*s* orbital before removing electrons from the 3*d* orbitals.
Co^{3+} $[Ar]4s^03d^6$

e) Cu^{2+} Begin by writing the electron configuration of the neutral atom. Remember, Cu is one of our exceptions.
Cu $[Ar]4s^13d^{10}$

Since this ion has a 2+ charge, remove two electrons to write the electron configuration of the ion. Since it is a transition metal, remove the electrons from the 4*s* orbital before removing electrons from the 3*d* orbitals.

Cu^{2+} $[Ar]4s^0 3d^9$

25. a) V^{5+} Begin by writing the electron configuration of the neutral atom.

V $[Ar]4s^2 3d^3$

Since this ion has a 5+ charge, remove five electrons to write the electron configuration of the ion. Since it is a transition metal, remove the electrons from the 4*s* orbital before removing electrons from the 3*d* orbitals.

V^{5+} $[Ar]4s^0 3d^0 = [Ne]3s^2 3p^6$

[Ne] [⇂↾] [⇂↾ | ⇂↾ | ⇂↾]

3*s* 3*p*

V^{5+} is diamagnetic

b) Cr^{3+} Begin by writing the electron configuration of the neutral atom. Remember, Cr is one of our exceptions.

Cr $[Ar]4s^1 3d^5$

Since this ion has a 3+ charge, remove three electrons to write the electron configuration of the ion. Since it is a transition metal, remove the electrons from the 4*s* orbital before removing electrons from the 3*d* orbitals.

Cr^{3+} $[Ar]4s^0 3d^3$

[Ar] [ ] [↑ | ↑ | ↑ | |]

4*s* 3*d*

Cr^{3+} is paramagnetic

c) Ni^{2+} Begin by writing the electron configuration of the neutral atom.

Ni $[Ar]4s^2 3d^8$

Since this ion has a 2+ charge, remove two electrons to write the electron configuration of the ion. Since it is a transition metal, remove the electrons from the 4*s* orbital before removing electrons from the 3*d* orbitals.

Ni^{2+} $[Ar]4s^0 3d^8$

[Ar] [ ] [⇂↾ | ⇂↾ | ⇂↾ | ↑ | ↑]

4*s* 3*d*

Ni^{2+} is paramagnetic

d) Fe^{3+} Begin by writing the electron configuration of the neutral atom.

Fe $[Ar]4s^2 3d^6$

Since this ion has a 3+ charge, remove three electrons to write the electron configuration of the ion. Since it is a transition metal, remove the electrons from the 4s orbital before removing electrons from the 3d orbitals.

Fe^{3+} $[Ar]4s^0 3d^5$

[Ar] [ ] [↑ | ↑ | ↑ | ↑ | ↑]

4s 3d

Fe^{3+} is paramagnetic

27. a) Li or Li^+ A Li atom is larger than Li^+ because cations are smaller than the atoms from which they are formed.

b) I^- or Cs^+ An I^- ion is larger than a Cs^+ ion because, although they are isoelectronic, I^- has two fewer protons than Cs^+, resulting in a lesser pull on the electrons and therefore larger radius.

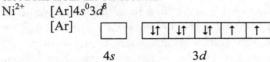

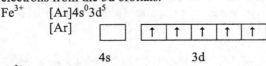

c) Cr or Cr^{3+} — A Cr atom is larger than Cr^{3+} because cations are smaller than the atoms from which they are formed.

d) O or O^{2-} — An O^{2-} ion is larger than an O atom because anions are larger than the atoms from which they are formed.

29. Since all the species are isoelectronic, the radius will depend on the number of protons in each species. The fewer the protons the larger the radius.
F: Z = 9; Ne: Z = 10; O: Z = 8; Mg: Z = 12; Na: Z = 11
So: $O^{2-} > F^- > Ne > Na^+ > Mg^{2+}$

31. a) Br or Bi Br has a higher ionization energy than Bi because, as you trace the path between Br and Bi on the periodic table, you move down a column (ionization energy decreases) and then to the left across a period (ionization energy decreases). These effects sum together for an overall decrease.

b) Na or Rb Na has a higher ionization energy than Rb because, as you trace a path between Na and Rb on the periodic table, you move down a column. Ionization energy decreases as you go down a column because of the increasing size of orbitals with increasing n.

c) As or At Based on periodic trends alone, it is impossible to tell which has a higher ionization energy because, as you trace the path between As and At you go to the right across a period (ionization energy increases) and then down a column (ionization energy decreases). These effects tend to oppose each other, and it is not obvious which will dominate.

d) P or Sn P has a higher ionization energy than Sn because, as you trace the path between P and Sn on the periodic table, you move down a column (ionization energy decreases) and then to the left across a period (ionization energy decreases). These effects sum together for an overall decrease.

33. Since ionization energy increases as you move to the right across a period and increases as you move up a column, the element with the smallest first ionization energy would be the element farthest to the left and lowest down on the periodic table. So, In has the smallest ionization energy, as you trace a path to the right and up on the periodic table, the next element reached is Si, continuing up and to the right you reach N and then continuing to the right you reach F. So, in the order of increasing first ionization energy the elements are:
In < Si < N < F.

35. The jump in ionization energy occurs when you change from removing a valence electron to removing a core electron. To determine where this jump occurs you need to look at the electron configuration of the atom.

a) Be — $1s^2 2s^2$ — The first and second ionization energies involve removing 2*s* electrons, the third ionization energy removes a core electron, so the jump will occur between the second and third ionization energies.

b) N — $1s^2 2s^2 2p^3$ — The first five ionization energies involve removing the 2*p* and 2*s* electrons, the sixth ionization energy removes a core electron, so the jump will occur between the fifth and sixth ionization energies.

c) O — $1s^2 2s^2 2p^4$ — The first six ionization energies involve removing the 2*p* and 2*s* electrons, the seventh ionization energy removes a core electron, so the jump will occur between the sixth and seventh ionization energies.

d) Li — $1s^2 2s^1$ — The first ionization energy involve removing a 2*s* electron, the second ionization energy removes a core electron, so the jump will occur between the first and second ionization energies.

37. a) Na or Rb — Na has a more negative electron affinity than Rb. In column 1A electron affinity becomes less negative as you go down the column.

b) B or S — S has a more negative electron affinity than B. As you trace from B to S in the periodic table, you move to the right which shows the value of the electron affinity becoming more negative. Also, as you move from period 2 to period 3, the value of the electron affinity becomes more negative. Both of these trends sum together for the value of the electron affinity to become more negative.

c) C or N C has the more negative electron affinity. As you trace from C to N across the periodic table, you would normally expect N to have the more negative electron affinity. However, N has a half-filled p sublevel which lends it extra stability, therefore it is harder to add an electron.

d) Li or F F has the more negative electron affinity. As you trace from Li to F on the periodic table you move to the right in the period. As you go to the right across a period, the value of the electron affinity generally becomes more negative.

39. a) Sr or Sb Sr is more metallic than Sb because, as we trace the path between Sr and Sb on the periodic table, we move to the right within the same period. Metallic character decreases as you go to the right.

b) As or Bi Bi is more metallic because, as we trace a path between As and Bi on the periodic table, we move down a column in the same family (metallic character increases).

c) Cl or O Based on periodic trends alone, we cannot tell which is more metallic because, as we trace the path between O and Cl, we go to the right across a period (metallic character decreases) and then down a column (metallic character increases). These effects tend to oppose each other, and it is not easy to tell which will predominate.

d) S or As As is more metallic than S because, as we trace the path between S and As on the periodic table, we move down a column (metallic character increases) and then to the left across a period (metallic character increases). These effects add together for an overall increase.

41. The order of increasing metallic character is: S < Se < Sb < In < Ba < Fr.
Metallic character decreases as you move left to right across a period and decreases as you move up a column, therefore, the element with the least metallic character will be to the top right of the periodic table. So, of these elements, S has the least metallic character. As you move down the column, the next element is Se, as you continue down and then to the right, you reach Sb, continuing to the right goes to In, going down the column and then to the right comes to Ba and then down the column and to the right is Fr.

43. Br: $1s^2 2s^2 2p^6 3s^2 3p^6 4s^2 3d^{10} 4p^5$
Kr: $1s^2 2s^2 2p^6 3s^2 3p^6 4s^2 3d^{10} 4p^6$
Krypton has a completely filled *p* sublevel giving it chemical stability. Bromine needs one electron to achieve a completely filled *p* sublevel and therefore has a highly negative electron affinity. It therefore easily takes on an electron and is reduced to the bromide ion, giving it the added stability of the filled *p* sublevel.

45. Write the electron configuration of vanadium
V: [Ar] $4s^2 3d^3$
Since this ion has a 3+ charge, remove three electrons to write the electron configuration of the ion. Since it is a transition metal, remove the electrons from the 4*s* orbital before removing electrons from the 3*d* orbitals.
V^{3+}: [Ar] $4s^0 3d^2$
Both vanadium and the V^{3+} ion have unpaired electrons and are paramagnetic.

47. Since K^+ has a 1+ charge you would need a cation with a similar size and a 1+ charge. Looking at the ions in the same family, Na^+ would be too small and Rb^+ would be too large. If we then consider Ar^+ and Ca^+ we would have ions of similar size and charge. Between these two Ca^+ would be the easier to achieve because the first ionization energy of Ca is similar to that of K while the first ionization energy of Ar is much larger. However, the second ionization energy of Ca is relatively low, making it easy to lose the second electron.

49. C has an outer shell electron configuration of ns^2np^2; based on this you would expect Si and Ge, which are in the same family, to be most like carbon. Ionization energies for both Si and Ge are similar and tend to be slightly lower than C but all are intermediate in the range of first ionization energies. The electron affinities of Si and Ge are close to that of C.

51. a) N: $[He]2s^2 2p^3$ Mg: $[Ne]3s^2$ O: $[He]2s^2 2p^4$
F: $[He]2s^2 2p^5$ Al: $[Ne]3s^2 3p^1$

b) Mg > Al > N > O > F

c) Al < Mg < O < N < F (from the table)

d) Mg and Al would have the largest radius because they are in period $n = 3$, Al is smaller than Mg because radius decreases as you move to the right across the period. F is smaller than O is smaller than N because as you move to the right across the period, radius decreases.
The first ionization energy of Al is smaller than the first ionization energy of Mg because Al loses the electron from the $3p$ orbital which is shielded by the electrons in the $3s$ orbital, while Mg loses the electron from the filled $3s$ orbital which has added stability because it is a filled orbital. The first ionization energy of O is lower than the first ionization energy of N because N has a half-filled $2p$ orbitals which adds extra stability, thus making it harder to remove the electron and the fourth electron in the O $2p$ orbitals experiences added electron-electron repulsion because it must pair with another electron in the same $2p$ orbital, thus making it easier to remove.

53. As you move to the right across a row in the periodic table for the main-group elements, the effective nuclear charge (Z_{eff}) experienced by the electrons in the outermost principal energy level increases, resulting in a stronger attraction between the outermost electrons and the nucleus and therefore, smaller atomic radii.
Across the row of transition elements, the number of electrons in the outermost principal energy level (highest n value) is nearly constant. As another proton is added to the nucleus with each successive element, another electron is added, but that electron goes into an $n_{highest} - 1$ orbital (a core level). The number of outermost electrons stays constant and they experience a roughly constant effective nuclear charge, keeping the radius approximately constant after the first couple of elements in the series.

55. The noble gases all have a filled outer quantum level, very high first ionization energies, and positive values for the electron affinity and are thus particularly unreactive. The lighter noble gases will not form any compounds because the ionization energies of He and Ne are both over 2000 kJ/mol. Since ionization energy decreases as you move down a column we find that the heavier noble gases, Ar, Kr, and Xe do form some compounds, they have ionization energies that are close to the ionization energy of H and can thus be forced to lose an electron.

57. Group 6A: ns^2np^4 Group 7A: ns^2np^5
The electron affinity of the group 7A elements are more negative than the group 6A elements in the same period because group 7A requires only one electron to achieve the noble gas configuration ns^2np^6 while the group 6A elements require 2 electrons. Adding one electron to the group 6A element will not give them any added stability and leads to extra electron-electron repulsions so the value of the electron affinity is less negative than that for group 7A.

59. 35 = Br = $[Ar]4s^23d^{10}4p^5$ 53 = I = $[Kr]5s^24d^{10}5p^6$
Br and I are both halogens with an outermost electron configuration of ns^2np^5, the next element with the same outermost electron configuration is 85, At.

61. a) Using Excel, make a table of radius, atomic number, and density. Using xy scatter, make a chart of radius vs. density. With an exponential trendline, estimate the density of argon and xenon. Also, make a chart or atomic number vs. density. With a linear trendline, estimate the density of argon and xenon.

element	radius(pm)	atomic number	density
He	32	2	0.18
Ne	70	10	0.90
Ar	98	18	
Kr	112	36	3.75
Xe	130	54	
Rn		86	9.73
		118	

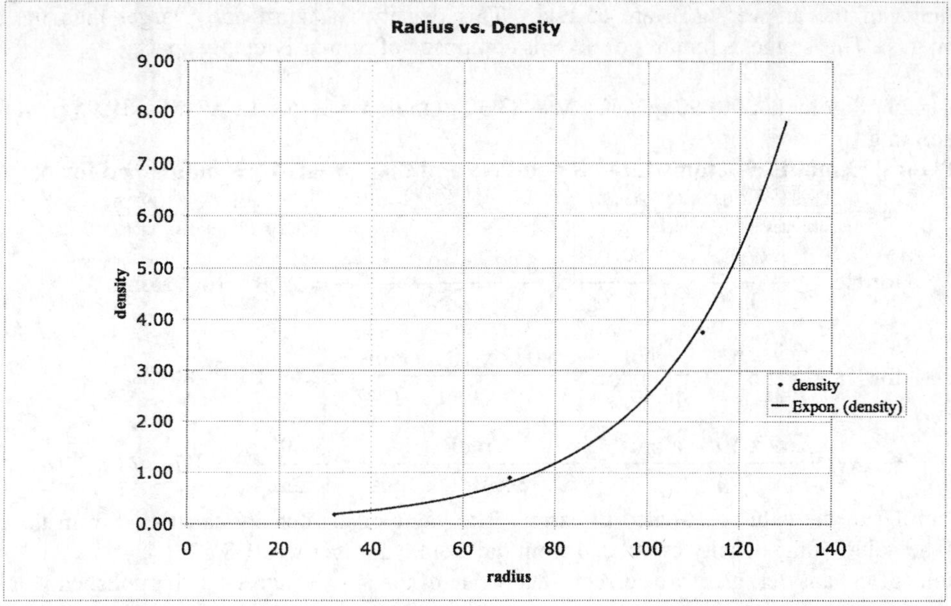

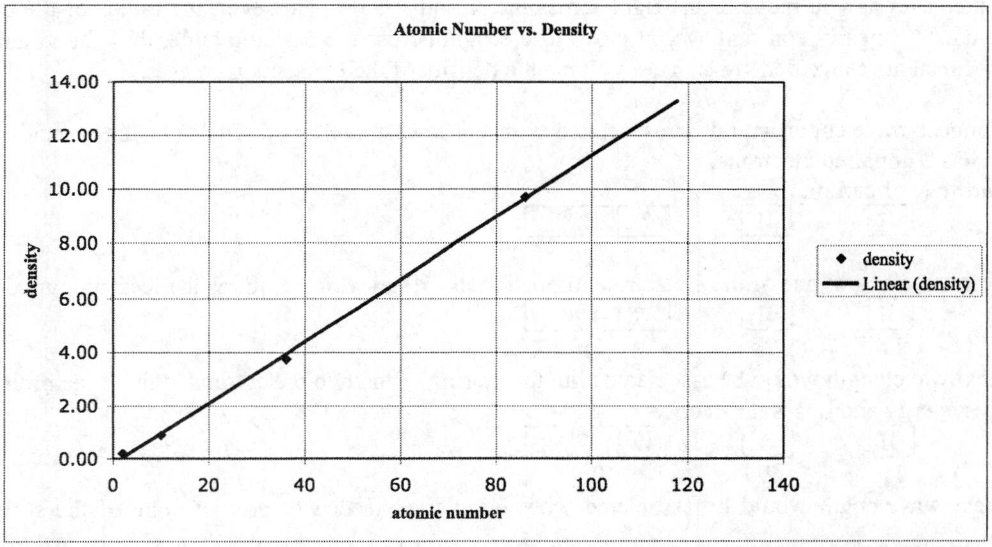

From the radius vs. density chart: Ar has a density of ~ 2 g/L and Xe has a density of ~ 7.7 g/L. From the atomic number vs. density chart: Ar has a density of ~1.8 g/L and Xe has a density of ~6 g/L.

b) Using the chart of atomic number vs. density, element 118 would be predicted to have a density of ~ 13 g/L.

c) **Given:** Ne: M = 20.18 g/mol; r = 70 pm **Find:** mass of neon; d neon

Conceptual Plan: M → m_{atom} and then r → vol_{atom} and then → d

$$\frac{6.022 \times 10^{23} \text{ atoms}}{\text{mol}} \qquad V = \frac{4}{3}\pi r^3 \qquad d = \frac{\text{mass}}{\text{vol}}$$

Solution: $\frac{20.18 \text{ g}}{\cancel{\text{mol}}} \times \frac{\cancel{\text{mol}}}{6.022 \times 10^{23} \text{ atoms}} = 3.35 \times 10^{-23} \text{ g/atom}$

$$V = \frac{4}{3} \times 3.14 \times (70 \cancel{\text{pm}})^3 \times \left(\frac{\cancel{\text{m}}}{10^{12} \cancel{\text{pm}}}\right)^3 \times \frac{\text{L}}{0.0010 \cancel{\text{m}^3}} = 1.\underline{4}4 \times 10^{-27} \text{ L}$$

$$d = \frac{3.35 \times 10^{-23} \text{ g}}{1.\underline{4}4 \times 10^{-27} \text{ L}} = 2.\underline{3}3 \times 10^4 = 2.3 \times 10^4 \text{ g/L}$$

Check: The units of the answer, g/L, are correct. This density is significantly larger than the actual density of neon gas. This suggests that a L of neon is composed of primarily empty space.

d) **Given:** Ne: M = 20.18 g/ mol, d = 0.90 g/L; Kr: M = 83.30 g/ mol, d = 3.75 g/L; Ar: M = 39.95 g/ mol
Find: d of argon in g/L
Conceptual Plan: d → mol/L → atoms/L for Kr and Ne and then atoms/L → mol/L → d for Ar

$$\text{mol} = \frac{\text{mass}}{\text{molar mass}} \qquad \frac{6.022 \times 10^{23}\ \text{atoms}}{\text{mol}} \qquad \frac{\text{mol}}{6.022 \times 10^{23}\ \text{atoms}} \qquad \frac{39.95\ \text{g}}{\text{mol}}$$

Solution: for Ne: $$\frac{0.90\ \cancel{\text{g}}}{\text{L}} \times \frac{\cancel{\text{mol}}}{20.18\ \cancel{\text{g}}} \times \frac{6.022 \times 10^{23}\ \text{atoms}}{\cancel{\text{mol}}} = 2.69 \times 10^{22}\ \text{atoms/ L}$$

for Kr: $$\frac{3.75\ \cancel{\text{g}}}{\text{L}} \times \frac{\cancel{\text{mol}}}{83.80\ \cancel{\text{g}}} \times \frac{6.022 \times 10^{23}\ \text{atoms}}{\cancel{\text{mol}}} = 2.69 \times 10^{22}\ \text{atoms/ L}$$

for Ar: $$\frac{2.69 \times 10^{22}\ \cancel{\text{atoms}}}{\text{L}} \times \frac{\cancel{\text{mol}}}{6.022 \times 10^{23}\ \cancel{\text{atoms}}} \times \frac{39.95\ \text{g}}{\cancel{\text{mol}}} = 1.78\ \text{g/ L}$$

This value is similar to the value calculated in part a. The value of the density calculated from the radius was 2 g/L and the value of the density calculated from the atomic number was 1.8 g/L.
Check: The units of the answer, g/L , are correct. The value of the answer agrees with published value.

63. The density increases as you move to the right across the first transition series. For the first transition series, the mass increases as you move to the right across the periodic table. However, the radius of the transition series elements stays nearly constant as you move to the right across the periodic table, thus the volume will remain nearly constant. Since density is mass/ volume the density of the elements increases.

65. The longest wavelength would be associated with the lowest energy state next to the ground state of carbon which has 2 unpaired electrons:
Ground state of carbon:

[↓↑] 1*s* [↓↑] 2*s* [↑ | ↑ |] 2*p*

Longest wavelength: one of the *p* electrons flipped in its orbital which requires the least amount of energy.

[↓↑] 1*s* [↓↑] 2*s* [↑ | ↓ |] 2*p*

The next wavelength would be associated with the pairing of the two *p* electrons in the same orbital because this requires energy and raises the energy.

[↓↑] 1*s* [↓↑] 2*s* [↓↑ | |] 2*p*

The next wavelength would be associated with the energy needed to promote one of the *s* electrons to a *p* orbital.

[↓↑] 1*s* [↑] 2*s* [↑ | ↑ | ↑] 2*p*

67. The element that would fill the 8*s* and 8*p* orbitals would have atomic number 168. The element is in the noble gas family and would have the properties of noble gases. It would have the electron configuration of $[118]8s^2 5g^{18} 6f^{14} 7d^{10} 8p^6$. The outer shell electron (highest *n* level) configuration would be $8s^2 8p^6$. The element would be relatively inert, have a first ionization energy less than 1037 kJ/mol (the first ionization energy of Rn), and have a positive electron affinity. It would be difficult to form compounds with other elements but would be able to form compounds with fluorine.

69. When you move down the column from Al to Ga, the size of the atom actually decreases because not much shielding is contributed by the 3d electrons in the Ga atom while there is a large increase in the nuclear charge, therefore, the effective nuclear charge is greater for Ga than for Al, so the ionization energy does not decrease. As you go from In to Tl, the ionization energy actually increases because the 4f electrons do not contribute to the shielding of the outermost electrons and there is a large increase in the effective nuclear charge.

71. The second electron is added to an ion with a 1 – charge, so there is a large repulsive force that has to be overcome to add the second electron. Thus, it will require energy to add the second electron and the second electron affinity will have a positive value.

73. If six electrons rather than eight electrons led to a stable configuration, the electron configuration of the stable configuration would be: ns^2np^4

a) A noble gas would have the electron configuration ns^2np^4. This could correspond to the O atom.

b) A reactive nonmetal would have one less electron than the stable configuration. This would have the electron configuration ns^2np^3. This could correspond to the N atom

c) A reactive metal would have one have one more electron than the stable configuration. This would have the electron configuration of ns^1. This could correspond to the Li atom.

75. An electron in a 5*p* orbital could have any one of the following combinations of quantum numbers.

5,1, -1,+1/2	5,1,-1,-1/2	5,1,0,+1/2	5,1,0,-1/2	5,1,1,+1/2	5,1,1,-1/2

An electron in a 6*d* orbital could have any one of the following combinations of quantum numbers.

6,2,-2,+1/2	6,2,-2,-1/2	6,2,-1,+1/2	6,2,-1,-1/2	6,2,0,+1/2	6,2,0,-1/2
6,2,1,+1/2	6,2,1,-1/2	6,2,2,+1/2	6,2,2,-1/2		

Chemical Bonding I: Lewis Theory

1. N : $1s^22s^22p^3$  The electrons included in the Lewis structure are: $2s^22p^3$.

3. a) Al: $1s^22s^22p^63s^23p^1$

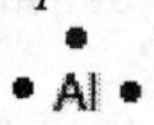

b) Na^+: $1s^22s^22p^6$

c) Cl: $1s^22s^22p^63s^23p^5$

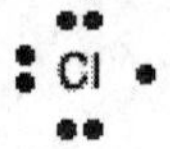

d) Cl^-: $1s^22s^22p^63s^23p^6$

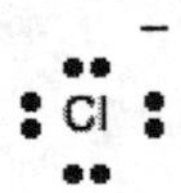

5. a) NaF: Draw the Lewis structures for Na and F based on their valence electrons. Na: $3s^1$ F: $2s^22p^5$

Sodium must lose one electron and be left with the octet from the previous shell, while fluorine needs to gain one electron to get an octet.

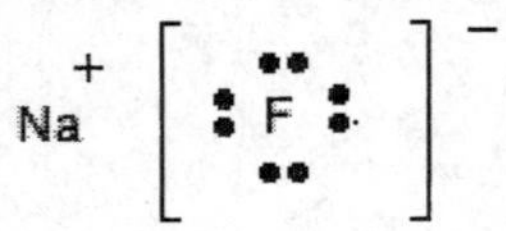

b) CaO: Draw the Lewis structures for Ca and O based on their valence electrons. Ca: $4s^2$ O: $2s^22p^4$

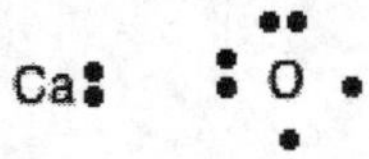

Calcium must lose two electrons and be left with the octet from the previous shell, while oxygen needs to gain two electrons to get an octet.

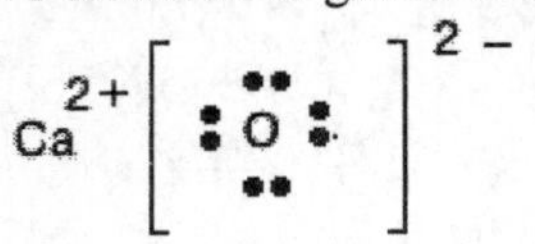

c) $SrBr_2$: Draw the Lewis structures for Sr and Br based on their valence electrons. Sr: $5s^2$ Br: $4s^24p^5$

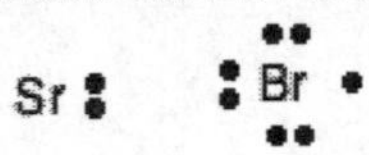

Strontium must lose two electrons and be left with the octet from the previous shell, while bromine needs to gain one electron to get an octet.

$$Sr^{2+} \quad 2 \left[\begin{array}{c} \bullet\bullet \\ \bullet\, Br \,\bullet \\ \bullet\bullet \end{array} \right]^{-}$$

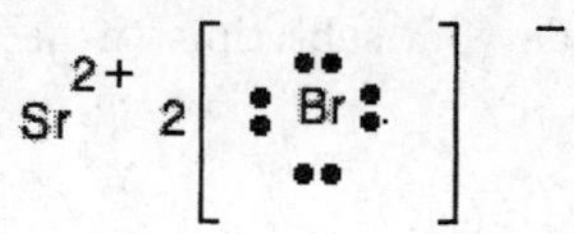

d) K_2O: Draw the Lewis structures for K and O based on their valence electrons. K: $4s^1$ O: $2s^2 2p^4$

$$K \bullet \qquad \overset{\bullet\bullet}{\underset{\bullet}{\,:\, O \,\bullet}}$$

Potassium must lose one electron and be left with the octet from the previous shell, while oxygen needs to gain two electrons to get an octet.

$$2K \quad^{+} \left[\begin{array}{c} \bullet\bullet \\ \bullet\, O \,\bullet \\ \bullet\bullet \end{array} \right]^{2-}$$

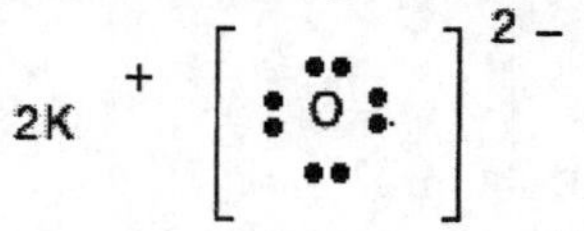

7. a) Sr and Se: Draw the Lewis structures for Sr and Se based on their valence electrons.
   Sr: $5s^2$ Se: $4s^2 4p^4$

$$Sr\overset{\bullet}{\underset{\bullet}{\,}} \qquad \overset{\bullet\bullet}{\underset{\bullet}{\,:\, Se \,\bullet}}$$

Strontium must lose two electrons and be left with the octet from the previous shell, while selenium needs to gain two electrons to get an octet.

$$Sr^{2+} \left[\begin{array}{c} \bullet\bullet \\ \bullet\, Se \,\bullet \\ \bullet\bullet \end{array} \right]^{2-}$$

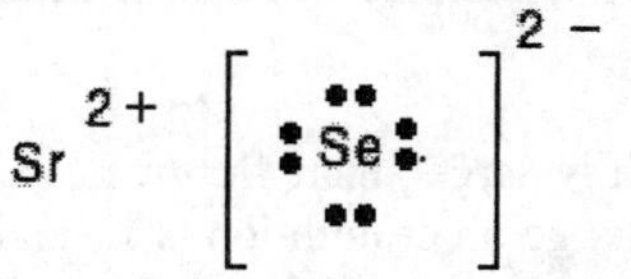

Thus, we need one Sr^{2+} and one Se^{2-}. Write the formula with subscripts (if necessary) to indicate the number of atoms.
   SrSe

b) Ba and Cl Draw the Lewis structures for Ba and Cl based on their valence electrons.
   Ba: $6s^2$ Cl: $3s^2 3p^5$

$$Ba\overset{\bullet}{\underset{\bullet}{\,}} \qquad \overset{\bullet\bullet}{\underset{\bullet\bullet}{\,:\, Cl \,\bullet}}$$

Barium must lose two electrons and be left with the octet from the previous shell, while chlorine needs to gain one electron to get an octet.

$$Ba^{2+} \quad 2 \left[\begin{array}{c} \bullet\bullet \\ \bullet\, Cl \,\bullet \\ \bullet\bullet \end{array} \right]^{-}$$

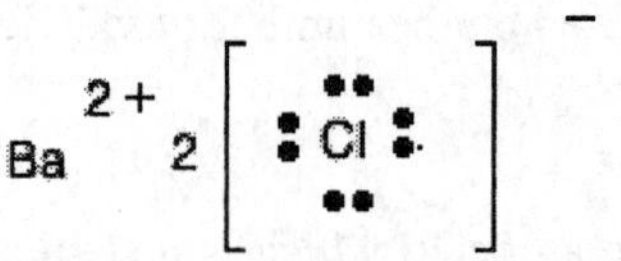

Thus, we need one Ba^{2+} and two Cl^{-}. Write the formula with subscripts (if necessary) to indicate the number of atoms.
   $BaCl_2$

c) Na and S Draw the Lewis structures for Na and S based on their valence electrons.
   Na: $3s^1$ S: $3s^2 3p^4$

$$Na \bullet \qquad \overset{\bullet\bullet}{\underset{\bullet}{\,:\, S \,\bullet}}$$

Sodium must lose one electron and be left with the octet from the previous shell, while sulfur needs to gain two electrons to get an octet.

$$2\,Na \quad^{+} \left[\begin{array}{c} \bullet\bullet \\ \bullet\, S \,\bullet \\ \bullet\bullet \end{array} \right]^{2-}$$

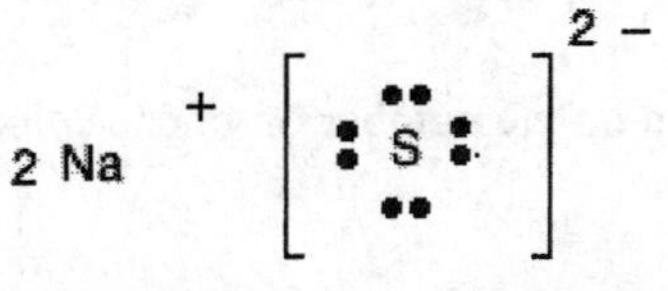

Thus, we need two Na^+ and one S^{2-}. Write the formula with subscripts (if necessary) to indicate the number of atoms.

Na_2S

d) Al and O Draw the Lewis structures for Al and O based on their valence electrons.

Al: $3s^2 3p^1$ O: $2s^2 2p^4$

Aluminum must lose three electrons and be left with the octet from the previous shell, while oxygen needs to gain two electrons to get an octet.

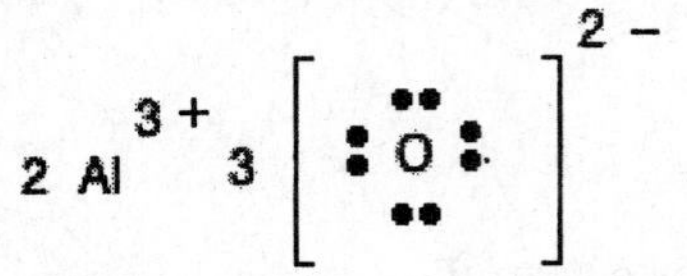

Thus, we need two Al^{3+} and three O^{2-} in order to lose and gain the same number of electrons. Write the formula with subscripts (if necessary) to indicate the number of atoms.

Al_2O_3

9. As the size of the alkaline metal ions increases down the column, so does the distance between the metal cation and the oxide anion. Therefore, the magnitude of the lattice energy of the oxides decreases, making the formation of the oxides less exothermic and the compounds less stable. Since the ions cannot get as close to each other, they therefore do not release as much energy.

11. Cesium is slightly larger than barium, but oxygen is slightly larger than fluorine, so we cannot use size to explain the difference in the lattice energy. However, the charge on cesium ion is 1+ and the charge on fluoride ion is 1 –, while the charge on barium ion is 2+ and the charge on oxide ion is 2 –. The coulombic equation states that the magnitude of the potential also depends on the product of the charges. Since the product of the charges for CsF = 1 –, and the product of the charges for BaO = 4 –, the stabilization for BaO relative to CsF should be about four times greater, which is what we see in its much more exothermic lattice energy.

13. a) Hydrogen: Write the Lewis structure of each atom based on the number of valence electrons.

When the two hydrogen atoms share their electrons, they each get a duet, which is a stable configuration for hydrogen

b) The halogens: Write the Lewis structure of each atom based on the number of valence electrons.

If the two halogens pair together they can each achieve an octet, which is a stable configuration. So, the halogens are predicted to exist as diatomic molecules.

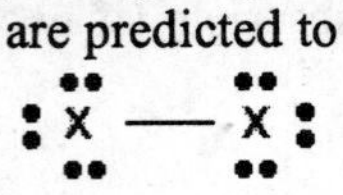

c.) Oxygen: Write the Lewis structure of each atom based on the number of valence electrons.

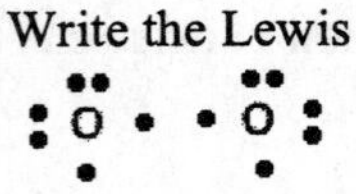

In order to achieve a stable octet on each oxygen, the oxygen atoms will need to share two electron pairs. So, oxygen is predicted to exist as a diatomic molecule with a double bond.

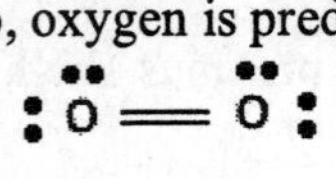

d) Nitrogen: Write the Lewis structure of each atom based on the number of valence electrons.

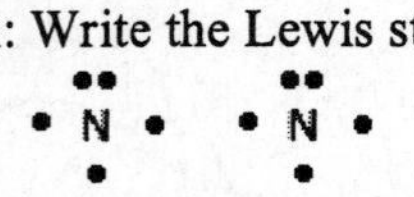

In order to achieve a stable octet on each nitrogen, the nitrogen atoms will need to share three electron pairs. So, nitrogen is predicted to exist as a diatomic molecule with a triple bond.

15. a) PH_3: Write the Lewis structure for each atom based on the valence electron.

The phosphorus will share an electron pair with each hydrogen in order to achieve a stable octet.

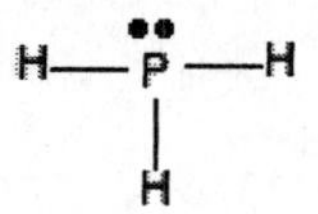

b) SCl_2: Write the Lewis structure for each atom based on the valence electron.

The sulfur will share an electron pair with each chlorine in order to achieve a stable octet.

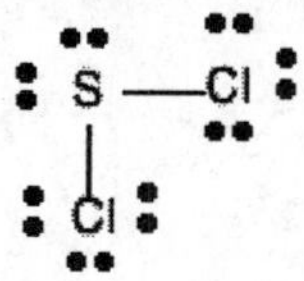

c) HI: Write the Lewis structure for each atom based on the valence electron.

The iodine will share an electron pair with hydrogen in order to achieve a stable octet.

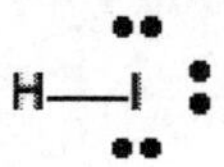

d) CH_4: Write the Lewis structure for each atom based on the valence electron.

The carbon will share an electron pair with each hydrogen in order to achieve a stable octet.

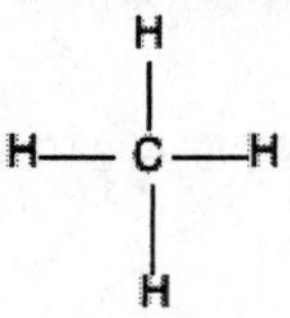

17. a) Br and Br: pure covalent From Figure 9.9 we find the electronegativity of Br is 2.5. Since both atoms are the same, the electronegativity difference (ΔEN) = 0, and using Table 9.1 we classify this bond as pure covalent.

b) C and Cl: polar covalent From Figure 9.9 we find the electronegativity of C is 2.5 and Cl is 3.0. The electronegativity difference (ΔEN) is $\Delta EN = 3.0 - 2.5 = 0.5$. Using Table 9.1 we classify this bond as polar covalent.

c) C and S: pure covalent From Figure 9.9 we find the electronegativity of C is 2.5 and S is 2.5. The electronegativity difference (ΔEN) is $\Delta EN = 2.5 - 2.5 = 0$. Using Table 9.1 we classify this bond as pure covalent.

d) Sr and O: ionic From Figure 9.9 we find the electronegativity of Sr is 1.0 and O is 3.5. The electronegativity difference (ΔEN) is $\Delta EN = 3.5 - 1.0 = 2.5$. Using Table 9.1 we classify this bond as ionic.

19. CO: Write the Lewis structure for each atom based on the valence electron.

The carbon will share three electron pairS with oxygen in order to achieve a stable octet.
The oxygen atom is more electronegative than the carbon atom, so the oxygen will have a partial negative charge and the carbon will have a partial positive charge.

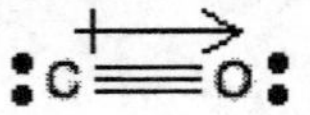

To estimate the percent ionic character, determine the difference in electronegativity between carbon and oxygen.
From figure 9.9 we find the electronegativity of C is 2.5 and O is 3.5. The electronegativity difference (ΔEN) is $\Delta EN = 3.5 - 2.5 = 1.0$.
From figure 9.11, we can estimate a percent ionic character of 25%

21. a.) CI_4: Write the correct skeletal structure for the molecule.

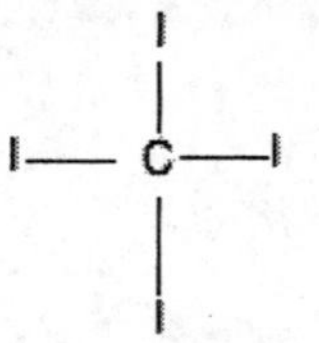

Calculate the total number of electrons for the Lewis structure by summing the valence electron of each atom in the molecule.

(number of valence e^- for C) + 4(number of valence e^- for I) = 4 + 4(7) = 32

Distribute the electrons among the atoms, giving octets (or duets for H) to as many atoms as possible. Begin with the bonding electrons, and then proceed to lone pairs on terminal atoms and finally to lone pairs of the central atom.

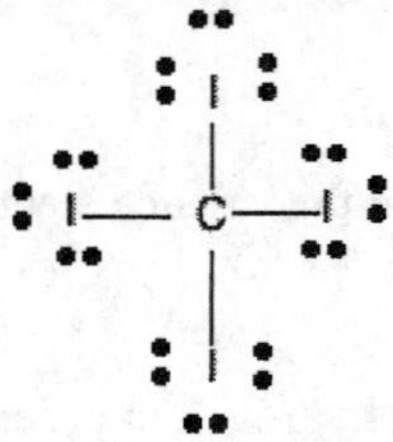

All 32 valence electrons are used

If any atom lacks an octet, form double or triple bonds as necessary to give them octets.
All atoms have octets, duets for H, structure is complete.

b) N_2O: Write the correct skeletal structure for the molecule

N is the less electronegative, so it is central

Calculate the total number of electrons for the Lewis structure by summing the valence electron of each atom in the molecule

2(number of valence e^- for N) + (number of valence e^- for O) = 2(5) + 6 = 16

Distribute the electrons among the atoms, giving octets (or duets for H) to as many atoms as possible. Begin with the bonding electrons, and then proceed to lone pairs on terminal atoms and finally to lone pairs of the central atom.

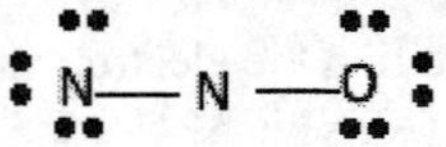

All 16 valence electrons are used

If any atom lacks an octet, form double or triple bonds as necessary to give them octets.

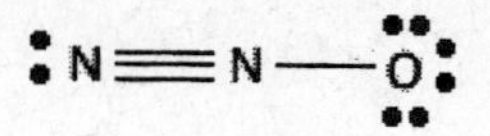

All atoms have octets, duets for H, structure is complete

c) SiH_4: Write the correct skeletal structure for the molecule
H is always terminal so Si is the central atom

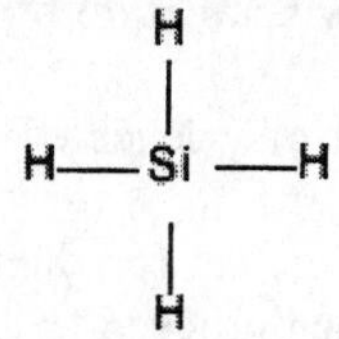

Calculate the total number of electrons for the Lewis structure by summing the valence electrons of each atom in the molecule

(number of valence e^- for Si) + 4(number of valence e^- for H) = 4 + 4(1) = 8

Distribute the electrons among the atoms, giving octets (or duets for H) to as many atoms as possible. Begin with the bonding electrons, and then proceed to lone pairs on terminal atoms and finally to lone pairs of the central atom.

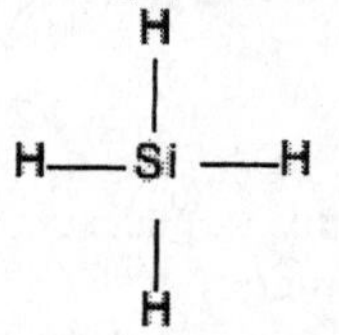

All 8 valence electrons are used.

If any atom lacks an octet, form double or triple bonds as necessary to give them octets.
All atoms have octets, duets for H, structure is complete.

d) Cl_2CO: Write the correct skeletal structure for the molecule
C is the least electronegative, so it is the central atom

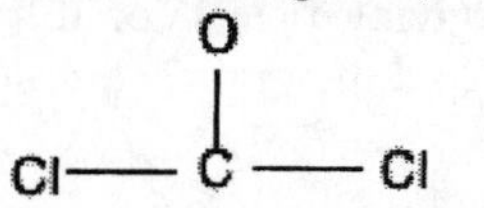

Calculate the total number of electrons for the Lewis structure by summing the valence electrons of each atom in the molecule

(number of valence e^- for C) + 2(number of valence e^- for Cl) + (number of valence e^- for O) = 4 + 2(7) + 6 = 24

Distribute the electrons among the atoms, giving octets (or duets for H) to as many atoms as possible. Begin with the bonding electrons, and then proceed to lone pairs on terminal atoms and finally to lone pairs of the central atom.

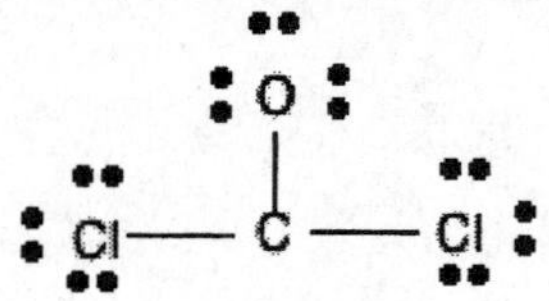

All 24 valence electrons are used.

If any atom lacks an octet, form double or triple bonds as necessary to give them octets.

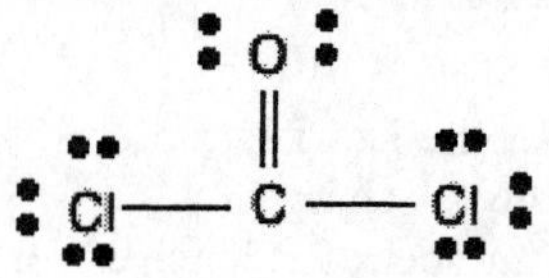

All atoms have octets, duets for H, structure is complete.

e) H_3COH: Write the correct skeletal structure for the molecule
C is less electronegative and H is terminal

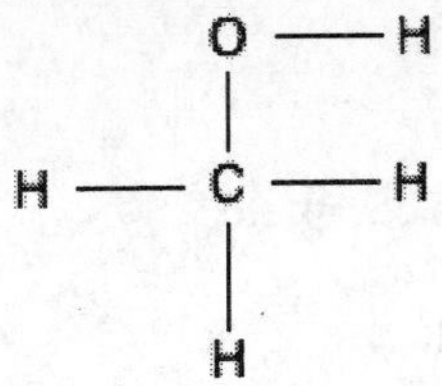

Calculate the total number of electrons for the Lewis structure by summing the valence electrons of each atom in the molecule

(number of valence e^- for C) + 4(number of valence e^- for H) + (number of valence e^- for O) = $4 + 4(1) + 6 = 14$

Distribute the electrons among the atoms, giving octets (or duets for H) to as many atoms as possible. Begin with the bonding electrons, and then proceed to lone pairs on terminal atoms and finally to lone pairs of the central atom.

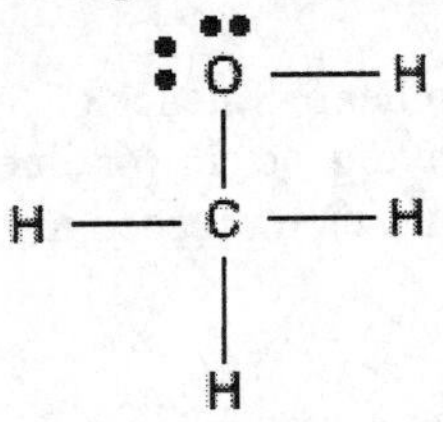

All 14 valence electrons are used.

If any atom lacks an octet, form double or triple bonds as necessary to give them octets.

All atoms have octets, duets for H, structure is complete.

f) OH^-: Write the correct skeletal structure for the ion

Calculate the total number of electrons for the Lewis structure by summing the valence electrons of each atom in the ion and adding 1 for the 1 – charge.

(number of valence e^- for O) + (number of valence e^- for H) + 1 = $6 + 1 + 1 = 8$

Distribute the electrons among the atoms, giving octets (or duets for H) to as many atoms as possible. Begin with the bonding electrons, and then proceed to lone pairs on terminal atoms and finally to lone pairs of the central atom.

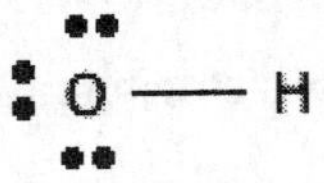

All 8 valence electrons are used.

If any atom lacks an octet, form double or triple bonds as necessary to give them octets.

Lastly, write the Lewis structure in brackets with the charge of the ion in the upper right-hand corner.

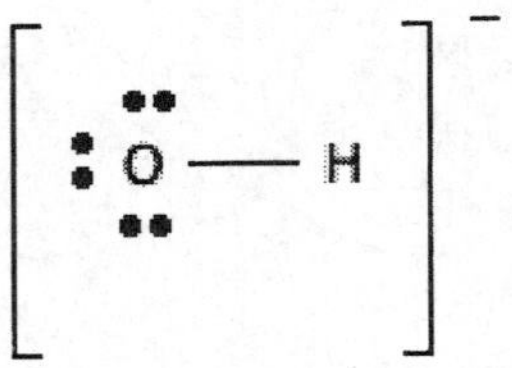

g) BrO^-: Write the correct skeletal structure for the ion

Calculate the total number of electrons for the Lewis structure by summing the valence electrons of each atom in the ion and adding 1 for the 1 – charge.

(number of valence e^- for O) + (number of valence e^- for Br) + 1 = $6 + 7 + 1 = 14$

Distribute the electrons among the atoms, giving octets (or duets for H) to as many atoms as possible. Begin with the bonding electrons, and then proceed to lone pairs on terminal atoms and finally to lone pairs of the central atom.

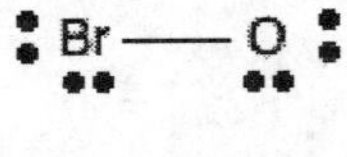

All 14 valence electrons are used.

If any atom lacks an octet, form double or triple bonds as necessary to give them octets.

Lastly, write the Lewis structure in brackets with the charge of the ion in the upper right-hand corner.

$$\left[\ddot{\underset{..}{Br}} - \ddot{\underset{..}{O}} \colon \right]^{-}$$

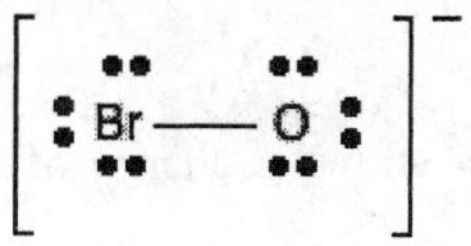

23. a) SeO_2: Write the correct skeletal structure for the molecule

Se is the less electronegative, so it is central

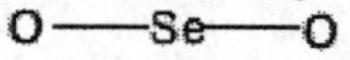

O —— Se —— O

Calculate the total number of electrons for the Lewis structure by summing the valence electrons of each atom in the molecule

(number of valence e⁻ for Se) + 2(number of valence e⁻ for O) = 6 + 2(6) = 18

Distribute the electrons among the atoms, giving octets (or duets for H) to as many atoms as possible. Begin with the bonding electrons, and then proceed to lone pairs on terminal atoms and finally to lone pairs of the central atom.

$$\colon \ddot{\underset{..}{O}} - \ddot{Se} - \ddot{\underset{..}{O}} \colon$$

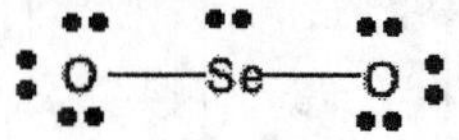

All 18 valence electrons are used.

If any atom lacks an octet, form double or triple bonds as necessary to give them octets.

$$\colon \ddot{\underset{..}{O}} - \ddot{Se} = \ddot{O} \colon$$

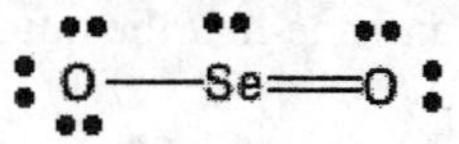

All atoms have octets, duets for H, structure is complete. However, the double bond can form from either oxygen atom, so there are two resonance forms.

$$\colon \ddot{\underset{..}{O}} - \ddot{Se} = \ddot{O} \colon \longleftrightarrow \colon \ddot{O} = \ddot{Se} - \ddot{\underset{..}{O}} \colon$$

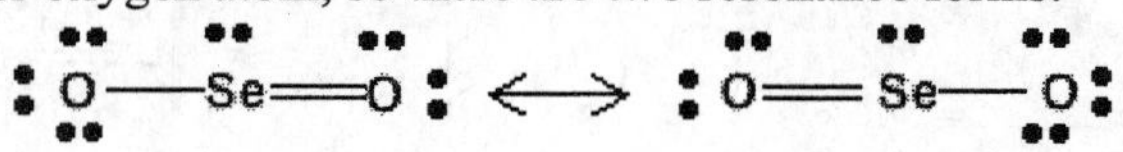

Calculate the formal charge on each atom by finding the number of valence electrons and subtracting the number of lone pair electrons and one-half the number of bonding electrons.

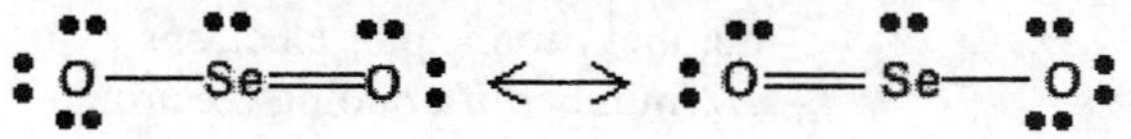

$$\colon \ddot{\underset{..}{O}} - \ddot{Se} = \ddot{O} \colon \longleftrightarrow \colon \ddot{O} = \ddot{Se} - \ddot{\underset{..}{O}} \colon$$

number of valence electrons	6	6	6	6	6	6
- number of lone pair electrons	6	2	4	4	2	6
- 1/2(number of bonding electrons)	1	3	2	2	3	1
Formal charge	−1	+1	0	0	+1	−1

b) $CO_3{}^{2-}$: Write the correct skeletal structure for the ion

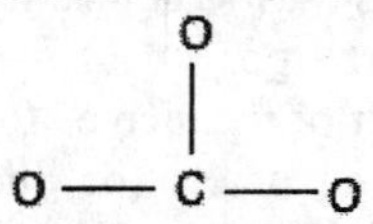

```
          O
          |
   O —— C —— O
```

Calculate the total number of electrons for the Lewis structure by summing the valence electrons of each atom in the ion and adding 2 for the 2 – charge.

3(number of valence e⁻ for O) + (number of valence e⁻ for C) + 2 = 3(6) + 4 + 2 = 24

Distribute the electrons among the atoms, giving octets (or duets for H) to as many atoms as possible. Begin with the bonding electrons, and then proceed to lone pairs on terminal atoms and finally to lone pairs of the central atom.

$$\colon \ddot{\underset{..}{O}} \colon$$
$$|$$
$$\colon \ddot{\underset{..}{O}} - C - \ddot{\underset{..}{O}} \colon$$

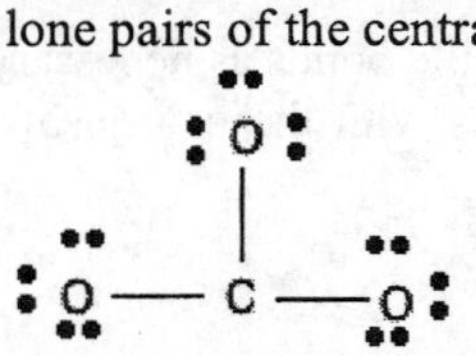

All 24 valence electrons are used.

If any atom lacks an octet, form double or triple bonds as necessary to give them octets.

$$\begin{array}{c}
\ddot{:}\ddot{O}\ddot{:}\\
|\\
\ddot{:}\underset{..}{\ddot{O}}-C=\ddot{O}\ddot{:}
\end{array}$$

Lastly, write the Lewis structure in brackets with the charge of the ion in the upper right-hand corner.

$$\left[\begin{array}{c}
\ddot{:}\ddot{O}\ddot{:}\\
|\\
\ddot{:}\underset{..}{\ddot{O}}-C=\ddot{O}\ddot{:}
\end{array}\right]^{2-}$$

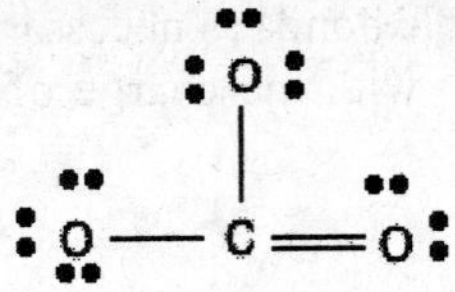

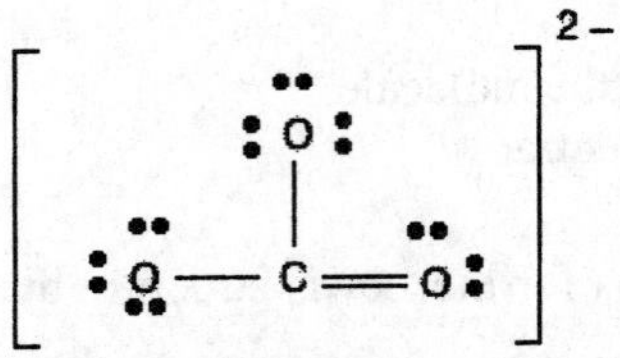

All atoms have octets, duets for H, structure is complete. However, the double bond can form from any oxygen atom, so there are three resonance forms.

$$\left[\begin{array}{c}
\ddot{:}\ddot{O}\ddot{:}\\
|\\
\ddot{:}\ddot{O}-C=\ddot{O}\ddot{:}
\end{array}\right]^{2-} \longleftrightarrow \left[\begin{array}{c}
\ddot{:}\ddot{O}\ddot{:}\\
|\\
\ddot{:}\ddot{O}=C-\ddot{O}\ddot{:}
\end{array}\right]^{2-} \longleftrightarrow \left[\begin{array}{c}
\ddot{:}\ddot{O}\ddot{:}\\
||\\
\ddot{:}\ddot{O}-C-\ddot{O}\ddot{:}
\end{array}\right]^{2-}$$

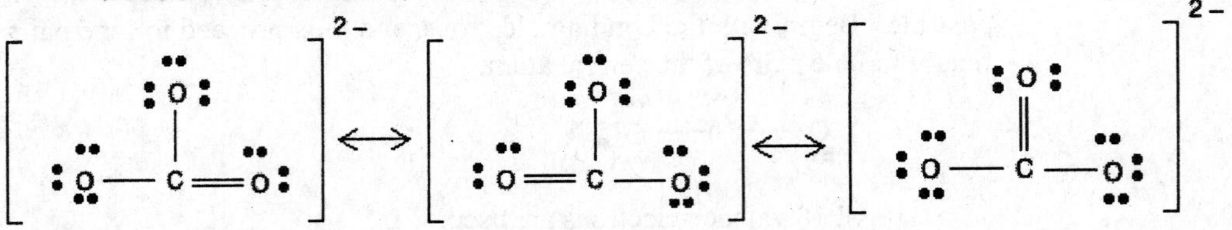

Calculate the formal charge on each atom by finding the number of valence electrons and subtracting the number of lone pair electrons and one-half the number of bonding electrons.

$$\left[\begin{array}{c}
\ddot{:}\ddot{O}\ddot{:}\\
|\\
\ddot{:}\underset{..}{\ddot{O}}-C=\ddot{O}\ddot{:}
\end{array}\right]^{2-}$$

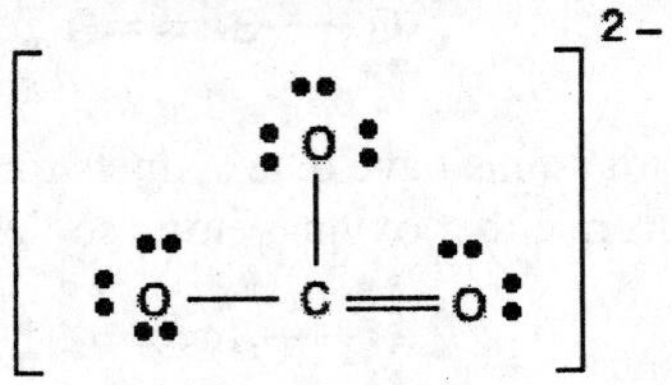

	O_{left}	O_{top}	O_{right}	C
number of valence electrons	6	6	6	4
- number of lone pair electrons	6	6	4	0
- 1/2(number of bonding electrons)	1	1	2	4
Formal charge	−1	−1	0	0

The sum of the formal charges is − 2 which is the overall charge of the ion. The other resonance forms would be the same.

c) ClO⁻ : Write the correct skeletal structure for the ion

$$Cl - O$$

Calculate the total number of electrons for the Lewis structure by summing the valence electrons of each atom in the ion and adding 1 for the 1 − charge.

(number of valence e⁻ for O) + (number of valence e⁻ for Cl) + 1 = 6 + 7 + 1 = 14

Distribute the electrons among the atoms, giving octets (or duets for H) to as many atoms as possible. Begin with the bonding electrons, and then proceed to lone pairs on terminal atoms and finally to lone pairs of the central atom.

$$:\ddot{Cl}-\ddot{O}:$$

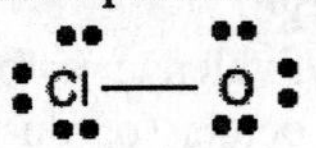

All 14 valence electrons are used.

If any atom lacks an octet, form double or triple bonds as necessary to give them octets.

Lastly, write the Lewis structure in brackets with the charge of the ion in the upper right-hand corner.

$$\left[:\ddot{Cl}-\ddot{O}:\right]^{-}$$

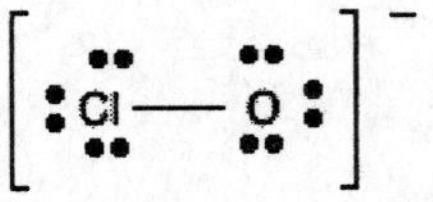

All atoms have octets, duets for H, structure is complete.

Calculate the formal charge on each atom by finding the number of valence electrons and subtracting the number of lone pair electrons and one-half the number of bonding electrons.

	Cl	O
number of valence electrons	7	6
- number of lone pair electrons	6	6
- 1/2(number of bonding electrons)	1	1
Formal charge	0	–1

The sum of the formal charges is – 1 which is the overall charge of the ion.

d) NO_2^-: Write the correct skeletal structure for the ion

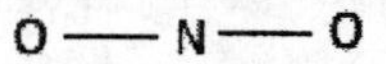

Calculate the total number of electrons for the Lewis structure by summing the valence electrons of each atom in the ion and adding 1 for the 1 – charge.

2(number of valence e^- for O) + (number of valence e^- for N) + 1 = 2(6) + 5 + 1 = 18

Distribute the electrons among the atoms, giving octets (or duets for H) to as many atoms as possible. Begin with the bonding electrons, and then proceed to lone pairs on terminal atoms and finally to lone pairs of the central atom.

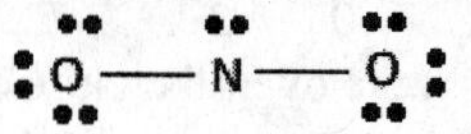

All 14 valence electrons are used.

If any atom lacks an octet, form double or triple bonds as necessary to give them octets.

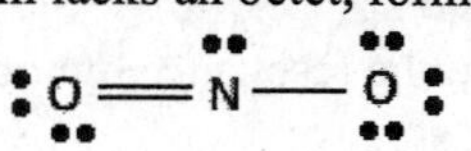

Lastly, write the Lewis structure in brackets with the charge of the ion in the upper right-hand corner.

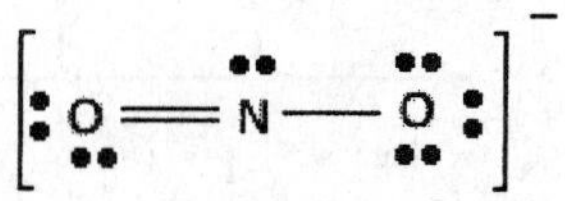

All atoms have octets, duets for H, structure is complete. However, the double bond can form from either oxygen atom, so there are two resonance forms.

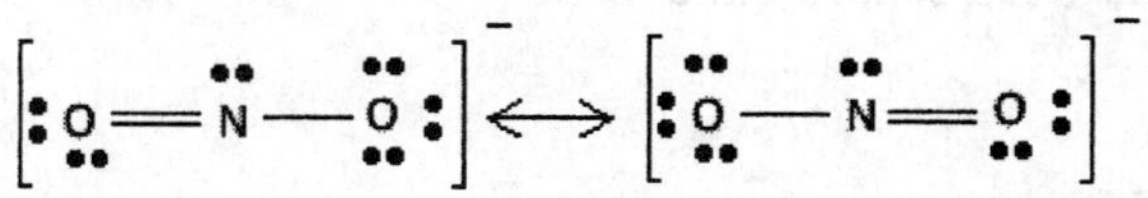

Calculate the formal charge on each atom by finding the number of valence electrons and subtracting the number of lone pair electrons and one-half the number of bonding electrons. Using the left side structure:

	O	N	O
number of valence electrons	6	5	6
- number of lone pair electrons	4	2	6
- 1/2(number of bonding electrons)	2	3	1
Formal charge	0	0	– 1

The sum of the formal charges is – 1 which is the overall charge of the ion.

25.

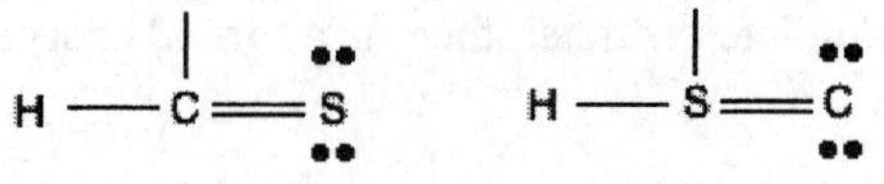

I II

Calculate the formal charge on each atom in structure I by finding the number of valence electrons and subtracting the number of lone pair electrons and one-half the number of bonding electrons.

	H_{left}	H_{top}	C	S
number of valence electrons	1	1	4	6
- number of lone pair electrons	0	0	0	4
- 1/2(number of bonding electrons)	1	1	4	2
Formal charge	0	0	0	0

The sum of the formal charges is 0 which is the overall charge of the molecule.
Calculate the formal charge on each atom in structure II by finding the number of valence electrons and subtracting the number of lone pair electrons and one-half the number of bonding electrons

	H_{left}	H_{top}	S	C
number of valence electrons	1	1	6	4
- number of lone pair electrons	0	0	0	4
- 1/2(number of bonding electrons)	1	1	4	2
Formal charge	0	0	+2	−2

The sum of the formal charges is 0 which is the overall charge of the molecule.
Structure I is the better Lewis structure because it has the least amount of formal charge on each atom.

27. 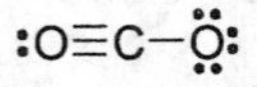 does not provide a significant contribution to the resonance hybrid as it has a +1 formal charge on a very electronegative oxygen.

	O_{left}	O_{right}	C
number of valence electrons	6	6	4
- number of lone pair electrons	2	6	0
- 1/2(number of bonding electrons)	3	1	4
Formal charge	+1	−1	0

29. a) BCl_3: Write the correct skeletal structure for the molecule.
B is the less electronegative, so it is central

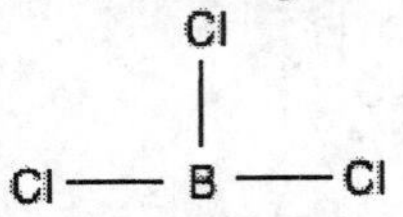

Calculate the total number of electrons for the Lewis structure by summing the valence electrons of each atom in the molecule

(number of valence e^- for B) + 3(number of valence e^- for Cl) = 3 +3(7) = 24

Distribute the electrons among the atoms, giving octets (or duets for H) to as many atoms as possible. Begin with the bonding electrons, and then proceed to lone pairs on terminal atoms and finally to lone pairs of the central atom.

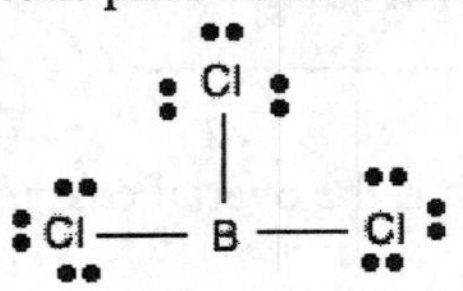

All 24 valence electrons are used.

B has an incomplete octet. If we complete the octet, there is a formal charge of – 1 on the B which is less electronegative than Cl.

b) NO_2: Write the correct skeletal structure for the molecule.
N is the less electronegative, so it is central

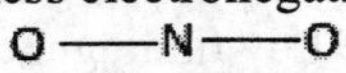

Calculate the total number of electrons for the Lewis structure by summing the valence electrons of each atom in the molecule

(number of valence e^- for N) + 2(number of valence e^- for O) = 5 +2(6) = 17

Distribute the electrons among the atoms, giving octets (or duets for H) to as many atoms as possible. Begin with the bonding electrons, and then proceed to lone pairs on terminal atoms and finally to lone pairs of the central atom.

All 17 valence electrons are used.

N has an incomplete octet. It has 7 electrons because we have an odd number of valence electrons.

c) BH_3: Write the correct skeletal structure for the molecule.

B is the less electronegative, so it is central

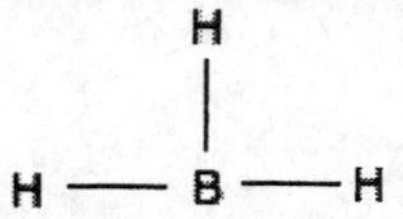

Calculate the total number of electrons for the Lewis structure by summing the valence electrons of each atom in the molecule

(number of valence e^- for B) + 3(number of valence e^- for H) = 3 +3(1) = 6

Distribute the electrons among the atoms, giving octets (or duets for H) to as many atoms as possible. Begin with the bonding electrons, and then proceed to lone pairs on terminal atoms and finally to lone pairs of the central atom.

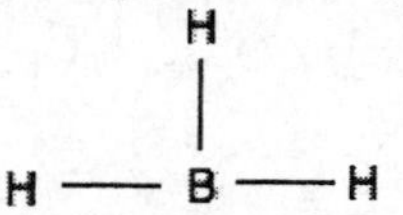

All 6 valence electrons are used.

B has an incomplete octet. H cannot double bond, so it is not possible to complete the octet on B with a double bond.

31. a) PO_4^{3-}: Write the correct skeletal structure for the ion.

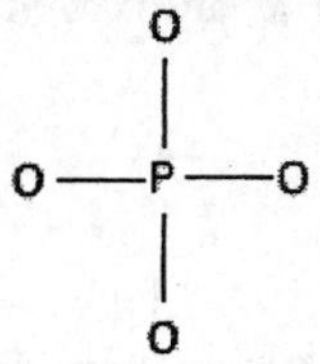

Calculate the total number of electrons for the Lewis structure by summing the valence electrons of each atom in the ion and adding 3 for the 3 – charge.

4(number of valence e^- for O) + (number of valence e^- for P) + 3 = 4(6) + 5 + 3 = 32

Distribute the electrons among the atoms, giving octets (or duets for H) to as many atoms as possible. Begin with the bonding electrons, and then proceed to lone pairs on terminal atoms and finally to lone pairs of the central atom.

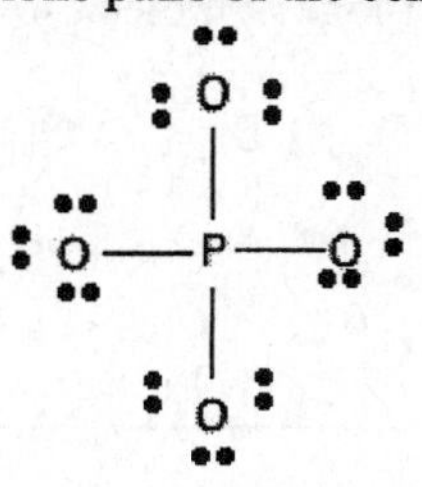

All 32 valence electrons are used.

Lastly, write the Lewis structure in brackets with the charge of the ion in the upper right-hand corner.

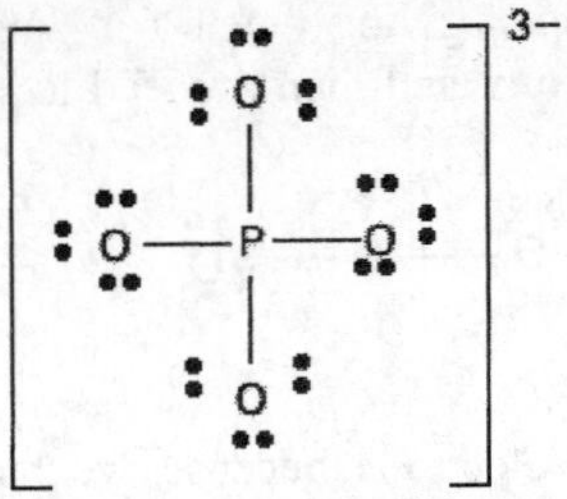

All atoms have octets, duets for H, structure is complete.
Calculate the formal charge on each atom by finding the number of valence electrons and subtracting the number of lone pair electrons and one-half the number of bonding electrons.

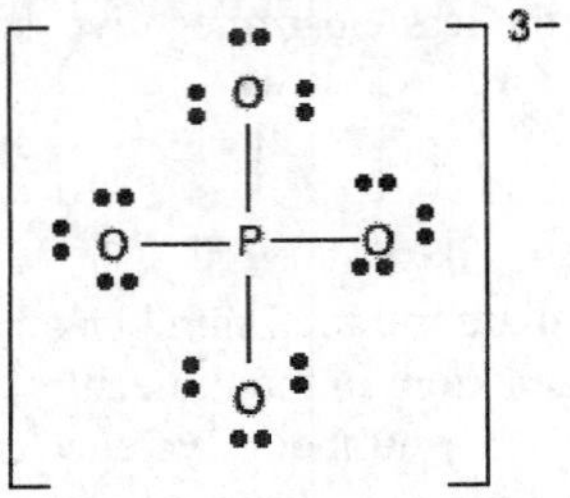

	O_{left}	O_{top}	O_{right}	O_{bottom}	P
number of valence electrons	6	6	6	6	5
- number of lone pair electrons	6	6	6	6	0
- 1/2(number of bonding electrons)	1	1	1	1	4
Formal charge	–1	–1	–1	–1	+1

The sum of the formal charges is –3, which is the overall charge of the ion. However, we can write a resonance structure with a double bond to an oxygen because P can expand its octet. This leads to lower formal charges on P and O.

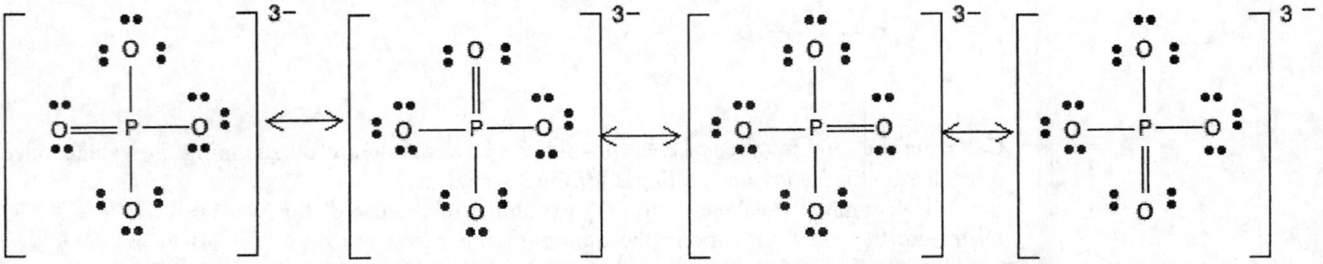

Using the leftmost structure calculate the formal charge on each atom by finding the number of valence electrons and subtracting the number of lone pair electrons and one-half the number of bonding electrons.

	O_{left}	O_{top}	O_{right}	O_{bottom}	P
number of valence electrons	6	6	6	6	5
- number of lone pair electrons	4	6	6	6	0
- 1/2(number of bonding electrons)	2	1	1	1	5
Formal charge	0	–1	–1	–1	0

The sum of the formal charges is –3, which is the overall charge of the ion. These resonance forms would all have the lower formal charges associated with the double bonded O and P.

b) CN^-: Write the correct skeletal structure for the ion.

C — N

Calculate the total number of electrons for the Lewis structure by summing the valence electrons of each atom in the ion and adding 1 for the 1 – charge.

(number of valence e^- for C) + (number of valence e^- for N) + 1 = 4 + 5 + 1 = 10

Distribute the electrons among the atoms, giving octets (or duets for H) to as many atoms as possible. Begin with the bonding electrons, and then proceed to lone pairs on terminal atoms and finally to lone pairs of the central atom.

$$:\!C\!-\!\overset{\bullet\bullet}{\underset{\bullet\bullet}{N}}\!:$$

All 10 valence electrons are used.

If any atom lacks an octet, form double or triple bonds as necessary to give them octets.

$$:\!C\!\equiv\!N\!:$$

Lastly, write the Lewis structure in brackets with the charge of the ion in the upper right-hand corner.

$$\left[:\!C\!\equiv\!N\!:\right]^{-}$$

All atoms have octets, duets for H, structure is complete.

Calculate the formal charge on each atom by finding the number of valence electrons and subtracting the number of lone pair electrons and one-half the number of bonding electrons.

$$\left[:\!C\!\equiv\!N\!:\right]^{-}$$

	C	N
number of valence electrons	4	5
- number of lone pair electrons	2	2
- 1/2(number of bonding electrons)	3	3
Formal charge	−1	0

The sum of the formal charges is − 1, which is the overall charge of the ion

c) SO_3^{2-}: Write the correct skeletal structure for the ion.

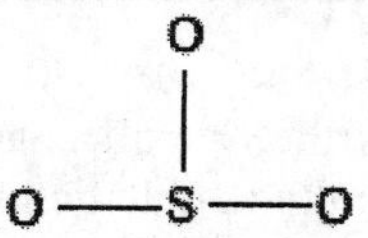

Calculate the total number of electrons for the Lewis structure by summing the valence electrons of each atom in the ion and adding 2 for the 2 – charge.

3(number of valence e⁻ for O) + (number of valence e⁻ for S) + 2 = 3(6) + 6 + 2 = 26

Distribute the electrons among the atoms, giving octets (or duets for H) to as many atoms as possible. Begin with the bonding electrons, and then proceed to lone pairs on terminal atoms and finally to lone pairs of the central atom.

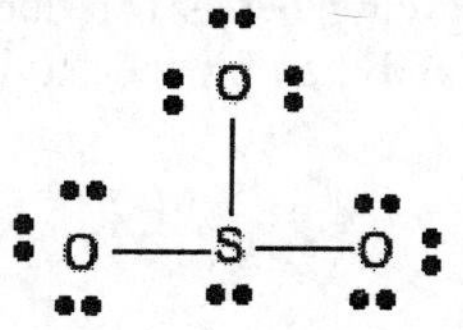

All 26 valence electrons are used.

Lastly, write the Lewis structure in brackets with the charge of the ion in the upper right-hand corner.

$$\left[\begin{array}{c}:\!\overset{\bullet\bullet}{O}\!: \\ | \\ :\!\overset{\bullet\bullet}{\underset{\bullet\bullet}{O}}\!-\!\overset{\bullet\bullet}{\underset{\bullet\bullet}{S}}\!-\!\overset{\bullet\bullet}{\underset{\bullet\bullet}{O}}\!:\end{array}\right]^{2-}$$

Calculate the formal charge on each atom by finding the number of valence electrons and subtracting the number of lone pair electrons and one-half the number of bonding electrons.

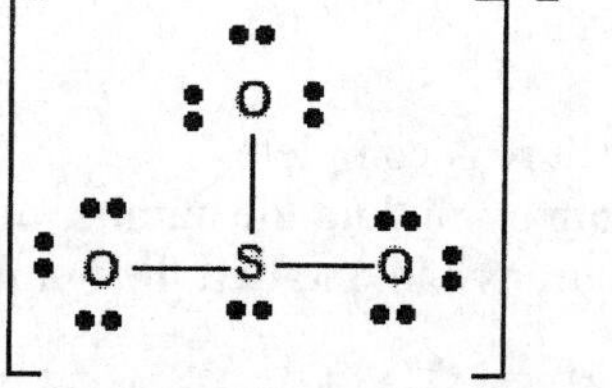

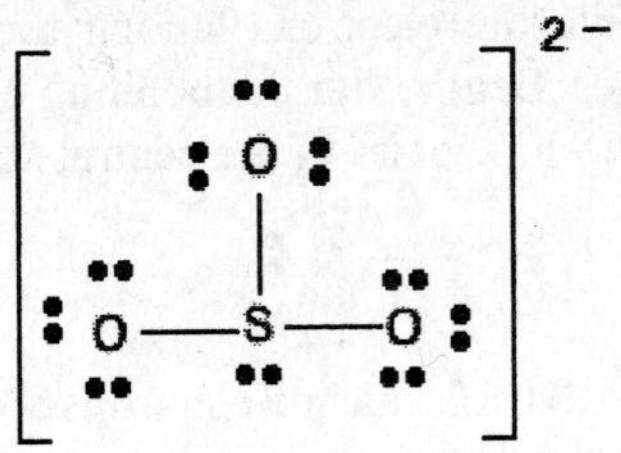

	O_{left}	O_{top}	O_{right}	S
number of valence electrons	6	6	6	6
- number of lone pair electrons	6	6	6	2
- 1/2(number of bonding electrons)	1	1	1	3
Formal charge	–1	–1	–1	+1

The sum of the formal charges is –2, which is the overall charge of the ion. However, we can write a resonance structure with a double bond to an oxygen because S can expand its octet. This leads to a lower formal charge

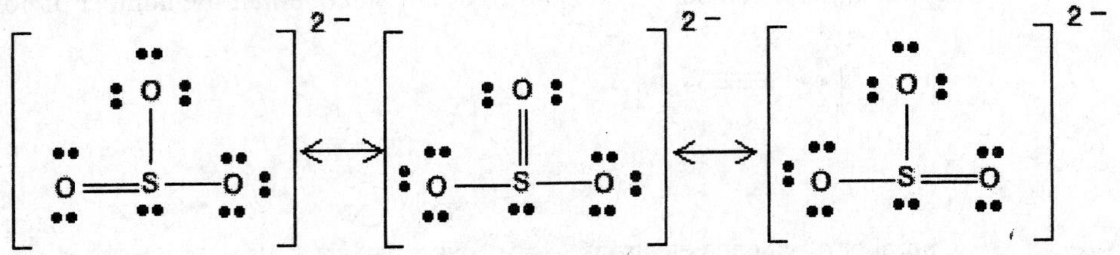

Using the leftmost resonance form: Calculate the formal charge on each atom by finding the number of valence electrons and subtracting the number of lone pair electrons and one-half the number of bonding electrons.

	O_{left}	O_{top}	O_{right}	S
number of valence electrons	6	6	6	6
- number of lone pair electrons	4	6	6	2
- 1/2(number of bonding electrons)	2	1	1	4
Formal charge	0	–1	–1	0

The sum of the formal charges is –2, which is the overall charge of the ion. These resonance forms would all have the lower formal charge on the double bonded O and S.

d) ClO_2^-: Write the correct skeletal structure for the ion.

O — Cl — O

Calculate the total number of electrons for the Lewis structure by summing the valence electrons of each atom in the ion and adding 1 for the 1 – charge.

2(number of valence e^- for O) + (number of valence e^- for Cl) + 1 = 2(6) + 7 + 1 = 20

Distribute the electrons among the atoms, giving octets (or duets for H) to as many atoms as possible. Begin with the bonding electrons, and then proceed to lone pairs on terminal atoms and finally to lone pairs of the central atom.

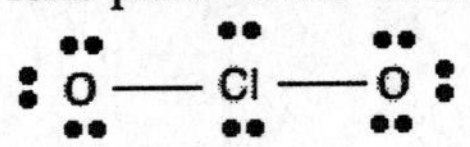

All 20 valence electrons are used.

Lastly, write the Lewis structure in brackets with the charge of the ion in the upper right-hand corner.

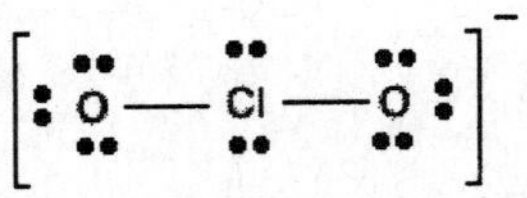

All atoms have octets, duets for H, structure is complete.

Calculate the formal charge on each atom by finding the number of valence electrons and subtracting the number of lone pair electrons and one-half the number of bonding electrons.

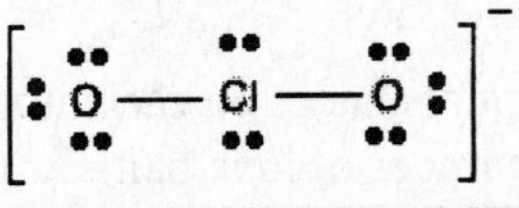

	O_{left}	O_{right}	Cl
number of valence electrons	6	6	7
- number of lone pair electrons	6	6	4
- 1/2(number of bonding electrons)	1	1	2
Formal charge	−1	−1	+1

The sum of the formal charges is −1, which is the overall charge of the ion. However, we can write a resonance structure with a double bond to an oxygen because Cl can expand its octet. This leads to a lower formal charge

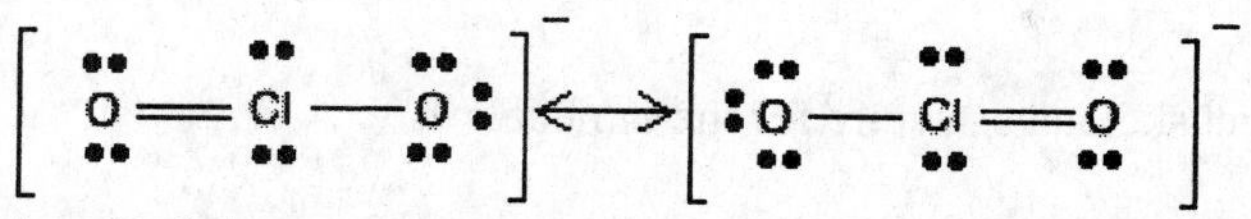

Using the leftmost resonance form: Calculate the formal charge on each atom by finding the number of valence electrons and subtracting the number of lone pair electrons and one – half the number of bonding electrons.

	O_{left}	O_{right}	Cl
number of valence electrons	6	6	7
- number of lone pair electrons	4	6	4
- 1/2(number of bonding electrons)	2	1	3
Formal charge	0	−1	0

The sum of the formal charges is −1, which is the overall charge of the ion. These resonance forms would all have the lower formal charge on the double bonded O and Cl.

33. a) PF_5: Write the correct skeletal structure for the molecule

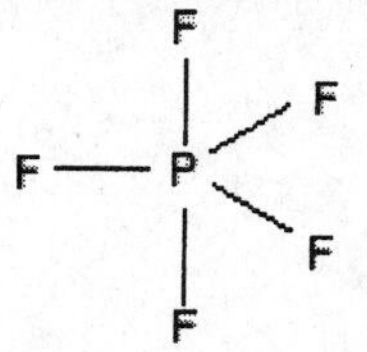

Calculate the total number of electrons for the Lewis structure by summing the valence electrons of each atom in the molecule

(number of valence e^- for P) + 5(number of valence e^- for F) = 5 +5(7) = 40

Distribute the electrons among the atoms, giving octets (or duets for H) to as many atoms as possible. Begin with the bonding electrons, and then proceed to lone pairs on terminal atoms and finally to lone pairs of the central atom. Arrange additional electrons around the central atom, giving it an expanded octet of up to 12 electrons.

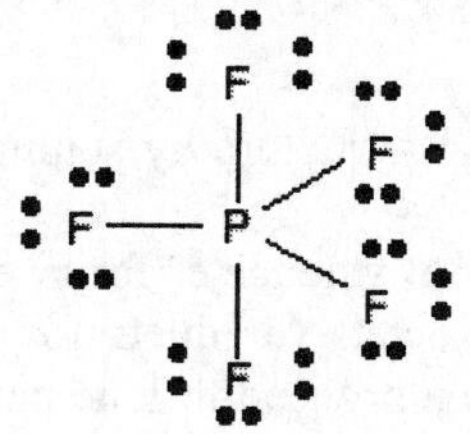

b) I_3^-: Write the correct skeletal structure for the ion

I —— I —— I

Calculate the total number of electrons for the Lewis structure by summing the valence electrons of each atom in the ion and adding 1 for the 1 – charge.

3(number of valence e^- for I) + 1 = 3(7) + 1 = 22

Distribute the electrons among the atoms, giving octets (or duets for H) to as many atoms as possible. Begin with the bonding electrons, and then proceed to lone pairs on terminal atoms and finally to lone pairs of the central atom. Arrange additional electrons around the central atom, giving it an expanded octet of up to 12 electrons.

Lastly, write the Lewis structure in brackets with the charge of the ion in the upper right-hand corner.

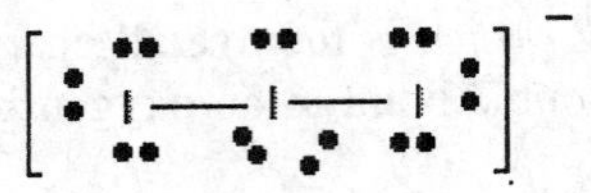

c) SF_4: Write the correct skeletal structure for the molecule

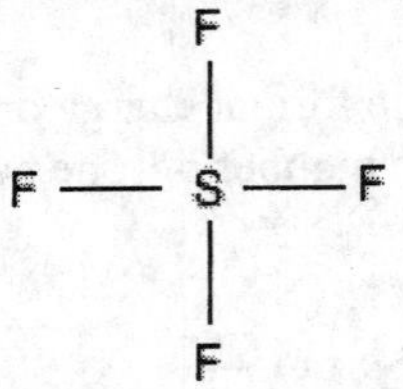

Calculate the total number of electrons for the Lewis structure by summing the valence electrons of each atom in the molecule

(number of valence e^- for S) + 4(number of valence e^- for F) = 6 + 4(7) = 34

Distribute the electrons among the atoms, giving octets (or duets for H) to as many atoms as possible. Begin with the bonding electrons, and then proceed to lone pairs on terminal atoms and finally to lone pairs of the central atom. Arrange additional electrons around the central atom, giving it an expanded octet of up to 12 electrons.

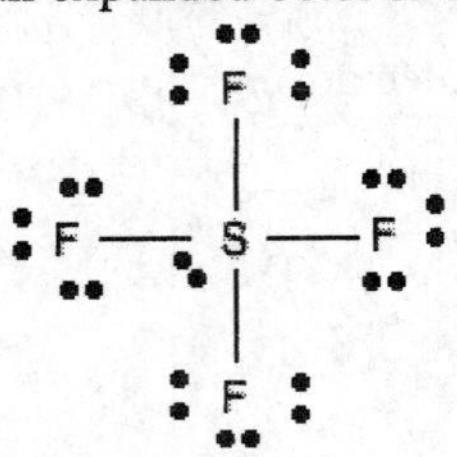

d) GeF_4: Write the correct skeletal structure for the molecule

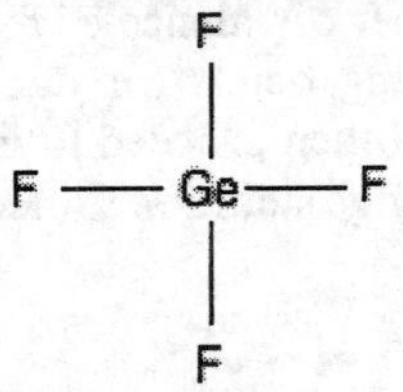

Calculate the total number of electrons for the Lewis structure by summing the valence electrons of each atom in the molecule

(number of valence e^- for Ge) + 4(number of valence e^- for F) = 4 + 4(7) = 32

Distribute the electrons among the atoms, giving octets (or duets for H) to as many atoms as possible. Begin with the bonding electrons, and then proceed to lone pairs on terminal atoms and finally to lone pairs of the central atom. Arrange additional electrons around the central atom, giving it an expanded octet of up to 12 electrons.

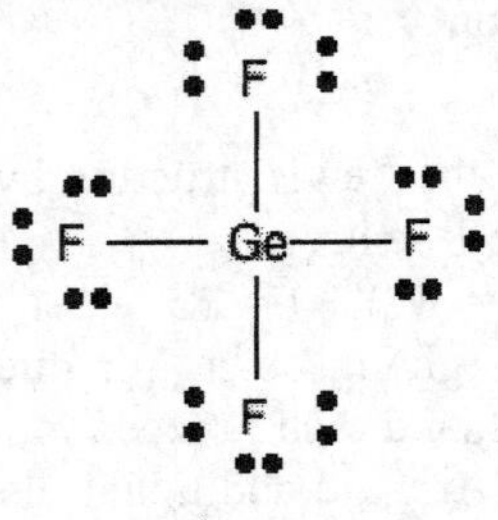

35. Bond strength: $H_3CCH_3 < H_2CCH_2 < HCCH$
Bond length: $H_3CCH_3 > H_2CCH_2 > HCCH$

Write the Lewis structures for the three compounds. Compare the C – C bonds. Triple bonds are stronger than double bonds which are stronger than single bonds. Also, single bonds are longer than double bonds are longer than triple bonds.

HCCH (10 e^-) H_2CCH_2 (12 e^-) H_3CCH_3(14 e^-)

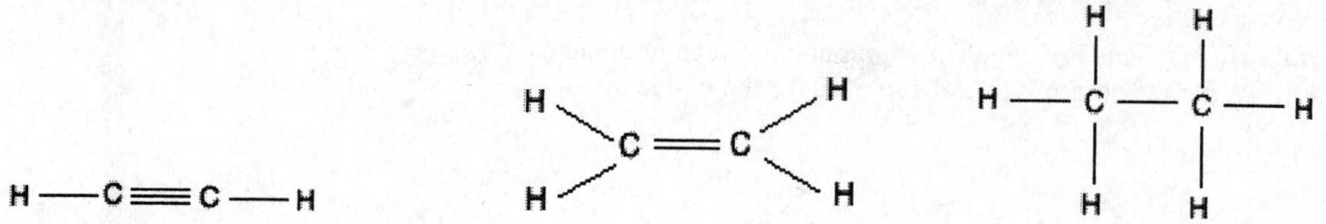

37. Rewrite the reaction using the Lewis structures of the molecules involved.

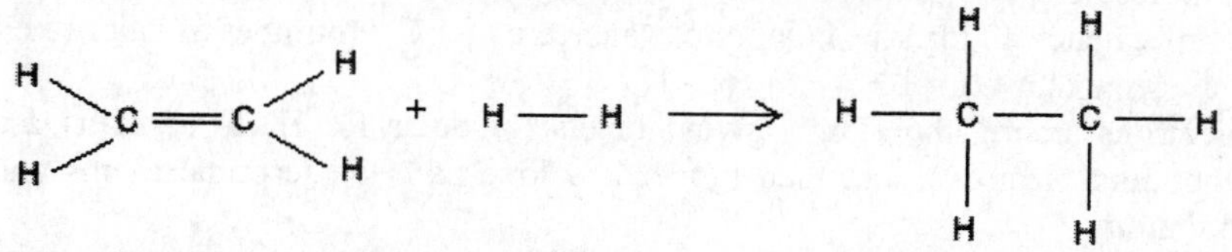

Determine which bonds are broken in the reaction and sum the bond energies of these

$\Sigma(\Delta H\text{'s bonds broken})$
$= 4\text{mol}(C - H) + 1\text{mol}(C = C) + 1\text{mol}(H - H)$
$= 4\text{mol}(414\text{ kJ/mol}) + 1\text{mol}(611\text{ kJ/mol}) + 1\text{mol}(436\text{ kJ/mol})$
$= 2703\text{ kJ/mol}$

Determine which bonds are formed in the reaction and sum the negatives of the bond energies of these

$\Sigma(-\Delta H\text{'s of bonds formed})$
$= -\ 6\text{mol}(C - H) - 1\text{mol}(C - C)$
$= -\ 6\text{mol}(414\text{ kJ/mol}) - 1\text{mol}\ (347\text{ kJ/mol})$
$= -2831\text{ kJ/mol}$

Find ΔH_{rxn} by summing the results of the two steps.

$\Delta H_{rxn} = \Sigma(\Delta H\text{'s bonds broken}) + \Sigma(-\Delta H\text{'s of bonds formed})$
$= 2703\text{ kJ/mol} - 2831\text{ kJ/mol}$
$= -\ 128\text{ kJ/mol}$

39. a) BI_3: This is a covalent compound between two nonmetals.
Write the correct skeletal structure for the molecule

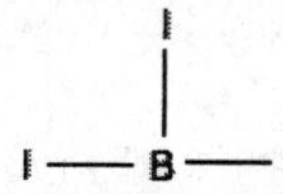

Calculate the total number of electrons for the Lewis structure by summing the valence electrons of each atom in the molecule

(number of valence e^- for B) + (number of valence e^- for I) = 3 +3(7) = 24

Distribute the electrons among the atoms, giving octets (or duets for H) to as many atoms as possible. Begin with the bonding electrons, and then proceed to lone pairs on terminal atoms and finally to lone pairs of the central atom.

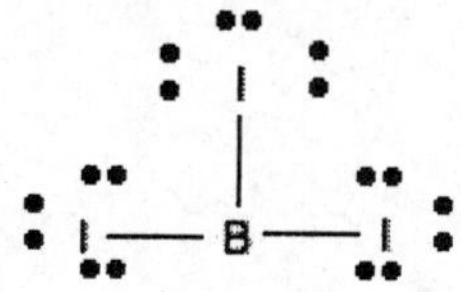

b) K_2S: This is an ionic compound between a metal and nonmetal
Draw the Lewis structures for K and S based on their valence electrons. K: $4s^1$ S: $3s^23p^4$

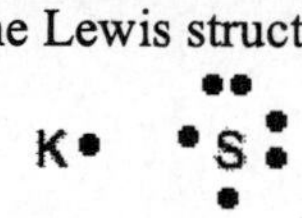

Potassium must lose one electron and be left with the octet from the previous shell, while sulfur needs to gain two electrons to get an octet.

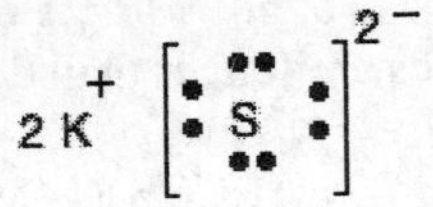

c) HCFO: This is a covalent compound between nonmetals.
Write the correct skeletal structure for the molecule

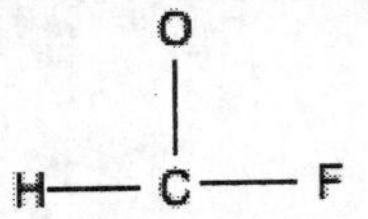

Calculate the total number of electrons for the Lewis structure by summing the valence electrons of each atom in the molecule

(number of valence e^- for H)+(number of valence e^- for C)+(number of valence e^- for F)+(number of valence e^- for O) = 1 + 4 + 7 +6 = 18

Distribute the electrons among the atoms, giving octets (or duets for H) to as many atoms as possible. Begin with the bonding electrons, and then proceed to lone pairs on terminal atoms and finally to lone pairs of the central atom.

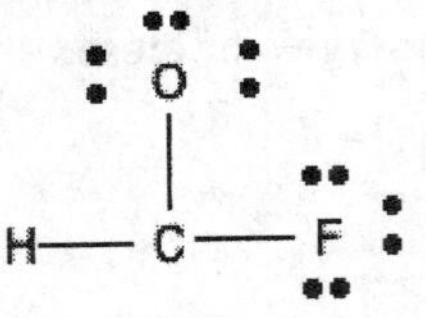

If any atom lacks an octet, form double or triple bonds as necessary to give them octets.

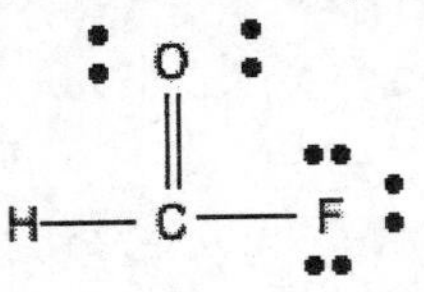

d) PBr_3: This is a covalent compound between two nonmetals.
Write the correct skeletal structure for the molecule

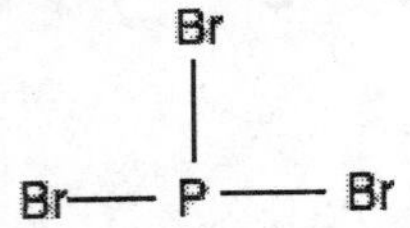

Calculate the total number of electrons for the Lewis structure by summing the valence electrons of each atom in the molecule

(number of valence e^- for P) + 3(number of valence e^- for Br) = 5 + 3(7) = 26

Distribute the electrons among the atoms, giving octets (or duets for H) to as many atoms as possible. Begin with the bonding electrons, and then proceed to lone pairs on terminal atoms and finally to lone pairs of the central atom.

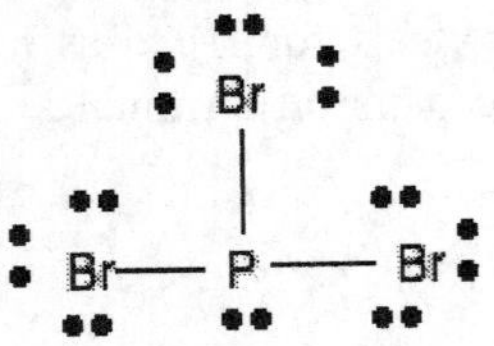

41. a) $BaCO_3$: Ba^{2+}

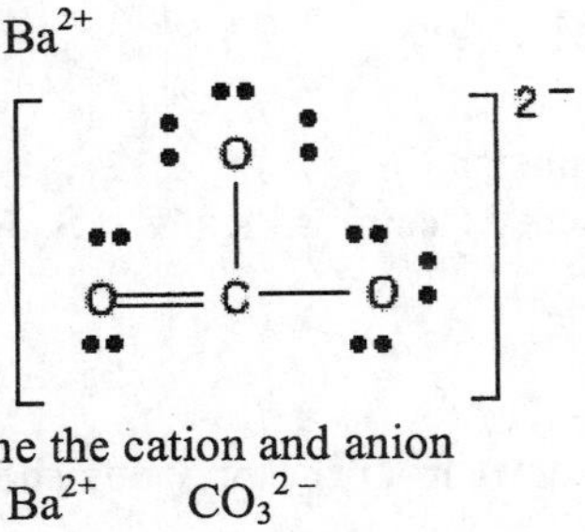

Determine the cation and anion

Ba^{2+} CO_3^{2-}

Write the Lewis structure for the barium cation based on the valence electrons

Ba $5s^2$ Ba^{2+} $5s^0$

Ba must lose two electrons and be left with the octet from the previous shell

Write the Lewis structure for the covalent anion.

Write the correct skeletal structure for the ion

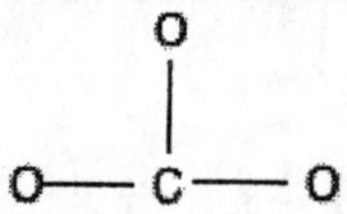

Calculate the total number of electrons for the Lewis structure by summing the valence electrons of each atom in the ion and adding two for the 2 – charge.

(number of valence e^- for C) + 3(number of valence e^- for O) = 4 + 3(6) + 2 = 24

Distribute the electrons among the atoms, giving octets (or duets for H) to as many atoms as possible. Begin with the bonding electrons, and then proceed to lone pairs on terminal atoms and finally to lone pairs of the central atom.

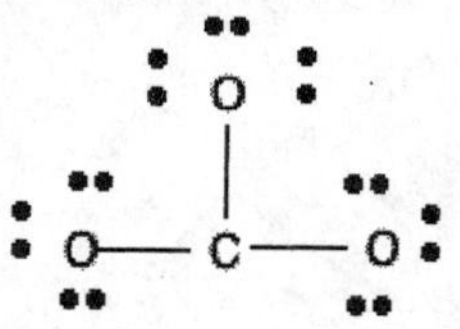

If any atom lacks an octet, form double or triple bonds as necessary to give them octets.

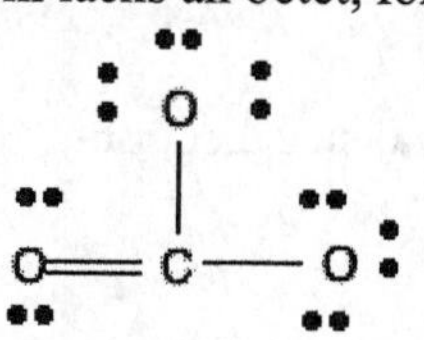

Lastly, write the Lewis structure in brackets with the charge of the ion in the upper right-hand corner.

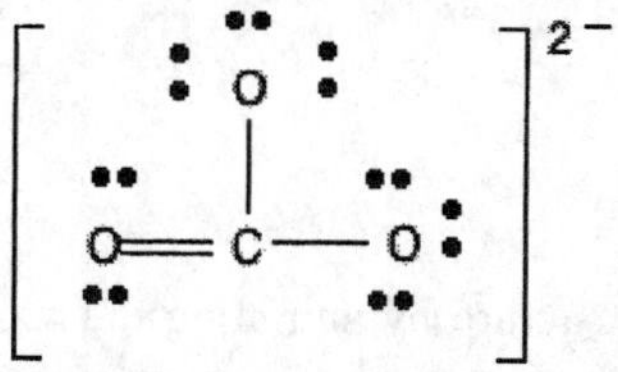

The double bond can be between the C and any of the oxygen atoms, so there are resonance structures

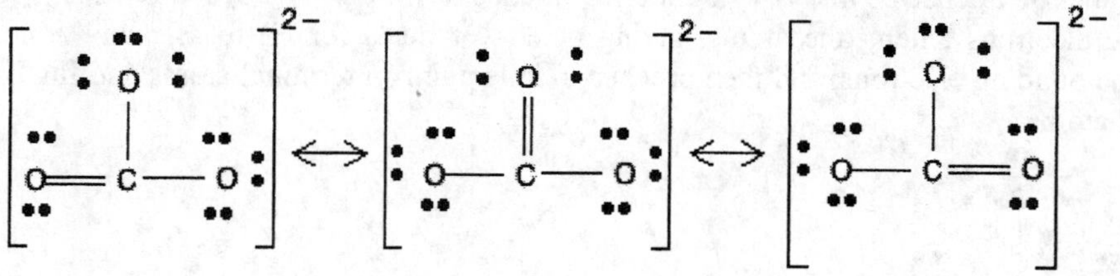

b) $Ca(OH)_2$: Ca^{2+}

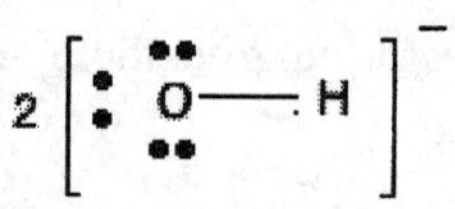

Determine the cation and anion

Ca^{2+} OH^-

Write the Lewis structure for the calcium cation based on the valence electrons

Ca $4s^2$ Ca^{2+} $4s^0$

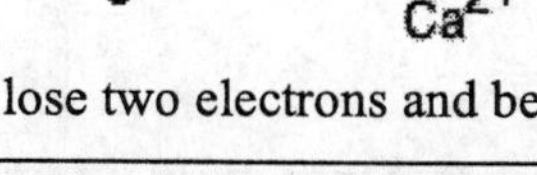

Ca must lose two electrons and be left with the octet from the previous shell

Write the Lewis structure for the covalent anion.
Write the correct skeletal structure for the ion

Calculate the total number of electrons for the Lewis structure by summing the valence electrons of each atom in the ion and adding one for the 1 – charge.

(number of valence e^- for H) + (number of valence e^- for O) +1 = 1 + 6 +1 = 8

Distribute the electrons among the atoms, giving octets (or duets for H) to as many atoms as possible. Begin with the bonding electrons, and then proceed to lone pairs on terminal atoms and finally to lone pairs of the central atom.

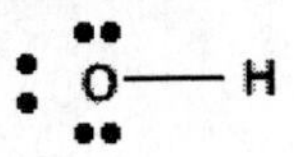

Lastly, write the Lewis structure in brackets with the charge of the ion in the upper right-hand corner.

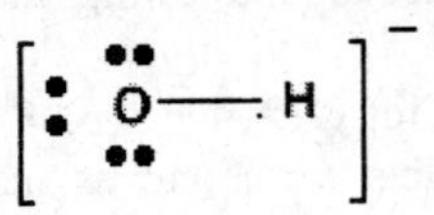

c) KNO_3: K^+

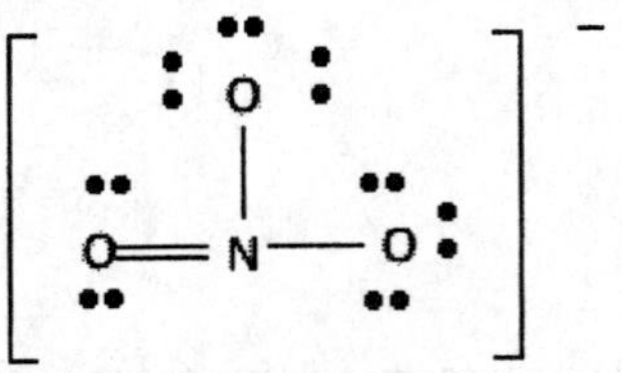

Determine the cation and anion

K^+ NO_3^-

Write the Lewis structure for the potassium cation based on the valence electrons

K $4s^1$ K^+ $4s^0$

K•

K^+

K must lose one electron and be left with the octet from the previous shell
Write the Lewis structure for the covalent anion.
Write the correct skeletal structure for the ion

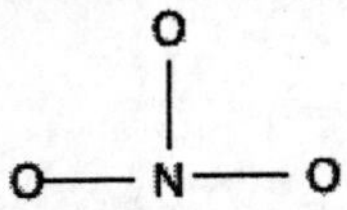

Calculate the total number of electrons for the Lewis structure by summing the valence electrons of each atom in the ion and adding one for the 1 – charge.

(number of valence e^- for N) + (number of valence e^- for O) = 5 + 3(6) +1 = 24

Distribute the electrons among the atoms, giving octets (or duets for H) to as many atoms as possible. Begin with the bonding electrons, and then proceed to lone pairs on terminal atoms and finally to lone pairs of the central atom.

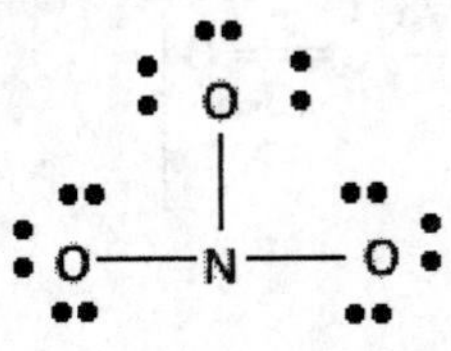

If any atom lacks an octet, form double or triple bonds as necessary to give them octets.

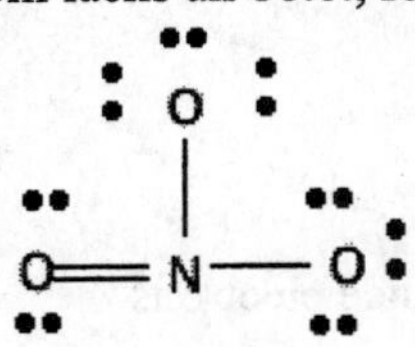

Lastly, write the Lewis structure in brackets with the charge of the ion in the upper right-hand corner.

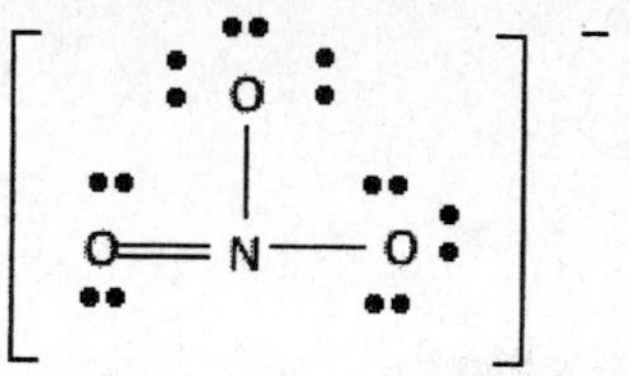

The double bond can be between the N and any of the oxygen atoms, so there are resonance structures

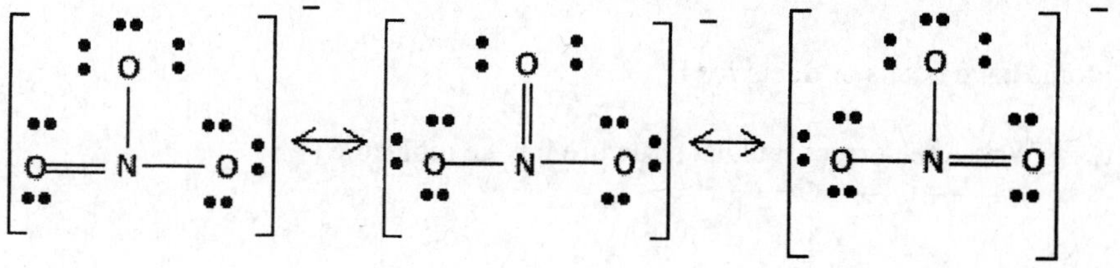

d) LiIO: Li^+

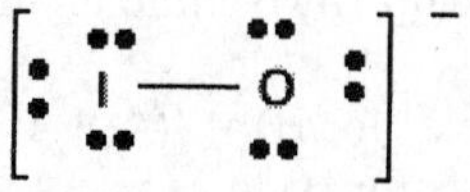

Determine the cation and anion

Li^+ IO^-

Write the Lewis structure for the lithium cation based on the valence electrons

Li $2s^1$ Li^+ $2s^0$

Li•

Li^+

Li must lose one electrons and be left with the octet from the previous shell

Write the Lewis structure for the covalent anion.

Write the correct skeletal structure for the ion

I — O

Calculate the total number of electrons for the Lewis structure by summing the valence electrons of each atom in the ion and adding one for the 1 – charge.

(number of valence e^- for I) + (number of valence e^- for O) = 7 +6 + 1 = 14

Distribute the electrons among the atoms, giving octets (or duets for H) to as many atoms as possible. Begin with the bonding electrons, and then proceed to lone pairs on terminal atoms and finally to lone pairs of the central atom.

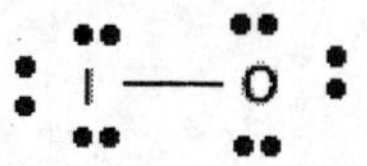

Lastly, write the Lewis structure in brackets with the charge of the ion in the upper right-hand corner.

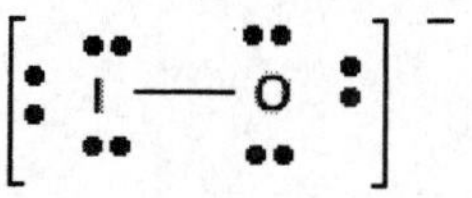

43. a) C_4H_8: Write the correct skeletal structure for the molecule

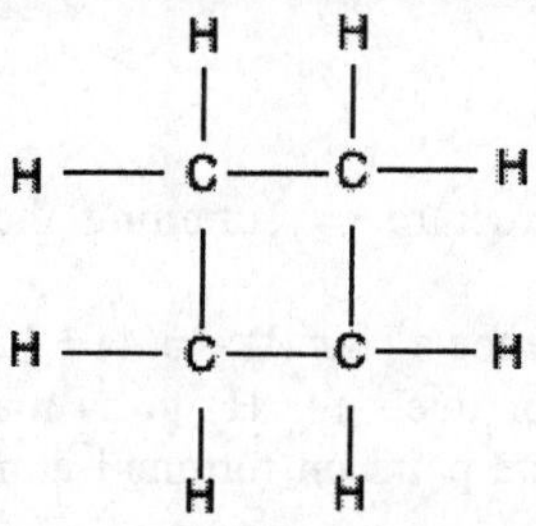

Calculate the total number of electrons for the Lewis structure by summing the valence electrons of each atom in the molecule.

4 (number of valence e^- for C) + 8(number of valence e^- for H) = 4(4) + 8(1) = 24

Distribute the electrons among the atoms, giving octets (or duets for H) to as many atoms as possible.

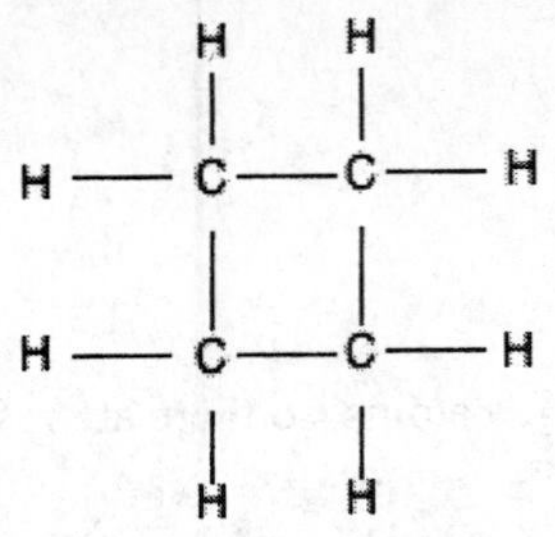

All atoms have octets or duets for H,

b) C_4H_4: Write the correct skeletal structure for the molecule

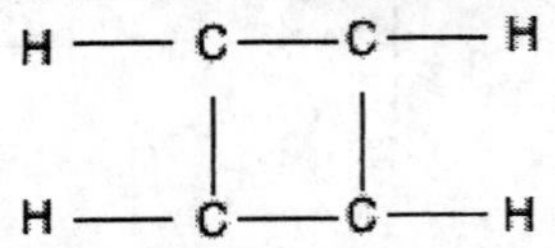

Calculate the total number of electrons for the Lewis structure by summing the valence electrons of each atom in the molecule.

4 (number of valence e^- for C) + 4(number of valence e^- for H) = 4(4) + 4(1) = 20

Distribute the electrons among the atoms, giving octets (or duets for H) to as many atoms as possible.

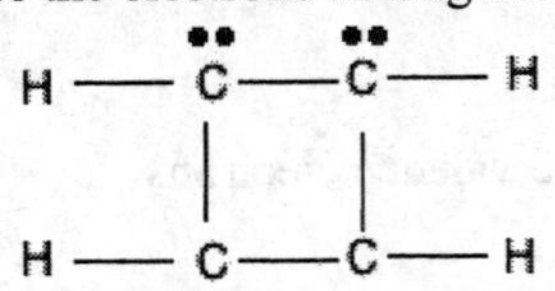

Complete octets by forming double bonds on alternating carbons, draw resonance structures.

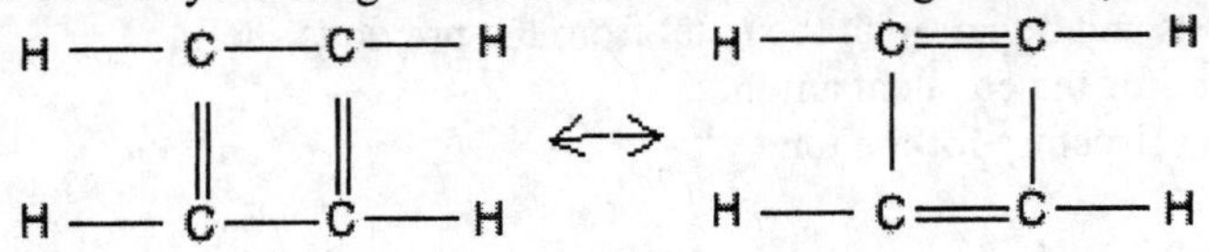

c) C_6H_{12}: Write the correct skeletal structure for the molecule.

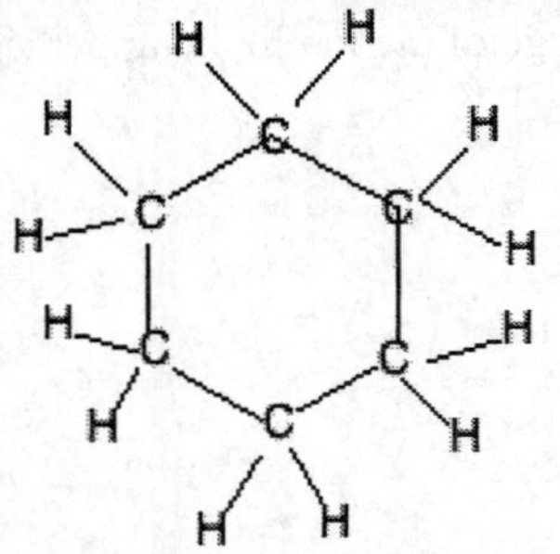

Calculate the total number of electrons for the Lewis structure by summing the valence electrons of each atom in the molecule.

6 (number of valence e^- for C) + 12(number of valence e^- for H) = 6(4) + 12(1) = 36

Distribute the electrons among the atoms, giving octets (or duets for H) to as many atoms as possible. Begin with the bonding electrons, and then proceed to lone pairs on terminal atoms and finally to lone pairs of the central atom.

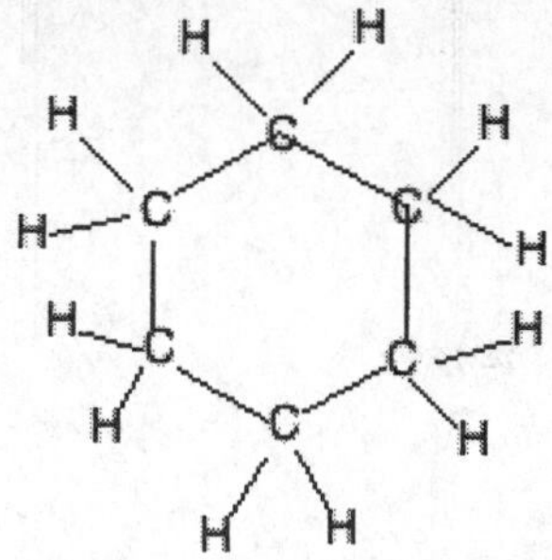

All 36 electrons are used and all atoms have octets or duets for H.

d) C_6H_6: Write the correct skeletal structure for the molecule.

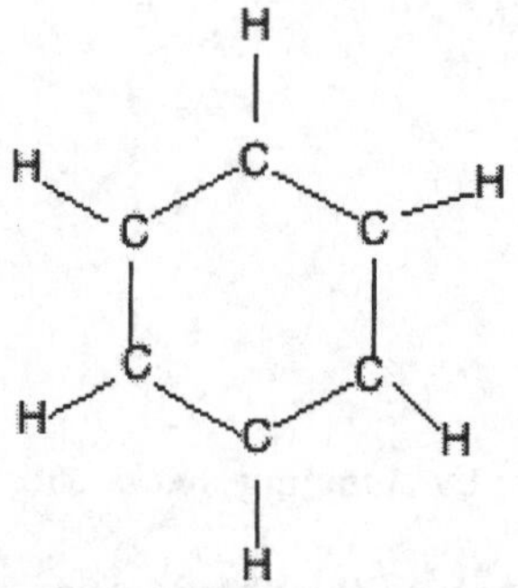

Calculate the total number of electrons for the Lewis structure by summing the valence electrons of each atom in the molecule.

6 (number of valence e^- for C) + 6(number of valence e^- for H) = 6(4) + 6(1) = 30

Distribute the electrons among the atoms, giving octets (or duets for H) to as many atoms as possible.

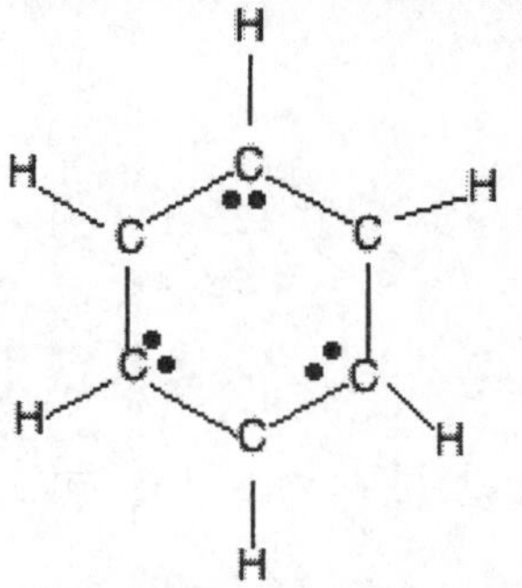

Complete octets by forming double bonds on alternating carbons, draw resonance structures.

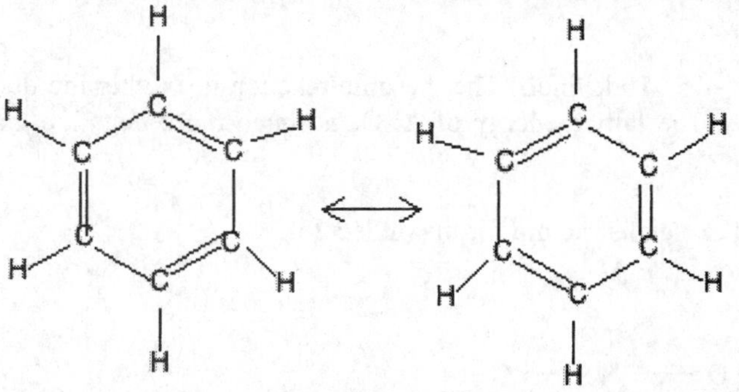

45. **Given:** 26.01% C; 4.38 % H; 69.52 % O; molar mass = 46.02 g/mol
Find: molecular formula and Lewis structure
Conceptual Plan:
convert mass to mol of each element → pseudoformula → empirical formula

$\frac{1 \text{ mol C}}{12.01 \text{ g C}}$ $\frac{1 \text{ mol H}}{1.008 \text{ g H}}$ $\frac{1 \text{ mol O}}{16.00 \text{ g O}}$ divide by smallest number

→ molecular formula → Lewis structure
empirical formula x n

Solution: $26.01\ \text{g C} \times \dfrac{1\ \text{mol C}}{12.01\ \text{g C}} = 2.166\ \text{mol C}$

$4.38\ \text{g H} \times \dfrac{1\ \text{mol H}}{1.008\ \text{g H}} = 4.345\ \text{mol H}$

$69.52\ \text{g O} \times \dfrac{1\ \text{mol O}}{16.00\ \text{g O}} = 4.345\ \text{mol O}$

$C_{2.166}H_{4.345}O_{4.345}$

$C_{\frac{2.166}{2.166}}H_{\frac{4.345}{2.166}}O_{\frac{4.345}{2.166}} \rightarrow CH_2O_2$

The correct empirical formula is CH_2O_2

empirical formula mass = (12.01 g/mol) + 2(1.008 g/mol) + 2(16.00 g/mol) = 46.03 g/mol

$n = \dfrac{\text{molar mass}}{\text{formula molar mass}} = \dfrac{46.02\ \text{g/mol}}{46.03\ \text{g/mol}} = 1$

molecular formula $= CH_2O_2 \times 1$

$= CH_2O_2$

Write the correct skeletal structure for the molecule

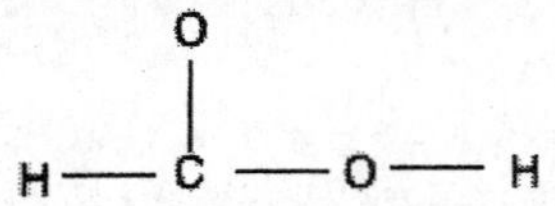

Calculate the total number of electrons for the Lewis structure by summing the valence electrons of each atom in the molecule.

(number of valence e^- for C) + 2(number of valence e^- for O) + 2(number of valence e^- for H)
= 4 + 2(6) + 2(1) = 18

Distribute the electrons among the atoms, giving octets (or duets for H) to as many atoms as possible. Begin with the bonding electrons, and then proceed to lone pairs on terminal atoms and finally to lone pairs of the central atom.

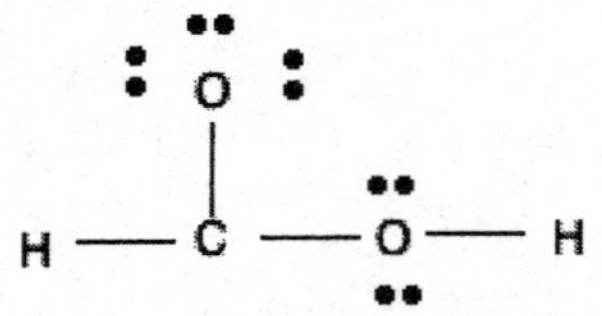

Complete the octet on C by forming a double bond

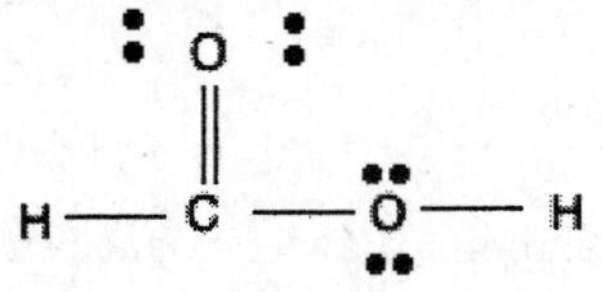

47. The lattice energy of Al_2O_3 is –15,916 kJ/mol. The thermite reaction is exothermic due to the energy released when the Al_2O_3 lattice forms. The lattice energy of Al_2O_3 is much more negative than the lattice energy of Fe_2O_3.

49. HNO_3 Write the correct skeletal structure for the molecule

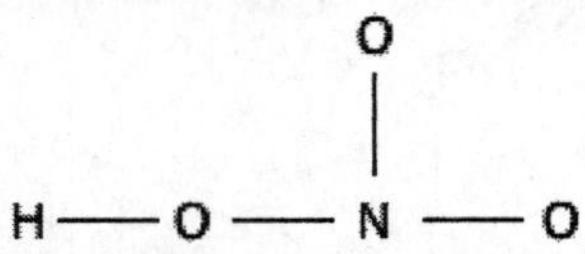

Calculate the total number of electrons for the Lewis structure by summing the valence electrons of each atom in the molecule.

3(number of valence e^- for O) + (number of valence e^- for N) + (number of valence e^- for H) = 3(6) + 5 +1 = 24

Distribute the electrons among the atoms, giving octets (or duets for H) to as many atoms as possible. Begin with the bonding electrons, and then proceed to lone pairs on terminal atoms and finally to lone pairs of the central atom.

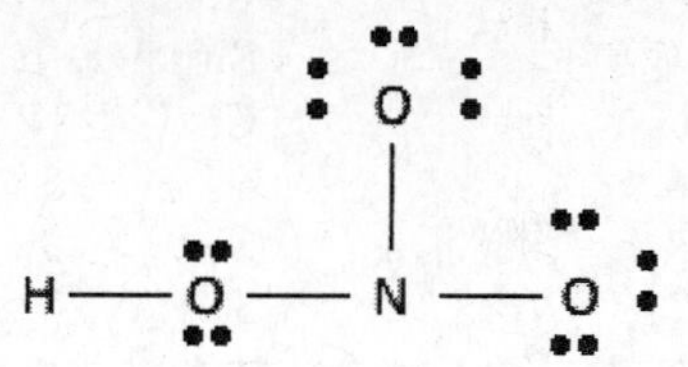

All 24 valence electrons are used

If any atom lacks an octet, form double or triple bonds as necessary to give them octets. The double bond can be formed to any of the three oxygen atoms, so there are three resonance forms.

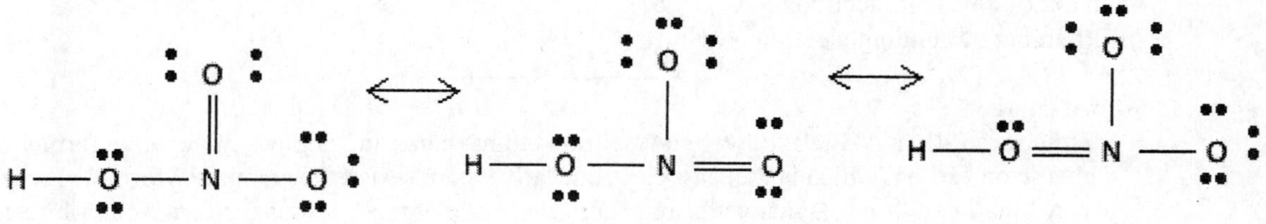

I II III

All atoms have octets, duets for H, structures is complete.
To determine which resonance hybrid(s) is most important, calculate the formal charge on each atom in each structure by finding the number of valence electrons and subtracting the number of lone pair electrons and one-half the number of bonding electrons.

	Structure I					Structure II				
	O_{left}	O_{top}	O_{right}	N	H	O_{left}	O_{top}	O_{right}	N	H
number of valence electrons	6	6	6	5	1	6	6	6	5	1
- number of lone pair electrons	4	4	6	0	0	4	6	4	0	0
- 1/2(number of bonding electrons)	2	2	1	4	1	2	1	2	4	1
Formal charge	0	0	−1	+1	0	0	−1	0	+1	0

	Structure III				
	O_{left}	O_{top}	O_{right}	N	H
number of valence electrons	6	6	6	5	1
- number of lone pair electrons	2	6	6	0	0
- 1/2(number of bonding electrons)	3	1	1	4	1
Formal charge	+1	−1	−1	+1	0

The sum of the formal charges is 0 for each structure, which is the overall charge of the molecule. However, in structures I and II, the individual formal charges are lower and these two forms would contribute equally to the structure of HNO_3 and structure III would be less important since the individual formal charges are higher.

51. CNO^- Write the skeletal structure:

C – N – O

Determine the number of valence electrons.

(valence e^- from C) + (valence e^- from N) + (valence e^- from O) +1(from the negative charge)

4 + 5 + 6 + 1 = 16

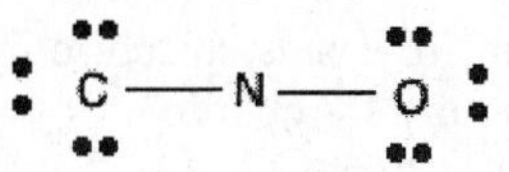

Distribute the electrons to complete octets if possible.

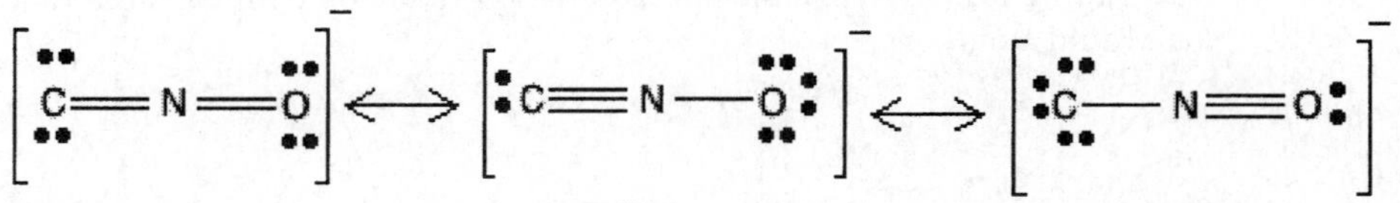

I II III

Determine the formal charge on each atom for each structure

	Structure I C	Structure I N	Structure I O	Structure II C	Structure II N	Structure II O
number of valence electrons	4	5	6	4	5	6
- number of lone pair electrons	4	0	4	2	0	6
- 1/2(number of bonding electrons)	2	4	2	3	4	1
Formal charge	–2	+1	0	–1	+1	–1

	Structure III C	Structure III N	Structure III O
number of valence electrons	4	5	6
- number of lone pair electrons	6	0	2
- 1/2(number of bonding electrons)	1	4	3
Formal charge	–3	+1	+1

Structures I, II, and III all follow the octet rule but have varying degrees of negative formal charge on carbon, which is the least electronegative atom. Also the amount of formal charge is very high in all three resonance forms. Therefore, none of these resonance forms contribute to the stability of the fulminate ion and the ion is not very stable.

53. $HCSNH_2$: Write the correct skeletal structure for the molecule.

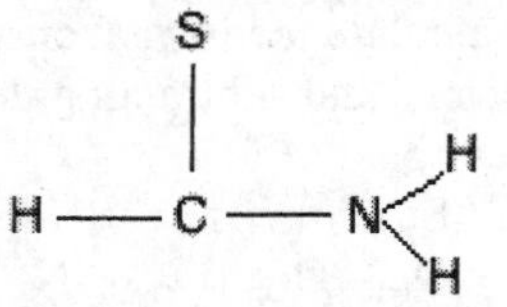

Calculate the total number of electrons for the Lewis structure by summing the valence electrons of each atom in the molecule.

(number of valence e^- for N) + (number of valence e^- for S) + (number of valence e^- for C) + 3(number of valence e^- for H) = 5 + 6 + 4 + 3(1) = 18

Distribute the electrons among the atoms, giving octets (or duets for H) to as many atoms as possible. Begin with the bonding electrons, and then proceed to lone pairs on terminal atoms and finally to lone pairs of the central atom.

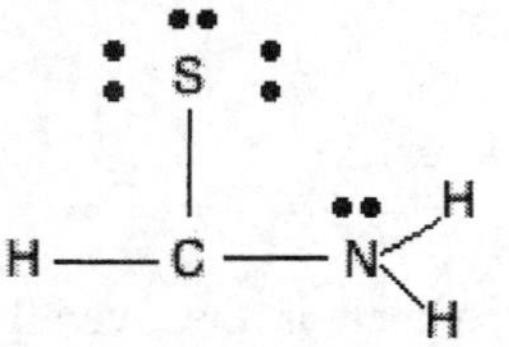

Complete the octet on C by forming a double bond.

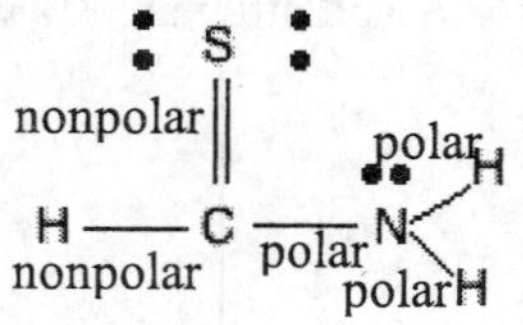

55. a) O_2^-: Write the correct skeletal structure for the radical

Calculate the total number of electrons for the Lewis structure by summing the valence electrons of each atom in the radical and adding 1 for the 1 – charge.

2(number of valence e^- for O) + 1 = 2(6) + 1 = 13

Distribute the electrons among the atoms, giving octets to as many atoms as possible. Begin with the bonding electrons, and then proceed to lone pairs on terminal atoms and finally to lone pairs of the central atom.

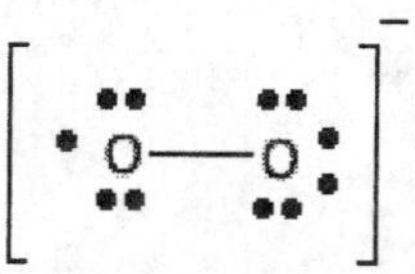

All 13 valence electrons are used.

O has an incomplete octet. It has 7 electrons because we have an odd number of valence electrons.

b) O⁻:   Write the Lewis structure based on the valence electrons: $2s^2 2p^5$.

$$\left[\; \cdot \ddot{\underset{\displaystyle \cdot\cdot}{\overset{\displaystyle \cdot\cdot}{O}}} \! : \; \right]^{-}$$

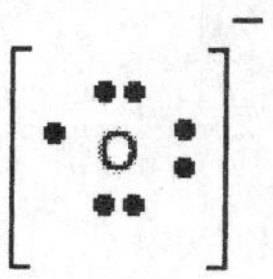

c) OH: Write the correct skeletal structure for the molecule

H —— O

Calculate the total number of electrons for the Lewis structure by summing the valence electrons of each atom in the molecule

(number of valence e⁻ for O) + (number of valence e⁻ for H) = 6 + 1 = 7

Distribute the electrons among the atoms, giving octets (or duets for H) to as many atoms as possible. Begin with the bonding electrons, and then proceed to lone pairs on terminal atoms and finally to lone pairs of the central atom.

All 7 valence electrons are used.

O has an incomplete octet. It has 7 electrons because we have an odd number of valence electrons.

d) CH₃OO: Write the correct skeletal structure for the radical

C is the less electronegative, so it is central

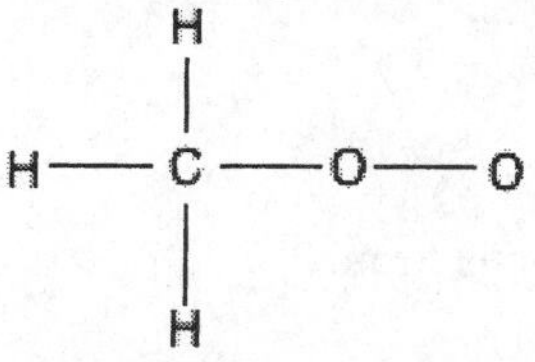

Calculate the total number of electrons for the Lewis structure by summing the valence electrons of each atom in the molecule

3(number of valence e⁻ for H) + (number of valence e⁻ for C) + 2(number of valence e⁻ for O) = 3(1) + 4 + 2(6) = 19

Distribute the electrons among the atoms, giving octets (or duets for H) to as many atoms as possible. Begin with the bonding electrons, and then proceed to lone pairs on terminal atoms and finally to lone pairs of the central atom.

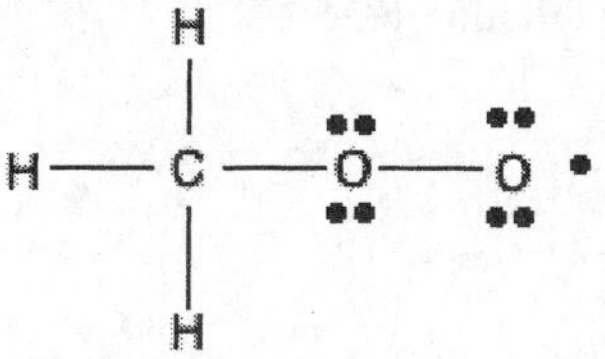

All 19 valence electrons are used.

O has an incomplete octet. It has 7 electrons because we have an odd number of valence electrons.

57. Rewrite the reaction using the Lewis structures of the molecules involved.

$$\text{H} - \text{H (g)} + 1/2\;\text{O} = \text{O (g)} \rightarrow \text{H} - \overset{\cdot\cdot}{\underset{\cdot\cdot}{\text{O}}} - \text{H}$$

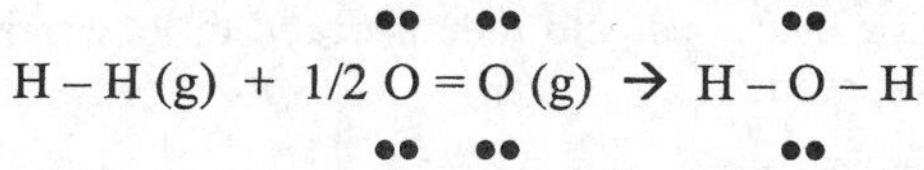

Determine which bonds are broken in the reaction and sum the bond energies of these

Σ(ΔH's bonds broken)

= 1mol(H – H) + 1/2mol(O = O)

= 1mol(436 kJ/mol) + 1/2mol(498)

= 685 kJ/mol

Determine which bonds are formed in the reaction and sum the negatives of the bond energies of these

Σ(-ΔH's of bonds formed)

= – 2mol(O – H)

= – 2mol(464 kJ/mol)

= –928 kJ/mol

Find ΔH_{rxn} by summing the results of the two steps.

ΔH_{rxn} = Σ(ΔH's bonds broken) + Σ(-ΔH's of bonds formed)

= 685 kJ/mol – 928 kJ/mol

= – 243 kJ/mol

$CH_4(g) + 2O_2(g) \rightarrow CO_2(g) + 2H_2O(g)$

Rewrite the reaction using the Lewis structures of the molecules involved.

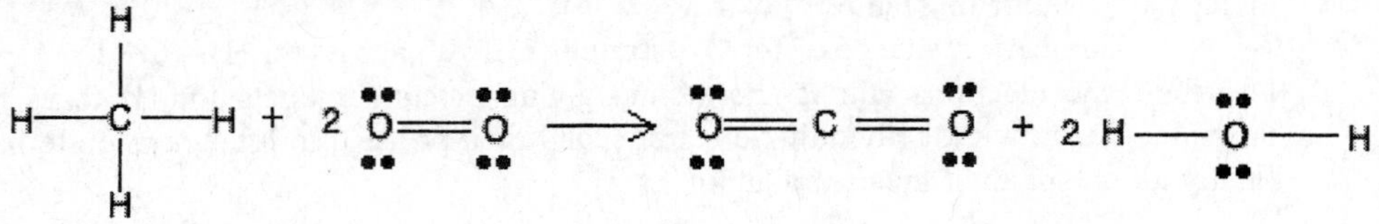

Determine which bonds are broken in the reaction and sum the bond energies of these

Σ(ΔH's bonds broken)

= 4mol(C – H) + 2mol(O = O)

= 4mol(414 kJ/mol) + 2mol(498)

= 2652 kJ/mol

Determine which bonds are formed in the reaction and sum the negatives of the bond energies of these

Σ(-ΔH's of bonds formed)

= – 2mol(C = O) – 4mol(O – H)

= – 2mol(799 kJ/mol) – 4mol(464 kJ/mol)

= –3454 kJ/mol

Find ΔH_{rxn} by summing the results of the two steps.

ΔH_{rxn} = Σ(ΔH's bonds broken) + Σ(-ΔH's of bonds formed)

= 2653 kJ/mol – 3454 kJ/mol

= – 802 kJ/mol

Compare

	kJ/mol	kJ/g
H_2	–243	–120
CH_4	–802	–50.1

So, methane yields more energy per mole but hydrogen yields more energy per gram.

59. a) Cl_2O_7: Write the correct skeletal structure for the molecule

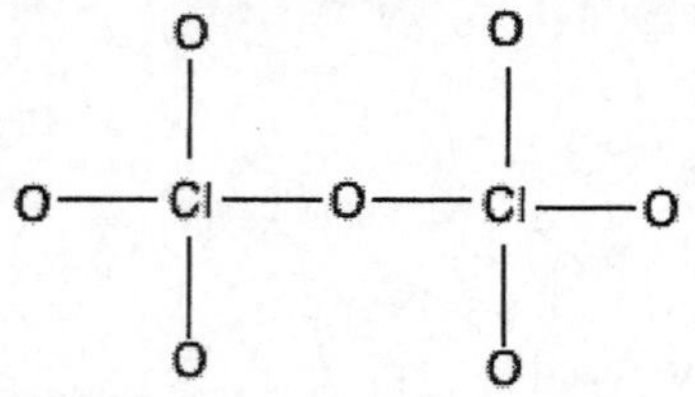

Calculate the total number of electrons for the Lewis structure by summing the valence electrons of each atom in the molecule

2(number of valence e^- for Cl) + 7(number of valence e^- for O) = 2(7)5 + 7(6) = 56

Distribute the electrons among the atoms, giving octets (or duets for H) to as many atoms as possible. Begin with the bonding electrons, and then proceed to lone pairs on terminal atoms and finally to lone pairs of the central atom.

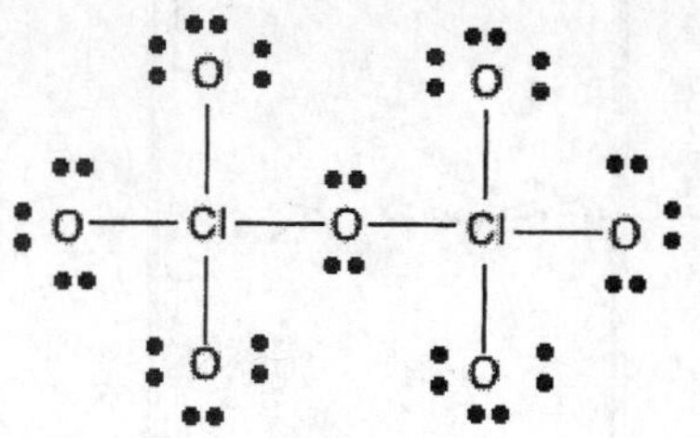

Form double bonds to minimize formal charge.

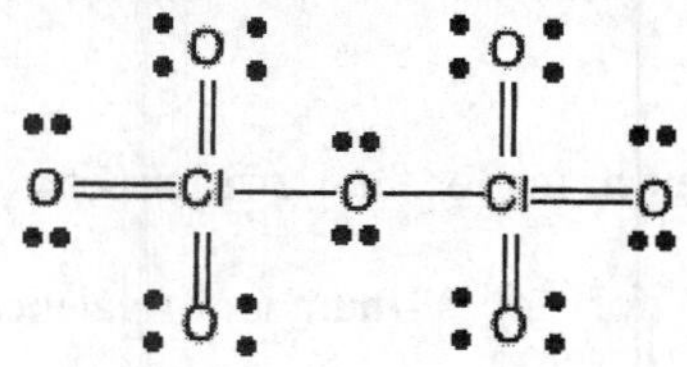

b) H_3PO_3: Write the correct skeletal structure for the molecule

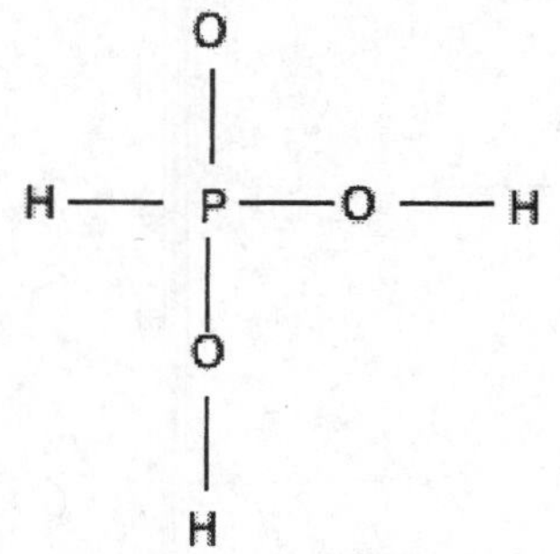

Calculate the total number of electrons for the Lewis structure by summing the valence electrons of each atom in the molecule

(number of valence e^- for P) + 3(number of valence e^- for O) + 3(number of valence e^- for H) = 5 + 3(6) + 3(1) = 26

Distribute the electrons among the atoms, giving octets (or duets for H) to as many atoms as possible. Begin with the bonding electrons, and then proceed to lone pairs on terminal atoms and finally to lone pairs of the central atom.

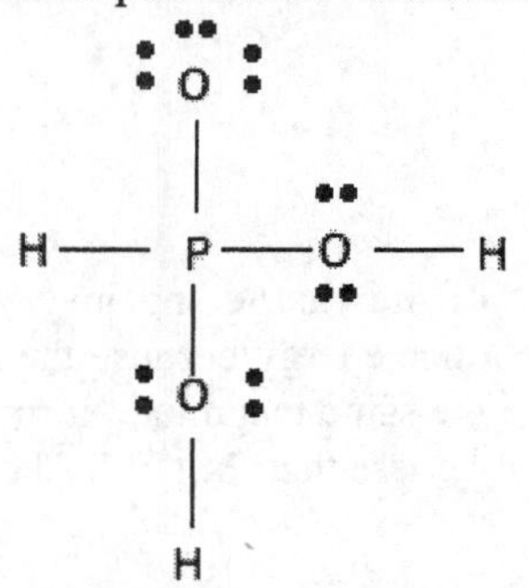

Form double bond to minimize formal charge.

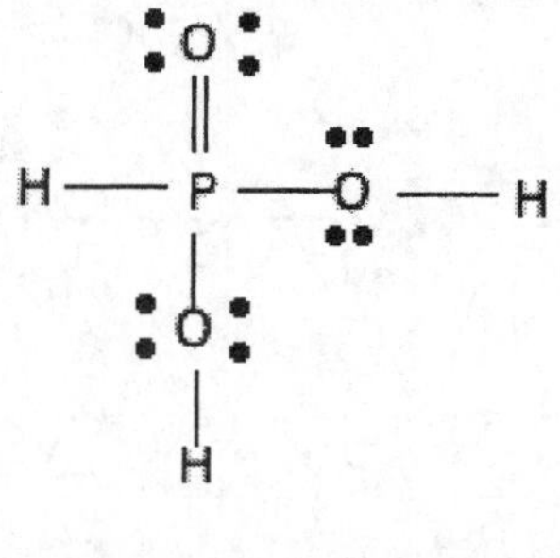

c) H_3AsO_4: Write the correct skeletal structure for the radical

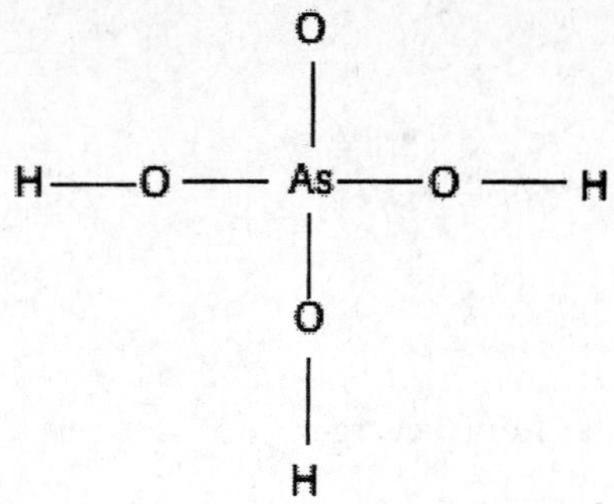

Calculate the total number of electrons for the Lewis structure by summing the valence electrons of each atom in the molecule

(number of valence e^- for As) + 4(number of valence e^- for O) + 3(number of valence e^- for H) = 5 + 4(6) + 3(1) = 32

Distribute the electrons among the atoms, giving octets (or duets for H) to as many atoms as possible. Begin with the bonding electrons, and then proceed to lone pairs on terminal atoms and finally to lone pairs of the central atom.

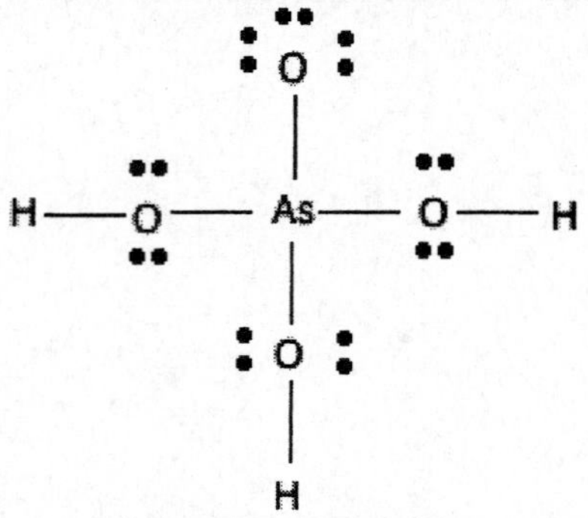

Form double bond to minimize formal charge.

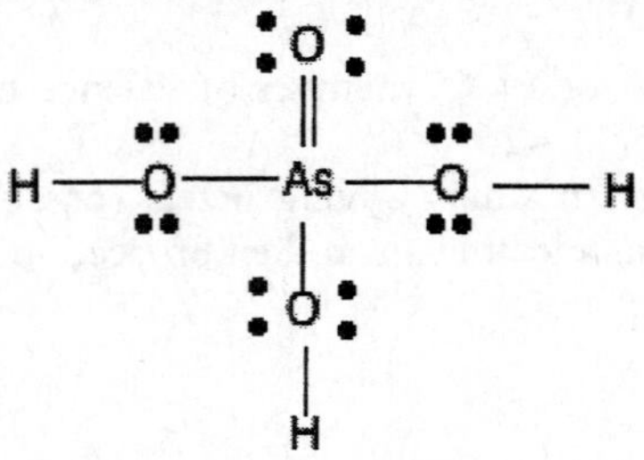

61. $Na^+F^- < Na^+O^{2-} < Mg^{2+}F^- < Mg^{2+}O^{2-} < Al^{3+}O^{2-}$

The lattice energy is proportional to the magnitude of the charge and inversely proportional to the distance between the atoms. Na^+F^- would have the smallest lattice energy because the magnitude of the charges on Na and F are the smallest. $Mg^{2+}F^-$ and Na^+O^{2-} both have the same magnitude formal charge, the O^{2-} is larger than F^- in size, Na^+ is larger than Mg^{2+}, so Na^+O^{2-} should be less than $Mg^{2+}F^-$. The magnitude of the charge makes $Mg^{2+}O^{2-} < Al^{3+}O^{2-}$.

63.

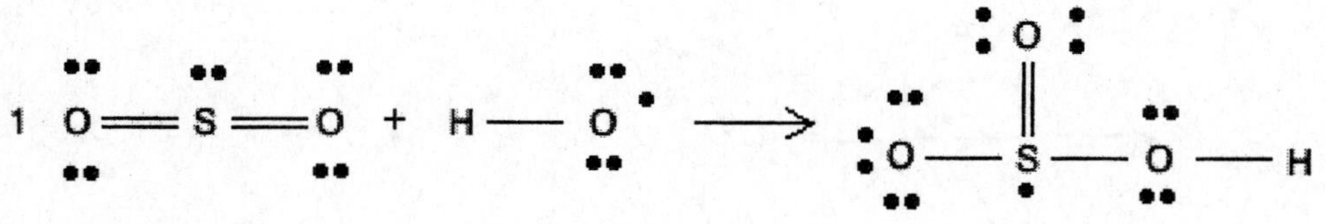

2 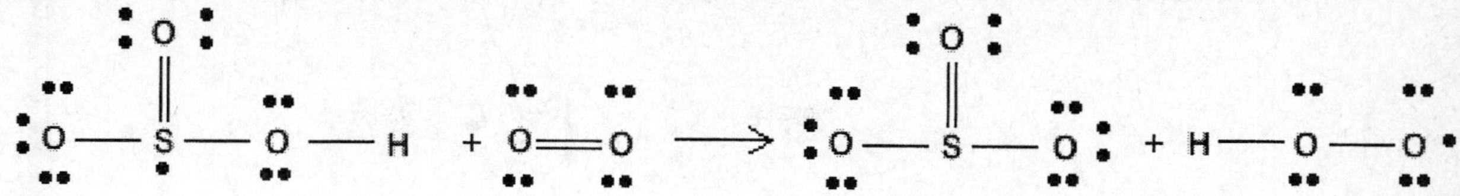

3 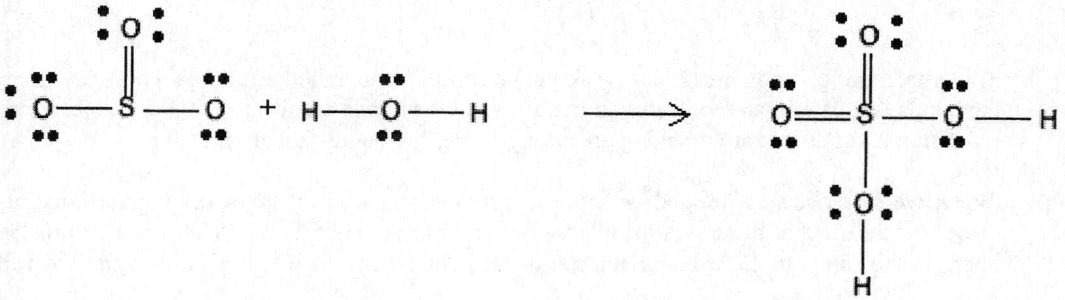

Step 1:

Bonds broken: 2mol(S = O) +1mol(H – O) = 2mol(523kJ/mol) + 1mol(464kJ/mol) = 1510 kJ/mol

Bonds formed: –2mol(S – O) –1mol(S = O) –1mol(O – H) =
–2mol(265 kJ/mol) –1mol(523 kJ/mol) –1mol(464 kJ/mol) = –1517 kJ/mol

ΔH_{step} = –7 kJ/mol

Step 2:

Bonds broken: 2mol(S – O) + 1mol(S = O) + 1mol(O – H) + 1mol(O = O) =
2mol(265 kJ/mol)+1mol(523 kJ/mol)+1mol(464 kJ/mol)+1mol(498 kJ/mol) =
2015 kJ/mol

Bonds formed: : –2mol(S – O) –1mol(S = O) –1mol(O – H) –1mol(O – O) =
–2mol(265 kJ/mol) –1mol(523 kJ/mol) –1mol(464 kJ/mol) –1mol(142 kJ/mol)=
–1659 kJ/mol

ΔH_{step}= +356 kJ/mol

Step 3:

Bonds broken: 2mol(S – O) + 1mol(S = O) + 2mol(O – H) =
2mol(265 kJ/mol)+1mol(523 kJ/mol)+2mol(464 kJ/mol) = 1981 kJ/mol

Bonds formed: –2mol(S – O) – 2mol(S = O) – 2mol(O – H) =
–2mol(265 kJ/mol) + –2mol(523 kJ/mol) + –2mol(464 kJ/mol) = –2504 kJ/mol

ΔH_{step}= – 523 kJ/mol

Hess's law states that ΔH for the reaction is the sum of ΔH of the steps:
ΔH_{rxn} = (– 7 kJ/mol) + (+356 kJ/mol) + (– 523 kJ/mol) = – 174 kJ/mol

65. **Given:** μ = 1.08 D HCl, 20% ionic and μ = 1.82 D HF, 45% ionic **Find:** r
Conceptual Plan: $\mu \rightarrow \mu_{calc} \rightarrow r$

$$\%\text{ ionic character} = \frac{\mu}{\mu_{calc}} \qquad \mu_{calc} = qr$$

Solution: For HCl $\mu_{calc} = \frac{1.08}{0.20} = 5.4\text{ D}$

$$\frac{5.4\ \cancel{\text{D}} \times \frac{3.34 \times 10^{-30}\ \cancel{\text{C}}\cdot\cancel{\text{m}}}{\cancel{\text{D}}} \times \frac{10^{12}\ \text{pm}}{\cancel{\text{m}}}}{1.6 \times 10^{-19}\ \cancel{\text{C}}} = 113\ \text{pm}$$

For HF $\mu_{calc} = \frac{1.82}{0.45} = 4.0\underline{4}\text{ D}$

$$\frac{4.0\underline{4}\ \cancel{\text{D}} \times \frac{3.34 \times 10^{-30}\ \cancel{\text{C}}\cdot\cancel{\text{m}}}{\cancel{\text{D}}} \times \frac{10^{12}\ \text{pm}}{\cancel{\text{m}}}}{1.6 \times 10^{-19}\ \cancel{\text{C}}} = 84\ \text{pm}$$

From Table 9.4, the bond length of HCl = 127 pm, and HF = 92 pm. Both of these values are slightly higher than the calculated values.

67. In order for the four P atoms to be equivalent, they must all be in the same electronic environment. That is, they must all see the same number of bonds and lone pair electrons. The only way to achieve this is with a

tetrahedral configuration where the P atoms are at the four points of the tetrahedron.

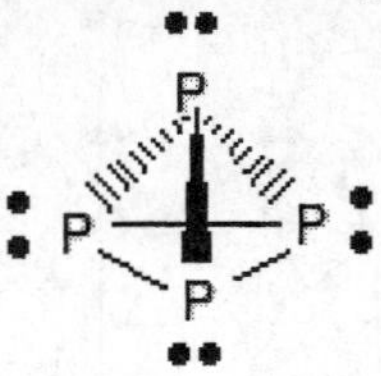

68. a) is true: Strong bonds break and weak bonds form. In an endothermic reaction, the energy required to break the bonds is greater than the energy given off when the bonds are formed ($\Delta H > 0$), therefore, in an endothermic reaction the bonds that are breaking are stronger than the bonds that are forming.

69. When we say that a compound is "energy rich" we mean that it gives off a great amount of energy when it reacts. It means that there is a lot of energy stored in the compound. This energy is released when the weak bonds in the compound break and much stronger bonds are formed in the product, thereby releasing energy.

70. In solid covalent compounds, the electrons in the bonds are shared directly between the atoms involved in the molecule. Each molecule is a distinct unit. Ionic compounds, on the other hand, are not distinct units. Rather, they are composed of alternating positive and negative ions in a three-dimensional crystalline array.

71. Lewis theory is successful because it allows us to understand and predict many chemical observations. We can use it to determine the formula of ionic compounds, to account for low melting points and boiling points of molecular compounds compared to ionic compounds. Lewis theory allows us to predict what molecules or ions will be stable, which will be more reactive, and which will not exist. Lewis theory, however, does not really tell us anything about how the bonds in the molecules and ions form. It does not give us a way to account for the paramagnetism of oxygen. And, by itself, Lewis theory does not really tell us anything about the shape of the molecule or ion.

Chapter 10
Chemical Bonding II: Molecular Shapes, Valence Bond Theory, and Molecular Orbital Theory

1. 4 electron pairs: A trigonal pyramidal molecular geometry has three bonding groups and one lone pair of electrons, so there are four electron pairs on atom A.

3. a) 4 total electron groups, 4 bonding groups, 0 lone pair
A tetrahedral molecular geometry has four bonding groups and no lone pair. So, there are four total electron groups, four bonding groups, and no lone pair.

b) 5 total electron groups, 3 bonding groups, 2 lone pairs
A T-shaped molecular geometry has three bonding groups and two lone pairs. So, there are five total electron groups, three bonding groups, and two lone pairs.

c) 6 total electron groups, 5 bonding groups, 1 lone pair
A square pyramidal molecular geometry has five bonding groups and one lone pair. So, there are six total electron groups, five bonding groups, and one lone pair.

5. a) PF_3: Electron geometry – tetrahedral; molecular geometry – trigonal pyramidal; bond angle = 109.5°
Because of the lone pair, the bond angle will be less than 109.5°.
Draw a Lewis structure for the molecule:
PF_3 has 26 valence electrons

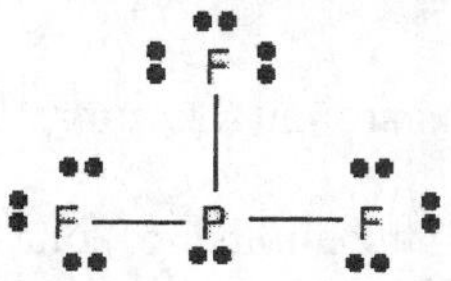

Determine the total number of electron groups around the central atom:
There are four electron groups on P

Determine the number of bonding groups and the number of lone pairs around the central atom:
There are three bonding groups and one lone pair

Use Table 10.1 to determine the electron geometry and molecular geometry and bond angles:
Four electron groups is tetrahedral electron geometry, three bonding groups and one lone pair is trigonal pyramidal molecular geometry, the idealized bond angles for tetrahedral are 109.5°; however, the lone pair will make the bond angle less than idealized.

b) SBr_2: Electron geometry – tetrahedral; molecular geometry – bent; bond angle = 109.5°
Because of the lone pairs, the bond angle will be less than 109.5°.
Draw a Lewis structure for the molecule:
SBr_3 has 20 valence electrons

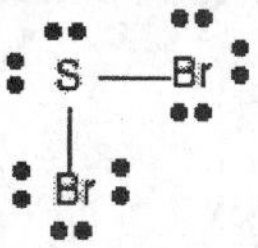

Determine the total number of electron groups around the central atom:
There are four electron groups on S
Determine the number of bonding groups and the number of lone pairs around the central atom:
There are two bonding groups and two lone pair
Use Table 10.1 to determine the electron geometry and molecular geometry and bond angles:
Four electron groups is tetrahedral electron geometry, two bonding groups and two lone pair is a bent molecular geometry, the idealized bond angles for tetrahedral are 109.5°; however, the lone pairs will make the bond angle less than idealized.

c) $CHCl_3$: Electron geometry – tetrahedral; molecular geometry – tetrahedral; bond angle = 109.5°
Because there are no lone pairs, the bond angle will be 109.5°.

Draw a Lewis structure for the molecule:

$CHCl_3$ has 26 valence electrons

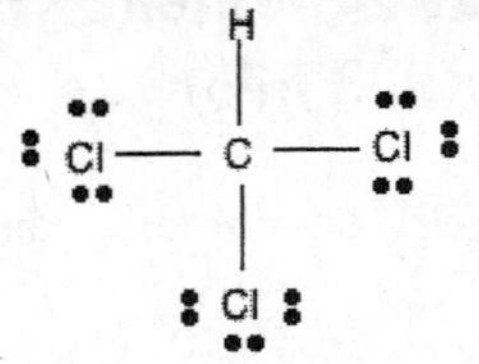

Determine the total number of electron groups around the central atom:
There are four electron groups on C
Determine the number of bonding groups and the number of lone pairs around the central atom:
There are four bonding groups and no lone pair
Use Table 10.1 to determine the electron geometry and molecular geometry and bond angles:
Four electron groups is tetrahedral electron geometry, four bonding groups and no lone pair is a tetrahedral molecular geometry, the idealized bond angles for tetrahedral are 109.5°; however, because the attached atoms have different electronegativities the bond angles are less than idealized.

d) CS_2: Electron geometry – linear; molecular geometry – linear; bond angle = 180°
Because there are no lone pairs, the bond angle will 180°.
Draw a Lewis structure for the molecule:

CS_2 has 16 valence electrons

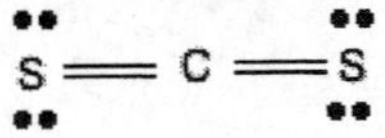

Determine the total number of electron groups around the central atom:
There are two electron groups on C
Determine the number of bonding groups and the number of lone pairs around the central atom:
There are two bonding groups and no lone pair
Use Table 10.1 to determine the electron geometry and molecular geometry and bond angles:
Two electron groups is linear geometry, two bonding groups and no lone pair is linear molecular geometry, the idealized bond angle is 180°

7. H_2O will have the smaller bond angle because lone pair–lone pair repulsions are greater than lone pair–bonding pair repulsions.

Draw the Lewis structures for both structures:

H_3O^+ has 8 valence electrons | H_2O has 8 valence electrons

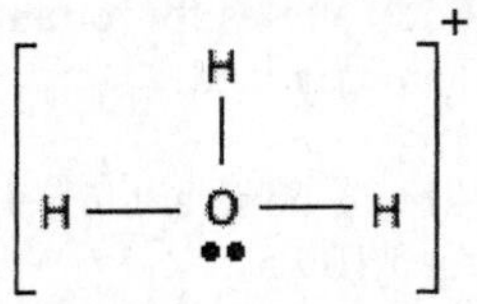

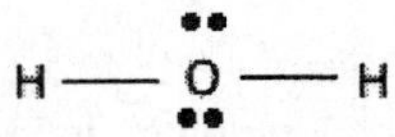

3 bonding groups and 1 lone pair | 2 bonding groups and 2 lone pairs

Both have 4 electron groups, but the 2 lone pairs in H_2O will cause the bond angle to be smaller because of the lone pair–lone pair repulsions.

9. a) SF_4 Draw a Lewis structure for the molecule:

SF_4 has 34 valence electrons

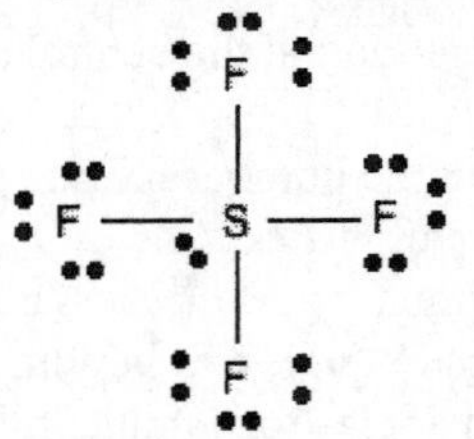

Determine the total number of electron groups around the central atom:
There are five electron groups on S
Determine the number of bonding groups and the number of lone pairs around the central atom:
There are four bonding groups and one lone pair

Use Table 10.1 to determine the electron geometry and molecular geometry:
The electron geometry is trigonal bipyramidal so the molecular geometry is seesaw
Sketch the molecule:

b) ClF_3 Draw a Lewis structure for the molecule:
ClF_3 has 28 valence electrons

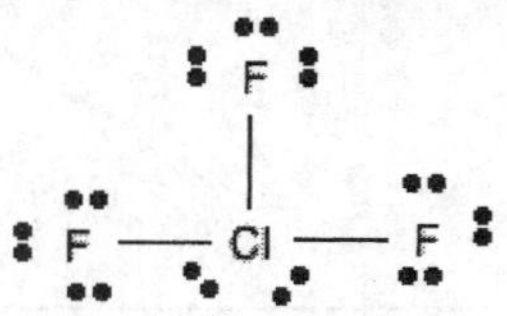

Determine the total number of electron groups around the central atom:
There are five electron groups on Cl
Determine the number of bonding groups and the number of lone pairs around the central atom:
There are three bonding groups and two lone pairs.
Use Table 10.1 to determine the electron geometry and molecular geometry:
The electron geometry is trigonal bipyramidal so the molecular geometry is T-shape
Sketch the molecule:

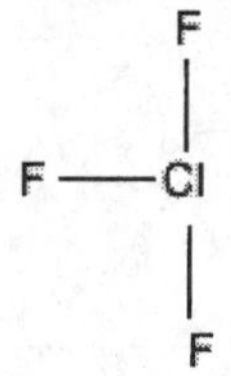

c) IF_2^- Draw a Lewis structure for the ion:
IF_2^- has 22 valence electrons

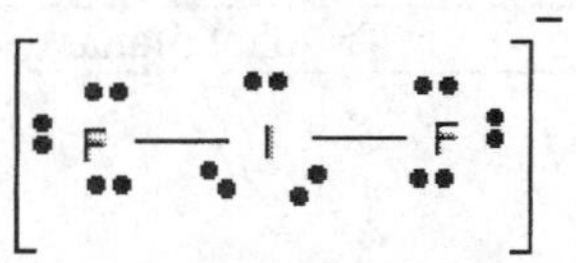

Determine the total number of electron groups around the central atom:
There are five electron groups on I
Determine the number of bonding groups and the number of lone pairs around the central atom:
There are two bonding groups and three lone pairs.
Use Table 10.1 to determine the electron geometry and molecular geometry:
The electron geometry is trigonal bipyramidal so the molecular geometry is linear
Sketch the ion:

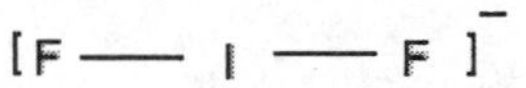

d) IBr_4^- Draw a Lewis structure for the ion:
IBr_4^- has 36 valence electrons

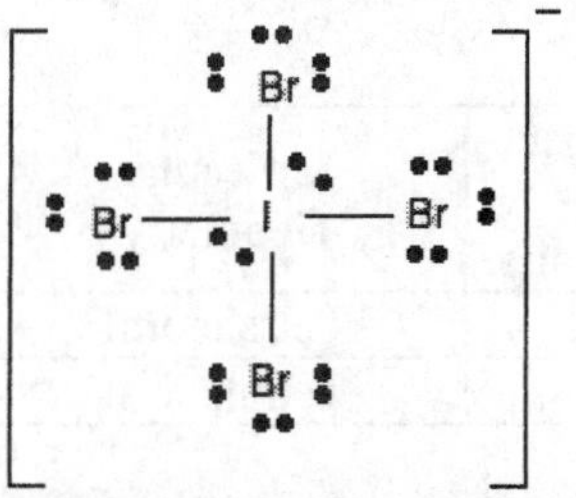

Determine the total number of electron groups around the central atom:

There are six electron groups on I
Determine the number of bonding groups and the number of lone pairs around the central atom:
There are four bonding groups and two lone pairs.
Use Table 10.1 to determine the electron geometry and molecular geometry:
The electron geometry is octahedral so the molecular geometry is square planar
Sketch the ion:

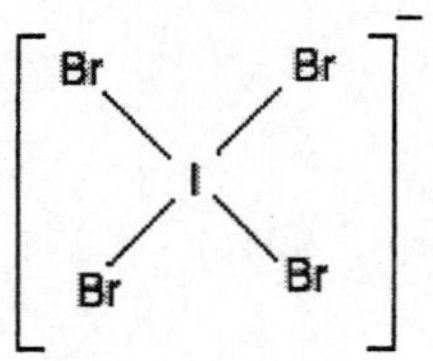

11. a) C_2H_2 Draw the Lewis structure:

Atom	Number of Electron Groups	Number of Lone Pairs	Molecular Geometry
Left C	2	0	Linear
Right C	2	0	Linear

Sketch the molecule:

b) C_2H_4 Draw the Lewis structure:

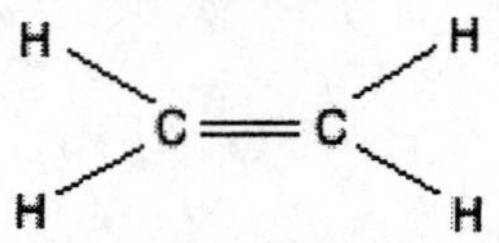

Atom	Number of Electron Groups	Number of Lone Pairs	Molecular Geometry
Left C	3	0	Trigonal planar
Right C	3	0	Trigonal planar

Sketch the molecule:

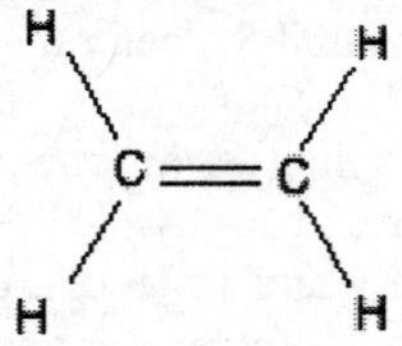

c) C_2H_6 Draw the Lewis structure:

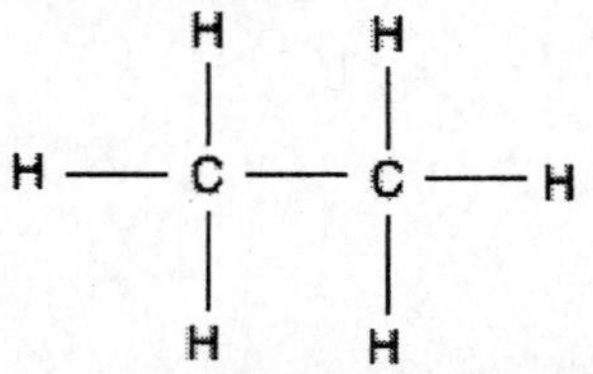

Atom	Number of Electron Groups	Number of Lone Pairs	Molecular Geometry
Left C	4	0	Tetrahedral
Right C	4	0	Tetrahedral

Sketch the molecule:

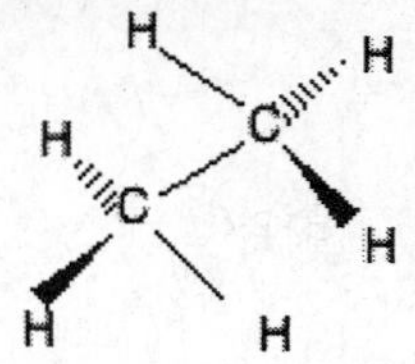

13. a) Four pairs of electrons gives a tetrahedral electron geometry, the lone pair would cause lone pair–bonded pair repulsions and would have a trigonal pyramidal molecular geometry.

b) Five pairs of electrons gives a trigonal bipyramidal electron geometry, the lone pair occupies an equatorial position in order to minimize lone pair–bonded pair repulsions and the molecule would have a seesaw molecular geometry.

c) Six pairs of electrons gives an octahedral electron geometry, the two lone pair would occupy opposite position in order to minimize lone pair–lone pair repulsions. The molecular geometry would be square planar.

15. a) CH_3OH Draw the Lewis structure, determine the geometry about each interior atom:

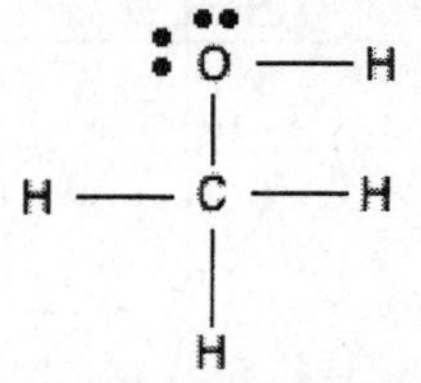

Atom	Number of Electron Groups	Number of Lone Pairs	Molecular Geometry
C	4	0	Tetrahedral
O	4	2	Bent

Sketch the molecule:

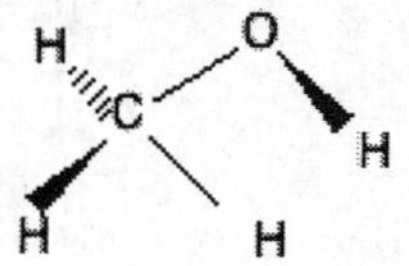

b) CH_3OCH_3 Draw the Lewis structure, determine the geometry about each interior atom:

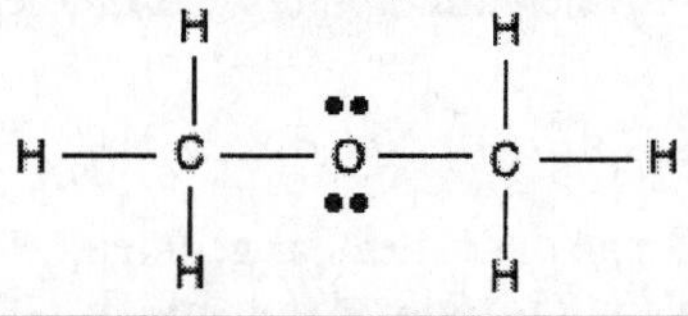

Atom	Number of Electron Groups	Number of Lone Pairs	Molecular Geometry
C	4	0	Tetrahedral
O	4	2	Bent
C	4	0	Tetrahedral

Sketch the molecule:

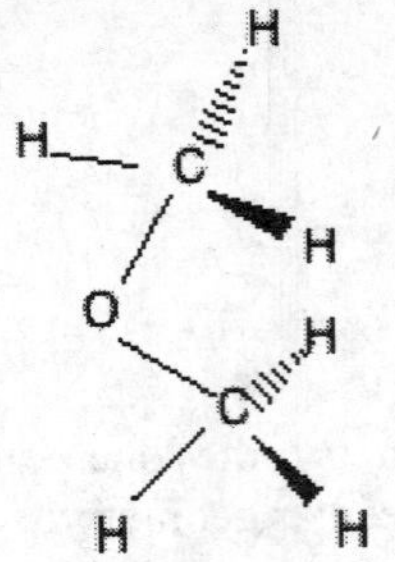

c) H_2O_2 Draw the Lewis structure, determine the geometry about each interior atom:

Atom	Number of Electron Groups	Number of Lone Pairs	Molecular Geometry
O	4	2	Bent
O	4	2	Bent

Sketch the molecule:

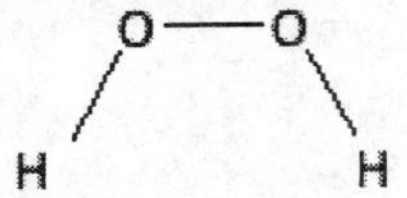

17. Draw the Lewis structure for CO_2 and CCl_4 and determine the molecular geometry.

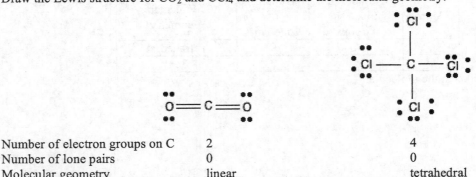

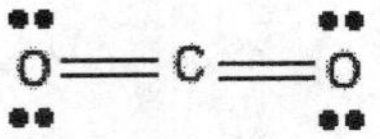

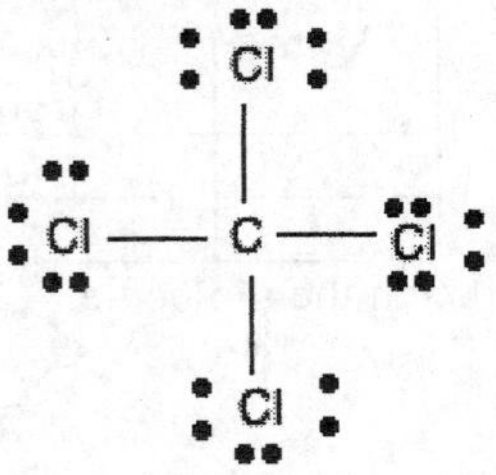

	CO_2	CCl_4
Number of electron groups on C	2	4
Number of lone pairs	0	0
Molecular geometry	linear	tetrahedral

Even though each molecule contains polar bonds, the sum of the bond dipoles gives a net dipole of zero for each molecule.

19. a) PF_3 – polar

Draw the Lewis structure and determine the molecular geometry:
The molecular geometry from problem 5 is trigonal pyramidal.

Determine if the molecule contains polar bonds:
The electronegativities of P = 2.1 and F = 4. Therefore the bonds are polar.

Determine whether the polar bonds add together to form a net dipole:
Because the molecule is trigonal pyramidal, the three dipole moments sum to a nonzero net dipole moment. The molecule is polar.

b) SBr_2 – polar

Draw the Lewis structure and determine the molecular geometry:
The molecular geometry from problem 5 is bent.

Determine if the molecule contains polar bonds:
The electronegativities of S = 2.5 and Br = 2.0. Therefore the bonds are polar.

Determine whether the polar bonds add together to form a net dipole:
Because the molecule is bent, the two dipole moments sum to a nonzero net dipole moment. The molecule is polar.

c) $CHCl_3$ – polar
Draw the Lewis structure and determine the molecular geometry:
The molecular geometry from problem 5 is tetrahedral.

Determine if the molecule contains polar bonds:
The electronegativities of C = 2.5, H = 2.1, and Cl = 3.0. Therefore the bonds are polar.

Determine whether the polar bonds add together to form a net dipole:
Because the bonds have different dipole moments because of the different atoms involved, the four dipole moments sum to a nonzero net dipole moment. The molecule is polar.

d) CS_2 – nonpolar
Draw the Lewis structure and determine the molecular geometry:
The molecular geometry from problem 5 is linear.

Determine if the molecule contains polar bonds:
The electronegativities of C = 2.5 and S = 2.5. Therefore the bonds are nonpolar. Also, the molecule is linear, which would result in a zero net dipole even if the bonds were polar.
The molecule is nonpolar.

21. a) ClO_3^- – polar
Draw the Lewis structure and determine the molecular geometry:

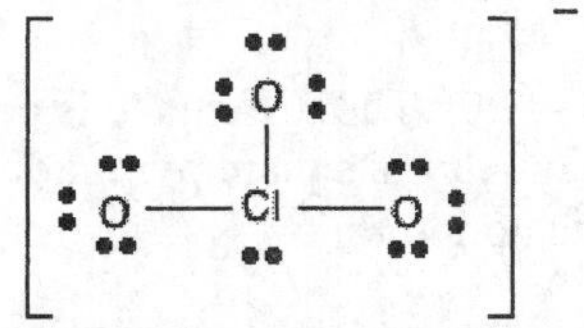

Four electron pairs, with one lone pair gives a trigonal pyramidal molecular geometry

Determine if the molecule contains polar bonds:
The electronegativities of Cl = 3.0 and O = 3.5. Therefore the bonds are polar.

Determine whether the polar bonds add together to form a net dipole:
Because the molecular geometry is trigonal pyramidal, the three dipole moments sum to a nonzero net dipole moment. The molecule is polar.

b) SCl_2 – polar
Draw the Lewis structure and determine the molecular geometry:

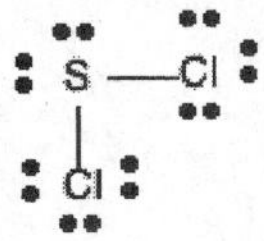

Four electron pairs with two lone pairs gives a bent molecular geometry

Determine if the molecule contains polar bonds:
The electronegativities of S = 2.5 and Cl = 3.0. Therefore the bonds are polar.

Determine whether the polar bonds add together to form a net dipole:
Because the molecular geometry is bent, the two dipole moments sum to a nonzero net dipole moment. The molecule is polar.

c) SCl_4 – polar
Draw the Lewis structure and determine the molecular geometry:

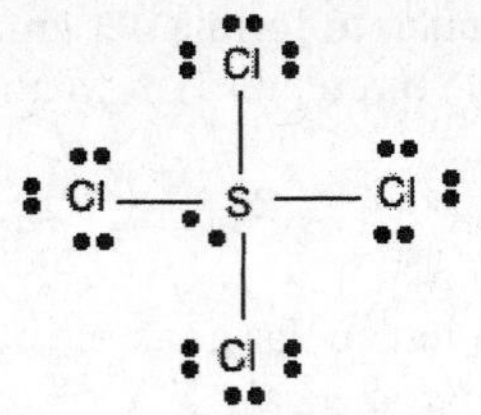

Five electron pairs with one lone pair give a seesaw molecular geometry

Determine if the molecule contains polar bonds:
The electronegativities of S = 2.5 and Cl = 3.0. Therefore the bonds are polar.

Determine whether the polar bonds add together to form a net dipole:
Because the molecular geometry is seesaw, the four dipole moments sum to a nonzero net dipole moment. The molecule is polar.

d) $BrCl_5$ – nonpolar
Draw the Lewis structure and determine the molecular geometry:

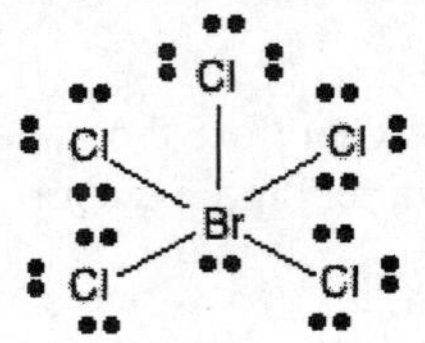

Six electron pairs with one lone pair give square pyramidal molecular geometry

Determine if the molecule contains polar bonds:
The electronegativity of Br = 2.8 and Cl = 3.0. The difference is only 0.2, therefore the bonds are nonpolar. Even though the molecular geometry is square pyramidal, the five bonds are nonpolar so there is no net dipole. The molecule is nonpolar.

23. a) Be $2s^2$ — 0 bonds can form. Beryllium contains no unpaired electrons, so no bonds can form without hybridization.

b) P $3s^2 3p^3$ — 3 bonds can form. Phosphorus contains 3 unpaired electrons, so 3 bonds can form without hybridization.

c) F $2s^2 2p^5$ — 1 bond can form. Fluorine contains 1 unpaired electron, so 1 bond can form without hybridization.

25. PH_3

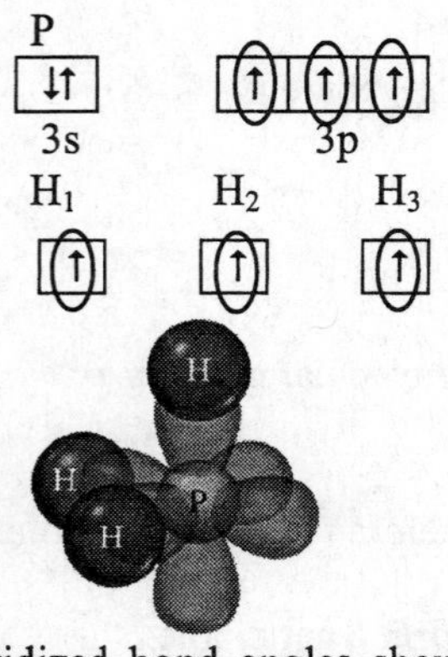

The unhybridized bond angles should be 90°. So, without hybridization, there is good agreement between valence bond theory and the actual bond angle of 93.3°

27. C $2s^2 2p^2$

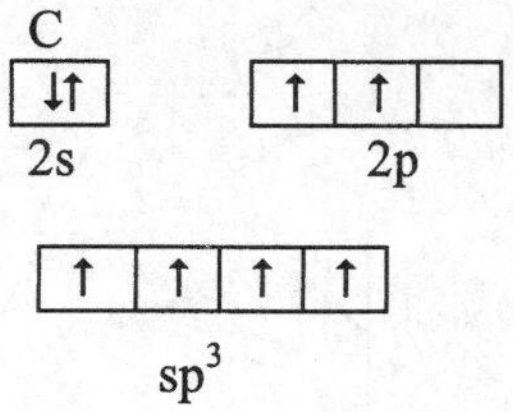

29. sp^2 Only sp^2 hybridization of this set of orbitals has a remaining p orbital to form a π bond.
sp^3 hybridization utilizes all 3 p orbitals
sp^3d^2 hybridization utilizes all 3 p orbitals and 2 d orbitals

31. a) CCl_4 Write the Lewis structure for the molecule:

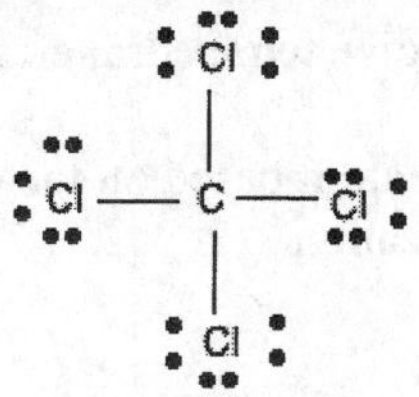

Use VSEPR to predict the electron geometry:
Four electron groups around the central atom give tetrahedral electron geometry

Select the correct hybridization for the central atom based on the electron geometry:
Tetrahedral electron geometry has sp^3 hybridization

Sketch the molecule and label the bonds:

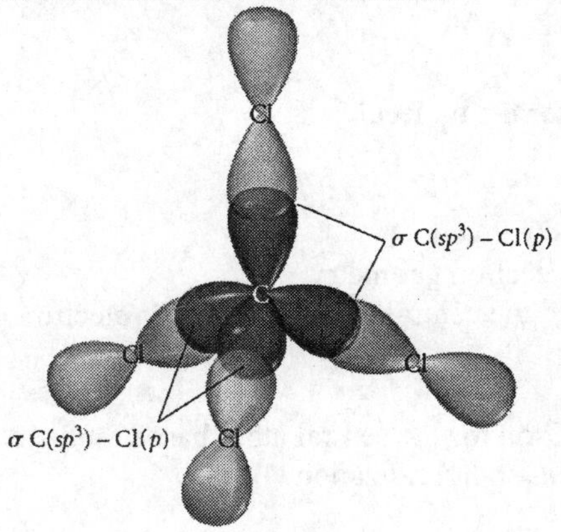

b) NH_3 Write the Lewis structure for the molecule:

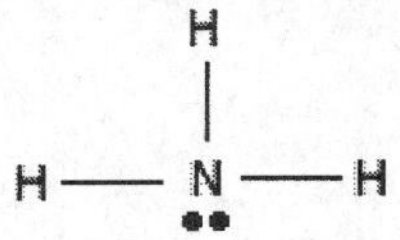

Use VSEPR to predict the electron geometry:
Four electron groups around the central atom give tetrahedral electron geometry

Select the correct hybridization for the central atom based on the electron geometry:
Tetrahedral electron geometry has sp^3 hybridization

Sketch the molecule and label the bonds:

Lone pair in N sp^3

$\sigma\ N(sp^3) - H(s)$ $\sigma\ N(sp^3) - H(s)$

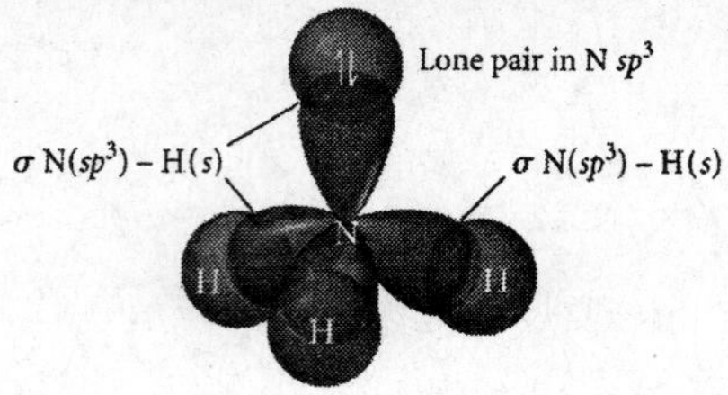

c) OF_2 Write the Lewis structure for the molecule:

Use VSEPR to predict the electron geometry:
Four electron groups around the central atom give tetrahedral electron geometry

Select the correct hybridization for the central atom based on the electron geometry:
Tetrahedral electron geometry has sp^3 hybridization

Sketch the molecule and label the bonds:

Lone pair in O sp^3

$\sigma\ O(sp^3) - F(p)$

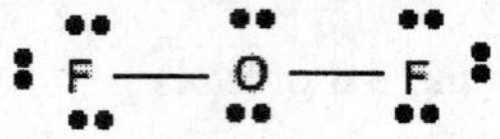

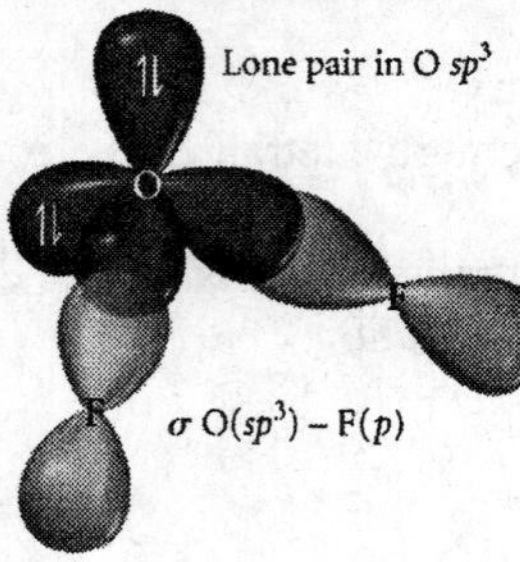

d) CO_2 Write the Lewis structure for the molecule:

$$\overset{\bullet\bullet}{\underset{\bullet\bullet}{O}} = C = \overset{\bullet\bullet}{\underset{\bullet\bullet}{O}}$$

Use VSEPR to predict the electron geometry:
Two electron groups around the central atom gives linear electron geometry

Select the correct hybridization for the central atom based on the electron geometry:
Linear electron geometry has sp hybridization

Sketch the molecule and label the bonds:

$\pi\ C(p_y) - O(p_y)$ $\pi\ C(p_z) - O(p_z)$

$\sigma\ C(sp) - O(p)$

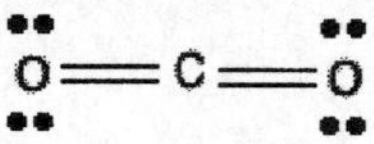

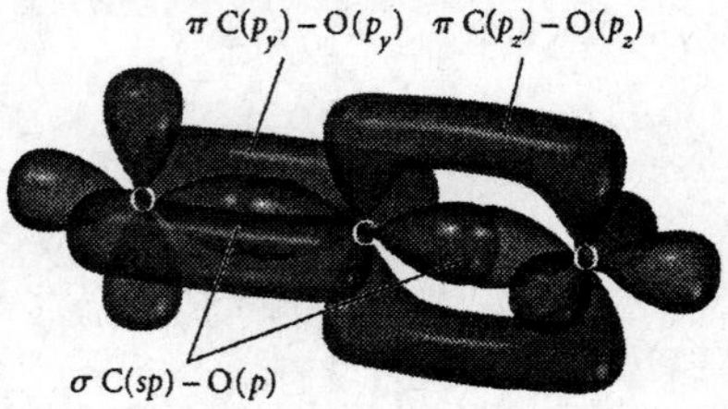

33. a) $COCl_2$ Write the Lewis structure for the molecule:

$$\overset{\bullet\ \bullet}{\underset{}{O}}$$
$$\|$$
$$\overset{\bullet\bullet}{\underset{\bullet\bullet}{Cl}} - C - \overset{\bullet\bullet}{\underset{\bullet\bullet}{Cl}}$$

Use VSEPR to predict the electron geometry:
Three electron groups around the central atom gives trigonal planar electron geometry

Select the correct hybridization for the central atom based on the electron geometry:

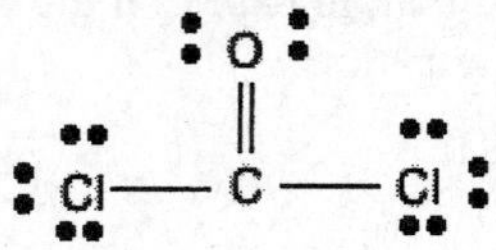

Trigonal planar electron geometry has sp^2 hybridization

Sketch the molecule and label the bonds:

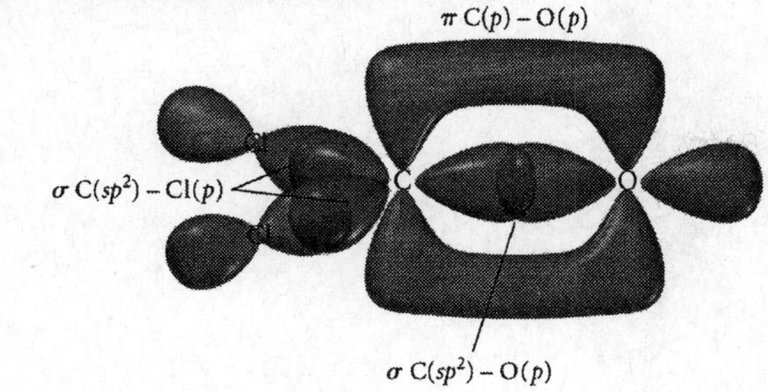

b) BrF_5 Write the Lewis structure for the molecule:

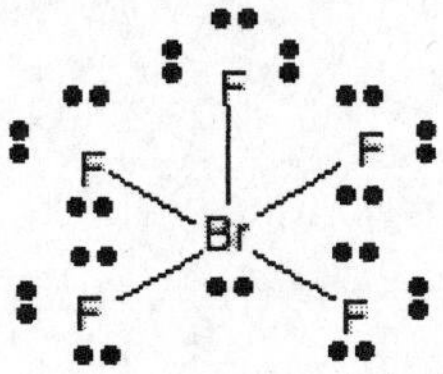

Use VSEPR to predict the electron geometry:
Six electron pairs around the central atoms gives octahedral electron geometry

Select the correct hybridization for the central atom based on the electron geometry:
Octahedral electron geometry has sp^3d^2 hybridization

Sketch the molecule and label the bonds:

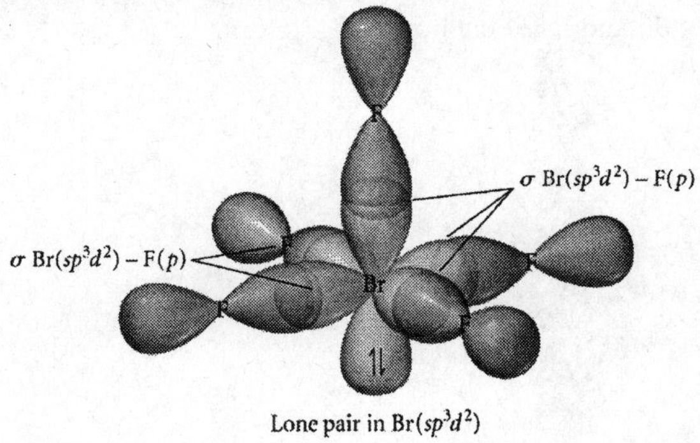

c) XeF_2 Write the Lewis structure for the molecule:

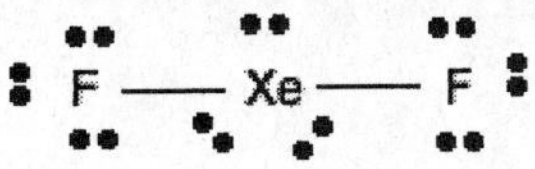

Use VSEPR to predict the electron geometry:
Five electron groups around the central atom gives trigonal bipyramidal geometry

Select the correct hybridization for the central atom based on the electron geometry:
Trigonal bipyramidal geometry has sp^3d hybridization

Sketch the molecule and label the bonds:

3 lone pair in Xe
sp^3d equatorial

$\sigma\ \text{Xe}(sp^3d) - \text{F}(p)$

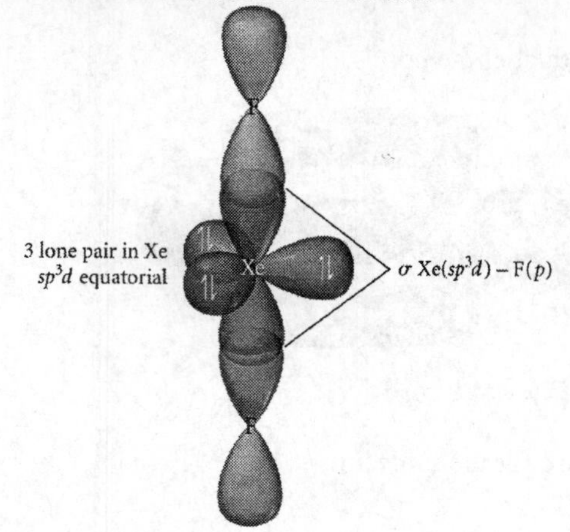

d) I_3^- Write the Lewis structure for the molecule:

$$\left[\ :\ddot{\text{I}}\ \overset{\cdot\cdot}{-}\ \ddot{\text{I}}\ \overset{\cdot\cdot}{-}\ \ddot{\text{I}}\ :\right]^-$$

Use VSEPR to predict the electron geometry:
Five electron groups around the central atom gives trigonal bipyramidal geometry

Select the correct hybridization for the central atom based on the electron geometry:
Trigonal bipyramidal geometry has sp^3d hybridization

Sketch the molecule and label the bonds:

3 lone pair in I
sp^3d equatorial

$\sigma\ \text{I}(sp^3d) - \text{I}(p)$

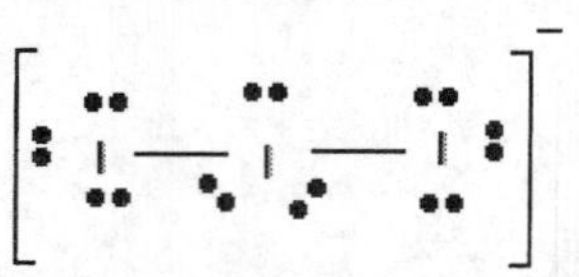

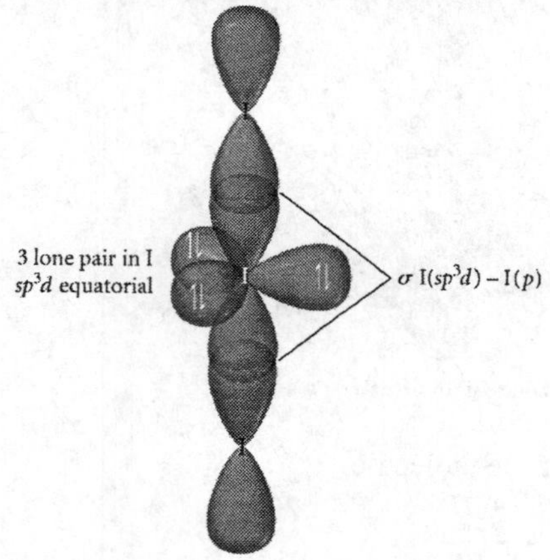

35. a) N_2H_2 Write the Lewis structure for the molecule:

$$\text{H} \overset{\ }{\text{---}} \overset{\cdot\cdot}{\text{N}} \text{===} \overset{\cdot\cdot}{\text{N}} \text{---} \text{H}$$

Use VSEPR to predict the electron geometry:
Three electron groups around each interior atom gives trigonal planar electron geometry

Select the correct hybridization for the central atom based on the electron geometry:
Trigonal planar electron geometry has sp^2 hybridization

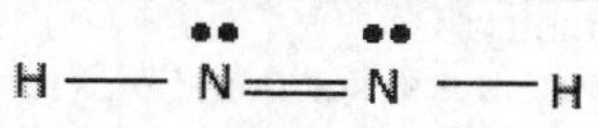

Sketch the molecule and label the bonds:

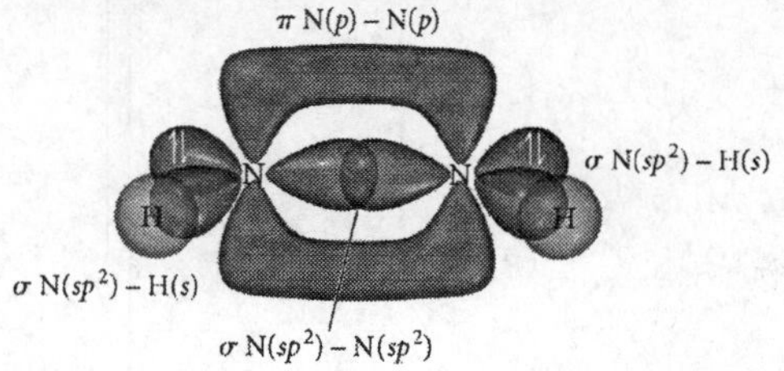

b) N_2H_4 Write the Lewis structure for the molecule:

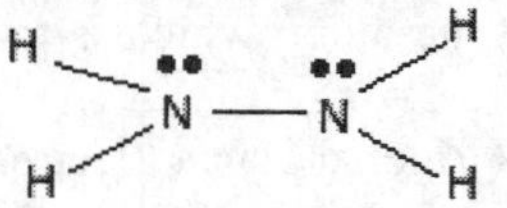

Use VSEPR to predict the electron geometry:
Four electron groups around each interior atom gives tetrahedral electron geometry

Select the correct hybridization for the central atom based on the electron geometry:
Tetrahedral electron geometry has sp^3 hybridization

Sketch the molecule and label the bonds:

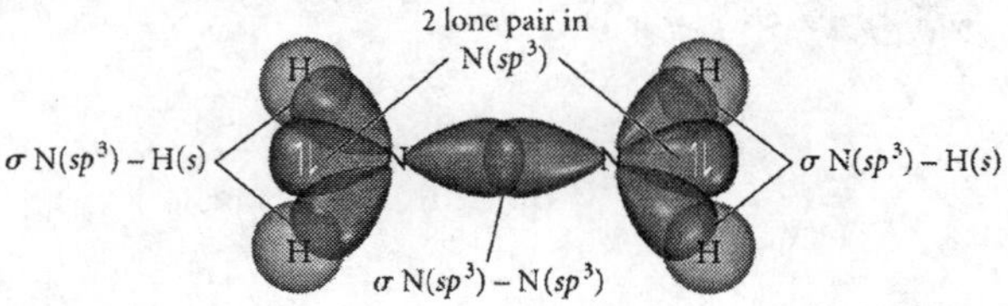

c) CH_3NH_2 Write the Lewis structure for the molecule:

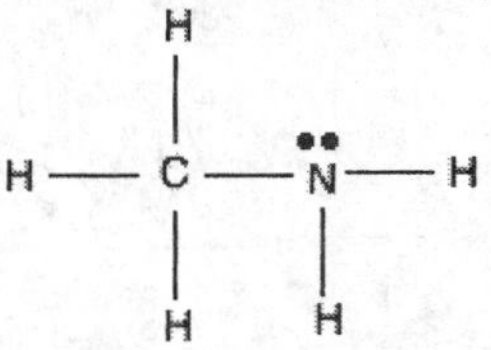

Use VSEPR to predict the electron geometry:
Four electron groups around the C gives tetrahedral electron geometry, four electron groups around the N gives tetrahedral geometry.

Select the correct hybridization for the central atom based on the electron geometry:
Tetrahedral electron geometry has sp^3 hybridization of both C and N

Sketch the molecule and label the bonds:

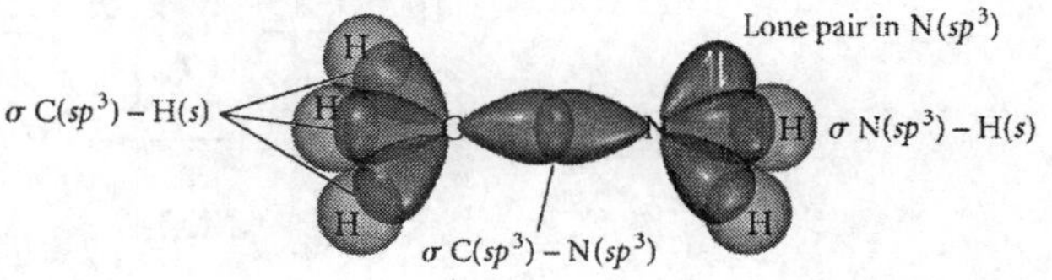

37. 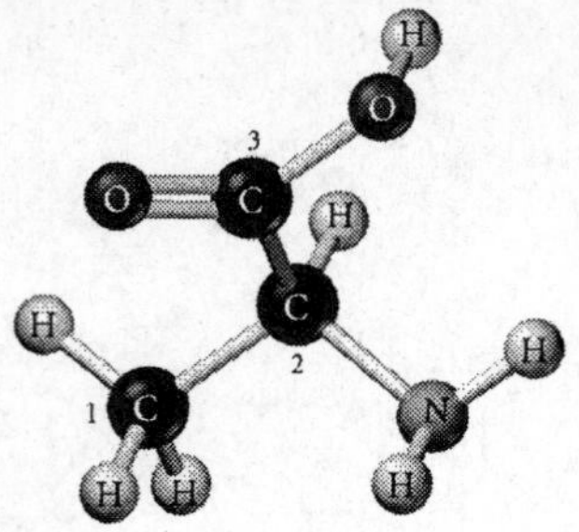

C – 1 and C – 2 each have four electron pairs around the atom, which is tetrahedral electron pair geometry. Tetrahedral electron pair geometry is sp^3 hybridization.
C – 3 has three electron pairs around the atom, which is trigonal planar electron pair geometry. Trigonal planar electron pair geometry is sp^2 hybridization.
O has four electron pairs around the atom, which is tetrahedral electron pair geometry. Tetrahedral electron pair geometry is sp^3 hybridization.
N has four electron pairs around the atom, which is tetrahedral electron pair geometry. Tetrahedral electron pair geometry is sp^3 hybridization.

39. 1s + 1s constructive interference results in a bonding orbital

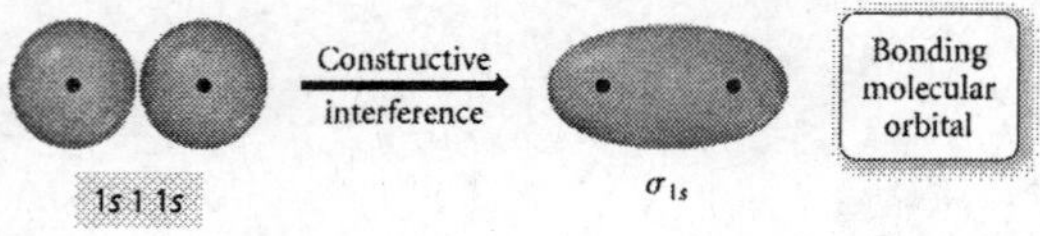

41. Be_2^+ has 7 electrons

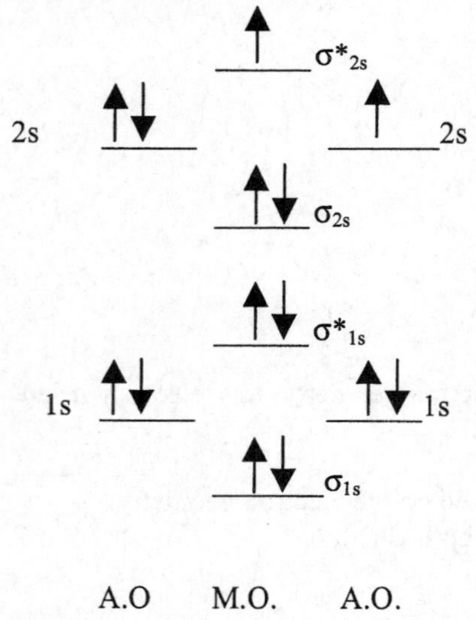

Be_2^- has 9 electrons

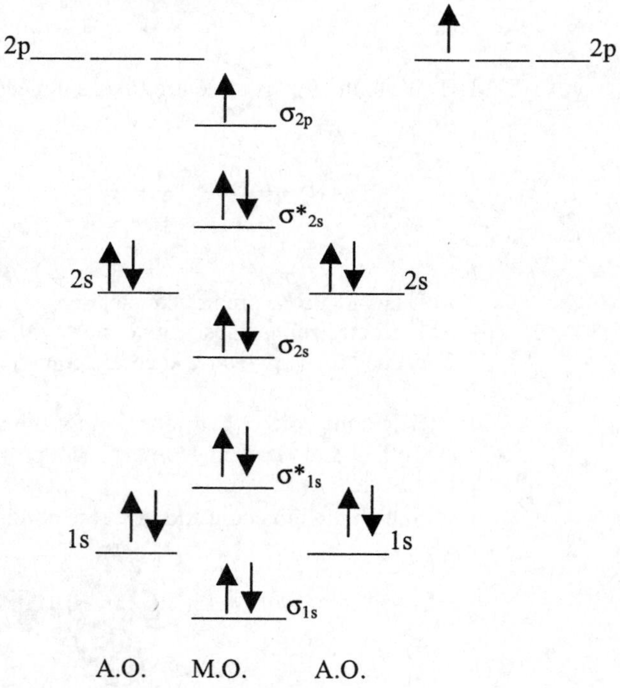

A.O.= Atomic Orbital; M.O. = Molecular Orbital

Bond order = $\frac{4-3}{2} = \frac{1}{2}$ stable

Bond order = $\frac{5-4}{2} = \frac{1}{2}$ stable

43. The bonding and antibonding molecular orbitals from the combination of p_x and p_x atomic orbitals lie along the internuclear axis.

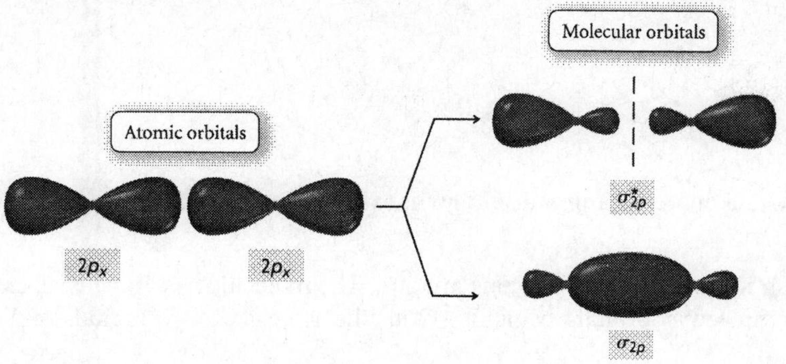

45. a) 4 valence electrons b) 6 valence electrons c) 8 valence electrons d) 9 valence electrons

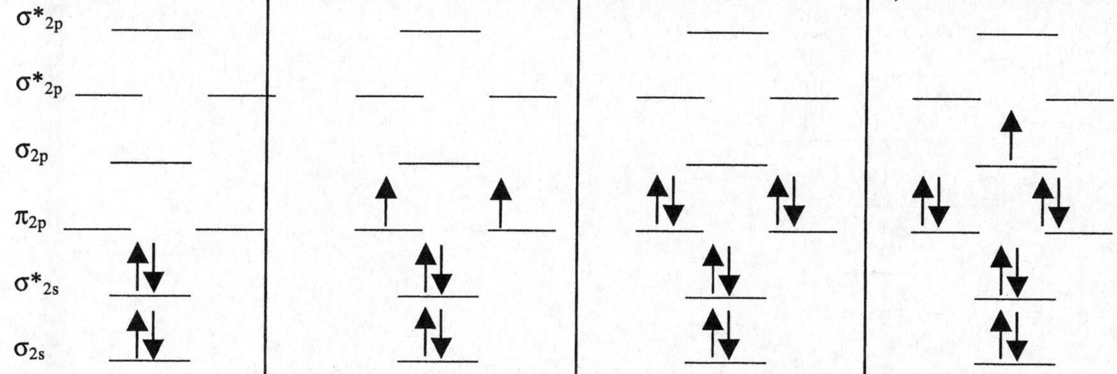

a) Bond order = $\frac{2-2}{2}=0$ diamagnetic

b) Bond order = $\frac{4-2}{2}=1$ paramagnetic

c) Bond order = $\frac{6-2}{2}=2$ diamagnetic

d) Bond order = $\frac{7-2}{2}=2.5$ paramagnetic

47. a) Write an energy level diagram for the molecular orbitals in H_2^{2-}. The ion has 4 valence electrons. Assign the electrons to the molecular orbitals beginning with the lowest energy orbitals and following Hund's rule.

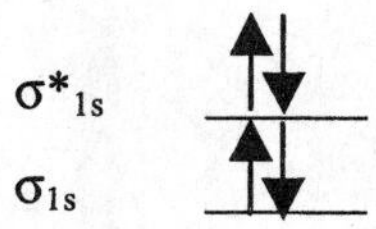

Bond order = $\frac{2-2}{2}=0$ With a bond order of 0, the ion will not exist.

b) Write an energy level diagram for the molecular orbitals in Ne_2. The molecule has 16 valence electrons. Assign the electrons to the molecular orbitals beginning with the lowest energy orbitals and following Hund's rule.

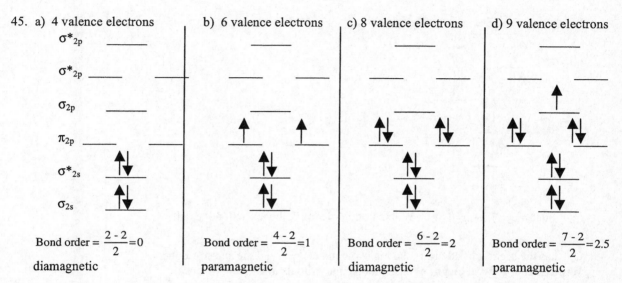

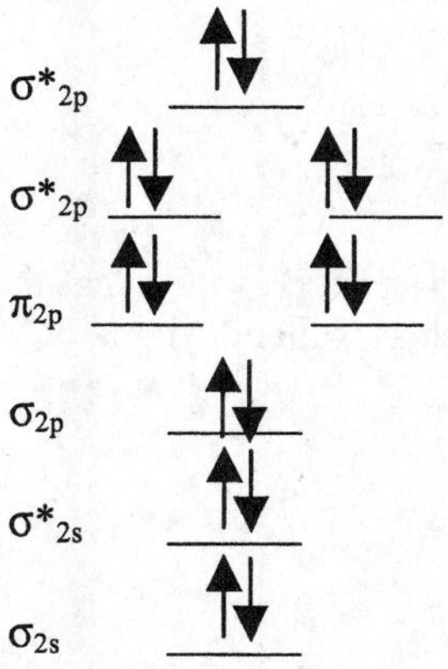

Bond order = $\frac{8-8}{2}=0$ With a bond order of 0, the molecule will not exist.

c) Write an energy level diagram for the molecular orbitals in He_2^{2+}. The ion has 2 valence electrons. Assign the electrons to the molecular orbitals beginning with the lowest energy orbitals and following Hund's rule.

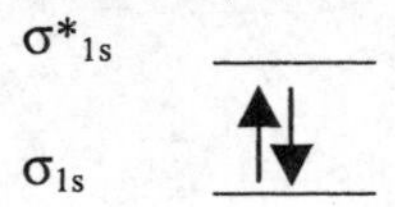

Bond order = $\frac{2-0}{2}=1$ With a bond order of 1, the ion will exist.

d) Write an energy level diagram for the molecular orbitals in F_2^{2-}. The molecule has 16 valence electrons. Assign the electrons to the molecular orbitals beginning with the lowest energy orbitals and following Hund's rule.

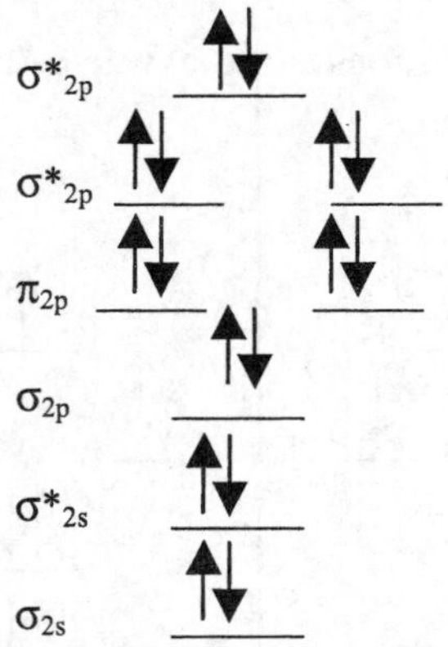

Bond order = $\frac{8-8}{2}=0$ With a bond order of 0, the ion will not exist.

49. C_2^- has the highest bond order, the highest bond energy, and the shortest bond.
Write an energy level diagram for the molecular orbitals in C_2.
Assign the electrons to the molecular orbitals beginning with the lowest energy orbitals and following Hund's rule for each of the species.
C_2 (8 valence electrons); C_2^+ (7 valence electrons): C_2^- (9 valence electrons)

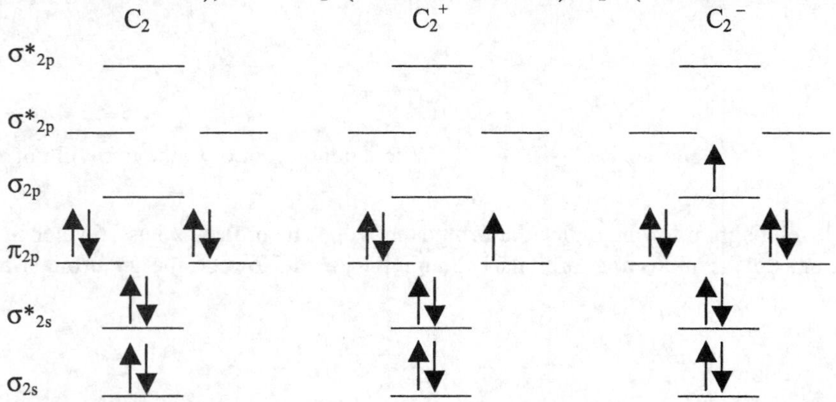

Bond order = $\frac{6-2}{2}=2$ Bond order = $\frac{5-2}{2}=1.5$ Bond order = $\frac{7-2}{2}=2.5$

C_2^- has the highest bond order at 2.5. Bond order is directly related to bond energy, so C_2^- has the largest bond energy and bond order is inversely related to bond length, so C_2^- has the shortest bond length.

51. a) COF_2 Write the Lewis structure for the molecule:

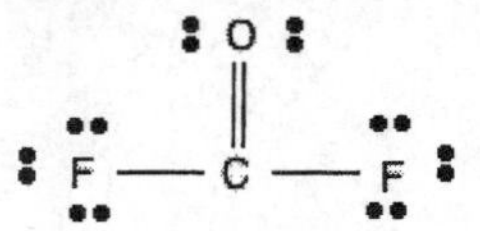

Use VSEPR to predict the electron geometry:

Three electron groups around the central atom gives trigonal planar electron geometry. Three bonding pairs of electrons give trigonal planar molecular geometry.

Determine if the molecule contains polar bonds:
The electronegativities of C = 2.5, O = 3.5, and F = 4.0 Therefore the bonds are polar.

Determine whether the polar bonds add together to form a net dipole:
Even though a trigonal planar molecular geometry normally is nonpolar, because the bonds have different dipole moments, the sum of the dipole moments is not zero. The molecule is polar.

Select the correct hybridization for the central atom based on the electron geometry:
Trigonal planar geometry has sp^2 hybridization.

Sketch the molecule and label the bonds:

$\pi\ C(p) - O(p)$

$\sigma\ C(sp^3) - F(p)$

$\sigma\ C(sp^2) - O(p)$

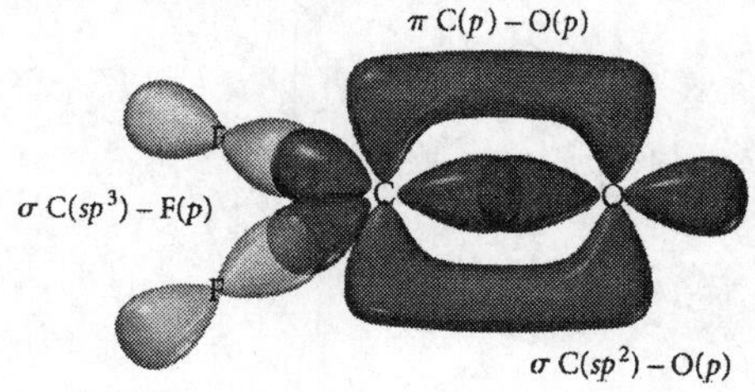

b) S_2Cl_2 Write the Lewis structure for the molecule:

$$\overset{\cdot\cdot}{\underset{\cdot\cdot}{Cl}} - \overset{\cdot\cdot}{\underset{\cdot\cdot}{S}} - \overset{\cdot\cdot}{\underset{\cdot\cdot}{S}} - \overset{\cdot\cdot}{\underset{\cdot\cdot}{Cl}}$$

Use VSEPR to predict the electron geometry:
Four electron groups around the central atom gives tetrahedral electron geometry. Two bonding pairs and two lone pairs of electrons gives bent molecular geometry.

Determine if the molecule contains polar bonds:
The electronegativities of S = 2.5 and Cl = 3.0. Therefore the bonds are polar.

Determine whether the polar bonds add together to form a net dipole:
In a bent molecular geometry the sum of the dipole moments is not zero. The molecule is polar.

Select the correct hybridization for the central atom based on the electron geometry:
Tetrahedral geometry has sp^3 hybridization.

Sketch the molecule and label the bonds:

$\sigma\ S(sp^3) - S(sp^3)$

$\sigma\ S(sp^3) - Cl(p)$

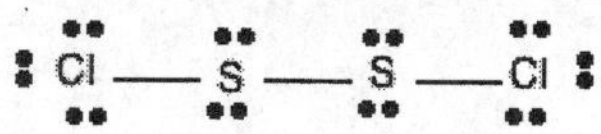

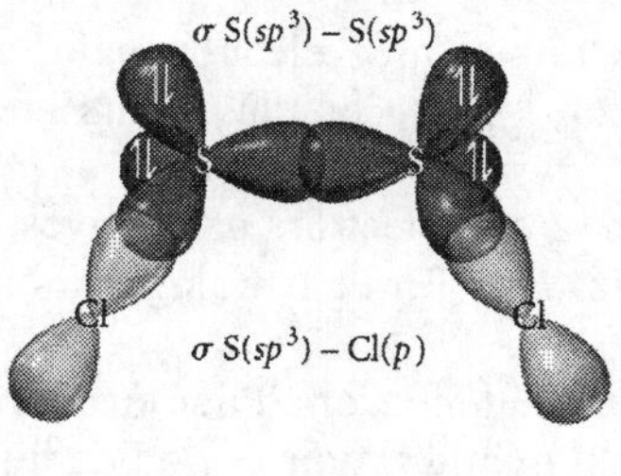

c) SF_4 Write the Lewis structure for the molecule:

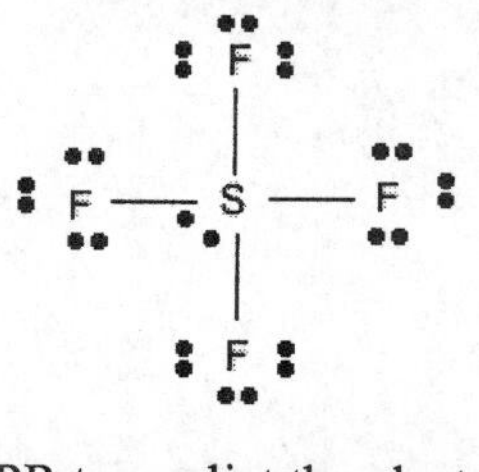

Use VSEPR to predict the electron geometry:

Five electron groups around the central atom gives trigonal bipyramidal electron geometry. Four bonding pairs and one lone pair of electrons gives seesaw molecular geometry.

Determine if the molecule contains polar bonds:
The electronegativities of S = 2.5 and F = 4.0. Therefore the bonds are polar.

Determine whether the polar bonds add together to form a net dipole:
In a seesaw molecular geometry the sum of the dipole moments is not zero. The molecule is polar.

Select the correct hybridization for the central atom based on the electron geometry:
Trigonal bipyramidal electron geometry has sp^3d hybridization.

Sketch the molecule and label the bonds:

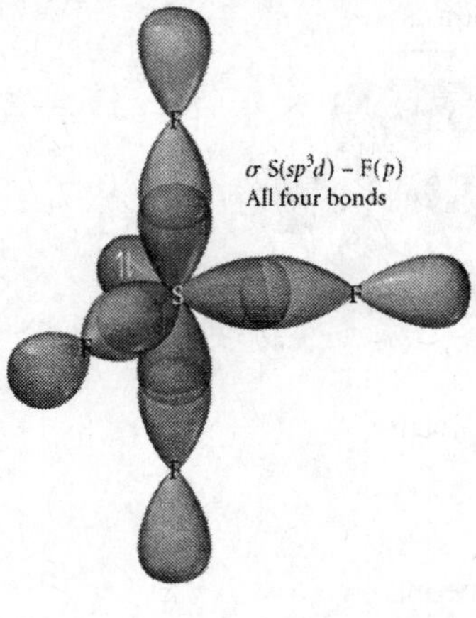

53. a) serine

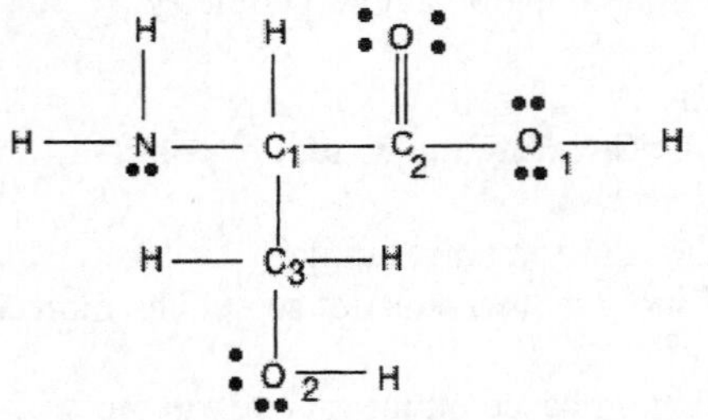

C – 1 and C – 3 each have four electron pairs around the atom. Four electron pairs gives tetrahedral electron geometry, tetrahedral electron geometry has sp^3 hybridization. Four bonding pairs and zero lone pairs gives tetrahedral molecular geometry.
C – 2 has three electron pairs around the atom. Three electron pairs gives trigonal planar geometry, trigonal planar geometry has sp^2 hybridization. Three bonding pairs and zero lone pair gives trigonal planar molecular geometry.
N has four electron pairs around the atom. Four electron pairs gives tetrahedral electron geometry, tetrahedral electron geometry has sp^3 hybridization. Three bonding pairs and one lone pair gives trigonal pyramidal molecular geometry.
O – 1 and O – 2 each has four electron pairs around the atom. Four electron pairs gives tetrahedral electron geometry, tetrahedral electron geometry has sp^3 hybridization. Two bonding pairs and two lone pairs gives bent molecular geometry.

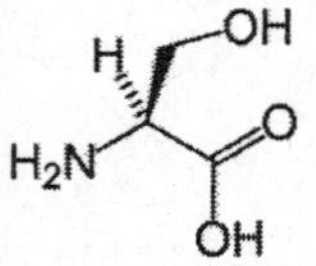

b) asparagine

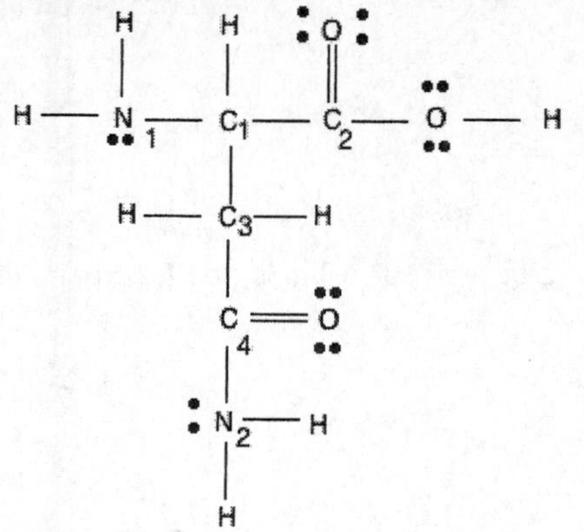

C – 1 and C – 3 each have four electron pairs around the atom. Four electron pairs gives tetrahedral electron geometry, tetrahedral electron geometry has sp^3 hybridization. Four bonding pairs and zero lone pairs gives tetrahedral molecular geometry.
C – 2 and C – 4 each have three electron pairs around the atom. Three electron pairs gives trigonal planar geometry, trigonal planar geometry has sp^2 hybridization. Three bonding pairs and zero lone pairs gives trigonal planar molecular geometry.
N – 1 and N – 2 each have four electron pairs around the atom. Four electron pairs gives tetrahedral electron geometry, tetrahedral electron geometry has sp^3 hybridization. Three bonding pairs and one lone pair gives trigonal pyramidal molecular geometry.
O has four electron pairs around the atom. Four electron pairs gives tetrahedral electron geometry, tetrahedral electron geometry has sp^3 hybridization. Two bonding pairs and two lone pairs gives bent molecular geometry.

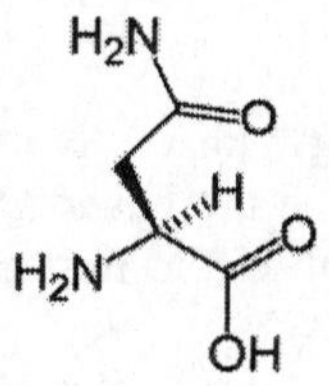

c) cysteine

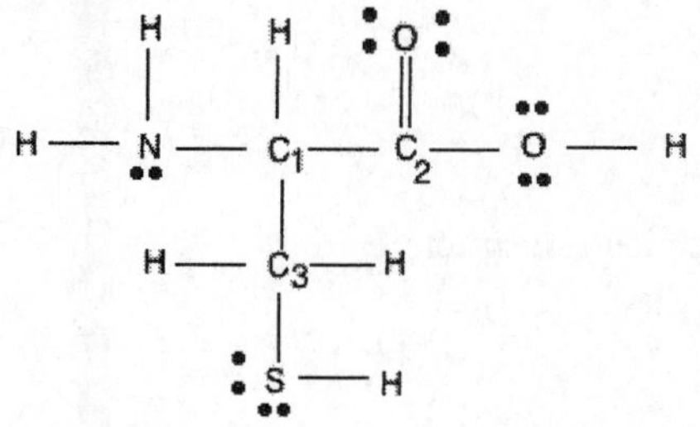

C – 1 and C – 3 each have four electron pairs around the atom. Four electron pairs gives tetrahedral electron geometry, tetrahedral electron geometry has sp^3 hybridization. Four bonding pairs and zero lone pairs gives tetrahedral molecular geometry.
C – 2 has three electron pairs around the atom. Three electron pairs gives trigonal planar geometry, trigonal planar geometry has sp^2 hybridization. Three bonding pairs and zero lone pairs gives trigonal planar molecular geometry.
N has four electron pairs around the atom. Four electron pairs gives tetrahedral electron geometry, tetrahedral electron geometry has sp^3 hybridization. Three bonding pairs and one lone pair gives trigonal pyramidal molecular geometry.
O and S have four electron pairs around the atom. Four electron pairs gives tetrahedral electron geometry, tetrahedral electron geometry has sp^3 hybridization. Two bonding pairs and two lone pairs gives bent molecular geometry.

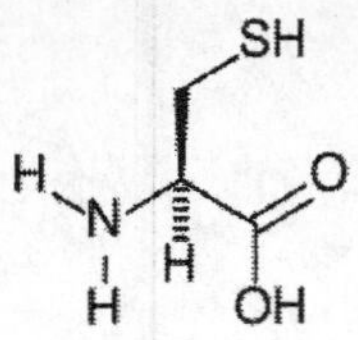

55. 4 π bonds; 25 σ bonds; the lone pair on the O's and N – 2 occupy sp^2 orbitals, the lone pairs on N – 1, N – 3, and N – 4 occupy sp^3 orbitals.

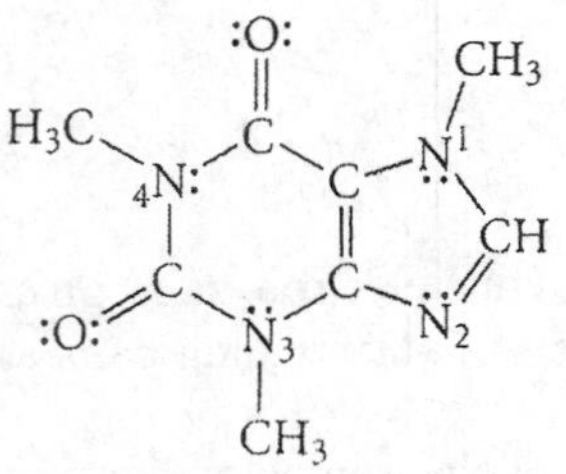

57. a) water soluble – the 4 C – OH bonds, the C = O bond and the C – O bonds in the ring, make the molecule polar. Because of the large electronegativity difference between the C and O, each of the bonds will have a dipole moment. The sum of the dipole moments does NOT give a net zero dipole moment, so the molecule is polar. Since it is polar, it will be water soluble.

b) fat soluble – There is only one C – O bond in the molecule. The dipole moment from this bond is not enough to make the molecule polar because of all of the nonpolar components of the molecule. The C – H bonds in the structure lead to a net dipole of zero for most of the sites in the molecule. Since the molecule is nonpolar, it is fat soluble.

c) water soluble – the carboxylic acid function (COOH group) along with the N atom in the ring make the molecule polar. Because of the electronegativity difference between the C and O and the C and N atoms, the bonds will have a dipole moment and the net dipole moment of the molecule is NOT zero, so the molecule is polar. Since the molecule is polar, it is water soluble.

d) fat soluble – The two O atoms in the structure contribute a very small amount to the net dipole moment of this molecule. The majority of the molecule is nonpolar because there is no net dipole moment at the interior C atoms. Because the molecule is nonpolar it is fat soluble.

59. BrF (14 valence electrons)

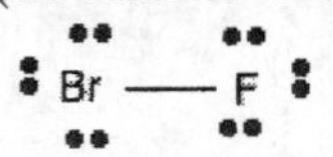

no central atom, no hybridization, no electron structure

BrF_2^- (22 valence electrons)

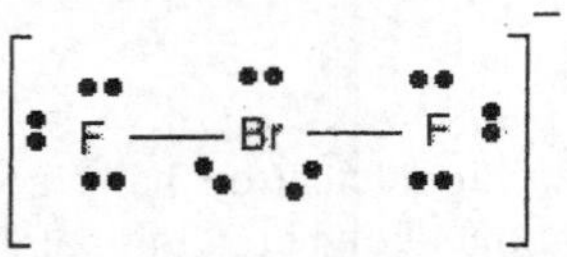

Five electron pairs on the central atom, electron geometry is trigonal bipyramidal, two bonding pairs and three lone pairs gives a linear molecular geometry. An electron geometry of trigonal bipyramidal has sp^3d hybridization.

BrF_3 (28 valence electrons)

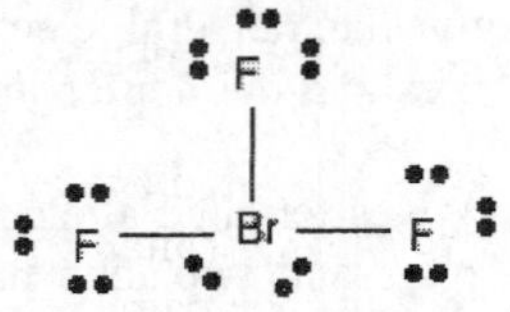

Five electron pairs on the central atom, electron geometry is trigonal bipyramidal, three bonding pairs and two lone pairs gives a T-shaped molecular geometry. An electron geometry of trigonal bipyramidal has sp^3d hybridization.

BrF_4^- (36 valence electrons)

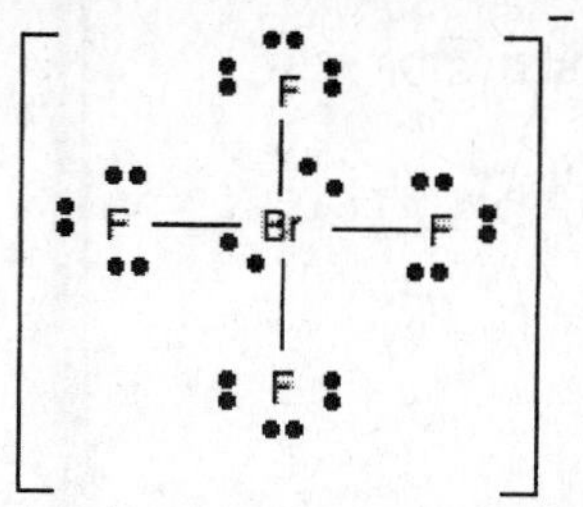

Six electron pairs on the central atom, electron geometry is octahedral, four bonding pairs and two lone pairs gives a square planar molecular geometry. An electron geometry of octahedral has sp^3d^2 hybridization.

BrF_5 (42 valence electrons)

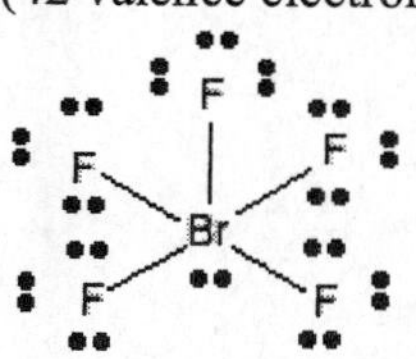

Six electron pairs on the central atom, electron geometry is octahedral, five bonding pairs and one lone pair gives a square pyramidal molecular geometry. An electron geometry of octahedral has sp^3d^2 hybridization.

61. According to valence bond theory, CH_4, NH_3, and H_2O are all sp^3 hybridized. This hybridization results in a tetrahedral electron group configuration with a 109.5° bond angle. NH_3 and H_2O deviate from this idealized bond angle because their lone electron pairs exist in their own sp^3 orbitals. The presence of lone pairs lowers the tendency for the central atom's orbitals to hybridize. As a result, as lone pairs are added, the bond angle moves further from the 109.5° hybrid angle to the 90° unhybridized angle.

63. Write the Lewis structure for :
Determine electron pair geometry around each central atom:
Determine the molecular geometry, determine idealized bond angles and predict actual bond angles:
NO_2

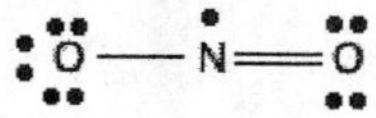

2 bonding pairs and a lone electron gives trigonal planar electron geometry, the molecular geometry will be bent. Trigonal planar electron geometry has idealized bond angles of 120°. The bond angle is expected to be slightly less than 120° because of the lone electron occupying the third sp^2 orbital.

NO_2^+

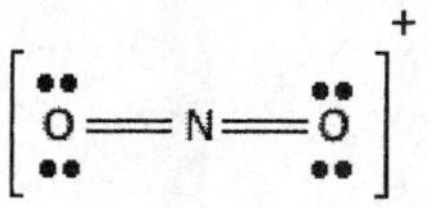

Two bonding pairs of electrons and no lone pairs gives linear electron geometry and molecular geometry. Linear electron geometry has a bond angle of 180°.

NO_2^-

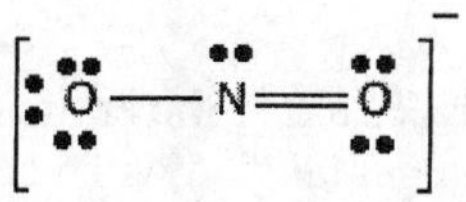

Two bonding pairs of electrons and one lone pair gives trigonal planar electron geometry, the molecular geometry will be bent.
Trigonal planar electron geometry has idealized bond angles of 120°. The bond angle is expected to be less than 120° because of the lone pair electrons occupying the third sp^2 orbital. Further, the bond angle should be less than the bond angle in NO_2 because the presence of lone pairs lowers the tendency for the central atom's orbitals to hybridize. As a result, as lone pairs are added, the bond angle moves further from the 120° hybrid angle to the 90° unhybridized angle and the two electrons will increase this tendency.

Chapter 11
Liquids, Solids, and Intermolecular Forces

1. a) dispersion forces

   b) dispersion forces and dipole–dipole forces

   c) dispersion forces

   d) dispersion forces, dipole–dipole forces, and hydrogen bonding

   e) dispersion forces

   f) dispersion forces, dipole–dipole forces and hydrogen bonding

   g) dispersion forces and dipole–dipole forces

   h) dispersion forces

3. a) CH_4 < b) CH_3CH_3 < c) CH_3CH_2Cl < d) CH_3CH_2OH. The first two molecules only exhibit dispersion forces, so the boiling point increases with increasing molar mass. The third molecule also exhibits dipole–dipole forces, which are stronger than dispersion forces. The last molecule exhibits hydrogen bonding. Since these are the strongest intermolecular forces in this group, the last molecule has the highest boiling point.

5. a) CH_3OH has the higher boiling point since it exhibits hydrogen bonding.

   b) CH_3CH_2OH has the higher boiling point since it exhibits hydrogen bonding.

   c) CH_3CH_3 has the higher boiling point since it has the larger molar mass.

7. a) Br_2 has the higher vapor pressure since it has the smaller molar mass.

   b) H_2S has the higher vapor pressure since it does not exhibit hydrogen bonding.
   c) PH_3 has the higher vapor pressure since it does not exhibit hydrogen bonding.

9. a) This will not form a homogeneous solution, since one is polar and one is nonpolar.

   b) This will form a homogeneous solution. There will be ion–dipole interactions between the K^+ and Cl^- ions and the water molecules. There will also be dispersion forces, dipole–dipole forces, and hydrogen bonding between the water molecules.

   c) This will form a homogeneous solution. There will be dispersion forces present.

   d) This will form a homogeneous solution. There will be dispersion forces, dipole–dipole forces, and hydrogen bonding.

11. Water will have the higher surface tension since it exhibits hydrogen bonding, a strong intermolecular force. Acetone cannot form hydrogen bonds.

13. Compound A will have the higher viscosity since it can interact with other molecules along the entire molecule, not just at a single point. Also the molecule is very flexible and the molecules can get tangled with each other.

15. In a clean glass tube the water can generate strong adhesive interactions with the glass (due to the dipoles at the surface of the glass). Water experiences adhesive forces with glass that are stronger than its cohesive forces, causing it to climb the surface of a glass tube. When grease or oil coats the glass this interferes with the formation of these adhesive interactions with the glass, since oils are nonpolar and cannot interact strongly with the dipoles in the water. Without experiencing these strong intermolecular forces with oil, the water's cohesive forces will be greater and it will be drawn away from the surface of the tube.

17. The water in the 12 cm diameter beaker will evaporate more quickly because there is more surface area for the molecules to evaporate from. The vapor pressure will be the same in the two containers because the vapor pressure is the pressure of the gas when it is in dynamic equilibrium with the liquid (evaporation rate = condensation rate). The vapor pressure is dependent only on the substance and the temperature. The 12 cm diameter container will reach this dynamic equilibrium faster.

19. The boiling point and higher heat of vaporization of oil are much higher than that of water, so it will not vaporize as quickly as the water. The evaporation of water cools your skin because evaporation is an endothermic process.

21. **Given:** 955 kJ from candy bar, water $d = 1.00$ g/ml **Find:** L(H_2O) vaporized at 100.0 °C
Other: $\Delta H^\circ_{vap} = 40.7$ kJ/mol
Conceptual Plan: $q \rightarrow$ **mol H_2O** $\rightarrow$ **g H_2O** $\rightarrow$ **mL H_2O** $\rightarrow$ **L H_2O**

$$\frac{1\,\text{mol}}{40.7\,\text{kJ}} \quad \frac{18.01\,\text{g}}{1\,\text{mol}} \quad \frac{1.00\,\text{mL}}{1.00\,\text{g}} \quad \frac{1\,\text{L}}{1000\,\text{mL}}$$

Solution: $$955\,\text{kJ} \times \frac{1\,\text{mol}}{40.7\,\text{kJ}} \times \frac{18.01\,\text{g}}{1\,\text{mol}} \times \frac{1.00\,\text{mL}}{1\,\text{g}} \times \frac{1\,\text{L}}{1000\,\text{mL}} = 0.423\,\text{L}\,H_2O$$

Check: The units (L) are correct. The magnitude of the answer (< 1 L) makes physical sense because we are vaporizing about 20 moles of water.

23. **Given:** 0.88 g water condenses on iron block 75.0 g at $T_i = 22$ °C **Find:** T_f(iron block)
Other: $\Delta H^\circ_{vap} = 44.0$ kJ/mol; $C_{Fe} = 0.449$ J/g · °C from text
Conceptual Plan: **g H_2O** $\rightarrow$ **mol H_2O** $\rightarrow$ q_{H2O} **(kJ)** $\rightarrow$ q_{H2O} **(J)** $\rightarrow$ q_{Fe} **then** q_{Fe}, m_{Fe}, $T_i \rightarrow T_f$

$$\frac{1\,\text{mol}}{18.01\,\text{g}} \quad \frac{-44.0\,\text{kJ}}{1\,\text{mol}} \quad \frac{1000\,\text{J}}{1\,\text{kJ}} \quad -q_{H2O} = q_{Fe} \quad q = m\,C_s\left(T_f - T_i\right)$$

Solution: $$0.88\,\text{g} \times \frac{1\,\text{mol}}{18.01\,\text{g}} \times \frac{-44.0\,\text{kJ}}{1\,\text{mol}} \times \frac{1000\,\text{J}}{1\,\text{kJ}} = -2\underline{1}49.92\,\text{J} \text{ then } -q_{H2O} = q_{Fe} = 2\underline{1}49.92\,\text{J} \text{ then}$$

$q = m\,C_s\left(T_f - T_i\right)$ Rearrange to solve for T_f.

$$T_f = \frac{m\,C_s\,T_i + q}{m\,C_s} = \frac{\left(75.0\,\text{g} \times 0.449\,\frac{\text{J}}{\text{g}\cdot{}^\circ\text{C}} \times 22\,{}^\circ\text{C}\right) + 2\underline{1}49.92\,\text{J}}{75.0\,\text{g} \times 0.449\,\frac{\text{J}}{\text{g}\cdot{}^\circ\text{C}}} = 86\,{}^\circ\text{C}$$

Check: The units (°C) are correct. The temperature rose, which is consistent with heat being added to the block. The magnitude of the answer (86 °C) makes physical sense because even though we have ~ 1/20 th of a mole, the energy involved in condensation is very large.

25. **Given:** **Find:** ΔH°_{vap} (NH_3) and normal boiling point

Temperature (K)	Vapor Pressure (torr)
200	65.3
210	134.3
220	255.7
230	456.0
235	597.0

Conceptual Plan: To find the heat of vaporization, use Excel or similar software to make a plot of the natural log of vapor pressure (ln *P*) as a function of the inverse of the temperature in K (1/*T*). Then fit the points to a line and determine the slope of the line. Since the slope = – $\Delta H_{vap}/R$, we find the heat of vaporization as follows:
slope = – $\Delta H_{vap}/R$ $\rightarrow$ ΔH_{vap} = – slope × *R* then J $\rightarrow$ kJ.

$$\frac{1\,\text{kJ}}{1000\,\text{J}}$$

For the normal boiling point, use the equation of the best fit line, substitute 760 torr for the pressure and calculate the temperature.

Solution: Data was plotted in Excel.
The slope of the best fitting line is – 2969.9 K.

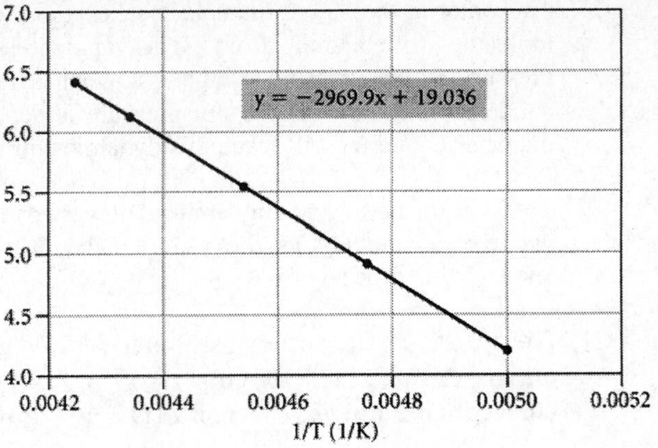

$$\Delta H_{vap} = -\,slope \times R = -\,-2969.9\ \cancel{K} \times \frac{8.314\ J}{\cancel{K}\ mol} =$$

$$= \frac{2.46917 \times 10^{4}\ \cancel{J}}{mol} \times \frac{1\ kJ}{1000\ \cancel{J}} = 24.7\ \frac{kJ}{mol}$$

$$\ln P = -\,2969.9K\left(\frac{1}{T}\right) + 19.036 \rightarrow$$

$$\ln 760 = -\,2969.9K\left(\frac{1}{T}\right) + 19.036 \rightarrow$$

$$2969.9K\left(\frac{1}{T}\right) = 19.036 - 6.63332 \rightarrow$$

$$T = \frac{2969.9K}{12.40268} = 239\ K$$

Check: The units (kJ/mol) are correct. The magnitude of the answer (25) is consistent with other values in the text.

27. **Given:** ethanol, ΔH°_{vap} = 38.56 kJ/mol; normal boiling point = 78.4 °C **Find:** $P_{Ethanol}$ at 15 °C
Conceptual Plan: **°C → K and kJ → J then ΔH°_{vap}, T_1, P_1, T_2 → P_2**

$K = °C + 273.15$ $\qquad \frac{1000\ J}{1\ kJ}$ $\qquad \ln\frac{P_2}{P_1} = \frac{-\Delta H_{vap}}{R}\left(\frac{1}{T_2} - \frac{1}{T_1}\right)$

Solution: T_1 = 78.4 °C + 273.15 = 351.6 K; T_2 = 15 °C + 273.15 = 288 K;

$\frac{38.56\ \cancel{kJ}}{mol} \times \frac{1000\ J}{1\ \cancel{kJ}} = 3.856 \times 10^{4}\ \frac{J}{mol}$ $\quad P_1$ = 760 torr $\quad \ln\frac{P_2}{P_1} = \frac{-\Delta H_{vap}}{R}\left(\frac{1}{T_2} - \frac{1}{T_1}\right)$ Substitute values in equation.

$$\ln\frac{P_2}{760\ torr} = \frac{-3.856 \times 10^{4}\ \frac{\cancel{J}}{\cancel{mol}}}{8.314\ \frac{\cancel{J}}{\cancel{K} \cdot \cancel{mol}}}\left(\frac{1}{288\ \cancel{K}} - \frac{1}{351.6\ \cancel{K}}\right) = -2.91302 \rightarrow$$

$$\frac{P_2}{760\ torr} = e^{-2.91302} = 0.054311 \rightarrow P_2 = 0.054311 \times 760\ torr = 41\ torr$$

Check: The units (torr) are correct. Since 15 °C is significantly below the boiling point, we expect the answer to be much less than 760 torr.

29. **Given:** 47.5 g water freezes **Find:** energy released **Other:** ΔH°_{fus} = 6.02 kJ/mol from text
Conceptual Plan: **g H_2O → mol H_2O → q_{H_2O} (kJ) → q_{H_2O} (J)**

$\frac{1\ mol}{18.01\ g}$ $\qquad \frac{-6.02\ kJ}{1\ mol}$ $\qquad \frac{1000\ J}{1\ kJ}$

Solution: $47.5\ \cancel{g} \times \frac{1\ \cancel{mol}}{18.01\ \cancel{g}} \times \frac{-6.02\ \cancel{kJ}}{1\ \cancel{mol}} \times \frac{1000\ J}{1\ \cancel{kJ}} = -\,15900\ J$ or 15900 J released or 15.9 kJ released

Check: The units (J) are correct. The magnitude (15900 J) makes sense since we are freezing about 3 moles of water. Freezing is exothermic, so heat is released.

31. **Given:** 8.5 g ice; 255 g water **Find:** ΔT of water
Other: ΔH°_{fus} = 6.0 kJ/mol; C_{H_2O} = 4.18 J/g · °C from text
Conceptual Plan: The first step is to calculate how much heat is removed from the water to melt the ice.
$q_{ice} = -\ q_{water}$ so g (ice) → mol (ice) → q_{fus}(kJ) → q_{fus} (J) → q_{water} (J) then q, m, C_s → ΔT_1

$\frac{1\ mol}{18.01 g}$ $\qquad \frac{6.0\ kJ}{1\ mol}$ $\qquad \frac{1000\ J}{1\ kJ}$ $\qquad q_{water} = -\ q_{ice}$ $\qquad q = m\,C_S \Delta T_1$

Now we have slightly cooled water (at a temperature of T_1) in contact with 0.0 °C water, and we can

calculate a second temperature drop of the water due to mixing of the water that was ice with the initially room temperature water. so $q_{ice} = -q_{water}$ with $m, C_s \rightarrow \Delta T_2$ with $\Delta T_1 \, \Delta T_2 \rightarrow \Delta T_{Total}$

$q = m\,C_s\,\Delta T_2$ *then set* $q_{ice} = -q_{H2O}$ $\quad \Delta T_{Total} = \Delta T_1 + \Delta T_2$

Solution: $8.5\ \text{g} \times \dfrac{1\ \text{mol}}{18.01\ \text{g}} \times \dfrac{6.0\ \text{kJ}}{1\ \text{mol}} \times \dfrac{1000\ \text{J}}{1\ \text{kJ}} = 2.\underline{8}3176 \times 10^3\ \text{J}$, $q_{water} = -\,q_{ice} = -\,2.\underline{8}3176 \times 10^3\ \text{J}$

$q = m C_S \Delta T$ Rearrange to solve for ΔT. $\Delta T_1 = \dfrac{q}{m C_S} = \dfrac{-\,2.\underline{8}3176 \times 10^3\ \text{J}}{255\ \text{g} \times 4.18\ \dfrac{\text{J}}{\text{g}\cdot{}^\circ\text{C}}} = -\,2.\underline{6}567\ {}^\circ\text{C}$.

$q = m C_S \Delta T$ substitute in values and set $q_{ice} = -\,q_{water}$.

$q_{ice} = m_{ice}\, C_{ice}\left(T_f - T_{icei}\right) = 8.5\ \text{g} \times 4.18\ \dfrac{\text{J}}{\text{g}\cdot{}^\circ\text{C}} \times \left(T_f - 0.0\ {}^\circ\text{C}\right) =$

$\rightarrow$

$-\,q_{water} = -\,m_{water}\, C_{water}\, \Delta T_{water2} = -\,255\ \text{g} \times 4.18\ \dfrac{\text{J}}{\text{g}\cdot{}^\circ\text{C}} \times \Delta T_{water2}$

$8.5\,T_f = -\,255\ \Delta T_{water2} = -\,255\ (T_f - T_{f1})$ Rearrange to solve for T_f. $8.5\,T_f + 255\ T_f = 255\ T_{f1} \rightarrow$

$26\underline{3}.5\ T_f = 255\ T_{f1} \rightarrow T_f = 0.96\underline{7}74\, T_{f1}$ but $\Delta T_1 = (T_{f1} - T_{i1}) = -\,2.\underline{6}567\ {}^\circ\text{C}$ which says that

$T_{f1} = T_{i1} - 2.\underline{6}567\ {}^\circ\text{C}$ and $\Delta T_{Total} = (T_f - T_{i1})$ so

$\Delta T_{Total} = 0.96\underline{7}74\ T_{f1} - T_{i1} = 0.96\underline{7}74(T_{i1} - 2.\underline{6}567\ {}^\circ\text{C}) - T_{i1} = -\,2.\underline{6}567\ {}^\circ\text{C} - 0.03\underline{2}26\,T_{i1}$.

This implies that the larger the initial temperature of the water, the larger the temperature drop. If the initial temperature was 90 °C, the temperature drop would be 5.6 °C. If the initial temperature was 25 °C, the temperature drop would be 3.5°C. If the initial temperature was 5 °C, the temperature drop would be 2.8 °C. This makes physical sense because the lower the initial temperature of the water, the less kinetic energy it initially has and the smaller the heat transfer from the water to the melted ice will be.

Check: The units (°C) are correct. The temperature drop from the melting of the ice is only 2.7 °C because the mass of the water is so much larger than the ice.

33. Given: 10.0 g ice $T_i = -10.0\ {}^\circ\text{C}$ to steam at $T_f = 110.0\ {}^\circ\text{C}$ **Find:** heat required (kJ)

Other: $\Delta H^\circ_{fus} = 6.02$ kJ/mol; $\Delta H^\circ_{vap} = 40.7$ kJ/mol; $C_{ice} = 2.09$ J/g·°C; $C_{H2O} = 4.18$ J/g·°C; $C_{steam} = 2.01$ J/g·°C

Conceptual Plan: Follow the heating curve in Figure 11.36. $q_{Total} = q_1 + q_2 + q_3 + q_4 + q_5$ where q_1, q_3, and q_5 are heating of a single phase then J $\rightarrow$ kJ and q_2 and q_4 are phase transitions.

$q = m C_S (T_f - T_i) \qquad \dfrac{1\ \text{kJ}}{1000\ \text{J}} \qquad q = m \times \dfrac{1\ \text{mol}}{18.01\ \text{g}} \times \dfrac{\Delta H}{1\ \text{mol}}$

Solution:

$q_1 = m_{ice}\, C_{ice}\left(T_{icef} - T_{icei}\right) = 10.0\ \text{g} \times 2.09\ \dfrac{\text{J}}{\text{g}\cdot{}^\circ\text{C}} \times (0.0\ {}^\circ\text{C} - -10.0\ {}^\circ\text{C}) = 209\ \text{J} \times \dfrac{1\ \text{kJ}}{1000\ \text{J}} = 0.209\ \text{kJ}$,

$q_2 = m \times \dfrac{1\ \text{mol}}{18.01\text{g}} \times \dfrac{\Delta H_{fus}}{1\ \text{mol}} = 10.0\ \text{g} \times \dfrac{1\ \text{mol}}{18.01\ \text{g}} \times \dfrac{6.02\ \text{kJ}}{1\ \text{mol}} = 3.3\underline{4}3\ \text{kJ}$,

$q_3 = m_{water}\, C_{water}\left(T_{waterf} - T_{wateri}\right) = 10.0\ \text{g} \times 4.18\ \dfrac{\text{J}}{\text{g}\cdot{}^\circ\text{C}} \times (100.0\ {}^\circ\text{C} - 0.0\ {}^\circ\text{C}) = 4180\ \text{J} \times \dfrac{1\ \text{kJ}}{1000\ \text{J}} = 4.18\ \text{kJ}$,

$q_4 = m \times \dfrac{1\ \text{mol}}{18.01\text{g}} \times \dfrac{\Delta H_{vap}}{1\ \text{mol}} = 10.0\ \text{g} \times \dfrac{1\ \text{mol}}{18.01\ \text{g}} \times \dfrac{40.7\ \text{kJ}}{1\ \text{mol}} = 22.\underline{5}99\ \text{kJ}$,

$q_5 = m_{steam}\, C_{steam}\left(T_{steamf} - T_{steami}\right) = 10.0\ \text{g} \times 2.01\ \dfrac{\text{J}}{\text{g}\cdot{}^\circ\text{C}} \times (110.0\ {}^\circ\text{C} - 100.0\ {}^\circ\text{C})$

$= 201\ \text{J} \times \dfrac{1\ \text{kJ}}{1000\ \text{J}} = 0.201\ \text{kJ}.$

$q_{Total} = q_1 + q_2 + q_3 + q_4 + q_5 = 0.209\ \text{kJ} + 3.3\underline{4}3\ \text{kJ} + 4.18\ \text{kJ} + 22.\underline{5}99\ \text{kJ} + 0.201\ \text{kJ} = 30.5\ \text{kJ}$

Check: The units (kJ) are correct. The amount of heat is dominated the vaporization step. Since we have less than 1 mole we expect less than 41 kJ.

35. a) solid

b) liquid

c) gas

d) supercritical fluid

e) solid/liquid equilibrium

f) liquid/gas equilibrium

g) solid/liquid/gas equilibrium

37. **Given:** nitrogen, normal boiling point = 77.3 K, normal melting point = 63.1 K, critical temperature = 126.2 K, critical pressure = 2.55×10^4 torr, triple point at 63.1 K and 94.0 torr
Find: Sketch phase diagram. Does nitrogen have a stable liquid phase at 1 atm?

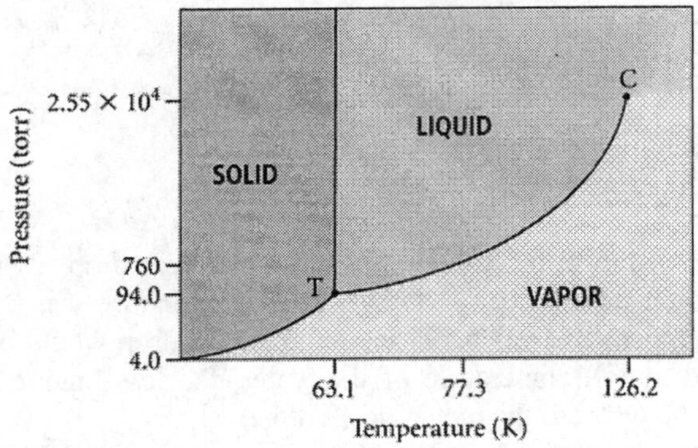

Nitrogen has a stable liquid phase at 1 atm.

39. a) 0.027 mmHg, the higher of the two triple points

b) The Rhombic phase is denser because if we start in the Monoclinic phase at 100 °C and increase the pressure, we will cross into the Rhombic phase.

41. Water has a low molar mass (18.01 g/mol), yet it is a liquid at room temperature. Water's high boiling point for its molar mass can be understood by examining the structure of the water molecule. The bent geometry of the water molecule and the highly polar nature of the O–H bonds result in a molecule with a significant dipole moment. Water's two O–H bonds (hydrogen directly bonded to oxygen) allow a water molecule to form strong hydrogen bonds with four other water molecules, resulting in a relatively high boiling point.

43. Water has an exceptionally high specific heat capacity, which has a moderating effect on the climate of coastal cities. Also, its high ΔH_{vap} causes water evaporation and condensation to have a strong effect on temperature. A tremendous amount of heat can be stored in large bodies of water. The heat will be absorbed or released from the large bodies of water preferentially over the land around it. In some cities, such as San Francisco, for example, the daily fluctuation in temperature can be less than 10 °C. This same moderating effect occurs over the entire planet, two-thirds of which is covered by water. In other words, without water, the daily temperature fluctuations on our planet might be more like those on Mars, where temperature fluctuations of 63 °C (113 °F) have been measured between midday and early morning.

45. a) 8 corner atoms x (1/8 atom / unit cell) = 1 atom / unit cell

b) 8 corner atoms x (1/8 atom / unit cell) + 1 atom in center = (1 + 1) atoms / unit cell = 2 atoms / unit cell

c) 8 corner atoms x (1/8 atom / unit cell) + 6 face-centered atoms x (1/2 atom / unit cell) = (1 + 3) atoms / unit cell = 4 atoms / unit cell

47. **Given:** platinum, face-centered cubic structure, $r = 139$ pm **Find:** edge length of unit cell and density (g/cm^3)
Conceptual Plan:
$r \rightarrow l$ and $l \rightarrow V(\text{pm}^3) \rightarrow V(\text{cm}^3)$ and $\mathfrak{M}$, FCC structure → m then $m, V \rightarrow d$

$l = 2\sqrt{2}\,r$ $\quad V = l^3 \quad \dfrac{(1\text{ cm})^3}{(10^{10}\text{ pm})^3} \quad m = \dfrac{4\text{ atoms}}{\text{unit cell}} \times \dfrac{\mathfrak{M}}{N_A} \quad d = m/V$

Solution: $l = 2\sqrt{2}\,r = 2\sqrt{2} \times 139\text{ pm} = 39\underline{3}.151\text{ pm} = 393\text{ pm}$ and

$$V = l^3 = (39\underline{3}.151\text{ pm})^3 \times \frac{(1\text{ cm})^3}{(10^{10}\text{ pm})^3} = 6.0\underline{7}682 \times 10^{-23}\text{ cm}^3 \text{ and}$$

$$m = \frac{4\text{ atoms}}{\text{unit cell}} \times \frac{\mathfrak{M}}{N_A} = \frac{4\text{ atoms}}{\text{unit cell}} \times \frac{195.09\text{ g}}{1\text{ mol}} \times \frac{1\text{ mol}}{6.022 \times 10^{23}\text{ atoms}} = 1.29\underline{5}848 \times 10^{-21}\frac{\text{g}}{\text{unit cell}} \text{ then}$$

$$d = \frac{m}{V} = \frac{1.29\underline{5}848 \times 10^{-21}\dfrac{\text{g}}{\text{unit cell}}}{6.0\underline{7}682 \times 10^{-23}\dfrac{\text{cm}^3}{\text{unit cell}}} = 21.3\frac{\text{g}}{\text{cm}^3}$$

Check: The units (pm and g/cm^3) are correct. The magnitude (393 pm) makes sense because it must be larger than the radius of an atom. The magnitude (21 g/ cm^3) is consistent for Pt from Chapter 1.

49. **Given:** rhodium, face-centered cubic structure, $d = 12.41$ g/cm^3 **Find:** r (Rh)
Conceptual Plan:
$\mathfrak{M}$, FCC structure → m then $m, V \rightarrow d$ then $V(\text{cm}^3) \rightarrow l(\text{cm}) \rightarrow l(\text{pm})$ then $l \rightarrow r$

$m = \dfrac{4\text{ atoms}}{\text{unit cell}} \times \dfrac{\mathfrak{M}}{N_A} \quad d = m/V \quad V = l^3 \quad \dfrac{10^{10}\text{ pm}}{1\text{ cm}} \quad l = 2\sqrt{2}\,r$

Solution: $m = \dfrac{4\text{ atoms}}{\text{unit cell}} \times \dfrac{\mathfrak{M}}{N_A} = \dfrac{4\text{ atoms}}{\text{unit cell}} \times \dfrac{102.905\text{ g}}{1\text{ mol}} \times \dfrac{1\text{ mol}}{6.022 \times 10^{23}\text{ atoms}} = 6.83\underline{5}271 \times 10^{-22}\dfrac{\text{g}}{\text{unit cell}}$

then $d = \dfrac{m}{V}$ Rearrange to solve for V. $V = \dfrac{m}{d} = \dfrac{6.83\underline{5}271 \times 10^{-22}\dfrac{\text{g}}{\text{unit cell}}}{12.41\dfrac{\text{g}}{\text{cm}^3}} = 5.50\underline{7}873 \times 10^{-23}\dfrac{\text{cm}^3}{\text{unit cell}}$ then

$V = l^3$ Rearrange to solve for l.

$l = \sqrt[3]{V} = \sqrt[3]{5.50\underline{7}873 \times 10^{-23}\text{ cm}^3} = 3.80\underline{4}831 \times 10^{-8}\text{ cm} \times \dfrac{10^{10}\text{ pm}}{1\text{ cm}} = 380.\underline{4}831\text{ pm}$ then $l = 2\sqrt{2}\,r$ Rearrange

to solve for r. $r = \dfrac{l}{2\sqrt{2}} = \dfrac{380.\underline{4}831\text{ pm}}{2\sqrt{2}} = 134.5\text{ pm}$

Check: The units (pm) are correct. The magnitude (135 pm) is consistent with atom diameters.

51. **Given:** polonium, simple cubic structure, $d = 9.3$ g/cm^3; $r = 167$ pm; $\mathfrak{M} = 209$ g/mol **Find:** estimate N_A
Conceptual Plan:
$r \rightarrow l$ and $l \rightarrow V(\text{pm}^3) \rightarrow V(\text{cm}^3)$ then $d, V \rightarrow m$ then $\mathfrak{M}$, SC structure → m

$l = 2r \quad V = l^3 \quad \dfrac{(1\text{ cm})^3}{(10^{10}\text{ pm})^3} \quad d = m/V \quad m = \dfrac{1\text{ atom}}{\text{unit cell}} \times \dfrac{\mathfrak{M}}{N_A}$

Solution: $l = 2r = 2 \times 167\text{ pm} = 334\text{ pm}$ and $V = l^3 = (334\text{ pm})^3 \times \dfrac{(1\text{ cm})^3}{(10^{10}\text{ pm})^3} = 3.7\underline{2}597 \times 10^{-23}\text{ cm}^3$ then

$d = \dfrac{m}{V}$ Rearrange to solve for m. $m = d\,V = 9.3\dfrac{\text{g}}{\text{cm}^3} \times \dfrac{3.7\underline{2}597 \times 10^{-23}\text{ cm}^3}{\text{unit cell}} = 3.4\underline{6}515 \times 10^{-22}\dfrac{\text{g}}{\text{unit cell}}$

then $m = \dfrac{1\text{ atom}}{\text{unit cell}} \times \dfrac{\mathfrak{M}}{N_A}$ Rearrange to solve for N_A.

$$N_A = \frac{1 \text{ atom}}{\text{unit cell}} \times \frac{\mathfrak{M}}{m} = \frac{1 \text{ atom}}{\cancel{\text{unit cell}}} \times \frac{209 \cancel{\text{g}}}{1 \text{ mol}} \times \frac{1 \cancel{\text{unit cell}}}{3.4\underline{6}515 \times 10^{-22} \cancel{\text{g}}} = 6.03 \times 10^{23} \frac{\text{atom}}{\text{mol}}$$

Check: The units (atoms/mol) are correct. The magnitude (6×10^{23}) is consistent with Avogadro's number.

53. a) atomic, since Ar is an atom

    b) molecular, since water is a molecule

    c) ionic, since K_2O is an ionic solid

    d) atomic, since iron is an atom

55. LiCl has the highest melting point since it is the only ionic solid in the group. The other three solids are held together by intermolecular forces while LiCl is held together by stronger coulombic interactions between the cations and anions of the crystal lattice.

57. a) TiO_2 because it is an ionic solid

    b) $SiCl_4$ because it has a higher molar mass, and therefore, has stronger dispersion forces

    c) Xe because it has a higher molar mass, and therefore, has stronger dispersion forces

    d) CaO because the ions have greater charge, stronger dipole–dipole interactions

59. The Ti atoms occupy the corner positions and the center of the unit cell – 8 corner atoms x (1/8 atom / unit cell) + 1 atom in center = (1 + 1) Ti atoms / unit cell = 2 Ti atoms / unit cell. The O atoms occupy four positions on the top and bottom faces and two positions inside the unit cell – 4 face-centered atoms x (1/2 atom / unit cell) + 2 atoms in the interior = (2 + 2) O atoms / unit cell = 4 O atoms / unit cell. Therefore there are 2 Ti atoms / unit cell and 4 O atoms / unit cell, so the ratio Ti:O is 2:4 or 1:2. The formula for the compound is TiO_2.

61. In CsCl: The Cs atoms occupy the center of the unit cell – 1 atom in center = 1 Cs atom / unit cell. The Cl atoms occupy corner positions of the unit cell – 8 corner atoms x (1/8 atom / unit cell) = 1 Cl atom / unit cell. Therefore there are 1 Cl atom / unit cell and 1 Cl atom / unit cell, so the ratio Cs:Cl is 1:1. The formula for the compound is CsCl, as expected.

    In $BaCl_2$: The Ba atoms occupy the corner positions and the face-centered positions of the unit cell – 8 corner atoms x (1/8 atom / unit cell) + 6 face-centered atoms x (1/2 atom / unit cell) = (1 + 3) Ba atoms / unit cell = 4 Ba atoms / unit cell. The Cl atoms occupy eight positions inside the unit cell – 8 Cl atoms / unit cell. Therefore there are 4 Ba atoms / unit cell and 8 Cl atoms / unit cell, so the ratio Ba:Cl is 4:8 or 1:2. The formula for the compound is $BaCl_2$, as expected.

63. a) Zn should have little or no band gap because it is the only metal in the group.

65. The general trend is that melting point increases with increasing mass. This is due to the fact that the electrons of the larger molecules are held more loosely and a stronger dipole moment can be induced more easily. HF is the exception to the rule. It has a relatively high melting point due to hydrogen bonding.

67. **Given:** $P_{H2O} = 23.76$ torr at 25 °C; 1.25 g water in 1.5 L container **Find:** m (H_2O) as liquid

    Conceptual Plan:

    °C → K and torr → atm then *P, V, T* → mol (*g*) → g (*g*) then g (*g*), g (*l*)$_i$ → g (*l*)$_f$

    $K = °C + 273.15$ $\quad \frac{1 \text{ atm}}{760 \text{ torr}}$ $\quad PV = nRT$ $\quad \frac{18.01 \text{ g}}{1 \text{ mol}}$ $\quad g\,(l)_f = g\,(l)_i - g\,(g)$

    Solution: $T = 25\ °C + 273.15 = 298$ K, $23.76 \cancel{\text{torr}} \times \frac{1 \text{ atm}}{760 \cancel{\text{torr}}} = 0.0312\underline{6}32$ atm then $PV = nRT$

    Rearrange to solve for n. $n = \frac{PV}{RT} = \frac{0.0312\underline{6}32 \cancel{\text{atm}} \times 1.5 \cancel{\text{L}}}{0.08206 \frac{\cancel{\text{L}} \cdot \cancel{\text{atm}}}{\cancel{\text{K}} \cdot \text{mol}} \times 298 \cancel{\text{K}}} = 0.001\underline{9}1768$ mol then

$0.00191768 \text{ mol} \times \frac{18.01 \text{ g}}{1 \text{ mol}} = 0.0345375 \text{ g}$ in gas phase then

$g(l)_f = g(l)_i - g(g) = 1.25 \text{ g} - 0.0345375 \text{ g} = 1.22 \text{ g}$ remaining as liquid Yes, there is 1.22 g of liquid.

Check: The units (g) are correct. The magnitude (1.2 g) is expected since very little material is expected to be in the gas phase.

69. Since we are starting at a temperature that is higher and a pressure that is lower than the triple point, the phase transitions will be gas → liquid → solid or condensation followed by freezing.

71. **Given:** Ice: $T_1 = 0$ °C exactly, $m = 53.5$ g; Water: $T_1 = 75$ °C, $m = 115$ g **Find:** T_f
Other: $\Delta H°_{fus} = 6.0$ kJ/mol; $C_{H2O} = 4.18$ J/g·°C
Conceptual Plan: $q_{ice} = -q_{water}$ **so g (ice) → mol (ice) →** q_{fus}**(kJ) →** q_{fus} **(J) →** q_{water} **(J) then**

$\frac{1 \text{ mol}}{18.01 \text{ g}}$ $\frac{6.02 \text{ kJ}}{1 \text{ mol}}$ $\frac{1000 \text{ J}}{1 \text{ kJ}}$ $q_{water} = -q_{ice}$

q, m, C_s → ΔT then $T_i, \Delta T$ → T_2 now we have slightly cooled water in contact with 0.0 °C water

$q = mC_S\Delta T$ $\Delta T = T_2 - T_i$

so $q_{ice} = -q_{water}$ **with m, C_s, T_i → T_f**

$q = m\,C_s(T_f - T_i)$ *then set* $q_{ice} = -q_{water}$

Solution: $53.5 \text{ g} \times \frac{1 \text{ mol}}{18.01 \text{ g}} \times \frac{6.02 \text{ kJ}}{1 \text{ mol}} \times \frac{1000 \text{ J}}{1 \text{ kJ}} = 1.78828 \times 10^4 \text{ J}$, $q_{water} = -q_{ice} = -1.78828 \times 10^4$ J

$q = mC_S\Delta T$ Rearrange to solve for ΔT. $\Delta T = \frac{q}{mC_S} = \frac{-1.78828 \times 10^4 \text{ J}}{115 \text{ g} \times 4.18 \frac{\text{J}}{\text{g} \cdot °\text{C}}} = -37.2017$ °C then

$\Delta T = T_2 - T_i$. Rearrange to solve for T_2. $T_2 = \Delta T + T_i = -37.2017 \text{ °C} + 75 \text{ °C} = 37.798 \text{ °C}$

$q = m\,C_s(T_f - T_i)$ substitute in values and set $q_{ice} = -q_{water}$.

$q_{ice} = m_{ice}\,C_{ice}(T_f - T_{icei}) = 53.5 \text{ g} \times 4.18 \frac{\text{J}}{\text{g} \cdot °\text{C}} \times (T_f - 0.0 \text{ °C}) =$

$-q_{water} = -m_{water}\,C_{water}(T_f - T_{water2}) = -115 \text{ g} \times 4.18 \frac{\text{J}}{\text{g} \cdot °\text{C}} \times (T_f - 37.798 \text{ °C})$

Rearrange to solve for T_f.

$53.5\,T_f = -115(T_f - 37.798 \text{ °C}) \rightarrow 53.5\,T_f = -115\,T_f + 4346.8 \text{ °C} \rightarrow -4346.8 \text{ °C} = -168.5\,T_f$

$\rightarrow T_f = \frac{-4346.8 \text{ °C}}{-168.5} = 25.8 \text{ °C} = 26 \text{ °C}$

Check: The units (°C) are correct. The temperature is between the two initial temperatures. Since the ice mass is about half the water mass, we are not surprised that the temperature is closer to the original ice temperature.

73. **Given:** Home: 6.0 m x 10.0 m x 2.2 m; $T = 30$ °C, $P_{H2O} = 85$ % of $P°_{H2O}$ **Find:** m (H_2O) removed
Other: $P°_{H2O} = 31.86$ mm Hg from text
Conceptual Plan:
l, w, h → V(m³) → V(cm³) → V(L) and $P°_{H2O}$ → P_{H2O} (mm Hg) → P_{H2O} (atm) and

$V = l\,w\,h$ $\frac{(100 \text{ cm})^3}{(1 \text{ m})^3}$ $\frac{1 \text{ L}}{1000 \text{ cm}^3}$ $P_{H2O} = 0.85\,P°_{H2O}$ $\frac{1 \text{ atm}}{760 \text{ mmHg}}$

°C → K then P, V, T → mol (H_2O) → g (H_2O)

$K = °C + 273.15$ $PV = nRT$ $\frac{18.01 \text{ g}}{1 \text{ mol}}$

Solution: $V = l\,wh = 6.0 \text{ m} \times 10.0 \text{ m} \times 2.2 \text{ m} = 132 \text{ m}^3 \times \frac{(100 \text{ cm})^3}{(1 \text{ m})^3} \times \frac{1 \text{ L}}{1000 \text{ cm}^3} = 1.32 \times 10^5 \text{ L}$,

$P_{H2O} = 0.85\ P^0_{H2O} = 0.85 \text{ x } 31.86 \text{ mmHg x } \dfrac{1 \text{ atm}}{760 \text{ mmHg}} = 0.035633 \text{ atm}$, $T = 30\ °C + 273.15 = 303$ K,

then $PV = nRT$ Rearrange to solve for n.

$$n = \frac{PV}{RT} = \frac{0.035633 \text{ atm x } 1.32 \text{ x } 10^5 \text{ L}}{0.08206 \dfrac{\text{L} \cdot \text{atm}}{\text{K} \cdot \text{mol}} \text{ x } 303 \text{ K}} = 189.17 \text{ mol} \text{ then } 189.17 \text{ mol x } \frac{18.01 \text{ g}}{1 \text{ mol}} = 3400 \text{ g to remove}$$

Check: The units (g) are correct. The magnitude of the answer (3400 g) makes sense since the volume of the house is so large. We are removing almost 200 moles of water.

75. CsCl has a higher melting point than AgI because of its higher coordination number. In CsCl, one anion bonds to 8 cations (and vice versa) while in AgI, one anion bonds only to 4 cations.

77. a) atoms are connected across the face diagonal (c), so $c = 4r$

b) From the Pythagorean Theorem $c^2 = a^2 + b^2$, from part a) $c = 4r$ and for a cubic structure $a = l, b = l$ so

$(4r)^2 = l^2 + l^2 \rightarrow 16r^2 = 2l^2 \rightarrow 8r^2 = l^2 \rightarrow l = \sqrt{8r^2} \rightarrow l = 2\sqrt{2}\,r$

79. **Given:** diamond, V (unit cell) = 0.0454 nm^3; d = 3.52 g/cm^3 **Find:** number of carbon atoms / unit cell
Conceptual Plan: V**(nm^3)** → V**(cm^3) then d, V → m → mol → atoms**

$\dfrac{(1 \text{ cm})^3}{(10^7 \text{ nm})^3}$ $\quad d = m/V \quad$ $\dfrac{1 \text{ mol}}{12.01 \text{ g}}$ $\quad \dfrac{6.022 \text{ x } 10^{23} \text{ atoms}}{1 \text{ mol}}$

Solution: $0.0454 \text{ nm}^3 \text{ x } \dfrac{(1 \text{ cm})^3}{(10^7 \text{ nm})^3} = 4.54 \text{ x } 10^{-23} \text{ cm}^3$ then $d = \dfrac{m}{V}$ Rearrange to solve for m.

$m = d\,V = 3.52 \dfrac{\text{g}}{\text{cm}^3} \text{ x } 4.54 \text{ x } 10^{-23} \text{ cm}^3 = 1.59808 \text{ x } 10^{-22} \text{ g}$ then

$$\frac{1.59808 \text{ x } 10^{-22} \text{ g}}{\text{unit cell}} \text{ x } \frac{1 \text{ mol}}{12.01 \text{ g}} \text{ x } \frac{6.022 \text{ x } 10^{23} \text{ atoms}}{1 \text{ mol}} = 8.01 \frac{\text{C atoms}}{\text{unit cell}} = 8 \frac{\text{C atoms}}{\text{unit cell}}$$

Check: The units (atoms) are correct. The magnitude (8) makes sense because it is a fairly small number and our answer within our calculation error of an integer.

81. a) $CO_2\,(s) \rightarrow CO_2\,(g)$ at 194.7 K

b) $CO_2\,(s) \rightarrow$ triple point at 216.5 K $\rightarrow CO_2\,(g)$ just above 216.5 K

c) $CO_2\,(s) \rightarrow CO_2\,(l)$ at somewhat above 216 K $\rightarrow CO_2\,(g)$ at around 250 K

d) CO_2 (s) $\rightarrow CO_2$ above the critical point where there is no distinction between liquid and gas. This change occurs at about 300 K.

83. **Given:** KCl, rock salt structure **Find:** density (g/cm^3) **Other:** r (K$^+$) = 133 pm; r (Cl$^-$) = 181 pm from Chapter 8
Conceptual Plan: Rock salt structure is a face-centered cubic structure with anions at the lattice points and cations in the holes between lattice sites → assume $r = r$(Cl$^-$), but $\mathfrak{M} = \mathfrak{M}$(KCl)
r**(K$^+$),** r**(Cl$^-$) →** l **and** l **→** V**(pm^3) →** V**(cm^3) and , FCC structure →** m **then** m, V **→** d

from Figure 11.52 $l = 2r(Cl^-) + 2r(K^+)$ $\quad V = l^3 \quad$ $\dfrac{(1 \text{ cm})^3}{(10^{10} \text{ pm})^3}$ $\quad m = \dfrac{4 \text{ formula units}}{\text{unit cell}} \text{ x } \dfrac{\mathfrak{M}}{N_A}$ $\quad d = m/V$

Solution: $l = 2r(Cl^-) + 2r(K^+) = 2(181 \text{ pm}) + 2(133 \text{ pm}) = 628 \text{ pm}$ and

$V = l^3 = (628 \text{ pm})^3 \text{ x } \dfrac{(1 \text{ cm})^3}{(10^{10} \text{ pm})^3} = 2.47673 \text{ x } 10^{-22} \text{ cm}^3$ and

$$m = \frac{4 \text{ formula units}}{\text{unit cell}} \times \frac{\mathfrak{M}}{N_A} = \frac{4 \text{ formula units}}{\text{unit cell}} \times \frac{74.55 \text{ g}}{1 \text{ mol}} \times \frac{1 \text{ mol}}{6.022 \times 10^{23} \text{ formula units}}$$

$$= 4.95\underline{1}976 \times 10^{-22} \frac{\text{g}}{\text{unit cell}}$$

then $$d = \frac{m}{V} = \frac{4.95\underline{1}976 \times 10^{-22} \frac{\text{g}}{\text{unit cell}}}{2.4\underline{7}673 \times 10^{-22} \frac{\text{cm}^3}{\text{unit cell}}} = 1.99\underline{9}40 \frac{\text{g}}{\text{cm}^3} = 2.00 \frac{\text{g}}{\text{cm}^3}$$

Check: The units (g/cm^3) are correct. The magnitude (2 g/ cm^3) is reasonable for a salt density. The published value is 1.98 g/cm^3. This method of estimating the density gives a value that is close to the experimentally measured density.

85. Decreasing the pressure will decrease the temperature of liquid nitrogen. Because the nitrogen is boiling, its temperature must be constant at a given pressure. As the pressure decreases, the boiling point decreases, and therefore so does the temperature. Remember that vaporization is an endothermic process, so as the nitrogen vaporizes it will remove heat from the liquid, dropping its temperature. If the pressure drops below the pressure of the triple point, the phase change will shift from vaporization to sublimation and the liquid nitrogen will become solid.

87. **Given:** cubic closest packing structure = cube with touching spheres of radius = r on alternating corners of a cube **Find:** body diagonal of cube and radius of tetrahedral hole

Solution: The cell edge length = l and $l^2 + l^2 = (2r)^2 \rightarrow 2\,l^2 = 4\,r^2 \rightarrow l^2 = 2\,r^2$. Since body diagonal = BD is the hypotenuse of the right triangle formed by the face diagonal and the cell edge we have $(BD)^2 = l^2 + (2r)^2 = 2\,r^2 + 4r^2 = 6\,r^2 \rightarrow BD = \sqrt{6}\,r$. The radius of the tetrahedral hole = r_T is half the body diagonal minus the radius of the sphere or

$$r_T = \frac{BD}{2} - r = \frac{\sqrt{6}\,r}{2} - r = \left(\frac{\sqrt{6}}{2} - 1\right) r = \left(\frac{\sqrt{6} - 2}{2}\right) r = \left(\frac{\sqrt{3}\sqrt{2} - \sqrt{2}\sqrt{2}}{\sqrt{2}\sqrt{2}}\right) r = \left(\frac{\sqrt{3} - \sqrt{2}}{\sqrt{2}}\right) r \approx 0.22474\,r$$

89. Melting of an ice cube in a glass of water will not raise or lower the level of the liquid in the glass as long as the ice is always floating in the liquid. This is because the ice will displace a volume of water based on its mass. By the same logic, melting floating icebergs will not raise the ocean levels (assuming that the dissolved solids content, and thus the density, will not change when the icebergs melt). Dissolving ice formations that are supported by land will raise the ocean levels, just as pouring more water into the glass will raise the liquid level in the glass.

91. Substance A will have the larger change in vapor pressure with the same temperature change. To understand this consider the Clausius–Clapeyron Equation: $\ln \frac{P_2}{P_1} = \frac{-\Delta H_{vap}}{R}\left(\frac{1}{T_2} - \frac{1}{T_1}\right)$, if we use the same temperatures we see that $\frac{P_2}{P_1} \;\alpha\; e^{-\Delta H_{vap}}$. So the smaller the heat of vaporization, the larger the final vapor pressure. We can also consider that the lower the heat of vaporization, the easier it is to convert the substance from a liquid to a gas. This again leads to Substance A having the larger change in vapor pressure.

93. $\Delta H_{sub} = \Delta H_{fus} + \Delta H_{vap}$ as long as the heats of fusion and vaporization are measured at the same temperatures.

95. Water has an exceptionally high specific heat capacity, which has a moderating effect on the temperature of the root cellar. A large amount of heat can be stored in a large vat of water. The heat will be absorbed or released from the large bodies of water preferentially over the area around it. As the temperature of the air drops, the water will release heat, keeping the temperature more constant. If the temperature of the cellar falls enough to begin to freeze the water, the heat given off during the freezing will further protect the food in the cellar.

Chapter 12
Solutions

1. a) hexane, toluene, or CCl_4; dispersion forces

   b) water, methanol, acetone; dispersion, dipole–dipole, hydrogen bonding

   c) hexane, toluene, or CCl_4; dispersion forces

   d) water, acetone, methanol, ethanol; dispersion, ion–dipole

3. $HOCH_2CH_2CH_2OH$ would be more soluble in water because it has –OH groups on both ends of the molecule, so it can hydrogen bond on both ends, not just one end.

5. a) water; dispersion, dipole–dipole, hydrogen bonding

   b) hexane; dispersion forces

   c) water; dispersion, dipole–dipole

   d) water; dispersion, dipole–dipole, hydrogen bonding

7. a) endothermic

   b) The lattice energy is greater in magnitude than the heat of hydration.

   c)

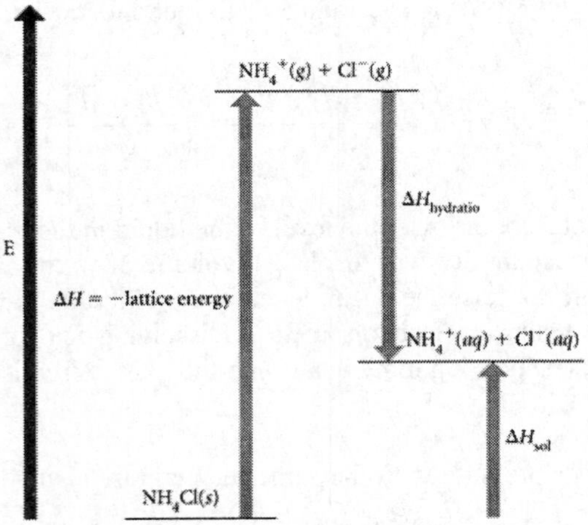

   d) The solution forms because chemical systems tend towards greater entropy.

9. **Given:** $AgNO_3$: Lattice Energy = – 820. kJ/mol, ΔH_{soln} = + 22.6 kJ/mol **Find:** $\Delta H_{hydration}$

   Conceptual Plan: Lattice Energy, ΔH_{soln} → $\Delta H_{hydration}$

   $\Delta H_{soln} = \Delta H_{solute} + \Delta H_{hydration}$ where $\Delta H_{solute} = -\Delta H_{lattice}$

   Solution: $\Delta H_{soln} = \Delta H_{solute} + \Delta H_{hydration}$ where $\Delta H_{solute} = -\Delta H_{lattice}$ so $\Delta H_{hydration} = \Delta H_{soln} + \Delta H_{lattice}$

   $\Delta H_{hydration} = 22.6\text{kJ/mol} - 820.\ \text{kJ/mol} = -797\ \text{kJ/mol}$

   Check: The units (kJ/mol) are correct. The magnitude of the answer (– 800) makes physical sense because the lattice energy is so negative and so it dominates the calculation.

11. **Given:** LiI: Lattice Energy = – 7.3 x 10^2 kJ/mol, $\Delta H_{hydration}$ = – 793 kJ/mol; 15.0 g LiI

   Find: ΔH_{soln} and heat evolved

   Conceptual Plan: Lattice Energy, $\Delta H_{hydration}$ → ΔH_{soln} and g → mol then mol, ΔH_{soln} → q

   $\Delta H_{soln} = \Delta H_{solute} + \Delta H_{hydration}$ where $\Delta H_{solute} = -\Delta H_{lattice}$ $\quad \frac{1\ \text{mol}}{133.843\ \text{g}}$ $\quad q = n\,\Delta H_{soln}$

   Solution: $\Delta H_{soln} = \Delta H_{solute} + \Delta H_{hydration}$ where $\Delta H_{solute} = -\Delta H_{lattice}$ so $\Delta H_{soln} = \Delta H_{hydration} - \Delta H_{lattice}$

$\Delta H_{soln} = -793$ kJ/mol – (– 730 kJ/mol) = – 60 kJ/mol = – 6 x 10^1 kJ/mol and

$$15.0 \text{ g} \times \frac{1 \text{ mol}}{133.843 \text{ g}} = 0.112072 \text{ mol}$$ then

$$q = n\,\Delta H_{soln} = 0.112072 \text{ mol} \times -6 \times 10^1 \frac{\text{kJ}}{\text{mol}} = -7 \text{ kJ or } 7 \text{ kJ released}$$

Check: The units (kJ/mol and kJ) are correct. The magnitude of the answer (– 60) makes physical sense because the lattice energy and the heat of hydration are about the same. The magnitude of the heat (7) makes physical sense since 15 g is much less than a mole, and so the amount of heat released is going to be small.

13. The solution is unsaturated since we are dissolving 25g of NaCl per 100 g of water and the solubility from the figure is ~ 35 g NaCl per 100 g of water.

15. At 40 °C the solution has 45 g of KNO_3 per 100 g of water and it can contain up to 63 g of KNO_3 per 100 g of water. At 0 °C the solubility from the figure is ~ 14 g KNO_3 per 100 g of water, so ~ 31 g KNO_3 per 100 g of water will precipitate out of solution.

17. Since the solubility of gases decrease as the temperature increased, dissolved oxygen will be removed from the solution.

19. Henry's Law says that as pressure increases, nitrogen will more easily dissolve in blood. To reverse this process, divers should ascend to lower pressures.

21. **Given:** room temperature, 80.0 L aquarium, $P_{Total} = 1.0$ atm; $\chi_{N2} = 0.78$ **Find:** m (N_2)
Other: $k_H(N_2) = 6.1 \times 10^{-4}$ M/L at 25 °C
Conceptual Plan: $P_{Total}, \chi_{N2} \rightarrow P_{N2}$ **then** $P_{N2}, k_H(N_2) \rightarrow S_{N2}$ **then** L → mol → g

$$P_{N2} = \chi_{N2}\,P_{Total} \qquad S_{N2} = k_H(N_2)P_{N2} \qquad S_{N2} \qquad \frac{28.01 \text{ g}}{1 \text{ mol}}$$

Solution: $P_{N2} = \chi_{N2}\,P_{Total} = 0.78 \times 1.0 \text{ atm} = 0.78 \text{ atm}$ then

$$S_{N2} = k_H(N_2)P_{N2} = 6.1 \times 10^{-4} \frac{\text{M}}{\text{atm}} \times 0.78 \text{ atm} = 4.758 \times 10^{-4} \text{ M}$$ then

$$80.0 \text{ L} \times 4.758 \times 10^{-4} \frac{\text{mol}}{\text{L}} \times \frac{28.01 \text{ g}}{1 \text{ mol}} = 1.1 \text{ g}$$

Check: The units (g) are correct. The magnitude of the answer (1) seems reasonable since we have 80 L of water and expect much less than a mole of nitrogen.

23. **Given:** NaCl and water; 133 g NaCl in 1.00 L solution **Find:** M, *m*, and mass percent
Other: $d = 1.08$ g/mL
Conceptual Plan: $g_{NaCl} \rightarrow$ mol **and** L → mL → g_{soln} **and** $g_{soln}\ g_{NaCl} \rightarrow g_{H2O} \rightarrow kg_{H2O}$ **then**

$$\frac{1 \text{ mol NaCl}}{58.44 \text{ g NaCl}} \qquad \frac{1000 \text{ mL}}{1 \text{ L}} \quad \frac{1.08 \text{ g}}{1 \text{ mL}} \qquad g_{H2O} = g_{soln} - g_{NaCl} \qquad \frac{1 \text{ kg}}{1000 \text{ g}}$$

mol, V → M **and** **mol, kg_{H2O} → m** **and** **$g_{soln}\ g_{NaCl}$ → mass percent**

$$\text{M} = \frac{\text{amount solute (moles)}}{\text{volume solution (L)}} \qquad m = \frac{\text{amount solute (moles)}}{\text{mass solvent (kg)}} \qquad \textit{mass percent} = \frac{\textit{mass solute}}{\textit{mass solution}} \times 100\ \%$$

Solution: $$133 \text{ g NaCl} \times \frac{1 \text{ mol NaCl}}{58.44 \text{ g NaCl}} = 2.27584 \text{ mol NaCl}$$ and

$$1.00 \text{ L} \times \frac{1000 \text{ mL}}{1 \text{ L}} \times \frac{1.08 \text{ g}}{1 \text{ mL}} = 1080 \text{ g soln}$$ and

$$g_{H2O} = g_{soln} - g_{NaCl} = 1080 \text{ g} - 133 \text{ g} = 947 \text{ g } H_2O \times \frac{1 \text{ kg}}{1000 \text{ g}} = 0.947 \text{ kg } H_2O$$ then

$$M = \frac{\text{amount solute (moles)}}{\text{volume solution (L)}} = \frac{2.2\underline{7}584 \text{ mol NaCl}}{1.00 \text{ L soln}} = 2.28 \text{ M} \quad \text{and}$$

$$m = \frac{\text{amount solute (moles)}}{\text{mass solvent (kg)}} = \frac{2.2\underline{7}584 \text{ mol NaCl}}{0.9\underline{4}7 \text{ kg } H_2O} = 2.4\ m \quad \text{and}$$

$$\textit{mass percent} = \frac{\textit{mass solute}}{\textit{mass solution}} \times 100\ \% = \frac{133 \text{ g NaCl}}{10\underline{8}0 \text{ g soln}} \times 100\ \% = 12.3\ \% \text{ by mass}.$$

Check: The units (M, m, and percent by mass) are correct. The magnitude of the answer (2.28 M) seems reasonable since we have 133 g NaCl, which is a couple of moles and we have 1 L. The magnitude of the answer (2.4 *m*) seems reasonable since it is a little higher than the molarity, which we expect since we only use the solvent weight in the denominator. The magnitude of the answer (12 %) seems reasonable since we have 133 g NaCl and just over 1000 g of solution.

25. **Given:** initial solution: 50.0 mL of 5.00 M KI; final solution contains: 3.25 g KI in 25.0 mL
Find: final volume to dilute initial solution to
Conceptual Plan:
final solution: $g_{KI} \rightarrow$ **mol and mL → L then mol,** $V \rightarrow M_2$ **then** $M_1, V_1, M_2 \rightarrow V_2$

$$\frac{1 \text{ mol KI}}{166.006 \text{ g KI}} \qquad \frac{1 \text{ L}}{1000 \text{ mL}} \qquad M = \frac{\text{amount solute (moles)}}{\text{volume solution (L)}} \qquad M_1 V_1 = M_2 V_2$$

Solution: $3.25 \text{ g KI} \times \frac{1 \text{ mol KI}}{166.006 \text{ g KI}} = 0.019\underline{5}776 \text{ mol KI}$ and $25.0 \text{ mL} \times \frac{1 \text{ L}}{1000 \text{ mL}} = 0.0250 \text{ mL}$

then $M = \frac{\text{amount solute (moles)}}{\text{volume solution (L)}} = \frac{0.019\underline{5}776 \text{ mol KI}}{0.0250 \text{ L soln}} = 0.78\underline{3}104 \text{ M}$ then $M_1 V_1 = M_2 V_2$.

Rearrange to solve for V_2. $V_2 = \frac{M_1}{M_2} \times V_1 = \frac{5.00 \text{ M}}{0.78\underline{3}104 \text{ M}} \times 50.0 \text{ mL} = 319 \text{ mL diluted volume}$

Check: The units (mL) are correct. The magnitude of the answer (319 mL) seems reasonable since we are starting with a concentration of 5 M and ending with a concentration of less than 1 M.

27. **Given:** $AgNO_3$ and water; 3.4 % Ag by mass, 4.8 L solution **Find:** m (Ag) **Other:** $d = 1.01$ g/mL
Conceptual Plan: $\textbf{L} \rightarrow \textbf{mL} \rightarrow g_{soln} \rightarrow g_{Ag}$

$$\frac{1000 \text{ mL}}{1 \text{ L}} \qquad \frac{1.01 \text{ g}}{1 \text{ mL}} \qquad \frac{3.4 \text{ g Ag}}{100 \text{ g soln}}$$

Solution: $4.8 \text{ L} \times \frac{1000 \text{ mL}}{1 \text{ L}} \times \frac{1.01 \text{ g}}{1 \text{ mL}} = 4\underline{8}48 \text{ g soln}$ then

$4\underline{8}48 \text{ g soln} \times \frac{3.4 \text{ g Ag}}{100 \text{ g soln}} = 160 \text{ g Ag} = 1.6 \times 10^2 \text{ g Ag}$.

Check: The units (g) are correct. The magnitude of the answer (160 g) seems reasonable since we have almost 5000 g solution.

29. **Given:** Ca^{2+} and water; 0.0085 % Ca^{2+} by mass, 1.2 g Ca **Find:** *m* (water)
Conceptual Plan: $g_{Ca} \rightarrow g_{soln} \rightarrow g_{H2O}$

$$\frac{100 \text{ g soln}}{0.0085 \text{ g Ca}} \qquad g_{H2O} = g_{soln} - g_{Ca}$$

Solution: $1.2 \text{ g Ca} \times \frac{100 \text{ g soln}}{0.0085 \text{ g Ca}} = 1\underline{4}118 \text{ g soln}$ then

$g_{H2O} = g_{soln} - g_{Ca} = 1\underline{4}118 \text{ g} - 1.2 \text{ g} = 1.4 \times 10^4 \text{ g water}$.

Check: The units (g) are correct. The magnitude of the answer (10^4 g) seems reasonable since we have such a low concentration of Ca.

31. **Given:** concentrated HNO_3: 70.3 % HNO_3 by mass, $d = 1.41$ g/mL; final solution: 1.15 L of 0.100 M HNO_3
Find: describe final solution preparation
Conceptual Plan:
$M_2, V_2 \rightarrow mol_{HNO3} \rightarrow g_{HNO3} \rightarrow g_{conc\ acid} \rightarrow mL_{conc\ acid}$ then describe method

$$mol = M\,V \qquad \frac{63.02\text{ g HNO}_3}{1\text{ mol HNO}_3} \qquad \frac{100\text{ g conc acid}}{70.3\text{ g HNO}_3} \qquad \frac{1\text{ mL}}{1.41\text{ g}}$$

Solution: $mol = M\,V = 0.100\,\frac{\text{mol HNO}_3}{1\text{ L soln}} \times 1.15\text{ L soln} = 0.115\text{ mol HNO}_3$ then

$$0.115\text{ mol HNO}_3 \times \frac{63.02\text{ g HNO}_3}{1\text{ mol HNO}_3} \times \frac{100\text{ g conc acid}}{70.3\text{ g HNO}_3} \times \frac{1\text{ mL conc acid}}{1.41\text{ g conc acid}} = 7.31\text{ mL conc acid}$$

. Prepare the solution by putting 1.00 L of distilled water in a container. Carefully pour in the 7.31 mL of the concentrated acid, mix the solution, and allow it to cool. Finally add enough water to generate a total volume of solution (1.15 L). It is important to add acid to water, and not the reverse, since there is such a large amount of heat released upon mixing.
Check: The units (mL) are correct. The magnitude of the answer (7 g) seems reasonable since we are starting with such a very concentrated solution and diluting it to a low concentration.

33. a) **Given:** 1.00×10^2 mL of 0.500 M KCl **Find:** describe final solution preparation
Conceptual Plan: mL $\rightarrow$ L then $M, V \rightarrow mol_{KCl} \rightarrow g_{KCl}$ then describe method

$$\frac{1\text{ L}}{1000\text{ mL}} \qquad mol = M\,V \qquad \frac{74.56\text{ g KCl}}{1\text{ mol KCl}}$$

Solution: $$1.00 \times 10^2\text{ mL} \times \frac{1\text{ L}}{1000\text{ mL}} = 0.100\text{ L}$$

$$mol = M\,V = 0.500\,\frac{\text{mol KCl}}{1\text{ L soln}} \times 0.100\text{ L soln} = 0.0500\text{ mol KCl}$$

then $0.0500\text{ mol KCl} \times \frac{74.56\text{ g KCl}}{1\text{ mol KCl}} = 3.73\text{ g KCl}$.

Prepare the solution by carefully adding 3.73 g KCl to a 100-mL volumetric flask. Add ~ 75 mL of distilled water and agitate the solution until the salt dissolves completely. Finally add enough water to generate a total volume of solution (add water to the mark on the flask).
Check: The units (g) are correct. The magnitude of the answer (4 g) seems reasonable since we are making a small volume of solution and the formula weight of KCl is ~ 75 g/mol.

b) **Given:** 1.00×10^2 g of 0.500 m KCl **Find:** describe final solution preparation
Conceptual Plan:
$m \rightarrow mol_{KCl}$ /1 kg solvent $\rightarrow g_{KCl}$/1 kg solvent then g_{KCl}/1 kg solvent, $g_{soln} \rightarrow g_{KCl}, g_{H2O}$

$$m = \frac{\text{amount solute (moles)}}{\text{mass solvent (kg)}} \qquad \frac{74.56\text{ g KCl}}{1\text{ mol KCl}} \qquad g_{soln} = g_{KCl} + g_{H2O}$$

then describe method

Solution: $m = \frac{\text{amount solute (moles)}}{\text{mass solvent (kg)}}$ so $0.500\ m = \frac{0.500\text{ mol KCl}}{1\text{ kg H}_2\text{O}}$ so

$$\frac{0.500\text{ mol KCl}}{1\text{ kg H}_2\text{O}} \times \frac{74.56\text{ g KCl}}{1\text{ mol KCl}} = \frac{37.2\underline{8}\text{ g KCl}}{1000\text{ g H}_2\text{O}} \qquad g_{soln} = g_{KCl} + g_{H2O} \text{ so } g_{soln} - g_{KCl} = g_{H2O}$$

substitute into ratio $\frac{0.037\underline{2}8\text{ g KCl}}{1\text{ g H}_2\text{O}} = \frac{x\text{ g KCl}}{100\text{ g soln} - x\text{ g KCl}}$ Rearrange and solve for x g KCl

$$0.037\underline{2}8\,(100\text{ g soln} - x\text{ g KCl}) = x\text{ g KCl} \rightarrow 3.7\underline{2}8 - 0.037\underline{2}8\,(x\text{ g KCl}) = x\text{ g KCl} \rightarrow$$

$$3.7\underline{2}8 = 1.037\underline{2}8\,(x\text{ g KCl}) \rightarrow \frac{3.7\underline{2}8}{1.037\underline{2}8} = x\text{ g KCl} = 3.59\text{ g KCl then}$$

$$g_{H2O} = g_{soln} - g_{KCl} = 100.\text{ g} - 3.59\text{ g} = 96.41\text{ g H}_2\text{O}.$$

Prepare the solution by carefully adding 3.59 g KCl to a container with 96.41 g of distilled water and agitate the solution until the salt dissolves completely.
Check: The units (g) are correct. The magnitude of the answer (3.6 g) seems reasonable since we are making a small volume of solution and the formula weight of KCl is ~ 75 g/mol.

c) **Given:** 1.00×10^2 g of 5.0 % KCl by mass **Find:** describe final solution preparation
Conceptual Plan: $g_{soln} \rightarrow g_{KCl}$ **then** $g_{KCl}, g_{soln} \rightarrow g_{H2O}$

$$\frac{5.0 \text{ g KCl}}{100 \text{ g soln}} \qquad g_{soln} = g_{KCl} + g_{H2O}$$

then describe method

Solution: $1.00 \times 10^2 \text{ g soln} \times \dfrac{5.0 \text{g KCl}}{100 \text{ g soln}} = 5.0 \text{ g KCl}$ then $g_{soln} = g_{KCl} + g_{H2O}$.

So $g_{H2O} = g_{soln} - g_{KCl} = 100. \text{ g} - 5.0 \text{ g} = 95 \text{ g } H_2O$
Prepare the solution by carefully adding 5.0 g KCl to a container with 95 g of distilled water and agitate the solution until the salt dissolves completely.
Check: The units (g) are correct. The magnitude of the answer (5 g) seems reasonable since we are making a small volume of solution and the solution is 5 % by mass KCl.

35. a) **Given:** 28.4 g of glucose ($C_6H_{12}O_6$) in 355 g water; final volume = 378 mL **Find:** molarity
Conceptual Plan: **mL → L and** $g_{C6H12O6} \rightarrow mol_{C6H12O6}$ **then** $mol_{C6H12O6}, V \rightarrow M$

$$\frac{1 \text{ L}}{1000 \text{ mL}} \qquad \frac{1 \text{ mol } C_6H_{12}O_6}{180.16 \text{ g } C_6H_{12}O_6} \qquad M = \frac{\text{amount solute (moles)}}{\text{volume solution (L)}}$$

Solution: $378 \text{ mL} \times \dfrac{1 \text{L}}{1000 \text{ mL}} = 0.378 \text{ L}$ and

$$28.4 \text{ g } C_6H_{12}O_6 \times \frac{1 \text{ mol } C_6H_{12}O_6}{180.16 \text{ g } C_6H_{12}O_6} = 0.15\underline{7}638 \text{ mol } C_6H_{12}O_6$$

$$M = \frac{\text{amount solute (moles)}}{\text{volume solution (L)}} = \frac{0.15\underline{7}638 \text{ mol } C_6H_{12}O_6}{0.378 \text{ L}} = 0.417 \text{ M}$$

Check: The units (M) are correct. The magnitude of the answer (0.4 M) seems reasonable since we have 1/8 mole in about 1/3 L.

b) **Given:** 28.4 g of glucose ($C_6H_{12}O_6$) in 355 g water; final volume = 378 mL **Find:** molality
Conceptual Plan: $g_{H2O} \rightarrow kg_{H2O}$ **and** $g_{C6H12O6} \rightarrow mol_{C6H12O6}$ **then** $mol_{C6H12O6}, kg_{H2O} \rightarrow m$

$$\frac{1 \text{kg}}{1000 \text{ g}} \qquad \frac{1 \text{ mol } C_6H_{12}O_6}{180.16 \text{ g } C_6H_{12}O_6} \qquad m = \frac{\text{amount solute (moles)}}{\text{mass solvent (kg)}}$$

Solution: $355 \text{ g} \times \dfrac{1 \text{kg}}{1000 \text{ g}} = 0.355 \text{ kg}$ and

$$28.4 \text{ g } C_6H_{12}O_6 \times \frac{1 \text{ mol } C_6H_{12}O_6}{180.16 \text{ g } C_6H_{12}O_6} = 0.15\underline{7}638 \text{ mol } C_6H_{12}O_6$$

$$m = \frac{\text{amount solute (moles)}}{\text{mass solvent (kg)}} = \frac{0.15\underline{7}638 \text{ mol } C_6H_{12}O_6}{0.355 \text{ kg}} = 0.444 \, m$$

Check: The units (m) are correct. The magnitude of the answer (0.4 m) seems reasonable since we have 1/8 mole in about 1/3 kg.

c) **Given:** 28.4 g of glucose ($C_6H_{12}O_6$) in 355 g water; final volume = 378 mL **Find:** percent by mass
Conceptual Plan: $g_{C6H12O6}, g_{H2O} \rightarrow g_{soln}$ **then** $g_{C6H12O6}, g_{soln} \rightarrow$ **percent by mass**

$$g_{soln} = g_{C6H12O6} + g_{H2O} \qquad \text{mass percent} = \frac{\text{mass solute}}{\text{mass solution}} \times 100\ \%$$

Solution: $g_{soln} = g_{C6H12O6} + g_{H2O} = 28.4\text{ g} + 355\text{ g} = 38\underline{3}.4\text{ g soln}$ then

$$mass\ percent = \frac{mass\ solute}{mass\ solution} \times 100\,\% = \frac{28.4\text{ g } C_6H_{12}O_6}{38\underline{3}.4\text{ g soln}} \times 100\,\% = 7.41\text{ percent by mass}$$

Check: The units (percent by mass) are correct. The magnitude of the answer (7 %) seems reasonable since we are dissolving 28 g in 355 g.

d) **Given:** 28.4 g of glucose ($C_6H_{12}O_6$) in 355 g water; final volume = 378 mL **Find:** mole fraction
Conceptual Plan:
$g_{C6H12O6} \rightarrow mol_{C6H12O6}$ and $g_{H2O} \rightarrow mol_{H2O}$ then $mol_{C6H12O6}, mol_{H2O} \rightarrow \chi_{C6H12O6}$

$$\frac{1\text{ mol } C_6H_{12}O_6}{180.16\text{ g } C_6H_{12}O_6} \qquad \frac{1\text{ mol } H_2O}{18.02\text{ g } H_2O} \qquad \chi = \frac{\text{amount solute (in moles)}}{\text{total amount of solute and solvent (in moles)}}$$

Solution: $$28.4\text{ g } C_6H_{12}O_6 \times \frac{1\text{ mol } C_6H_{12}O_6}{180.16\text{ g } C_6H_{12}O_6} = 0.15\underline{7}638\text{ mol } C_6H_{12}O_6$$ and

$$355\text{ g } H_2O \times \frac{1\text{ mol } H_2O}{18.02\text{ g } H_2O} = 19.\underline{7}003\text{ mol } H_2O$$ then

$$\chi = \frac{\text{amount solute (in moles)}}{\text{total amount of solute and solvent (in moles)}} = \frac{0.15\underline{7}638\text{ mol}}{0.15\underline{7}638\text{ mol} + 19.\underline{7}003\text{ mol}} = 0.00794$$

Check: The units (none) are correct. The magnitude of the answer (0.008) seems reasonable since we have many more grams of water and water has a much lower molecular weight.

e) **Given:** 28.4 g of glucose ($C_6H_{12}O_6$) in 355 g water; final volume = 378 mL **Find:** mole percent
Conceptual Plan: use answer from part d) then $\chi_{C6H12O6} \rightarrow$ mole percent

$$\chi \times 100\,\%$$

Solution: $mole\ percent = \chi \times 100\,\% = 0.00794 \times 100\,\% = 0.794$ mole percent

Check: The units (%) are correct. The magnitude of the answer (0.8) seems reasonable since we have many more grams of water and water has a much lower molecular weight and we are just increasing the answer from part d) by a factor of 100.

37. **Given:** 3.0 % H_2O_2 by mass, $d = 1.01$ g/mL **Find:** molarity
Conceptual Plan:
Assume exactly 100 g of solution; $g_{Solution} \rightarrow g_{H2O2} \rightarrow mol_{H2O2}$ **and** $g_{Solution} \rightarrow mL_{Solution} \rightarrow L_{Solution}$

$$\frac{3.0\text{ g } H_2O_2}{100\text{ g Solution}} \qquad \frac{1\text{ mol } H_2O_2}{34.02\text{ g } H_2O_2} \qquad \frac{1\text{ mL}}{1.01\text{ g}} \qquad \frac{1\text{ L}}{1000\text{ mL}}$$

then $mol_{H2O2}, L_{Solution} \rightarrow M$

$$M = \frac{\text{amount solute (moles)}}{\text{volume solution (L)}}$$

Solution: $$100\text{ g Solution} \times \frac{3.0\text{ g } H_2O_2}{100\text{ g Solution}} \times \frac{1\text{ mol } H_2O_2}{34.02\text{ g } H_2O_2} = 0.08\underline{8}1834\text{ mol } H_2O_2$$ and

$$100\text{ g Solution} \times \frac{1\text{ mL Solution}}{1.01\text{ g Solution}} \times \frac{1\text{ L Solution}}{1000\text{ mL Solution}} = 0.099\underline{0}099\text{ L Solution}$$ then

$$M = \frac{\text{amount solute (moles)}}{\text{volume solution (L)}} = \frac{0.08\underline{8}1834\text{ mol } H_2O_2}{0.099\underline{0}099\text{ L Solution}} = 0.89\text{ M } H_2O_2.$$

Check: The units (M) are correct. The magnitude of the answer (1) seems reasonable since we are starting with a low concentration solution and pure water is ~ 55.5 M.

39. **Given:** 36 % HCl by mass **Find:** molality and mole fraction

Conceptual Plan:

Assume exactly 100 g of solution; $g_{Solution} \rightarrow g_{HCl} \rightarrow mol_{HCl}$ **and** $g_{HCl}, g_{Solution} \rightarrow g_{Solvent} \rightarrow kg_{Solvent}$

$$\frac{36 \text{ g HCl}}{100 \text{ g Solution}} \qquad \frac{1 \text{ mol HCl}}{36.46 \text{ g HCl}} \qquad g_{soln} = g_{HCl} + g_{H2O} \qquad \frac{1 \text{ kg}}{1000 \text{ g}}$$

then $mol_{HCl}, kg_{Solvent} \rightarrow m$ **and** $g_{Solvent} \rightarrow mol_{Solvent}$ **then** $mol_{HCl}, mol_{Solvent} \rightarrow \chi_{HCl}$

$$m = \frac{\text{amount solute (moles)}}{\text{mass solvent (kg)}} \qquad \frac{1 \text{ mol } H_2O}{18.02 \text{ g } H_2O} \qquad \chi = \frac{\text{amount solute (in moles)}}{\text{total amount of solute and solvent (in moles)}}$$

Solution: $100 \text{ g Solution} \times \frac{36 \text{ g HCl}}{100 \text{ g Solution}} = 36 \text{ g HCl} \times \frac{1 \text{ mol HCl}}{36.46 \text{ g HCl}} = 0.9\underline{8}7383 \text{ mol HCl}$ and

$g_{soln} = g_{HCl} + g_{H2O}$ Rearrange to solve for $g_{Solvent}$. $g_{H2O} = g_{soln} - g_{HCl} = 100\text{ g} - 36\text{ g} = 64\text{ g } H_2O$

$64 \text{ g } H_2O \times \frac{1 \text{ kg } H_2O}{1000 \text{ g } H_2O} = 0.064 \text{ kg } H_2O$ then

$$m = \frac{\text{amount solute (moles)}}{\text{mass solvent (kg)}} = \frac{0.9\underline{8}7383 \text{ mol HCl}}{0.064 \text{ kg}} = 15\ m \text{ HCl}$$ and

$64 \text{ g } H_2O \times \frac{1 \text{ mol } H_2O}{18.02 \text{ g } H_2O} = 3.\underline{5}5161 \text{ mol } H_2O$ then

$$\chi = \frac{\text{amount solvent (in moles)}}{\text{total amount of solute and solvent (in moles)}} = \frac{0.9\underline{8}7383 \text{ mol}}{0.9\underline{8}7383 \text{ mol} + 3.\underline{5}5161 \text{ mol}} = 0.22.$$

Check: The units (m and unitless) are correct. The magnitudes of the answers (15 and 0.2) seem reasonable since we are starting with a high concentration solution and the molar mass of water is much less than that of HCl.

41. The level has decreased more in the beaker filled with pure water. The dissolved salt in the seawater decreases the vapor pressure and subsequently lowers the rate of vaporization.

43. **Given:** 28.5 g of glycerin ($C_3H_8O_3$) in 125 mL water at 30 °C; $P°_{H2O}$ = 31.8 torr **Find:** P_{H2O}

Other: d (H_2O) = 1.00 g/mL; glycerin is not ionic solid

Conceptual Plan:

$g_{C3H8O3} \rightarrow mol_{C3H8O3}$ **and** $mL_{H2O} \rightarrow g_{H2O} \rightarrow mol_{H2O}$ **then** $mol_{C3H8O3}, mol_{H2O} \rightarrow \chi_{H2O}$

$$\frac{1 \text{ mol } C_3H_8O_3}{92.09 \text{ g } C_3H_8O_3} \qquad \frac{1.00 \text{ g}}{1 \text{ mL}} \quad \frac{1 \text{ mol } H_2O}{18.01 \text{ g } H_2O} \qquad \chi = \frac{\text{amount solute (in moles)}}{\text{total amount of solute and solvent (in moles)}}$$

then $\chi_{H2O}, P°_{H2O} \rightarrow P_{H2O}$

$$P_{solution} = \chi_{solvent} P^o_{solvent}$$

Solution: $28.5 \text{ g } C_3H_8O_3 \times \frac{1 \text{ mol} C_3H_8O_3}{92.09 \text{ g } C_3H_8O_3} = 0.30\underline{9}466 \text{ mol} C_3H_8O_3$ and

$125 \text{ mL} \times \frac{1.00 \text{ g}}{1 \text{ mL}} \times \frac{1 \text{ mol} H_2O}{18.01 \text{ g } H_2O} = 6.9\underline{4}059 \text{ mol} H_2O$ then

$$\chi = \frac{\text{amount solvent (in moles)}}{\text{total amount of solute and solvent (in moles)}} = \frac{6.9\underline{4}059 \text{ mol}}{0.30\underline{9}466 \text{ mol} + 6.9\underline{4}059 \text{ mol}} = 0.95\underline{7}316$$ then

$P_{solution} = \chi_{solvent} P^o_{solvent} = 0.95\underline{7}316 \times 31.8 \text{ torr} = 30.4 \text{ torr}$

Check: The units (torr) are correct. The magnitude of the answer (30 torr) seems reasonable since it is a drop from the pure vapor pressure. Very few moles of glycerin are added, so the pressure will not drop much.

45. **Given:** 5.50 % NaCl by mass in water at 25 °C; **Find:** P_{H2O} **Other:** $P°_{H2O}$ = 23.78 torr, i_{NaCl} = 1.9
Conceptual Plan: % NaCl by mass → g_{NaCl}, g_{H2O} then g_{NaCl} → mol_{NaCl} and

$$\frac{5.50 \text{ g NaCl}}{100 \text{ g (NaCl + H}_2\text{O)}} \qquad \frac{1 \text{ mol NaCl}}{58.44 \text{ g NaCl}}$$

g_{H2O} → mol_{H2O} then mol_{NaCl}, mol_{H2O} → χ_{H2O} then χ_{H2O}, $P°_{H2O}$ → P_{H2O}

$$\frac{1 \text{ mol H}_2\text{O}}{18.01 \text{ g H}_2\text{O}} \qquad \chi = \frac{\text{amount solute (in moles)}}{\text{total amount of solute and solvent (in moles)}} \qquad P_{solution} = \chi_{solvent} P^{\circ}_{solvent}$$

Solution: $\frac{5.50 \text{ g NaCl}}{100 \text{ g (NaCl + H}_2\text{O)}}$ means 5.50 g NaCl and (100 g – 5.50 g) = 94.5 g H_2O then

$$5.50 \text{ g NaCl} \times \frac{1 \text{ mol NaCl}}{58.44 \text{ g NaCl}} = 0.094\underline{1}136 \text{ mol NaCl} \text{ and } 94.5 \text{ g H}_2\text{O} \times \frac{1 \text{ mol H}_2\text{O}}{18.01 \text{ g H}_2\text{O}} = 5.2\underline{4}708 \text{ mol H}_2\text{O}$$

the number of moles of solute = i_{NaCl} x n_{NaCl} so

$$\chi_{solv} = \frac{\text{amount solvent (in moles)}}{\text{total amount solute and solvent particles (in moles)}} = \frac{5.2\underline{4}708 \text{ mol}}{5.2\underline{4}708 \text{ mol} + 1.9(0.094\underline{1}136 \text{ mol})} = 0.96\underline{7}04$$

then $P_{soln} = \chi_{solv} P^{\circ}_{solv} = 0.96\underline{7}04 \times 23.78 \text{ torr} = 23.0 \text{ torr}$

Check: The units (torr) are correct. The magnitude of the answer (23 torr) seems reasonable since it is a drop from the pure vapor pressure. Only a fraction of a mole of NaCl is added, so the pressure will not drop much.

47. **Given:** 50.0 g of heptane (C_7H_{16}) and 50.0 g of octane (C_8H_{18}) at 25 °C; $P°_{C7H16}$ = 45.8 torr; $P°_{C8H18}$ = 10.9 torr

a) **Find:** P_{C7H16}, P_{C8H18}

Conceptual Plan:

g_{C7H16} → mol_{C7H16} and g_{C8H18} → mol_{C8H18} then mol_{C7H16}, mol_{C8H18} → χ_{C7H16}, χ_{C8H18}

$$\frac{1 \text{ mol C}_7\text{H}_{16}}{100.20 \text{ g C}_7\text{H}_{16}} \qquad \frac{1 \text{ mol C}_8\text{H}_{18}}{114.22 \text{ g C}_8\text{H}_{18}} \qquad \chi_{C7H16} = \frac{\text{amount C}_7\text{H}_{16} \text{ (in moles)}}{\text{total amount (in moles)}} \qquad \chi_{C8H18} = 1 - \chi_{C7H16}$$

then χ_{C7H16}, $P°_{C7H16}$ → P_{C7H16} and χ_{C8H18}, $P°_{C8H18}$ → P_{C8H18}

$$P_{C7H16} = \chi_{C7H16} P^{o}_{C7H16} \qquad P_{C8H18} = \chi_{C8H18} P^{o}_{C8H18}$$

Solution: $50.0 \text{ g C}_7\text{H}_{16} \times \frac{1 \text{ mol C}_7\text{H}_{16}}{100.20 \text{ g C}_7\text{H}_{16}} = 0.49\underline{9}002 \text{ mol C}_7\text{H}_{16}$ and

$50.0 \text{ g C}_8\text{H}_{18} \times \frac{1 \text{ mol C}_8\text{H}_{18}}{114.22 \text{ g C}_8\text{H}_{18}} = 0.43\underline{7}752 \text{ mol C}_8\text{H}_{18}$ then

$$\chi_{C7H16} = \frac{\text{amount C}_7\text{H}_{16} \text{ (in moles)}}{\text{total amount (in moles)}} = \frac{0.49\underline{9}002 \text{ mol}}{0.49\underline{9}002 \text{ mol} + 0.43\underline{7}752 \text{ mol}} = 0.53\underline{2}693 \text{ and}$$

$\chi_{C8H18} = 1 - \chi_{C7H16} = 1 - 0.53\underline{2}693 = 0.46\underline{7}307$ then

$P_{C7H16} = \chi_{C7H16} P^{o}_{C7H16} = 0.53\underline{2}693 \times 45.8 \text{ torr} = 24.4 \text{ torr}$ and

$P_{C8H18} = \chi_{C8H18} P^{o}_{C8H18} = 0.46\underline{7}307 \times 10.9 \text{ torr} = 5.09 \text{ torr}$

Check: The units (torr) are correct. The magnitude of the answer (24 and 5 torr) seems reasonable since it we expect a drop in half from the pure vapor pressures since we have roughly a 50:50 mole ratio of the two components.

b) **Find:** P_{Total}

Conceptual Plan: P_{C7H16}, P_{C8H18} → P_{Total}

$$P_{Total} = P_{C7H16} + P_{C8H18}$$

Solution: $P_{Total} = P_{C7H16} + P_{C8H18} = 24.4 \text{ torr} + 5.09 \text{ torr} = 29.5 \text{ torr}$

Check: The units (torr) are correct. The magnitude of the answer (30 torr) seems reasonable considering the two pressures.

c) **Find:** mass percent composition of the gas phase

Conceptual Plan: since n α P and we are calculating a mass percent, which is a ratio of masses, we can simply convert 1 torr to 1 mole so

P_{C7H16}, P_{C8H18} → n_{C7H16}, n_{C8H18} **then** mol_{C7H16} → g_{C7H16} **and** mol_{C8H18} → g_{C8H18}

$$\frac{100.20 \text{ g } C_7H_{16}}{1 \text{ mol } C_7H_{16}} \qquad \frac{114.22 \text{ g } C_8H_{18}}{1 \text{ mol } C_8H_{18}}$$

then g_{C7H16}, g_{C8H18} → **mass percents**

$$\textit{mass percent} = \frac{\textit{mass solute}}{\textit{mass solution}} \times 100\ \%$$

Solution: so $n_{C7H16} = 24.4$ mol and $n_{C8H18} = 5.09$ mol then

$$24.4 \text{ mol } C_7H_{16} \times \frac{100.20 \text{ g } C_7H_{16}}{1 \text{ mol } C_7H_{16}} = 24\underline{4}4.88 \text{ g } C_7H_{16} \text{ and}$$

$$5.09 \text{ mol } C_8H_{18} \times \frac{114.22 \text{ g } C_8H_{18}}{1 \text{ mol } C_8H_{18}} = 58\underline{1}.380 \text{ g } C_8H_{18} \text{ then}$$

$$\textit{mass percent} = \frac{\textit{mass solute}}{\textit{mass solution}} \times 100\ \% = \frac{24\underline{4}4.88 \text{ g } C_7H_{16}}{24\underline{4}4.88 \text{ g } C_7H_{16} + 58\underline{1}.380 \text{ g } C_8H_{18}} \times 100\ \% =$$

$= 80.8$ percent by mass C_7H_{16}
then $100\ \% - 80.8\ \% = 19.2$ percent by mass C_8H_{18}
Check: The units (%) are correct. The magnitudes of the answers (81 % and 19 %) seem reasonable considering the two pressures.

d) The two mass percents are different because the vapor is richer in the more volatile component (the lighter molecule).

49. **Given:** 55.8 g of glucose ($C_6H_{12}O_6$) in 455 g water **Find:** T_f and T_b
Other: $K_f = 1.86$ °C/m; $K_b = 0.512$ °C/m;
Conceptual Plan: g_{H2O} → kg_{H2O} **and** $g_{C6H12O6}$ → $mol_{C6H12O6}$ **then** $mol_{C6H12O6}$, kg_{H2O} → m

$$\frac{1 \text{ kg}}{1000 \text{ g}} \qquad \frac{1 \text{ mol } C_6H_{12}O_6}{180.16 \text{ g } C_6H_{12}O_6} \qquad m = \frac{\text{amount solute (moles)}}{\text{mass solvent (kg)}}$$

m, K_f → ΔT_f → T_f **and** m, K_b → ΔT_b → T_b

$$\Delta T_f = K_f\, m \quad T_f = T_f^o - \Delta T_f \qquad \Delta T_b = K_b\, m \quad \Delta T_b = T_b - T_b^o$$

Solution: $455 \text{ g} \times \frac{1 \text{ kg}}{1000 \text{ g}} = 0.455 \text{ kg}$ and $55.8 \text{ g } C_6H_{12}O_6 \times \frac{1 \text{ mol } C_6H_{12}O_6}{180.16 \text{ g } C_6H_{12}O_6} = 0.30\underline{9}725 \text{ mol } C_6H_{12}O_6$

then $m = \frac{\text{amount solute (moles)}}{\text{mass solvent (kg)}} = \frac{0.30\underline{9}725 \text{ mol } C_6H_{12}O_6}{0.455 \text{ kg}} = 0.68\underline{0}714\ m$ then

$$\Delta T_f = K_f\, m = 1.86 \frac{°C}{m} \times 0.68\underline{0}714\ m = 1.27\ °C \text{ then } T_f = T_f^o - \Delta T_f = 0.00\ °C - 1.27\ °C = -1.27\ °C \text{ and}$$

$$\Delta T_b = K_b\, m = 0.512 \frac{°C}{m} \times 0.68\underline{0}714\ m = 0.349\ °C \text{ and } \Delta T_b = T_b - T_b^o \text{ so}$$

$$T_b = T_b^o + \Delta T_b = 100.000\ °C + 0.349\ °C = 100.349\ °C$$

Check: The units (°C) are correct. The magnitudes of the answers seem reasonable since the molality is ~ 2/3. The shift in boiling point is less than the shift in freezing point because the constant is smaller for freezing.

51. **Given:** 17.5 g of unknown nonelectrolyte in 100.0 g water, $T_f = -1.8$ °C **Find:** $\mathfrak{M}$
Other: $K_f = 1.86$ °C/m
Conceptual Plan: g_{H2O} → kg_{H2O} **and** T_f → ΔT_f **then** ΔT_f, K_f → m **then** m, kg_{H2O} → mol_{Unk}

$$\frac{1 \text{ kg}}{1000 \text{ g}} \qquad T_f = T_f^o - \Delta T_f \qquad \Delta T_f = K_f\, m \qquad m = \frac{\text{amount solute (moles)}}{\text{mass solvent (kg)}}$$

then g_{Unk}, **mol** $_{Unk}$ → $\mathfrak{M}$

$$\mathfrak{M} = \frac{g_{Unk}}{mol_{Unk}}$$

Solution: $100.0\ \cancel{g} \times \frac{1\text{kg}}{1000\ \cancel{g}} = 0.1000\ \text{kg}$ and $T_f = T_f^o - \Delta T_f$ so

$\Delta T_f = T_f^o - T_f = 0.00\ °C - (-1.8\ °C) = +1.8\ °C$ $\quad \Delta T_f = K_f m$ Rearrange to solve for *m*.

$$m = \frac{\Delta T_f}{K_f} = \frac{1.8\ \cancel{°C}}{1.86\ \frac{\cancel{°C}}{m}} = 0.9\underline{6}774\ m \quad \text{then} \quad m = \frac{\text{amount solute (moles)}}{\text{mass solvent (kg)}} \quad \text{so}$$

$$mol_{Unk} = m_{Unk} \times kg_{H2O} = 0.9\underline{6}774 \frac{\text{mol Unk}}{\cancel{\text{kg}}} \times 0.1000\ \cancel{\text{kg}} = 0.09\underline{6}774\ \text{mol Unk} \quad \text{then}$$

$$\mathfrak{M} = \frac{g_{Unk}}{mol_{Unk}} = \frac{17.5\ \text{g}}{0.09\underline{6}774\ \text{mol}} = 180\ \frac{\text{g}}{\text{mol}} = 1.8 \times 10^2\ \frac{\text{g}}{\text{mol}}$$

Check: The units (g/mol) are correct. The magnitude of the answer (180 g/mol) seems reasonable since the molality is ~ 0.1 and we have ~18 g. It is a reasonable molecular weight for a solid or liquid.

53. **Given:** 24.6 g of glycerin ($C_3H_8O_3$) in 250.0 mL of solution at 298 K **Find:** Π
Conceptual Plan: mL_{soln} → L_{soln} **and** g_{C3H8O3} → mol_{C3H8O3} **then** mol_{C3H8O3}, L_{soln} → **M then**

$$\frac{1\ \text{L}}{1000\ \text{mL}} \qquad \frac{1\ \text{mol}\ C_3H_8O_3}{92.09\ \text{g}\ C_3H_8O_3} \qquad M = \frac{\text{amount solute (moles)}}{\text{volume solution (L)}}$$

M, *T* → Π

$\Pi = MRT$

Solution:

$$250.0\ \cancel{\text{mL}} \times \frac{1\ \text{L}}{1000\ \cancel{\text{mL}}} = 0.2500\ \text{L} \quad \text{and} \quad 24.6\ \cancel{\text{g}\ C_3H_8O_3} \times \frac{1\ \text{mol}\ C_3H_8O_3}{92.09\ \cancel{\text{g}\ C_3H_8O_3}} = 0.26\underline{7}130\ \text{mol}\ C_3H_8O_3 \quad \text{then}$$

$$M = \frac{\text{amount solute (moles)}}{\text{volume solution (L)}} = \frac{0.26\underline{7}130\ \text{mol}\ C_3H_8O_3}{0.2500\ \text{L}} = 1.0\underline{6}852\ \text{M} \quad \text{then}$$

$$\Pi = MRT = 1.0\underline{6}852\ \frac{\cancel{\text{mol}}}{\cancel{\text{L}}} \times 0.08206\ \frac{\cancel{\text{L}} \cdot \text{atm}}{\cancel{\text{K}} \cdot \cancel{\text{mol}}} \times 298\ \cancel{\text{K}} = 26.1\ \text{atm}$$

Check: The units (atm) are correct. The magnitude of the answer (26 atm) seems reasonable since the molarity is ~ 1.

55. **Given:** 27.55 mg unknown protein in 25.0 mL solution; $\Pi = 3.22$ torr at 25 °C **Find:** $\mathfrak{M}_{\text{unknown protein}}$
Conceptual Plan:
°C → K and torr → atm then Π, *T* → M then mL_{soln} → L_{soln} **then**

$$K = °C + 273.15 \qquad \frac{1\ \text{atm}}{760\ \text{torr}} \qquad \Pi = MRT \qquad \frac{1\ \text{L}}{1000\ \text{mL}}$$

L_{soln}, **M →** $\text{mol}_{\text{unknown protein}}$ **and mg → g then** $g_{\text{unknown protein}}$, $\text{mol}_{\text{unknown protein}}$ → $\mathfrak{M}_{\text{unknown protein}}$

$$M = \frac{\text{amount solute (moles)}}{\text{volume solution (L)}} \qquad \frac{1\ \text{g}}{1000\ \text{mg}} \qquad \mathfrak{M} = \frac{g_{\text{unknown protein}}}{\text{mol}_{\text{unknown protein}}}$$

Solution: $25\ °C + 273.15 = 298\ \text{K}$ and $\quad 3.22\ \cancel{\text{torr}} \times \frac{1\ \text{atm}}{760\ \cancel{\text{torr}}} = 0.0042\underline{3}684\ \text{atm}$ $\quad \Pi = MRT$ for M.

$$M = \frac{\Pi}{RT} = \frac{0.0042\underline{3}684\ \cancel{\text{atm}}}{0.08206\ \frac{\text{L} \cdot \cancel{\text{atm}}}{\cancel{\text{K}} \cdot \text{mol}} \times 298\ \cancel{\text{K}}} = 1.7\underline{3}258 \times 10^{-4}\ \frac{\text{mol}}{\text{L}} \quad \text{then} \quad 25.0\ \cancel{\text{mL}} \times \frac{1\ \text{L}}{1000\ \cancel{\text{mL}}} = 0.0250\ \text{L}$$

then $M = \dfrac{\text{amount solute (moles)}}{\text{volume solution (L)}}$ Rearrange to solve for mol $_{\text{unknown protein}}$.

$$\text{mol}_{\text{unknown protein}} = M \times L = 1.73\underline{2}58 \times 10^{-4} \frac{\text{mol}}{\cancel{\text{L}}} \times 0.0250 \cancel{\text{L}} = 4.33\underline{1}46 \times 10^{-6} \text{ mol}$$ and

$$27.55 \cancel{\text{mg}} \times \frac{1\text{g}}{1000 \cancel{\text{mg}}} = 0.02755 \text{ g}$$ then

$$\mathfrak{M} = \frac{\text{g}_{\text{unknown protein}}}{\text{mol}_{\text{unknown protein}}} = \frac{0.02755 \text{ g}}{4.33\underline{1}46 \times 10^{-6} \text{ mol}} = 6.36 \times 10^{3} \frac{\text{g}}{\text{mol}}$$

Check: The units (g/mol) are correct. The magnitude of the answer (6400 g/mol) seems reasonable for a large biological molecule. A small amount of material is put into 0.025 L, so the concentration is very small and the molecular weight is large.

57. a) **Given:** 0.100 m of K_2S, completely dissociated **Find:** T_f, T_b

Other: $K_f = 1.86$ °C/m; $K_b = 0.512$ °C/m;

Conceptual Plan: $m, i, K_f \rightarrow \Delta T_f$ then $\Delta T_f \rightarrow T_f$ and $m, i, K_b \rightarrow \Delta T_b$ then $\Delta T_b \rightarrow T_b$

$\Delta T_f = K_f i m$ $i = 3$ $\quad T_f = T_f^\circ - \Delta T_f$ $\quad \Delta T_b = K_b i m$ $i = 3$ $\quad T_b = T_b^\circ + \Delta T_b$

Solution: $\Delta T_f = K_f i m = 1.86 \frac{°\text{C}}{\cancel{m}} \times 3 \times 0.100 \cancel{m} = 0.558$ °C then

$T_f = T_f^o - \Delta T_f = 0.000°\text{C} - 0.558\ °\text{C} = -0.558\ °\text{C}$ and

$\Delta T_b = K_b i m = 0.512 \frac{°\text{C}}{\cancel{m}} \times 3 \times 0.100 \cancel{m} = 0.154$ °C then

$T_b = T_b^o - \Delta T_b = 100.000°\text{C} + 0.154\ °\text{C} = 100.154\ °\text{C}$.

Check: The units (°C) are correct. The magnitude of the answer (– 0.6 °C and 100.2 °C) seems reasonable since the molality of the particles is 0.3. The shift in boiling point is less than the shift in freezing point because the constant is smaller for freezing.

b) **Given:** 21.5 g $CuCl_2$ in 4.50 x 10^2 g water, completely dissociated **Find:** T_f, T_b

Other: $K_f = 1.86$ °C/m; $K_b = 0.512$ °C/m;

Conceptual Plan: $\text{g}_{H2O} \rightarrow \text{kg}_{H2O}$ and $\text{g}_{CuCl2} \rightarrow \text{mol}_{CuCl2}$ then $\text{mol}_{CuCl2}, \text{kg}_{H2O} \rightarrow m$

$\frac{1\text{kg}}{1000\text{ g}}$ $\quad \frac{1 \text{ mol } CuCl_2}{134.46 \text{ g } CuCl_2}$ $\quad m = \frac{\text{amount solute (moles)}}{\text{mass solvent (kg)}}$

$m, i, K_f \rightarrow \Delta T_f \rightarrow T_f$ and $m, i, K_b \rightarrow \Delta T_b \rightarrow T_b$

$\Delta T_f = K_f i m$ $i = 3$ $\quad T_f = T_f^\circ - \Delta T_f$ $\quad \Delta T_b = K_b i m$ $i = 3$ $\quad T_b = T_b^\circ + \Delta T_b$

Solution:

$4.5 \times 10^2 \cancel{\text{g}} \times \frac{1\text{kg}}{1000 \cancel{\text{g}}} = 0.450 \text{ kg}$ and $21.5 \cancel{\text{g } CuCl_2} \times \frac{1 \text{ mol } CuCl_2}{134.46 \cancel{\text{g } CuCl_2}} = 0.15\underline{9}904 \text{ mol } CuCl_2$ then

$m = \frac{\text{amount solute (moles)}}{\text{mass solvent (kg)}} = \frac{0.15\underline{9}904 \text{ mol } CuCl_2}{0.450 \text{ kg}} = 0.35\underline{5}341\ m$ then

$\Delta T_f = K_f i m = 1.86 \frac{°\text{C}}{\cancel{m}} \times 3 \times 0.35\underline{5}341 \cancel{m} = 1.98$ °C then

$T_f = T_f^o - \Delta T_f = 0.000°\text{C} - 1.98\ °\text{C} = -1.98\ °\text{C}$ and

$\Delta T_b = K_b i m = 0.512 \frac{°\text{C}}{\cancel{m}} \times 3 \times 0.35\underline{5}341 \cancel{m} = 0.546$ °C then

$T_b = T_b^o - \Delta T_b = 100.000°\text{C} + 0.546\ °\text{C} = 100.546\ °\text{C}$.

Check: The units (°C) are correct. The magnitude of the answer (– 2 °C and 100.5 °C) seems reasonable since the molality of the particles is ~ 1. The shift in boiling point is less than the shift in freezing point because the constant is smaller for freezing.

c) **Given:** 5.5 % by mass $NaNO_3$, completely dissociated **Find:** T_f, T_b
Other: $K_f = 1.86$ °C/m; $K_b = 0.512$ °C/m;
Conceptual Plan:
percent by mass → g_{NaNO3}, g_{H2O} then g_{H2O} → kg_{H2O} and g_{NaNO3} → mol_{NaNO3} then

$$mass\ percent = \frac{mass\ solute}{mass\ solution} \times 100\ \% \qquad \frac{1\,kg}{1000\ g} \qquad \frac{1\ mol\ NaNO_3}{84.99\ g\ NaNO_3}$$

mol_{NaNO3}, kg_{H2O} → m then m, i, K_f → ΔT_f → T_f and m, i, K_b → ΔT_b → T_b

$$m = \frac{\text{amount solute (moles)}}{\text{mass solvent (kg)}} \qquad \Delta T_f = K_f i m \quad i = 2 \quad T_f = T_f^{\circ} - \Delta T_f \qquad \Delta T_b = K_b i m \quad i = 2 \quad T_b = T_b^{\circ} + \Delta T_b$$

Solution: $mass\ percent = \dfrac{mass\ solute}{mass\ solution} \times 100\ \%$ so 5.5 % by mass $NaNO_3$ means 5.5 g $NaNO_3$ and

100.0 g – 5.5 g = 94.5 g water. Then $94.5\ \cancel{g} \times \dfrac{1\,kg}{1000\ \cancel{g}} = 0.0945$ kg and

$$5.5\ \cancel{g\ NaNO_3} \times \frac{1\ mol\ NaNO_3}{84.99\ \cancel{g\ NaNO_3}} = 0.06\underline{4}713\ mol\ NaNO_3 \text{ then}$$

$$m = \frac{\text{amount solute (moles)}}{\text{mass solvent (kg)}} = \frac{0.06\underline{4}713\ mol\ NaNO_3}{0.0945\ kg} = 0.6\underline{8}480\ m \text{ then}$$

$$\Delta T_f = K_f i m = 1.86 \frac{°C}{\cancel{m}} \times 2 \times 0.6\underline{8}480\ \cancel{m} = 2.5\ °C \text{ then}$$

$$T_f = T_f^{o} - \Delta T_f = 0.000°C - 2.3\ °C = -2.5\ °C \quad \text{and}$$

$$\Delta T_b = K_b i m = 0.512 \frac{°C}{\cancel{m}} \times 2 \times 0.6\underline{8}480\ \cancel{m} = 0.70\ °C \text{ then}$$

$$T_b = T_b^{o} - \Delta T_b = 100.000°C + 0.64\ °C = 100.70\ °C\,.$$

Check: The units (°C) are correct. The magnitude of the answer (– 2.5 °C and 100.7 °C) seems reasonable since the molality of the particles is ~ 1. The shift in boiling point is less than the shift in freezing
point because the constant is smaller for freezing.

59. a) **Given:** 0.100 m of $FeCl_3$ **Find:** T_f **Other:** $K_f = 1.86$ °C/m; $i_{measured} = 3.4$
Conceptual Plan: **m, i, K_f → ΔT_f then ΔT_f → T_f**

$$\Delta T_f = K_f i m \qquad T_f = T_f^{\circ} - \Delta T_f$$

Solution: $\Delta T_f = K_f i m = 1.86 \dfrac{°C}{\cancel{m}} \times 3.4 \times 0.100\ \cancel{m} = 0.632\ °C$ then

$$T_f = T_f^{o} - \Delta T_f = 0.000°C - 0.632\ °C = -0.632\ °C\,.$$

Check: The units (°C) are correct. The magnitude of the answer (– 0.6 °C) seems reasonable since the molality of the particles is 0.3.

b) **Given:** 0.085 M of K_2SO_4 at 298 K **Find:** Π **Other:** $i_{measured} = 2.6$
Conceptual Plan: **M, i, T → Π**

$$\Pi = iMRT$$

Solution: $\Pi = iMRT = 2.6 \times 0.085 \dfrac{\cancel{mol}}{\cancel{L}} \times 0.08206 \dfrac{\cancel{L} \cdot atm}{\cancel{K} \cdot \cancel{mol}} \times 298\ \cancel{K} = 5.4$ atm

Check: The units (atm) are correct. The magnitude of the answer (5 atm) seems reasonable since the molarity is ~ 0.2.

c) **Given:** 1.22 % by mass $MgCl_2$ **Find:** T_b **Other:** $K_b = 0.512\ °C/m$; $i_{measured} = 2.7$

Conceptual Plan:

percent by mass → g_{MgCl2}, g_{H2O} then g_{H2O} → kg_{H2O} and g_{MgCl2} → mol_{MgCl2} then

$$mass\ percent = \frac{mass\ solute}{mass\ solution} \times 100\ \% \qquad \frac{1\,kg}{1000\ g} \qquad \frac{1\ mol\,MgCl_2}{95.22\ g\ MgCl_2}$$

mol_{MgCl2}, kg_{H2O} → m then m, i, K_b → ΔT_b → T_b

$$m = \frac{\text{amount solute (moles)}}{\text{mass solvent (kg)}} \qquad \Delta T_b = K_b\, i\, m \qquad T_b = T_b^{\circ} + \Delta T_b$$

Solution: $mass\ percent = \dfrac{mass\ solute}{mass\ solution} \times 100\,\%$ so 1.22 % by mass $MgCl_2$ means 1.22 g $MgCl_2$ and

100.00 g – 1.22 g = 98.78 g water. Then $98.78\ \cancel{g} \times \dfrac{1\,kg}{1000\ \cancel{g}} = 0.09878\ kg$ and

$$1.22\ \cancel{g\,MgCl_2} \times \frac{1\ mol\,MgCl_2}{95.22\ \cancel{g\,MgCl_2}} = 0.01\underline{2}8124\ mol\ MgCl_2 \text{ then}$$

$$m = \frac{\text{amount solute (moles)}}{\text{mass solvent (kg)}} = \frac{0.01\underline{2}8124\ mol\ MgCl_2}{0.09878\ kg} = 0.12\underline{9}706\ m \text{ then}$$

$$\Delta T_b = K_b\, i\, m = 0.512 \frac{°C}{\cancel{m}} \times 2.7 \times 0.12\underline{9}706\ \cancel{m} = 0.18\ °C \text{ then}$$

$T_b = T_b^{\circ} - \Delta T_b = 100.000°C + 0.18\ °C = 100.18\ °C$.

Check: The units (°C) are correct. The magnitude of the answer (100.2 °C) seems reasonable since the molality of the particles is ~ 1/3.

61. **Given:** 0.100 M of ionic solution, Π = 8.3 atm at 25 °C **Find:** $i_{measured}$

Conceptual Plan: °C → K then Π, M, T → i

$$K = °C + 273.15 \qquad \Pi = i\,M\,R\,T$$

Solution: 25 °C + 273.15 = 298 K then $\Pi = i\,M\,R\,T$ Rearrange to solve for i.

$$i = \frac{\Pi}{MRT} = \frac{8.3\ \cancel{atm}}{0.100\ \dfrac{\cancel{mol}}{\cancel{L}} \times 0.08206\ \dfrac{\cancel{L} \cdot \cancel{atm}}{\cancel{K} \cdot \cancel{mol}} \times 298\ \cancel{K}} = 3.4$$

Check: The units (none) are correct. The magnitude of the answer (3) seems reasonable for an ionic solution with a high osmotic pressure.

63. Chloroform is polar and has stronger solute–solvent interactions than nonpolar carbon tetrachloride.

65. **Given:** $KClO_4$: Lattice Energy = – 599 kJ/mol, $\Delta H_{hydration}$ = – 548 kJ/mol; 10.0 g $KClO_4$ in 100.00 mL solution **Find:** ΔH_{soln} and ΔT **Other:** C_s = 4.05 J/g °C; d = 1.05 g/mL

Conceptual Plan:

Lattice Energy, $\Delta H_{hydration}$ → ΔH_{soln} and g → mol then mol, ΔH_{soln} → q(kJ) → q(J)

$$\Delta H_{soln} = \Delta H_{solute} + \Delta H_{hydration} \text{ where } \Delta H_{solute} = -\Delta H_{lattice} \qquad \frac{1\ mol}{138.56\ g} \qquad q = n\,\Delta H_{soln} \qquad \frac{1000\ J}{1\ kJ}$$

then mL_{soln} → g_{soln} then q, g_{soln}, C_s → ΔT

$$\frac{1.05\ g}{1\ mL} \qquad q = m\,C_s\,\Delta T$$

Solution: $\Delta H_{soln} = \Delta H_{solute} + \Delta H_{hydration}$ where $\Delta H_{solute} = -\Delta H_{lattice}$ so $\Delta H_{soln} = \Delta H_{hydration} - \Delta H_{lattice}$

ΔH_{soln} = – 548 kJ/mol – (– 599 kJ/mol) = +51 kJ/mol and $10.0\ \cancel{g} \times \dfrac{1\ mol}{138.56\ \cancel{g}} = 0.07\underline{2}1709\ mol$ then

$$q = n\,\Delta H_{soln} = 0.07\underline{2}1709\ \cancel{mol} \times 51 \frac{kJ}{\cancel{mol}} = +3.\underline{6}807\ \cancel{kJ} \times \frac{1000\ J}{1\ \cancel{kJ}} = +3\underline{6}80.7\ J \text{ absorbed then}$$

$$100.0 \text{ mL} \times \frac{1.05 \text{ g}}{1 \text{ mL}} = 105 \text{ g}$$ Since heat is absorbed when $KClO_4$ dissolves, the temperature will drop or $q = -3\underline{6}80.7$ J and $q = m\,C_s\,\Delta T$ Rearrange to solve for ΔT.

$$\Delta T = \frac{q}{m\,C_s} = \frac{-3\underline{6}80.7 \text{ J}}{105 \text{ g} \times 4.05 \frac{\text{J}}{\text{g}\cdot{}^\circ\text{C}}} = -8.7\ {}^\circ\text{C}$$

Check: The units (kJ/mol and °C) are correct. The magnitude of the answer (51 kJ/mol) makes physical sense because the lattice energy is larger than the heat of hydration. The magnitude of the temperature change (– 9 °C) makes physical sense since heat is absorbed and the heat of solution is fairly small.

67. **Given:** Argon, 0.0537 L; 25 °C, P_{Ar} = 1.0 atm to make 1.0 L saturated solution **Find:** k_H(Ar)
Conceptual Plan: **°C → K** and **P_{Ar}, V, T → mol$_{He}$** then **mol$_{He}$, V_{soln}, P_{Ar} → k_H(He)**

$K = {}^\circ C + 273.15$ $\quad PV = nRT \quad S_{Ar} = k_H(Ar)P_{Ar}$ with $S_{Ar} = \frac{mol_{Ar}}{L_{soln}}$

Solution: 25 °C + 273.15 = 298 K and $PV = nRT$ Rearrange to solve for n.

$$n = \frac{PV}{RT} = \frac{1.0 \text{ atm} \times 0.0537 \text{ L}}{0.08206 \frac{\text{L}\cdot\text{atm}}{\text{K}\cdot\text{mol}} \times 298 \text{ K}} = 0.002\underline{1}9597 \text{ mol}$$ then $S_{Ar} = k_H(Ar)P_{Ar}$ with $S_{Ar} = \frac{mol_{Ar}}{L_{soln}}$

Substitute in values and rearrange to solve for k_H.

$$k_H(Ar) = \frac{mol_{Ar}}{L_{soln}P_{He}} = \frac{0.002\underline{1}9597 \text{ mol}}{1.0 \text{ L}_{soln} \times 1.0 \text{ atm}} = 2.2 \times 10^{-3} \frac{\text{M}}{\text{atm}}$$

Check: The units (M/atm) are correct. The magnitude of the answer (10^{-3}) seems reasonable since it is consistent with other values in the text.

69. **Given:** 0.0020 ppm by mass Hg = legal limit; 0.0040 ppm by mass Hg = contaminated water; 50.0 mg Hg ingested **Find:** volume of contaminated water
Conceptual Plan: **mg$_{Hg}$ → g$_{Hg}$ → g$_{H2O}$ → mL$_{H2O}$ → L$_{H2O}$**

$$\frac{1 \text{ g}}{1000 \text{ mg}} \quad \frac{10^6 \text{ g water}}{0.0040 \text{ g Hg}} \quad \frac{1 \text{ mL}}{1.00 \text{ g}} \quad \frac{1 \text{ L}}{1000 \text{ mL}}$$

Solution:

$$50.0 \text{ mg Hg} \times \frac{1 \text{ g Hg}}{1000 \text{ mg Hg}} \times \frac{10^6 \text{ g water}}{0.0040 \text{ g Hg}} \times \frac{1 \text{ mL water}}{1.00 \text{ g water}} \times \frac{1 \text{ L water}}{1000 \text{ mL water}} = 1.3 \times 10^4 \text{ L water}$$

Check: The units (L) are correct. The magnitude of the answer (10^4 L) seems reasonable since the concentration is so low.

71. **Given:** 12.5 % NaCl by mass in water at 55 °C; 2.5 L vapor **Find:** g$_{H2O}$ in vapor
Other: P°_{H2O} = 118 torr, i_{NaCl} = 2.0 (complete dissociation)
Conceptual Plan: **% NaCl by mass → g$_{NaCl}$, g$_{H2O}$** then **g$_{NaCl}$ → mol$_{NaCl}$** and

$$\frac{12.5 \text{ g NaCl}}{100 \text{ g } (NaCl + H_2O)} \qquad \frac{1 \text{ mol NaCl}}{58.44 \text{ g NaCl}}$$

g$_{H2O}$ → mol$_{H2O}$ then **mol$_{NaCl}$, mol$_{H2O}$ → χ_{NaCl} → χ_{H2O}** then **χ_{H2O}, P°_{H2O} → P_{H2O}**

$$\frac{1 \text{ mol } H_2O}{18.02 \text{ g } H_2O} \quad \chi = \frac{\text{amount solute (in moles)}}{\text{total amount of solute and solvent (in moles)}} \quad \chi_{H2O} = 1 - i_{NaCl}\chi_{NaCl} \quad P_{solution} = \chi_{solvent}P^o_{solvent}$$

then **torr → atm** and **°C → K** **P, V, T → mol$_{H2O}$ → g$_{H2O}$**

$$\frac{1 \text{ atm}}{760 \text{ torr}} \quad K = {}^\circ C + 273.15 \quad PV = nRT \quad \frac{18.02 \text{ g } H_2O}{1 \text{ mol } H_2O}$$

Solution: $\frac{12.5 \text{ g NaCl}}{100 \text{ g } (NaCl + H_2O)}$ means 12.5 g NaCl and (100 g – 12.5 g) = 87.5 g H_2O then

$$12.5\ \text{g NaCl} \times \frac{1\ \text{mol NaCl}}{58.44\ \text{g NaCl}} = 0.21\underline{3}895\ \text{mol NaCl} \text{ and } 87.5\ \text{g H}_2\text{O} \times \frac{1\ \text{mol H}_2\text{O}}{18.02\ \text{g H}_2\text{O}} = 4.8\underline{5}572\ \text{mol H}_2\text{O}$$

$$\text{then } \chi = \frac{\text{amount solute (in moles)}}{\text{total amount of solute and solvent (in moles)}} = \frac{0.21\underline{3}895\ \text{mol}}{0.21\underline{3}895\ \text{mol} + 4.8\underline{5}572\ \text{mol}} = 0.042\underline{1}916$$

then $\chi_{H2O} = 1 - i_{NaCl}\chi_{NaCl} = 1 - (2.0 \times 0.042\underline{1}916) = 0.91\underline{5}617$ then

$P_{solution} = \chi_{solvent} P^{o}_{solvent} = 0.91\underline{5}617 \times 118\ \text{torr} = 10\underline{8}.043\ \text{torr H}_2\text{O}$ then

$$10\underline{8}.043\ \text{torr H}_2\text{O} \times \frac{1\ \text{atm}}{760\ \text{torr}} = 0.14\underline{2}162\ \text{atm}$$

and 55 °C + 273.15 = 328 K then $PV = nRT$ Rearrange to solve for n.

$$n = \frac{PV}{RT} = \frac{0.14\underline{2}162\ \text{atm} \times 2.5\ \text{L}}{0.08206 \frac{\text{L} \cdot \text{atm}}{\text{K} \cdot \text{mol}} \times 328\ \text{K}} = 0.01\underline{3}204\ \text{mol} \text{ then}$$

$$0.01\underline{3}204\ \text{mol H}_2\text{O} \times \frac{18.02\ \text{g H}_2\text{O}}{1\ \text{mol H}_2\text{O}} = 0.24\ \text{g H}_2\text{O}$$

Check: The units (g) are correct. The magnitude of the answer (0.2 g) seems reasonable since there is very little mass in a vapor.

73. **Given:** T_b = 106.5 °C aqueous solution **Find:** T_f **Other:** K_f = 1.86 °C/m; K_b = 0.512 °C/m
Conceptual Plan: $T_b \rightarrow \Delta T_b$ **then** $\Delta T_b, K_b \rightarrow m$ **then** $m, K_f \rightarrow \Delta T_f \rightarrow T_f$

$$T_b = T_b^o + \Delta T_b \qquad \Delta T_b = K_b m \qquad \Delta T_f = K_f m \quad T_f = T_f^\circ - \Delta T_f$$

Solution: $T_b = T_b^o + \Delta T_b$ so $\Delta T_b = T_b - T_b^o = 106.5°\text{C} - 100.0\ °\text{C} = 6.5\ °\text{C}$ then $\Delta T_b = K_b m$

Rearrange to solve for m. $m = \frac{\Delta T_b}{K_b} = \frac{6.5\ °\text{C}}{0.512 \frac{°\text{C}}{m}} = 1\underline{2}.695\ m$ then

$\Delta T_f = K_f m = 1.86 \frac{°\text{C}}{m} \times 1\underline{2}.695\ m = 2\underline{3}.6\ °\text{C}$ then $T_f = T_f^o - \Delta T_f = 0.000°\text{C} - 2\underline{3}.6\ °\text{C}\ °\text{C} = -24\ °\text{C}$.

Check: The units (°C) are correct. The magnitude of the answer (– 24 °C) seems reasonable since the shift in boiling point is less than the shift in freezing point because the constant is smaller for freezing.

75. a) **Given:** 0.90 % NaCl by mass per volume; isotonic aqueous solution at 25 °C; KCl; i = 1.9
Find: % KCl by mass per volume
Conceptual Plan: Isotonic solutions will have the same number of particles. Since i is the same,

$$\frac{1\ \text{mol KCl}}{1\ \text{mol NaCl}}$$

The new % mass per volume will be the mass ratio of the two salts.

$$\textit{percent by mass per volume} = \frac{\textit{mass solute}}{\text{V}} \times 100\ \% \qquad \frac{1\ \text{mol NaCl}}{58.44\ \text{g NaCl}} \text{ and } \frac{74.56\ \text{g KCl}}{1\ \text{mol KCl}}$$

Solution:

$$\textit{percent by mass per volume} = \frac{\textit{mass solute}}{\text{V}} \times 100\ \% =$$

$$= \frac{0.0090\ \text{g NaCl}}{\text{V}} \times \frac{1\ \text{mol NaCl}}{58.44\ \text{g NaCl}} \times \frac{1\ \text{mol KCl}}{1\ \text{mol NaCl}} \times \frac{74.56\ \text{g KCl}}{1\ \text{mol KCl}} \times 100\ \% = 1.1\ \%\ \text{KCl by mass per volume}$$

Check: The units (% KCl by mass per volume) are correct. The magnitude of the answer (1.1) seems reasonable since the molar mass of KCl is larger than the molar mass of NaCl.

b) **Given:** 0.90 % NaCl by mass per volume; isotonic aqueous solution at 25 °C; NaBr; i = 1.9
Find: % NaBr by mass per volume

Conceptual Plan: Isotonic solutions will have the same number of particles. Since *i* is the same, $\frac{1 \text{ mol NaBr}}{1 \text{ mol NaCl}}$ **the new % mass per volume will be the mass ratio of the two salts.**

$$\text{percent by mass per volume} = \frac{\text{mass solute}}{V} \times 100\ \% \quad \frac{1 \text{ mol NaCl}}{58.44 \text{ g NaCl}} \text{ and } \frac{102.90 \text{ g NaBr}}{1 \text{ mol NaBr}}$$

Solution:

$$\text{percent by mass per volume} = \frac{\text{mass solute}}{V} \times 100\ \% =$$

$$= \frac{0.0090 \text{ g NaCl}}{V} \times \frac{1 \text{ mol NaCl}}{58.44 \text{ g NaCl}} \times \frac{1 \text{ mol NaBr}}{1 \text{ mol NaCl}} \times \frac{102.90 \text{ g NaBr}}{1 \text{ mol NaBr}} \times 100\ \%$$

= 1.6 % NaBr by mass per volume

Check: The units (% NaBr by mass per volume) are correct. The magnitude of the answer (1.6) seems reasonable since the molar mass of NaBr is larger than the molar mass of NaCl.

c) **Given:** 0.90 % NaCl by mass per volume; isotonic aqueous solution at 25 °C; glucose ($C_6H_{12}O_6$); $i = 1.9$
Find: % glucose by mass per volume
Conceptual Plan: Isotonic solutions will have the same number of particles. Since glucose is a nonelectrolyte, the *i* is not the same, then use the mass ratio of the two compounds.

$$\frac{1.9 \text{ mol } C_6H_{12}O_6}{1 \text{ mol NaCl}} \quad \text{percent by mass per volume} = \frac{\text{mass solute}}{V} \times 100\ \% \quad \frac{1 \text{ mol NaCl}}{58.44 \text{ g NaCl}} \text{ and } \frac{180.16 \text{ g } C_6H_{12}O_6}{1 \text{ mol } C_6H_{12}O_6}$$

Solution:

$$\text{percent by mass per volume} = \frac{\text{mass solute}}{V} \times 100\ \% =$$

$$= \frac{0.0090 \text{ g NaCl}}{V} \times \frac{1 \text{ mol NaCl}}{58.44 \text{ g NaCl}} \times \frac{1.9 \text{ mol } C_6H_{12}O_6}{1 \text{ mol NaCl}} \times \frac{180.16 \text{ g } C_6H_{12}O_6}{1 \text{ mol } C_6H_{12}O_6} \times 100\ \% =$$

= 5.3 % $C_6H_{12}O_6$ by mass per volume

Check: The units (% $C_6H_{12}O_6$ by mass per volume) are correct. The magnitude of the answer (1.6) seems reasonable since the molar mass of $C_6H_{12}O_6$ is larger than the molar mass of NaCl and we need more moles of $C_6H_{12}O_6$ since it is a nonelectrolyte.

77. **Given:** 4.5701 g of $MgCl_2$ and 43.238 g water, $P_{soln} = 0.3624$ atm, $P^o_{soln} = 0.3804$ atm at 348.0 K
Find: $i_{measured}$
Conceptual Plan: $\mathbf{g_{MgCl2} \rightarrow mol_{MgCl2}}$ **and** $\mathbf{g_{H2O} \rightarrow mol_{H2O}}$ **then** $\mathbf{P_{soln}, P^o_{soln} \rightarrow \chi_{MgCl2}}$

$$\frac{1 \text{ mol } MgCl_2}{95.218 \text{ g } MgCl_2} \qquad \frac{1 \text{ mol } H_2O}{18.015 \text{ g } H_2O} \qquad P_{Soln} = (1 - \chi_{MgCl2})\, P^o_{H2O}$$

then $\mathbf{mol_{MgCl2}, mol_{H2O}, \chi_{MgCl2} \rightarrow \mathit{i}}$

$$\chi_{MgCl2} = \frac{i\,(\text{moles } MgCl_2)}{\text{moles } H_2O + i\,(\text{moles } MgCl_2)}$$

Solution: $4.5701 \text{ g } MgCl_2 \times \frac{1 \text{ mol } MgCl_2}{95.218 \text{ g } MgCl_2} = 0.04799\underline{6}177 \text{ mol } MgCl_2$ and

$$43.238 \text{ g } H_2O \times \frac{1 \text{ mol } H_2O}{18.015 \text{ g } H_2O} = 2.400\underline{1}110 \text{ mol } H_2O \text{ then } P_{Soln} = (1 - \chi_{MgCl2})\, P^o_{H2O} \text{ so}$$

$$\chi_{MgCl2} = 1 - \frac{P_{Soln}}{P^o_{H2O}} = 1 - \frac{0.3624 \text{ atm}}{0.3804 \text{ atm}} = 0.0473\underline{1}861 \text{ Solve for } i.$$

$$i\,(0.04799\underline{6}177) = 0.0473\underline{1}861\,(2.400\underline{1}110 + i\,(0.04799\underline{6}177)) \rightarrow$$

$$i\,(0.04799\underline{6}177 - 0.00227\underline{1}112) = 0.113\underline{5}699 \rightarrow i = \frac{0.113\underline{5}699}{0.0457\underline{2}506} = 2.484.$$

Check: The units (none) are correct. The magnitude of the answer (2.5) seems reasonable for $MgCl_2$ since we expect *i* to be 3 if it completely dissociates. Since Mg is small and doubly charges, we expect a significant drop from 3.

79. **Given:** $T_b = 375.5$ K aqueous solution **Find:** P_{H2O} **Other:** $P°_{H2O} = 0.2467$ atm; $K_b = 0.512$ °C/*m*

Conceptual Plan: $T_b \rightarrow \Delta T_b$ **then** $\Delta T_b, K_b \rightarrow m$ **assume 1kg water** $kg_{H2O} \rightarrow mol_{H2O}$ **then**

$$T_b = T_b^{\circ} + \Delta T_b \qquad \Delta T_b = K_b\, m \qquad \frac{1\ \text{mol H}_2\text{O}}{18.02\ \text{g H}_2\text{O}}$$

$m \rightarrow mol_{Solute}$ **then** $mol_{H2O}, mol_{Solute} \rightarrow \chi_{H2O}$ **then** $\chi_{H2O}, P°_{H2O} \rightarrow P_{H2O}$

$$m = \frac{\text{amount solute (moles)}}{\text{mass solvent (kg)}} \qquad \chi_{H2O} = \frac{\text{moles H}_2\text{O}}{\text{moles H}_2\text{O} + \text{moles solute}} \qquad P_{H2O} = \chi_{H2O}\, P^{o}_{H2O}$$

Solution: $T_b = T_b^o + \Delta T_b$ so $\Delta T_b = T_b - T_b^o = 375.5\ \text{K} - 373.15\ \text{K} = 2.4\ \text{K} = 2.4\ °\text{C}$ then

$\Delta T_b = K_b\, m$ Rearrange to solve for *m*. $m = \dfrac{\Delta T_b}{K_b} = \dfrac{2.4\ \cancel{°C}}{0.512\ \frac{\cancel{°C}}{m}} = 4.6\underline{8}75\ m$ then

$$1000\ \cancel{\text{g H}_2\text{O}} \times \frac{1\ \text{mol H}_2\text{O}}{18.02\ \cancel{\text{g H}_2\text{O}}} = 55.4\underline{9}390\ \text{mol H}_2\text{O} \text{ then}$$

$$m = \frac{\text{amount solute (moles)}}{\text{mass solvent (kg)}} = \frac{x\ \text{mol}}{1\ \text{kg}} = 4.6\underline{8}75\ m \text{ so}$$

$$x = 4.6\underline{8}75\ \text{mol then } \chi_{H2O} = \frac{\text{moles H}_2\text{O}}{\text{moles H}_2\text{O} + \text{moles solute}} = \frac{55.4\underline{9}390\ \cancel{\text{mol}}}{55.4\underline{9}390\ \cancel{\text{mol}} + 4.\underline{6}875\ \cancel{\text{mol}}} = 0.92\underline{2}1106$$

then $P_{H2O} = \chi_{H2O}\, P^o_{H2O} = 0.92\underline{2}1106 \times 0.2467\ \text{atm} = 0.227\ \text{atm}$

Check: The units (atm) are correct. The magnitude of the answer (0.227 atm) seems reasonable since the mole fraction is lowered by ~ 8%.

81. **Given:** equal masses of carbon tetrachloride (CCl_4) and chloroform ($CHCl_3$) at 316 K; $P°_{CCl4} = 0.354$ atm; $P°_{CHCl3} = 0.526$ atm **Find:** χ_{CCl4}, χ_{CHCl3} in vapor; and P_{CHCl3} in flask of condensed vapor

Conceptual Plan: assume 100 grams of each $g_{CCl4} \rightarrow mol_{CCl4}$ **and** $g_{CHCl3} \rightarrow mol_{CHCl3}$ **then**

$$\frac{1\ \text{mol CCl}_4}{153.82\ \text{g CCl}_4} \qquad \frac{1\ \text{mol CHCl}_3}{119.38\ \text{g CHCl}_3}$$

$mol_{CCl4}, mol_{CHCl3} \rightarrow \chi_{CCl4}, \chi_{CHCl3}$ **then** $\chi_{CCl4}, P°_{CCl4} \rightarrow P_{CCl4}$ **and** $\chi_{CHCl3}, P°_{CHCl3} \rightarrow P_{CHCl3}$ **then**

$$\chi_{CCl4} = \frac{\text{amount CCl}_4\ \text{(in moles)}}{\text{total amount (in moles)}} \qquad \chi_{CHCl3} = 1 - \chi_{CCl4} \qquad P_{CCl4} = \chi_{CCl4}\, P^o_{CCl4} \qquad P_{CHCl3} = \chi_{CHCl3}\, P^o_{CHCl3}$$

$P_{CCl4}, P_{CHCl3} \rightarrow P_{Total}$ **then since** $n \propto P$ **and we are calculating a mass percent, which is a ratio of masses,**

$$P_{Total} = P_{CCl4} + P_{CHCl3}$$

we can simply convert 1 atm to 1 mole so $P_{CCl4}, P_{CHCl3} \rightarrow n_{CCl4}, n_{CHCl3}$ **then**

$$\chi_{CCl4} = \frac{\text{amount CCl}_4\ \text{(in moles)}}{\text{total amount (in moles)}}$$

$mol_{CCl4}, mol_{CHCl3} \rightarrow \chi_{CCl4}, \chi_{CHCl3}$ **then for the second vapor** $\chi_{CHCl3}, P°_{CHCl3} \rightarrow P_{CHCl3}$

$$\chi_{CHCl3} = 1 - \chi_{CCl4} \qquad P_{CHCl3} = \chi_{CHCl3}\, P^o_{CHCl3}$$

Solution: $100.00\ \cancel{\text{g CCl}_4} \times \dfrac{1\ \text{mol CCl}_4}{153.82\ \cancel{\text{g CCl}_4}} = 0.6501\underline{1}051\ \text{mol CCl}_4$ and

$$100.00\ \cancel{\text{g CHCl}_3} \times \frac{1\ \text{mol CHCl}_3}{119.38\ \cancel{\text{g CHCl}_3}} = 0.837\underline{6}6125\ \text{mol CHCl}_3 \text{ then}$$

$$\chi_{CCl4} = \frac{\text{amount CCl}_4\ \text{(in moles)}}{\text{total amount (in moles)}} = \frac{0.6501\underline{1}051\ \cancel{\text{mol}}}{0.6501\underline{1}051\ \cancel{\text{mol}} + 0.837\underline{6}6125\ \cancel{\text{mol}}} = 0.4369\underline{6}925 \text{ and}$$

$\chi_{CHCl3} = 1 - \chi_{CCl4} = 1 - 0.43696925 = 0.56303075$ then

$P_{CCl4} = \chi_{CCl4}\, P^o_{CCl4} =$ 0.43696925 x 0.354 atm = 0.154687 atm and

$P_{CHCl3} = \chi_{CHCl3}\, P^o_{CHCl3} =$ 0.56303075 x 0.526 atm = 0.296154 atm then

$P_{Total} = P_{CCl4} + P_{CHCl3} =$ 0.154687 atm + 0.296154 atm = 0.450841 atm then

mol $_{CCl4}$ = 0.154687 mol and mol $_{CHCl3}$ = 0.296154 mol then

$$\chi_{CCl4} = \frac{\text{amount CCl}_4\text{ (in moles)}}{\text{total amount (in moles)}} = \frac{0.154687 \text{ mol}}{0.154687 \text{ mol} + 0.296154 \text{ mol}} = 0.343108 = 0.343 \text{ in the first vapor}$$

and $\chi_{CHCl3} = 1 - \chi_{CCl4} = 1 - 0.343108 = 0.656892 = 0.657$ in the first vapor then in the second vapor

$P_{CHCl3} = \chi_{CHCl3}\, P^o_{CHCl3} =$ 0.656892 x 0.526 atm = 0.345525 atm = 0.346 atm

Check: The units (none and atm) are correct. The magnitudes of the answers seem reasonable since it we expect the lighter component to be found preferentially in the vapor phase. This effect is magnified in the second vapor.

83. **Given:** N_2: $k_H(N_2) = 6.1 \times 10^{-4}$ M/L at 25 °C; 14.6 mg/L at 50 °C and 1.00 atm; $P_{N2} = 0.78$ atm; O_2: $k_H(O_2) = 1.3 \times 10^{-3}$ M/L at 25 °C; 27.8 mg/L at 50 °C and 1.00 atm; $P_{O2} = 0.21$ atm; and 1.5 L water
Find: $V(N_2)$ and $V(O_2)$
Conceptual Plan: at 25 °C: $P_{Total}, \chi_{N2} \rightarrow P_{N2}$ then $P_{N2}, k_H(N_2) \rightarrow S_{N2}$ then L → mol

$P_{N2} = \chi_{N2}\, P_{Total}$ $S_{N2} = k_H(N_2) P_{N2}$ S_{N2}

at 50 °C: L → mL → mg → g → mol then $mol_{25\,°C}, mol_{25\,°C} \rightarrow mol_{removed}$ then °C → K

$\frac{1000\text{ mL}}{1\text{ L}}$ $\frac{14.6\text{ mg}}{1\text{ L}}$ $\frac{1\text{ g}}{1000\text{ mg}}$ $\frac{1\text{ mol}}{28.01\text{ g}}$ $mol_{removed} = mol_{25°C} - mol_{50°C}$ K = °C + 273.15

then $P, n, T \rightarrow V$

$PV = nRT$

at 25 °C: $P_{Total}, \chi_{O2} \rightarrow P_{O2}$ then $P_{O2}, k_H(O_2) \rightarrow S_{O2}$ then L → mol

$P_{O2} = \chi_{O2}\, P_{Total}$ $S_{O2} = k_H(O_2) P_{O2}$ S_{O2}

at 50 °C: L → mL → mg → g → mol then $mol_{25\,°C}, mol_{25\,°C} \rightarrow mol_{removed}$ then °C → K

$\frac{1000\text{ mL}}{1\text{ L}}$ $\frac{27.8\text{ mg}}{1\text{ L}}$ $\frac{1\text{ g}}{1000\text{ mg}}$ $\frac{1\text{ mol}}{32.00\text{ g}}$ $mol_{removed} = mol_{25°C} - mol_{50°C}$ K = °C + 273.15

then $P, n, T \rightarrow V$

$PV = nRT$

Solution:

at 25 °C: $P_{N2} = \chi_{N2}\, P_{Total} = 0.78 \times 1.0 \text{ atm} = 0.78 \text{ atm}$ then

$$S_{N2} = k_H(N_2)P_{N2} = 6.1 \times 10^{-4} \frac{\text{M}}{\text{atm}} \times 0.78 \text{ atm} = 4.758 \times 10^{-4} \text{ M then}$$

$$1.5 \text{ L} \times 4.758 \times 10^{-4} \frac{\text{mol}}{\text{L}} = 0.00071371 \text{ mol}$$

$$\text{at 50 °C:}\quad 1.5 \text{ L} \times \frac{14.6 \text{ mg}}{1 \text{ L} \cdot \text{atm}} \times 0.78 \text{ atm} \times \frac{1 \text{ g}}{1000 \text{ mg}} \times \frac{1 \text{ mol}}{28.01 \text{ g}} = 0.00060985 \text{ mol} \quad \text{then}$$

$mol_{removed} = mol_{25°C} - mol_{50°C} = 0.00071371 \text{ mol} - 0.00060985 \text{ mol} = 1.039 \times 10^{-4} \text{ mol}$.

then 50 °C + 273.15 = 323 K then $PV = nRT$ Rearrange to solve for V.

$$V = \frac{nRT}{P} = \frac{1.039 \times 10^{-4} \text{ mol} \times 0.08206 \frac{\text{L} \cdot \text{atm}}{\text{K} \cdot \text{mol}} \times 323 \text{ K}}{1.00 \text{ atm}} = 0.0027539 \text{ L}$$

at 25 °C: $P_{O2} = \chi_{O2}\, P_{Total} = 0.21 \times 1.0 \text{ atm} = 0.21 \text{ atm}$ then

$$S_{O2} = k_H(O_2)P_{O2} = 1.3 \times 10^{-3} \frac{\text{M}}{\text{atm}} \times 0.21 \text{ atm} = 2.73 \times 10^{-4} \text{ M then}$$

$$1.5 \text{ L} \times 2.73 \times 10^{-4} \frac{\text{mol}}{\text{L}} = 0.0004095 \text{ mol}$$

at 50 °C: $1.5\ \cancel{L} \times \frac{27.8\ \cancel{mg}}{1\ \cancel{L}\cdot\cancel{atm}} \times 0.21\ \cancel{atm} \times \frac{1\ \cancel{g}}{1000\ \cancel{mg}} \times \frac{1\ mol}{32.00\ \cancel{g}} = 0.00027366\ g$ then

$mol_{removed} = mol_{25°C} - mol_{50°C} = 0.0004095\ mol - 0.00027366\ mol = 1.358 \times 10^{-4}\ mol$

then $50\ °C + 273.15 = 323\ K$ then $PV = nRT$ Rearrange to solve for V.

$$V = \frac{nRT}{P} = \frac{1.358 \times 10^{-4}\ \cancel{mol} \times 0.08206 \frac{L\cdot\cancel{atm}}{\cancel{K}\cdot\cancel{mol}} \times 323\ \cancel{K}}{1.00\ \cancel{atm}} = 0.0035994\ L$$ finally

$V_{Total} = V_{N2} + V_{O2} = 0.0027526\ L + 0.0035994\ L = 0.0064\ L$

Check: The units (L) are correct. The magnitude of the answer (0.006 L) seems reasonable since we have so little dissolved gas at room temperature and most is still soluble at 50 °C.

85. **Given:** 1.10 g glucose ($C_6H_{12}O_6$) and sucrose ($C_{12}H_{22}O_{11}$) mixture in 25.0 mL solution and $\Pi = 3.78$ atm at 298 K **Find:** percent composition of mixture

Conceptual Plan: **$\Pi, T \rightarrow$ M then $mL_{soln} \rightarrow L_{soln}$ then L_{soln}, M $\rightarrow$ $mol_{mixture}$ then**

$\Pi = MRT$ $\frac{1\ L}{1000\ mL}$ $M = \frac{\text{amount solute (moles)}}{\text{volume solution (L)}}$

$mol_{mixture}$, $g_{mixture}$ $\rightarrow$ $mol_{C6H12O6}$, $mol_{C12H22O11}$ then

$g_{mixture} = mol\ C_6H_{12}O_6 \times \frac{180.16\ g\ C_6H_{12}O_6}{1\ mol\ C_6H_{12}O_6} + mol\ C_{12}H_{22}O_{11} \times \frac{342.30\ g\ C_{12}H_{22}O_{11}}{1\ mol\ C_{12}H_{22}O_{11}}$ with $mol_{mixture} = mol\ C_6H_{12}O_6 + mol\ C_{12}H_{22}O_{11}$

$mol_{C6H12O6}$ $\rightarrow$ $g_{C6H12O6}$ and $mol_{C12H22O11}$ $\rightarrow$ $g_{C12H22O11}$ and $g_{C6H12O6}$, $g_{C12H22O11}$ $\rightarrow$ mass percents

$\frac{180.16\ g\ C_6H_{12}O_6}{1\ mol\ C_6H_{12}O_6}$ $\frac{342.30\ g\ C_{12}H_{22}O_{11}}{1\ mol\ C_{12}H_{22}O_{11}}$ $mass\ percent = \frac{mass\ solute}{mass\ solution} \times 100\ \%$

Solution: $\Pi = MRT$ Rearrange to solve for M.

$$M = \frac{\Pi}{RT} = \frac{3.78\ \cancel{atm}}{0.08206 \frac{L\cdot\cancel{atm}}{\cancel{K}\cdot mol} \times 298\ \cancel{K}} = 0.154577 \frac{mol\ mixture}{L}$$ then $25.0\ \cancel{mL} \times \frac{1\ L}{1000\ \cancel{mL}} = 0.0250\ L$

then $M = \frac{\text{amount solute (moles)}}{\text{volume solution (L)}}$ so

$mol_{mixture} = M \times L_{soln} = 0.154577 \frac{mol\ mixture}{\cancel{L}} \times 0.0250\ \cancel{L} = 0.00386442\ mol\ mixture$ then

$g_{mixture} = \cancel{mol\ C_6H_{12}O_6} \times \frac{180.16\ g\ C_6H_{12}O_6}{1\ \cancel{mol\ C_6H_{12}O_6}} + \cancel{mol\ C_{12}H_{22}O_{11}} \times \frac{342.30\ g\ C_{12}H_{22}O_{11}}{1\ \cancel{mol\ C_{12}H_{22}O_{11}}}$ with

$mol_{mixture} = mol\ C_6H_{12}O_6 + mol\ C_{12}H_{22}O_{11}$ so

$1.10\ g = mol\ C_6H_{12}O_6 \times \frac{180.16\ g\ C_6H_{12}O_6}{1\ mol\ C_6H_{12}O_6} + (0.00386442\ mol - mol\ C_6H_{12}O_6) \times \frac{342.30\ g\ C_{12}H_{22}O_{11}}{1\ mol\ C_{12}H_{22}O_{11}} \rightarrow$

$1.10 = 180.16 \times mol\ C_6H_{12}O_6 + 1.32228 - 342.30 \times mol\ C_6H_{12}O_6 \rightarrow 162.14 \times mol\ C_6H_{12}O_6 = 0.22228 \rightarrow$

$x\ mol\ C_6H_{12}O_6 = \frac{0.22228}{162.14} = 0.00137091\ mol\ C_6H_{12}O_6$ then

$mol\ C_{12}H_{22}O_{11} = mol_{mixture} - mol\ C_6H_{12}O_6 = 0.00386442\ mol - 0.00137091\ mol = 0.0024935\ mol\ C_{12}H_{22}O_{11}$ then

$0.00137091\ \cancel{mol\ C_6H_{12}O_6} \times \frac{180.16\ g\ C_6H_{12}O_6}{1\ \cancel{mol\ C_6H_{12}O_6}} = 0.24698\ g\ C_6H_{12}O_6$ and

$0.0024935\ \cancel{mol\ C_{12}H_{22}O_{11}} \times \frac{342.30\ g\ C_{12}H_{22}O_{11}}{1\ \cancel{mol\ C_{12}H_{22}O_{11}}} = 0.85353\ g\ C_{12}H_{22}O_{11}$ finally

$$mass\ percent = \frac{mass\ solute}{mass\ solution} \times 100\% = \frac{0.24698\ \text{g}\ C_6H_{12}O_6}{0.24698\ \text{g}\ C_6H_{12}O_6 + 0.85353\ \text{g}\ C_{12}H_{22}O_{11}} \times 100\%$$

$= 22.44\ \%\ C_6H_{12}O_6$ by mass

and $100.00\% - 22.44\% = 77.56\ \%\ C_{12}H_{22}O_{11}$ by mass

Check: The units (% by mass) are correct. We expect the percent by $C_6H_{12}O_6$ to be larger than that for $C_{12}H_{22}O_{11}$ since the $\text{g}_{\text{mixture}}/\text{mol}_{\text{mixture}} = 285$ g/mol, which is closer to $C_{12}H_{22}O_{11}$ than $C_6H_{12}O_6$ and the molar mass of $C_{12}H_{22}O_{11}$ is larger than the molar mass of $C_6H_{12}O_6$.

87. **Given:** isopropyl alcohol ($(CH_3)_2CHOH$) and propyl alcohol ($CH_3CH_2CH_2OH$) at 313 K; solution 2/3 by mass isopropyl alcohol $P_{2/3} = 0.110$ atm; solution 1/3 by mass isopropyl alcohol $P_{1/3} = 0.089$ atm; **Find:** $P°_{iso}$ and $P°_{pro}$ and explain why they are different

Conceptual Plan: since these are isomers, they have the same molar mass and so the fraction by mass is the same as the mole fraction so mole fractions, P_{soln}'s → $P°$'s

$$\chi_{iso} = \frac{\text{amount iso (in moles)}}{\text{total amount (in moles)}} \quad \chi_{pro} = 1 - \chi_{iso} \quad P_{iso} = \chi_{iso} P^{\circ}_{iso} \quad P_{pro} = \chi_{pro} P^{\circ}_{pro} \text{ and } P_{soln} = P_{iso} + P_{pro}$$

Solution:

Solution 1: $\chi_{iso} = 2/3$ and $\chi_{iso} = 1/3$ $P_{soln} = P_{iso} + P_{pro}$ so $0.110\ \text{atm} = 2/3 P^{\circ}_{iso} + 1/3 P^{\circ}_{pro}$ and

Solution 2: $\chi_{iso} = 1/3$ and $\chi_{iso} = 2/3$ $P_{soln} = P_{iso} + P_{pro}$ so $0.089\ \text{atm} = 1/3 P^{\circ}_{iso} + 2/3 P^{\circ}_{pro}$. We now have 2 equations and 2 unknowns and a number of ways to solve this. One way is to rearrange the first equation for $P°_{iso}$ and then substitute into the other equation. Thus, $P^{\circ}_{iso} = 3/2\,(0.110\ \text{atm} - 1/3 P^{\circ}_{pro})$ and

$$0.089\ \text{atm} = \frac{1}{3}\,\frac{3}{2}(0.110\ \text{atm} - 1/3 P^{\circ}_{pro}) + \frac{2}{3}P^{\circ}_{pro} \rightarrow 0.089\ \text{atm} = 0.0550\ \text{atm} - \frac{1}{6}P^{\circ}_{pro} + \frac{2}{3}P^{\circ}_{pro} \rightarrow$$

$$\frac{1}{2}P^{\circ}_{pro} = 0.0340\ \text{atm} \rightarrow P^{\circ}_{pro} = 0.0680\ \text{atm} = 0.068\ \text{atm} \text{ and then}$$

$$P^{\circ}_{iso} = 3/2\,(0.110\ \text{atm} - 1/3 P^{\circ}_{pro}) = 3/2\,(0.110\ \text{atm} - 1/3(0.0680\ \text{atm})) = 0.131\ \text{atm}$$

The major intermolecular attractions are between the OH groups. The OH group at the end of the chain in propyl alcohol is more accessible than the one in the middle of the chain in isopropyl alcohol. In addition, the molecular shape of propyl alcohol is a straight chain of carbon atoms, while that of isopropyl alcohol has a branched chain and is more like a ball. The contact area between two ball-like objects is smaller than that of two chain-like objects. The smaller contact area in isopropyl alcohol means the molecules do not attract each other as strongly as do those of propyl alcohol. As a result of both of these factors, the vapor pressure of isopropyl alcohol is higher.

Check: The units (atm) are correct. The magnitude of the answers seems reasonable since the solution partial pressures are both ~0.1 atm.

89. a) The two substances mix because their intermolecular forces between themselves are roughly equal to the forces between each other and there is a pervasive tendency to increase randomness, which happens when the two substances mix.

b) $\Delta H_{soln} \approx 0$, since the intermolecular forces between themselves are roughly equal to the forces between each other.

c) ΔH_{solute} and $\Delta H_{solvent}$ are positive, ΔH_{mix} is negative and equals the sum of ΔH_{solute} and $\Delta H_{solvent}$.

91. d) More solute particles are found in an ionic solution, since the solute breaks apart into its ions. The vapor pressure is lowered the more solute particle are present.

93. The balloon not only loses He, it also takes in N_2 and O_2 from the air surrounding the balloon (due to the tendency for mixing), increasing the density of the balloon.

Chapter 13
Chemical Kinetics

1. a) $\text{Rate} = -\frac{1}{2}\frac{\Delta[\text{HBr}]}{\Delta t} = \frac{\Delta[\text{H}_2]}{\Delta t} = \frac{\Delta[\text{Br}_2]}{\Delta t}$

b) **Given:** first 15.0 s; 0.500 M to 0.455 M **Find:** average rate
Conceptual Plan: **t_1, t_2, $[\text{HBr}]_1$, $[\text{HBr}]_2$ → average rate**

$$\text{Rate} = -\frac{1}{2}\frac{\Delta[\text{HBr}]}{\Delta t}$$

Solution: $\text{Rate} = -\frac{1}{2}\frac{[\text{HBr}]_{t_2} - [\text{HBr}]_{t_1}}{t_2 - t_1} = -\frac{1}{2}\frac{0.455\ \text{M} - 0.500\ \text{M}}{15.0\,\text{s} - 0.0\ \text{s}} = 1.5 \times 10^{-3}\ \text{M s}^{-1}$

Check: The units (M s^{-1}) are correct. The magnitude of the answer (10^{-3}M s^{-1}) makes physical sense because rates are always positive and we are not changing the concentration much in 15 s.

c) **Given:** 0.500 L vessel and part b) data **Find:** mol_{Br2} formed
Conceptual Plan: **average rate, t_1, t_2, → Δ [Br_2] then Δ [Br_2], L → mol_{Br2} formed**

$$\text{Rate} = \frac{\Delta[\text{Br}_2]}{\Delta t} \qquad M = \frac{mol_{Br2}}{L}$$

Solution: $\text{Rate} = 1.5 \times 10^{-3}\ \text{M s}^{-1} = \frac{\Delta[\text{Br}_2]}{\Delta t} = \frac{\Delta[\text{Br}_2]}{15.0\,\text{s} - 0.0\ \text{s}}$ Rearrange to solve for Δ [Br_2].

$\Delta[\text{Br}_2] = 1.5 \times 10^{-3}\ \frac{\text{M}}{\text{s}} \times 15.0\ \text{s} = 0.02\underline{2}5\ \text{M}$ then $M = \frac{mol_{Br2}}{L}$. Rearrange to solve for mol_{Br2}.

$0.2\underline{2}5\ \frac{\text{mol Br}_2}{\text{L}} \times 0.500\ \text{L} = 0.011\ \text{mol Br}_2$.

Check: The units (mol) are correct. The magnitude of the answer (0.01 mol) makes physical sense because the times are the same as in part b) and we need to divide by 2 twice because of the stoichiometric coefficient difference and the volume of the vessel.

3. a) $\text{Rate} = -\frac{1}{2}\frac{\Delta[\text{A}]}{\Delta t} = -\frac{\Delta[\text{B}]}{\Delta t} = \frac{1}{3}\frac{\Delta[\text{C}]}{\Delta t}$

b) **Given:** $\frac{\Delta[\text{A}]}{\Delta t} = -0.100\ \text{M/s}$ **Find:** $\frac{\Delta[\text{B}]}{\Delta t}$, and $\frac{\Delta[\text{C}]}{\Delta t}$

Conceptual Plan: $\frac{\Delta[\text{A}]}{\Delta t}$ → $\frac{\Delta[\text{B}]}{\Delta t}$, **and** $\frac{\Delta[\text{C}]}{\Delta t}$

$$\text{Rate} = -\frac{1}{2}\frac{\Delta[\text{A}]}{\Delta t} = -\frac{\Delta[\text{B}]}{\Delta t} = \frac{1}{3}\frac{\Delta[\text{C}]}{\Delta t}$$

Solution: $\text{Rate} = -\frac{1}{2}\frac{\Delta[\text{A}]}{\Delta t} = -\frac{\Delta[\text{B}]}{\Delta t} = \frac{1}{3}\frac{\Delta[\text{C}]}{\Delta t}$ Substitute in value and solve for the two desired values.

$-\frac{1}{2}\frac{-0.100\ \text{M}}{\text{s}} = -\frac{\Delta[\text{B}]}{\Delta t}$ so $\frac{\Delta[\text{B}]}{\Delta t} = -0.0500\ \text{M s}^{-1}$ and $-\frac{1}{2}\frac{-0.100\ \text{M}}{\text{s}} = \frac{1}{3}\frac{\Delta[\text{C}]}{\Delta t}$ so

$\frac{\Delta[\text{C}]}{\Delta t} = 0.150\ \text{M s}^{-1}$

Check: The units (M s^{-1}) are correct. The magnitude of the answer ($-0.05\ \text{M s}^{-1}$) makes physical sense because fewer moles of B are reacting for every mole of A and the change in concentration with time is negative since this is a reactant. The magnitude of the answer ($0.15\ \text{M s}^{-1}$) makes physical sense because more moles of C are being formed for every mole of A reacting and the change in concentration with time is positive since this is a product.

5. a) **Given:** [C_4H_8] versus time data **Find:** average rate between 0 and 10 s, and between 40 and 50 s
Conceptual Plan: **t_1, t_2, [C_4H_8]$_1$, [C_4H_8]$_2$ → average rate**

$$\text{Rate} = -\frac{\Delta[C_4H_8]}{\Delta t}$$

Solution: for 0 to 10 s: $\text{Rate} = -\frac{[C_4H_8]_{t_2} - [C_4H_8]_{t_1}}{t_2 - t_1} = -\frac{0.913\text{ M} - 1.000\text{ M}}{10.\text{ s} - 0.\text{ s}} = 8.7 \times 10^{-3}\text{ M s}^{-1}$ and

for 40 to 50 s: $\text{Rate} = -\frac{[C_4H_8]_{t_2} - [C_4H_8]_{t_1}}{t_2 - t_1} = -\frac{0.637\text{ M} - 0.697\text{ M}}{50.\text{ s} - 40.\text{ s}} = 6.0 \times 10^{-3}\text{ M s}^{-1}$

Check: The units (M s^{-1}) are correct. The magnitude of the answer (10^{-3} M s^{-1}) makes physical sense because rates are always positive and we are not changing the concentration much in 10 s. Also reactions slow as they proceed because the concentration of the reactants is decreasing.

b) **Given:** [C_4H_8] versus time data **Find:** $\frac{\Delta[C_2H_4]}{\Delta t}$ between 20 and 30 s

Conceptual Plan: **t_1, t_2, [C_4H_8]$_1$, [C_4H_8]$_2$ →** $\frac{\Delta[C_2H_4]}{\Delta t}$

$$\text{Rate} = -\frac{\Delta[C_4H_8]}{\Delta t} = \frac{1}{2}\frac{\Delta[C_2H_4]}{\Delta t}$$

Solution: $\text{Rate} = -\frac{[C_4H_8]_{t_2} - [C_4H_8]_{t_1}}{t_2 - t_1} = -\frac{0.763\text{ M} - 0.835\text{ M}}{30.\text{ s} - 20.\text{ s}} = 7.2 \times 10^{-3}\text{ M s}^{-1} = \frac{1}{2}\frac{\Delta[C_2H_4]}{\Delta t}$

Rearrange to solve for $\frac{\Delta[C_2H_4]}{\Delta t}$. So $\frac{\Delta[C_2H_4]}{\Delta t} = 2(7.2 \times 10^{-3}\text{ M s}^{-1}) = 1.4 \times 10^{-2}\text{ M s}^{-1}$.

Check: The units (M s^{-1}) are correct. The magnitude of the answer (10^{-2} M s^{-1}) makes physical sense because rate of product formation is always positive and we are not changing the concentration much in 10 s. The rate of change of the product is faster than the decline of the reactant because of the stoichiometric coefficients.

7. a) **Given:** [Br_2] versus time plot
Find: (i) average rate between 0 and 25 s; (ii) instantaneous rate at 25 s; and (iii) instantaneous rate of HBr formation at 50 s
Conceptual Plan: (i) t_1, t_2, [Br_2]$_1$, [Br_2]$_2$ → average rate then

$$\text{Rate} = -\frac{\Delta[Br_2]}{\Delta t}$$

(ii) draw tangent at 25 s and determine slope → instantaneous rate then

$$\text{Rate} = -\frac{\Delta[Br_2]}{\Delta t}$$

(iii) draw tangent at 50 s and determine slope → instantaneous rate → $\frac{\Delta[HBr]}{\Delta t}$

$$\text{Rate} = -\frac{\Delta[Br_2]}{\Delta t} \qquad \text{Rate} = \frac{1}{2}\frac{\Delta[HBr]}{\Delta t}$$

Solution:

(i) $\text{Rate} = -\frac{[Br_2]_{t_2} - [Br_2]_{t_1}}{t_2 - t_1} = -\frac{0.75\text{ M} - 1.00\text{ M}}{25\text{ s} - 0.\text{ s}} = 1.0 \times 10^{-2}\text{ M s}^{-1}$

and (ii) at 25 s:

$\text{Slope} = \frac{\Delta y}{\Delta x} = \frac{0.68\text{ M} - 0.85\text{ M}}{35\text{ s} - 15\text{ s}} = -8.5 \times 10^{-3}\text{ M s}^{-1}$

since the slope $= \frac{\Delta[Br_2]}{\Delta t}$ and $\text{Rate} = -\frac{\Delta[Br_2]}{\Delta t}$,

then Rate $= -(-8.5 \times 10^{-3}\text{ M s}^{-1}) = 8.5 \times 10^{-3}\text{ M s}^{-1}$

(iii) at 50 s:

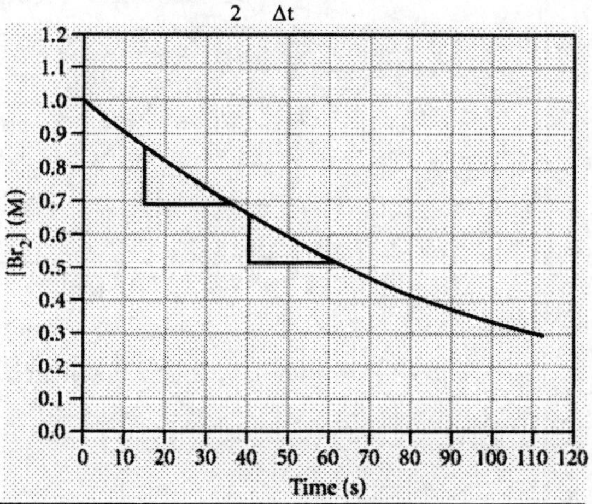

$$\text{Slope} = \frac{\Delta y}{\Delta x} = \frac{0.53\text{ M} - 0.66\text{ M}}{60.\text{ s} - 40.\text{ s}} = -6.5 \times 10^{-3}\text{ M s}^{-1}$$

since the slope $= \frac{\Delta[Br_2]}{\Delta t}$ and Rate $= -\frac{\Delta[Br_2]}{\Delta t} = \frac{1}{2}\frac{\Delta[HBr]}{\Delta t}$ then

$$\frac{\Delta[HBr]}{\Delta t} = -2\frac{\Delta[Br_2]}{\Delta t} = -2\,(-6.5 \times 10^{-3}\text{ M s}^{-1}) = 1.3 \times 10^{-2}\text{ M s}^{-1}$$

Check: The units (M s^{-1}) are correct. The magnitude of the first answer is larger than the second answer because the rate is slowing down and the first answer includes the initial portion of the data. The magnitudes of the answers (10^{-3} M s^{-1}) make physical sense because rates are always positive and we are not changing the concentration much.

b) **Given:** $[Br_2]$ versus time data; and $[HBr]_0 = 0$ M
Find: plot [HBr] with time

Conceptual Plan: Since Rate $= -\frac{\Delta[Br_2]}{\Delta t} = \frac{1}{2}\frac{\Delta[HBr]}{\Delta t}$ **. The rate of change of [HBr] will be twice that of $[Br_2]$. The plot will start at the origin.**
Solution:

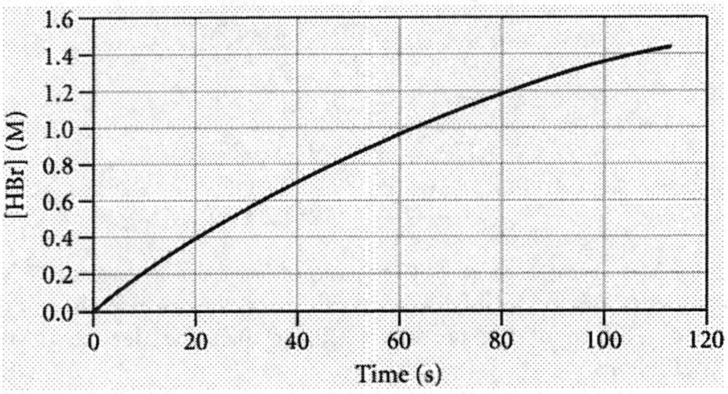

Check: The units (M versus s) are correct. The plot makes sense because the plot has the same general shape of the original plot, only we are increasing instead of decreasing and our concentration axis has changed by a factor of two (to account for the difference in stoichiometric coefficients).

9. a) **Given:** Rate versus [A] plot **Find:** reaction order
Conceptual Plan: Look at shape of plot and match to possibilities.
Solution: The plot is a linear plot, so Rate α [A] or the reaction is first order.
Check: The order of the reaction is a common reaction order.

b) **Given:** part a) **Find:** sketch plot of [A] versus time
Conceptual Plan: Using result from part a), shape plot of [A] versus time should be curved with [A] decreasing. Use 1.0 M as initial concentration.
Solution:

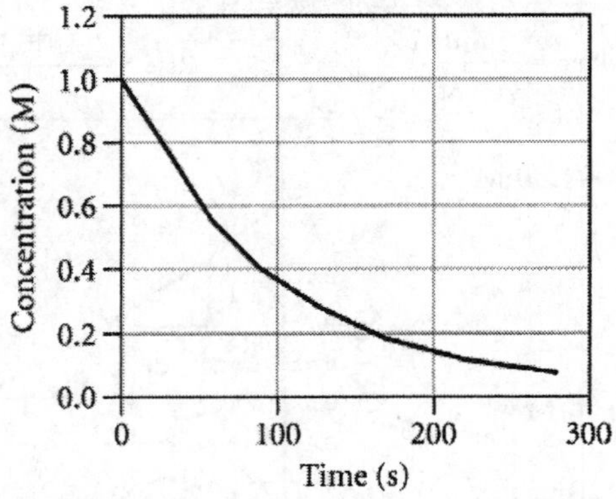

Check: The plot has a shape that matches the one in the text for first order plots.

c) **Given: part a)** **Find:** write a rate law and estimate k
Conceptual Plan: **Using result from part a), the slope of the plot is the rate constant.**

Solution: $\text{Slope} = \dfrac{\Delta y}{\Delta x} = \dfrac{0.010\ \frac{\cancel{M}}{s} - 0.00\ \frac{\cancel{M}}{s}}{1.0\ \cancel{M} - 0.0\ \cancel{M}} = 0.010\ s^{-1}$ so Rate = $k\,[A]^1$ or Rate = $k\,[A]$ or Rate = $0.010\ s^{-1}\,[A]$

Check: The units (s^{-1}) are correct. The magnitude of the answer ($10^{-2}\ s^{-1}$) makes physical sense because, rate and concentration data. Remember that concentration is in units of M, so plugging the rate constant into the equation has the units of the rate as M s^{-1}, which is correct.

11. **Given:** reaction order: a) first-order; b) second-order; and c) zero-order **Find:** units of k
Conceptual Plan: **Using rate law, rearrange to solve for k.**

$$\text{Rate} = k\,[A]^n, \text{ where n = reaction order}$$

Solution: for all cases Rate has units of M s^{-1} and [A] has units of M

a) Rate = $k\,[A]^1 = k\,[A]$ so $k = \dfrac{\text{Rate}}{[A]} = \dfrac{\frac{\cancel{M}}{s}}{\cancel{M}} = s^{-1}$;

b) Rate = $k\,[A]^2$ so $k = \dfrac{\text{Rate}}{[A]^2} = \dfrac{\frac{\cancel{M}}{s}}{\cancel{M}M} = M^{-1}\,s^{-1}$;

c) Rate = $k\,[A]^0 = k = M\,s^{-1}$.

Check: The units (s^{-1}, $M^{-1}s^{-1}$ and Ms^{-1}) are correct. The units for k change with the reaction order so that the units on the rate remain as M s^{-1}.

13. **Given:** A, B, and C react to form products. Reaction is first order in A, second order in B, and zero order in C
Find: a) rate law; b) overall order of reaction; c) factor change in rate if [A] doubled; d) factor change in rate if [B] doubled; e) factor change in rate if [C] doubled; and f) factor change in rate if [A], [B], and [C] doubled.
Conceptual Plan:
a) Using general rate law form, substitute in values for orders.

$$\text{Rate} = k\,[A]^m\,[B]^n\,[C]^p, \text{ where m, n, and p = reaction orders}$$

b) Using rate law in part a) add up all reaction orders.

$$\textit{overall reaction order} = m + n + p$$

c) through f) Using rate law from part a) substitute in concentration changes.

$$\frac{\text{Rate 2}}{\text{Rate 1}} = \frac{k\,[A]_2^1\,[B]_2^2}{k\,[A]_1^1\,[B]_1^2}$$

Solution:

a) m = 1, n = 2, and p = 0 so Rate = $k\,[A]^1[B]^2[C]^0$ or Rate = $k\,[A][B]^2$.

b) $\textit{overall reaction order} = m + n + p = 1 + 2 + 0 = 3$ so it is a third order reaction overall.

c) $\dfrac{\text{Rate 2}}{\text{Rate 1}} = \dfrac{k\,[A]_2^1\,[B]_2^2}{k\,[A]_1^1\,[B]_1^2}$ and $[A]_2 = 2\,[A]_1$, $[B]_2 = [B]_1$, $[C]_2 = [C]_1$, so $\dfrac{\text{Rate 2}}{\text{Rate 1}} = \dfrac{\cancel{k}\,(2\,\cancel{[A]_1})^1\,\cancel{[B]_1^2}}{\cancel{k}\,\cancel{[A]_1^1}\,\cancel{[B]_1^2}} = 2$ so the reaction rate doubles (factor of 2).

d) $\dfrac{\text{Rate 2}}{\text{Rate 1}} = \dfrac{k\,[A]_2^1\,[B]_2^2}{k\,[A]_1^1\,[B]_1^2}$ and $[A]_2 = [A]_1$, $[B]_2 = 2\,[B]_1$, $[C]_2 = [C]_1$, so $\dfrac{\text{Rate 2}}{\text{Rate 1}} = \dfrac{\cancel{k}\,\cancel{[A]_1}\,(2\,\cancel{[B]_1})^2}{\cancel{k}\,\cancel{[A]_1^1}\,\cancel{[B]_1^2}} = 2^2 = 4$ so the reaction rate quadruples (factor of 4).

e) $\dfrac{\text{Rate 2}}{\text{Rate 1}} = \dfrac{k\,[A]_2^1\,[B]_2^2}{k\,[A]_1^1\,[B]_1^2}$ and $[A]_2 = [A]_1$, $[B]_2 = [B]_1$, $[C]_2 = 2\,[C]_1$, so $\dfrac{\text{Rate 2}}{\text{Rate 1}} = \dfrac{\cancel{k}\,\cancel{[A]_1^1}\,\cancel{[B]_1^2}}{\cancel{k}\,\cancel{[A]_1^1}\,\cancel{[B]_1^2}} = 1$ so the reaction rate is unchanged (factor of 1).

f) $\dfrac{\text{Rate 2}}{\text{Rate 1}} = \dfrac{k\,[A]_2^1\,[B]_2^2}{k\,[A]_1^1\,[B]_1^2}$ and $[A]_2 = 2\,[A]_1$, $[B]_2 = 2\,[B]_1$, $[C]_2 = 2\,[C]_1$, so

$\dfrac{\text{Rate 2}}{\text{Rate 1}} = \dfrac{\cancel{k}\,(2\,\cancel{[A]_1})^1\,(2\,\cancel{[B]_1})^2}{\cancel{k}\,\cancel{[A]_1^1}\,\cancel{[B]_1^2}} = 2 \times 2^2 = 8$ so the reaction rate goes up by a factor of 8.

Check: The units (none) are correct. The rate law is consistent with the orders given and the overall order is larger that any of the individual orders. The factors are consistent with the reaction orders. The larger the order, the larger the factor. When all concentrations are changed the rate changes the most. If a reactant is not in the rate law, then changing its concentration has no effect on the reaction rate.

15. **Given:** table of [A] versus initial rate **Find:** rate law and k
Conceptual Plan: Using general rate law form, compare rate ratios to determine reaction order.

$$\frac{\text{Rate 2}}{\text{Rate 1}} = \frac{k\,[A]_2^n}{k\,[A]_1^n}$$

Then use one of the concentration/ initial rate pairs to determine *k*.

$$\text{Rate} = k[A]^n$$

Solution: $\dfrac{\text{Rate 2}}{\text{Rate 1}} = \dfrac{k\,[A]_2^n}{k\,[A]_1^n}$ Comparing the first two sets of data: $\dfrac{0.210\ \cancel{\text{M/s}}}{0.053\ \cancel{\text{M/s}}} = \dfrac{\cancel{k}\,(0.200\ \cancel{\text{M}})^n}{\cancel{k}\,(0.100\ \cancel{\text{M}})^n}$ and $3.\underline{9}623 = 2^n$ so n = 2. If we compare the first and the last data sets: $\dfrac{0.473\ \cancel{\text{M/s}}}{0.053\ \cancel{\text{M/s}}} = \dfrac{\cancel{k}\,(0.300\ \cancel{\text{M}})^n}{\cancel{k}\,(0.100\ \cancel{\text{M}})^n}$ and $8.\underline{9}245 = 3^n$ so n = 2. This second comparison is not necessary, but it increases our confidence in the reaction order. So $\text{Rate} = k[A]^2$. Selecting the second data set and rearranging the rate equation

$$k = \frac{\text{Rate}}{[A]^2} = \frac{0.210\frac{\text{M}}{\text{s}}}{(0.200\,\text{M})^2} = 5.25\ \text{M}^{-1}\ \text{s}^{-1} \text{ so Rate} = 5.25\ \text{M}^{-1}\ \text{s}^{-1}[A]^2$$

Check: The units (none and $M^{-1}\ s^{-1}$) are correct. The rate law is a common form. The rate is changing more rapidly than the concentration, so second order is consistent. The rate constant is consistent with the units necessary to get rate as M/s and the magnitude is reasonable since we have a second order reaction.

17. **Given:** table of $[NO_2]$ and $[F_2]$ versus initial rate **Find:** rate law, k, and overall order
Conceptual Plan: Using general rate law form, compare rate ratios to determine reaction order of each reactant. Be sure to choose data that changes only one concentration at a time.

$$\frac{\text{Rate 2}}{\text{Rate 1}} = \frac{k\,[NO_2]_2^m\,[F_2]_2^n}{k\,[NO_2]_1^m\,[F_2]_1^n}$$

Then use one of the concentration/ initial rate pairs to determine *k*.

$$\text{Rate} = k\,[NO_2]^m\,[F_2]^n$$

Solution: $\dfrac{\text{Rate 2}}{\text{Rate 1}} = \dfrac{k\,[NO_2]_2^m\,[F_2]_2^n}{k\,[NO_2]_1^m\,[F_2]_1^n}$ Comparing the first two sets of data:

$\dfrac{0.051\ \cancel{\text{M/s}}}{0.026\ \cancel{\text{M/s}}} = \dfrac{\cancel{k}\,(0.200\ \cancel{\text{M}})^m\,(\cancel{0.100\ \text{M}})^n}{\cancel{k}\,(0.100\ \cancel{\text{M}})]^m\,(\cancel{0.100\ \text{M}})^n}$ and $1.\underline{9}615 = 2^m$ so m = 1. If we compare the second and the third data sets: $\dfrac{0.103\ \cancel{\text{M/s}}}{0.051\ \cancel{\text{M/s}}} = \dfrac{\cancel{k}\,(\cancel{0.200\ \text{M}})^m\,(0.200\ \cancel{\text{M}})^n}{\cancel{k}\,(\cancel{0.200\ \text{M}})^m\,(0.100\ \cancel{\text{M}})^n}$ and $2 = 2^n$ so n = 1. Other comparisons can be made, but are not necessary. They should reinforce these values of the reaction orders. So $\text{Rate} = k\,[NO_2]\,[F_2]$.

Selecting the last data set and rearranging the rate equation $k = \frac{\text{Rate}}{[NO_2][F_2]} = \frac{0.411\frac{\cancel{M}}{s}}{(0.400\ \cancel{M})(0.400\ M)} = 2.57\ M^{-1}s^{-1}$

so Rate = $2.57\ M^{-1}s^{-1}[NO_2][F_2]$ and the reaction is second order overall.

Check: The units (none and $M^{-1}s^{-1}$) are correct. The rate law is a common form. The rate is changing as rapidly as each concentration is changing, which is consistent with first order in each reactant. The rate constant is consistent with the units necessary to get rate as M/s and the magnitude is reasonable since we have a second order reaction.

19. a) The reaction is zero order. Since the slope of the plot is independent of the concentration, there is no dependence of the concentration of the reactant in the rate law.

b) The reaction is first order. The expression for the half-life of a first order reaction is $t_{1/2} = \frac{0.693}{k}$, which is independent of the reactant concentration.

c) The reaction is second order. The integrated rate expression for a second order reaction is $\frac{1}{[A]_t} = kt + \frac{1}{[A]_0}$, which is linear when the inverse of the concentration is plotted versus time.

21. **Given:** table of [AB] versus time **Find:** reaction order, *k*, and [AB] at 25 s

Conceptual Plan: Look at the data and see if any common reaction orders can be eliminated. If the data does not show an equal concentration drop with time, then zero order can be eliminated. Look for changes in the half-life (compare time for concentration to drop to one half of any value). If the half-life is not constant, then the first order can be eliminated. If the half-life is getting longer as the concentration drops, this might suggest second order. Plot the data as indicated by the appropriate rate law. Determine *k* from the slope of the plot. Finally calculate the [AB] at 25 s by using the appropriate integrated rate expression.

Solution: By the above logic, we can eliminate both the zero order and the first order reactions. (Alternatively, you could make all three plots and only one should be linear.) This suggests that we should have a second order reaction. Plot 1/[AB] versus time. Since $\frac{1}{[AB]_t} = kt + \frac{1}{[AB]_0}$, the slope will be the rate constant. The slope can be determined by measuring Δy/Δx on the plot or by using functions, such as "add trendline" in Excel. Thus the rate constant is $0.0225\ M^{-1}\ s^{-1}$ and the rate law is Rate = $0.0225\ M^{-1}s^{-1}[AB]^2$.

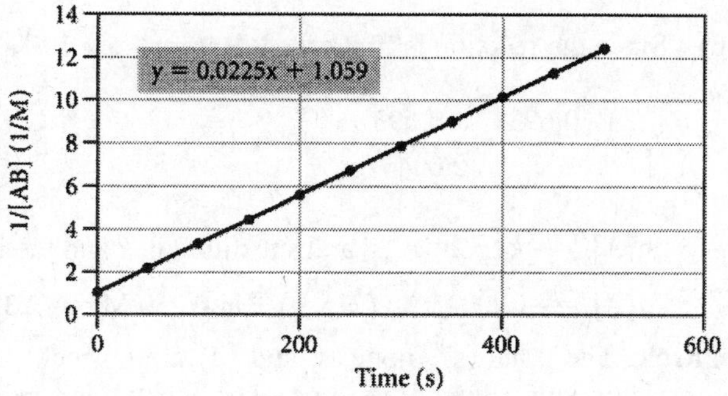

Finally, use $\frac{1}{[AB]_t} = kt + \frac{1}{[AB]_0}$, substitute in the values of $[AB]_0$, 25 s and *k* and rearrange to solve for [AB] at 25 s. $[AB]_t = \frac{1}{kt + \frac{1}{[AB]_0}} = \frac{1}{(0.0225\ M^{-1}\cancel{s^{-1}})(25\ \cancel{s}) + \left(\frac{1}{0.950\ M}\right)} = 0.619\ M$

Check: The units (none, $M^{-1}s^{-1}$, and M) are correct. The rate law is a common form. The plot was extremely linear, confirming second order kinetics. The rate constant is consistent with the units necessary to get rate as M/s and the magnitude is reasonable since we have a second order reaction. The [AB] at 25 s is in between the values at 0 s and 50 s.

23. **Given:** table of $[C_4H_8]$ versus time **Find:** reaction order, *k*, and reaction rate when $[C_4H_8] = 0.25$ M

Conceptual Plan: Look at the data and see if any common reaction orders can be eliminated. If the data does not show an equal concentration drop with time, then zero order can be eliminated. Look for changes in half-life (compare time for concentration to drop to one half of any value). If the half-life is not constant, then the first order can be eliminated. If the half-life is getting longer as the concentration drops, this might suggest second order. Plot the data as indicated by the appropriate rate law.

Determine k from the slope of the plot. Finally calculate the reaction rate when $[C_4H_8] = 0.25$ M by using the rate law.

Solution: By the above logic, we can see that the reaction is most likely first order. It takes about 60 s for the concentration to be cut in half for any concentration. Plot ln $[C_4H_8]$ versus time. Since $\ln[A]_t = -kt + \ln[A]_0$, the negative of the slope will be the rate constant. The slope can be determined by measuring Δy/Δx on the plot or by using functions, such as "add trendline" in Excel. Thus the rate constant is 0.0112 s^{-1} and the rate law is $\text{Rate} = 0.0112\ s^{-1}[C_4H_8]$. Finally, use $\text{Rate} = 0.0112\ s^{-1}[C_4H_8]$, substitute in the values of $[C_4H_8]$. $\text{Rate} = 0.0112\ s^{-1}[0.25\ M] = 2.8 \times 10^{-3}\ M\ s^{-1}$.

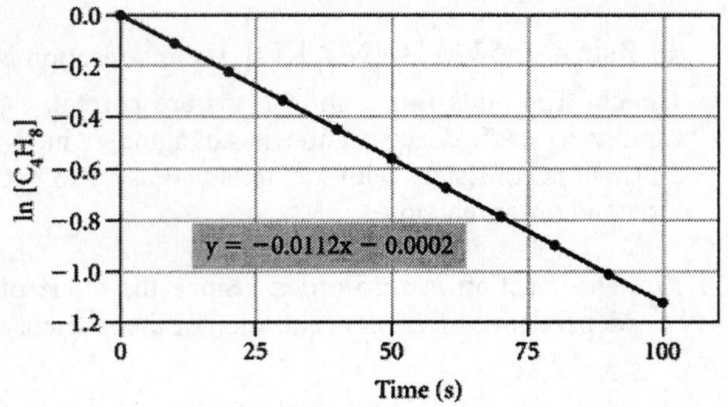

Check: The units (none, s^{-1}, and M s^{-1}) are correct. The rate law is a common form. The plot was extremely linear, confirming first order kinetics. The rate constant is consistent with the units necessary to get rate as M/s and the magnitude is reasonable since we have a first order reaction. The rate when $[C_4H_8]$ = 0. 25 M is consistent with the average rate using 90 s and 100 s.

25. **Given:** plot of ln [A] versus time has slope = – 0.0045/s; $[A]_0 = 0.250$ M
Find: a) k; b) rate law; c) $t_{1/2}$; and d) [A] after 225 s
Conceptual Plan:
a) **A plot of ln [A] versus time is linear for a first order reaction. Using** $\ln[A]_t = -kt + \ln[A]_0$**, the rate constant is the negative of the slope.**
b) **Rate law is first order. Add rate constant from part a).**
c) **For a first order reaction,** $t_{1/2} = \dfrac{0.693}{k}$**. Substitute in k from part a).**
d) **Use the integrated rate law,** $\ln[A]_t = -kt + \ln[A]_0$**, and substitute in k and the initial concentration.**
Solution:
a) Since the rate constant is the negative of the slope, $k = 4.5 \times 10^{-3}\ s^{-1}$.
b) Since the reaction is first order, Rate = $4.5 \times 10^{-3}\ s^{-1}$ [A].
c) $t_{1/2} = \dfrac{0.693}{k} = \dfrac{0.693}{0.0045/s} = 1.5 \times 10^2$ s.
d) $\ln[A]_t = -kt + \ln[A]_0$, and substitute in k and the initial concentration. So
$\ln[A]_t = -(0.0045/\cancel{s})\ (225\ \cancel{s}) + \ln 0.250\ M = -2.39879$ and $[A]_{250s} = e^{-2.39879} = 0.0908$ M
Check: The units (s^{-1}, none, s, and M) are correct. The rate law is a common form. The rate constant is consistent with value of the slope. The half-life is consistent with a small value of k. The concentration at 225 s is consistent with being between one and two half-lives.

27. **Given:** decomposition of SO_2Cl_2, first order; $k = 1.42 \times 10^{-4}\ s^{-1}$
Find: a) $t_{1/2}$; b) t to decrease to 25 % of $[SO_2Cl_2]_0$; c) t to 0.78 M when $[SO_2Cl_2]_0 = 1.00$ M; and d) $[SO_2Cl_2]$ after 2.00×10^2 s and 5.00×10^2 s when $[SO_2Cl_2]_0 = 0.150$ M
Conceptual Plan:
a) $k \rightarrow t_{1/2}$
$$t_{1/2} = \frac{0.693}{k}$$
b) $[SO_2Cl_2]_0$, 25 % of $[SO_2Cl_2]_0$, $k \rightarrow$ t
$$\ln[A]_t = -kt + \ln[A]_0$$
c) $[SO_2Cl_2]_0$, $[SO_2Cl_2]_t$, $k \rightarrow$ t
$$\ln[A]_t = -kt + \ln[A]_0$$

d) $[SO_2Cl_2]_0$, t, k → $[SO_2Cl_2]_t$

$$\ln[A]_t = -kt + \ln[A]_0$$

Solution:

a) $t_{1/2} = \frac{0.693}{k} = \frac{0.693}{1.42 \times 10^{-4}\ s^{-1}} = 4.88 \times 10^3\ s$

b) $[SO_2Cl_2]_t = 0.25\ [SO_2Cl_2]_0$. Since $\ln[SO_2Cl_2]_t = -kt + \ln[SO_2Cl_2]_0$ rearrange to solve for t

$$t = -\frac{1}{k}\ln\frac{[SO_2Cl_2]_t}{[SO_2Cl_2]_0} = -\frac{1}{1.42 \times 10^{-4}\ s^{-1}}\ln\frac{0.25\ \cancel{[SO_2Cl_2]_0}}{\cancel{[SO_2Cl_2]_0}} = 9.8 \times 10^3\ s$$

c) $[SO_2Cl_2]_t = 0.78$ M; $[SO_2Cl_2]_0 = 1.00$ M. Since $\ln[SO_2Cl_2]_t = -kt + \ln[SO_2Cl_2]_0$ rearrange to solve for t

$$t = -\frac{1}{k}\ln\frac{[SO_2Cl_2]_t}{[SO_2Cl_2]_0} = -\frac{1}{1.42 \times 10^{-4}\ s^{-1}}\ln\frac{0.78\ \cancel{M}}{1.00\ \cancel{M}} = 1.7 \times 10^3\ s$$

d) $[SO_2Cl_2]_0 = 0.150$ M and 2.00×10^2 s in

$\ln[SO_2Cl_2]_t = -(1.42 \times 10^{-4}\ \cancel{s^{-1}})(2.00 \times 10^2\ \cancel{s}) + \ln 0.150\ M = -1.9\underline{2}552$ →

$[SO_2Cl_2]_t = e^{-1.9\underline{2}552} = 0.146$ M and $[SO_2Cl_2]_0 = 0.150$ M and 5.00×10^2 s in

$\ln[SO_2Cl_2]_t = -(1.42 \times 10^{-4}\ \cancel{s^{-1}})(5.00 \times 10^2\ \cancel{s}) + \ln 0.150\ M = -1.9\underline{6}812$ → $[SO_2Cl_2]_t = e^{-1.9\underline{6}812} = 0.140$ M

Check: The units (s, s, s, and M) are correct. The rate law is a common form. The half-life is consistent with a small value of k. The time to 25% is consistent with two half-lives. The time to 0.78 M is consistent with being less than one half-life. The final concentrations are consistent with the time being less than one half-life.

29. **Given:** $t_{1/2}$ for radioactive decay of U-238 = 4.5 billion years and independent of $[U\text{-}238]_0$ **Find:** t to decrease by 10 %; number U-238 atoms today, when 1.5×10^{18} atoms formed 13.8 billion years ago

Conceptual Plan: $t_{1/2}$ independent of concentration implies first order kinetics, $t_{1/2}$ → k then

$$t_{1/2} = \frac{0.693}{k}$$

90 % of $[U\text{-}238]_0$, k → t and $[U\text{-}238]_0$, t, k → $[U\text{-}238]_t$

$$\ln[A]_t = -kt + \ln[A]_0 \qquad \ln[A]_t = -kt + \ln[A]_0$$

Solution: $t_{1/2} = \frac{0.693}{k}$ rearrange to solve for k. $k = \frac{0.693}{t_{1/2}} = \frac{0.693}{4.5 \times 10^9\ yr} = 1.\underline{5}4 \times 10^{-10}\ yr^{-1}$ then

$[U\text{-}238]_t = 0.10\ [U\text{-}238]_0$. Since $\ln[U\text{-}238]_t = -kt + \ln[U\text{-}238]_0$ rearrange to solve for t

$t = -\frac{1}{k}\ln\frac{[U\text{-}238]_t}{[U\text{-}238]_0} = -\frac{1}{1.\underline{5}4 \times 10^{-10}\ yr^{-1}}\ln\frac{0.90\ \cancel{[U\text{-}238]_0}}{\cancel{[U\text{-}238]_0}} = 6.8 \times 10^8$ yr and $[U\text{-}238]_0 = 1.5 \times 10^{18}$ atoms;

$t = 13.8 \times 10^9$ yr

$\ln[U\text{-}238]_t = -kt + \ln[U\text{-}238]_0 = -(1.\underline{5}4 \times 10^{-10}\ \cancel{yr^{-1}})(13.8 \times 10^9\ \cancel{yr}) + \ln(1.5 \times 10^{18}\ atoms) = 39.7\underline{2}6797$ →

$[U\text{-}238]_t = e^{39.7\underline{2}6797} = 1.8 \times 10^{17}$ atoms

Check: The units (yr and atoms) are correct. The time to 10 % decay is consistent with less than one half-life. The final concentration is consistent with the time being about three half-lives.

31.

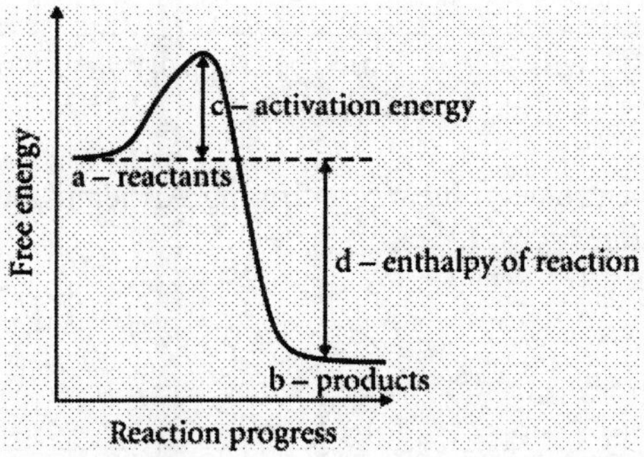

33. **Given:** activation energy = 56.8 kJ/mol, frequency factor = 1.5 x 10^{11} /s, 25 °C **Find:** rate constant
Conceptual Plan: **°C → K and kJ/mol → J/mol then E_a, T, A → k**

$$K = °C + 273.15 \qquad \frac{1000\ J}{1\ kJ} \qquad k = A\, e^{-E_a/RT}$$

Solution: $T = 25°C + 273.15 = 298$ K and $\dfrac{56.8\ \text{kJ}}{\text{mol}} \times \dfrac{1000\ \text{J}}{1\ \text{kJ}} = 5.68 \times 10^4\ \dfrac{\text{J}}{\text{mol}}$ then

$$k = A\, e^{-E_a/RT} = (1.5 \times 10^{11}\ s^{-1})\, e^{\frac{-5.68 \times 10^4 \frac{J}{mol}}{\left(8.314 \frac{J}{K\ mol}\right) 298\ K}} = 17\ s^{-1}$$

Check: The units (s^{-1}) are correct. The rate constant is consistent with a large activation energy and a large frequency factor.

35. **Given:** table of rate constant versus T **Find:** E_a, and A

Conceptual Plan: **Since** $\ln k = \dfrac{-E_a}{R}\left(\dfrac{1}{T}\right) + \ln A$ **a plot of ln k versus 1/T will have a slope = $-E_a/R$ and an intercept = ln A.**

Solution: The slope can be determined by measuring Δy/Δx on the plot or by using functions, such as "add trendline" in Excel. Since the slope = – 30189 K = $-E_a/R$ then

$$E_a = -(slope)R = -(-30189\ K)\left(8.314 \frac{J}{K\ mol}\right)\left(\frac{1\ kJ}{1000\ J}\right) = 251\ \frac{kJ}{mol}$$

and intercept = 27.399 = ln A then $A = e^{intercept} = e^{27.399} = 7.93 \times 10^{11}\ s^{-1}$.

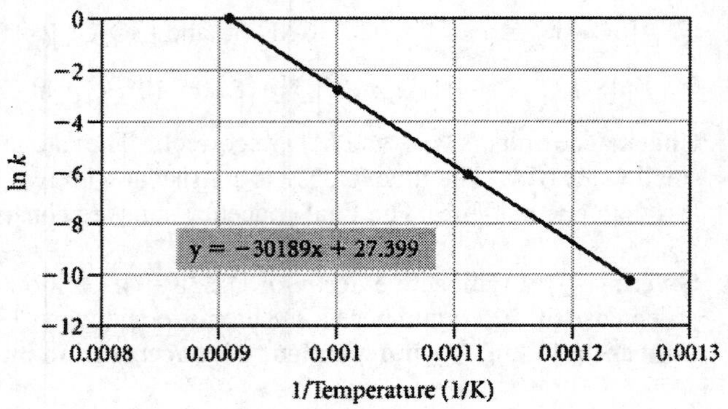

Check: The units (kJ/mol and s^{-1}) are correct. The plot was extremely linear, confirming Arrhenius behavior. The activation and frequency factor are typical for many reactions.

37. **Given:** table of rate constant versus T **Find:** E_a, and A

Conceptual Plan: **Since** $\ln k = \dfrac{-E_a}{R}\left(\dfrac{1}{T}\right) + \ln A$ **a plot of ln k versus 1/T will have a slope = $-E_a/R$ and an intercept = ln A.**

Solution: The slope can be determined by measuring Δy/Δx on the plot or by using functions, such as "add trendline" in Excel. Since the slope = – 2767.2 K = $-E_a/R$ then

$$E_a = -(slope)R = -(-2767.2\ K)\left(8.314 \frac{J}{K\ mol}\right)\left(\frac{1\ kJ}{1000\ J}\right) = 23.0\ \frac{kJ}{mol}$$

and intercept = 25.112 = ln A then $A = e^{intercept} = e^{25.112} = 8.05 \times 10^{10}\ s^{-1}$.

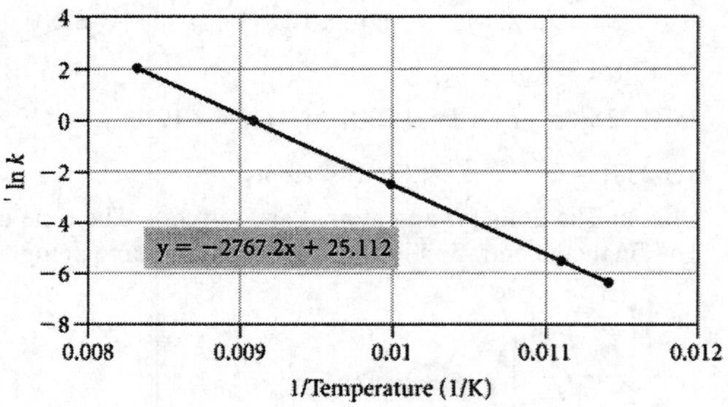

Check: The units (kJ/mol and s^{-1}) are correct. The plot was extremely linear, confirming Arrhenius behavior. The activation and frequency factor are typical for many reactions.

39. **Given:** rate constant = 0.0117/s at 400. K, and 0.689/s at 450. K **Find:** a) E_a and b) rate constant at 425 K
Conceptual Plan: a) $k_1, T_1, k_2, T_2 \rightarrow E_a$ **then J/mol → kJ/mol b)** $E_a, k_1, T_1, T_2 \rightarrow k_2$

$$\ln\left(\frac{k_2}{k_1}\right)=\frac{E_a}{R}\left(\frac{1}{T_1}-\frac{1}{T_2}\right) \qquad \frac{1\text{ kJ}}{1000\text{ J}} \qquad \ln\left(\frac{k_2}{k_1}\right)=\frac{E_a}{R}\left(\frac{1}{T_1}-\frac{1}{T_2}\right)$$

Solution: a) $\ln\left(\frac{k_2}{k_1}\right)=\frac{E_a}{R}\left(\frac{1}{T_1}-\frac{1}{T_2}\right)$ Rearrange to solve for E_a.

$$E_a=\frac{R\ln\left(\frac{k_2}{k_1}\right)}{\left(\frac{1}{T_1}-\frac{1}{T_2}\right)}=\frac{8.314\frac{\text{J}}{\text{K mol}}\ln\left(\frac{0.689\ \text{s}^{-1}}{0.0117\ \text{s}^{-1}}\right)}{\left(\frac{1}{400.\ \text{K}}-\frac{1}{450.\ \text{K}}\right)}=1.22\times10^5\frac{\text{J}}{\text{mol}}\times\frac{1\text{ kJ}}{1000\text{ J}}=122\frac{\text{kJ}}{\text{mol}}\text{ and}$$

b) $\ln\left(\frac{k_2}{k_1}\right)=\frac{E_a}{R}\left(\frac{1}{T_1}-\frac{1}{T_2}\right)$ with $k_1 = 0.0117/\text{s}$, $T_1 = 400.$ K, $T_2 = 425$ K Rearrange to solve for k_2.

$$\ln k_2=\frac{E_a}{R}\left(\frac{1}{T_1}-\frac{1}{T_2}\right)+\ln k_1=\frac{1.22\times10^5\frac{\text{J}}{\text{mol}}}{8.314\frac{\text{J}}{\text{K mol}}}\left(\frac{1}{400.\ \text{K}}-\frac{1}{425\ \text{K}}\right)+\ln 0.0117\ s^{-1}=-2.29\underline{0}2 \rightarrow$$

$$k_2=e^{-2.29\underline{0}2}=0.101\ \text{s}^{-1}$$

Check: The units (kJ/mol and s^{-1}) are correct. The activation energy is typical for a reaction. The rate constant at 425 K is in between the values given at 400 K and 450 K.

41. **Given:** rate constant doubles from at 10.0 °C to 20.0 °C **Find:** E_a
Conceptual Plan: °C → K then $k_1, T_1, k_2, T_2 \rightarrow E_a$ **then J/mol → kJ/mol;**

$$\text{K} = {}^\circ\text{C} + 273.15 \qquad \ln\left(\frac{k_2}{k_1}\right)=\frac{E_a}{R}\left(\frac{1}{T_1}-\frac{1}{T_2}\right) \qquad \frac{1\text{ kJ}}{1000\text{ J}}$$

Solution: T_1 = 10.0 °C + 273.15 = 283.2 K and T_2 = 20.0 °C + 273.15 = 293.2 K and $k_2 = 2\,k_1$ then

$\ln\left(\frac{k_2}{k_1}\right)=\frac{E_a}{R}\left(\frac{1}{T_1}-\frac{1}{T_2}\right)$ Rearrange to solve for E_a.

$$E_a=\frac{R\ln\left(\frac{k_2}{k_1}\right)}{\left(\frac{1}{T_1}-\frac{1}{T_2}\right)}=\frac{8.314\frac{\text{J}}{\text{K mol}}\ln\left(\frac{2\,k_1}{k_1}\right)}{\left(\frac{1}{283.2\ \text{K}}-\frac{1}{293.2\ \text{K}}\right)}=4.785\times10^4\frac{\text{J}}{\text{mol}}\times\frac{1\text{ kJ}}{1000\text{ J}}=47.85\frac{\text{kJ}}{\text{mol}}$$

Check: The units (kJ/mol) are correct. The activation energy is typical for a reaction.

43. Reaction a. would have the faster rate because the orientation factor, *p*, would be larger for this reaction since the reactants are symmetrical.

45. Since the first reaction is the slow step, it is the rate determining step. Using this first step to determine the rate law, Rate = $k_1\,[AB]^2$. Since this is the observed rate law, this mechanism is consistent with the experimental data.

47. a) The overall reaction is the sum of the steps in the mechanism:

$$Cl_2\,(g) \underset{k_2}{\overset{k_1}{\rightleftharpoons}} \cancel{2\,Cl\,(g)}$$

$$\cancel{Cl\,(g)} + CHCl_3\,(g) \xrightarrow{k_3} HCl\,(g) + \cancel{CCl_3\,(g)}$$

$$\cancel{Cl\,(g)} + \cancel{CCl_3\,(g)} \xrightarrow{k_4} CCl_4\,(g)$$

$$Cl_2\,(g) + CHCl_3\,(g) \rightarrow HCl\,(g) + CCl_4\,(g)$$

b) The intermediates are the species that are generated by one step and consumed by other steps. These are a Cl (*g*) and CCl_3 (*g*).

c) Since the second step is the rate determining step, Rate = k_3 [Cl] [$CHCl_3$]. Since Cl is an intermediate, its concentration cannot appear in the rate law. Using the fast equilibrium in the first step, we see that $k_1[Cl_2] = k_2[Cl]^2$ or $[Cl] = \sqrt{\frac{k_1}{k_2}[Cl_2]}$. Substituting this into the first rate expression we get that Rate = $k_3\sqrt{\frac{k_1}{k_2}}[Cl_2]^{1/2}[CHCl_3]$. Simplifying this expression we see Rate = $k\,[Cl_2]^{1/2}\,[CHCl_3]$.

49. Heterogeneous catalysts require a large surface area because catalysis can only happen at the active sites on the surface. A greater surface area means greater opportunity for the substrate to react, which results in a speedier reaction.

51. Assume Rate ratio α k ratio (since concentration terms will cancel each other) and $k = A\,e^{-E_a/RT}$. T = 25 °C + 273.15 = 298 K, E_{a1} = 1.25 x 10^5 J/mol, and E_{a2} = 5.5 x 10^4 J/mol. Ratio of rates will be

$$\frac{k_2}{k_1} = \frac{\cancel{A}\,e^{-E_{a2}/RT}}{\cancel{A}\,e^{-E_{a1}/RT}} = \frac{e^{\frac{-5.5\times10^4\,\frac{\cancel{J}}{\cancel{mol}}}{\left(8.314\,\frac{\cancel{J}}{\cancel{K}\,\cancel{mol}}\right)298\,\cancel{K}}}}{e^{\frac{-1.25\times10^5\,\frac{\cancel{J}}{\cancel{mol}}}{\left(8.314\,\frac{\cancel{J}}{\cancel{K}\,\cancel{mol}}\right)298\,\cancel{K}}}} = \frac{e^{-22.199}}{e^{-50.453}} = 10^{12}$$

53. **Given:** table of [CH_3CN] versus time **Find:** a) reaction order, k; b) $t_{1/2}$; and c) t for 90 % conversion
Conceptual Plan: a) and b) Look at the data and see if any common reaction orders can be eliminated. If the data does not show an equal concentration drop with time, then zero order can be eliminated. Look for changes in the half-life (compare time for concentration to drop to one half of any value). If the half-life is not constant, then the first order can be eliminated. If the half-life is getting longer as the concentration drops, this might suggest second order. Plot the data as indicated by the appropriate rate law or if it is first order and there is an obvious half-life in the date, a plot is not necessary. Determine k from the slope of the plot (or using half-life equation for first order). c) Finally calculate the time to 90 % conversion using the appropriate integrated rate equation.
Solution: a and b) By the above logic, we can see that the reaction is first order. It takes 15.0 h for the concentration to be cut in half for any concentration (1.000 M to 0.501 M; 0.794 M to 0.398 M; and 0.631 M to 0.316 M), so $t_{1/2}$ = 15.0 h. Then use $t_{1/2} = \frac{0.693}{k}$ and rearrange to solve for k.

$$k = \frac{0.693}{t_{1/2}} = \frac{0.693}{15.0\text{ h}} = 0.0462\text{ h}^{-1}.$$

c) $[CH_3CN]_t$ = 0.10 $[CH_3CN]_0$. Since $\ln[CH_3CN]_t = -kt + \ln[CH_3CN]_0$ rearrange to solve for t

$$t = -\frac{1}{k}\ln\frac{[CH_3CN]_t}{[CH_3CN]_0} = -\frac{1}{0.0462\text{ hr}^{-1}}\ln\frac{0.10\,\cancel{[CH_3CN]_0}}{\cancel{[CH_3CN]_0}} = 49.8\text{ hr}.$$

Check: The units (none, h^{-1}, h, and h) are correct. The rate law is a common form. The data showed a constant half-life very clearly. The rate constant is consistent with the units necessary to get rate as M/s and the magnitude is reasonable since we have a first order reaction. The time to 90 % conversion is consistent with a time between three and four half-lives.

55. **Given:** $\text{Rate} = k\frac{[A][C]^2}{[B]^{1/2}}$ = 0.0115 M/s at certain initial concentrations of A, B and C; double A and C concentration and triple B concentration **Find:** reaction rate
Conceptual Plan: $[A]_1$, $[B]_1$, $[C]_1$, Rate 1, $[A]_2$, $[B]_2$, $[C]_2$ → Rate 2

$$\frac{\text{Rate 2}}{\text{Rate 1}} = \frac{k\frac{[A]_2[C]_2^2}{[B]_2^{1/2}}}{k\frac{[A]_1[C]_1^2}{[B]_1^{1/2}}}$$

Solution: $\frac{\text{Rate 2}}{\text{Rate 1}} = \frac{k\frac{[A]_2[C]_2^2}{[B]_2^{1/2}}}{k\frac{[A]_1[C]_1^2}{[B]_1^{1/2}}}$ Rearrange to solve for Rate 2. $\text{Rate 2} = \frac{k\frac{[A]_2[C]_2^2}{[B]_2^{1/2}}}{k\frac{[A]_1[C]_1^2}{[B]_1^{1/2}}}\text{Rate 1}$ $[A]_2 = 2[A]_1$, $[B]_2 = 3[B]_1$, $[C]_2 = 2[C]_1$ and Rate 1 = 0.0115 M/s so

$$\text{Rate 2} = \frac{k\frac{2[A]_1(2[C]_1)^2}{(3[B]_1)^{1/2}}}{k\frac{[A]_1[C]_1^2}{[B]_1^{1/2}}}0.0115\ \frac{\text{M}}{\text{s}} = \frac{2^3}{3^{1/2}}0.0115\ \frac{\text{M}}{\text{s}} = 0.0531\frac{\text{M}}{\text{s}}$$

Check: The units (M s^{-1}) are correct. The should increase because we have a factor of 8 (2^3) divided by the square root of three.

57. **Given:** N_2O_5 decomposes to NO_2 and O_2, first order in $[N_2O_5]$; $t_{1/2}$ = 2.81 h at 25 °C; V = 1.5 L, $P°_{N2O5}$ = 745 torr **Find:** P_{O2} after 215 minutes

Conceptual Plan:

Write a balanced reaction. then $t_{1/2} \rightarrow k$ then °C → K and torr → atm then

$N_2O_5 \rightarrow 2\ NO_2 + ½\ O_2$ $\quad t_{1/2} = \frac{0.693}{k}$ $\quad K = °C + 273.15$ $\quad \frac{1\text{ atm}}{760\text{ torr}}$

$P°_{N2O5}$, V, T → n/V then min → h $[N_2O_5]_0$, t, k → $[N_2O_5]_t$ then $[N_2O_5]_0$, $[N_2O_5]_t$ → $[O_2]_t$

$PV = nRT$ $\quad \frac{1\text{ hr}}{60\text{ min}}$ $\quad \ln[A]_t = -kt + \ln[A]_0$ $\quad [O_2]_t = ([N_2O_5]_0 - [N_2O_5]_t) \times \frac{1/2\text{ mol }O_2}{1\text{ mol }N_2O_5}$

then $[O_2]_t$, V, T → $P°_{O2}$ then finally atm → torr

$PV = nRT$ $\quad \frac{760\text{ torr}}{1\text{ atm}}$

Solution: $t_{1/2} = \frac{0.693}{k}$ and rearrange to solve for k. $k = \frac{0.693}{t_{1/2}} = \frac{0.693}{2.81\text{hr}} = 0.24\underline{6}619\text{ hr}^{-1}$. Then

T = 25 °C + 273.15 = 298 K. $745\text{ torr} \times \frac{1\text{ atm}}{760\text{ torr}} = 0.98\underline{0}263$ atm then $PV = nRT$ Rearrange to solve for

n/V. $\frac{n}{V} = \frac{P}{RT} = \frac{0.98\underline{0}263\text{ atm}}{0.08206\frac{\text{L}\cdot\text{atm}}{\text{K}\cdot\text{mol}} \times 298\text{ K}} = 0.040\underline{0}862$ M then $215\text{ min} \times \frac{1\text{ hr}}{60\text{ min}} = 3.5\underline{8}333$ hr .

Since $\ln[N_2O_5]_t = -kt + \ln[N_2O_5]_0 = -(0.24\underline{6}619\text{ hr}^{-1})(3.5\underline{8}333\text{ hr}) + \ln(0.040\underline{0}862\text{ M}) = -4.1\underline{0}044$ → $[N_2O_5]_t = e^{-4.1\underline{0}044} = 0.016\underline{5}653$ M then

$[O_2]_t = ([N_2O_5]_0 - [N_2O_5]_t) \times \frac{1/2\text{ mol }O_2}{1\text{ mol }N_2O_5} = (0.040\underline{0}862\frac{\text{mol }N_2O_5}{\text{L}} - 0.016\underline{5}653\frac{\text{mol }N_2O_5}{\text{L}}) \times \frac{1/2\text{ mol }O_2}{1\text{ mol }N_2O_5} =$

$= 0.011\underline{7}605\text{ M }O_2$

then finally $PV = nRT$ rearrange to solve for P.

$$P = \frac{n}{V}RT = 0.011\underline{7}605\frac{\text{mol}}{\text{L}} \times 0.08206\frac{\text{L atm}}{\text{K mol}} \times 298\text{ K} = 0.287589\text{ atm} \times \frac{760\text{ torr}}{1\text{ atm}} = 219\text{ torr}$$

Check: The units (torr) are correct. The pressure is reasonable because it must be less than one half of the original pressure.

59. **Given:** I_2 formation from I atoms, second order in I; $k = 1.5 \times 10^{10}\ M^{-1}\ s^{-1}$, $[I]_0 = 0.0100$ M
Find: t to decrease by 95 %
Conceptual Plan: $[I]_0$, $[I]_t$, k → t

$$\frac{1}{[A]_t} = kt + \frac{1}{[A]_0}$$

Solution: $[I]_t = 0.05\,[I]_0 = 0.05 \times 0.0100\text{ M} = 0.0005\text{ M}$. Since $\frac{1}{[I]_t} = kt + \frac{1}{[I]_0}$ rearrange to solve for t

$$t = \frac{1}{k}\left(\frac{1}{[I]_t} - \frac{1}{[I]_0}\right) = \frac{1}{(1.5 \times 10^{10}\ \cancel{\text{M}^{-1}}\,\text{s}^{-1})}\left(\frac{1}{0.0005\ \cancel{\text{M}}} - \frac{1}{0.0100\ \cancel{\text{M}}}\right) = \underline{1}.267 \times 10^{-7}\text{ s} = 1 \times 10^{-7}\text{ s}.$$

Check: The units (s) are correct. We expect the time to be extremely small because the rate constant is so large.

61. a) There are two elementary steps in the reaction mechanism because there are two peaks in the reaction progress diagram.

b)

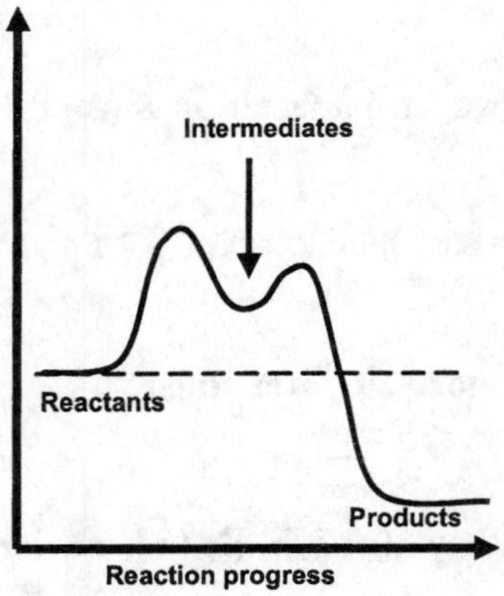

c) The first step is the rate limiting step because it has the higher activation energy.

d) The overall reaction is exothermic because the products are at a lower energy than the reactants.

63. **Given:** n-butane desorption from single crystal aluminum oxide, first order; $k = 0.128\text{ s}^{-1}$ at 150 K; initially completely covered

Find: a) $t_{½}$; b) t for 25 % and for 50 % to desorb; c) fraction remaining after 10 s and 20 s

Conceptual Plan: a) $k \rightarrow t_{½}$ **b)** $[C_4H_{10}]_0, [C_4H_{10}]_t, k \rightarrow t$ **c)** $[C_4H_{10}]_0, t, k \rightarrow [C_4H_{10}]_t$

$$t_{1/2} = \frac{0.693}{k} \qquad \ln[A]_t = -kt + \ln[A]_0 \qquad \ln[A]_t = -kt + \ln[A]_0$$

Solution:

a) $$t_{1/2} = \frac{0.693}{k} = \frac{0.693}{0.128\text{ s}^{-1}} = 5.41\text{ s}.$$

b) $\ln[C_4H_{10}]_t = -kt + \ln[C_4H_{10}]_0$ Rearrange to solve for t. For 25 % desorbed $[C_4H_{10}]_t = 0.75\,[C_4H_{10}]_0$

and $$t = -\frac{1}{k}\ln\frac{[C_4H_{10}]_t}{[C_4H_{10}]_0} = -\frac{1}{0.128\text{ s}^{-1}}\ln\frac{0.75\,\cancel{[C_4H_{10}]_0}}{\cancel{[C_4H_{10}]_0}} = 2.2\text{ s}.$$ For 50 % desorbed

$[C_4H_{10}]_t = 0.50\,[C_4H_{10}]_0$ and $$t = -\frac{1}{k}\ln\frac{[C_4H_{10}]_t}{[C_4H_{10}]_0} = -\frac{1}{0.128\text{ s}^{-1}}\ln\frac{0.50\,\cancel{[C_4H_{10}]_0}}{\cancel{[C_4H_{10}]_0}} = 5.4\text{ s}.$$

c) For 10 s $\ln[C_4H_{10}]_t = -kt + \ln[C_4H_{10}]_0 = -(0.128\ \cancel{\text{s}^{-1}})(10\ \cancel{\text{s}}) + \cancel{\ln(1.00)} = -1.\underline{2}8 \rightarrow$

$[C_4H_{10}]_t = e^{-1.\underline{2}8} = 0.28 =$ fraction covered and for 20 s

$\ln[C_4H_{10}]_t = -kt + \ln[C_4H_{10}]_0 = -(0.128\ \cancel{\text{s}^{-1}})(20\ \cancel{\text{s}}) + \cancel{\ln(1.00)} = -2.\underline{5}6 \rightarrow$

$[C_4H_{10}]_t = e^{-2.\underline{5}6} = 0.077 =$ fraction covered.

Check: The units (s, s, s, none, and none) are correct. The half-life is reasonable considering the size of the rate constant. The time to 25 % desorbed is less than one half-life. The time to 50 % desorbed is the half-life. The fraction at 10 s is consistent with about two half-lives. The fraction covered at 20 s is consistent with about four half-lives.

65. a) **Given:** table of rate constant versus T **Find:** E_a, and A

Conceptual Plan: **First convert temperature data into Kelvins (°C + 273.15 = K). Since** $\ln k = \frac{-E_a}{R}\left(\frac{1}{T}\right) + \ln A$ **a plot of ln k versus $1/T$ will have a slope = $-E_a/R$ and an intercept = ln A.**

Solution: The slope can be determined by measuring Δy/Δx on the plot or by using functions, such as "add trendline" in Excel.

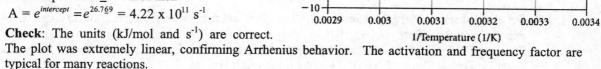

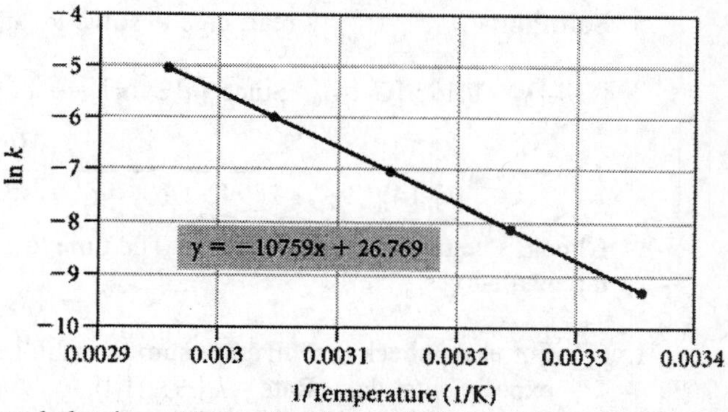

Since the slope = − 10759 K = − E_a/R then

$E_a = -(slope)R =$

$$= -(-10\underline{7}59\ \text{K})\left(8.314\ \frac{\text{J}}{\text{K mol}}\right)\left(\frac{1\ \text{kJ}}{1000\ \text{J}}\right)$$ an

$$= 89.5\ \frac{\text{kJ}}{\text{mol}}$$

d intercept = 26.7$\underline{6}$9 = ln A then

$A = e^{intercept} = e^{26.7\underline{6}9} = 4.22 \times 10^{11}\ \text{s}^{-1}$.

Check: The units (kJ/mol and s^{-1}) are correct.

The plot was extremely linear, confirming Arrhenius behavior. The activation and frequency factor are typical for many reactions.

b) **Given:** part a) results **Find:** k at 15 °C

Conceptual Plan: **°C → K then T, E_a, A → k**

°C + 273.15 = K $\quad \ln k = \frac{-E_a}{R}\left(\frac{1}{T}\right) + \ln A$

Solution: 15 °C + 273.15 = 288 K then

$$\ln k = \frac{-E_a}{R}\left(\frac{1}{T}\right) + \ln A = \frac{-89.5\ \frac{\text{kJ}}{\text{mol}} \times \frac{1000\ \text{J}}{1\ \text{kJ}}\ E_a}{8.314\ \frac{\text{J}}{\text{K mol}}}\left(\frac{1}{288\ \text{K}}\right) + \ln(4.22 \times 10^{11}\ \text{s}^{-1}) = -1\underline{0}.610 \rightarrow$$

$k = e^{-1\underline{0}.610} = 2.5 \times 10^{-5}\ \text{M}^{-1}\text{s}^{-1}$.

Check: The units ($M^{-1}\ s^{-1}$) are correct. The value of the rate constant is less than the value at 25 °C.

c) **Given:** part a) results, 0.155 M C_2H_5Br and 0.250 M OH^- at 75 °C **Find:** initial reaction rate

Conceptual Plan:

°C → K then T, E_a, A → k then k, [C_2H_5Br], [OH^-] → initial reaction rate

°C + 273.15 = K $\quad \ln k = \frac{-E_a}{R}\left(\frac{1}{T}\right) + \ln A \quad$ Rate = $k\,[C_2H_5Br][OH^-]$

Solution: 75 °C + 273.15 = 348 K then

$$\ln k = \frac{-E_a}{R}\left(\frac{1}{T}\right) + \ln A = \frac{-89.5\ \frac{\text{kJ}}{\text{mol}} \times \frac{1000\ \text{J}}{1\ \text{kJ}}\ E_a}{8.314\ \frac{\text{J}}{\text{K mol}}}\left(\frac{1}{348\ \text{K}}\right) + \ln(4.22 \times 10^{11}\ \text{s}^{-1}) = -4.\underline{1}656 \rightarrow$$

$k = e^{-4.\underline{1}656} = 1.\underline{5}521 \times 10^{-2}\ \text{M}^{-1}\text{s}^{-1}$.

Rate = $k\,[C_2H_5Br][OH^-] = (1.\underline{5}521 \times 10^{-2}\ \text{M}^{-1}\text{s}^{-1})(0.155\ \text{M})(0.250\ \text{M}) = 6.0 \times 10^{-4}\ \text{M s}^{-1}$

Check: The units ($M\ s^{-1}$) are correct. The value of the rate is reasonable considering the value of the rate constant (larger than in the table) and the fact that the concentrations are less than 1 M.

67. a) No, because the activation energy is zero. This means that the rate constant ($k = A\,e^{-E_a/RT}$) will be independent of temperature.

b) No bond is broken and the two radicals attract each other.

d) Formation of diatomic gases from atomic gases.

69. **Given:** $t_{1/2}$ for radioactive decay of C-14 = 5730 years; bone has 19.5 % C-14 in living bone
Find: age of bone
Conceptual Plan:
radioactive decay implies first order kinetics, $t_{1/2}$ → k then 19.5 % of $[C\text{-}14]_0$, k → t

$$t_{1/2} = \frac{0.693}{k} \qquad ln[A]_t = -kt + ln[A]_0$$

Solution: $t_{1/2} = \frac{0.693}{k}$ rearrange to solve for k. $k = \frac{0.693}{t_{1/2}} = \frac{0.693}{5730\ \text{yr}} = 1.2\underline{0}942 \times 10^{-4}\ \text{yr}^{-1}$ then

$[C\text{-}14]_t = 0.195\ [C\text{-}14]_0$. Since $ln[C\text{-}14]_t = -kt + ln[C\text{-}14]_0$ rearrange to solve for t

$$t = -\frac{1}{k}\ln\frac{[C\text{-}14]_t}{[C\text{-}14]_0} = -\frac{1}{1.2\underline{0}942 \times 10^{-4}\ \text{yr}^{-1}}\ln\frac{0.195\ \cancel{[C\text{-}14]_0}}{\cancel{[C\text{-}14]_0}} = 1.35 \times 10^4\ \text{yr}.$$

Check: The units (yr) are correct. The time to 19.5 % decay is consistent with the time being between two and three half-lives.

71. a) For each, check that all steps sum to overall reaction and that the predicted rate law is consistent with experimental data (Rate = $k\ [H_2]\ [I_2]$).
For the first mechanism, the single step is the overall reaction. The rate law is determined by the stoichiometry, so Rate = $k\ [H_2]\ [I_2]$ and the mechanism is valid.
For the second mechanism, the overall reaction is the sum of the steps in the mechanism:

$$I_2\ (g) \underset{k_2}{\overset{k_1}{\rightleftharpoons}} \cancel{2\ I\ (g)}$$

$$H_2\ (g) + \cancel{2\ I\ (g)} \xrightarrow{k_3} 2\ HI\ (g)$$ So the sum matches the overall reaction.

$$H_2\ (g) + I_2\ (g) \rightarrow 2\ HI\ (g)$$

Since the second step is the rate determining step, Rate = $k_3\ [H_2]\ [I]^2$. Since I is an intermediate, its concentration cannot appear in the rate law. Using the fast equilibrium in the first step, we see that $k_1[I_2] = k_2\ [I]^2$ or $[I]^2 = \frac{k_1}{k_2}[I_2]$. Substituting this into the first rate expression, we get that

Rate = $k_3 \frac{k_1}{k_2} [H_2]\ [I_2]$ and the mechanism is valid.

For the third mechanism, the overall reaction is the sum of the steps in the mechanism:

$$I_2\ (g) \underset{k_2}{\overset{k_1}{\rightleftharpoons}} \cancel{2\ I\ (g)}$$

$$H_2\ (g) + \cancel{I\ (g)} \underset{k_4}{\overset{k_3}{\rightleftharpoons}} \cancel{H_2I\ (g)}$$

$$\cancel{H_2I\ (g)} + \cancel{I\ (g)} \xrightarrow{k_5} 2\ HI\ (g)$$ So the sum matches the overall reaction.

$$H_2\ (g) + I_2\ (g) \rightarrow 2\ HI\ (g)$$

Since the third step is the rate determining step, Rate = $k_5\ [H_2I]\ [I]$. Since H_2I and I are intermediates, their concentrations cannot appear in the rate law. Using the fast equilibrium in the first step, we see that $k_1[I_2] = k_2\ [I]^2$ or $[I] = \sqrt{\frac{k_1}{k_2}[I_2]}$. Using the fast equilibrium in the second step, we see that $k_3[H_2]\ [I] = k_4\ [H_2I]$ or $[H_2I] = \frac{k_3}{k_4}[H_2]\ [I]$. Substituting both of these expressions for the intermediates into the first rate expression, we get that Rate = $k_5\ [H_2I]\ [I] = k_5 \frac{k_3}{k_4}[H_2]\ [I]\ \sqrt{\frac{k_1}{k_2}[I_2]}$. Substituting again for [I] we get Rate = $k_5 \frac{k_3}{k_4}[H_2] \sqrt{\frac{k_1}{k_2}[I_2]}\ \sqrt{\frac{k_1}{k_2}[I_2]}$. Simplifying this expression we see

Rate = $k_5 \frac{k_3}{k_4} \frac{k_1}{k_2} [H_2]\ [I_2]$ and the mechanism is valid.

b) To distinguish between mechanisms, you could look for the buildup of I(g) and/or H_2I(g), the intermediates.

73. a) For a zero order reaction, the rate is independent of the concentration. If the first half goes in the first 100 minutes, the second half will go in the second 100 minutes. This means that there will be none or 0 % left at 200 minutes.

b) For a first order reaction, the half-life is independent of concentration. This means that if half of the reactant decomposes in the first 100 minutes, then half of this (or another 25 % of the original amount) will decompose in the second 100 minutes. This means that at 200 minutes 50 % + 25 % = 75 % has decomposed or 25 % remains.

c) For a second order reaction, $t_{1/2} = \dfrac{1}{k[A]_0} = 100 \text{ min}$ and the integrated rate expression is

$\dfrac{1}{[A]_t} = kt + \dfrac{1}{[A]_0}$. We can rearrange the first expression to solve for k as $k = \dfrac{1}{100 \text{ min } [A]_0}$.

Substituting this and 200 minutes into the integrated rate expression, we get $\dfrac{1}{[A]_t} = \dfrac{200 \cancel{\text{min}}}{100 \cancel{\text{min}} [A]_0} + \dfrac{1}{[A]_0}$

$\rightarrow \dfrac{1}{[A]_t} = \dfrac{3}{[A]_0} \rightarrow \dfrac{[A]_t}{[A]_0} = \dfrac{1}{3}$ or 33 % remains.

75. Using the energy diagram show and using Hess' Law, we can see that the activation energy for the decomposition is equal to the activation energy for the formation reaction plus the heat of formation of 2 moles of HI or $E_{a\,formation} = E_{a\,decomposition} + 2\,\Delta H_f^0(\text{HI})$. So $E_{a\,formation} = 185 \text{ kJ} + 2 \text{mol}(-5.65 \text{ kJ/mol}) = 174 \text{ kJ}$.
Check: Since the reaction is endothermic, we expect the activation energy in the reverse direction to be less in the forward direction.

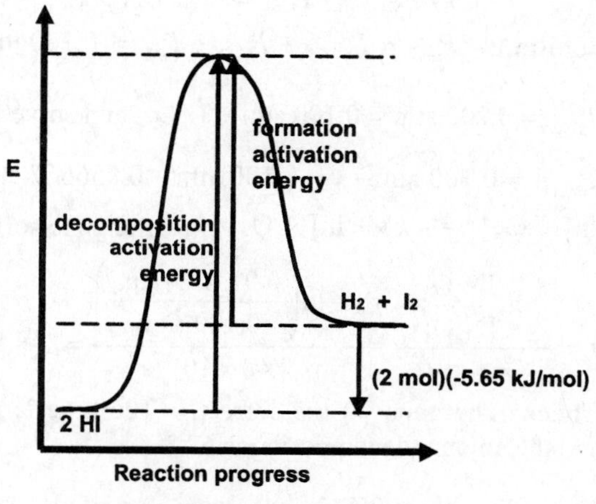

Note: energy axis is not to scale.

77. a) Since the rate determining step involves the collision of two molecules, the expected reaction order would be second order.

b) The proposed mechanism is:

$$CH_3NC + \cancel{CH_3NC} \underset{k_2}{\overset{k_1}{\rightleftharpoons}} \cancel{CH_3NC^*} + \cancel{CH_3NC} \quad \text{(fast)}$$

$$\cancel{CH_3NC^*} \xrightarrow{k_3} CH_3CN \quad \text{(slow)}$$ So the sum matches the overall reaction.

$$CH_3NC \rightarrow CH_3CN$$

Since the second step is the rate determining step, Rate = $k_3[CH_3NC^*]$. Since CH_3NC^* is an intermediate, its concentration cannot appear in the rate law. Using the fast equilibrium in the first step, we see that $k_1[CH_3NC]^2 = k_2[CH_3NC^*][CH_3NC]$ or $[CH_3NC^*] = \dfrac{k_1}{k_2}[CH_3NC]$. Substituting this into the first rate expression we get that Rate = $k_3\dfrac{k_1}{k_2}[CH_3NC]$, which simplifies to Rate = $k\,[CH_3NC]$. This matches the experimental observation of first order and the mechanism is valid.

79. a) **Given:** N_2O_5 decomposes to NO_2 and O_2, first order in $[N_2O_5]$; $k = 7.48 \times 10^{-3}\ s^{-1}$; $P°_{N2O5} = 0.100$ atm
Find: t to $P_{Total} = 0.145$ atm
Conceptual Plan: **Write a balanced reaction. then**

$$N_2O_5 \rightarrow 2\,NO_2 + \tfrac{1}{2}\,O_2$$

Write expression for P_{Total} in terms of amount reacted then $x, P°_{N2O5}, k \rightarrow t$

let $x = P_{N2O5reacted}$ $P_{Total} = P_{N2O5} + P_{NO2} + P_{O2}$ $\qquad \ln[A]_t = -kt + \ln[A]_0$

Solution: $P_{Total} = P_{N2O5} + P_{NO2} + P_{O2} = (0.100\text{ atm} - x) + (2x) + (1/2x) = 0.100\text{ atm} + 1.5\,x$ Set $P_{Total} = 0.145\text{ atm} = 0.100\text{ atm} + 1.5\,x$ and solve for x. $x = \dfrac{0.145\text{ atm} - 0.100\text{ atm}}{1.5} = 0.030\text{ atm}$ then $P_{N2O5} = 0.100\text{ atm} - x = 0.100\text{ atm} - 0.030\text{ atm} = 0.070\text{ atm}$. Since $P \alpha n/V$ or M

$\ln[N_2O_5]_t = -kt + \ln[N_2O_5]_0$ rearrange to solve for t.

$$t = -\frac{\ln\dfrac{[N_2O_5]_t}{[N_2O_5]_0}}{k} = -\frac{\ln\left(\dfrac{0.070\text{ atm}}{0.100\text{ atm}}\right)}{7.48\times10^{-3}\text{ s}^{-1}} = 4\underline{7}.684\text{ s} = 48\text{ s}$$

Check: The units (s) are correct. The time is reasonable because it is less than one half-life and the amount decomposing is less that 50 %.

b) **Given:** N_2O_5 decomposes to NO_2 and O_2, first order in $[N_2O_5]$; $k = 7.48 \times 10^{-3}\text{ s}^{-1}$; $P°_{N2O5} = 0.100$ atm
Find: t to $P_{Total} = 0.200$ atm
Conceptual Plan: Write a balanced reaction. then

$N_2O_5 \rightarrow 2\,NO_2 + ½\,O_2$

Write expression for P_{Total} in terms of amount reacted then $x, P°_{N2O5}, k \rightarrow t$

let $x = P_{N2O5reacted}$ $P_{Total} = P_{N2O5} + P_{NO2} + P_{O2}$ $\qquad \ln[A]_t = -kt + \ln[A]_0$

Solution: $P_{Total} = P_{N2O5} + P_{NO2} + P_{O2} = (0.100\text{ atm} - x) + (2x) + (1/2x) = 0.100\text{ atm} + 1.5\,x$ Set $P_{Total} = 0.200\text{ atm} = 0.100\text{ atm} + 1.5\,x$ and solve for x. $x = \dfrac{0.200\text{ atm} - 0.100\text{ atm}}{1.5} = 0.06\underline{6}667\text{ atm}$ then $P_{N2O5} = 0.100\text{ atm} - x = 0.100\text{ atm} - 0.06\underline{6}667\text{ atm} = 0.03\underline{3}333\text{ atm}$. Since $P \alpha n/V$ or M

$\ln[N_2O_5]_t = -kt + \ln[N_2O_5]_0$ rearrange to solve for t.

$$t = -\frac{\ln\dfrac{[N_2O_5]_t}{[N_2O_5]_0}}{k} = -\frac{\ln\left(\dfrac{0.03\underline{3}333\text{ atm}}{0.100\text{ atm}}\right)}{7.48\times10^{-3}\text{ s}^{-1}} = 1\underline{4}6.873\text{ s} = 150\text{ s} = 1.5\times10^2\text{ s}.$$

Check: The units (s) are correct. The time is reasonable because it is between one and two half-lives and the amount decomposing is 67 %.

c) **Given:** N_2O_5 decomposes to NO_2 and O_2, first order in $[N_2O_5]$; $k = 7.48 \times 10^{-3}\text{ s}^{-1}$; $P°_{N2O5} = 0.100$ atm
Find: P_{Total} after 100 s
Conceptual Plan:
$P°_{N2O5}, k, t \rightarrow P_{N2O5}$ then Write a balanced reaction. then $P°_{N2O5}, P_{N2O5} \rightarrow x$ then

$\ln[A]_t = -kt + \ln[A]_0$ $\qquad N_2O_5 \rightarrow 2\,NO_2 + ½\,O_2$ $\qquad x = P^0_{N2O5} - P_{N2O5}$

Write expression for P_{Total} in terms of amount reacted.

let $x = P_{N2O5reacted}$ $P_{Total} = P_{N2O5} + P_{NO2} + P_{O2}$

Solution: Since $P \alpha n/V$ or M

$\ln[N_2O_5]_t = -kt + \ln[N_2O_5]_0 = -(7.48\times10^{-3}\text{ s}^{-1})(100\text{ s}) + \ln(0.100\text{ atm}) = -3.0\underline{5}059$

$P_{N2O5} = e^{-3.0\underline{5}059} = 0.047\underline{3}312\text{ atm}$ so $x = P^0_{N2O5} - P_{N2O5} = 0.100\text{ atm} - 0.047\underline{3}312\text{ atm} = 0.05\underline{2}669\text{ atm}$

finally $P_{Total} = P_{N2O5} + P_{NO2} + P_{O2} = (0.100\text{ atm} - x) + (2x) + (1/2x) = 0.100\text{ atm} + 1.5\,x =$

$= 0.100\text{ atm} + 1.5(0.05\underline{2}669\text{ atm}) = 0.179\text{ atm}$

Check: The units (atm) are correct. The pressure is reasonable because the time is between those for parts a and b.

81. **Given:** N_2O_3 decomposes to NO_2 and NO, first order in $[N_2O_3]$, table of $[NO_2]$ versus time, at 50,000 s = $[N_2O_3] = 0$ **Find:** k
Conceptual Plan: Write a balanced reaction. then write expression for P_{NO2} in terms P_{N2O3}, then

$N_2O_3 \rightarrow NO_2 + NO$ $\qquad P_{N2O3} = 0.784\text{ atm} - P_{NO2}$

Plot ln P_{N2O3} versus time (this will have the same slope as ln $[N_2O_3]$ versus time). Since $\ln[A]_t = -kt + \ln[A]_0$, the negative of the slope will be the rate constant.

Solution: $P_{N2O3} = 0.784$ atm – P_{NO2} Plot ln P_{N2O3} versus time (this will have the same slope as ln $[N_2O_3]$ versus time). Since $\ln[A]_t = -kt + \ln[A]_0$, the negative of the slope will be the rate constant. The slope can be determined by measuring Δy/Δx on the plot or by using functions, such as "add trendline" in Excel. The last point (50,000 s) cannot be plotted because the concentration is 0 and the ln 0 is undefined,

Thus the rate constant is 3.20×10^{-4} s^{-1} and the rate law is Rate = $3.20\ 10^{-4}\ s^{-1}[N_2O_3]$.

Check: The units (s^{-1}) are correct. The rate constant is typical for a reaction.

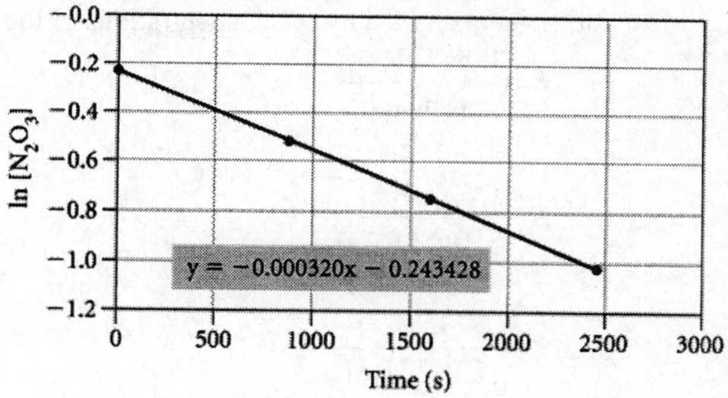

83. Reaction A is second order. A plot of 1/[A] versus time will be linear $\left(\frac{1}{[A]_t} = kt + \frac{1}{[A]_0}\right)$. Reaction B is first order. A plot of ln [A] versus time will be linear ($\ln[A]_t = -kt + \ln[A]_0$). Reactant concentrations drop more quickly for first order reactions than for second order reactions.

Chapter 14
Chemical Equilibrium

1. The equilibrium constant is defined as the concentrations of the products raised to their stoichiometric coefficients divided by the concentrations of the reactants raised to their stoichiometric coefficients.

   a) $K = \dfrac{[SbCl_3][Cl_2]}{[SbCl_5]}$

   b) $K = \dfrac{[NO]^2[Br_2]}{[BrNO]^2}$

   c) $K = \dfrac{[CS_2][H_2]^4}{[CH_4][H_2S]^2}$

   d) $K = \dfrac{[CO_2]^2}{[CO]^2[O_2]}$

3. With an equilibrium constant of 1.4×10^{-5}, the value of the equilibrium constant is small, therefore, the concentration of reactants will be greater than the concentration of products. This is independent of the initial concentration of the reactants and products.

5. (i) has 10 H_2 and 10 I_2
   (ii) has 7 H_2 and 7 I_2 and 6 HI
   (iii) has 5 H_2 and 5 I_2 and 10 HI
   (iv) has 4 H_2 and 4 I_2 and 12 HI
   (v) has 3 H_2 and 3 I_2 and 14 HI
   (vi) has 3 H_2 and 3 I_2 and 14 HI

   a) Concentration of (v) and (vi) are the same so the system reached equilibrium at (v).

   b) If a catalyst was added to the system, the system would reach the conditions at (v) sooner since a catalyst speeds up the reaction but does not change the equilibrium conditions.

   c) The final figure (vi) would have the same amount of reactants and products since a catalyst speeds up the reaction, but does not change the equilibrium concentrations.

7. a) If you reverse the reaction, invert the equilibrium constant. So, $K' = \dfrac{1}{K_p} = \dfrac{1}{2.26x10^4} = 4.42 \times 10^{-5}$. The reactants will be favored.

   b) If you multiply the coefficients in the equation by a factor, raise the equilibrium constant to the same factor.
   So, $K' = (K_p)^{1/2} = (2.26 \times 10^4)^{1/2} = 1.50 \times 10^2$. The products will be favored.

   c) Begin with the reverse of the reaction and invert the equilibrium constant.
   $K_{reverse} = \dfrac{1}{K_p} = \dfrac{1}{2.26x10^4} = 4.42 \times 10^{-5}$
   Then, multiply the reaction by 2 and raise the value of $K_{reverse}$ to the 2 power.
   $K' = (K_{reverse})^2 = (4.42 \times 10^{-5})^2 = 1.96 \times 10^{-9}$. The reactants will be favored.

9. To find the equilibrium constant for reaction 3, you need to combine reactions 1 and 2 to get reaction 3. Begin by reversing reaction 2, then multiply reaction 1 by 2 and add the two new reactions. When you add reactions you multiply the values of *K*.

$$N_2(g) + O_2(g) \rightleftharpoons \cancel{2NO}(g) \qquad K_1 = \frac{1}{K_p} = \frac{1}{2.1 \times 10^{30}} = 4.\underline{7}6 \times 10^{-31}$$

$$\cancel{2NO}(g) + Br_2(g) \rightleftharpoons 2\,NOBr(g) \qquad K_2 = (K_p)^2 = (5.3)^2 = 2\underline{8}.09$$

$$N_2(g) + O_2(g) + Br_2(g) \rightleftharpoons 2\,NOBr(g) \qquad K_3 = K_1K_2 = (4.\underline{7}6 \times 10^{-31})(2\underline{8}.09) = 1.3 \times 10^{-29}$$

11. a) **Given:** $K_p = 6.26 \times 10^{-22}$ T = 298K **Find:** K_c

Conceptual Plan: $K_p \rightarrow K_c$

$$K_p = K_c(RT)^{\Delta n}$$

Solution: Δn = mol product gas – mol reactant gas = 2 – 1 = 1

$$K_c = \frac{K_p}{(RT)^{\Delta n}} = \frac{6.26 \times 10^{-22}}{\left(0.08206 \frac{L \cdot atm}{mol \cdot \cancel{K}} \times 298\ \cancel{K}\right)^1} = 2.56 \times 10^{-23}$$

Check: Substitute into the equation and confirm that you get the original value of K_p.

$$K_p = K_c(RT)^{\Delta n} = (2.56 \times 10^{-23})\left(0.08206 \frac{L \cdot atm}{mol \cdot K} \times 298\right)^1 = 6.26 \times 10^{-22}$$

b) **Given:** $K_p = 7.7 \times 10^{24}$ T = 298K **Find:** K_c

Conceptual Plan: $K_p \rightarrow K_c$

$$K_p = K_c(RT)^{\Delta n}$$

Solution: Δn = mol product gas – mol reactant gas = 4 – 2 = 2

$$K_c = \frac{K_p}{(RT)^{\Delta n}} = \frac{7.7 \times 10^{24}}{\left(0.08206 \frac{L \cdot atm}{mol \cdot \cancel{K}} \times 298\ \cancel{K}\right)^2} = 1.3 \times 10^{22}$$

Check: Substitute into the equation and confirm that you get the original value of K_p.

$$K_p = K_c(RT)^{\Delta n} = (1.3 \times 10^{22})\left(0.08206 \frac{L \cdot atm}{mol \cdot K} \times 298\right)^2 = 7.7 \times 10^{24}$$

c) **Given:** $K_p = 81.9$ T = 298K **Find:** K_c

Conceptual Plan: $K_p \rightarrow K_c$

$$K_p = K_c(RT)^{\Delta n}$$

Solution: Δn = mol product gas – mol reactant gas = 2 – 2 = 0

$$K_c = \frac{K_p}{(RT)^{\Delta n}} = \frac{81.9}{\left(0.08206 \frac{L \cdot atm}{mol \cdot \cancel{K}} \times 298\ \cancel{K}\right)^0} = 81.9$$

Check: Substitute into the equation and confirm that you get the original value of K_p.

$$K_p = K_c(RT)^{\Delta n} = (81.9)\left(0.08206 \frac{L \cdot atm}{mol \cdot K} \times 298\right)^0 = 81.9$$

13. a) Since H_2O is a liquid, it is omitted from the equilibrium expression. $K_{eq} = \frac{[HCO_3^-][OH^-]}{[CO_3^{2-}]}$

b) Since $KClO_3$ and KCl are both solids, they are omitted from the equilibrium expression. $K_{eq} = [O_2]^3$

c) Since H_2O is a liquid, it is omitted from the equilibrium expression. $K_{eq} = \frac{[H_3O^+][F^-]}{[HF]}$

d) Since H_2O is a liquid, it is omitted from the equilibrium expression. $K_{eq} = \frac{[NH_4^+][OH^-]}{[NH_3]}$

15. **Given:** At equilibrium: [CO] = 0.105M, $[H_2]$ = 0.114M, $[CH_3OH]$ = 0.185M **Find:** K_c

Conceptual Plan: Balanced reaction → equilibrium expression → K_c

Solution: $K_c = \dfrac{[CH_3OH]}{[CO][H_2]^2} = \dfrac{(0.185)}{(0.105)(0.114)^2} = 136$

Check: The answer is reasonable since the concentration of products is greater than the concentration of reactants and the equilibrium constant should be greater than 1.

17. At 500K: **Given:** At equilibrium: $[N_2]$ = 0.115M, $[H_2]$ = 0.105M, and $[NH_3]$ = 0.439 **Find:** K_c

Conceptual Plan: Balanced reaction → equilibrium expression → K_c

Solution: $K_c = \dfrac{[NH_3]^2}{[N_2][H_2]^3} = \dfrac{(0.439)^2}{(0.115)(0.105)^3} = 1.45 \text{ x } 10^3$

Check: The value is reasonable since the concentration of products is greater than the concentration of reactants.

At 575K: **Given:** At equilibrium: $[N_2]$ = 0.110M, $[NH_3]$ = 0.128, K_c = 9.6 **Find:** $[H_2]$

Conceptual Plan: Balanced reaction → equilibrium expression → $[H_2]$

Solution: $K_c = \dfrac{[NH_3]^2}{[N_2][H_2]^3}$ $\quad 9.6 = \dfrac{(0.128)^2}{(0.110)(x)^3}$ $\quad x = 0.249$

Check: Plug the value for x back into the equilibrium expression, and check the value.

$$9.6 = \frac{(0.128)^2}{(0.110)(0.249)^3}$$

At 775K: **Given:** At equilibrium: $[N_2]$ = 0.120M, $[H_2]$ = 0.140, K_c = 0.0584 **Find:** $[NH_3]$

Conceptual Plan: Balanced reaction → equilibrium expression → $[NH_3]$

Solution: $K_c = \dfrac{[NH_3]^2}{[N_2][H_2]^3}$ $\quad 0.0584 = \dfrac{(x)^2}{(0.120)(0.140)^3}$ $\quad x = 0.00439$

Check: Plug the value for x back into the equilibrium expression, and check the value.

$$0.0584 = \frac{(0.00439)^2}{(0.120)(0.140)^3}$$

19. **Given:** P_{NO} = 118 torr; P_{Br2} = 176 torr, K_p = 28.4 **Find:** P_{NOBr}

Conceptual Plan: **torr → atm and then balanced reation → equilibrium expression → P_{NOBr}**

$\dfrac{1 \text{ atm}}{760 \text{ torr}}$

Solution: $P_{NO} = 118 \text{ torr x } \dfrac{1 \text{ atm}}{760 \text{ torr}} = 0.15\underline{5}3 \text{ atm}$ $\quad P_{Br_2} = 176 \text{ torr x } \dfrac{1 \text{ atm}}{760 \text{ torr}} = 0.23\underline{1}6$

$K_p = \dfrac{P^2_{NOBr}}{P^2_{NO}P_{Br_2}}$ $\quad 28.4 = \dfrac{x^2}{(0.15\underline{5}3)^2(0.23\underline{1}6)}$ $\quad x = 0.398 \text{ atm} = 303 \text{ torr}$

Check: Plug the value for x back into the equilibrium expression, and check the value.

$$28.4 = \frac{(0.398)^2}{(0.15\underline{5}3)^2(0.23\underline{1}6)}$$

21. **Given:** $[Fe^{3+}]_{initial} = 1.0 \text{ x } 10^{-3}$M; $[SCN^-]_{initial} = 8.0 \text{ x } 10^{-4}$M; $[FeSCN^{2+}]_{eq} = 1.7 \text{ x } 10^{-4}$M **Find:** K_c

Conceptual Plan:
1. **Prepare ICE table**
2. **Calculate concentration change for known value**
3. **Calculate concentration changes for other reactants/products**
4. **Determine equilibrium concentration**
5. **Write the equilibrium expression and determine K_c**

Solution: $Fe^{3+}(aq) + SCN^-(aq) \leftrightarrows FeSCN^{3+}(aq)$

	$[Fe^{3+}]$	$[SCN^-]$	$[FeSCN^{3+}]$
I	$1.0\text{x}10^{-3}$	$8.0\text{x}10^{-4}$	0.00
C	$-1.7\text{x}10^{-4}$	$-1.7\text{x}10^{-4}$	$+1.7\text{x}10^{-4}$
E	$8.3\text{x}10^{-4}$	$6.3\text{x}10^{-4}$	$1.7\text{x}10^{-4}$

$$K_c = \frac{[FeSCN^{2+}]}{[Fe^{3+}][SCN^-]} = \frac{(1.7 \times 10^{-4})}{(8.3 \times 10^{-4})(6.3 \times 10^{-4})} = 3.3 \times 10^2$$

23. **Given:** 3.67 L flask; 0.763 g H_2 initial; 96.9 g I_2 initial; 90.4 g HI equilibrium   **Find:** K_c
    Conceptual Plan: g → mol → M and then

$$n = \frac{g}{\text{molar mass}} \qquad M = \frac{n}{V}$$

        1. Prepare ICE table
        2. Calculate concentration change for known value
        3. Calculate concentration changes for other reactants/products
        4. Determine equilibrium concentration
        5. Write the equilibrium expression and determine K_c

    Solution: $0.763 \text{ g } H_2 \times \frac{1 \text{ mol } H_2}{2.016 \text{ g } H_2} = 0.37\underline{8}5 \text{ mol } H_2$ $\frac{0.37\underline{8}5 \text{ mol } H_2}{3.67 \text{ L}} = 0.103 \text{ M}$

             $96.9 \text{ g } I_2 \times \frac{1 \text{ mol } I_2}{253.8 \text{ g } I_2} = 0.38\underline{1}8 \text{ mol } I_2$ $\frac{0.38\underline{1}8 \text{ mol } I_2}{3.67 \text{ L}} = 0.104 \text{ M}$

             $90.4 \text{ g HI} \times \frac{1 \text{ mol HI}}{127.9 \text{ g } I_2} = 0.70\underline{6}8 \text{ mol HI}$ $\frac{0.70\underline{6}8 \text{ mol HI}}{3.67 \text{ L}} = 0.193 \text{ M}$

	$H_2(g)$	$+ I_2 (g)$	$\rightleftharpoons 2HI(g)$
	$[H_2]$	$[I_2]$	$[HI]$
I	0.103	0.104	0.00
C	- 0.0965	- 0.0965	+0.193
E	0.0065	0.0075	0.193

$$K_c = \frac{[HI]^2}{[H_2][I_2]} = \frac{(0.193)^2}{(0.0065)(0.0075)} = 764$$

25. **Given:** $K_c = 8.5 \times 10^{-3}$; $[NH_3] = 0.166M$; $[H_2S] = 0.166M$ **Find:** Will solid form or decompose?
    Conceptual Plan: Calculate Q → compare Q and K_c
    Solution: $Q = [NH_3][H_2S] = (0.166)(0.166) = 0.0276$
    $Q = 0.0276$ and $K_c = 8.5 \times 10^{-3}$ so $Q > K_c$ and the reaction will shift to the left, so more solid will form.

27. **Given:** 6.55 g Ag_2SO_4, 1.5 L solution, $K_c = 1.1 \times 10^{-5}$ **Find:** will more solid dissolve
    Conceptual Plan:
        g Ag_2SO_4 → mol Ag_2SO_4 → $[Ag_2SO_4]$ → $[Ag^+],[SO_4^{-2}]$ → calculate Q and compare to K_c.

        $\frac{1 \text{ mol } Ag_2SO_4}{311.81 \text{ g}}$ $[\ ] = \frac{\text{mol } Ag_2SO_4}{\text{vol solution}}$ $Q = [Ag^+]^2[SO_4^{2-}]$

    Solution: $6.55 \text{ g } Ag_2SO_4 \left(\frac{1 \text{ mol } Ag_2SO_4}{311.81 \text{ g } Ag_2SO_4} \right) = 0.0210 \text{ mol } Ag_2SO_4$

    $\frac{0.0210 \text{ mol } Ag_2SO_4}{1.5 \text{ L solution}} = 0.01\underline{4}0 \text{ M } Ag_2SO_4$

    $[Ag^+] = 2[Ag_2SO_4] = 2(0.01\underline{4}0 \text{ M}) = 0.02\underline{8}0 \text{ M}$ $[SO_4^{2-}] = [Ag_2SO_4] = 0.01\underline{4}0 \text{ M}$

    $Q = [Ag^+]^2[SO_4^{2-}] = (0.02\underline{8}0)^2(0.01\underline{4}0) = 1.1 \times 10^{-5}$

    $Q = K_c$ so the system is at equilibrium and is a saturated solution. Therefore, if more solid is
    added it will not dissolve.

29. a)   **Given:** $[A] = 1.0$ M, $[B] = 0.0$ $K_c = 2.0$; $a = 1$, $b = 1$ **Find:** [A], [B] at equilibrium
    **Conceptual Plan: Prepare an ICE table; and then calculate Q; compare Q and K_c and predict the
    direction of the reaction; represent the change with x; sum the table and determine the
    equilibrium values; put the equilibrium values in the equilibrium expression and solve for x.
    Determine [A] and [B].**

Solution:

$A(g) \leftrightarrows B(g)$

	[A]	[B]
I	1.0	0.00
C	- x	x
E	1 – x	x

$Q = \frac{[B]}{[A]} = \frac{0}{1} = 0$ $Q < K$ therefore, the reaction will proceed to the right by x

$K_c = \frac{[B]}{[A]} = \frac{(x)}{(1 - x)} = 2.0$ x = 0.67

[A] = 1 - 0.67 = 0.33M [B] = 0.67M

Check: Plug the values into the equilibrium expression: $K_c = \frac{0.67}{0.33} = 2.0$

b) **Given:** [A] = 1.0 M, [B] = 0.0 K_c = 2.0; a = 2, b = 2 **Find:** [A], [B] at equilibrium
Conceptual Plan: Prepare an ICE table; and then calculate Q; compare Q and K_c and predict the direction of the reaction; represent the change with x; sum the table and determine the equilibrium values; put the equilibrium values in the equilibrium expression and solve for x. Determine [A] and [B].
Solution:

$2A(g) \leftrightarrows 2B(g)$

	[A]	[B]
I	1.0	0.00
C	- 2x	2x
E	1 – 2x	2x

$Q = \frac{[B]^2}{[A]^2} = \frac{0}{1} = 0$ $Q < K$ therefore, the reaction will proceed to the right by x

$K_c = \frac{[B]^2}{[A]^2} = \frac{(2x)^2}{(1 - 2x)^2} = 2.0$ x = 0.2$\underline{9}$3

[A] = 1 - 2(0.2$\underline{9}$3) = 0.414 = 0.41M [B] = 2(0.2$\underline{9}$3) = 0.586 = 0.59M

Check: Plug the values into the equilibrium expression: $K_c = \frac{(0.59)^2}{(0.41)^2} = 2.0$

c) **Given:** [A] = 1.0 M, [B] = 0.0 K_c = 2.0; a = 1, b = 2 **Find:** [A], [B] at equilibrium
Conceptual Plan: Prepare an ICE table; and then calculate Q; compare Q and K_c and predict the direction of the reaction; represent the change with x; sum the table and determine the equilibrium values; put the equilibrium values in the equilibrium expression and solve for x. Determine [A] and [B].
Solution:

$A(g) \leftrightarrows 2B(g)$

	[A]	[B]
I	1.0	0.00
C	- x	2x
E	1 – x	2x

$Q = \frac{[B]^2}{[A]} = \frac{0}{1} = 0$ $Q < K$ therefore, the reaction will proceed to the right by x

$K_c = \frac{[B]^2}{[A]} = \frac{(2x)^2}{(1 - x)} = 2.0$ $4x^2 + 2x - 2 = 0$

$(4x - 2)(x + 1) = 0$; x = - 1 or x = 0.5, therefore, x = 0.5

[A] = 1 - 0.50 = 0.50M [B] = 2x = 2(0.50) = 1.0M

Check: Plug the values into the equilibrium expression: $K_c = \frac{(1.0)^2}{0.50} = 2.0$

31. **Given:** $[N_2O_4] = 0.0500$ M, $[NO_2] = 0.0$ $K_c = 0.513$ **Find:** $[N_2O_4]$, $[NO_2]$ at equilibrium
Conceptual Plan: Prepare an ICE table; and then calculate Q; compare Q and K_c and predict the direction of the reaction; represent the change with x; sum the table and determine the equilibrium values; put the equilibrium values in the equilibrium expression and solve for x. Determine $[N_2O_4]$ and $[NO_2]$.
Solution:

	$N_2O_4(g)$ ⇆	$2\,NO_2(g)$
	$[N_2O_4]$	$[NO_2]$
I	0.0500	0.00
C	- x	2x
E	0.0500 – x	2x

$$Q = \frac{[NO_2]^2}{[N_2O_4]} = \frac{0}{0.0500} = 0 \qquad Q < K \text{ therefore, the reaction will proceed to the right by x}$$

$$K_c = \frac{[NO_2]^2}{[N_2O_4]} = \frac{(2x)^2}{(0.0500 - x)} = 0.513 \qquad 4x^2 + 0.513x - 0.02565 = 0$$

$$\frac{-b \pm \sqrt{b^2 - 4ac}}{2a} = \frac{-0.513 \pm \sqrt{(0.513)^2 - 4(4)(-0.02565)}}{2(4)} = \frac{-0.513 \pm \sqrt{0.6735}}{2(4)}$$

x = - 0.1667 or x = 0.0385, therefore, x = 0.0385

$[A] = 0.0500 - 0.03850 = 0.0115M \qquad [B] = 2x = 2(0.03850) = 0.0770M$

Check: Plug the values into the equilibrium expression: $K_c = \dfrac{(0.0770)^2}{0.0115} = 0.515$

33. **Given:** $[CO] = 0.10$ M, $[CO_2] = 0.0$ $K_c = 4.0 \times 10^3$ **Find:** $[CO_2]$ at equilibrium
Conceptual Plan: Prepare an ICE table; and then calculate Q; compare Q and K_c and predict the direction of the reaction; represent the change with x; sum the table and determine the equilibrium values; put the equilibrium values in the equilibrium expression and solve for x. Determine [CO], $[Cl_2]$, and $[COCl_2]$.
Solution:

	$NiO(s)$	$+CO(g)$	⇆ $Ni(s)$	$+ CO_2(g)$
		[CO]		$[CO_2]$
I		0.10		0.0
C		- x		x
E		0.10 –x		x

$$Q = \frac{[CO_2]}{[CO]} = \frac{0}{(0.10)} = 0 \qquad Q < K \text{ therefore, the reaction will proceed to the right by x}$$

$$K_c = \frac{[CO_2]}{[CO]} = \frac{x}{(0.10 - x)} = 4.0 \times 10^3 \qquad 4.0 \times 10^3(0.10 - x) = x$$

x = 0.10

$[CO_2] = 0.10$ M

Check: since the equilibrium constant is so large, the reaction goes essentially to completion, therefore, it is reasonable that the concentration of product is 0.10 M

35. **Given:** $[HC_2H_3O_2] = 0.210$M M, $[H_3O^+] = 0.0$, $[C_2H_3O_2^-] = 0.0$, $K_c = 1.8 \times 10^{-5}$
Find: $[HC_2H_3O2]$, $[H_2O^+]$, $[C_2H_3O_2^-]$ at equilibrium
Conceptual Plan: Prepare an ICE table; and then calculate Q; compare Q and K_c and predict the direction of the reaction; represent the change with x; sum the table and determine the equilibrium values; put the equilibrium values in the equilibrium expression and solve for x. Determine [CO], $[Cl_2]$, and $[COCl_2]$.
Solution:

	$HC_2H_3O_2(aq)$	$+H_2O(l)$	⇆ $H_3O^+(aq)$ +	$C_2H_3O_2^-(aq)$
	$[HC_2H_3O_2]$	$[H_2O]$	$[H_3O^+]$	$[C_2H_3O_2^-]$
I	0.210		0.0	0.0
C	- x		x	x

E	0.210–x	x	x

$$Q = \frac{[H_3O^+][C_2H_3O_2^-]}{[HC_2H_3O_2]} = \frac{0}{(0.210)} = 0$$ $Q < K$ therefore, the reaction will proceed to the right by x

$$K_c = \frac{[H_3O^+][C_2H_3O_2^-]}{[HC_2H_3O_2]} = \frac{(x)(x)}{(0.210 - x)} = 1.8 \times 10^{-5}$$

assume x is small compared to 0.210

$x^2 = 0.210(1.8 \times 10^{-5})$

$x = 0.00194$ check assumption: $\frac{0.00194}{0.210} \times 100 = 0.92$ %; assumption valid

$[H_3O^+] = [C_2H_3O_2^-] = 0.00194M$

$[HC_2H_3O_2] = 0.210 - 0.00194 = 0.20\underline{8}1 = 0.208M$

Check: Plug the values into the equilibrium expression: $K_c = \frac{(0.00194)(0.00194)}{(0.208)} = 1.81 \times 10^{-5}$; within 1 significant figure of true value, answers are valid.

37. **Given:** $P_{Br2} = 755$ torr, $P_{Cl2} = 735$ torr, $P_{BrCl} = 0.0$, $K_p = 1.11 \times 10^{-4}$ **Find:** P_{BrCl} at equilibrium
Conceptual Plan: Torr → atm and then: Prepare an ICE table; and then calculate Q; compare Q and K_c and predict the direction of the reaction; represent the change with x; sum the table and determine the equilibrium values; put the equilibrium values in the equilibrium expression and solve for x. Determine P_{BrCl}
Solution:

$$P_{Br_2} = 755 \text{ torr} \times \frac{1 \text{ atm}}{760 \text{ torr}} = 0.99\underline{3}4 \text{ atm} \qquad P_{Cl_2} = 735 \text{ torr} \times \frac{1 \text{ atm}}{760 \text{ torr}} = 0.96\underline{7}1 \text{ atm}$$

	Br_2 (g)	$+Cl_2(g)$	⇆ 2BrCl (g)
	P_{Br2}	P_{Cl2}	P_{BrCl}
I	0.9934	0.9671	0.0
C	- x	-x	2x
E	0.9934–x	0.9671– x	2x

$$Q = \frac{P^2_{BrCl}}{P_{Br_2}P_{Cl_2}} = \frac{0}{(0.9934)(0.9671)} = 0$$ $Q < K$ therefore, the reaction will proceed to the right by x

$$K_p = \frac{P^2_{BrCl}}{P_{Br_2}P_{Cl_2}} = \frac{(2x)^2}{(0.9934 - x)(0.9671 - x)} = 1.11 \times 10^{-4}$$

asusme x is small compared to 0.9934 and 0.9671

$$\frac{(2x)^2}{(0.9934)(0.9671)} = 1.11 \times 10^{-4} \qquad 4x^2 = 1.066 \times 10^{-4}$$

$x = 0.00516 = 3.92$ torr

$P_{BrCl} = 2x = 2(3.92 \text{ torr}) = 7.84$ torr

Check: Plug the values into the equilibrium expression:

$$K_c = \frac{(2(0.00516))^2}{(0.9934 - 0.00516)(0.99671 - 0.00516)} = 1.065 \times 10^{-4} = 1.12 \times 10^{-4};$$

within 1 significant figure of the true value, therefore, answers are valid.

39. a) **Given:** [A] = 1.0 M, [B] = [C] = 0.0, $K_c = 1.0$ **Find:** [A], [B], [C] at equilibrium
Conceptual Plan: Prepare an ICE table; and then calculate Q; compare Q and K_c and predict the direction of the reaction; represent the change with x; sum the table and determine the equilibrium values; put the equilibrium values in the equilibrium expression and solve for x. Determine [A], [B], and [C].
Solution:

	A (g)	⇆	B (g) +	C(g)
	[A]		[B]	[C]
I	1.0		0.0	0.0

	$A(g)$	$\leftrightarrows$	$B(g)$	$C(g)$
C	- x		x	x
E	1.0–x		x	x

$Q = \frac{[B][C]}{[A]} = \frac{0}{(1.0)} = 0$ $\quad Q < K$ therefore, the reaction will proceed to the right by x

$K_c = \frac{[B][C]}{[A]} = \frac{(x)(x)}{(1.0 - x)} = 1.0$

$x^2 = 1.0(1.0 - x)$

$x^2 + x - 1 = 0$

$\frac{-b \pm \sqrt{b^2 - 4ac}}{2a} \qquad \frac{-1 \pm \sqrt{1^2 - 4(1)(-1)}}{2(1)}$

x = 0.6$\underline{1}$8 or x = - 1.618, therefore, x = 0.6$\underline{1}$8

[B] = [C] = x = 0.6$\underline{1}$8 = 0.62M

[A] = 1.0 - 0.6$\underline{1}$8 = 0.382 = 0.38M

Check: Plug the values into the equilibrium expression:

$K_c = \frac{(0.62)(0.62)}{(0.38)} = 1.01 = 1.0$; which is the equilibrium constant so the values are correct.

b) **Given:** [A] = 1.0 M, [B] = [C] = 0.0, K_c = 0.010 **Find:** [A], [B], [C] at equilibrium
Conceptual Plan: Prepare an ICE table; and then calculate Q; compare Q and K_c and predict the direction of the reaction; represent the change with x; sum the table and determine the equilibrium values; put the equilibrium values in the equilibrium expression and solve for x. Determine [A], [B], and [C].
Solution:

	$A(g)$	$\leftrightarrows$	$B(g)$	$+ C(g)$
	[A]		[B]	[C]
I	1.0		0.0	0.0
C	- x		x	x
E	1.0–x		x	x

$Q = \frac{[B][C]}{[A]} = \frac{0}{(1.0)} = 0$ $\quad Q < K$ therefore, the reaction will proceed to the right by x

$K_c = \frac{[B][C]}{[A]} = \frac{(x)(x)}{(1.0 - x)} = 0.010$

$x^2 = 0.010(1.0 - x)$

$x^2 + 0.010x - 0.010 = 0$

$\frac{-b \pm \sqrt{b^2 - 4ac}}{2a} \qquad \frac{-(0.010) \pm \sqrt{(0.010)^2 - 4(1)(-0.010)}}{2(1)}$

x = 0.09$\underline{5}$12 or x = - 0.1015, therefore, x = 0.09$\underline{5}$12

[B] = [C] = x = 0.09$\underline{5}$12 = 0.095M

[A] = 1.0 - 0.09$\underline{5}$12 = 0.90488 = 0.90M

Check: Plug the values into the equilibrium expression:

$K_c = \frac{(0.095)(0.095)}{(0.90)} = 0.01002 = 0.010$; which is the equilibrium constant so the values are correct.

c) **Given:** [A] = 1.0 M, [B] = [C] = 0.0, $K_c = 1.0 \times 10^{-5}$ **Find:** [A], [B], [C] at equilibrium
Conceptual Plan: Prepare an ICE table; and then calculate Q; compare Q and K_c and predict the direction of the reaction; represent the change with x; sum the table and determine the equilibrium values; put the equilibrium values in the equilibrium expression and solve for x. Determine [A], [B], and [C].
Solution:

	$A(g)$	$\leftrightarrows$	$B(g)$	$+ C(g)$
	[A]		[B]	[C]
I	1.0		0.0	0.0

C	- x	x	x
E	1.0–x	x	x

$Q = \frac{[B][C]}{[A]} = \frac{0}{(1.0)} = 0$ $\quad Q < K$ therefore, the reaction will proceed to the right by x

$K_c = \frac{[B][C]}{[A]} = \frac{(x)(x)}{(1.0 - x)} = 1.0 \times 10^{-5}$

assume x is small compared to 1.0

$x^2 = 1.0(1.0 \times 10^{-5})$

$x = 0.00316$ check assumption: $\frac{0.00316}{1.0} \times 100 = 0.32\%$, assumption valid

$[B] = [C] = x = 0.00316 = 0.0032M$

$[A] = 1.0 - 0.00316 = 0.9968 = 1.0M$

Check: Plug the values into the equilibrium expression:

$K_c = \frac{(0.0032)(0.0032)}{(1.0)} = 1.024 \times 10^{-5} = 1.0 \times 10^{-5}$;

which is the equilibrium constant so the values are correct

41. **Given:** $CO(g) + Cl_2(g) \leftrightarrows COCl_2(g)$ at equilibrium **Find:** what is the effect of each of the following
    a) $COCl_2$ is added to the reaction mixture: Adding $COCl_2$ increases the concentration of $COCl_2$ and causes the reaction to shift to the left.

    b) Cl_2 is added to the reaction mixture: Adding Cl_2 increases the concentration of Cl_2 and causes the reaction to shift to the right.

    c) $COCl_2$ is removed from the reaction mixture: removing the $COCl_2$ decreases the concentration of $COCl_2$ and causes the reaction to shift to the right.

43. **Given:** $2KClO_3(s) \leftrightarrows 2KCl(s) + 3O_2(g)$ at equilibrium **Find:** what is the effect of each of the following
    a) O_2 is removed from the reaction mixture: Removing the O_2 decreases the concentration of O_2 and causes the reaction to shift to the right.

    b) KCl is added to the reaction mixture: Adding KCl does not cause any change in the reaction. KCl is a solid and the concentration remains constant so the addition of more solid does not change the equilibrium concentration.

    c) $KClO_3$ is added to the reaction mixture: Adding $KClO_3$ does not cause any change in the reaction. $KClO_3$ is a solid and the concentration remains constant so the addition of more solid does not change the equilibrium concentration.

    d) O_2 is added to the reaction mixture: Adding O_2 increases the concentration of O_2 and causes the reaction to shift to the left.

45. a) **Given:** $I_2(g) \leftrightarrows 2I(g)$ at equilibrium **Find:** the effect of increasing the volume
    The chemical equation has 2 moles of gas on the right and 1 mole of gas on the left. Increasing the volume of the reaction mixture decreases the pressure and causes the reaction to shift to the right (toward the side with more moles of gas particles).

    b) **Given:** $2H_2S(g) \leftrightarrows 2H_2(g) + S_2(g)$ **Find:** the effect of decreasing the volume
    The chemical equation has 3 moles of gas on the right and 2 moles of gas on the left. Decreasing the volume of the reaction mixture increases the pressure and causes the reaction to shift to the left (toward the side with fewer moles of gas particles).

    c) **Given:** $I_2(g) + Cl_2(g) \leftrightarrows 2ICl(g)$ **Find:** the effect of decreasing the volume
    The chemical equation has 2 moles of gas on the right and 2 moles of gas on the left. Decreasing the volume of the reaction mixture increases the pressure but causes no shift in the reaction because the moles are equal on both sides.

47. **Given:** $C(s) + CO_2(g) \leftrightarrows 2CO(g)$ is endothermic **Find:** the effect of increasing the temperature
Since the reaction is endothermic we can think of the heat as a reactant, increasing the temperature is equivalent to adding a reactant causing the reaction to shift to the right. This will cause an increase in the concentration of products and a decrease in the concentration of reactant, therefore, the value of K will increase.
Find: the effect of decreasing the temperature
Since the reaction is endothermic we can think of the heat as a reactant, decreasing the temperature is equivalent to removing a reactant causing the reaction to shift to the left. This will cause a decrease in the concentration of products and an increase in the concentration of reactants, therefore, the value of K will decrease.

49. **Given:** $C(s) + 2H_2(g) \leftrightarrows CH_4(g)$ is exothermic **Find:** determine which will favor CH_4
   a) adding more C to the reaction mixture: does NOT favor CH_4. Adding C does not cause any change in the reaction. C is a solid and the concentration remains constant so the addition of more solid does not change the equilibrium concentration.
   b) adding more H_2 to the reaction mixture: favors CH_4. Adding H_2 increases the concentration of H_2 causing the reaction to shift to the right.

   c) raising the temperature of the reaction mixture: does NOT favor CH_4. Since the reaction is exothermic we can think of heat as a product, raising the temperature is equivalent to adding a product causing the reaction to shift to the left.

   d) lowering the volume of the reaction mixture: favors CH_4. The chemical equation has 1 mole of gas on the right and 2 moles of gas on the left. Decreasing the volume of the reaction mixture increases the pressure and causes the reaction to shift to the right (toward the side with fewer moles of gas particles).

   e) adding a catalyst to the reaction mixture: does NOT favor CH_4. A catalyst added to the reaction mixture only speeds up the reaction, it does not change the equilibrium concentration.

   f) adding neon gas to the reaction mixture: does NOT favor CH_4. Adding an inert gas to a reaction mixture at a fixed volume has no effect on the equilibrium

51. a) To find the value of K for the new equation, combine the two given equations to yield the new equation. Reverse equation 1, and use $1/K_1$ and then add to equation 2. To find K for equation 3 use $(1/K_1)(K_2)$.

$HbO_2(aq)$	$\leftrightarrows$	~~Hb(aq)~~ + $O_2(aq)$	$K_1 = 1/1.8$
~~Hb(aq)~~ + $CO(aq)$	$\leftrightarrows$	$HbCO(aq)$	$K_2 = 306$
$HbO_2(aq) + CO(aq)$	$\leftrightarrows$	$HbCO(aq) + O_2(aq)$	$K_3 = K_1K_2 = (1/1.8)(306) = 170$

   b) **Given:** $O_2 = 20\%$, $CO = 0.10\%$ **Find:** The ratio $\frac{[HbCO]}{[HbO_2]}$

   Conceptual Plan: Determine the equilibrium expression and then determine $\frac{[HbCO]}{[HbO_2]}$

   Solution: $K = \frac{[HbCO][O_2]}{[HbO_2][CO]}$ $\quad 170 = \frac{[HbCO](20.0)}{[HbO_2](0.10)}$ $\quad \frac{[HbCO]}{[HbO_2]} = 170\left(\frac{0.10}{20.0}\right) = \frac{0.85}{1.0}$

   Since the ratio is almost 1:1, 0.10% CO will replace about 50% of the O_2 in the blood. The CO blocks the uptake of O_2 by the blood and is therefore highly toxic.

53. **Given:** $C_2H_4(g) + Cl_2(g) \leftrightarrows C_2H_4Cl_2(g)$ is exothermic
Find: Which of the following will maximize $C_2H_4Cl_2$
   a) increasing the reaction volume. Will not maximize $C_2H_4Cl_2$. The chemical equation has 1 mole of gas on the right and 2 moles of gas on the left. Increasing the volume of the reaction mixture decreases the pressure and causes the reaction to shift to the left (toward the side with more moles of gas particles).

   b) removing $C_2H_4Cl_2$ as it forms. Will maximize $C_2H_4Cl_2$. Removing the $C_2H_4Cl_2$ will decrease the concentration of $C_2H_4Cl_2$ and will cause the reaction to shift to the right, producing more $C_2H_4Cl_2$.

   c) lowering the reaction temperature. Will maximize $C_2H_4Cl_2$. The reaction is exothermic so we can think of heat as a product. Lowering the temperature will cause the reaction to shift to the right, producing more $C_2H_4Cl_2$.

d) adding Cl_2. Will maximize $C_2H_4Cl_2$. Adding Cl_2 increases the concentration of Cl_2 so the reaction shifts to the right, which will produce more $C_2H_4Cl_2$.

55. **Given:** Reaction 1 at equilibrium: $P_{H2} = 0.958$ atm; $P_{I2} = 0.877$ atm; $P_{HI} = 0.0200$ atm: reaction 2: $P_{H2} = P_{I2} = 0.621$ atm; $P_{HI} = 0.101$ atm **Find:** Is reaction 2 at equilibrium, if not, what is the P_{HI} at equilibrium.
Conceptual Plan: Use equilibrium partial pressures to determine K_p. Use K_p to determine if reaction 2 is at equilibrium. Prepare an ICE table; and then calculate Q; compare Q and K_p and predict the direction of the reaction; represent the change with x; sum the table and determine the equilibrium values; put the equilibrium values in the equilibrium expression and solve for x. Determine P_{HI}.
Solution: $H_2(g) + I_2(g) \leftrightarrows 2HI(g)$

	P_{H2}	P_{I2}	P_{HI}
Reaction 1:	0.958	0.877	0.020

$$K_p = \frac{P_{HI}^2}{P_{H_2}P_{I_2}} = \frac{(0.020)^2}{(0.958)(0.877)} = 4.7\underline{6}10 \times 10^{-4}$$

	$H_2(g)$ +	$I_2(g) \leftrightarrows$	$2HI(g)$
Reaction 2:	P_{H2}	P_{I2}	P_{HI}
I	0.621	0.621	0.101
C	x	x	-2x
E	0.621+x	0.621+x	0.101 -2x

$$Q = \frac{P_{HI}^2}{P_{H_2}P_{I_2}} = \frac{0.101^2}{(0.621)(0.621)} = 0.0264\text{: } Q > K \text{ so the reaction shifts to the left}$$

$$K_p = \frac{P_{HI}^2}{P_{H_2}P_{I_2}} = \frac{(0.101 - x)^2}{(0.621+x)(0.621+x)} = 4.7\underline{6}10 \times 10^{-4}$$

$$\sqrt{\frac{(0.101 - 2x)^2}{(0.621+x)(0.621+x)}} = \sqrt{4.7\underline{6}10 \times 10^{-4}}$$

$$\frac{(0.101 - 2x)}{(0.621+x)} = 2.1\underline{8}2 \times 10^{-2}$$

$x = 0.04325 = 0.0433$

$P_{H_2} = P_{I_2} = 0.621+x = 0.621 + 0.0433 = 0.664$ atm; $P_{HI} = 0.101 - 2x = 0.101 - 2(0.0433) = 0.0144$ atm

Check: Plug the values into the equilibrium expression:

$$K_p = \frac{(0.0144)^2}{(0.664)} = 4.703 \times 10^{-4} = 4.70 \times 10^{-4}$$; this value is close to the original equilibrium constant

57. **Given:** 200.0 L container; 1.27 kg N_2; 0.310 kg H_2; 725K; $K_p = 5.3 \times 10^{-5}$
Find: mass in g of NH_3 and % yield
Conceptual Plan:
$K_p \rightarrow K_c$ and then kg → g → mol → M and then prepare an ICE table; represent the change

$K_p = K_c(RT)^{\Delta n}$ $\quad \frac{1000 \text{ g}}{\text{kg}} \quad \frac{\text{g}}{\text{molar mass}} \quad \frac{\text{mol}}{\text{vol}}$

with x; sum the table and determine the equilibrium values; put the equilibrium values in the equilibrium expression and solve for x. Determine $[NH_3]$. Then M → mol → g and then determine

M x vol $\quad$ mol x molar mass

theoretical yield NH_3 → % yield

determine limiting reactant $\quad \frac{\text{actual yield}}{\text{theoretical yield}}$

Solution:

$$K_p = K_c(RT)^{\Delta n} \qquad K_c = \frac{K_p}{(RT)^{\Delta n}} = \frac{5.3 \times 10^{-5}}{\left((0.0821 \frac{\text{L}\cdot\text{atm}}{\text{mol}\cdot\text{K}})(725\text{K})\right)^{-2}} = 0.1\underline{8}77$$

$$n_{N_2} = 1.27\ \text{kg N}_2 \times \frac{1000\ \text{g}}{\text{kg}} \times \frac{1\ \text{mol N}_2}{28.00\ \text{g N}_2} = 45.\underline{3}57\ \text{mol N}_2 \qquad [N_2] = \frac{45.\underline{3}57\ \text{mol}}{200.0\ \text{L}} = 0.22\underline{6}78$$

$$n_{H_2} = 0.310\ \text{kg H}_2 \times \frac{1000\ \text{g}}{\text{kg}} \times \frac{1\ \text{mol H}_2}{2.016\ \text{g H}_2} = 15\underline{3}.77\ \text{mol H}_2 \qquad [H_2] = \frac{15\underline{3}.77\ \text{mol}}{200.0\ \text{L}} = 0.76\underline{8}85$$

$N_2(g) + 3H_2(g) \leftrightarrows 2NH_3(g)$

Reaction 1:	$[N_2]$	$[H_2]$	$[NH_3]$
I	0.22$\underline{6}$8	0.76$\underline{8}$9	0.0
C	-x	-3x	2x
E	0.2268-x	0.7689-3x	2x

Reaction shifts to the right

$$K_c = \frac{[NH_3]^2}{[N_2][H_2]^3} = \frac{(2x)^2}{(0.2268\text{-}x)(0.7689\text{-}3x)^3} = 0.1877$$

Assume x is small compared to 0.2268 and 3x is small compared to 0.7689

$$\frac{(2x)^2}{(0.2268)(0.7689)^3} = 0.1877 \qquad x = 0.06956$$

check assumptions: $\frac{0.06956}{0.2268} \times 100\% = 30.7$ not valid and $\frac{3(0.06956)}{0.7689} \times 100\% = 27.1\%$

Use method of successive substitution to solve for x. This yields

$x = 0.0461$

$[NH_3] = 2x = 2(0.0461) = 0.0922M$

Check: Plug the values into the equilibrium expression:

$$K_c = \frac{(0.0922)^2}{(0.2268 - 0.0461)(0.7689 - 3(0.0461))^3} = 0.1876\ ;$$

this value is close to the original equilibrium constant

Determine grams NH_3: $0.0922 \frac{\text{mol NH}_3}{\text{L}} \times \frac{200.0\ \text{L}}{} \times \frac{17.02\ \text{g NH}_3}{\text{mol NH}_3} = 3\underline{1}3.8\ \text{g} = 3.1 \times 10^2\ \text{g}$

Determine the theoretical yield: Determine the limiting reactant:

$$1.27\ \text{kg N}_2 \times \frac{1000\ \text{g}}{\text{kg}} \times \frac{1\ \text{mol N}_2}{28.0\ \text{g N}_2} \times \frac{2\ \text{mol NH}_3}{1\ \text{mol N}_2} \times \frac{17.02\ \text{g NH}_3}{\text{mol NH}_3} = 15\underline{4}4\ \text{g NH}_3$$

$$0.310\ \text{kg H}_2 \times \frac{1000\ \text{g}}{\text{kg}} \times \frac{1\ \text{mol H}_2}{2.016\ \text{g H}_2} \times \frac{2\ \text{mol NH}_3}{3\ \text{mol H}_2} \times \frac{17.02\ \text{g NH}_3}{\text{mol NH}_3} = 17\underline{4}5\ \text{g NH}_3$$

N_2 produces the least amount of NH_3 therefore, it is the limiting reactant and the theoretical yield is 1.54×10^3 g NH_3.

$$\%\ \text{yield} = \frac{3.1 \times 10^2\ \text{g}}{1.54 \times 10^3\ \text{g}} \times 100 = 20.\%$$

59. **Given:** At equilibrium: $P_{CO} = 0.30$ atm; $P_{Cl2} = 0.10$ atm; $P_{COCl2} = 0.60$ atm, add 0.40 atm Cl_2
Find: P_{CO} when system returns to equilibrium.
Conceptual Plan: Use equilibrium partial pressures to determine K_p. For the new conditions: prepare an ICE table; represent the change with x; sum the table and determine the equilibrium values; put the equilibrium values in the equilibrium expression and solve for x. Determine P_{CO}.
Solution: $CO(g) + Cl_2(g) \leftrightarrows COCl_2(g)$

Condition 1:	P_{CO}	P_{Cl2}	P_{COCl2}
	0.30	0.10	0.60

$$K_p = \frac{P_{COCl_2}}{P_{CO}P_{Cl_2}} = \frac{(0.60)}{(0.30)(0.10)} = 20.$$

$CO(g) + Cl_2(g) \leftrightarrows COCl_2(g)$

Condition 2:	P_{CO}	P_{Cl2}	P_{COCl2}
I	0.30	0.10+0.40	0.60
C	-x	-x	+x
E	0.30-x	0.50-x	0.60 + x

Reaction shifts to the right because the concentration of Cl_2 was increased

$$K_p = \frac{P_{COCl_2}}{P_{CO}P_{Cl_2}} = \frac{(0.60+x)}{(0.30-x)(0.50-x)} = 20$$

$$20x^2 - 17x + 2.4 = 0$$

$$\frac{-b \pm \sqrt{b^2 - 4ac}}{2a} = \frac{-(-17) \pm \sqrt{(-17)^2 - 4(20)(2.4)}}{2(20)}$$

x = 0.67 or 0.18 So, x = 0.18

$P_{CO} = 0.30 - 0.18 = 0.12$ atm; $P_{Cl_2} = 0.50 - 0.18 = 0.32$; $P_{COCl_2} = 0.60 + 0.18 = 0.78$ atm

Check: Plug the values into the equilibrium expression:

$$K_p = \frac{(0.78)}{(0.12)(0.32)} = 20.3 = 20.$$; this is the same as the original equilibrium constant.

61. **Given:** $K_p = 0.76$; P_{total} at equilibrium = 1.00 **Find:** $P_{initial}$ CCl_4
Conceptual Plan: Prepare an ICE table; represent the P(CCl_4) with A and the change with x; sum the table and determine the equilibrium values; use the total pressure and solve for A in terms of x. Determine partial pressure of each at equilibrium, use the equilibrium expression to determine x, and then determine A.

Solution: $CCl_4(g) \leftrightarrows C(s) + 2Cl_2(g)$

	P_{CCl4}	P_C	P_{Cl2}
I	A	constant	0.00
C	-x		+2x
E	A-x		2x

$P_{Total} = P_{CCl_4} + P_{Cl_2}$ 1.0 = A -x + 2x A = 1 - x

$P_{CCl_4} = (A - x) = (1-x) - x = 1 - 2x$; $P_{Cl_2} = (2x)$

$$K_p = \frac{P^2_{Cl_2}}{P_{CCl_4}} = \frac{(2x)^2}{(1-2x)} = 0.76$$

$4x^2 + 1.52x - 0.76 = 0$ x = 0.285 or -0.665 so x = $0.2\underline{8}5$

A = 1 - x = 1.0 - 0.285 = 0.715 = 0.72 atm

Check: Plug values into equilibrium expression:

$$K_p = \frac{P^2_{Cl_2}}{P_{CCl_4}} = \frac{(2x)^2}{(A-x)} = \frac{(2(0.285))^2}{(0.715-0.285)} = 0.755 = 0.76;$$ the original equilibrium constant

63. **Given:** V = 0.654 L, T = 1000 K, $K_p = 3.9 \times 10^{-2}$ **Find:** mass CaO as equilibrium
Conceptual Plan: $K_p \rightarrow P_{CO2} \rightarrow n(CO_2) \rightarrow n(CaO) \rightarrow g$
PV = nRT stoichiometry g = n(molar mass)

Solution: Since $CaCO_3$ and CaO are solids, they are not included in the equilibrium expression.

$$K_p = P_{CO_2} = 3.9 \times 10^{-2} \qquad n = \frac{PV}{RT} = \frac{(3.9 \times 10^{-2}\ \text{atm})(0.654\ \text{L})}{(0.0821\ \frac{\text{L} \cdot \text{atm}}{\text{mol} \cdot \text{K}})(1000\ \text{K})} = 3.\underline{1}06 \times 10^{-4}\ \text{mol } CO_2$$

$$3.\underline{1}06 \times 10^{-4}\ \text{mol } CO_2 \times \frac{1\ \text{mol CaO}}{1\ \text{mol } CO_2} \times \frac{56.1\ \text{g CaO}}{1\ \text{mol CaO}} = 0.0174\ \text{g} = 0.017\ \text{g CaO}$$

Check: The small value of *K* would give a small amount of products, so we would not expect to have a large mass of CaO formed.

65. a) **Given:** NO (P = 522 torr), O_2 (P = 421 torr); at equilibrium, P_{total} = 748 torr **Find:** K_p

Conceptual Plan: Prepare an ICE table; represent the change with x; sum the table and determine the equilibrium values; use the total pressure and solve for x. torr → atm → K_p

Solution: $2NO(g) + O_2(g) \leftrightarrows 2NO_2(g)$

	P_{NO}	P_{O2}	P_{NO2}
I	522 torr	421 torr	0.00
C	-2x	-x	+2x
E	522-2x	421-x	2x

$P_{Total} = P_{NO} + P_{O_2} + P_{NO_2}$ 748 = 522 -2x + (421 - x) + 2x

x = 195 torr P_{NO} = (522 - 2(195) = 132 torr; P_{O_2} = (421 - 195) = 226 torr; P_{NO_2} = 2(195) = 390 torr

$P_{NO} = 132 \text{ torr} \times \frac{1 \text{ atm}}{760 \text{ torr}} = 0.1737 \text{ atm}$; $P_{O_2} = 226 \text{ torr} \times \frac{1 \text{ atm}}{760 \text{ torr}} = 0.2974 \text{ atm}$;

$P_{NO_2} = 390 \text{ torr} \times \frac{1 \text{ atm}}{760 \text{ torr}} = 0.5132 \text{ atm}$

$$K_p = \frac{P^2_{NO_2}}{P^2_{NO}P_{O_2}} = \frac{(0.5132)^2}{(0.1737)^2(0.2974)} = 29.34 = 29.3$$

b) **Given:** NO (P = 255 torr), O_2 (P = 185 torr), K_p = 29.3 **Find:** equilibrium P_{NO2}

Conceptual Plan:

torr → atm and then prepare an ICE table; represent the change with x; sum the table and

$\frac{\text{atm}}{760 \text{ torr}}$

determine the equilibrium values; put the equilibrium values in the equilibrium expression and solve for x. Determine P N_2O_4.

Solution: $P_{NO} = 255 \text{ torr} \times \frac{1 \text{ atm}}{760 \text{ torr}} = 0.3355 \text{ atm}$ $P_{O_2} = 185 \text{ torr} \times \frac{1 \text{ atm}}{760 \text{ torr}} = 0.2434 \text{ atm}$

$2NO(g) + O_2(g) \leftrightarrows 2NO_2(g)$

	P_{NO}	P_{O2}	P_{NO2}
I	0.3355	0.2434	0.00
C	-2x	-x	+2x
E	0.3355-2x	0.2434-x	2x

$$K_p = \frac{P^2_{NO_2}}{P^2_{NO}P_{O_2}} = \frac{(2x)^2}{(0.3355 - 2x)^2(0.2434\text{-}x)} = 29.3$$

$-117.2x^3 + 63.847x^2 - 12.869x + 0.80272 = 0$

x = 0.1112 P_{NO_2} = 2x = 2(0.1112) = 0.2224 atm

$0.2224 \text{ atm} \times \frac{760 \text{ torr}}{1 \text{ atm}} = 169.1 \text{ torr} = 169 \text{ torr}$

Check: Plug the values into the equilibrium expression:

$K_p = \frac{(0.2224)^2}{(0.1131)^2(0.1322)} = 29.249 = 29.2$; this is within 1 significant figure of the original equilibrium constant.

67. **Given:** P_{NOCl} at equilibrium = 115 torr; K_p = 0.27, T = 700 K, **Find:** initial pressure NO, Cl_2

Conceptual Plan:

torr → atm and then prepare an ICE table; represent the change with x; sum the table and

$\frac{\text{atm}}{760 \text{ torr}}$

determine the equilibrium values; put the equilibrium values in the equilibrium expression and determine initial pressure.

Solution: $115 \text{ torr} \times \frac{1 \text{ atm}}{760 \text{ torr}} = 0.1513 \text{ atm}$

$2NO(g) + Cl_2(g) \leftrightarrows 2NOCl(g)$

	P_{NO}	P_{Cl2}	P_{NOCl}
I	A	A	0.00

Let A = initial pressure of NO and Cl_2

C	-2x	-x	+2x	
E	A-2x	A-x	0.151	2x = 0.151, so, x = 0.0756

So, at equilibrium: P_{NO} = A-0.151 P_{Cl2} = A-0.0756 P_{NOCl} = 0.151

$$K_p = \frac{P^2_{NOCL}}{P^2_{NO}P_{Cl_2}} = \frac{(0.151)^2}{(A - 0.151)^2(A - 0.0756)} = 0.27$$

$$0.27A^3 - 0.1019A^2 + 0.01231A - 0.022336 = 0$$

$$A = 0.565$$

$$P_{NO} = P_{Cl_2} = A = 0.565\text{atm} = 429 \text{ torr}$$

Check: Plug the values into the equilibrium expression:

$$K_p = \frac{P^2_{NOCL}}{P^2_{NO}P_{Cl_2}} = \frac{(0.151)^2}{(0.565 - 0.151)^2(0.565 - 0.0756)} \; 0.2\underline{7}2 \;;$$

this is within 1 digit of the significant figure of the original equilibrium constant

69. **Given:** P= 0.750 atm, density = 0.520 g/L, T = 337°C **Find:** K_c

Conceptual Plan: Prepare an ICE table; represent the P(CCl_4) with A and the change with x; sum the table and determine the equilibrium values; use the total pressure and solve for A in terms of x. Determine partial pressure of each at equilibrium in terms of x and then use the density to determine

$$d = \frac{PM}{RT}$$

the apparent molar mass and then use the mole fraction (in terms of P) and the molar mass of each

$$\chi_A = \frac{P_A}{P_{Total}}$$

gas to determine x.

Solution: $2NO_2(g) \leftrightarrows 2NO(g) + O_2(g)$

	P_{NO2}	P_{NO}	P_{O2}
I	A	0.00	0.00
C	-2x	+2x	+x
E	A-2x	2x	x

$P_{Total} = P_{NO_2} + P_{NO} + P_{O_2}$ 0.750 = A -2x + 2x + x A = 0.750 - x

$P_{NO_2} = (A - 2x) = (0.750\text{-}x) - 2x = (0.750 - 3x)$; $P_{NO} = (2x)$; $P_{O_2} = x$

$$d = \frac{PM}{RT} \quad M = \frac{dRT}{P} = \frac{(0.520\frac{g}{L})(0.0821\frac{L \cdot atm}{mol \cdot K})(610\ K)}{0.750\ atm} = 34.72 \text{ g/mol}$$

$$M = \chi_{NO_2}M_{NO_2} + \chi_{NO}M_{NO} + \chi_{O_2}M_{O_2} = \frac{P_{NO_2}}{P_{total}}M_{NO_2} + \frac{P_{NO}}{P_{total}}M_{NO} + \frac{P_{O_2}}{P_{total}}M_{O_2}$$

$$P_{Total}M = P_{NO_2}M_{NO_2} + P_{NO}M_{NO} + P_{O_2}M_{O_2}$$

$$(0.750)(34.7) = (0.750 - 3x)(46.0) + 2x(30.0) + x(32.0)$$

$$x = 0.184$$

$$K_p = \frac{P^2_{NO}P_{O_2}}{P^2_{NO_2}} = \frac{(2x)^2(x)}{(0.750\text{-}3x)} = \frac{(2(0.184))^2(0.184)}{(0.750\text{-}3(0.184))^2} = 0.63\underline{5}6$$

$$K_p = K_c(RT)^{\Delta n} \qquad 0.63\underline{5}6 = K_c((0.0821\frac{L \cdot atm}{mol \cdot K})(610K))^1$$

$$K_c = 1.27 \times 10^{-2}$$

71. Yes, the direction will depend on the volume. If the initial moles of A and B are equal, the initial concentrations of A and B are equal regardless of the volume. Since $K_c = \frac{[B]^2}{[A]} = 1$, if the volume is such that the [A] = [B] < 1.0, then $Q < K$ and the reaction goes to the right to reach equilibrium, but, if the volume is such that the [A] = [B] > 1.0, then $Q > K$ and the reaction goes to the left to reach equilibrium.

73. An examination of the data shows when $P_A = 1.0$ then $P_B = 1.0$, therefore, $K_p = \frac{P_B^b}{P_A^a} = \frac{(1.0)^b}{(1.0)^a} = 1.0$.

Therefore, the value of the numerator and denominator must be equal. We see from the data that $P_B = \sqrt{P_A}$, so $P_B^2 = P_A$. Since, the stoichiometric coefficients become exponents in the equilibrium expression, $a = 1$ and $b = 2$.

Chapter 15
Acids and Bases

1. a) acid $HNO_3(aq) \rightarrow H^+(aq) + NO_3^-(aq)$

   b) acid $NH_4^+(aq) \rightarrow H^+(aq) + NH_3(aq)$

   c) base $KOH(aq) \rightarrow K^+(aq) + OH^-(aq)$

   d) acid $HC_2H_3O_2(aq) \rightarrow H^+(aq) + C_2H_3O_2^-(aq)$

3. a) Since H_2CO_3 donates a proton to H_2O, it is the acid. After H_2CO_3 donates the proton, it becomes HCO_3^-, the conjugate base. Since H_2O accepts a proton, it is the base. After H_2O accepts the proton, it becomes H_3O^+, the conjugate acid.

   b) Since H_2O donates a proton to NH_3, it is the acid. After H_2O donates the proton, it becomes OH^-, the conjugate base. Since NH_3 accepts a proton, it is the base. After NH_3 accepts the proton, it becomes NH_4^+, the conjugate acid.

   c) Since HNO_3 donates a proton to H_2O, it is the acid. After HNO_3 donates the proton, it becomes NO_3^-, the conjugate base. Since H_2O accepts a proton, it is the base. After H_2O accepts the proton, it becomes H_3O^+, the conjugate acid.

   d) Since H_2O donates a proton to C_5H_5N, it is the acid. After H_2O donates the proton, it becomes OH^-, the conjugate base. Since C_5H_5N accepts a proton, it is the base. After C_5H_5N accepts the proton, it becomes $C_5H_5NH^+$, the conjugate acid.

5. a) Cl^- $HCl(aq) + H_2O(l) \rightarrow H_3O^+(aq) + Cl^-(aq)$

   b) HSO_3^- $H_2SO_3(aq) + H_2O(l) \leftrightharpoons H_3O^+(aq) + HSO_3^-(aq)$

   c) CHO_2^- $HCHO_2(aq) + H_2O(l) \leftrightharpoons H_3O^+(aq) + CHO_2^-(aq)$

   d) F^- $HF(aq) + H_2O(l) \leftrightharpoons H_3O^+(aq) + F^-(aq)$

7. $H_2PO_4^- + H_2O(l) \leftrightharpoons H_3O^+(aq) + HPO_4^{2-}(aq)$
   $H_2PO_4^- + H_2O(l) \leftrightharpoons H_3PO_4(aq) + OH^-(aq)$

9. a) HNO_3 is a strong acid

   b) HCl is a strong acid

   c) HBr is a strong acid

   d) H_2SO_3 is a weak acid $H_2SO_3(aq) + H_2O(l) \leftrightharpoons H_3O^+(aq) + HSO_3^-(aq)$

   $$K_a = \frac{[H_3O^+][HSO_3^-]}{[H_2SO_3]}$$

11. a) contains no HA, 10 H^+, and 10 A^-

    b) contains 3 HA, 3 H^+, and 7 A^-

    c) contains 9 HA, 1 H^+, and 1 A^-

    So, solution a > solution b > solution c

13. a) F^- is a stronger base than Cl^-.

F^- is the conjugate base of HF (a weak acid), Cl^- is the conjugate base of HCl (a strong acid), the weaker the acid, the stronger the conjugate base.

b) NO_2^- is a stronger base than NO_3^-.
NO_2^- is the conjugate base of HNO_2 (a weak acid), NO_3^- is the conjugate base of HNO_3 (a strong acid), the weaker the acid, the stronger the conjugate base.

c) ClO^- is a stronger base than F^-.
F^- is the conjugate base of HF ($K_a = 3.5 \times 10^{-4}$), ClO^- is the conjugate base of HClO ($K_a = 2.9 \times 10^{-8}$)
HClO is the weaker acid, the weaker the acid, the stronger the conjugate base.

15. a) **Given:** $K_w = 1.0 \times 10^{-14}$, $[H_3O^+] = 9.7 \times 10^{-9}$ M **Find:** $[OH^-]$
Conceptual Plan: $[H_3O^+] \rightarrow [OH^-]$
$K_w = 1 \times 10^{-14} = [H_3O^+][OH^-]$
Solution:
$K_w = 1 \times 10^{-14} = (9.7 \times 10^{-9})[OH^-]$
$[OH^-] = 1.0 \times 10^{-6}$ M
$[OH^-] > [H_3O^+]$ so the solution is basic

b) **Given:** $K_w = 1.0 \times 10^{-14}$, $[H_3O^+] = 2.2 \times 10^{-6}$ M **Find:** $[OH^-]$
Conceptual Plan: $[H_3O^+] \rightarrow [OH^-]$
$K_w = 1 \times 10^{-14} = [H_3O^+][OH^-]$
Solution:
$K_w = 1 \times 10^{-14} = (2.2 \times 10^{-6})[OH^-]$
$[OH^-] = 4.5 \times 10^{-9}$ M
$[H_3O^+] > [OH^-]$ so the solution is acidic

c) **Given:** $K_w = 1.0 \times 10^{-14}$, $[H_3O^+] = 1.2 \times 10^{-9}$ M **Find:** $[OH^-]$
Conceptual Plan: $[H_3O^+] \rightarrow [OH^-]$
$K_w = 1 \times 10^{-14} = [H_3O^+][OH^-]$
Solution:
$K_w = 1 \times 10^{-14} = (1.2 \times 10^{-9})[OH^-]$
$[OH^-] = 8.3 \times 10^{-6}$ M
$[OH^-] > [H_3O^+]$ so the solution is basic

17. a) **Given:** $[H_3O^+] = 1.7 \times 10^{-8}$ M **Find:** pH and pOH
Conceptual Plan: $[H_3O^+] \rightarrow$ **pH** $\rightarrow$ **pOH**
pH = -log$[H_3O^+]$ pH + pOH = 14
Solution: pH = -log(1.7×10^{-8}) = 7.77 pOH = 14.00 - 7.77 = 6.23
pH > 7 so the solution is basic.

b) **Given:** $[H_3O^+] = 1. \times 10^{-7}$ M **Find:** pH and pOH
Conceptual Plan: $[H_3O^+] \rightarrow$ **pH** $\rightarrow$ **pOH**
pH = -log$[H_3O^+]$ pH + pOH = 14
Solution: pH = -log(1.0×10^{-7}) = 7.00 pOH = 14.00 - 7.00 = 7.00
pH = 7 so the solution is neutral.

c) **Given:** $[H_3O^+] = 2.2 \times 10^{-6}$ M **Find:** pH and pOH
Conceptual Plan: $[H_3O^+] \rightarrow$ **pH** $\rightarrow$ **pOH**
pH = -log$[H_3O^+]$ pH + pOH = 14
Solution: pH = -log(2.2×10^{-6}) = 5.66 pOH = 14.00 - 5.66 = 8.34
pH < 7 so the solution is acidic.

19. pH = -log[H_3O^+] K_w = 1 x 10^{-14} = [H_3O^+][OH^-]

[H_3O^+]	[OH^-]	pH	Acidic or basic
7.1 x 10^{-4}	1.4 x 10^{-11}	**3.15**	acidic
3.7 x 10^{-9}	2.7 x 10^{-6}	8.43	basic
8 x 10^{-12}	1 x 10^{-3}	**11.1**	basic
6.2 x 10^{-4}	**1.6 x 10^{-11}**	3.20	acidic

$[H_3O^+] = 10^{-3.15} = 7.1 \times 10^{-4}$ $\qquad [OH^-] = \dfrac{1 \times 10^{-14}}{7.1 \times 10^{-4}} = 1.4 \times 10^{-11}$

$[OH^-] = \dfrac{1 \times 10^{-14}}{3.7 \times 10^{-9}} = 2.7 \times 10^{-6}$ $\qquad pH = -\log(3.7 \times 10^{-9}) = 8.43$

$[H_3O^+] = 10^{-11.1} = 8 \times 10^{-12}$ $\qquad [OH^-] = \dfrac{1 \times 10^{-14}}{8 \times 10^{-12}} = 1 \times 10^{-3}$

$[H_3O^+] = \dfrac{1 \times 10^{-14}}{1.6 \times 10^{-11}} = 6.2 \times 10^{-4}$ $\qquad pH = -\log(6.2 \times 10^{-4}) = 3.20$

21. **Given:** $K_w = 2.4 \times 10^{-14}$ at 37°C **Find:** [H_3O^+], pH
Conceptual Plan: K_w → [H_3O^+] → pH
$K_w = [H_3O^+][OH^-]$ pH = -log[H_3O^+]
Solution: $H_2O(l) + H_2O(l) \leftrightarrows H_3O^+(aq) + OH^-(aq)$
$K_w = [H_3O^+][OH^-]$
$[H_3O^+] = [OH^-] = \sqrt{K_w} = \sqrt{2.4 \times 10^{-14}} = 1.5 \times 10^{-7}$
$pH = -\log[H_3O^+] = -\log(1.5 \times 10^{-7}) = 6.81$
Check: The value of K_w increased indicating more products formed, so the [H_3O^+] increases and the pH decreases from the values at 25°C

23. a) **Given:** 0.15M HCl (strong acid) **Find:** [H_3O^+],[OH^-], pH
Conceptual Plan: [HCl] → [H_3O^+] → pH and then [H_3O^+] → [OH^-]
[HCl] → [H_3O^+] pH = -log[H_3O^+] [H_3O^+][OH^-] = 1 x 10^{-14}
Solution: 0.15 M HCl = 0.15 M H_3O^+ pH = −log(0.15) = 0.82
[OH^-] = 1 x 10^{-14} / 0.15 M = 6.67 x 10^{-14}
Check: HCl is a strong acid with a relatively high concentration, so we expect the pH to be low and the [OH^-] to be small.

b) **Given:** 0.025M HNO_3 (strong acid) **Find:** [H_3O^+],[OH^-], pH
Conceptual Plan: [HNO_3] → [H_3O^+] → pH and then [H_3O^+] → [OH^-]
[HNO_3] → [H_3O^+] pH = -log[H_3O^+] [H_3O^+][OH^-] = 1 x 10^{-14}
Solution: 0.025 M HNO_3 = 0.025 M H_3O^+ pH = −log(0.025) = 1.60
[OH^-] = 1.0 x 10^{-14} / 0.025 M = 4.0 x 10^{-13}
Check: HNO_3 is a strong acid, so we expect the pH to be low and the [OH^-] to be small.

c) **Given:** 0.072M HBr and 0.015M HNO_3 (strong acids) **Find:** [H_3O^+],[OH^-], pH
Conceptual Plan: [HBr]+ [HNO_3] → [H_3O^+] → pH and then [H_3O^+] → [OH^-]
[HBr] + [HNO_3] → [H_3O^+] pH = -log[H_3O^+] [H_3O^+][OH^-] = 1 x 10^{-14}
Solution: 0.072 M HBr = 0.072 M H_3O^+ and 0.015M HNO_3 = 0.015M H_3O^+
Total H_3O^+ = 0.072M + 0.015M = 0.087M pH = −log(0.087) = 1.06
[OH^-] = 1.0 x 10^{-14} / 0.087 M = 1.1 x 10^{-13}
Check: HBr and HNO_3 are both strong acids and completely dissociate, this gives a relatively high concentration, so we expect the pH to be low and the [OH^-] to be small.

d) **Given:** HNO_3 = 0.855% by mass, $d_{solution}$ = 1.01 g/mL **Find:** [H_3O^+],[OH^-], pH
Conceptual Plan:
% mass HNO_3 → g HNO_3 → mol HNO_3 and then g soln → mL soln → L soln → M HNO_3
$\dfrac{\%}{100}$ $\quad \dfrac{\text{mol HNO}_3}{63.018 \text{ g HNO}_3}$ $\quad \dfrac{1.01 \text{ g soln}}{\text{mL soln}}$ $\quad \dfrac{1000 \text{ mL soln}}{\text{L soln}}$ $\quad \dfrac{\text{mol HNO}_3}{\text{L soln}}$
→ M H_3O^+ → pH and then [H_3O^+] → [OH^-]
[HNO_3] → [H_3O^+] pH = -log[H_3O^+] [H_3O^+][OH^-] = 1 x 10^{-14}

Solution: $\dfrac{0.855 \text{ g HNO}_3}{100 \text{ g soln}}$ x $\dfrac{1 \text{ mol HNO}_3}{63.018 \text{ g HNO}_3}$ x $\dfrac{1.01 \text{ g soln}}{\text{mL soln}}$ x $\dfrac{1000 \text{ mL soln}}{\text{L soln}}$ = 0.137 M HNO$_3$

0.137 M HNO$_3$ = 0.137 M H$_3$O$^+$ pH = $-\log(0.137)$ = 0.863

[OH$^-$] = 1.00 x 10^{-14} / 0.137 M = 7.30 x 10^{-14}

Check: HNO$_3$ is a strong acid and completely dissociates, this gives a relatively high concentration, so we expect the pH to be low and the [OH$^-$] to be small.

25. a) **Given:** pH = 1.25, 0.250 L   **Find:** g HI

   Conceptual Plan: pH → [H$_3$O$^+$] → [HI] → mol HI → g HI
   pH = -log[H$_3$O$^+$]   [H$_3$O$^+$]→ [HI] mol = MV g = mol(127.9 g/mol)

   Solution: [H$_3$O$^+$] = 10$^{-1.25}$ = 0.056M = [HI] $\dfrac{0.056 \text{ mol HI}}{\text{L}}$ x 0.250 L x $\dfrac{127.90 \text{ g HI}}{\text{mol HI}}$ = 1.8 g HI

   b) **Given:** pH = 1.75, 0.250 L   **Find:** g HI

   Conceptual Plan: pH → [H$_3$O$^+$] → [HI] → mol HI → g HI
   pH = -log[H$_3$O$^+$]   [H$_3$O$^+$]→ [HI]   mol = MV   g = mol(127.9 g/mol)

   Solution [H$_3$O$^+$] = 10$^{-1.75}$ = 0.0178M = [HI] $\dfrac{0.0178 \text{ mol HI}}{\text{L}}$ x 0.250 L x $\dfrac{127.90 \text{ g HI}}{\text{mol HI}}$ = 0.57 g HI

   c) **Given:** pH = 2.85, 0.250 L   **Find:** g HI

   Conceptual Plan: pH → [H$_3$O$^+$] → [HI] → mol HI → g HI
   pH = -log[H$_3$O$^+$]   [H$_3$O$^+$]→ [HI]   mol = MV   g = mol(127.9 g/mol)

   Solution [H$_3$O$^+$] = 10$^{-2.85}$ = 0.0014M = [HI] $\dfrac{0.0014 \text{ mol HI}}{\text{L}}$ x 0.250 L x $\dfrac{127.90 \text{ g HI}}{\text{mol HI}}$ = 0.045 g HI

27. **Given:** 0.100M benzoic acid. K_a = 6.5 x 10^{-5} **Find:** [H$_3$O$^+$], pH

   Conceptual Plan: Write a balanced reaction. Prepare an ICE table; represent the change with x; sum the table and determine the equilibrium values; put the equilibrium values in the equilibrium expression and solve for x. Determine [H$_3$O$^+$] and pH

   Solution: HC$_7$H$_5$O$_2$(aq) + H$_2$O(l) ⇆ H$_3$O$^+$(aq) + C$_7$H$_5$O$_2$$^-$ (aq)

I	0.100M	0.0	0.0
C	- x	x	x
E	0.100 – x	x	x

   $K_a = \dfrac{[\text{H}_3\text{O}^+][\text{C}_7\text{H}_5\text{O}_2^-]}{[\text{HC}_7\text{H}_5\text{O}_2]} = \dfrac{(x)(x)}{(0.100 - x)}$ = 6.5 x 10^{-5}

   Assume x is small compared to 0.100

   x^2 = (6.5 x 10^{-5})(0.100)   x = 2.5 x 10^{-3} M = [H$_3$O$^+$]

   check assumption: $\dfrac{2.5 \times 10^{-3}}{0.100}$ x 100% = 2.5%; assumption valid

   pH = -log(2.5 x 10^{-3}) = 2.60

29. a) **Given:** 0.500M HNO$_2$. K_a = 4.6 x 10^{-4} **Find:** pH

   Conceptual Plan: Write a balanced reaction. Prepare an ICE table; represent the change with x; sum the table and determine the equilibrium values; put the equilibrium values in the equilibrium expression and solve for x. Determine [H$_3$O$^+$] and pH

   Solution: HNO$_2$(aq) + H$_2$O(l) ⇆ H$_3$O$^+$(aq) + NO$_2$$^-$ (aq)

I	0.500M	0.0	0.0
C	- x	x	x
E	0.500 – x	x	x

   $K_a = \dfrac{[\text{H}_3\text{O}^+][\text{NO}_2^-]}{[\text{HNO}_2]} = \dfrac{(x)(x)}{(0.500 - x)}$ = 4.6 x 10^{-4}

   Assume x is small compared to 0.500

   x^2 = (4.6 x 10^{-4})(0.500)   x = 0.015M = [H$_3$O$^+$]

Check assumption: $\frac{0.015}{0.500}$ x 100% =3.0% assumption valid

pH = -log(0.015) = 1.82

b) **Given:** 0.100M HNO_2. $K_a = 4.6 \times 10^{-4}$ **Find:** pH
Conceptual Plan: Write a balanced reaction. Prepare an ICE table; represent the change with x; sum the table and determine the equilibrium values; put the equilibrium values in the equilibrium expression and solve for x. Determine $[H_3O^+]$ and pH
Solution: $HNO_2(aq) + H_2O(l) \leftrightarrows H_3O^+(aq) + NO_2^-(aq)$

I	0.100M	0.0	0.0
C	- x	x	x
E	0.100 – x	x	x

$$K_a = \frac{[H_3O^+][NO_2^-]}{[HNO_2]} = \frac{(x)(x)}{(0.100 - x)} = 4.6 \times 10^{-4}$$

Assume x is small compared to 0.100

$x^2 = (4.6 \times 10^{-4})(0.100)$ $x = 0.0068M = [H_3O^+]$

Check assumption: $\frac{0.0068}{0.100}$ x 100% =6.8% assumption not valid

$x^2 = (4.6 \times 10^{-4})(0.100 - x)$ $x^2 + 4.6 \times 10^{-4}x - 4.6 \times 10^{-5} = 0$

x = 0.00656

pH = -log(0.00656) = 2.18

c) **Given:** 0.100M HNO_2. $K_a = 4.6 \times 10^{-4}$ **Find:** pH
Conceptual Plan: Write a balanced reaction. Prepare an ICE table; represent the change with x; sum the table and determine the equilibrium values; put the equilibrium values in the equilibrium expression and solve for x. Determine $[H_3O^+]$ and pH
Solution: $HNO_2(aq) + H_2O(l) \leftrightarrows H_3O^+(aq) + NO_2^-(aq)$

I	0.0100M	0.0	0.0
C	- x	x	x
E	0.0100 – x	x	x

$$K_a = \frac{[H_3O^+][NO_2^-]}{[HNO_2]} = \frac{(x)(x)}{(0.0100 - x)} = 4.6 \times 10^{-4}$$

Assume x is small compared to 0.100

$x^2 = (4.6 \times 10^{-4})(0.0100)$ $x = 0.0021M = [H_3O^+]$

Check assumption: $\frac{0.0021}{0.0100}$ x 100% =21% assumption not valid

$x^2 = (4.6 \times 10^{-4})(0.0100 - x)$ $x^2 + 4.6 \times 10^{-4}x - 4.6 \times 10^{-6} = 0$

x = 0.0019

pH = -log(0.0019) = 2.72

31. **Given:** 15.0 mL glacial acetic, d = 1.05 g/mL, dilute to 1.50 L, $K_a = 1.8 \times 10^{-5}$ **Find:** pH
Conceptual Plan: mL acetic acid → g acetic acid → mol acetic acid → M and then write a balanced reaction. XXX $\frac{1.05 \text{ g}}{\text{mL}}$ XXX $\frac{\text{mol acetic acid}}{60.05 \text{ g}}$ XXX $M = \frac{\text{mol}}{\text{L}}$
Prepare an ICE table; represent the change with x; sum the table and determine the equilibrium values; put the equilibrium values in the equilibrium expression and solve for x. Determine $[H_3O^+]$ and pH

Solution: $15.0 \text{ mL} \times \frac{1.05 \text{ g}}{\text{mL}} \times \frac{1 \text{ mol}}{60.05 \text{ g}} \times \frac{1}{1.50 \text{ L}} = 0.17\underline{4}8 \text{ M}$

$HC_2H_3O_2(aq) + H_2O(l) \leftrightarrows H_3O^+(aq) + C_2H_3O_2^-(aq)$

I	0.1748M	0.0	0.0
C	- x	x	x
E	0.1748 – x	x	x

$$K_a = \frac{[H_3O^+][C_2H_3O_2^-]}{[HC_2H_3O_2]} = \frac{(x)(x)}{(0.1748 - x)} = 1.8 \times 10^{-5}$$

Assume x is small compared to 0.1748

$x^2 = (1.8 \times 10^{-5})(0.1748)$ $\quad x = 0.00177M = [H_3O^+]$

Check assumption: $\frac{0.00177}{0.1748} \times 1005 = 1.0\%$ assumption valid

$pH = -\log(0.00177) = 2.75$

33. **Given:** 0.185 M HA, pH = 2.95 **Find:** K_a

Conceptual Plan: pH → $[H_3O^+]$ and then write a balanced reaction, prepare an ICE table, calculate equilibrium concentrations, and then plug into the equilibrium expression to solve for K_a.

Solution: $[H_3O^+] = 10^{-2.95} = 0.0012\ M = [A^-]$

	$HA(aq) + H_2O(l) \leftrightharpoons$	$H_3O^+(aq)$ +	$A^-(aq)$
I	0.185M	0.0	0.0
C	- x	x	x
E	0.185 – 0.0012	0.0012	0.0012

$$K_a = \frac{[H_3O^+][A^-]}{[HA]} = \frac{(0.0012)(0.0012)}{(0.185 - 0.0012)} = 7.8 \times 10^{-6}$$

35. **Given:** 0.125 HCN $K_a = 4.9 \times 10^{-10}$ **Find:** % ionization

Conceptual Plan: Write a balanced reaction. Prepare an ICE table; represent the change with x; sum the table and determine the equilibrium values; put the equilibrium values in the equilibrium expression and solve for x and then x → % ionization

$$\% \text{ ionization} = \frac{x}{[HCN]_{original}} \times 100$$

Solution:	$HCN(aq) + H_2O(l) \leftrightharpoons$	$H_3O^+(aq)$ +	$CN^-(aq)$
I	0.125M	0.0	0.0
C	- x	x	x
E	0.125 – x	x	x

$$K_a = \frac{[H_3O^+][CN^-]}{[HCN]} = \frac{(x)(x)}{(0.125 - x)} = 4.9 \times 10^{-10}$$

Assume x is small compared to 0.125

$x^2 = (4.9 \times 10^{-10})(0.125)$ $\quad x = 7.83 \times 10^{-6}$

$$\% \text{ ionization} = \frac{7.83 \times 10^{-6}}{0.125} \times 100 = 0.0063\% \text{ ionized}$$

37. a) **Given:** 1.00M $HC_2H_3O_2$ $K_a = 1.8 \times 10^{-5}$ **Find:** % ionization

Conceptual Plan: Write a balanced reaction. Prepare an ICE table; represent the change with x; sum the table and determine the equilibrium values; put the equilibrium values in the equilibrium expression and solve for x and then x → % ionization

$$\% \text{ ionization} = \frac{x}{[HC_2H_3O_2]_{original}} \times 100$$

Solution: $HC_2H_3O_2(aq) + H_2O(l) \leftrightharpoons H_3O^+(aq) + C_2H_3O_2^-(aq)$

		H_3O^+	$C_2H_3O_2^-$
I	1.0 M	0.0	0.0
C	- x	x	x
E	1.00 – x	x	x

$$K_a = \frac{[H_3O^+][C_2H_3O_2^-]}{[HC_2H_3O_2]} = \frac{(x)(x)}{(1.00 - x)} = 1.8 \times 10^{-5}$$

Assume x is small compared to1.00

$x^2 = (1.8 \times 10^{-5})(1.00)$ $\quad x = 0.00424$

$$\% \text{ ionization} = \frac{0.00424}{1.00} \times 100 = 0.42\% \text{ ionized}$$

b) **Given:** 0.500 M $HC_2H_3O_2$ $K_a = 1.8 \times 10^{-5}$ **Find:** % ionization
Conceptual Plan: Write a balanced reaction. Prepare an ICE table; represent the change with x; sum the table and determine the equilibrium values; put the equilibrium values in the equilibrium expression and solve for x and then x → % ionization

$$\% \text{ ionization} = \frac{x}{[HC_2H_3O_2]_{original}} \times 100$$

Solution: $HC_2H_3O_2 \text{ (aq)} + H_2O\text{(l)} \leftrightharpoons H_3O^+\text{(aq)} + C_2H_3O_2^- \text{ (aq)}$

I	0.500 M	0.0	0.0
C	- x	x	x
E	0.500 – x	x	x

$$K_a = \frac{[H_3O^+][C_2H_3O_2^-]}{[HC_2H_3O_2]} = \frac{(x)(x)}{(0.500 - x)} = 1.8 \times 10^{-5}$$

Assume x is small compared to 0.500

$$x^2 = (1.8 \times 10^{-5})(0.500) \qquad x = 0.00300$$

$$\% \text{ ionization} = \frac{0.00300}{0.500} \times 100 = 0.60\% \text{ ionized}$$

c) **Given:** 0.100M $HC_2H_3O_2$ $K_a = 1.8 \times 10^{-5}$ **Find:** % ionization
Conceptual Plan: Write a balanced reaction. Prepare an ICE table; represent the change with x; sum the table and determine the equilibrium values; put the equilibrium values in the equilibrium expression and solve for x and then x → % ionization

$$\% \text{ ionization} = \frac{x}{[HC_2H_3O_2]_{original}} \times 100$$

Solution: $HC_2H_3O_2 \text{ (aq)} + H_2O\text{(l)} \leftrightharpoons H_3O^+\text{(aq)} + C_2H_3O_2^- \text{ (aq)}$

I	0.100 M	0.0	0.0
C	- x	x	x
E	0.100 – x	x	x

$$K_a = \frac{[H_3O^+][C_2H_3O_2^-]}{[HC_2H_3O_2]} = \frac{(x)(x)}{(0.100 - x)} = 1.8 \times 10^{-5}$$

Assume x is small compared to 1.00

$$x^2 = (1.8 \times 10^{-5})(0.100) \qquad x = 0.00134$$

$$\% \text{ ionization} = \frac{0.00134}{0.100} \times 100 = 1.3\% \text{ ionized}$$

d) **Given:** 0.0500M $HC_2H_3O_2$ $K_a = 1.8 \times 10^{-5}$ **Find:** % ionization
Conceptual Plan: Write a balanced reaction. Prepare an ICE table; represent the change with x; sum the table and determine the equilibrium values; put the equilibrium values in the equilibrium expression and solve for x and then x → % ionization

$$\% \text{ ionization} = \frac{x}{[HC_2H_3O_2]_{original}} \times 100$$

Solution: $HC_2H_3O_2 \text{ (aq)} + H_2O\text{(l)} \leftrightharpoons \qquad H_3O^+\text{(aq)} + \qquad C_2H_3O_2^- \text{ (aq)}$

I	0.0500 M	0.0	0.0
C	- x	x	x
E	0.0500 – x	x	x

$$K_a = \frac{[H_3O^+][C_2H_3O_2^-]}{[HC_2H_3O_2]} = \frac{(x)(x)}{(0.0500 - x)} = 1.8 \times 10^{-5}$$

Assume x is small compared to 0.0500

$$x^2 = (1.8 \times 10^{-5})(0.0500) \qquad x = 9.49 \times 10^{-4}$$

$$\% \text{ ionization} = \frac{9.49 \times 10^{-4}}{0.0500} \times 100 = 1.9\% \text{ ionized}$$

39. **Given:** 0.148 M HA 1.55% dissociation **Find:** K_a
Conceptual Plan: M → $[H_3O^+]$ → K_a and then write a balanced reaction, determine equilibrium concentration and plug into the equilibrium expression.

Solution: $(0.148\text{ M HA})(0.0155) = 0.002294\ [H_3O^+] = [A^-]$

	$HA(aq) + H_2O(l) \leftrightarrows$	$H_3O^+(aq) +$	$A^-(aq)$
I	0. 148M	0.0	0.0
C	- x	x	x
E	0.148 – 0.002294	0.002294	0.002294

$$K_a = \frac{[H_3O^+][A^-]}{[HA]} = \frac{(0.002294)(0.002294)}{(0.148 - 0.002294)} = 3.61 \times 10^{-5}$$

41. a) **Given:** 0.250 M HF $K_a = 3.5 \times 10^{-4}$ **Find:** pH, % dissociation

Conceptual Plan: Write a balanced reaction. Prepare an ICE table; represent the change with x; sum the table and determine the equilibrium values; put the equilibrium values in the equilibrium expression and solve for x and then x → % ionization

$$\% \text{ ionization} = \frac{x}{[HF]_{original}} \times 100$$

Solution:

	$HF(aq) + H_2O(l) \leftrightarrows$	$H_3O^+(aq) +$	$F^-(aq)$
I	0.250M	0.0	0.0
C	- x	x	x
E	0.250 – x	x	x

$$K_a = \frac{[H_3O^+][F^-]}{[HF]} = \frac{(x)(x)}{(0.250 - x)} = 3.5 \times 10^{-4}$$

Assume x is small compared to 0.250

$x^2 = (3.5 \times 10^{-4})(0.250)$ $x = 0.00935M = [H_3O^+]$

$$\frac{0.00935}{0.250} \times 100\% = 3.7\%$$

$pH = -\log(0.00935) = 2.03$

b) **Given:** 0.100 M HF $K_a = 3.5 \times 10^{-4}$ **Find:** pH, % dissociation

Conceptual Plan: Write a balanced reaction. Prepare an ICE table; represent the change with x; sum the table and determine the equilibrium values; put the equilibrium values in the equilibrium expression and solve for x and then x → % ionization

$$\% \text{ ionization} = \frac{x}{[HF]_{original}} \times 100$$

Solution:

	$HF(aq) + H_2O(l) \leftrightarrows$	$H_3O^+(aq) +$	$F^-(aq)$
I	0.1000M	0.0	0.0
C	- x	x	x
E	0.100 – x	x	x

$$K_a = \frac{[H_3O^+][F^-]}{[HF]} = \frac{(x)(x)}{(0.100 - x)} = 3.5 \times 10^{-4}$$

Assume x is small compared to 0.100

$x^2 = (3.5 \times 10^{-4})(0.100)$ $x = 0.00592M = [H_3O^+]$

$$\frac{0.00592}{0.100} \times 100\% = 5.9\%$$

x > 5.0 % therefore, assumption invalid

$x^2 + 3.5 \times 10^{-4}x - 3.5 \times 10^{-5} = 0$

x = 0.00574 or -0.00609

$pH = -\log(0.00574) = 2.24$

$$\% \text{ dissociation} = \frac{0.00574}{0.100} \times 100 = 5.7\%$$

c) **Given:** 0.050 M HF $K_a = 3.5 \times 10^{-4}$ **Find:** pH, % dissociation
Conceptual Plan: Write a balanced reaction. Prepare an ICE table; represent the change with x; sum the table and determine the equilibrium values; put the equilibrium values in the equilibrium expression and solve for x and then x → % ionization

$$\% \text{ ionization} = \frac{x}{[HF]_{original}} \times 100$$

Solution: $HF(aq) + H_2O(l) \leftrightarrows H_3O^+(aq) + F^-(aq)$

	HF	H_3O^+	F^-
I	0.050M	0.0	0.0
C	- x	x	x
E	0.050 – x	x	x

$$K_a = \frac{[H_3O^+][F^-]}{[HF]} = \frac{(x)(x)}{(0.050 - x)} = 3.5 \times 10^{-4}$$

Assume x is small compared to 0.050

$x^2 = (3.5 \times 10^{-4})(0.050)$ $x = 0.00418M = [H_3O^+]$

$$\frac{0.00418}{0.050} \times 100\% = 8.4\%$$

Assumption invalid.

$x^2 + 3.5 \times 10^{-4}x - 1.75 \times 10^{-5} = 0$

$x = 0.00401$ or -0.00436

$pH = -\log(0.00401) = 2.40$

$$\% \text{ dissociation} = \frac{0.00401}{0.050} \times 100\% = 8.0\%$$

43. $H_3PO_4(aq) + H_2O(l) \leftrightarrows H_3O^+(aq) + H_2PO_4^-(aq)$ $K_{a1} = \frac{[H_3O^+][H_2PO_4^-]}{[H_3PO_4]}$

$H_2PO_4^-(aq) + H_2O(l) \leftrightarrows H_3O^+(aq) + HPO_4^{2-}(aq)$ $K_{a2} = \frac{[H_3O^+][HPO_4^{2-}]}{[H_2PO_4^-]}$

$HPO_4^{2-}(aq) + H_2O(l) \leftrightarrows H_3O^+(aq) + PO_4^{3-}(aq)$ $K_{a3} = \frac{[H_3O^+][PO_4^{3-}]}{[HPO_4^{2-}]}$

45. a) **Given:** 0.350 M H_3PO_4 $K_{a1} = 7.5 \times 10^{-3}$, $K_{a2} = 6.2 \times 10^{-8}$ **Find:** $[H_3O^+]$, pH
Conceptual Plan: K_{a1} much larger than K_{a2}, so use K_{a1} to calculate $[H_3O^+]$. Write a balanced reaction. Prepare an ICE table; represent the change with x; sum the table and determine the equilibrium values; put the equilibrium values in the equilibrium expression and solve for x.
Solution: $H_3PO_4(aq) + H_2O(l) \leftrightarrows H_3O^+(aq) + H_2PO_4^-(aq)$

	H_3PO_4	H_3O^+	$H_2PO_4^-$
I	0.350M	0.0	0.0
C	- x	x	x
E	0.350 – x	x	x

$$K_a = \frac{[H_3O^+][H_2PO_4^-]}{[H_3PO_4]} = \frac{(x)(x)}{(0.350 - x)} = 7.3 \times 10^{-3}$$

Assume x is small compared to 0.350

$x^2 = (7.3 \times 10^{-3})(0.350)$ $x = 0.0512M = [H_3O^+]$

Check assumption: $\frac{0.0512}{0.350} \times 100\% = 14.6\%$ assumption not valid

$x^2 + 7.5 \times 10^{-3}\ x - 0.002625 = 0$ $x = 0.04762 = [H_3O^+]$

$pH = -\log(0.04762) = 1.32$

b) **Given:** 0.350 M $H_2C_2O_4$ $K_{a1} = 6.0 \times 10^{-2}$, $K_{a2} = 6.0 \times 10^{-5}$ **Find:** $[H_3O^+]$, pH
Conceptual Plan: K_{a1} much larger than K_{a2}, so use K_{a1} to calculate $[H_3O^+]$. Write a balanced reaction. Prepare an ICE table; represent the change with x; sum the table and determine the equilibrium values; put the equilibrium values in the equilibrium expression and solve for x.

Solution: $H_2C_2O_4(aq) + H_2O(l) \leftrightarrows H_3O^+(aq) + HC_2O_4^-(aq)$

I	0.350M	0.0	0.0
C	- x	x	x
E	0.350 – x	x	x

$$K_a = \frac{[H_3O^+][HC_2O_4^-]}{[H_2C_2O_4]} = \frac{(x)(x)}{(0.350 - x)} = 6.0 \times 10^{-2}$$

$x^2 + 6.0 \times 10^{-2}\ x - 0.021 = 0$ $\qquad x = 0.1\underline{1}79 = 0.12\text{ M }[H_3O^+]$

$pH = -\log(0.1\underline{1}79) = 0.93$

47. a) **Given:** 0.15 M NaOH **Find:** $[OH^-]$, $[H_3O^+]$, pH, pOH

Conceptual Plan: [NaOH] → $[OH^-]$ → $[H_3O^+]$ → pH → pOH

$K_w = [H_3O^+][OH^-]$ $\quad pH = -\log[H_3O^+]$ $\quad pH + pOH = 14$

Solution: $[OH^-] = [NaOH] = 0.15M$

$$[H_3O^+] = \frac{K_w}{[OH^-]} = \frac{1 \times 10^{-14}}{0.15\text{ M}} = 6.7 \times 10^{-14}\text{M}$$

$pH = -\log(6.7 \times 10^{-14}) = 13.18$

$pOH = 14.00 - 13.18 = 0.82$

b) **Given:** 1.5×10^{-3} M $Ca(OH)_2$ **Find:** $[OH^-]$, $[H_3O^+]$, pH, pOH

Conceptual Plan: $[Ca(OH)_2]$ → $[OH^-]$ → $[H_3O^+]$ → pH → pOH

$K_w = [H_3O^+][OH^-]$ $\quad pH = -\log[H_3O^+]$ $\quad pH + pOH = 14$

Solution: $[OH^-] = 2[Ca(OH)_2] = 2(1.5 \times 10^{-3}) = 0.0030$ M

$$[H_3O^+] = \frac{K_w}{[OH^-]} = \frac{1 \times 10^{-14}}{0.0030\text{ M}} = 3.\underline{3}3 \times 10^{-12}\text{M}$$

$pH = -\log(3.\underline{3}3 \times 10^{-12}) = 11.48$

$pOH = 14.00 - 11.48 = 2.52$

c) **Given:** 4.8×10^{-4} M $Sr(OH)_2$ **Find:** $[OH^-]$, $[H_3O^+]$, pH, pOH

Conceptual Plan: $[Sr(OH)_2]$ → $[OH^-]$ → $[H_3O^+]$ → pH → pOH

$K_w = [H_3O^+][OH^-]$ $\quad pH = -\log[H_3O^+]$ $\quad pH + pOH = 14$

Solution: $[OH^-] = [Sr(OH)_2] = 2(4.8 \times 10^{-4}) = 9.6 \times 10^{-4}$ M

$$[H_3O^+] = \frac{K_w}{[OH^-]} = \frac{1 \times 10^{-14}}{9.6 \times 10^{-4}\text{ M}} = 1.\underline{0}4 \times 10^{-11}\text{M}$$

$pH = -\log(1.\underline{0}4 \times 10^{-11}) = 10.98$

$pOH = 14.00 - 10.98 = 3.02$

d) **Given:** 8.7×10^{-5} M KOH **Find:** $[OH^-]$, $[H_3O^+]$, pH, pOH

Conceptual Plan: [KOH] → $[OH^-]$ → $[H_3O^+]$ → pH → pOH

$K_w = [H_3O^+][OH^-]$ $\quad pH = -\log[H_3O^+]$ $\quad pH + pOH = 14$

Solution: $[OH^-] = [KOH] = 8.7 \times 10^{-5}$ M

$$[H_3O^+] = \frac{K_w}{[OH^-]} = \frac{1 \times 10^{-14}}{8.7 \times 10^{-5}\text{M}} = 1.\underline{1} \times 10^{-10}\text{M}$$

$pH = -\log(1.\underline{1} \times 10^{-10}) = 9.94$

$pOH = 14.00 - 9.94 = 4.06$

49. **Given:** 3.85% KOH by mass, d = 1.01 g/mL **Find:** pH

Conceptual Plan:

% mass → g KOH → mol KOH and mass soln → mL soln → L soln → M KOH → $[OH^{-1}]$

$\frac{1\text{ mol KOH}}{56.01\text{ g KOH}}$ $\qquad \frac{1.01\text{ g soln}}{\text{mL soln}}$ $\qquad \frac{1000\text{ mL soln}}{\text{L soln}}$ $\qquad \frac{\text{mol KOH}}{\text{L soln}}$

→ pOH → pH

$pOH = -\log[OH^-]$ $\quad pH + pOH = 14$

Solution: $\frac{3.85 \text{ g KOH}}{100.0 \text{ g soln}} \times \frac{1 \text{ mol KOH}}{56.01 \text{ g KOH}} \times \frac{1.01 \text{ g soln}}{\text{mL soln}} \times \frac{1000 \text{ mL soln}}{\text{L soln}} = 0.69\underline{4}2 \text{ M KOH}$

$[OH^-] = [KOH] = 0.69\underline{4}2$ M pOH = -log(0.69$\underline{4}$2) = 0.159 pH = 14.000 - 0.159 = 13.841

51. a) $NH_3(aq) + H_2O(l) \leftrightarrows NH_4^+(aq) + OH^-(aq)$ $K_b = \frac{[NH_4^+][OH^-]}{[NH_3]}$

b) $HCO_3^-(aq) + H_2O(l) \leftrightarrows H_2CO_3(aq) + OH^-(aq)$ $K_b = \frac{[H_2CO_3][OH^-]}{[HCO_3^-]}$

c) $CH_3NH_2(aq) + H_2O(l) \leftrightarrows CH_3NH_3^+(aq) + OH^-(aq)$ $K_b = \frac{[CH_3NH_3^+][OH^-]}{[CH_3NH_2]}$

53. **Given:** 0.15M NH_3 $K_b = 1.76 \times 10^{-5}$ **Find:** $[OH^-]$, pH, pOH

Conceptual Plan: Write a balanced reaction. Prepare an ICE table; represent the change with x; sum the table and determine the equilibrium values; put the equilibrium values in the equilibrium expression and solve for x.

x = [OH⁻] → pOH → pH

pOH = -log[OH⁻] pH + pOH = 14

Solution: $NH_3(aq) + H_2O(l) \leftrightarrows NH_4^+(aq) + OH^-(aq)$

	NH_3	NH_4^+	OH^-
I	0.15	0.0	0.0
C	-x	x	x
E	0.15 – x	x	x

$K_b = \frac{[NH_4^+][OH^-]}{[NH_3]} = \frac{(x)(x)}{(0.15 - x)} = 1.76 \times 10^{-5}$

Assume x is small

$x^2 = (1.76 \times 10^{-5})(0.15)$ $x = [OH^-] = 0.001\underline{6}2$ M

pOH = -log(0.001$\underline{6}$2) = 2.79

pH = 14.00 - 2.79 = 11.21

55. **Given:** pK_b = 10.4, 455 mg/L caffeine **Find:** pH

Conceptual Plan: pK_b → K_b and then mg/L → g/L → mol/L and then write a balanced reaction. Prepare an ICE table; represent the change with x; sum the table and determine the equilibrium values; put the equilibrium values in the equilibrium expression and solve for x.

x = [OH⁻] → pOH → pH

pOH = -log[OH⁻] pH + pOH = 14

Solution: $K_b = 10^{-10.4} = \underline{3}.98 \times 10^{-11}$

$\frac{455 \text{ mg caffeine}}{\text{L soln}} \times \frac{\text{g caffeine}}{1000 \text{ mg caffeine}} \times \frac{1 \text{ mol caffeine}}{194.19 \text{ g}} = 0.0023\underline{4}2 \text{ M caffeine}$

$C_8H_{10}N_4O_2(aq) + H_2O(l) \leftrightarrows HC_8H_{10}N_4O_2^+(aq) + OH^-(aq)$

	$C_8H_{10}N_4O_2$	$HC_8H_{10}N_4O_2^+$	OH^-
I	0.0023$\underline{4}$2	0.0	0.0
C	-x	x	x
E	0.0023$\underline{4}$2– x	x	x

$K_b = \frac{[HC_8H_{10}N_4O_2^+][OH^-]}{[C_8H_{10}N_4O_2]} = \frac{(x)(x)}{(0.0023\underline{4}2 - x)} = \underline{3}.98 \times 10^{-11}$

Assume x is small

$x^2 = (\underline{3}.98 \times 10^{-11})(0.0023\underline{4}2)$ $x = [OH^-] = \underline{3}.05 \times 10^{-7}$ M

$\frac{\underline{3}.05 \times 10^{-7} \text{ M}}{0.0023\underline{4}2} \times 100\% = 0.013\%$; assumption is valid

pOH = -log($\underline{3}.05 \times 10^{-7}$) = 6.5   pH = 14.00 - 6.5 = 7.5

57. **Given:** 0.150 M morphine, pH = 10.5 **Find:** K_b

Conceptual Plan:

pH → pOH → [OH⁻] and then write a balanced equation, prepare an ICE table, determine equilibrium concentrations → K_b

$pH = pOH = 14 \quad pOH = -\log[OH^-]$

Solution: pOH = 14.0 – 10.5 = 3.5 $[OH^-] = 10^{-3.5} = 3.16 \times 10^{-4} = [Hmorphine^+]$

	morphine(aq) + $H_2O(l)$ ⇆	$Hmorphine^+(aq)$+	$OH^-(aq)$
I	0.150	0.0	0.0
C	-x	x	x
E	0.150– x	3.16×10^{-4}	3.16×10^{-4}

$$K_b = \frac{[Hmorphine^+][OH^-]}{[morphine]} = \frac{(3.16 \times 10^{-4})(3.16 \times 10^{-4})}{(0.150 - 3.16 \times 10^{-4})} = 6.68 \times 10^{-7} = 7 \times 10^{-7}$$

59. a) pH neutral: Br^- is the conjugate base of a strong acid, therefore, it is pH neutral.

b) weak base: ClO^- is the conjugate base of a weak acid, therefore, it is a weak base.
$ClO^-(aq) + H_2O(l) \leftrightarrows HClO(aq) + OH^-(aq)$

c) weak base: CN^- is the conjugate base of a weak acid, therefore, it is a weak base.
$CN^-(aq) + H_2O(l) \leftrightarrows HCN(aq) + OH^-(aq)$

d) pH neutral: Cl^- is the conjugate base of a strong acid, therefore, it is pH neutral.

61. **Given:** $[F^-] = 0.140$ M, $K_a(HF) = 3.5 \times 10^{-4}$ **Find:** $[OH^-]$, pOH

Conceptual Plan:

Determine K_b. Write a balanced reaction. Prepare an ICE table; represent the change with x;

$K_b = \frac{K_w}{K_a}$

sum the table and determine the equilibrium values; put the equilibrium values in the equilibrium expression and solve for x. Determine [OH⁻] → pOH → pH

$pOH = -\log[OH^-] \quad pH + pOH = 14$

Solution:

	$F^-(aq)$ + $H_2O(l)$ ⇆	$HF(aq)$+	$OH^-(aq)$
I	0.140	0.0	0.0
C	-x	x	x
E	0.140– x	x	x

$$K_b = \frac{K_w}{K_a} = \frac{1 \times 10^{-14}}{3.5 \times 10^{-4}} = \frac{(x)(x)}{(0.140 - x)}$$

Assume x is small

$x = 2.0 \times 10^{-6} = [OH^-]$ $pOH = -\log(2.0 \times 10^{-6}) = 5.70$

$pH = 14.00 - 5.70 = 8.30$

63. a) weak acid: NH_4^+ is the conjugate acid of a weak base, therefore, it is a weak acid.
$NH_4^+(aq) + H_2O(l) \leftrightarrows H_3O^+(aq) + NH_3(aq)$

b) pH neutral: Na^+ is the counterion of a strong base, therefore, it is pH neutral.

c) weak acid: The Co^{3+} cation is a small, highly charged metal cation, therefore, it is a weak acid.
$Co(H_2O)_6^{3+}(aq) + H_2O(l) \leftrightarrows Co(H_2O)_5(OH)^{2+}(aq) + H_3O^+(aq)$

d) weak acid: $CH_2NH_3^+$ is the conjugate acid of a weak base, therefore, it is a weak acid.
$CH_2NH_3^+(aq) + H_2O(l) \leftrightarrows H_3O^+(aq) + CH_2NH_2(aq)$

65. a) acidic: $FeCl_3$ Fe^{3+} is a small, highly charged metal cation and is therefore, acidic. Cl^- is the conjugate base of a strong acid, therefore, it is pH neutral.

b) basic: NaF Na^+ is the counterion of a strong base, therefore, it is pH neutral. F^- is the conjugate base of a weak acid, therefore, it is basic.

c) pH neutral: $CaBr_2$ Ca^{2+} is the counterion of a strong base, therefore, it is pH neutral. Br^- is the conjugate base of a strong acid, therefore, it is pH neutral.

d) acidic: NH_4Br NH_4^+ is the conjugate acid of a weak base, therefore, it is acidic. Br^- is the conjugate base of a strong acid, therefore, it is pH neutral.

e) acidic: $C_6H_5NH_3NO_2$ $C_6H_5NH_3^+$ is the conjugate acid of a weak base, therefore, it is a weak acid. NO_2^- is the conjugate base of a weak acid, therefore, it is a weak base. To determine pH, compare K values. $K_a(C_6H_5NH_3^+) = \dfrac{1 \times 10^{-14}}{3.9 \times 10^{-10}} = 2.6 \times 10^{-5}$ $K_b(NO_2^-) = \dfrac{1 \times 10^{-14}}{4.6 \times 10^{-4}} = 2.2 \times 10^{-11}$

$K_a > K_b$ therefore, the solution is acidic.

67. Identify each species and determine acid, base, neutral.
$NaCl$ pH neutral: Na^+ is the counterion of a strong base, therefore, it is pH neutral. Cl^- is the conjugate base of a strong acid, therefore, it is pH neutral.
NH_4Cl acidic: NH_4^+ is the conjugate acid of a weak base, therefore, it is acidic. Cl^- is the conjugate base of a strong acid, therefore, it is pH neutral.
$NaHCO_3$ basic: Na^+ is the counterion of a strong base, therefore, it is pH neutral. HCO_3^- is the conjugate base of a weak acid, therefore, it is basic.
NH_4ClO_2 acidic: NH_4ClO_2 NH_4^+ is the conjugate acid of a weak base, therefore, it is a weak acid. ClO_2^- is the conjugate base of a weak acid, therefore, it is a weak base. $K_a(NH_4^+) = 5.6 \times 10^{-10}$ $K_b(ClO_2^-) = 9.1 \times 10^{-13}$
$NaOH$ strong base
Increasing acidity: $NaOH < NaHCO_3 < NaCl < NH_4ClO_2 < NH_4Cl$

69. a) **Given:** 0.10 M NH_4Cl **Find:** pH
Conceptual Plan: Identify each species and determine which will contribute to pH. Write a balanced reaction. Prepare an ICE table; represent the change with x; sum the table and determine the equilibrium values; put the equilibrium values in the equilibrium expression and solve for x. Determine $[H_3O^+]$ → pH
Solution: NH_4^+ is the conjugate acid of a weak base, therefore, it is acidic. Cl^- is the conjugate base of a strong acid, therefore, it is pH neutral.

	$NH_4^+(aq)$ + $H_2O(l)$ ⇆	$NH_3(aq)$ +	$H_3O^+(aq)$
I	0.10	0.0	0.0
C	-x	x	x
E	0.10– x	x	x

$$K_a = \frac{K_w}{K_b} = \frac{1 \times 10^{-14}}{1.8 \times 10^{-5}} = 5.\underline{5}6 \times 10^{-10} = \frac{(x)(x)}{(0.10 - x)}$$

Assume x is small

$x = 7.\underline{4}5 \times 10^{-6} = [H_3O^+]$ $pH = -\log(7.\underline{4}5 \times 10^{-6}) = 5.13$

b) **Given:** 0.10 M $NaC_2H_3O_2$ **Find:** pH
Conceptual Plan: Identify each species and determine which will contribute to pH. Write a balanced reaction. Prepare an ICE table; represent the change with x; sum the table and determine the equilibrium values; put the equilibrium values in the equilibrium expression and solve for x. Determine $[OH^-]$ → pOH → pH
$pOH = -\log[OH^-]$ $pH + pOH = 14$
Solution: Na^+ is the counterion of a strong base, therefore, it is pH neutral. $C_2H_3O_2^-$ is the conjugate base of a weak acid, therefore, it is basic.

	$C_2H_3O_2^-(aq)$ + $H_2O(l)$ ⇆	$HC_2H_3O_2(aq)$+	$OH^-(aq)$
I	0.10	0.0	0.0
C	-x	x	x
E	0.10– x	x	x

$$K_b = \frac{K_w}{K_a} = \frac{1 \times 10^{-14}}{1.8 \times 10^{-5}} = 5.\underline{5}6 \times 10^{-10} = \frac{(x)(x)}{(0.10 - x)}$$

Assume x is small

$x = 7.\underline{4}5 \times 10^{-6} = [OH^-]$ $pOH = -\log(7.\underline{4}5 \times 10^{-6}) = 5.13$

$pH = 14.00 - 5.13 = 8.87$

c) **Given:** 0.10 M NaCl **Find:** pH
Conceptual Plan: Identify each species and determine which will contribute to pH.
Solution: Na^+ is the counterion of a strong base, therefore, it is pH neutral. Cl^- is the conjugate base of a strong acid, therefore, it is pH neutral.
pH = 7.0

71. **Given:** 0.15 M KF **Find:** concentration of all species
Conceptual Plan: Identify each species and determine which will contribute to pH. Write a balanced reaction. Prepare an ICE table; represent the change with x; sum the table and determine the equilibrium values; put the equilibrium values in the equilibrium expression and solve for x. And then, determine $[OH^-] \rightarrow [H_3O^+]$
$K_w = [H_3O^+][OH^-]$
Solution: K^+ is the counterion of a strong base, therefore, is pH neutral. F^- is the conjugate base of a weak acid, therefore, it is basic.

$F^-(aq)$ + $H_2O(l)$ ⇆	$HF(aq)$ +	$OH^-(aq)$	
I	0.15	0.0	0.0
C	-x	x	x
E	0.15 – x	x	x

$$K_b = \frac{K_w}{K_a} = \frac{1 \times 10^{-14}}{3.5 \times 10^{-4}} = \frac{(x)(x)}{(0.15 - x)}$$

Assume x is small

$x = 2.1 \times 10^{-6} = [OH^-] = [HF]$ $\qquad [H_3O^+] = \frac{K_w}{[OH^-]} = \frac{1 \times 10^{-14}}{2.1 \times 10^{-6}} = 4.8 \times 10^{-9}$

$[K^+] = 0.15$ M
$[F^-] = (0.15 - 2.1 \times 10^{-6}) = 0.15$M
$[HF] = 2.1 \times 10^{-6}$
$[OH^-] = 2.1 \times 10^{-6}$
$[H_3O^+] = 4.8 \times 10^{-9}$

73. a) HCl is the stronger acid. HCl is the weaker bond, therefore, it is more acidic.

b) HF is the stronger acid. F is more electronegative than O, so the bond is more polar and more acidic.

c) H_2Se is the stronger acid. The H – Se bond is weaker, therefore, it is more acidic.

75. a) H_2SO_4 is the stronger acid because it has more oxygen atoms.

b) $HClO_2$ is the stronger acid because it has more oxygen atoms.

c) HClO is the stronger acid because Cl is more electronegative than Br.

d) CCl_3COOH is the stronger acid because Cl is more electronegative than H.

77. S^{2-} is the stronger base. Base strength is determined from the corresponding acid. The weaker the acid, the stronger the base. H_2S is the weaker acid because it has a stronger bond.

79. a) Lewis acid: Fe^{3+} has an empty d orbital and can accept lone pair electrons.

b) Lewis acid: BH_3 has an empty p orbital to accept a lone pair of electrons.

c) Lewis base: NH_3 has a lone pair of electrons to donate.

d) Lewis base: F^- has lone pair electrons to donate.

81. a) Fe^{3+} accepts electron pair from H_2O, so Fe^{3+} is the Lewis acid and H_2O is the Lewis base.

b) Zn^{2+} accepts electron pair from NH_3, so Zn^{2+} is the Lewis acid and NH_3 is the Lewis base.

c) The empty p orbital on B accepts an electron pair from $(CH_3)_3N$, so BF_3 is the Lewis acid and $(CH_3)_3N$ is the Lewis base.

83. a) weak acid. The beaker contains 10 HF molecule, $2H_3O^+$ ions, and $2F^-$ ions. Since both the molecule and the ions exist in solution, the acid is a weak acid.

b) strong acid. The beaker contains $12H_3O^+$ ions and $2OH^-$ ions. Since the molecule is completely ionized in solution, the acid is a strong acid.

c) weak acid. The beaker contains 10 $HCHO_2$ molecules, $2H_3O^+$ ions, and $2CHO_2^-$ ions. Since both the molecule and the ions exist in solution, the acid is a weak acid.

d) strong acid. The beaker contains $12H_3O^+$ ions and $2NO_3^-$ ions. Since the molecule is completely ionized in solution, the acid is a strong acid.

85. $HbH^+(aq) + O_2(aq) \leftrightarrows HbO_2(aq) + H^+(aq)$
Using LeChatelier's Principle, if the $[H^+]$ increases, the reaction will shift left, if the $[H^+]$ decreases, the reaction will shift right. So, if the pH of blood is too acidic (low pH; $[H^+]$ increased), the reaction will shift to the left. This will cause less of the HbO_2 in the blood and decrease the oxygen-carrying capacity of the hemoglobin in the blood.

87. **Given:** 4.00×10^2 mg $Mg(OH)_2$, 2.00×10^2 mL HCl solution, pH = 1.3
Find: volume neutralized, % neutralized.
Conceptual Plan:
mg $Mg(OH)_2$ → g $Mg(OH)_2$ → mol $Mg(OH)_2$ and then pH → $[H_3O^+]$ and then mol $Mg(OH)_2$

$$\frac{\text{g Mg(OH)}_2}{1000\text{ mg}} \quad \frac{\text{mol Mg(OH)}_2}{58.326\text{ g}} \qquad \text{pH} = -\log[H_3O^+]$$

mol OH^- → mol H_3O^+ → vol H_3O^+ → % neutralized.

$$\frac{2\text{ OH}^-}{\text{Mg(OH)}_2} \quad \frac{H_3O^+}{OH^-} \qquad \frac{\text{mol }H_3O^+}{M(H_3O^+)} \qquad \frac{\text{vol HCl neutralized}}{\text{total vol HCl}} \times 100$$

Solution: $[H_3O^+] = 10^{-1.3} = 0.05011\text{ M} = 0.05\text{ M}$

$$4.00 \times 10^2\ \text{mg Mg(OH)}_2 \times \frac{\text{g Mg(OH)}_2}{1000\ \text{mg Mg(OH)}_2} \times \frac{1\ \text{mol Mg(OH)}_2}{58.326\ \text{g Mg(OH)}_2} \times \frac{2\ \text{mol OH}^-}{1\ \text{mol Mg(OH)}_2}$$

$$\times \frac{1\ \text{mol H}_3\text{O}^+}{1\ \text{mol OH}^-} \times \frac{1\ \text{L}}{0.0501\ \text{mol H}_3\text{O}^+} \times \frac{1000\ \text{mL}}{\text{L}} = 27\underline{3}.7\ \text{mL} = 274\ \text{mL neutralized}$$

Stomach contains 2.00×10^2 mL HCl at pH = 1.3 and 4.00×10^2 mg will neutralize 274 mL of pH 1.3 HCl, so all of the stomach acid will be neutralized.

89. **Given:** pH of Great Lakes acid rain = 4.5, West Coast = 5.4
Find: $[H_3O^+]$ and ratio of Great Lakes/ West Coast
Conceptual Plan: pH → $[H_3O^+]$, and then ratio of $[H_3O^+]$ Great Lakes to West Coast.

$$\text{pH} = -\log[H_3O^+]$$

Solution: Great Lakes: $[H_3O^+] = 10^{-4.5} = 3.16 \times 10^{-5}$M West Coast: $[H_3O^+] = 10^{-5.4} = 3.98 \times 10^{-6}$M

$$\frac{\text{Great Lakes}}{\text{West Coast}} = \frac{\underline{3}.16 \times 10^{-5}\text{M}}{\underline{3}.98 \times 10^{-6}\text{M}} = 7.94 = 8 \text{ times more acidic}$$

91. **Given:** 6.5×10^2 mg aspirin, 8 ounces water, $pK_a = 3.5$ **Find:** pH of solution
Conceptual Plan:
mg aspirin → g aspirin → mol aspirin and ounces → quart → L and then [aspirin] and pK_a

$$\frac{\text{g aspirin}}{1000\text{ mg}} \quad \frac{\text{mol aspirin}}{180.16\text{ g}} \qquad \frac{1\text{ qt}}{32\text{ ounces}} \quad \frac{1\text{ L}}{1.0567\text{ qt}} \qquad \frac{\text{mol aspirin}}{\text{L soln}} \quad pK_a = -\log K_a$$

→ K_a and then: Write a balanced reaction. Prepare an ICE table; represent the change with x; sum the table and determine the equilibrium values; put the equilibrium values in the equilibrium expression and solve for x. Determine $[H_3O^+]$ → pH

$$\text{pH} = -\log[H_3O^+]$$

Solution: $\frac{6.5 \times 10^2 \text{ mg aspirin}}{8 \text{ ounces}} \times \frac{1 \text{ gram aspirin}}{1000 \text{ mg aspirin}} \times \frac{\text{mol aspirin}}{180.16 \text{ g}} \times \frac{32 \text{ ounces}}{\text{qt}} \times \frac{1.0567 \text{ qt}}{1 \text{ L}} = 0.01\underline{5}2 \text{ M}$

$K_a = 10^{-3.5} = 3.\underline{1}6 \times 10^{-4}$

	aspirin(aq) + H_2O(l) $\leftrightarrows$	H_3O^+(aq) +	aspirin$^-$ (aq)
I	0.01$\underline{5}$2M	0.0	0.0
C	- x	x	x
E	0.01$\underline{5}$2– x	x	x

$$K_a = \frac{[H_3O^+][\text{aspirin}^-]}{[\text{aspirin}]} = \frac{(x)(x)}{(0.0152 - x)} = 3.\underline{1}6 \times 10^{-4}$$

$x^2 + 3.16 \times 10^{-4}x - 4.80 \times 10^{-6} = 0 \qquad x = 2.\underline{0}4 \times 10^{-3} \text{ M} = [H_3O^+]$

$pH = -\log(2.\underline{0}4 \times 10^{-3}) = 2.69 = 2.7$

93. a) **Given:** 0.0100M $HClO_4$ **Find:** pH

Conceptual Plan: $[HClO_4] \rightarrow [H_3O^+] \rightarrow pH$

$[HClO_4] \rightarrow [H_3O^+] \quad pH = -\log[H_3O^+]$

Solution: 0.0100 M $HClO_4$ = 0.0100 M H_3O^+ $\qquad pH = -\log(0.0100) = 2.000$

b) **Given:** 0.115M $HClO_2$, $K_a = 1.1 \times 10^{-2}$ **Find:** pH

Conceptual Plan: Write a balanced reaction. Prepare an ICE table; represent the change with x; sum the table and determine the equilibrium values; put the equilibrium values in the equilibrium expression and solve for x. Determine $[H_3O^+]$ and pH

Solution: $HClO_2$ (aq) + H_2O(l) $\leftrightarrows$ H_3O^+(aq) + ClO_2^- (aq)

I	0.115M	0.0	0.0
C	- x	x	x
E	0.115 – x	x	x

$$K_a = \frac{[H_3O^+][ClO_2^-]}{[HClO_2]} = \frac{(x)(x)}{(0.115 - x)} = 1.1 \times 10^{-2}$$

Assume x is small compared to 0.115

$x^2 = (1.1 \times 10^{-2})(0.115) \qquad x = 0.03\underline{5}6\text{M} = [H_3O^+]$

Check assumption: $\frac{0.03\underline{5}6}{0.115} \times 100 = 30.9\%$ assumption not valid

$x^2 + 1.1 \times 10^{-2}x - 0.001265 = 0$

$x = 0.03\underline{0}49$ or -0.0415

$pH = -\log(0.03\underline{0}49) = 1.52$

c) **Given:** 0.045M $Sr(OH)_2$ **Find:** pH

Conceptual Plan: $[Sr(OH)_2] \rightarrow [OH^-] \rightarrow [H_3O^+] \rightarrow pH$

$K_w = [H_3O^+][OH^-] \quad pH = -\log[H_3O^+] \quad pH + pOH = 14$

Solution: $[OH^-] = [Sr(OH)_2] = 2(0.045) = 0.090$ M

$$[H_3O^+] = \frac{K_w}{[OH^-]} = \frac{1 \times 10^{-14}}{0.090 \text{ M}} = 1.\underline{1}1 \times 10^{-13}\text{M}$$

$pH = -\log(1.\underline{1}1 \times 10^{-13}) = 12.95$

d) **Given:** 0.0852 KC_6H_5O, K_a (HC_6H_5O)= 4.9×10^{-10} **Find:** pH

Conceptual Plan: Identify each species and determine which will contribute to pH. Write a balanced reaction. Prepare an ICE table; represent the change with x; sum the table and determine the equilibrium values; put the equilibrium values in the equilibrium expression and solve for x. Determine $[OH^-] \rightarrow pOH \rightarrow pH$

$pOH = -\log[OH^-] \quad pH + pOH = 14$

Solution: K^+ is the counterion of a strong base, therefore, is pH neutral. CN^- is the conjugate base of a weak acid, therefore, it is basic.

	CN^- (aq) + $H_2O(l)$ ⇆	HCN (aq)+	OH^- (aq)
I	0.0852	0.0	0.0
C	-x	x	x
E	0.0852 – x	x	x

$$K_b = \frac{K_w}{K_a} = \frac{1 \times 10^{-14}}{4.9 \times 10^{-10}} = 2.\underline{0}4 \times 10^{-5} = \frac{(x)(x)}{(0.0852 - x)}$$

Assume x is small

$x = 1.\underline{3}2 \times 10^{-3} = [OH^-]$ $\quad$ $pOH = -\log(1.\underline{3}2 \times 10^{-3}) = 2.88$

$pH = 14.00 - 2.88 = 11.12$

e) **Given:** 0.155 NH_4Cl, K_b (NH_3)= 1.8×10^{-5} **Find:** pH
Conceptual Plan: Identify each species and determine which will contribute to pH. Write a balanced reaction. Prepare an ICE table; represent the change with x; sum the table and determine the equilibrium values; put the equilibrium values in the equilibrium expression and solve for x. Determine $[H_3O^+]$ → pH
Solution: NH_4^+ is the conjugate acid of a weak base, therefore, it is acidic. Cl^- is the conjugate base of a strong acid, therefore, it is pH neutral.

	NH_4^+ (aq) + $H_2O(l)$ ⇆	NH_3 (aq) +	H_3O^+ (aq)
I	0.155	0.0	0.0
C	-x	x	x
E	0.155– x	x	x

$$K_a = \frac{K_w}{K_b} = \frac{1 \times 10^{-14}}{1.8 \times 10^{-5}} = 5.\underline{5}6 \times 10^{-10} = \frac{(x)(x)}{(0.155 - x)}$$

Assume x is small

$x = 9.\underline{2}8 \times 10^{-6} = [H_3O^+]$ $\quad$ $pH = -\log(9.\underline{2}8 \times 10^{-6}) = 5.03$

95. a) sodium cyanide = NaCN $\quad$ nitric acid = HNO_3
$H^+(aq) + CN^-(aq)$ ⇆ $HCN(aq)$

b) ammonium chloride = NH_4Cl $\quad$ sodium hydroxide = NaOH
$NH_4^+(aq) + OH^-(aq)$ ⇆ $NH_3(aq) + H_2O(l)$

c) sodium cyanide = NaCN $\quad$ ammonium bromide = NH_4Br
$NH_4^+(aq) + CN^-(aq)$ ⇆ $NH_3(aq) + HCN(aq)$

d) potassium hydrogen sulfate = $KHSO_4$ lithium acetate = $LiC_2H_3O_2$
$HSO_4^-(aq) + C_2H_3O_2^-(aq)$ ⇆ $SO_4^{2-}(aq) + HC_2H_3O_2$ (aq)

e) sodium hypochlorite = NaClO $\quad$ ammonia = NH_3
No reaction, both are bases

97. **Given:** 1.0M urea, pH = 7.050 $\quad$ **Find:** K_a Hurea$^+$
Conceptual Plan: pH → pOH → $[OH^-]$ → K_b(urea) → K_a(Hurea$^+$). Write a balanced reaction.
pH + pOH = 14 $\quad$ pOH = -log[OH^-]
Prepare an ICE table; represent the change with x; sum the table and determine the equilibrium values; put the equilibrium values in the equilibrium expression and determine K_b.
Solution: pOH = 14.000 – 7.050 = 6.950 $\quad$ $[OH^{-1}] = 10^{-6.950} = 1.1\underline{2}2 \times 10^{-7}$M

	urea (aq) + $H_2O(l)$ ⇆	Hurea$^+$ (aq)+	OH^- (aq)
I	1.0	0.0	0.0
C	-x	x	x
E	1.0– $1.1\underline{2}2 \times 10^{-7}$	$1.1\underline{2}2 \times 10^{-7}$	$1.1\underline{2}2 \times 10^{-7}$

$$K_b = \frac{[Hurea^+][OH^-]}{[urea]} = \frac{(1.1\underline{2}2 \times 10^{-7})(1.1\underline{2}2 \times 10^{-7})}{(1.0 - 1.1\underline{2}2 \times 10^{-7})} = 1.2\underline{5}89 \times 10^{-14}$$

$$K_a = \frac{K_w}{K_b} = \frac{1.00 \times 10^{-14}}{1.2\underline{5}89 \times 10^{-14}} = 0.79\underline{4}3 = 0.794$$

99. The calculation is incorrect because it neglects the contribution from the autoionization of water.
HI is a strong acid, so $[H_3O^+]$ from HI $= 1.0 \times 10^{-7}$

$$H_2O(l) + H_2O(l) \leftrightarrows H_3O^+(aq) + OH^-(aq)$$

I		1×10^{-7}	0.0
C	-x	x	x
E		$1 \times 10^{-7} + x$	x

$K_w = [H_3O^+][OH^-] = 1.0 \times 10^{-14}$

$(1 \times 10^{-7} + x)(x) = 1.0 \times 10^{-14}$ $\qquad x^2 + 1 \times 10^{-7} x - 1.0 \times 10^{-14} = 0$

$x = 6.18 \times 10^{-8}$

$[H_3O^+] = (1 \times 10^{-7} + x) = (1 \times 10^{-7} + 6.18 \times 10^{-8}) = 1.618 \times 10^{-7}$

$pH = -\log(1.618 \times 10^{-7}) = 6.79$

101. **Given:** 0.00115M HCl, 0.01000M $HClO_2$ $K_a = 1.1 \times 10^{-2}$ $\qquad$ **Find:** pH
Conceptual Plan: Use HCl to determine $[H_3O^+]$, use $[H_3O^+]$ and $HClO_2$ to determine dissociation of $HClO_2$. Write a balanced reaction, prepare an ICE table, calculate equilibrium concentrations, and then plug into the equilibrium expression.
Solution: 0.00115M HCl = 0.00115M H_3O^+

$$H\,ClO_2(aq) + H_2O(l) \leftrightarrows H_3O^+(aq) + ClO_2^-(aq)$$

I	0.0100M	0.00115	0.0
C	- x	x	x
E	0.01000 – x	0.00115 + x	x

$K_a = \dfrac{[H_3O^+][ClO_2^-]}{[HClO_2]} = \dfrac{(0.00115 + x)(x)}{(0.0100 - x)} = 1.1 \times 10^{-2}$

$x^2 + 0.01215x - 1.1 \times 10^{-4} = 0$ $\qquad\qquad$ $x = 0.006045$

$[H_3O^+] = 0.00115 + 0.006045 = 0.007195$

$pH = -\log(0.007195) = 2.14$

103. **Given:** 1.0M HA, $K_a = 1.0 \times 10^{-8}$ $K = 4.0$ for reaction 2 **Find:** $[H^+]$, $[A^-]$, $[HA_2^-]$
Conceptual Plan: Combine reaction 1 and reaction 2 and determine equilibrium expression and the value of K. Prepare an ICE table, calculate equilibrium concentrations, and then plug into the equilibrium expression
Solution:

HA(aq)	$\leftrightarrows$	$H^+(aq) +$	$A^-(aq)$		$K = 1.0 \times 10^{-8}$
HA(aq) + $A^-(aq)$	$\leftrightarrows$	$HA_2^+(aq)$			$K = 4.0$
2HA(aq)	$\leftrightarrows$	$H^+(aq) +$	$HA_2^+(aq)$		$K = 4.0 \times 10^{-8}$

I	1.0	0.0	0.0
C	- 2x	x	x
E	1.0 – 2x	x	x

$K = \dfrac{[H^+][HA_2^-]}{[HA]^2} = 4.0 \times 10^{-8} = \dfrac{(x)(x)}{(1.0 - 2x)^2}$

$x^2 + 16 \times 10^{-8}x - 4 \times 10^{-8} = 0$ $\qquad\qquad$ $x = 1.9992 \times 10^{-4}$

	HA(aq)	$\leftrightarrows$	$H^+(aq) +$	$A^-(aq)$ $\quad K = 1.0 \times 10^{-8}$
I	1.0		1.9992×10^{-4}	0.0
C	- 2y		y	y
E	1.0 – 2y		$1.9992 \times 10^{-4} + y$	y

$K = \dfrac{[H^+][A^-]}{[HA]} = 1.0 \times 10^{-8} = \dfrac{(1.9992 \times 10^{-4} + y)(y)}{(1.0 - y)}$

$y^2 + 1.9992 \times 10^{-4}x - 1 \times 10^{-8} = 0$ $\qquad\qquad$ $y = 4.29 \times 10^{-5}$

$[H^+] = x + y = 1.9992 \times 10^{-4} + 4.29 \times 10^{-5} = 2.4 \times 10^{-4}$
$[A^-] = y = 4.29 \times 10^{-5}$
$[HA_2^-] = x = 2.0 \times 10^{-4}$

---

105. Solution b would be most acidic.

a) 0.0100 M HCl (strong acid) and 0.0100 M KOH(strong base), since the concentrations are equal, the acid and base will completely neutralize each other and the resulting solution will be pH neutral.

b) 0.0100 M HF (weak acid) and 0.0100 KBr (salt). K_a (HF) = 3.5 x 10^{-4}. The weak acid will produce an acidic solution. K^+ is the counterion of a strong base and is pH neutral. Br^- is the conjugate base of a strong acid and is pH neutral.

c) 0.0100 M NH_4Cl (salt) and 0.100 M CH_3NH_3Br. $K_b(NH_3)$ = 1.8 x 10^{-5}, $K_b(CH_3NH_2)$ = 4.4 x 10^{-4}. NH_4^+ is the conjugate acid of a weak base and $CH_3NH_3^+$ is the conjugate acid of a weak base. Cl^- and Br^- are the conjugate base of a strong acid and will be pH neutral. The solution will be acidic, however, K_a for the conjugate acids is much smaller K_a for HF, so the solution will be acidic, but not as acidic as HF.

d) 0.100 M NaCN (salt) and 0.100 M $CaCl_2$. Na^+ and Ca^{2+} ion are the counterion of a strong base, therefore, they are pH neutral. Cl^- is the conjugate base of a strong acid and is pH neutral. CN^- is the conjugate base of a weak acid and will produce a basic solution.

107. $CH_3COOH < CH_2ClCOOH < CHCl_2COOH < CCl_3COOH$

Since Cl is more electronegative than H, as you add Cl you increase the amount of electronegativity and pull electron density away from the O – H bond, making it more acidic.

Chapter 16
Aqueous Ionic Equilibrium

1. The only solution that HNO_2 will ionize less in is d) 0.10 M $NaNO_2$. It is the only solution that generates a common ion (NO_2^-) with nitrous acid.

3. a) **Given:** 0.15 M $HCHO_2$ and 0.10 M $NaCHO_2$ **Find:** pH **Other:** K_a ($HCHO_2$) = 1.8×10^{-4}

Conceptual Plan:

M $NaCHO_2$ → M CHO_2^- then M $HCHO_2$, M CHO_2^- → $[H_3O^+]$ → pH

$NaCHO_2(aq) \rightarrow Na^+(aq) + CHO_2^-(aq)$ ICE Chart $pH = -\log [H_3O^+]$

Solution: Since 1 CHO_2^- ion is generated for each $NaCHO_2$, $[CHO_2^-]$ = 0.10 M CHO_2^-.

$$HCHO_2(aq) + H_2O(l) \rightleftharpoons H_3O^+(aq) + CHO_2^-(aq)$$

	$[HCHO_2]$	$[H_3O^+]$	$[CHO_2^-]$
Initial	0.15	≈ 0.00	0.10
Change	$-x$	$+x$	$+x$
Equil	$0.15 - x$	$+x$	$0.10 + x$

$$K_a = \frac{[H_3O^+][CHO_2^-]}{[HCHO_2]} = 1.8 \times 10^{-4} = \frac{x(0.10 + x)}{0.15 - x}$$ Assume x is small ($x \ll 0.10 < 0.15$) so

$$\frac{x(0.10 + \cancel{x})}{0.15 - \cancel{x}} = 1.8 \times 10^{-4} = \frac{x(0.10)}{0.15}$$ and $x = 2.7 \times 10^{-4}$ M = $[H_3O^+]$. Confirm that assumption is valid

$$\frac{2.7 \times 10^{-4}}{0.10} \times 100\,\% = 0.27\,\%$$ so assumption is valid. Finally,

$pH = -\log [H_3O^+] = -\log (2.7 \times 10^{-4}) = 3.57$

Check: The units (none) are correct. The magnitude of the answer makes physical sense because pH should be greater than $-\log(0.15) = 0.82$ because this is a weak acid and there is a common ion effect.

b) **Given:** 0.12 M NH_3 and 0.18 M NH_4Cl **Find:** pH **Other:** K_b (NH_3) = 1.79×10^{-5}

Conceptual Plan:

M NH_4Cl → M NH_4^+ then M NH_3, M NH_4^+ → $[OH^-]$ → $[H_3O^+]$ → pH

$NH_4Cl(aq) \rightarrow NH_4^+(aq) + Cl^-(aq)$ ICE Chart $K_w = [H_3O^+][OH^-]$ $pH = -\log [H_3O^+]$

Solution: Since 1 NH_4^+ ion is generated for each NH_4Cl, $[NH_4^+]$ = 0.18 M NH_4^+.

$$NH_3(aq) + H_2O(l) \rightleftharpoons NH_4^+(aq) + OH^-(aq)$$

	$[NH_3]$	$[NH_4^+]$	$[OH^-]$
Initial	0.12	0.18	≈ 0.00
Change	$-x$	$+x$	$+x$
Equil	$0.12 - x$	$0.18 + x$	$+x$

$$K_b = \frac{[NH_4^+][OH^-]}{[NH_3]} = 1.79 \times 10^{-5} = \frac{(0.18 + x)x}{0.12 - x}$$

Assume x is small ($x \ll 0.12 < 0.18$) so $\frac{(0.18 + \cancel{x})x}{0.12 - \cancel{x}} = 1.79 \times 10^{-5} = \frac{(0.18)x}{0.12}$ and $x = 1.\underline{1}9333 \times 10^{-5}$ M =

$[OH^-]$. Confirm that assumption is valid $\frac{1.\underline{1}9333 \times 10^{-5}}{0.12} \times 100\,\% = 9.9 \times 10^{-3}\,\%$ so assumption is valid.

$$K_w = [H_3O^+][OH^-] \text{ so } [H_3O^+] = \frac{K_w}{[OH^-]} = \frac{1.0 \times 10^{-14}}{1.\underline{1}9333 \times 10^{-5}} = 8.\underline{3}799 \times 10^{-10}\text{ M}.$$ Finally,

$pH = -\log [H_3O^+] = -\log (8.\underline{3}799 \times 10^{-10}) = 9.08$

Check: The units (none) are correct. The magnitude of the answer makes physical sense because pH should be less than $14 + \log(0.12) = 13.1$ because this is a weak base and there is a common ion effect.

5. **Given:** 0.15 M $HC_7H_5O_2$ in pure water and in 0.10 M $NaC_7H_5O_2$

Find: % ionization in both solutions **Other:** K_a ($HC_7H_5O_2$) = 6.5×10^{-5}

Conceptual Plan:
pure water: M $HC_7H_5O_2$ → $[H_3O^+]$ → % ionization then in $NaC_7H_5O_2$ solution:

ICE Chart $\quad \% \text{ ionization} = \dfrac{[H_3O^+]_{equil}}{[HC_7H_5O_2]_0} \times 100\ \%$

M $NaC_7H_5O_2$ → M $C_7H_5O_2^-$ then M H $C_7H_5O_2$, M $C_7H_5O_2^-$ → $[H_3O^+]$ → % ionization

$NaC_7H_5O_2\,(aq) \rightarrow Na^+\,(aq) + C_7H_5O_2^-\,(aq)$ ICE Chart $\quad \% \text{ ionization} = \dfrac{[H_3O^+]_{equil}}{[HC_7H_5O_2]_0} \times 100\ \%$

Solution: in pure water:

$$HC_7H_5O_2\,(aq) + H_2O(l) \rightleftharpoons H_3O^+\,(aq) + C_7H_5O_2^-(aq)$$

	$[HC_7H_5O_2]$	$[H_3O^+]$	$[C_7H_5O_2^-]$
Initial	0.15	≈ 0.00	0.00
Change	$-x$	$+x$	$+x$
Equil	$0.15 - x$	$+x$	$+x$

$$K_a = \frac{[H_3O^+][C_7H_5O_2^-]}{[HC_7H_5O_2]} = 6.5 \times 10^{-5} = \frac{x^2}{0.15 - x}$$

Assume x is small ($x << 0.10$) so $\dfrac{x^2}{0.15 - \cancel{x}} = 6.5 \times 10^{-5} = \dfrac{x^2}{0.15}$ and $x = 3.\underline{1}225 \times 10^{-3}$ M$= [H_3O^+]$. Then

$\% \text{ ionization} = \dfrac{[H_3O^+]_{equil}}{[HC_7H_5O_2]_0} \times 100\ \% = \dfrac{3.\underline{1}225 \times 10^{-3}}{0.15} \times 100\ \% = 2.1\ \%$, which also confirms that the assumption is valid (since it is less than 5 %). In $NaC_7H_5O_2$ solution: Since 1 $C_7H_5O_2^-$ ion is generated for each $NaC_7H_5O_2$, $[C_7H_5O_2^-] = 0.10$ M $C_7H_5O_2^-$.

$$HC_7H_5O_2\,(aq) + H_2O(l) \rightleftharpoons H_3O^+\,(aq) + C_7H_5O_2^-(aq)$$

	$[HC_7H_5O_2]$	$[H_3O^+]$	$[C_7H_5O_2^-]$
Initial	0.15	≈ 0.00	0.10
Change	$-x$	$+x$	$+x$
Equil	$0.15 - x$	$+x$	$0.10 + x$

$K_a = \dfrac{[H_3O^+][C_7H_5O_2^-]}{[HC_7H_5O_2]} = 6.5 \times 10^{-5} = \dfrac{x(0.10 + x)}{0.15 - x}$ Assume x is small ($x << 0.10 < 0.15$) so

$\dfrac{x(0.10 + \cancel{x})}{0.15 - \cancel{x}} = 6.5 \times 10^{-5} = \dfrac{x(0.10)}{0.15}$ and $x = 9.\underline{7}5 \times 10^{-5}$ M$= [H_3O^+]$, then

$\% \text{ ionization} = \dfrac{[H_3O^+]_{equil}}{[HC_7H_5O_2]_0} \times 100\ \% = \dfrac{9.\underline{7}5 \times 10^{-5}}{0.15} \times 100\ \% = 0.065\ \%$, which also confirms that the assumption is valid (since it is less than 5 %). The percent ionization in the sodium benzoate solution is less than in pure water because of the common ion effect. An increase in one of the products (benzoate ion) shifts the equilibrium to the left, so less acid dissociates.

Check: The units (%) are correct. The magnitude of the answer makes physical sense because the acid is weak and so the percent ionization is low. With a common ion present, the percent ionization decreases.

7. a) **Given:** 0.15 M HF **Find:** pH **Other:** K_a (HF) = 3.5×10^{-4}

Conceptual Plan: M HF → $[H_3O^+]$ → pH

ICE Chart $\quad pH = -\log [H_3O^+]$

Solution:

$$HF(aq) + H_2O(l) \rightleftharpoons H_3O^+\,(aq) + F^-(aq)$$

	[HF]	$[H_3O^+]$	$[F^-]$
Initial	0.15	≈ 0.00	0.00
Change	$-x$	$+x$	$+x$
Equil	$0.15 - x$	$+x$	$+x$

$K_a = \dfrac{[H_3O^+][F^-]}{[HF]} = 3.5 \times 10^{-4} = \dfrac{x^2}{0.15 - x}$ Assume x is small ($x << 0.15$) so $\dfrac{x^2}{0.15 - \cancel{x}} = 3.5 \times 10^{-4} = \dfrac{x^2}{0.15}$ and $x = 7.\underline{2}457 \times 10^{-3}$ M$= [H_3O^+]$. Confirm that assumption is valid $\frac{7.2457 \times 10^{-3}}{0.15} \times 100\ \% = 4.8\ \% < 5\ \%$ so assumption is valid.

Finally, $pH = -\log[H_3O^+] = -\log(7.2457 \times 10^{-3}) = 2.14$.

Check: The units (none) are correct. The magnitude of the answer makes physical sense because pH should be greater than – log (0.15) = 0.82 because this is a weak acid.

b) **Given:** 0.15 M NaF **Find:** pH **Other:** K_a (HF) = 3.5×10^{-4}

Conceptual Plan:

M NaF → M F⁻ and $K_a \rightarrow K_b$ then M F⁻ → [OH⁻] → [H₃O⁺] → pH

$NaF\,(aq) \rightarrow Na^+\,(aq) + F^-\,(aq)$ $K_w = K_a\,K_b$ *ICE Chart* $K_w = [H_3O^+][OH^-]$ $pH = -\log[H_3O^+]$

Solution: Since 1 F^- ion is generated for each NaF, $[F^-] = 0.15\ M\ F^-$. Since $K_w = K_a\,K_b$, rearrange to solve for K_b. $K_b = \frac{K_w}{K_a} = \frac{1.00 \times 10^{-14}}{3.5 \times 10^{-4}} = 2.8571 \times 10^{-11}$

$$F^-(aq) + H_2O(l) \rightleftharpoons HF(aq) + OH^-(aq)$$

	$[F^-]$	$[HF]$	$[OH^-]$
Initial	0.15	0.00	≈ 0.00
Change	$-x$	$+x$	$+x$
Equil	$0.15 - x$	$+x$	$+x$

$$K_b = \frac{[HF][OH^-]}{[F^-]} = 2.8571 \times 10^{-11} = \frac{x^2}{0.15 - x}$$

Assume x is small ($x << 0.15$) so $\frac{x^2}{0.15 - x} = 2.8571 \times 10^{-11} = \frac{x^2}{0.15}$ and $x = 2.0702 \times 10^{-6}\ M = [OH^-]$.

Confirm that assumption is valid $\frac{2.0702 \times 10^{-6}}{0.15} \times 100\ \% = 0.0014\ \% < 5\ \%$ so assumption is valid.

$K_w = [H_3O^+][OH^-]$ so $[H_3O^+] = \frac{K_w}{[OH^-]} = \frac{1.0 \times 10^{-14}}{2.0702 \times 10^{-6}} = 4.8305 \times 10^{-9}\ M$. Finally,

$pH = -\log[H_3O^+] = -\log(4.8305 \times 10^{-9}) = 8.32$.

Check: The units (none) are correct. The magnitude of the answer makes physical sense because pH should be slightly basic, since the fluoride ion is a very weak base.

c) **Given:** 0.15 M HF and 0.15 M NaF **Find:** pH **Other:** K_a (HF) = 3.5×10^{-4}

Conceptual Plan: M NaF → M F⁻ then M HF, M F⁻ → [H₃O⁺] → pH

$NaF\,(aq) \rightarrow Na^+\,(aq) + F^-\,(aq)$ ICE Chart $pH = -\log[H_3O^+]$

Solution: Since 1 F^- ion is generated for each NaF, $[F^-] = 0.15\ M\ F^-$.

$$HF(aq) + H_2O(l) \rightleftharpoons H_3O^+(aq) + F^-(aq)$$

	$[HF]$	$[H_3O^+]$	$[F^-]$
Initial	0.15	≈ 0.00	0.15
Change	$-x$	$+x$	$+x$
Equil	$0.15 - x$	$+x$	$0.15 + x$

$$K_a = \frac{[H_3O^+][F^-]}{[HF]} = 3.5 \times 10^{-4} = \frac{x(0.15 + x)}{0.15 - x}$$

Assume x is small ($x << 0.15$) so $\frac{x(0.15 + x)}{0.15 - x} = 3.5 \times 10^{-4} = \frac{x(0.15)}{0.15}$ and $x = 3.5 \times 10^{-4}\ M = [H_3O^+]$.

Confirm that assumption is valid $\frac{3.5 \times 10^{-4}}{0.15} \times 100\ \% = 0.23\ \% < 5\ \%$ so assumption is valid.

Finally, $pH = -\log[H_3O^+] = -\log(3.5 \times 10^{-4}) = 3.46$.

Check: The units (none) are correct. The magnitude of the answer makes physical sense because pH should be greater than in part a (2.14) because of the common ion effect suppressing the dissociation of the weak acid.

9. When an acid (such as HCl) is added it will react with the conjugate base of the buffer system as follows: $HCl + NaC_2H_3O_2 \rightarrow HC_2H_3O_2 + NaCl$. When a base (such as NaOH) is added it will react with the weak acid of the buffer system as follows: $NaOH + HC_2H_3O_2 \rightarrow H_2O + NaC_2H_3O_2$. The reaction generates the other buffer system component.

11. a) **Given:** 0.15 M $HCHO_2$ and 0.10 M $NaCHO_2$ **Find:** pH **Other:** K_a ($HCHO_2$) = 1.8×10^{-4}
Conceptual Plan: identify acid and base components then M $NaCHO_2$ → M CHO_2^- then
acid = $HCHO_2$ base = CHO_2^- $NaCHO_2\ (aq) \rightarrow Na^+\ (aq) + CHO_2^-\ (aq)$
K_a, M $HCHO_2$, M CHO_2^- → pH

$$pH = pK_a + \log\frac{[base]}{[acid]}$$

Solution: Acid = $HCHO_2$, so [acid] = [$HCHO_2$] = 0.15 M. Base = CHO_2^-. Since 1 CHO_2^- ion is generated for each $NaCHO_2$, [CHO_2^-] = 0.10 M CHO_2^- = [base]. Then

$$pH = pK_a + \log\frac{[base]}{[acid]} = -\log(1.8 \times 10^{-4}) + \log\frac{0.10\ \cancel{M}}{0.15\ \cancel{M}} = 3.57\ .$$

Note that in order to use the Henderson–Hasselbalch Equation, the assumption that x is small must be valid. This was confirmed in problem #3.
Check: The units (none) are correct. The magnitude of the answer makes physical sense because pH should be less than the pK_a of the acid because there is more acid than base. The answer agrees with problem #3.

b) **Given:** 0.12 M NH_3 and 0.18 M NH_4Cl **Find:** pH **Other:** K_b (NH_3) = 1.79×10^{-5}
Conceptual Plan:
identify acid and base components then M NH_4Cl → M NH_4^+ and K_b → pK_b → pK_a
acid = NH_4^+ base = NH_3 $NH_4Cl\ (aq) \rightarrow NH_4^+\ (aq) + Cl^-\ (aq)$ $pK_b = -\log K_b$ $14 = pK_a + pK_b$
then pK_a, M NH_3, M NH_4^+ → pH

$$pH = pK_a + \log\frac{[base]}{[acid]}$$

Solution: Base = NH_3, [base] = [NH_3] = 0.12 M Acid = NH_4^+. Since 1 NH_4^+ ion is generated for each NH_4Cl, [NH_4^+] = 0.18 M NH_4^+ = [acid].
Since K_b (NH_3) = 1.79×10^{-5}, $pK_b = -\log K_b = -\log(1.79 \times 10^{-5}) = 4.75$. Since $14 = pK_a + pK_b$,

$$pK_a = 14 - pK_b = 14 - 4.75 = 9.25 \text{ then } pH = pK_a + \log\frac{[base]}{[acid]} = 9.25 + \log\frac{0.12\ \cancel{M}}{0.18\ \cancel{M}} = 9.07$$

Note that in order to use the Henderson–Hasselbalch Equation, the assumption that x is small must be valid. This was confirmed in problem #3.
Check: The units (none) are correct. The magnitude of the answer makes physical sense because pH should be less than the pK_a of the acid because there is more acid than base. The answer agrees with problem #3, within the error of the value.

13. a) **Given:** 0.125 M HClO and 0.150 M KClO **Find:** pH **Other:** K_a (HClO) = 2.9×10^{-8}
Conceptual Plan: identify acid and base components then M KClO → M ClO^- then
acid = HClO base = ClO^- $KClO\ (aq) \rightarrow K^+\ (aq) + ClO^-\ (aq)$
M HClO, M ClO^- → pH

$$pH = pK_a + \log\frac{[base]}{[acid]}$$

Solution: Acid = HClO, so [acid] = [HClO] = 0.125 M. Base = ClO^-. Since 1 ClO^- ion is generated for each KClO, [ClO^-] = 0.150 M ClO^- = [base]. Then

$$pH = pK_a + \log\frac{[base]}{[acid]} = -\log(2.9 \times 10^{-8}) + \log\frac{0.150\ \cancel{M}}{0.125\ \cancel{M}} = 7.62\ .$$

Check: The units (none) are correct. The magnitude of the answer makes physical sense because pH should be greater than the pK_a of the acid because there is more base than acid.

b) **Given:** 0.175 M $C_2H_5NH_2$ and 0.150 M $C_2H_5NH_3Br$ **Find:** pH **Other:** K_b ($C_2H_5NH_2$) = 5.6×10^{-4}
Conceptual Plan: identify acid and base components then M $C_2H_5NH_3Br$ → M $C_2H_5NH_3^+$ and
acid = $C_2H_5NH_3^+$ base = $C_2H_5NH_2$ $C_2H_5NH_3Br\ (aq) \rightarrow C_2H_5NH_3^+\ (aq) + Br^-\ (aq)$
K_b → pK_b → pK_a then pK_a, M $C_2H_5NH_2$, M $C_2H_5NH_3^+$ → pH
$pK_b = -\log K_b$ $14 = pK_a + pK_b$ $pH = pK_a + \log\frac{[base]}{[acid]}$

Solution: Base = $C_2H_5NH_2$, [base] = [$C_2H_5NH_2$] = 0.175 M Acid = $C_2H_5NH_3^+$. Since 1 $C_2H_5NH_3^+$ ion is generated for each $C_2H_5NH_3Br$, $[C_2H_5NH_3^+] = 0.150$ M $C_2H_5NH_3^+$ = [acid]. Since K_b ($C_2H_5NH_2$) = 5.6 x 10^{-4}, $pK_b = -\log K_b = -\log(5.6 \times 10^{-4}) = 3.25$. Since $14 = pK_a + pK_b$,

$$pK_a = 14 - pK_b = 14 - 3.25 = 10.75 \text{ then } pH = pK_a + \log\frac{[base]}{[acid]} = 10.75 + \log\frac{0.175\ \cancel{M}}{0.150\ \cancel{M}} = 10.82.$$

Check: The units (none) are correct. The magnitude of the answer makes physical sense because pH should be greater than the pK_a of the acid because there is more base than acid.

c) **Given:** 10.0 g $HC_2H_3O_2$ and 10.0 g $NaC_2H_3O_2$ in 150.0 mL solution **Find:** pH
Other: K_a ($HC_2H_3O_2$) = 1.8 x 10^{-5}
Conceptual Plan:
identify acid and base components then mL → L and g $HC_2H_3O_2$ → mol $HC_2H_3O_2$

acid = $HC_2H_3O_2$ base = $C_2H_3O_2^-$ $\qquad \frac{1\text{ L}}{1000\text{ mL}} \qquad \frac{1\text{ mol } HC_2H_3O_2}{60.05\text{ g } HC_2H_3O_2}$

then mol $HC_2H_3O_2$, L → M $HC_2H_3O_2$ and g $NaC_2H_3O_2$ → mol $NaC_2H_3O_2$ then

$M = \frac{mol}{L} \qquad \frac{1\text{ mol } NaC_2H_3O_2}{82.04\text{ g } NaC_2H_3O_2}$

mol $NaC_2H_3O_2$, L → M $NaC_2H_3O_2$ → M $C_2H_3O_2^-$ then M $HC_2H_3O_2$, M $C_2H_3O_2^-$ → pH

$M = \frac{mol}{L} \qquad NaC_2H_3O_2\,(aq) \rightarrow Na^+\,(aq) + C_2H_3O_2^-\,(aq) \qquad pH = pK_a + \log\frac{[base]}{[acid]}$

Solution: $150.0\ \cancel{mL} \times \frac{1\text{ L}}{1000\ \cancel{mL}} = 0.1500\text{ L}$ and

$$10.0\text{ g } \cancel{HC_2H_3O_2} \times \frac{1\text{ mol } HC_2H_3O_2}{60.05\text{ g } \cancel{HC_2H_3O_2}} = 0.16\underline{6}528\text{ mol } HC_2H_3O_2$$

$$\text{then } M = \frac{mol}{L} = \frac{0.16\underline{6}528\text{ mol } HC_2H_3O_2}{0.1500\text{ L}} = 1.1\underline{1}019\text{ M } HC_2H_3O_2 \text{ and}$$

$$10.0\text{ g } \cancel{NaC_2H_3O_2} \times \frac{1\text{ mol } NaC_2H_3O_2}{82.04\text{ g } \cancel{NaC_2H_3O_2}} = 0.12\underline{1}892\text{ mol } NaC_2H_3O_2 \text{ then}$$

$$M = \frac{mol}{L} = \frac{0.12\underline{1}892\text{ mol } NaC_2H_3O_2}{0.1500\text{ L}} = 0.81\underline{2}612\text{ M } NaC_2H_3O_2.$$ So Acid = $HC_2H_3O_2$, so [acid] = $[HC_2H_3O_2]$ = 1.1$\underline{1}$019 M and Base = $C_2H_3O_2^-$. Since 1 $C_2H_3O_2^-$ ion is generated for each $NaC_2H_3O_2$, $[C_2H_3O_2^-]$ = 0.81$\underline{2}$612 M $C_2H_3O_2^-$ = [base]. Then

$$pH = pK_a + \log\frac{[base]}{[acid]} = -\log(1.8 \times 10^{-5}) + \log\frac{0.81\underline{2}612\ \cancel{M}}{1.1\underline{1}019\ \cancel{M}} = 4.61.$$

Check: The units (none) are correct. The magnitude of the answer makes physical sense because pH should be less than the pK_a of the acid because there is more acid than base.

15. a) **Given:** 50.0 mL of 0.15 M $HCHO_2$ and 75.0 mL of 0.13 M $NaCHO_2$ **Find:** pH
Other: K_a ($HCHO_2$) = 1.8 x 10^{-4}
Conceptual Plan:
identify acid and base components then mL $HCHO_2$, mL $NaCHO_2$ → total mL then

acid = $HCHO_2$ base = CHO_2^- $\qquad$ total mL = mL $HCHO_2$ + mL $NaCHO_2$

mL $HCHO_2$, M $HCHO_2$, total mL → buffer M $HCHO_2$ and

$M_1 V_1 = M_2 V_2$

mL $NaCHO_2$, M $NaCHO_2$, total mL → buffer M $NaCHO_2$ → buffer M CHO_2^- then

$M_1 V_1 = M_2 V_2 \qquad NaCHO_2\,(aq) \rightarrow Na^+\,(aq) + CHO_2^-\,(aq)$

K_a, M $HCHO_2$, M CHO_2^- → pH

$pH = pK_a + \log\frac{[base]}{[acid]}$

Solution: total mL = mL $HCHO_2$ + mL $NaCHO_2$ = 50.0 mL + 75.0 mL = 125.0 mL. Then since $M_1 V_1 = M_2 V_2$ rearrange to solve for M_2. $M_2 = \frac{M_1 V_1}{V_2} = \frac{(0.15 \text{ M})(50.0 \text{ mL})}{125.0 \text{ mL}} = 0.060 \text{ M HCHO}_2$ and

$M_2 = \frac{M_1 V_1}{V_2} = \frac{(0.13 \text{ M})(75.0 \text{ mL})}{125.0 \text{ mL}} = 0.078 \text{ M NaCHO}_2$. Acid = $HCHO_2$, so [acid] = [$HCHO_2$] = 0.060 M. Base = CHO_2^-. Since 1 CHO_2^- ion is generated for each $NaCHO_2$, [CHO_2^-] = 0.078 M CHO_2^- = [base]. Then $\text{pH} = \text{p}K_a + \log\frac{[\text{base}]}{[\text{acid}]} = -\log(1.8 \times 10^{-4}) + \log\frac{0.078 \text{ M}}{0.060 \text{ M}} = 3.86$.

Check: The units (none) are correct. The magnitude of the answer makes physical sense because pH should be greater than the pK_a of the acid because there is more base than acid.

b) **Given:** 125.0 mL of 0.10 M NH_3 and 250.0 mL of 0.10 M NH_4Cl **Find:** pH
Other: K_b (NH_3) = 1.79 x 10^{-5}
Conceptual Plan:
identify acid and base components then mL NH_3, mL NH_4Cl → total mL then
acid = NH_4^+ base = NH_3 total mL = mL NH_3 + mL NH_4Cl

mL NH_3, M NH_3, total mL → buffer M NH_3 and
$M_1 V_1 = M_2 V_2$

mL NH_4Cl, M NH_4Cl, total mL → buffer M NH_4Cl → buffer M NH_4^+ and K_b → pK_b → pK_a then
$M_1 V_1 = M_2 V_2$ $NH_4Cl\,(aq) \rightarrow NH_4^+\,(aq) + Cl^-\,(aq)$ $\text{p}K_b = -\log K_b$ $14 = \text{p}K_a + \text{p}K_b$

then pK_a, M NH_3, M NH_4^+ → pH
$\text{pH} = \text{p}K_a + \log\frac{[\text{base}]}{[\text{acid}]}$

Solution: total mL = mL NH_3 + mL NH_4Cl = 125.0 mL + 250.0 mL = 375.0 mL. Then since $M_1 V_1 = M_2 V_2$ rearrange to solve for M_2. $M_2 = \frac{M_1 V_1}{V_2} = \frac{(0.10 \text{ M})(125.0 \text{ mL})}{375.0 \text{ mL}} = 0.03\underline{3}333 \text{ M NH}_3$ and

$M_2 = \frac{M_1 V_1}{V_2} = \frac{(0.10 \text{ M})(250.0 \text{ mL})}{375.0 \text{ mL}} = 0.06\underline{6}667 \text{ M NH}_4\text{Cl}$. Base = NH_3, [base] = [NH_3] = 0.03$\underline{3}$333 M
Acid = NH_4^+. Since 1 NH_4^+ ion is generated for each NH_4Cl, [NH_4^+] = 0.06$\underline{6}$6667 M NH_4^+ = [acid].
Since K_b (NH_3) = 1.79 x 10^{-5}, $\text{p}K_b = -\log K_b = -\log(1.79 \times 10^{-5}) = 4.75$. Since $14 = \text{p}K_a + \text{p}K_b$,

$\text{p}K_a = 14 - \text{p}K_b = 14 - 4.75 = 9.25$ then $\text{pH} = \text{p}K_a + \log\frac{[\text{base}]}{[\text{acid}]} = 9.25 + \log\frac{0.03\underline{3}333 \text{ M}}{0.06\underline{6}667 \text{ M}} = 8.95$.

Check: The units (none) are correct. The magnitude of the answer makes physical sense because pH should be less than the pK_a of the acid because there is more acid than base.

17. **Given:** NaF / HF buffer at pH = 4.00 **Find:** [NaF] / [HF] **Other:** K_a (HF) = 3.5 x 10^{-4}
Conceptual Plan: identify acid and base components then pH, K_a → [NaF] / [HF]
acid = HF base = F^- $\text{pH} = \text{p}K_a + \log\frac{[\text{base}]}{[\text{acid}]}$

Solution: $\text{pH} = \text{p}K_a + \log\frac{[\text{base}]}{[\text{acid}]} = -\log(3.5 \times 10^{-4}) + \log\frac{[\text{NaF}]}{[\text{HF}]} = 4.00$. Solve for [NaF] / [HF].

$\log\frac{[\text{NaF}]}{[\text{HF}]} = 4.00 - 3.46 = 0.54 \rightarrow \frac{[\text{NaF}]}{[\text{HF}]} = 10^{0.54} = 3.5$.

Check: The units (none) are correct. The magnitude of the answer makes physical sense because the pH is greater than the pK_a of the acid, so there needs to be more base than acid.

19. **Given:** 150.0 mL buffer of 0.15 M benzoic acid at pH = 4.25 **Find:** mass sodium benzoate
Other: K_a ($HC_7H_5O_2$) = 6.5 x 10^{-5}
Conceptual Plan: identify acid and base components then pH, K_a, [$HC_7H_5O_2$] → [$NaC_7H_5O_2$]
acid = $HC_7H_5O_2$ base = $C_7H_5O_2^-$ $\text{pH} = \text{p}K_a + \log\frac{[\text{base}]}{[\text{acid}]}$

mL → L then [$NaC_7H_5O_2$], L → mol $NaC_7H_5O_2$ → g $NaC_7H_5O_2$

$\frac{1\ L}{1000\ mL}$ $\qquad M = \frac{mol}{L}$ $\qquad \frac{144.11\ g\ NaC_7H_5O_2}{1\ mol\ NaC_7H_5O_2}$

Solution: $pH = pK_a + \log \frac{[base]}{[acid]} = -\log(6.5 \times 10^{-5}) + \log \frac{[NaC_7H_5O_2]}{0.15\ M} = 4.25$. Solve for [$NaC_7H_5O_2$].

$\log \frac{[NaC_7H_5O_2]}{0.15\ M} = 4.25 - 4.19 = 0.06291 \rightarrow \frac{[NaC_7H_5O_2]}{0.15\ M} = 10^{0.06291} = 1.1559 \rightarrow [NaC_7H_5O_2] = 0.17338\ M$.

Convert to moles using $M = \frac{mol}{L}$.

$\frac{0.17338\ mol\ NaC_7H_5O_2}{1\ L} \times 0.150\ L = 0.026007\ mol\ NaC_7H_5O_2 \times \frac{144.11\ g\ NaC_7H_5O_2}{1\ mol\ NaC_7H_5O_2} = 3.7\ g\ NaC_7H_5O_2$.

Check: The units (g) are correct. The magnitude of the answer makes physical sense because the volume of solution is small and the concentration is low, so much less than a mole is needed.

21. a) **Given:** 250.0 mL buffer 0.250 M $HC_2H_3O_2$ and 0.250 M $NaC_2H_3O_2$ **Find:** initial pH
Other: K_a ($HC_2H_3O_2$) = 1.8×10^{-5}
Conceptual Plan: identify acid and base components then M $NaC_2H_3O_2$ → M $C_2H_3O_2^-$ then
acid = $HC_2H_3O_2$ base = $C_2H_3O_2^-$ $\qquad NaC_2H_3O_2\ (aq) \rightarrow Na^+\ (aq) + C_2H_3O_2^-\ (aq)$
M $HC_2H_3O_2$, M $C_2H_3O_2^-$ → pH
$pH = pK_a + \log \frac{[base]}{[acid]}$

Solution: Acid = $HC_2H_3O_2$, so [acid] = [$HC_2H_3O_2$] = 0.250 M. Base = $C_2H_3O_2^-$. Since 1 $C_2H_3O_2^-$ ion is generated for each $NaC_2H_3O_2$, [$C_2H_3O_2^-$] = 0.250 M $C_2H_3O_2^-$ = [base]. Then

$pH = pK_a + \log \frac{[base]}{[acid]} = -\log(1.8 \times 10^{-5}) + \log \frac{0.250\ M}{0.250\ M} = 4.74$.

Check: The units (none) are correct. The magnitude of the answer makes physical sense because pH is equal to the pK_a of the acid because there are equal amounts of acid and base.

b) **Given:** 250.0 mL buffer 0.250 M $HC_2H_3O_2$ and 0.250 M $NaC_2H_3O_2$, add 0.0050 mol HCl
Find: pH **Other:** K_a ($HC_2H_3O_2$) = 1.8×10^{-5}
Conceptual Plan: Part I: Stoichiometry:
mL → L then [$NaC_2H_3O_2$], L → mol $NaC_2H_3O_2$ and [$HC_2H_3O_2$], L → mol $HC_2H_3O_2$
$\frac{1\ L}{1000\ mL}$ $\qquad M = \frac{mol}{L}$ $\qquad M = \frac{mol}{L}$

write balanced equation then
$HCl + NaC_2H_3O_2 \rightarrow HC_2H_3O_2 + NaCl$
mol $NaC_2H_3O_2$, mol $HC_2H_3O_2$, mol HCl → mol $NaC_2H_3O_2$, mol $HC_2H_3O_2$ then
set up stoichiometry table

Part II: Equilibrium:
mol $NaC_2H_3O_2$, mol $HC_2H_3O_2$, L, K_a → pH
$pH = pK_a + \log \frac{[base]}{[acid]}$

Solution: $250.0\ mL \times \frac{1\ L}{1000\ mL} = 0.2500\ L$ then

$\frac{0.250\ mol\ HC_2H_3O_2}{1\ L} \times 0.250\ L = 0.0625\ mol\ HC_2H_3O_2$ and

$\frac{0.250\ mol\ NaC_2H_3O_2}{1\ L} \times 0.250\ L = 0.0625\ mol\ NaC_2H_3O_2$ set up table to track changes:

	$HCl\ (aq)$ +	$NaC_2H_3O_2\ (aq)$ →	$HC_2H_3O_2\ (aq)$ +	$NaCl\ (aq)$
Before addition	≈ 0.00 mol	0.0625 mol	0.0625 mol	0.00 mol
Addition	0.0050 mol	–	–	–
After addition	≈ 0.00 mol	0.0575 mol	0.0675 mol	0.0050 mol

Since the amount of HCl is small, there are still significant amounts of both buffer components, so the Henderson–Hasselbalch Equation can be used to calculate the new pH.

$$pH = pK_a + \log\frac{[\text{base}]}{[\text{acid}]} = -\log(1.8 \times 10^{-5}) + \log\frac{\frac{0.0575\ \text{mol}}{0.250\ \text{L}}}{\frac{0.0675\ \text{mol}}{0.250\ \text{L}}} = 4.68.$$

Check: The units (none) are correct. The magnitude of the answer makes physical sense because the pH dropped slightly when acid was added.

c) **Given:** 250.0 mL buffer 0.250 M $HC_2H_3O_2$ and 0.250 M $NaC_2H_3O_2$, add 0.0050 mol NaOH
Find: pH **Other:** K_a ($HC_2H_3O_2$) = 1.8×10^{-5}
Conceptual Plan: Part I: Stoichiometry:
mL → L then [$NaC_2H_3O_2$], L → mol $NaC_2H_3O_2$ and [$HC_2H_3O_2$], L → mol $HC_2H_3O_2$

$\frac{1\ \text{L}}{1000\ \text{mL}}$ $M = \frac{\text{mol}}{\text{L}}$ $M = \frac{\text{mol}}{\text{L}}$

write balanced equation then
$NaOH + HC_2H_3O_2 \rightarrow H_2O + NaC_2H_3O_2$
mol $NaC_2H_3O_2$, mol $HC_2H_3O_2$, mol NaOH → mol $NaC_2H_3O_2$, mol $HC_2H_3O_2$ then
set up stoichiometry table
Part II: Equilibrium:
mol $NaC_2H_3O_2$, mol $HC_2H_3O_2$, L, K_a → pH

$$pH = pK_a + \log\frac{[\text{base}]}{[\text{acid}]}$$

Solution: $250.0\ \text{mL} \times \frac{1\ \text{L}}{1000\ \text{mL}} = 0.2500\ \text{L}$ then

$\frac{0.250\ \text{mol}\ HC_2H_3O_2}{1\ \text{L}} \times 0.2500\ \text{L} = 0.0625\ \text{mol}\ HC_2H_3O_2$ and

$\frac{0.250\ \text{mol}\ NaC_2H_3O_2}{1\ \text{L}} \times 0.2500\ \text{L} = 0.0625\ \text{mol}\ NaC_2H_3O_2$ set up table to track changes:

	$NaOH\ (aq)$ +	$HC_2H_3O_2\ (aq)$ →	$NaC_2H_3O_2\ (aq)$ +	$H_2O\ (l)$
Before addition	≈ 0.00 mol	0.0625 mol	0.0625 mol	–
Addition	0.0050 mol	–	–	–
After addition	≈ 0.00 mol	0.0575 mol	0.0675 mol	–

Since the amount of NaOH is small, there are still significant amounts of both buffer components, so the Henderson–Hasselbalch Equation can be used to calculate the new pH.

$$pH = pK_a + \log\frac{[\text{base}]}{[\text{acid}]} = -\log(1.8 \times 10^{-5}) + \log\frac{\frac{0.0675\ \text{mol}}{0.2500\ \text{L}}}{\frac{0.0575\ \text{mol}}{0.2500\ \text{L}}} = 4.81.$$

Check: The units (none) are correct. The magnitude of the answer makes physical sense because the pH rose slightly when base was added.

23. a) **Given:** 500.0 mL pure water **Find:** initial pH and after adding 0.010 mol HCl
Conceptual Plan:
pure water has a pH of 7.00 then mL → L then mol HCl, L → [H_3O^+] → pH

$\frac{1\ \text{L}}{1000\ \text{mL}}$ $M = \frac{\text{mol}}{\text{L}}$ $pH = -\log[H_3O^+]$

Solution: pure water has a pH of 7.00 so initial pH = 7.00 then $500.0\ \text{mL} \times \frac{1\ \text{L}}{1000\ \text{mL}} = 0.5000\ \text{L}$

then $M = \frac{\text{mol}}{\text{L}} = \frac{0.010\ \text{mol HCl}}{0.5000\ \text{L}} = 0.020\ \text{M HCl}$. Since HCl is a strong acid, it dissociates completely,

so $pH = -\log[H_3O^+] = -\log(0.020) = 1.70$.

Check: The units (none) are correct. The magnitudes of the answers make physical sense because the pH starts neutral and then dropps significantly when acid is added and there is no buffer present.

b) **Given:** 500.0 mL buffer 0.125 M $HC_2H_3O_2$ and 0.115 M $NaC_2H_3O_2$
Find: initial pH and after adding 0.010 mol HCl **Other:** K_a ($HC_2H_3O_2$) = 1.8×10^{-5}
Conceptual Plan: initial pH:
identify acid and base components then M $NaC_2H_3O_2$ → M $C_2H_3O_2^-$ then
acid = $HC_2H_3O_2$ base = $C_2H_3O_2^-$ $NaC_2H_3O_2\ (aq) \rightarrow Na^+\ (aq) + C_2H_3O_2^-\ (aq)$
M $HC_2H_3O_2$, M $C_2H_3O_2^-$ → pH

$$pH = pK_a + \log\frac{[base]}{[acid]}$$

pH after HCl addition: Part I: Stoichiometry:
mL → L then [$NaC_2H_3O_2$], L → mol $NaC_2H_3O_2$ and [$HC_2H_3O_2$], L → mol $HC_2H_3O_2$

$$\frac{1\ L}{1000\ mL} \qquad M = \frac{mol}{L} \qquad M = \frac{mol}{L}$$

write balanced equation then
$HCl + NaC_2H_3O_2 \rightarrow HC_2H_3O_2 + NaCl$
mol $NaC_2H_3O_2$, mol $HC_2H_3O_2$, mol HCl → mol $NaC_2H_3O_2$, mol $HC_2H_3O_2$ then
set up stoichiometry table
Part II: Equilibrium:
mol $NaC_2H_3O_2$, mol $HC_2H_3O_2$, L, K_a → pH

$$pH = pK_a + \log\frac{[base]}{[acid]}$$

Solution: initial pH: Acid = $HC_2H_3O_2$, so [acid] = [$HC_2H_3O_2$] = 0.125 M. Base = $C_2H_3O_2^-$. Since 1 $C_2H_3O_2^-$ ion is generated for each $NaC_2H_3O_2$, [$C_2H_3O_2^-$] = 0.115 M $C_2H_3O_2^-$ = [base]. Then

$$pH = pK_a + \log\frac{[base]}{[acid]} = -\log(1.8 \times 10^{-5}) + \log\frac{0.115\ \cancel{M}}{0.125\ \cancel{M}} = 4.71.$$

pH after HCl addition:

$$500.0\ \cancel{mL} \times \frac{1\ L}{1000\ \cancel{mL}} = 0.5000\ L \text{ then } \frac{0.125\ mol\ HC_2H_3O_2}{1\ \cancel{L}} \times 0.5000\ \cancel{L} = 0.0625\ mol\ HC_2H_3O_2 \text{ and}$$

$$\frac{0.115\ mol\ NaC_2H_3O_2}{1\ \cancel{L}} \times 0.5000\ \cancel{L} = 0.0575\ mol\ NaC_2H_3O_2 \text{ set up table to track changes:}$$

	$HCl\ (aq)$ +	$NaC_2H_3O_2\ (aq)$ →	$HC_2H_3O_2\ (aq)$ +	$NaCl\ (aq)$
Before addition	≈ 0.00 mol	0.0575 mol	0.0625 mol	0.00 mol
Addition	0.010 mol	–	–	–
After addition	≈ 0.00 mol	0.0475 mol	0.0725 mol	0.010 mol

Since the amount of HCl is small, there are still significant amounts of both buffer components, so the Henderson–Hasselbalch Equation can be used to calculate the new pH.

$$pH = pK_a + \log\frac{[base]}{[acid]} = -\log(1.8 \times 10^{-5}) + \log\frac{\frac{0.0475\ \cancel{mol}}{\cancel{0.5000\ L}}}{\frac{0.0725\ \cancel{mol}}{\cancel{0.5000\ L}}} = 4.56.$$

Check: The units (none) are correct. The magnitudes of the answers make physical sense because the pH started below the pK_a of the acid and it dropped slightly when acid was added.

c) **Given:** 500.0 mL buffer 0.155 M $CH_3CH_2NH_2$ and 0.145 M $CH_3CH_2NH_3Cl$
Find: initial pH and after adding 0.010 mol HCl **Other:** K_b ($CH_3CH_2NH_2$) = 5.6×10^{-4}
Conceptual Plan: initial pH:
identify acid and base components then M $CH_3CH_2NH_3Cl$ → M $CH_3CH_2NH_3^+$
acid = $C_2H_5NH_3^+$ base = $C_2H_5NH_2$ $C_2H_5NH_3Cl\ (aq) \rightarrow C_2H_5NH_3^+\ (aq) + Cl^-\ (aq)$

and K_b → pK_b → pK_a then pK_a, M $CH_3CH_2NH_2$, M $CH_3CH_2NH_3^+$ → pH

$$pK_b = -\log K_b \qquad 14 = pK_a + pK_b \qquad pH = pK_a + \log\frac{[base]}{[acid]}$$

pH after HCl addition: Part I: Stoichiometry:
mL → L then $[CH_3CH_2NH_2]$, L → mol $CH_3CH_2NH_2$ and

$$\frac{1\ L}{1000\ mL} \qquad M = \frac{mol}{L}$$

$[CH_3CH_2NH_3Cl]$, L → mol $CH_3CH_2NH_3Cl$ write balanced equation then

$$M = \frac{mol}{L} \qquad HCl + CH_3CH_2NH_2 \rightarrow CH_3CH_2NH_3Cl$$

mol $CH_3CH_2NH_2$, mol $CH_3CH_2NH_3Cl$, mol HCl → mol $CH_3CH_2NH_2$, mol $CH_3CH_2NH_3Cl$ then
set up stoichiometry table

Part II: Equilibrium:
mol $CH_3CH_2NH_2$, mol $CH_3CH_2NH_3Cl$, L, K_a → pH

$$pH = pK_a + \log\frac{[base]}{[acid]}$$

Solution: Base = $CH_3CH_2NH_2$, [base] = $[CH_3CH_2NH_2]$ = 0.155 M Acid = $CH_3CH_2NH_3^+$. Since 1 $CH_3CH_2NH_3^+$ ion is generated for each $CH_3CH_2NH_3Cl$, $[CH_3CH_2NH_3^+]$ = 0.145 M $CH_3CH_2NH_3^+$ = [acid]. Since K_b $(CH_3CH_2NH_2)$ = 5.6×10^{-4}, $pK_b = -\log K_b = -\log(5.6 \times 10^{-4}) = 3.25$. Since $14 = pK_a + pK_b$, $pK_a = 14 - pK_b = 14 - 3.25 = 10.75$ then

$$pH = pK_a + \log\frac{[base]}{[acid]} = 10.75 + \log\frac{0.155\ \cancel{M}}{0.145\ \cancel{M}} = 10.78\ .$$

pH after HCl addition: $500.0\ \cancel{mL} \times \frac{1\ L}{1000\ \cancel{mL}} = 0.5000\ L$ then

$$\frac{0.155\ mol\ CH_3CH_2NH_2}{1\ \cancel{L}} \times 0.5000\ \cancel{L} = 0.0775\ mol\ CH_3CH_2NH_2 \text{ and}$$

$$\frac{0.145\ mol\ CH_3CH_2NH_3Cl}{1\ \cancel{L}} \times 0.5000\ \cancel{L} = 0.0725\ mol\ CH_3CH_2NH_3Cl$$ set up table to track changes:

	$HCl\ (aq)$ +	$CH_3CH_2NH_2\ (aq)$ →	$CH_3CH_2NH_3Cl\ (aq)$
Before addition	≈ 0.00 mol	0.0775 mol	0.0725 mol
Addition	0.010 mol	–	–
After addition	≈ 0.00 mol	0.06$\underline{7}$5 mol	0.08$\underline{2}$5 mol

Since the amount of HCl is small, there are still significant amounts of both buffer components, so the Henderson–Hasselbalch Equation can be used to calculate the new pH.

$$pH = pK_a + \log\frac{[base]}{[acid]} = 10.75 + \log\frac{\dfrac{0.06\underline{7}5\ \cancel{mol}}{\cancel{0.5000\ L}}}{\dfrac{0.08\underline{2}5\ \cancel{mol}}{\cancel{0.5000\ L}}} = 10.66\ .$$

Check: The units (none) are correct. The magnitudes of the answers make physical sense because the initial pH should be greater than the pK_a of the acid because there is more base than acid and the pH drops slightly when acid is added.

25. **Given:** 350.00 mL 0.150 M HF and 0.150 M NaF buffer
Find: mass NaOH to raise pH to 4.00 and mass NaOH to raise pH to 4.00 with buffer concentrations raised to 0.350 M
Other: K_a (HF) = 3.5×10^{-4}
Conceptual Plan: identify acid and base components Since [NaF] = [HF] then initial pH = pK_a.
acid = HF base = F^- $\qquad$ pH = pK_a

final pH, pK_a → [NaF]/[HF] and mL → L then [HF], L → mol HF and [NaF], L → mol NaF

$$pH = pK_a + \log\frac{[base]}{[acid]} \qquad \frac{1\ L}{1000\ mL} \qquad M = \frac{mol}{L} \qquad M = \frac{mol}{L}$$

then write balanced equation then
$NaOH + HF \rightarrow NaF + H_2O$

mol HF, mol NaF, [NaF]/[HF] → mol NaOH → g NaOH

set up stoichiometry table $\dfrac{40.00 \text{ g NaOH}}{1 \text{ mol NaOH}}$

Finally, when the buffer concentrations are raised to 0.350 M, simply multiply the g NaOH by ratio of concentrations (0.350 M / 0.150 M).

Solution: initial $pH = pK_a = -\log(3.5 \times 10^{-4}) = 3.46$ then

$pH = pK_a + \log\dfrac{[\text{base}]}{[\text{acid}]} = -\log(3.5 \times 10^{-4}) + \log\dfrac{[NaF]}{[HF]} = 4.00$. Solve for [NaF] / [HF].

$\log\dfrac{[NaF]}{[HF]} = 4.00 - 3.46 = 0.54 \rightarrow \dfrac{[NaF]}{[HF]} = 10^{0.54} = 3.5$. $350.0 \text{ mL} \times \dfrac{1 \text{ L}}{1000 \text{ mL}} = 0.3500 \text{ L}$ then

$\dfrac{0.150 \text{ mol HF}}{1 \text{ L}} \times 0.3500 \text{ L} = 0.0525 \text{ mol HF}$ and $\dfrac{0.150 \text{ mol NaF}}{1 \text{ L}} \times 0.3500 \text{ L} = 0.0525 \text{ mol NaF}$

set up table to track changes:

	NaOH (*aq*)	+ HF (*aq*)	→ NaF (*aq*)	+ H_2O (*aq*)
Before addition	0.00 mol	0.0525 mol	0.0525 mol	–
Addition	x	–	–	–
After addition	≈ 0.00 mol	$(0.0525 - x)$ mol	$(0.0525 + x)$ mol	–

Since $\dfrac{[NaF]}{[HF]} = 3.5 = \dfrac{(0.0525 + x) \text{ mol}}{(0.0525 - x) \text{ mol}}$, solve for x. Note that the ratio of moles is the same as the ratio of concentrations, since the volume for both terms is the same. $3.5\,(0.0525 - x) = (0.0525 + x) \rightarrow$ $0.18375 - 3.5x = 0.0525 + x \rightarrow 0.13125 = 4.5x \rightarrow x = 0.029167$ mol NaOH then

$0.029167 \text{ mol NaOH} \times \dfrac{40.00 \text{ g NaOH}}{1 \text{ mol NaOH}} = 1.1667 \text{ g NaOH} = 1.2 \text{ g NaOH}$. Finally multiply the NaOH mass by the ratio of concentrations $1.1667 \text{ g NaOH} \times \dfrac{0.350 \text{ M}}{0.150 \text{ M}} = 2.7 \text{ g NaOH}$.

Check: The units (g) are correct. The magnitudes of the answers make physical sense because there is much less than a mole of each of the buffer components, so there must be much less than a mole of NaOH. The higher the buffer concentrations, the higher the buffer capacity and the mass of NaOH it can neutralize.

27. a) Yes, this will be a buffer because NH_3 is a weak base and NH_4^+ is its conjugate acid. The ratio of base to acid is 0.10/0.15 = 0.67, so the pH will be within 1 pH unit of the pK_a.

b) No, this will not be a buffer solution because HCl is a strong acid and NaOH is a strong base.

c) Yes, this will be a buffer because HF is a weak acid and the NaOH will convert 20.0/50.0 = 40 % of the acid to its conjugate base.

d) No, this will not be a buffer solution because both components are bases.

e) No, this will not be a buffer solution because both components are bases.

29. a) **Given:** blood buffer 0.024 M HCO_3^- and 0.0012 M H_2CO_3, $pK_a = 6.1$ **Find:** initial pH

Conceptual Plan: identify acid and base components then M HCO_3^-, M H_2CO_3 → pH

acid = H_2CO_3 base = HCO_3^- $pH = pK_a + \log\dfrac{[\text{base}]}{[\text{acid}]}$

Solution: Acid = H_2CO_3, so [acid] = $[H_2CO_3]$ = 0.0012 M. Base = HCO_3^-, so [base] = $[HCO_3^-]$ = 0.024 M HCO_3^-. Then $pH = pK_a + \log\dfrac{[\text{base}]}{[\text{acid}]} = 6.1 + \log\dfrac{0.024 \text{ M}}{0.0012 \text{ M}} = 7.4$.

Check: The units (none) are correct. The magnitude of the answer makes physical sense because pH is greater than the pK_a of the acid because there is more base than acid.

b) **Given:** 5.0 L of blood buffer **Find:** mass HCl to lower pH to 7.0

Conceptual Plan: final pH, pK_a → $[HCO_3^-]/[H_2CO_3]$ then $[HCO_3^-]$, L → mol HCO_3^- and

$$pH = pK_a + \log\frac{[base]}{[acid]} \qquad M = \frac{mol}{L}$$

$[H_2CO_3]$, L → mol H_2CO_3 then write balanced equation then

$$M = \frac{mol}{L} \qquad H^+ + HCO_3^- \rightarrow H_2CO_3$$

mol HCO_3^-, mol H_2CO_3, $[HCO_3^-]/[H_2CO_3]$ → mol HCl → g HCl

set up stoichiometry table $\frac{36.46 \text{ g HCl}}{1 \text{ mol HCl}}$

Solution: $pH = pK_a + \log\frac{[base]}{[acid]} = 6.1 + \log\frac{[HCO_3^-]}{[H_2CO_3]} = 7.0$. Solve for $[HCO_3^-]/[H_2CO_3]$.

$\log\frac{[HCO_3^-]}{[H_2CO_3]} = 7.0 - 6.1 = 0.9 \rightarrow \frac{[HCO_3^-]}{[H_2CO_3]} = 10^{0.9} = \underline{7}.9433$. Then

$\frac{0.024 \text{ mol } HCO_3^-}{1 \cancel{L}} \times 5.0 \cancel{L} = 0.12 \text{ mol } HCO_3^-$ and $\frac{0.0012 \text{ mol } H_2CO_3}{1 \cancel{L}} \times 5.0 \cancel{L} = 0.0060 \text{ mol } H_2CO_3$

Since HCl is a strong acid, $[HCl] = [H^+]$, and set up table to track changes:

	H^+ *(aq)*	+ HCO_3^- *(aq)*	→ H_2CO_3 *(aq)*
Before addition	≈ 0.00 mol	0.12 mol	0.0060 mol
Addition	*x*	–	–
After addition	≈ 0.00 mol	(0.12 – x) mol	(0.0060 + x) mol

Since $\frac{[HCO_3^-]}{[H_2CO_3]} = \underline{7}.9433 = \frac{(0.12 - x)\ \cancel{mol}}{(0.0060 + x)\ \cancel{mol}}$, solve for *x*. Note that the ratio of moles is the same as the ratio of concentrations, since the volume for both terms is the same. $\underline{7}.9433(0.0060 + x) = (0.12 - x) \rightarrow 0.04\underline{7}6598 + \underline{7}.9433x = 0.12 - x \rightarrow \underline{8}.9433x = 0.0\underline{7}234 \rightarrow x = 0.00\underline{8}0888$ mol HCl then

$0.00\underline{8}0888 \cancel{\text{mol HCl}} \times \frac{36.46 \text{ g HCl}}{1 \cancel{\text{mol HCl}}} = 0.\underline{2}9492 \text{ g HCl} = 0.3 \text{ g HCl}$.

Check: The units (g) are correct. The amount of acid needed is small because the concentrations of the buffer components are very low and the buffer starts only 0.4 pH units above the final pH.

c) **Given:** 5.0 L of blood buffer **Find:** mass NaOH to raise pH to 7.8

Conceptual Plan: final pH, pK_a → $[HCO_3^-]/[H_2CO_3]$ then $[HCO_3^-]$, L → mol HCO_3^- and

$$pH = pK_a + \log\frac{[base]}{[acid]} \qquad M = \frac{mol}{L}$$

$[H_2CO_3]$, L → mol H_2CO_3 then write balanced equation then

$$M = \frac{mol}{L} \qquad OH^- + H_2CO_3 \rightarrow HCO_3^- + H_2O$$

mol HCO_3^-, mol H_2CO_3, $[HCO_3^-]/[H_2CO_3]$ → mol NaOH → g NaOH

set up stoichiometry table $\frac{40.00 \text{ g NaOH}}{1 \text{ mol NaOH}}$

Solution: $pH = pK_a + \log\frac{[base]}{[acid]} = 6.1 + \log\frac{[HCO_3^-]}{[H_2CO_3]} = 7.8$.

Solve for $[HCO_3^-]/[H_2CO_3]$. $\log\frac{[HCO_3^-]}{[H_2CO_3]} = 7.8 - 6.1 = 1.7 \rightarrow \frac{[HCO_3^-]}{[H_2CO_3]} = 10^{1.7} = 50.\underline{1}1872$. Then

$\frac{0.024 \text{ mol } HCO_3^-}{1 \cancel{L}} \times 5.0 \cancel{L} = 0.12 \text{ mol } HCO_3^-$ and $\frac{0.0012 \text{ mol } H_2CO_3}{1 \cancel{L}} \times 5.0 \cancel{L} = 0.0060 \text{ mol } H_2CO_3$

Since NaOH is a strong base, $[NaOH] = [OH^-]$, and set up table to track changes:

$$OH^- (aq) + H_2CO_3 (aq) \rightarrow HCO_3^- (aq) + H_2O (l)$$

	OH^-	H_2CO_3	HCO_3^-	H_2O
Before addition	≈ 0.00 mol	0.0060 mol	0.12 mol	–
Addition	x	–	–	
After addition	≈ 0.00 mol	$(0.0060 - x)$ mol	$(0.12 + x)$ mol	

Since $\frac{[HCO_3^-]}{[H_2CO_3]} = 50.\underline{1}1872 = \frac{(0.12 + x)\ \text{mol}}{(0.0060 - x)\ \text{mol}}$, solve for x. Note that the ratio of moles is the same as the ratio of concentrations, since the volume for both terms is the same. $50.\underline{1}1872(0.0060 - x) = (0.12 + x)$ → $0.3\underline{0}071 - 50.\underline{1}1872x = 0.12 + x$ → $51.\underline{1}1872x = 0.1\underline{8}071$ → $x = 0.003\underline{5}351$ mol NaOH

then $0.003\underline{5}351\ \text{mol NaOH} \times \frac{40.00\ \text{g NaOH}}{1\ \text{mol NaOH}} = 0.1\underline{4}141\ \text{g NaOH} = 0.14\ \text{g NaOH}$.

Check: The units (g) are correct. The amount of base needed is small because the concentrations of the buffer components are very low.

31. **Given:** $HC_2H_3O_2/KC_2H_3O_2$, $HClO_2/KClO_2$, NH_3/NH_4Cl, and $HClO/KClO$ potential buffer systems to create buffer at pH = 7.20 **Find:** best buffer system and ratio of component masses
Other: K_a ($HC_2H_3O_2$) = 1.8×10^{-5}, K_a ($HClO_2$) = 1.8×10^{-4}, K_b (NH_3) = 1.79×10^{-5}, K_a ($HClO$) = 2.9×10^{-8}
Conceptual Plan:
calculate pK_a of all potential buffer acids for the base K_b → pK_b → pK_a then

$pK_a = -\log K_a$ $\qquad$ $pK_b = -\log K_b \quad 14 = pK_a + pK_b$

and choose the pK_a that is closest to 7.20. Then pH, K_a → [base]/[acid] → mass base/mass acid

$$\text{pH} = pK_a + \log\frac{[\text{base}]}{[\text{acid}]} \qquad \frac{\mathfrak{M}\ (\text{base})}{\mathfrak{M}\ (\text{acid})}$$

Solution: for $HC_2H_3O_2/KC_2H_3O_2$: $pK_a = -\log K_a = -\log(1.8 \times 10^{-5}) = 4.74$; for $HClO_2/KClO_2$: $pK_a = -\log K_a = -\log(1.8 \times 10^{-4}) = 3.74$; for NH_3/NH_4Cl: $pK_b = -\log K_b = -\log(1.79 \times 10^{-5}) = 4.75$. Since $14 = pK_a + pK_b$, $pK_a = 14 - pK_b = 14 - 4.75 = 9.25$; and for HClO/KClO: $pK_a = -\log K_a = -\log(2.9 \times 10^{-8}) = 7.54$. So the HClO/KClO buffer system has the pK_a that is the closest to 7.20. So, $\text{pH} = pK_a + \log\frac{[\text{base}]}{[\text{acid}]} = 7.54 + \log\frac{[\text{KClO}]}{[\text{HClO}]} = 7.1$. Solve for [KClO]/[HClO].

$\log\frac{[\text{KClO}]}{[\text{HClO}]} = 7.20 - 7.54 = -0.34$ → $\frac{[\text{KClO}]}{[\text{HClO}]} = 10^{-0.34} = 0.4\underline{5}7088$. Then convert to mass ratio using

$$\frac{\mathfrak{M}\ (\text{base})}{\mathfrak{M}\ (\text{acid})},\ 0.4\underline{5}7088\ \frac{\frac{\text{KClO mol}}{\text{L}}}{\frac{\text{HClO mol}}{\text{L}}} \times \frac{\frac{90.55\ \text{g KClO}}{\text{mol KClO}}}{\frac{52.46\ \text{g HClO}}{\text{mol HClO}}} = 0.79\ \frac{\text{g KClO}}{\text{g HClO}}$$

Check: The units (none and g base/g acid)) are correct. The buffer system with the K_a closest to 10^{-7} is the best choice. The magnitude of the answer makes physical sense because the buffer needs more acid than base (and this fact is not overcome by the heavier molar mass of the base).

33. **Given:** 500.0 mL of 0.100 M HNO_2 / 0.150 M KNO_2 buffer and a) 250 mg NaOH, b) 350 mg KOH, c) 1.25 g HBr and d) 1.35 g HI **Find:** if buffer capacity is exceeded
Conceptual Plan:
mL → L then [HNO_2], L → mol HNO_2 and [KNO_2], L → mol KNO_2 then

$\frac{1\ \text{L}}{1000\ \text{mL}}$ $\qquad$ $M = \frac{\text{mol}}{\text{L}}$ $\qquad$ $M = \frac{\text{mol}}{\text{L}}$

then calculate moles of acid or base to be added to the buffer mg → g → mol then

$\frac{1\ \text{g}}{1000\ \text{mg}}$ $\qquad$ $\mathfrak{M}$

compare the added amount to the buffer amount of the opposite component. Ratio of base/acid must be between 0.1 and 10 to maintain the buffer integrity.

Solution: $500.00\ \text{mL} \times \frac{1\ \text{L}}{1000\ \text{mL}} = 0.5000\ \text{L}$ then $\frac{0.100\ \text{mol HNO}_2}{1\ \text{L}} \times 0.5000\ \text{L} = 0.0500\ \text{mol HNO}_2$ and

$$\frac{0.150 \text{ mol } KNO_2}{1 \text{ L}} \times 0.5000 \text{ L} = 0.0750 \text{ mol } KNO_2$$

a) for NaOH: $250 \text{ mg NaOH} \times \frac{1 \text{ g NaOH}}{1000 \text{ mg NaOH}} \times \frac{1 \text{ mol NaOH}}{40.00 \text{ g NaOH}} = 0.006\underline{2}5 \text{ mol NaOH}$. Since the buffer contains 0.0500 mol acid, the amount of acid is reduced by 0.00625/0.0500 = 12.5 % and the ratio of base/acid is still between 0.1and 10. The buffer capacity is not exceeded.

b) for KOH: $350 \text{ mg KOH} \times \frac{1 \text{ g KOH}}{1000 \text{ mg KOH}} \times \frac{1 \text{ mol KOH}}{56.11 \text{ g KOH}} = 0.006\underline{2}4 \text{ mol KOH}$. Since the buffer contains 0.0500 mol acid, the amount of acid is reduced by 0.00624/0.0500 = 12.5 % and the ratio of base/acid is still between 0.1 and 10. The buffer capacity is not exceeded.

c) for HBr: $1.25 \text{ g HBr} \times \frac{1 \text{ mol HBr}}{80.91 \text{ g HBr}} = 0.015\underline{4}496 \text{ mol HBr}$. Since the buffer contains 0.0750 mol base, the amount of acid is reduced by 0.0154/0.0750 = 20.6 % and the ratio of base/acid is still between 0.1 and 10. The buffer capacity is not exceeded.

d) for HI: $1.35 \text{ g HI} \times \frac{1 \text{ mol HI}}{127.91 \text{ g HI}} = 0.010\underline{5}545 \text{ mol HI}$. Since the buffer contains 0.0750 mol base, the amount of acid is reduced by 0.0106/0.0750 = 14.1 % and the ratio of base/acid is still between 0.1 and 10. The buffer capacity is not exceeded.

35. (i) The equivalence point of a titration is where the pH rises sharply as base is added. The pH as the equivalence point is the midpoint of the sharp rise at ~ 50 mL added base. For (a) the pH = ~ 8 and for (b) the pH = ~ 7.

(ii) Graph (a) represents a weak acid and graph (b) represents a strong acid. A strong acid titration starts at a lower pH, has a flatter initial region and a sharper rise at the equivalence point than a weak acid. The pH at the equivalence point of a strong acid is neutral, while the pH at the equivalence point of a weak acid is basic.

37. **Given:** 20.0 mL 0.200 M KOH and 0.200 M CH_3NH_2 titrated with 0.100 M HI

a) **Find:** volume of base to reach equivalence point

Conceptual Plan: The answer for both titrations will be the same since the initial concentration and volumes of the bases are the same. Write balanced equation then mL → L then

$HI + KOH \rightarrow KI + H_2O$ and $HI + KOH \rightarrow CH_3NH_3I$ $\qquad \frac{1 \text{ L}}{1000 \text{ mL}}$

[base], L → mol base then set mol base = mol acid and [HI], mol HI → L HI → mL HI

$M = \frac{mol}{L}$ balanced equation has 1:1 stoichiometry $M = \frac{mol}{L}$ $\frac{1000 \text{ mL}}{1 \text{ L}}$

Solution: $20.0 \text{ mL base} \times \frac{1 \text{ L}}{1000 \text{ mL}} = 0.0200 \text{ L base}$ then

$\frac{0.200 \text{ mol base}}{1 \text{ L}} \times 0.0200 \text{ L} = 0.00400 \text{ mol base}$. So mol base = 0.00400 mol = mol HI then

$0.00400 \text{ mol HI} \times \frac{1 \text{ L HI}}{0.100 \text{ mol HI}} = 0.0400 \text{ L HI} \times \frac{1000 \text{ mL}}{1 \text{ L}} = 40.0 \text{ mL HI}$ for both titrations.

Check: The units (mL) are correct. The volume of acid is twice the volume of bases because the concentration of the base is twice that of the acid in each case. The answer for both titrations is the same because the stoichiometry is the same for both titration reactions.

b) The pH at the equivalence point will be neutral for KOH (since it is a strong base) and it will be acidic for CH_3NH_2 (since it is a weak base).

c) The initial pH will be lower for CH_3NH_2 (since it is a weak base and will only partially dissociate and not raise the pH as high as KOH (since it is a strong base and so it dissociates completely) at the same base concentration.

d) The titration curves will look like:

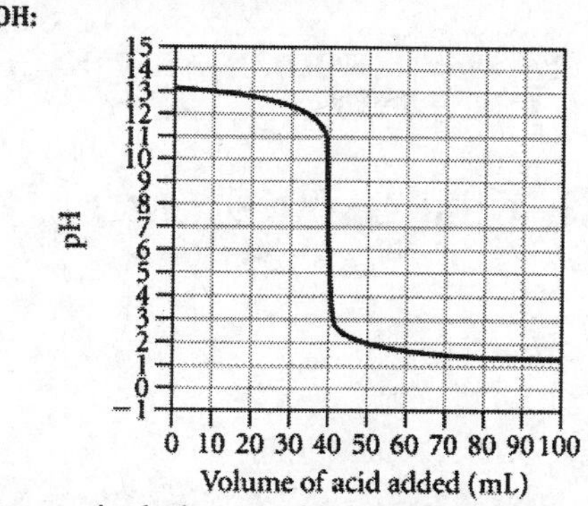

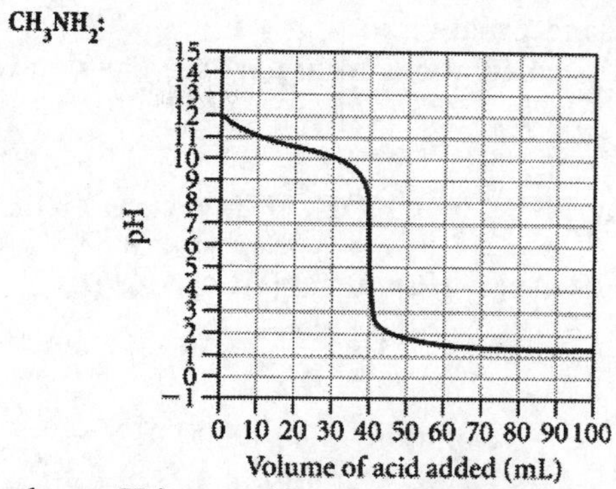

Important features to include are a high initial pH (if strong base pH is over 13 and lower for a weak base), flat initial region (very flat for strong base, not as flat for weak base where pH halfway to equivalence point is the pK_b of the base), sharp drop at equivalence point, pH at equivalence point (neutral for strong base and lower for weak base), and then flatten out at low pH.

39. a) The equivalence point of a titration is where the pH rises sharply as base is added. The volume at the equivalence point is ~ 30 mL. The pH as the equivalence point is the midpoint of the sharp rise at ~ 30 mL added base, which is a pH = ~ 9.

b) At 0 mL the pH is calculated by doing an equilibrium calculation of a weak acid in water (as done in Chapter 15).

c) The pH one-half way to the equivalence point is equal to the pK_a of the acid, or ~ 15 mL.

d) The pH at the equivalence point, or ~ 30 mL, is calculated by doing an equilibrium problem with the K_b of the acid. At the equivalence point, all of the acid has been converted to its conjugate base.

e) Beyond the equivalence point, or ~ 30 mL, there is excess base. All of the acid has been converted to its conjugate base and so the pH is calculated by focusing on this excess base concentration.

41. **Given:** 35.0 mL of 0.175 M HBr titrated with 0.200 M KOH

a) **Find:** initial pH

Conceptual Plan: Since HBr is a strong acid, it will dissociate completely, so initial pH = – log $[H_3O^+]$ = – log [HBr].

Solution: pH = – log [HBr] = – log 0.175 = 0.757

Check: The units (none) are correct. The pH is reasonable since the concentration is greater than 0.1 M and the acid dissociates completely, the pH is less than 1.

b) **Find:** volume of base to reach equivalence point

Conceptual Plan: Write balanced equation then mL → L then [HBr], L → mol HBr then

$HBr + KOH \rightarrow KBr + H_2O$ $\qquad \frac{1\ L}{1000\ mL} \qquad M = \frac{mol}{L}$

set mol acid (HBr) = mol base (KOH) and [KOH], mol KOH → L KOH → mL KOH

balanced equation has 1:1 stoichiometry $\qquad M = \frac{mol}{L} \qquad \frac{1000\ mL}{1\ L}$

Solution: $35.0\ \text{mL HBr} \times \frac{1\ L}{1000\ mL} = 0.0350\ L\ HBr$ then

$\frac{0.175\ mol\ HBr}{1\ L} \times 0.0350\ L = 0.006125\ mol\ HBr$.

So mol acid = mol HBr = 0.006125 mol = mol KOH then

$0.006125\ mol\ KOH \times \frac{1\ L}{0.200\ mol\ KOH} = 0.030625\ L\ KOH \times \frac{1000\ mL}{1\ L} = 30.6\ mL\ KOH$.

Check: The units (mL) are correct. The volume of base is a little less than the volume of acid because the concentration of the base is a little greater than that of the acid.

c) **Find:** pH after adding 10.0 mL of base
Conceptual Plan:
Use calculations from part b. Then mL → L then [KOH], L → mol KOH then

$\frac{1\text{ L}}{1000\text{ mL}}$ $\quad$ $M = \frac{\text{mol}}{\text{L}}$

mol HBr, mol KOH → mol excess HBr (set up stoichiometry table) **and L HBr, L KOH → total L** (L HBr + L KOH = total L) **then**

mol excess HBr, L → [HBr] → pH

$M = \frac{\text{mol}}{\text{L}}$ $\quad$ $\text{pH} = -\log[\text{HBr}]$

Solution: $10.0\ \text{mL KOH} \times \frac{1\text{ L}}{1000\text{ mL}} = 0.0100\text{ L KOH}$ then

$\frac{0.200\text{ mol KOH}}{1\text{ L}} \times 0.0100\text{ L} = 0.00200\text{ mol KOH}$.

Since KOH is a strong base, [KOH] = $[OH^-]$, and set up table to track changes:

	$KOH\,(aq)$	+ $HBr\,(aq)$	→ $KBr\,(aq)$	+ $H_2O\,(l)$
Before addition	0.00 mol	0.006125 mol	0.00 mol	–
Addition	0.00200 mol	–	–	–
After addition	≈ 0.00 mol	0.004125 mol	0.00200 mol	–

Then 0.0350 L HBr + 0.0100 L KOH = 0.0450 L total volume. So mol excess acid = mol HBr = 0.004125 mol in 0.0450 L so $[\text{HBr}] = \frac{0.004125\text{ mol HBr}}{0.0450\text{ L}} = 0.0916667\text{ M}$ and

$\text{pH} = -\log[\text{HBr}] = -\log 0.0916667 = 1.038$.

Check: The units (none) are correct. The pH is a little higher than the initial pH, which is expected since this is a strong acid.

d) **Find:** pH at equivalence point
Solution: Since this is a strong acid–strong base titration, the pH at the equivalence point is neutral or 7.

e) **Find:** pH after adding 5.0 mL of base beyond the equivalence point
Conceptual Plan: Use calculations from parts b & c. Then the pH is only dependent on the amount of excess base and the total solution volumes.
mL excess → L excess then [KOH], L excess → mol KOH excess

$\frac{1\text{ L}}{1000\text{ mL}}$ $\quad$ $M = \frac{\text{mol}}{\text{L}}$

then L HBr, L KOH to equivalence point, L KOH excess → total L then
(L HBr + L KOH to equivalence point + L KOH excess = total L)

mol excess KOH, total L → [KOH] = [OH⁻] → $[H_3O^+]$ → pH

$M = \frac{\text{mol}}{\text{L}}$ $\quad$ $K_w = [H_3O^+][OH^-]$ $\quad$ $\text{pH} = -\log[H_3O^+]$

Solution: $5.0\ \text{mL KOH} \times \frac{1\text{ L}}{1000\text{ mL}} = 0.0050\text{ L KOH excess}$ then

$\frac{0.200\text{ mol KOH}}{1\text{ L}} \times 0.0050\text{ L} = 0.0010\text{ mol KOH excess}$. Then 0.0350 L HBr + 0.0306 L KOH + 0.0050 L KOH = 0.0706 L total volume. $[\text{KOH excess}] = \frac{0.0010\text{ mol KOH excess}}{0.0706\text{ L}} = 0.014164\text{ M KOH excess}$

Since KOH is a strong base, [KOH] excess = $[OH^-]$. $K_w = [H_3O^+][OH^-]$ so

$[H_3O^+] = \frac{K_w}{[OH^-]} = \frac{1.0 \times 10^{-14}}{0.014164} = 7.06 \times 10^{-13}\text{ M}$. Finally, $\text{pH} = -\log[H_3O^+] = -\log(7.06 \times 10^{-13}) = 12.15$.

Check: The units (none) are correct. The pH is rising sharply at the equivalence point, so the pH after 5 mL past the equivalence point should be quite basic.

43. **Given:** 25.0 mL of 0.115 M RbOH titrated with 0.100 M HCl

a) **Find:** initial pH

Conceptual Plan: Since RbOH is a strong base, it will dissociate completely, so $[RbOH] = [OH^-] \rightarrow [H_3O^+] \rightarrow pH$

$K_w = [H_3O^+][OH^-]$ $\quad pH = -\log [H_3O^+]$

Solution: Since RbOH is a strong base, [RbOH] excess = $[OH^-]$. $K_w = [H_3O^+][OH^-]$ so

$$[H_3O^+] = \frac{K_w}{[OH^-]} = \frac{1.0 \times 10^{-14}}{0.115} = 8.\underline{6}9565 \times 10^{-14}\text{ M and } pH = -\log [H_3O^+] = -\log(8.\underline{6}9565 \times 10^{-14}) = 13.06$$

Check: The units (none) are correct. The pH is reasonable since the concentration is greater than 0.1 M and the base dissociates completely, the pH is greater than 13.

b) **Find:** volume of acid to reach equivalence point

Conceptual Plan: Write balanced equation then mL → L then [RbOH], L → mol RbOH then

$HCl + RbOH \rightarrow RbCl + H_2O$ $\quad \frac{1\text{ L}}{1000\text{ mL}}$ $\quad M = \frac{mol}{L}$

set mol base (RbOH) = mol acid (HCl) and [HCl], mol HCl → L HCl → mL HCl

balanced equation has 1:1 stoichiometry $\quad M = \frac{mol}{L}$ $\quad \frac{1000\text{ mL}}{1\text{ L}}$

Solution: $25.0\text{ mL RbOH} \times \frac{1\text{ L}}{1000\text{ mL}} = 0.0250\text{ L RbOH}$ then

$\frac{0.115\text{ mol RbOH}}{1\text{ L}} \times 0.0250\text{ L} = 0.0028\underline{7}5\text{ mol RbOH}$. So mol base = mol RbOH = $0.0028\underline{7}5$ mol = mol HCl then $0.0028\underline{7}5\text{ mol HCl} \times \frac{1\text{ L}}{0.100\text{ mol HCl}} = 0.028\underline{7}5\text{ L HCl} \times \frac{1000\text{ mL}}{1\text{ L}} = 28.8\text{ mL HCl}$.

Check: The units (mL) are correct. The volume of acid is greater than the volume of base because the concentration of the base is a little greater than that of the acid.

c) **Find:** pH after adding 5.0 mL of acid

Conceptual Plan: Use calculations from part b. Then mL → L then [HCl], L → mol HCl then

$\frac{1\text{ L}}{1000\text{ mL}}$ $\quad M = \frac{mol}{L}$

mol RbOH, mol HCl → mol excess RbOH and L RbOH, L HCl → total L then

set up stoichiometry table $\quad$ L RbOH + L HCl = total L

mol excess RbOH, L → $[RbOH] = [OH^-] \rightarrow [H_3O^+] \rightarrow pH$

$M = \frac{mol}{L}$ $\quad K_w = [H_3O^+][OH^-]$ $\quad pH = -\log [H_3O^+]$

Solution: $5.0\text{ mL HCl} \times \frac{1\text{ L}}{1000\text{ mL}} = 0.0050\text{ L HCl}$ then $\frac{0.100\text{ mol HCl}}{1\text{ L}} \times 0.0050\text{ L} = 0.00050\text{ mol HCl}$.

Since HCl is a strong acid, $[HCl] = [H_3O^+]$, and set up table to track changes:

	$HCl\ (aq)$	+ $RbOH\ (aq)$	→ $RbCl\ (aq)$	+ $H_2O\ (l)$
Before addition	0.00 mol	0.0028$\underline{7}$5 mol	0.00 mol	–
Addition	0.00050 mol	–	–	–
After addition	≈ 0.00 mol	0.0023$\underline{7}$5 mol	0.00050 mol	–

Then 0.0250 L RbOH + 0.0050 L HCl = 0.0300 L total volume. So mol excess base = mol RbOH = $0.0023\underline{7}5$ mol in 0.0300 L so $[RbOH] = \frac{0.0023\underline{7}5\text{ mol RbOH}}{0.0300\text{ L}} = 0.079\underline{1}667\text{ M}$. Since RbOH is a strong base, [RbOH] excess = $[OH^-]$. $K_w = [H_3O^+][OH^-]$ so $[H_3O^+] = \frac{K_w}{[OH^-]} = \frac{1.0 \times 10^{-14}}{0.079\underline{1}667} = 1.\underline{2}6316 \times 10^{-13}\text{ M}$

and $pH = -\log [H_3O^+] = -\log(1.\underline{2}6316 \times 10^{-13}) = 12.90$.

Check: The units (none) are correct. The pH is a little lower than the initial pH, which is expected since this is a strong base.

d) **Find:** pH at equivalence point
Solution: Since this is a strong acid–strong base titration, the pH at the equivalence point is neutral or 7.

e) **Find:** pH after adding 5.0 mL of acid beyond the equivalence point
Conceptual Plan: Use calculations from parts b & c. Then the pH is only dependent on the amount of excess acid and the total solution volumes. Then
mL excess → L excess then [HCl], L excess → mol HCl excess

$\frac{1\ L}{1000\ mL}$ $\qquad$ $M = \frac{mol}{L}$

then L RbOH, L HCl to equivalence point, L HCl excess → total L then

L RbOH + L HCl to equivalence point + L HCl excess = total L

mol excess HCl, total L → [HCl] = $[H_3O^+]$ → pH

$M = \frac{mol}{L}$ $\qquad$ $pH = -\log [H_3O^+]$

Solution: $5.0\ \text{mL HCl} \times \frac{1\ L}{1000\ \text{mL}} = 0.0050\ L$ HCl excess then

$\frac{0.100\ \text{mol HCl}}{1\ L} \times 0.0050\ L = 0.00050$ mol HCl excess. Then 0.0250 L RbOH + 0.0288 L HCl + 0.0050 L HCl = 0.0588 L total volume. $[\text{HCl excess}] = \frac{0.00050\ \text{mol HCl excess}}{0.0588\ L} = 0.0085\underline{0}34$ M HCl excess

Since HCl is a strong acid, [HCl] excess = $[H_3O^+]$. Finally, $pH = -\log [H_3O^+] = -\log (0.0085\underline{0}34) = 2.07$.
Check: The units (none) are correct. The pH is dropping sharply at the equivalence point, so the pH after 5 mL past the equivalence point should be quite acidic.

45. **Given:** 20.0 mL of 0.105 M $HC_2H_3O_2$ titrated with 0.125 M NaOH **Other:** K_a ($HC_2H_3O_2$) = 1.8×10^{-5}

a) **Find:** initial pH
Conceptual Plan:
Since $HC_2H_3O_2$ is a weak acid, set up an equilibrium problem using the initial concentration.
So M $HC_2H_3O_2$ → $[H_3O^+]$ → pH

ICE Chart $\qquad$ $pH = -\log [H_3O^+]$

Solution:

$$HC_2H_3O_2(aq) + H_2O(l) \rightleftharpoons H_3O^+(aq) + C_2H_3O_2^-(aq)$$

	$[HC_2H_3O_2]$	$[H_3O^+]$	$[C_2H_3O_2^-]$
Initial	0.105	≈ 0.00	0.00
Change	$-x$	$+x$	$+x$
Equil	$0.105 - x$	$+x$	$+x$

$K_a = \frac{[H_3O^+][C_2H_3O_2^-]}{[HC_2H_3O_2]} = 1.8 \times 10^{-5} = \frac{x^2}{0.105 - x}$ Assume x is small ($x << 0.105$) so

$\frac{x^2}{0.105 - x} = 1.8 \times 10^{-5} = \frac{x^2}{0.105}$ and $x = 1.\underline{3}748 \times 10^{-3}$ M = $[H_3O^+]$. Confirm that assumption is valid

$\frac{1.\underline{3}748 \times 10^{-3}}{0.105} \times 100\ \% = 1.3\ \% < 5\ \%$ so assumption is valid. Finally,

$pH = -\log [H_3O^+] = -\log (1.\underline{3}748 \times 10^{-3}) = 2.86$

Check: The units (none) are correct. The magnitude of the answer makes physical sense because pH should be greater than $-\log (0.105) = 0.98$ because this is a weak acid.

b) **Find:** volume of base to reach equivalence point
Conceptual Plan:
Write balanced equation then mL → L then [$HC_2H_3O_2$], L → mol $HC_2H_3O_2$ then

$HC_2H_3O_2 + NaOH \rightarrow NaC_2H_3O_2 + H_2O$ $\quad \frac{1\ L}{1000\ mL} \quad M = \frac{mol}{L}$

set mol acid($HC_2H_3O_2$) = mol base(NaOH) and [NaOH], mol NaOH → L NaOH → mL NaOH

balanced equation has 1:1 stoichiometry $\quad M = \frac{mol}{L} \quad \frac{1000\ mL}{1\ L}$

Solution: $20.0\ \text{mL}\ HC_2H_3O_2 \times \frac{1\ L}{1000\ \text{mL}} = 0.0200\ L\ HC_2H_3O_2$ then

$\frac{0.105\ mol\ HC_2H_3O_2}{1\ L} \times 0.0200\ L = 0.00210\ mol\ HC_2H_3O_2$. So mol acid = mol $HC_2H_3O_2$ = 0.00210 mol =

mol NaOH then $0.00210\ \text{mol NaOH} \times \frac{1\ L}{0.125\ \text{mol NaOH}} = 0.0168\ \text{L NaOH} \times \frac{1000\ mL}{1\ L} = 16.8\ mL\ NaOH$.

Check: The units (mL) are correct. The volume of base is a little less than the volume of acid because the concentration of the base is a little greater than that of the acid.

c) **Find:** pH after adding 5.0 mL of base
Conceptual Plan:
Use calculations from part b. Then mL → L then [NaOH], L → mol NaOH then

$\frac{1\ L}{1000\ mL} \quad M = \frac{mol}{L}$

mol $HC_2H_3O_2$, mol NaOH → mol excess $HC_2H_3O_2$, mol $C_2H_3O_2^-$ and

set up stoichiometry table

L $HC_2H_3O_2$, L NaOH → total L then

L $HC_2H_3O_2$ + L NaOH = total L

then mol excess $HC_2H_3O_2$, L → [$HC_2H_3O_2$] and mol excess $C_2H_3O_2^-$, L → [$C_2H_3O_2^-$] then

$M = \frac{mol}{L} \quad M = \frac{mol}{L}$

M $HC_2H_3O_2$, M $C_2H_3O_2^-$ → [H_3O^+] → pH

ICE Chart $\quad pH = -\log[H_3O^+]$

Solution: $5.0\ \text{mL NaOH} \times \frac{1\ L}{1000\ \text{mL}} = 0.0050\ L\ NaOH$ then

$\frac{0.125\ mol\ NaOH}{1\ L} \times 0.0050\ L = 0.00062\underline{5}\ mol\ NaOH$. Set up table to track changes:

	NaOH (*aq*)	+ $HC_2H_3O_2$ (*aq*)	→ $NaC_2H_3O_2$ (*aq*)	+ H_2O (*l*)
Before addition	0.00 mol	0.00210 mol	0.00 mol	–
Addition	0.00062$\underline{5}$ mol	–	–	–
After addition	≈ 0.00 mol	0.0014$\underline{7}$5 mol	0.00062$\underline{5}$ mol	–

Then 0.0200 L H $C_2H_3O_2$ + 0.0050 L NaOH = 0.0250 L total volume. Then

$[HC_2H_3O_2] = \frac{0.0014\underline{7}5\ mol\ HC_2H_3O_2}{0.0250\ L} = 0.0590\ M$ and

$[NaC_2H_3O_2] = \frac{0.00062\underline{5}\ mol\ C_2H_3O_2^-}{0.0250\ L} = 0.025\ M$. Since 1 $C_2H_3O_2^-$ ion is generated for each $NaC_2H_3O_2$,

$[C_2H_3O_2^-] = 0.025\ M\ C_2H_3O_2^-$.

$$HC_2H_3O_2\,(aq) + H_2O(l) \rightleftharpoons H_3O^+(aq) + C_2H_3O_2^-(aq)$$

	$[HC_2H_3O_2]$	$[H_3O^+]$	$[C_2H_3O_2^-]$
Initial	0.0590	≈ 0.00	0.025
Change	$-x$	$+x$	$+x$
Equil	$0.0590 - x$	$+x$	$0.025 + x$

$K_a = \frac{[H_3O^+][C_2H_3O_2^-]}{[HC_2H_3O_2]} = 1.8 \times 10^{-5} = \frac{x(0.025 + x)}{0.0590 - x}$ Assume x is small ($x \ll 0.025 < 0.0590$) so

$$\frac{x(0.025 + \cancel{x})}{0.0590 - \cancel{x}} = 1.8 \times 10^{-5} = \frac{x(0.025)}{0.0590}$$ and $x = 4.\underline{2}48 \times 10^{-5}$ M $= [H_3O^+]$. Confirm that assumption is valid

$$\frac{4.\underline{2}48 \times 10^{-5}}{0.025} \times 100\ \% = 0.17\ \% < 5\ \%$$ so assumption is valid. Finally,

$pH = -\log[H_3O^+] = -\log(4.\underline{2}48 \times 10^{-5}) = 4.37$.

Check: The units (none) are correct. The pH is a little higher than the initial pH, which is expected since some of the acid has been neutralized.

d) **Find:** pH at one-half of the equivalence point

Conceptual Plan: Since this is a weak acid–strong base titration, the pH at one-half the equivalence point is the pK_a of the weak acid.

Solution: $pH = pK_a = -\log K_a = -\log(1.8 \times 10^{-5}) = 4.74$.

Check: The units (none) are correct. Since this is a weak acid–strong base titration, the pH at one-half the equivalence point is the pK_a of the weak acid, so it should be a little below 5.

e) **Find:** pH at equivalence point

Conceptual Plan: Use calculations from parts b. Then since all of the weak acid has been converted to its conjugate base, the pH is only dependent on the hydrolysis reaction of the conjugate base. The mol $C_2H_3O_2^-$ = initial mol $HC_2H_3O_2$ and L $HC_2H_3O_2$, L NaOH to equivalence point → total L then

$L\ HC_2H_3O_2 + L\ NaOH = \text{total L}$

mol excess $C_2H_3O_2^-$, L → $[C_2H_3O_2^-]$ and K_a → K_b then do equilibrium calculation:

$M = \frac{mol}{L}$ $\qquad K_w = K_a K_b$

$[C_2H_3O_2^-]$, K_b → $[OH^-]$ → $[H_3O^+]$ → pH

set up ICE table $\qquad K_w = [H_3O^+][OH^-] \qquad pH = -\log[H_3O^+]$

Solution: mol $C_2H_3O_2^-$ = initial mol $HC_2H_3O_2$ = 0.00210 mol and total volume = L $HC_2H_3O_2$ + L NaOH = 0.020 L + 0.0168 L = 0.0368 L then $[C_2H_3O_2^-] = \frac{0.00210\ \text{mol}\ C_2H_3O_2^-}{0.0368\ \text{L}} = 0.057\underline{0}652$ M and

$K_w = K_a K_b$. Rearrange to solve for K_b. $K_b = \frac{K_w}{K_a} = \frac{1.0 \times 10^{-14}}{1.8 \times 10^{-5}} = 5.\underline{5}556 \times 10^{-10}$. Set up ICE table

$$C_2H_3O_2^-(aq) + H_2O(l) \rightleftharpoons HC_2H_3O_2(aq) + OH^-(aq)$$

	$[C_2H_3O_2^-]$	$[HC_2H_3O_2]$	$[OH^-]$
Initial	0.057$\underline{0}$652	≈ 0.00	≈ 0.00
Change	$-x$	$+x$	$+x$
Equil	0.057$\underline{0}$652 $- x$	$+x$	$+x$

$$K_b = \frac{[HC_2H_3O_2][OH^-]}{[C_2H_3O_2^-]} = 5.\underline{5}556 \times 10^{-10} = \frac{x^2}{0.057\underline{0}652 - x}$$ Assume x is small ($x << 0.057$) so

$$\frac{x^2}{0.057\underline{0}652 - \cancel{x}} = 5.\underline{5}556 \times 10^{-10} = \frac{x^2}{0.057\underline{0}652}$$ and $x = 5.\underline{6}305 \times 10^{-6}$ M $= [OH^-]$.

Confirm that assumption is valid $\frac{5.\underline{6}305 \times 10^{-6}}{0.057\underline{0}652} \times 100\ \% = 0.0099\ \% < 5\ \%$ so assumption is valid.

$K_w = [H_3O^+][OH^-]$ so $[H_3O^+] = \frac{K_w}{[OH^-]} = \frac{1.0 \times 10^{-14}}{5.\underline{6}305 \times 10^{-6}} = 1.\underline{7}760 \times 10^{-9}$ M. Finally,

$pH = -\log[H_3O^+] = -\log(1.\underline{7}760 \times 10^{-9}) = 8.75$.

Check: The units (none) are correct. Since this is a weak acid–strong base titration, the pH at the equivalence point is basic.

f) **Find:** pH after adding 5.0 mL of base beyond the equivalence point

Conceptual Plan: Use calculations from parts b & c. Then the pH is only dependent on the amount of excess base and the total solution volumes.

mL excess → L excess then [NaOH], L excess → mol NaOH excess

$$\frac{1\text{ L}}{1000\text{ mL}} \qquad M = \frac{\text{mol}}{\text{L}}$$

then L $HC_2H_3O_2$, L NaOH to equivalence point, L NaOH excess → total L then

L $HC_2H_3O_2$ + L NaOH to equivalence point + L NaOH excess = total L

mol excess NaOH, total L → [NaOH] = [OH⁻] → [H_3O^+] → pH

$$M = \frac{\text{mol}}{\text{L}} \qquad K_w = [H_3O^+][OH^-] \qquad pH = -\log[H_3O^+]$$

Solution: $5.0 \text{ mL NaOH} \times \dfrac{1\text{ L}}{1000\text{ mL}} = 0.0050\text{ L NaOH excess}$ then

$\dfrac{0.125\text{ mol NaOH}}{1\text{ L}} \times 0.0050\text{ L} = 0.000625\text{ mol NaOH excess}$. Then 0.0200 L $HC_2H_3O_2$ + 0.0168 L NaOH + 0.0050 L NaOH = 0.0418 L total volume.

$[\text{NaOH excess}] = \dfrac{0.000625\text{ mol NaOH excess}}{0.0418\text{ L}} = 0.014\underline{9}522\text{ M NaOH excess}$ Since NaOH is a strong base, [NaOH] excess = [OH^-]. The strong base overwhelms the weak base and is insignificant in the calculation. $K_w = [H_3O^+][OH^-]$ so $[H_3O^+] = \dfrac{K_w}{[OH^-]} = \dfrac{1.0 \times 10^{-14}}{0.014\underline{9}522} = 6.\underline{6}88 \times 10^{-13}\text{ M}$. Finally,

$pH = -\log[H_3O^+] = -\log(6.\underline{6}88 \times 10^{-13}) = 12.17$.

Check: The units (none) are correct. The pH is rising sharply at the equivalence point, so the pH after 5 mL past the equivalence point should be quite basic.

47. **Given:** 25.0 mL of 0.175 M CH_3NH_2 titrated with 0.150 M HBr **Other:** $K_b(CH_3NH_2) = 4.4 \times 10^{-4}$

a) **Find:** initial pH

Conceptual Plan: Conceptual Plan: Since CH_3NH_2 is a weak base, set up an equilibrium problem using the initial concentration, so M CH_3NH_2 → [OH^-] → [H_3O^+] → pH

ICE Chart $K_w = [H_3O^+][OH^-]$ $pH = -\log[H_3O^+]$

Solution:

$$CH_3NH_2(aq) + H_2O(l) \rightleftharpoons CH_3NH_3^+(aq) + OH^-(aq)$$

	$[CH_3NH_2]$	$[CH_3NH_3^+]$	$[OH^-]$
Initial	0.175	0.00	≈ 0.00
Change	$-x$	$+x$	$+x$
Equil	$0.175 - x$	$+x$	$+x$

$$K_b = \frac{[CH_3NH_3^+][OH^-]}{[CH_3NH_2]} = 4.4 \times 10^{-4} = \frac{x^2}{0.175 - x}$$

Assume x is small ($x << 0.175$) so $\dfrac{x^2}{0.175 - x} = 4.4 \times 10^{-4} = \dfrac{x^2}{0.175}$ and $x = 8.\underline{7}750 \times 10^{-3}\text{ M} = [OH^-]$.

Confirm that assumption is valid $\dfrac{8.\underline{7}750 \times 10^{-3}}{0.175} \times 100\,\% = 5.0\,\%$ so assumption is valid.

$K_w = [H_3O^+][OH^-]$ so $[H_3O^+] = \dfrac{K_w}{[OH^-]} = \dfrac{1.0 \times 10^{-14}}{8.\underline{7}750 \times 10^{-3}} = 1.\underline{1}396 \times 10^{-12}\text{ M}$. Finally,

$pH = -\log[H_3O^+] = -\log(1.\underline{1}396 \times 10^{-12}) = 11.94$.

Check: The units (none) are correct. The magnitude of the answer makes physical sense because pH should be less than 14 + log (0.175) = 13.2 because this is a weak base.

b) **Find:** volume of acid to reach equivalence point

Conceptual Plan: Write balanced equation then mL → L then [CH_3NH_2], L → mol CH_3NH_2

$HBr + CH_3NH_2 \rightarrow CH_3NH_3Br + H_2O$ $\qquad \dfrac{1\text{ L}}{1000\text{ mL}} \qquad M = \dfrac{\text{mol}}{\text{L}}$

then set mol base (CH_3NH_2) = mol acid (HBr) and [HBr], mol HBr → L HBr → mL HBr

balanced equation has 1:1 stoichiometry $\qquad M = \frac{mol}{L} \qquad \frac{1000\ mL}{1\ L}$

Solution: $25.0\ \text{mL}\ CH_3NH_2 \times \frac{1\ L}{1000\ mL} = 0.0250\ L\ CH_3NH_2$ then

$\frac{0.175\ mol\ CH_3NH_2}{1\ L} \times 0.0250\ L = 0.004375\ mol\ CH_3NH_2$. So mol base = mol CH_3NH_2 = 0.002875 mol

= mol HBr then $0.004375\ mol\ HBr \times \frac{1\ L}{0.150\ mol\ HBr} = 0.0291667\ L\ HBr \times \frac{1000\ mL}{1\ L} = 29.2\ mL\ HBr$.

Check: The units (mL) are correct. The volume of acid is a greater than the volume of base because the concentration of the base is a little greater than that of the acid.

c) **Find:** pH after adding 5.0 mL of acid
Conceptual Plan: Use calculations from part b. Then mL → L then [HBr], L → mol HBr

$\frac{1\ L}{1000\ mL} \qquad M = \frac{mol}{L}$

then mol CH_3NH_2, mol HBr → mol excess CH_3NH_2 and L CH_3NH_2, L HBr → total L then

set up stoichiometry table $\qquad$ L CH_3NH_2 + L HBr = total L

Since there are significant concentrations of both the acid and the conjugate base species, this is a buffer solution and so Henderson–Hasselbalch Equation $\left(pH = pK_a + \log\frac{[base]}{[acid]}\right)$ can be used. Convert K_b to K_a using $K_w = K_a\ K_b$. Also note that ratio of concentrations is the same as the ratio of moles, since the volume is the same for both species.

Solution: $5.0\ mL\ HBr \times \frac{1\ L}{1000\ mL} = 0.0050\ L\ HBr$ then $\frac{0.150\ mol\ HBr}{1\ L} \times 0.0050\ L = 0.00075\ mol\ HBr$.

Set up table to track changes:

	HBr (*aq*)	+ CH_3NH_2 (*aq*)	→ CH_3NH_3Br (*aq*)
Before addition	0.00 mol	0.004375 mol	0.00 mol
Addition	0.00075 mol	–	–
After addition	≈ 0.00 mol	0.003625 mol	0.000750 mol

then $K_w = K_a\ K_b$ so

$K_a = \frac{K_w}{K_b} = \frac{1.0 \times 10^{-14}}{4.4 \times 10^{-4}} = 2.2727 \times 10^{-11}\ M$ then use Henderson–Hasselbalch Equation, since the solution is a buffer. $pH = pK_a + \log\frac{[base]}{[acid]} = -\log(2.2727 \times 10^{-11}) + \log\frac{0.003625}{0.000750} = 11.33$

Check: The units (none) are correct. The pH is a little lower than the last pH, which is expected since some of the base has been neutralized.

d) **Find:** pH at one-half of the equivalence point
Conceptual Plan: Since this is a weak base–strong acid titration, the pH at one-half the equivalence point is the pK_a of the conjugate acid of weak base.
Solution: $pH = pK_a = -\log K_a = -\log(2.2727 \times 10^{-11}) = 10.64$.

Check: The units (none) are correct. Since this is a weak acid–strong base titration, the pH at one-half the equivalence point is the pK_a of the conjugate acid of the weak base, so it should be a little below 11.

e) **Find:** pH at equivalence point
Conceptual Plan: Use calculations from above. Since all of the weak base has been converted to its conjugate acid, the pH is only dependent on the hydrolysis reaction of the conjugate acid. The mol $CH_3NH_3^+$ = initial mol CH_3NH_2 and L CH_3NH_2, L HBr to equivalence point → total L then mol $CH_3NH_3^+$, L → [$CH_3NH_3^+$]

L CH_3NH_2 + L HBr = total L $\qquad M = \frac{mol}{L}$

then do equilibrium calculation: [$CH_3NH_3^+$], K_a → [H_3O^+] → pH

set up ICE table $\qquad pH = -\log[H_3O^+]$

Solution: mol base = mol acid = mol $CH_3NH_3^+$ = 0.004375 mol. Then total volume = L CH_3NH_2 + L

HBr = 0.0250 L + 0.0292 L = 0.0542 then $[CH_3NH_3^+] = \dfrac{0.004375 \text{ mol mol } CH_3NH_3^+}{0.0542 \text{ L}} = 0.080\underline{7}196$ M .

$$CH_3NH_3^+(aq) + H_2O(l) \rightleftharpoons CH_3NH_2(aq) + H_3O^+(aq)$$

		$[CH_3NH_3^+]$	$[CH_3NH_2]$	$[H_3O^+]$
Set up ICE table	*Initial*	0.080\underline{7}196	≈ 0.00	≈ 0.00
	Change	$-x$	$+x$	$+x$
	Equil	0.080\underline{7}196 $- x$	$+x$	$+x$

$K_a = \dfrac{[CH_3NH_2][H_3O^+]}{[CH_3NH_3^+]} = 2.\underline{2}727 \times 10^{-11} = \dfrac{x^2}{0.080\underline{7}196 - x}$ Assume x is small ($x << 0.0807$) so

$\dfrac{x^2}{0.080\underline{7}196 - \cancel{x}} = 2.\underline{2}727 \times 10^{-11} = \dfrac{x^2}{0.080\underline{7}196}$ and $x = 1.\underline{3}544 \times 10^{-6} = [H_3O^+]$.

Confirm that assumption is valid $\dfrac{1.\underline{3}544 \times 10^{-6}}{0.080\underline{7}106} \times 100\% = 0.0017\% < 5\%$ so assumption is valid.

Finally, $pH = -\log[H_3O^+] = -\log(1.\underline{3}544 \times 10^{-6}) = 5.87$.

Check: The units (none) are correct. Since this is a weak base–strong acid titration, the pH at the equivalence point is acidic.

f) **Find:** pH after adding 5.0 mL of acid beyond the equivalence point
 Conceptual Plan: Use calculations from parts b & c. Then the pH is only dependent on the amount of excess acid and the total solution volumes.
 mL excess → L excess   then   [HBr], L excess → mol HBr excess

$$\dfrac{1\text{ L}}{1000 \text{ mL}} \qquad\qquad M = \dfrac{mol}{L}$$

then   L CH_3NH_2, L HBr to equivalence point, L HBr excess → total L   then
 L CH_3NH_2 + L HBr to equivalence point + L HBr excess = total L
 mol excess HBr, total L → [HBr] = [H$_3$O$^+$] →   pH

$$M = \dfrac{mol}{L} \qquad\qquad pH = -\log[H_3O^+]$$

Solution: $5.0 \text{ m\cancel{L} HBr} \times \dfrac{1\text{ L}}{1000 \text{ m\cancel{L}}} = 0.0050$ L HBr excess   then

$\dfrac{0.150 \text{ mol HBr}}{1 \cancel{L}} \times 0.0050 \text{ }\cancel{L} = 0.00075$ mol HBr excess.   Then 0.0250 L $CH_3NH_2 + 0.0292$ L HBr +

0.0050 L HBr = 0.0592 L total volume.

$[HBr \text{ excess}] = \dfrac{0.00075 \text{ mol HBr excess}}{0.0592 \text{ L}} = 0.01\underline{2}669$ M HBr excess

Since HBr is a strong acid, [HBr] excess = [H$_3$O$^+$]. The strong acid overwhelms the weak acid and is insignificant in the calculation. Finally, $pH = -\log[H_3O^+] = -\log(0.01\underline{2}669) = 1.90$.

Check: The units (none) are correct. The pH is dropping sharply at the equivalence point, so the pH after 5 mL past the equivalence point should be quite acidic.

49. i) acid a is more concentrated, since the equivalence point (where sharp pH rise occurs) is at a higher volume of added base.

 ii) acid b has the larger K_a, since the pH at a volume of added base equal to half of the equivalence point volume is lower.

51. **Given:** 0.229 g unknown monoprotic acid titrated with 0.112 M NaOH and curve   **Find:** molar mass and pK_a of acid
 Conceptual Plan: The equivalence point is where sharp pH rise occurs. The pK_a is the pH at a volume of added base equal to half of the equivalence point volume. Then mL NaOH → L NaOH

$$\dfrac{1\text{ L}}{1000 \text{ mL}}$$

Chapter 16 – Aqueous Ionic Equilibria

then [NaOH], L NaOH → mol NaOH = mol acid   then mol NaOH, g acid → molar mass

$M = \frac{mol}{L}$ $\frac{g\ acid}{mol\ acid}$

Solution: The equivalence point is at 25 mL NaOH. The pH at 0.5 x 25 mL = 13 mL is ~3 = pK_a. then

$$25\ \text{mL NaOH} \times \frac{1\ \text{L}}{1000\ \text{mL}} = 0.025\ \text{L NaOH}$$ then

$$\frac{0.112\ \text{mol NaOH}}{1\ \text{L}} \times 0.025\ \text{L} = 0.0028\ \text{mol NaOH} = 0.0028\ \text{mol acid}$$ then

$$\text{Molar Mass} = \frac{0.229\ \text{g acid}}{0.0028\ \text{mol acid}} = 82\ \text{g/mol}.$$

Check: The units (none and g/mol) are correct. The pK_a is consistent with a weak acid. The molar mass is reasonable for an acid (>1 g/mol).

53. Since the exact conditions of the titration are not given, a rough calculation will suffice. Looking at the pattern of earlier problems, the pH at the equivalence point of a titration of a weak acid and a strong base is the hydrolysis of the conjugate base of the weak acid that has been diluted by a factor of roughly 2 with base. If it is assumed that the initial concentration of the weak acid is ~ 0.1 M, then the conjugate base concentration will be ~ 0.05 M. From earlier calculations it can be seen that the $K_b = \frac{K_w}{K_a} = \frac{[OH^-]^2}{0.05}$ thus

$$[OH^-] = \sqrt{\frac{0.05\,K_w}{K_a}} = \sqrt{\frac{5 \times 10^{-16}}{K_a}}$$ and the $\text{pH} = 14 + \log\sqrt{\frac{5 \times 10^{-16}}{K_a}}$.

a) for HF, the pK_a = 3.5 x 10^{-4} and so the above equation approximates the pH at the equivalence point of ~ 8.0. Looking at Table 16.1, phenol red or *m*-nitrophenol will change at the appropriate pH range.

b) for HCl, the pH at the equivalence point is 7, since HCl is a strong acid. Looking at Table 16.1, alizarin, bromthymol blue, or phenol red will change at the appropriate pH range.

c) for HCN, the pK_a = 4.9 x 10^{-10} and so the above equation approximates the pH at the equivalence point of ~ 11.0. Looking at Table 16.1, alizarin yellow R will change at the appropriate pH range.

55. For the dissolution reaction, start with the ionic compound as a solid and put it in equilibrium with the appropriate cation and anion, making sure to include the appropriate stoichiometric coefficients. The K_{sp} expression is the product of the concentrations of the cation and anion concentrations raised to their stoichiometric coefficients.

a) $BaSO_4\,(s) \rightleftharpoons Ba^{2+}\,(aq) + SO_4^{2-}\,(aq)$ and $K_{sp} = [Ba^{2+}][SO_4^{2-}]$.

b) $PbBr_2\,(s) \rightleftharpoons Pb^{2+}\,(aq) + 2\,Br^-\,(aq)$ and $K_{sp} = [Pb^{2+}][Br^-]^2$.

c) $Ag_2CrO_4\,(s) \rightleftharpoons 2\,Ag^+\,(aq) + CrO_4^{2-}\,(aq)$ and $K_{sp} = [Ag^+]^2[CrO_4^{2-}]$.

57. **Given:** ionic compound formula and Table 16.2 of K_{sp} values   **Find:** molar solubility (*S*)
Conceptual Plan: The expression of the solubility product constant of A_mX_n is: $K_{sp} = [A^{n+}]^m\,[X^{m-}]^n$. The molar solubility of a compound, A_mX_n, can be computed directly from K_{sp} by solving for *S* in the expression: $K_{sp} = (mS)^m\,(nS)^n = m^m\,n^n\,S^{m+n}$.

Solution:

a) for AgBr, K_{sp} = 5.35 x 10^{-13}, A = Ag^+, m = 1, X = Br^-, and n = 1 so K_{sp} = 5.35 x 10^{-13} = S^2. Rearrange to solve for *S*. $S = \sqrt{5.35 \times 10^{-13}} = 7.31 \times 10^{-7}$ M.

b) for $Mg(OH)_2$, K_{sp} = 2.06 x 10^{-13}, A = Mg^{2+}, m = 1, X = OH^-, and n = 2 so K_{sp} = 2.06 x 10^{-13} = 2^2S^3. Rearrange to solve for *S*. $S = \sqrt[3]{\frac{2.06 \times 10^{-13}}{4}} = 3.72 \times 10^{-5}$ M.

c) for CaF_2, K_{sp} = 1.46 x 10^{-10}, A = Ca^{2+}, m = 1, X = F^-, and n = 2 so K_{sp} = 1.46 x 10^{-10} = 2^2S^3. Rearrange to solve for S. $S = \sqrt[3]{\dfrac{1.46 \times 10^{-10}}{4}} = 3.32 \times 10^{-4}$ M .

Check: The units (M) are correct. The molar solubilities are much less than one and dependent not only on the value of the K_{sp}, but also the stoichiometry of the ionic compound. The more ions that are generated, the greater the molar solubility for the same value of the K_{sp}.

59. **Given:** ionic compound formula and molar solubility (S) **Find:** K_{sp}
Conceptual Plan: The expression of the solubility product constant of A_mX_n is: $K_{sp} = [A^{n+}]^m [X^{m-}]^n$. The molar solubility of a compound, A_mX_n, can be computed directly from K_{sp} by solving for S in the expression: $K_{sp} = (mS)^m (nS)^n = m^m n^n S^{m+n}$.
Solution:
a) for NiS, $S = 3.27 \times 10^{-11}$ M, A = Ni^{2+}, m = 1, X = S^{2-}, and n = 1 so $K_{sp} = S^2 = (3.27 \times 10^{-11})^2 = 1.07 \times 10^{-21}$.

b) for PbF_2, $S = 5.63 \times 10^{-3}$ M, A = Pb^{2+}, m = 1, X = F^-, and n = 2 so $K_{sp} = 2^2S^3 = 2^2 (5.63 \times 10^{-3})^3 = 7.14 \times 10^{-7}$.

c) for MgF_2, $S = 2.65 \times 10^{-4}$ M, A = Mg^{2+}, m = 1, X = F^-, and n = 2 so $K_{sp} = 2^2S^3 = 2^2 (2.65 \times 10^{-4})^3 = 7.44 \times 10^{-11}$.

Check: The units (none) are correct. The K_{sp} values are much less than one and dependent not only on the value of the solubility, but also the stoichiometry of the ionic compound. The more ions that are generated, the smaller the K_{sp} for the same value of the S.

61. **Given:** ionic compound formulas AX and AX_2 and $K_{sp} = 1.5 \times 10^{-5}$ **Find:** higher molar solubility (S)
Conceptual Plan: The expression of the solubility product constant of A_mX_n is: $K_{sp} = [A^{n+}]^m [X^{m-}]^n$. The molar solubility of a compound, A_mX_n, can be computed directly from K_{sp} by solving for S in the expression: $K_{sp} = (mS)^m (nS)^n = m^m n^n S^{m+n}$.
Solution: for AX, $K_{sp} = 1.5 \times 10^{-5}$, m = 1, and n = 1 so $K_{sp} = 1.5 \times 10^{-5} = S^2$. Rearrange to solve for S. $S = \sqrt{1.5 \times 10^{-5}} = 3.9 \times 10^{-3}$ M . For AX_2, $K_{sp} = 1.5 \times 10^{-5}$, m = 1, and n = 2 so $K_{sp} = 1.5 \times 10^{-5} = 2^2S^3$. Rearrange to solve for S. $S = \sqrt[3]{\dfrac{1.5 \times 10^{-5}}{4}} = 1.6 \times 10^{-2}$ M . Since 10^{-2} M > 10^{-3} M, AX_2 has a higher molar solubility.
Check: The units (M) are correct. The more ions that are generated, the greater the molar solubility for the same value of the K_{sp}.

63. **Given:** $Fe(OH)_2$ in 100.0 mL solution **Find:** grams of $Fe(OH)_2$ **Other:** $K_{sp} = 4.87 \times 10^{-17}$
Conceptual Plan: The expression of the solubility product constant of A_mX_n is: $K_{sp} = [A^{n+}]^m [X^{m-}]^n$. The molar solubility of a compound, A_mX_n, can be computed directly from K_{sp} by solving for S in the expression: $K_{sp} = (mS)^m (nS)^n = m^m n^n S^{m+n}$. Then solve for S, then mL → L then

$$\frac{1\ \text{L}}{1000\ \text{mL}}$$

S, L → mol $Fe(OH)_2$ → g $Fe(OH)_2$

$$M = \frac{\text{mol}}{\text{L}} \qquad \frac{89.87\ \text{g Fe(OH)}_2}{1\ \text{mol Fe(OH)}_2}$$

Solution: for $Fe(OH)_2$, $K_{sp} = 4.87 \times 10^{-17}$, A = Fe^{2+}, m = 1, X = OH^-, and n = 2 so $K_{sp} = 4.87 \times 10^{-17} = 2^2S^3$. Rearrange to solve for S.

$$S = \sqrt[3]{\frac{4.87 \times 10^{-17}}{4}} = 2.3\underline{0}050 \times 10^{-6}\ \text{M. Then } 100.0\ \cancel{\text{mL}} \times \frac{1\ \text{L}}{1000\ \cancel{\text{mL}}} = 0.1000\ \text{L then}$$

$$\frac{2.3\underline{0}050 \times 10^{-6}\ \text{mol Fe(OH)}_2}{1\ \cancel{\text{L}}} \times 0.1000\ \cancel{\text{L}} = 2.3\underline{0}050 \times 10^{-7}\ \cancel{\text{mol Fe(OH)}_2} \times \frac{89.87\ \text{g Fe(OH)}_2}{1\ \cancel{\text{mol Fe(OH)}_2}} = 2.07 \times 10^{-5}\ \text{g Fe(OH)}_2.$$

Check: The units (g) are correct. The solubility rules from Chapter 4 (most hydroxides are insoluble) suggest that very little $Fe(OH)_2$ will dissolve, so the magnitude of the answer is not surprising.

65. a) **Given:** BaF_2 **Find:** molar solubility (S) in pure water **Other:** K_{sp} (BaF_2) = 2.45×10^{-5}
Conceptual Plan: The expression of the solubility product constant of A_mX_n is: $K_{sp} = [A^{n+}]^m [X^{m-}]^n$. The molar solubility of a compound, A_mX_n, can be computed directly from K_{sp} by solving for S in the

expression: $K_{sp} = (mS)^m (nS)^n = m^m n^n S^{m+n}$.

Solution: BaF_2, K_{sp} = 2.45 x 10^{-5}, A = Ba^{2+}, m = 1, X = F^-, and n = 2 so K_{sp} = 2.45 x 10^{-5} = $2^2 S^3$.

Rearrange to solve for *S*. $S = \sqrt[3]{\frac{2.45 \times 10^{-5}}{4}} = 1.83 \times 10^{-2}$ M.

b) **Given:** BaF_2 **Find:** molar solubility (*S*) in 0.10 M $Ba(NO_3)_2$ **Other:** K_{sp} (BaF_2) = 2.45 x 10^{-5}

Conceptual Plan: M $Ba(NO_3)_2$ → M Ba^{2+} then M Ba^{2+}, K_{sp} → *S*

$Ba(NO_3)_2$ (*s*) → Ba^{2+} (*aq*) + 2 NO_3^- (*aq*) ICE Chart

Solution: Since 1 Ba^{2+} ion is generated for each $Ba(NO_3)_2$, [Ba^{2+}] = 0.10 M.

BaF_2 (*s*) ⇌	Ba^{2+} (*aq*) +	2 F^-(*aq*)
Initial	0.10	0.00
Change	*S*	2*S*
Equil	0.10 + *S*	2*S*

K_{sp} (BaF_2) = [Ba^{2+}] [F^-]^2 = 2.45 x 10^{-5} = (0.10 + *S*)(2 *S*)^2.

Assume *S* << 0.10, 2.45 x 10^{-5} = (0.10)(2 *S*)^2, and *S* = 7.83 x 10^{-3} M. Confirm that assumption is valid $\frac{7.83 \times 10^{-3}}{0.10}$ x 100 % = 7.8 % > 5 % so assumption is not valid. Since expanding the expression will give a third order polynomial, that is not easily solved directly. Solve by successive approximations. Substitute *S* = 7.83 x 10^{-3} M for the *S* term that is part of a sum (i.e., the one in (0.10 + *S*)). Thus, 2.45 x 10^{-5} = (0.10 + 7.83 x 10^{-3})(2 *S*)^2 and *S* = 7.53 x 10^{-3} M. Substitute this new *S* value again. Thus, 2.45 x 10^{-5} = (0.10 + 7.53 x 10^{-3})(2 *S*)^2 and *S* = 7.55 x 10^{-3} M. Substitute this new *S* value again. Thus, 2.45 x 10^{-5} = (0.10 + 7.55 x 10^{-3})(2 *S*)^2 and *S* = 7.55 x 10^{-3} M. So the solution has converged and *S* = 7.55 x 10^{-3} M.

c) **Given:** BaF_2 **Find:** molar solubility (*S*) in 0.15 M NaF **Other:** K_{sp} (BaF_2) = 2.45 x 10^{-5}

Conceptual Plan: M NaF → M F^- then M F^-, K_{sp} → *S*

NaF (*s*) → Na^+ (*aq*) + F^- (*aq*) ICE Chart

Solution: Since 1 F^- ion is generated for each NaF, [F^-] = 0.15 M.

BaF_2 (*s*) ⇌	Ba^{2+} (*aq*) +	2 F^-(*aq*)
Initial	0.00	0.15
Change	*S*	2*S*
Equil	*S*	0.15 + 2*S*

K_{sp} (BaF_2) = [Ba^{2+}] [F^-]^2 = 2.45 x 10^{-5} = (*S*)(0.15 + 2 *S*)^2.

Since 2 *S* << 0.15, 2.45 x 10^{-5} = (*S*)(0.15)^2, and *S* = 1.09 x 10^{-3} M. Confirm that assumption is valid $\frac{2\,(1.09 \times 10^{-3})}{0.10}$ x 100 % = 2.2 % < 5 % so assumption is valid.

Check: The units (M) are correct. The solubility of the BaF_2 decreases in the presence of a common ion. The effect of the anion is greater because the K_{sp} expression has the anion concentration squared.

67. **Given:** $Ca(OH)_2$ **Find:** molar solubility (*S*) in buffers at a) pH = 4, b) pH = 7, and c) pH = 9

Other: K_{sp} ($Ca(OH)_2$) = 4.68 x 10^{-6}

Conceptual Plan: pH → [H_3O^+] → [OH^-] then M OH^-, K_{sp} → *S*

$[H_3O^+] = 10^{-pH}$ $K_w = [H_3O^+][OH^-]$ set up ICE table

Solution:

a) pH = 4, so $[H_3O^+] = 10^{-pH} = 10^{-4} = 1 \times 10^{-4}$ M then $K_w = [H_3O^+][OH^-]$ so

$[OH^-] = \frac{K_w}{[H_3O^+]} = \frac{1.0 \times 10^{-14}}{1 \times 10^{-4}} = 1 \times 10^{-10}$ M then

$Ca(OH)_2$ (*s*) ⇌	Ca^{2+} (*aq*) +	2 OH^-(*aq*)
Initial	0.00	1 x 10^{-10}
Change	*S*	–
Equil	*S*	1 x 10^{-10}

K_{sp} ($Ca(OH)_2$) = [Ca^{2+}] [OH^-]^2 = 4.68 x 10^{-6} = *S*(1 x 10^{-10})^2.and *S* = 5 x 10^{14} M.

b) pH = 7, so $[H_3O^+] = 10^{-pH} = 10^{-7} = 1 \times 10^{-7}$ M then $K_w = [H_3O^+][OH^-]$ so

$[OH^-] = \frac{K_w}{[H_3O^+]} = \frac{1.0 \times 10^{-14}}{1 \times 10^{-7}} = 1 \times 10^{-7}$ M then

$Ca(OH)_2\,(s) \rightleftharpoons Ca^{2+}\,(aq) + 2OH^-(aq)$		
Initial	0.00	1×10^{-7}
Change	*S*	–
Equil	*S*	1×10^{-7}

$K_{sp}\,(Ca(OH)_2) = [Ca^{2+}]\,[OH^-]^2 = 4.68 \times 10^{-6} = S\,(1 \times 10^{-7})^2$ and $S = 5 \times 10^8$ M.

c) $pH = 9$, so $[H_3O^+] = 10^{-pH} = 10^{-9} = 1 \times 10^{-9}$ M then $K_w = [H_3O^+][OH^-]$ so

$[OH^-] = \dfrac{K_w}{[H_3O^+]} = \dfrac{1.0 \times 10^{-14}}{1 \times 10^{-9}} = 1 \times 10^{-5}$ M then

$Ca(OH)_2\,(s) \rightleftharpoons Ca^{2+}\,(aq) + 2\,OH^-(aq)$		
Initial	0.00	1×10^{-5}
Change	*S*	–
Equil	*S*	1×10^{-5}

$K_{sp}\,(Ca(OH)_2) = [Ca^{2+}]\,[OH^-]^2 = 4.68 \times 10^{-6} = S\,(1 \times 10^{-5})^2$. and $S = 5 \times 10^4$ M.

Check: The units (M) are correct. The solubility of the $Ca(OH)_2$ decreases as the pH increases (and the hydroxide ion concentration increases). Realize that these molar solubilities are not achievable because the density of pure $Ca(OH)_2$ is ~ 30 M. The bottom line is that as long as the hydroxide concentration can be controlled with a buffer, the $Ca(OH)_2$ will be very soluble.

69. a) $BaCO_3$ will be more soluble in acidic solutions because CO_3^{2-} is basic. In acidic solutions it can be converted to HCO_3^- and $H_2CO_3^{2-}$. These species are not CO_3^{2-} so they do not appear in the K_{sp} expression.

b) CuS will be more soluble in acidic solutions because S^{2-} is basic. In acidic solutions it can be converted to HS^- and H_2S^{2-}. These species are not S^{2-} so they do not appear in the K_{sp} expression.

c) AgCl will not be more soluble in acidic solutions because Cl^- will not react with acidic solutions, because HCl is a strong acid.

d) PbI_2 will not be more soluble in acidic solutions because I^- will not react with acidic solutions, because HI is a strong acid.

71. **Given:** 0.015 M NaF and 0.010 M $Ca(NO_3)_2$ **Find:** will a precipitate form, if so, identify it
Other: $K_{sp}\,(CaF_2) = 1.46 \times 10^{-10}$
Conceptual Plan: Look at all possible combinations and consider the solubility rules from Chapter 4. Salts of alkali metals (Na) are very soluble, so NaF and $NaNO_3$ will be very soluble. Nitrate compounds are very soluble so $NaNO_3$ will be very soluble. The only possibility for a precipitate is CaF_2. Determine if a precipitate will form by determining the concentration of the Ca^{2+} and F^- in solution. Then compute the reaction quotient, Q. If $Q > K_{sp}$ then a precipitate will form.
Solution: Since the only possible precipitate is CaF_2, calculate the concentrations of Ca^{2+} and F^-. $NaF\,(s) \rightarrow Na^+\,(aq) + F^-\,(aq)$. Since 1 F^- ion is generated for each NaF, $[F^-] = 0.015$ M. $Ca(NO_3)_2\,(s) \rightarrow Ca^{2+}\,(aq) + 2\,NO_3^-\,(aq)$. Since 1 Ca^{2+} ion is generated for each $Ca(NO_3)_2$, $[Ca^{2+}] = 0.010$ M. Then calculate Q (CaF_2), $A = Ca^{2+}$, $m = 1$, $X = F^-$, and $n = 2$ Since $Q = [A^{n+}]^m\,[X^{m-}]^n$, then
$Q\,(CaF_2) = [Ca^{2+}]\,[F^-]^2 = (0.010)\,(0.015)^2 = 2.3 \times 10^{-6} > 1.46 \times 10^{-10} = K_{sp}\,(CaF_2)$, so a precipitate will form.
Check: The units (none) are correct. The solubility of the CaF_2 is low, and the concentration of ions are extremely large compared to the K_{sp}, so a precipitate will form.

73. **Given:** 75.0 mL of NaOH with pOH = 2.58 and 125.0 mL of 0.0018 M $MgCl_2$
Find: will a precipitate form, if so, identify it **Other:** $K_{sp}\,(Mg(OH)_2) = 2.06 \times 10^{-13}$
Conceptual Plan: Look at all possible combinations and consider the solubility rules from Chapter 4. Salts of alkali metals (Na) are very soluble, so NaOH and NaCl will be very soluble. Chloride compounds are generally very soluble so $MgCl_2$ and NaCl will be very soluble. The only possibility for a precipitate is $Mg(OH)_2$. Determine if a precipitate will form by determining the concentration of the Mg^{2+} and OH^- in solution. Since pH, not NaOH concentration, is given pOH → $[OH^-]$ then

$[OH^-] = 10^{-pOH}$

mix solutions and calculate diluted concentrations mL NaOH, mL $MgCl_2$ → mL total then

mL NaOH + mL $MgCl_2$ = total mL

mL, initial M → final M then compute the reaction quotient, Q.

$M_1V_1 = M_2V_2$

If Q > K_{sp} then a precipitate will form.

Solution: Since the only possible precipitate is $Mg(OH)_2$, calculate the concentrations of Mg^{2+} and OH^-. for NaOH at pOH = 2.58, so $[OH^-] = 10^{-pOH} = 10^{-2.58} = 2.\underline{6}3027 \times 10^{-3}$ M and $MgCl_2$ (s) → Mg^{2+} (aq) + 2 Cl^- (aq). Since 1 Mg^{2+} ion is generated for each $MgCl_2$, $[Mg^{2+}] = 0.0018$ M. Then total mL = mL NaOH + mL $MgCl_2$ = 75.0 mL + 125.0 mL = 200.0 mL. Then $M_1V_1 = M_2V_2$,

rearrange to solve for M_2. $M_2 = M_1\frac{V_1}{V_2} = 2.\underline{6}3027 \times 10^{-3}\text{ M OH}^- \times \frac{75.0\text{ mL}}{200.0\text{ mL}} = 9.\underline{8}635 \times 10^{-4}\text{ M OH}^-$ and

$M_2 = M_1\frac{V_1}{V_2} = 0.0018\text{ M Mg}^+ \times \frac{125.0\text{ mL}}{200.0\text{ mL}} = 1.\underline{1}25 \times 10^{-3}\text{ M Mg}^{2+}$. Calculate Q ($Mg(OH)_2$), A = Mg^{2+}, m = 1, X = OH^-, and n = 2 Since Q= $[A^{n+}]^m [X^{m-}]^n$, then Q ($Mg(OH)_2$) = $[Mg^{2+}] [OH^-]^2$ = $(1.\underline{1}25 \times 10^{-3})(9.\underline{8}635 \times 10^{-4})^2 = 1.1 \times 10^{-9} > 2.06 \times 10^{-13} = K_{sp}$ ($Mg(OH)_2$), so a precipitate will form.

Check: The units (none) are correct. The solubility of the $Mg(OH)_2$ is low, and the NaOH (a base) is high enough that the product of the concentration of ions are large compared to the K_{sp}, so a precipitate will form.

75. **Given:** KOH as precipitation agent in a) 0.015 M $CaCl_2$, b) 0.0025 M $Fe(NO_3)_2$, and c) 0.0018 M $MgBr_2$
Find: concentration of KOH necessary to form a precipitate
Other: K_{sp} ($Ca(OH)_2$) = 4.68×10^{-6}, K_{sp} ($Fe(OH)_2$) = 4.87×10^{-17}, K_{sp} ($Mg(OH)_2$) = 2.06×10^{-13}
Conceptual Plan: The solubility rules from Chapter 4 state that most hydroxides are insoluble, so all precipitate will be hydroxides. Determine the concentration of the cation in solution. Since all metals have an oxidation state of +2 and [OH⁻] = [KOH], all of the K_{sp} = [cation] [KOH]² and so $[KOH] = \sqrt{\frac{K_{sp}}{[cation]}}$.

Solution:

a) $CaCl_2$ (*s*) → Ca^{2+} (*aq*) + 2 Cl^- (*aq*). Since 1 Ca^{2+} ion is generated for each $CaCl_2$, $[Ca^{2+}] = 0.015$ M.

Then $[KOH] = \sqrt{\frac{K_{sp}}{[cation]}} = \sqrt{\frac{4.68 \times 10^{-6}}{0.015}} = 0.018\text{ M KOH}$.

b) $Fe(NO_3)_2$ (*s*) → Fe^{2+} (*aq*) + 2 NO_3^- (*aq*). Since 1 Fe^{2+} ion is generated for each $Fe(NO_3)_2$, $[Fe^{2+}]$ = 0.0025 M. Then $[KOH] = \sqrt{\frac{K_{sp}}{[cation]}} = \sqrt{\frac{4.87 \times 10^{-17}}{0.0025}} = 1.4 \times 10^{-7}\text{ M KOH}$.

c) $MgBr_2$ (*s*) → Mg^{2+} (*aq*) + 2 Br^- (*aq*). Since 1 Mg^{2+} ion is generated for each $MgBr_2$, $[Mg^{2+}]$ = 0.0018 M. Then $[KOH] = \sqrt{\frac{K_{sp}}{[cation]}} = \sqrt{\frac{2.06 \times 10^{-13}}{0.0018}} = 1.1 \times 10^{-5}\text{ M KOH}$.

Check: The units (none) are correct. Since all cations have an oxidation state of +2, it can be seen that the [KOH] needed to precipitate the hydroxide is lower the smaller the K_{sp}.

77. **Given:** solution with 1.1×10^{-3} M $Zn(NO_3)_2$ and 0.150 M NH_3 **Find:** $[Zn^{2+}]$ at equilibrium
Other: K_f ($Zn(NH_3)_4^{2+}$) = 2.0×10^9
Conceptual Plan: Write balanced equation and expression for K_f. Use initial concentrations to set up ICE table. Since the K_f is so large, assume that reaction essentially goes to completion. Solve for $[Zn^{2+}]$ at equilibrium.
Solution: $Zn(NO_3)_2$ (*s*) → Zn^{2+} (*aq*) + 2 NO_3^- (*aq*). Since 1 Zn^{2+} ion is generated for each $Zn(NO_3)_2$, $[Zn^{2+}] = 1.1 \times 10^{-3}$ M. Balanced equation is:

$Zn^{2+}(aq) + 4\,NH_3(aq) \rightleftharpoons Zn(NH_3)_4^{2+}(aq)$ Set up ICE table with initial concentrations

	$[Zn^{2+}]$	$[NH_3]$	$[Zn(NH_3)_4^{2+}]$
Initial	1.1×10^{-3}	0.150	0.00
Change	$\approx 1.1 \times 10^{-3}$	$\approx -4(1.1 \times 10^{-3})$	$\approx 1.1 \times 10^{-3}$
Equil	*x*	$0.14\underline{5}6$	1.1×10^{-3}

Since K_f is so large and since initially $[NH_3] > 4[Zn^{2+}]$ the reaction essentially goes to completion then write equilibrium expression and solve for *x*.

$K_f = \dfrac{[Zn(NH_3)_4^{2+}]}{[Zn^{2+}][NH_3]^4} = 2.0 \times 10^9 = \dfrac{1.1 \times 10^{-3}}{x(0.14\underline{5}6)^4}$ So $x = 1.2 \times 10^{-9}$ M Zn^{2+}. Since x is insignificant compared to the initial concentration, the assumption is valid.

Check: The units (M) are correct. Since K_f is so large, the reaction essentially goes to completion and $[Zn^{2+}]$ is extremely small.

79. **Given:** 150.0 mL solution of 2.05 g sodium benzoate and 2.47 g benzoic acid **Find:** pH
Other: K_a ($HC_7H_5O_2$) = 6.5×10^{-5}
Conceptual Plan: **g $NaC_7H_5O_2$ → mol $NaC_7H_5O_2$ and g $HC_7H_5O_2$ → mol $HC_7H_5O_2$**

$$\frac{1 \text{ mol } NaC_7H_5O_2}{144.11 \text{ g } NaC_7H_5O_2} \qquad \frac{1 \text{ mol } HC_7H_5O_2}{122.13 \text{ g } HC_7H_5O_2}$$

Since the two components are in the same solution, the ratio of [base]/[acid] = (mol base)/(mol acid). Then K_a, mol $NaC_7H_5O_2$, mol $HC_7H_5O_2$ → pH

$$\text{pH} = \text{p}K_a + \log\frac{[\text{base}]}{[\text{acid}]}$$

Solution: $2.05 \text{ g } NaC_7H_5O_2 \times \dfrac{1 \text{ mol } NaC_7H_5O_2}{144.11 \text{ g } NaC_7H_5O_2} = 0.014\underline{2}252 \text{ mol } NaC_7H_5O_2$ and

$$2.47 \text{ g } HC_7H_5O_2 \times \frac{1 \text{ mol } HC_7H_5O_2}{122.13 \text{ g } HC_7H_5O_2} = 0.020\underline{2}244 \text{ mol } HC_7H_5O_2 \text{ then}$$

$$\text{pH} = \text{p}K_a + \log\frac{[\text{base}]}{[\text{acid}]} = \text{p}K_a + \log\frac{\text{mol base}}{\text{mol acid}} = -\log(6.5 \times 10^{-5}) + \log\frac{0.014\underline{2}252 \text{ mol}}{0.020\underline{2}244 \text{ mol}} = 4.03.$$

Check: The units (none) are correct. The magnitude of the answer makes physical sense because the pH is a little lower than the pK_a of the acid because there is more acid than base in the buffer solution.

81. **Given:** 150.0 mL of 0.25 M $HCHO_2$ and 75.0 ml of 0.20 M NaOH **Find:** pH **Other:** K_a ($HCHO_2$) = 1.8×10^{-4}
Conceptual Plan: In this buffer, the base is generated by converting some of the formic acid to the formate ion. Part I: Stoichiometry:
mL → L then L, initial $HCHO_2$ M → mol $HCHO_2$ then mL → L then

$$\frac{1 \text{ L}}{1000 \text{ mL}} \qquad M = \frac{\text{mol}}{\text{L}} \qquad \frac{1 \text{ L}}{1000 \text{ mL}}$$

L, initial NaOH M → mol NaOH then write balanced equation then

$$M = \frac{\text{mol}}{\text{L}} \qquad NaOH + HCHO_2 \rightarrow H_2O + NaCHO_2$$

mol $HCHO_2$, mol NaOH → mol $NaCHO_2$, mol $HCHO_2$ then
set up stoichiometry table
Part II: Equilibrium:
Since the two components are in the same solution, the ratio of [base]/[acid] = (mol base)/(mol acid). Then K_a, mol $NaCHO_2$, mol $HCHO_2$ → pH

$$\text{pH} = \text{p}K_a + \log\frac{[\text{base}]}{[\text{acid}]}$$

Solution: $150.0 \text{ mL} \times \dfrac{1 \text{ L}}{1000 \text{ mL}} = 0.1500 \text{ L}$ then

$$0.1500 \text{ L } HCHO_2 \times \frac{0.25 \text{ mol } HCHO_2}{1 \text{ L } HCHO_2} = 0.03\underline{7}5 \text{ mol } HCHO_2.$$

Then $75.0 \text{ mL} \times \dfrac{1 \text{ L}}{1000 \text{ mL}} = 0.0750 \text{ L}$ then $0.0750 \text{ L NaOH} \times \dfrac{0.20 \text{ mol NaOH}}{1 \text{ L NaOH}} = 0.015 \text{ mol NaOH}$ then

set up table to track changes:

	$NaOH\ (aq)$ +	$HCHO_2\ (aq)$ →	$NaCHO_2\ (aq)$ +	$H_2O\ (l)$
Before addition	0.00 mol	0.03$\underline{7}$5 mol	0.00 mol	–
Addition	0.015 mol	–	–	–
After addition	≈ 0.00 mol	0.02$\underline{2}$5 mol	0.015 mol	–

Since the amount of NaOH is small, there are significant amounts of both buffer components, so the

Henderson–Hasselbalch Equation can be used to calculate the pH.

$$\text{pH} = pK_a + \log\frac{[\text{base}]}{[\text{acid}]} = pK_a + \log\frac{\text{mol base}}{\text{mol acid}} = -\log(1.8 \times 10^{-4}) + \log\frac{0.015\ \cancel{\text{mol}}}{0.02\underline{2}5\ \cancel{\text{mol}}} = 3.57\,.$$

Check: The units (none) are correct. The magnitude of the answer makes physical sense because the pH is a little lower than the pK_a of the acid because there is more acid than base in the buffer solution.

83. **Given:** 1.0 L of buffer of 0.25 mol NH_3 and 0.25 mol NH_4Cl; adjust to pH = 8.75
Find: mass NaOH or HCl **Other:** K_b (NH_3) = 1.79×10^{-5}
Conceptual Plan: To decide which reagent needs to be added to adjust pH, calculate the initial pH. Since the mol NH_3 = mol NH_4Cl, the pH = pK_a so K_b → pK_b → pK_a then

acid = NH_4^+ base = NH_3 $\quad pK_b = -\log K_b \quad 14 = pK_a + pK_b$

final pH, pK_a → [NH_3]/[NH_4^+] then [NH_3], L → mol NH_3 and [NH_4^+], L → mol NH_4^+

$\text{pH} = pK_a + \log\frac{[\text{base}]}{[\text{acid}]} \qquad M = \frac{\text{mol}}{L} \qquad M = \frac{\text{mol}}{L}$

then write balanced equation then

$H^+ + NH_3 \rightarrow NH_4^+$

mol NH_3, mol NH_4^+, [NH_3]/[NH_4^+] → mol HCl → g HCl

set up stoichiometry table $\qquad \frac{36.46\text{ g HCl}}{1\text{ mol HCl}}$

Solution: Since K_b (NH_3) = 1.79×10^{-5}, $pK_b = -\log K_b = -\log(1.79 \times 10^{-5}) = 4.75$. Since $14 = pK_a + pK_b$, $pK_a = 14 - pK_b = 14 - 4.75 = 9.25$. Since the desired pH is lower (8.75) HCl (a strong acid) needs to be added. Then $\text{pH} = pK_a + \log\frac{[\text{base}]}{[\text{acid}]} = 9.25 + \log\frac{[NH_3]}{[NH_4^+]} = 8.75$. Solve for $\frac{[NH_3]}{[NH_4^+]}$.

$\log\frac{[NH_3]}{[NH_4^+]} = 8.75 - 9.25 = -0.50 \rightarrow \frac{[NH_3]}{[NH_4^+]} = 10^{-0.50} = 0.3\underline{1}623$. Then $\frac{0.25\text{ mol }NH_3}{1\ \cancel{L}} \times 1.0\ \cancel{L} = 0.25\text{ mol }NH_3$

and $\frac{0.25\text{ mol }NH_4Cl}{1\ \cancel{L}} \times 1.0\ \cancel{L} = 0.25\text{ mol }NH_4Cl = 0.25\text{ mol }NH_4^+$ Since HCl is a strong acid, [HCl] = [H^+], and set up table to track changes:

	H^+ (*aq*) +	NH_3 (*aq*) →	NH_4^+ (*aq*)
Before addition	≈ 0.00 mol	0.25 mol	0.25 mol
Addition	*x*	–	–
After addition	≈ 0.00 mol	(0.25 – *x*) mol	(0.25 + *x*) mol

Since $\frac{[NH_3]}{[NH_4^+]} = 0.3\underline{1}623 = \frac{(0.25 - x)\ \cancel{\text{mol}}}{(0.25 + x)\ \cancel{\text{mol}}}$, solve for *x*. Note that the ratio of moles is the same as the ratio of concentrations, since the volume for both terms is the same. $0.3\underline{1}623(0.25 + x) = (0.25 - x) \rightarrow 0.07\underline{9}0575 + 0.3\underline{1}623x = 0.25 - x \rightarrow 1.3\underline{1}623x = 0.1\underline{7}094 \rightarrow x = 0.1\underline{2}987$ mol HCl then

$0.1\underline{2}987\ \cancel{\text{mol HCl}} \times \frac{36.46\text{ g HCl}}{1\ \cancel{\text{mol HCl}}} = 4.7\text{ g HCl}$.

Check: The units (g) are correct. The magnitude of the answer makes physical sense because there is much less than a mole of each of the buffer components, so there must be much less than a mole of HCl.

85. a) **Given:** potassium hydrogen phthalate = KHP = $KHC_8H_4O_4$ titration with NaOH
Find: balanced equation
Conceptual Plan: The reaction will be a titration of the acid proton, leaving the phthalate ion intact. The K will not be titrated since it is basic.
Solution: $NaOH\,(aq) + KHC_8H_4O_4\,(aq) \rightarrow Na^+\,(aq) + K^+\,(aq) + C_8H_4O_4^{2-}\,(aq) + H_2O\,(l)$
Check: An acid–base reaction generates a salt (soluble here) and water. There is only one acidic proton in KHP.

b) **Given:** 0.5527 g KHP titrated with 25.87 mL of NaOH solution **Find:** [NaOH]

Conceptual Plan:

g KHP → mol KHP → mol NaOH and mL → L then mol NaOH and mL → M NaOH

$$\frac{1\ mol\ KHP}{204.22\ g\ KHP} \qquad \text{1:1 from balance equation} \qquad \frac{1\ L}{1000\ mL} \qquad M = \frac{mol}{L}$$

Solution: $0.5527\ \cancel{g\ KHP} \times \dfrac{1\ mol\ KHP}{204.22\ \cancel{g\ KHP}} = 0.002706395\ mol\ KHP$; mol KHP = mol acid = mol base =

0.002706395 mol NaOH then $25.87\ \cancel{mL} \times \dfrac{1\ L}{1000\ \cancel{mL}} = 0.02587\ L$ then

$$[NaOH] = \frac{0.002706395\ mol\ NaOH}{0.02587\ L} = 0.1046\ M\ NaOH\ .$$

Check: The units (M) are correct. The magnitude of the answer makes physical sense because there is much less than a mole of acid. The magnitude of the moles of acid and base are smaller than the volume of base in liters.

87. **Given:** 0.25 mol weak acid with 10.0 mL of 3.00 M KOH diluted to 1.5000 L has pH = 3.85 **Find:** pK_a of acid

Conceptual Plan: mL → L then M KOH, L → mol KOH then write balanced reaction

$$\frac{1\ L}{1000\ mL} \qquad\qquad M = \frac{mol}{L} \qquad\qquad KOH + HA \rightarrow NaA + H_2O$$

added mol KOH, initial mol acid → equil. mol KOH, equil. mol acid then

set up stoichiometry table

equil. mol KOH, equil. mol acid, pH → pK_a

$$pH = pK_a + \log\frac{[base]}{[acid]}$$

Solution: $10.00\ \cancel{mL} \times \dfrac{1\ L}{1000\ \cancel{mL}} = 0.01000\ L$ then $0.01000\ \cancel{L\ KOH} \times \dfrac{3.00\ mol\ KOH}{1\ \cancel{L\ KOH}} = 0.0300\ mol\ KOH$ then

Since KOH is a strong base, [KOH] = [OH⁻], and set up table to track changes:

$$KOH\ (aq)\ +\ HA\ (aq)\ \rightarrow\ KA\ (aq) + H_2O\ (l)$$

Before addition	≈ 0.00 mol	0.25 mol	0.00 mol	–
Addition	0.0300 mol	–	–	–
After addition	≈ 0.00 mol	0.23 mol	0.0300 mol	–

Since the ratio of base to acid is between 0.1 and 10, it is a buffer solution. Note that the ratio of moles is the same as the ratio of concentrations, since the volume for both terms is the same.

$$pH = pK_a + \log\frac{[base]}{[acid]} = pK_a + \log\frac{0.0300\ mol}{0.23\ mol} = 3.85 .\ \text{Solve for } pK_a.\quad pK_a = 3.85 - \log\frac{0.0300\ mol}{0.23\ mol} = 4.73 .$$

Check: The units (none) are correct. The magnitude of the answer makes physical sense because there is more acid than base at equilibrium, so the pK_a is higher than the pH of the solution.

89. **Given:** saturated $CaCO_3$ solution; precipitate 1.00×10^2 mg $CaCO_3$ **Find:** volume of solution evaporated

Other: $K_{sp}(CaCO_3) = 4.96 \times 10^{-9}$

Conceptual Plan: mg $CaCO_3$ → g $CaCO_3$ → mol $CaCO_3$

$$\frac{1\ g\ CaCO_3}{1000\ mg\ CaCO_3} \qquad \frac{1\ mol\ CaCO_3}{100.09\ g\ CaCO_3}$$

The expression of the solubility product constant of A_mX_n is: $K_{sp} = [A^{n+}]^m [X^{m-}]^n$. The molar solubility of a compound, A_mX_n, can be computed directly from K_{sp} by solving for S in the expression:

$$K_{sp} = (mS)^m (nS)^n = m^m\ n^n\ S^{m+n}.$$

Then mol $CaCO_3$, S → L

$$M = \frac{mol}{L}$$

Solution: $1.00 \times 10^2\ \cancel{mg\ CaCO_3} \times \dfrac{1\ \cancel{g\ CaCO_3}}{1000\ \cancel{mg\ CaCO_3}} \times \dfrac{1\ mol\ CaCO_3}{100.09\ \cancel{g\ CaCO_3}} = 9.99101 \times 10^{-4}\ mol\ CaCO_3$ then

$K_{sp} = 4.96 \times 10^{-9}$, A = Ca^{2+}, m = 1, X = CO_3^{2-}, and n = 1 so $K_{sp} = 4.96 \times 10^{-9} = S^2$. Rearrange to solve for S.

Chapter 16 – Aqueous Ionic Equilibria

$S = \sqrt{4.96 \times 10^{-9}} = 7.0\underline{4}273 \times 10^{-5}$ M. Finally, $9.9\underline{9}101 \times 10^{-4} \cancel{\text{mol CaCO}_3} \times \frac{1\text{ L}}{7.0\underline{4}273 \times 10^{-5} \cancel{\text{mol CaCO}_3}} = 14.2$ L.

Check: The units (L) are correct. The volume should be large since the solubility is low.

91. **Given:** $[Ca^{2+}] = 9.2$ mg/dL and K_{sp} ($Ca_2P_2O_7$) = 8.64×10^{-13} **Find:** $[P_2O_7^{4-}]$ to form precipitate
Conceptual Plan: mg Ca^{2+} /dL → g Ca^{2+} /dL → mol Ca^{2+}/dL → mol Ca^{2+}/L then

$\frac{1\text{ g Ca}^{2+}}{1000\text{ mg Ca}^{2+}}$ $\frac{1\text{ mol Ca}^{2+}}{40.08\text{ g Ca}^{2+}}$ $\frac{10\text{ dL}}{1\text{ L}}$

Write balanced equation and expression for K_{sp}. Then $[Ca^{2+}]$, K_{sp} → $[P_2O_7^{4-}]$

Solution: $9.2 \frac{\cancel{\text{mg Ca}^{2+}}}{\cancel{\text{dL}}} \times \frac{1\text{ g }\cancel{\text{Ca}^{2+}}}{1000\text{ mg }\cancel{\text{Ca}^{2+}}} \times \frac{1\text{ mol Ca}^{2+}}{40.08\text{ g }\cancel{\text{Ca}^{2+}}} \times \frac{10\text{ }\cancel{\text{dL}}}{1\text{ L}} = 2.\underline{2}9541 \times 10^{-3}\text{ M Ca}^{2+}$ then write equation $Ca_2P_2O_7$ (*s*) → 2 Ca^{2+} (*aq*) + $P_2O_7^{4-}$ (*aq*). So $K_{sp} = [Ca^{2+}]^2 [P_2O_7^{4-}] = 8.64 \times 10^{-13}$ = $= (2.\underline{2}9541 \times 10^{-3})^2 [P_2O_7^{4-}]$. Solve for $[P_2O_7^{4-}]$ then $[P_2O_7^{4-}] = 1.6 \times 10^{-7}$ M.
Check: The units (M) are correct. Since K_{sp} is so small and the sodium concentration is fairly high, the urate concentration is driven to a very low level.

93. **Given:** CuS in 0.150 M NaCN **Find:** molar solubility (*S*)
Other: K_f ($Cu(CN)_4^{2-}$) = 1.0×10^{25}, K_{sp} (CuS) = 1.27×10^{-36}
Conceptual Plan: Identify the appropriate solid and complex ion. Write balanced equations for dissolving the solid and forming the complex ion. Add these two reactions to get the desired overall reaction. Using the rules from Chapter 14, multiply the individual reaction *K*'s to get the overall *K* for the sum of these reactions. Then M NaCN, *K* → *S*

ICE Chart

Solution: Identify the solid as CuS and the complex ion as $Cu(CN)_4^{2-}$. Write the individual reactions and add them together.

$CuS\ (s) \rightleftharpoons \cancel{Cu^{2+}(aq)} + S^{2-}\ (aq) \qquad K_{sp} = 1.27 \times 10^{-36}$

$\cancel{Cu^{2+}(aq)} + 4\ CN^-(aq) \rightleftharpoons Cu(CN)_4^{2-}\ (aq) \qquad K_f = 1.0 \times 10^{25}$

$CuS\ (s) + 4\ CN^-(aq) \rightleftharpoons Cu(CN)_4^{2-}\ (aq) + S^{2-}\ (aq)$

Since the overall reaction is the simple sum of the two reactions, the overall reaction $K = K_f K_{sp} = (1.0 \times 10^{25}) \times (1.27 \times 10^{-36}) = 1.\underline{2}7 \times 10^{-11}$. NaCN (*s*) → Na^+ (*aq*) + CN^- (*aq*). Since 1 CN^- ion is generated for each NaCN, $[CN^-] = 0.150$ M. Set up ICE table.

$CuS\ (s) + 4\ CN^-(aq) \rightleftharpoons Cu(CN)_4^{2-}\ (aq) + S^{2-}\ (aq)$

	$[CN^-]$	$[Cu(CN)_4^{2-}]$	$[S^{2-}]$
Initial	0.150	0.00	0.00
Change	$-4S$	$+S$	$+S$
Equil	$0.150 - 4S$	$+S$	$+S$

$K = \frac{[Cu(CN)_4^{2-}][S^{2-}]}{[CN^-]^4} = 1.\underline{2}7 \times 10^{-11} = \frac{S^2}{(0.150 - 4S)^4}$.

Assume *S* is small ($4S << 0.150$) so $\frac{S^2}{(0.150 - \cancel{4S})^4} = 1.\underline{2}7 \times 10^{-11} = \frac{S^2}{(0.150)^4}$ and $S = 8.\underline{0}183 \times 10^{-8}$ =

$= 8.0 \times 10^{-8}$ M. Confirm that assumption is valid $\frac{4(8.\underline{0}183 \times 10^{-8})}{0.150} \times 100\ \% = 0.00021\ \% << 5\ \%$ so assumption is valid.
Check: The units (M) are correct. Since K_f is large, the overall *K* is larger than the original K_{sp} and the solubility of CuS increases over that of pure water $\left(\sqrt{1.27 \times 10^{-36}} = 1.13 \times 10^{-18}\text{ M}\right)$.

95. **Given:** 100.0 mL of 0.36 M NH_2OH and 50.0 mL of 0.26 M HCl and K_b (NH_2OH) = 1.10×10^{-8} **Find:** pH
Conceptual Plan: identify acid and base components mL → L then $[NH_2OH]$, L → mol NH_2OH

acid = NH_3OH^+ base = NH_2OH $\frac{1\text{ L}}{1000\text{ mL}}$ $M = \frac{\text{mol}}{\text{L}}$

then mL → L then [HCl], L → mol HCl then write balanced equation then

$\frac{1\text{ L}}{1000\text{ mL}}$ $M = \frac{\text{mol}}{\text{L}}$ $HCl + NH_2OH \rightarrow NH_3OHCl$

mol NH_2OH, mol HCl → mol excess NH_2OH, mol NH_3OH^+
set up stoichiometry table

Since there are significant amounts of both the acid and the conjugate base species, this is a buffer solution and so Henderson–Hasselbalch Equation $\left(pH = pK_a + \log\frac{[base]}{[acid]}\right)$ **can be used. Convert K_b to K_a using $K_w = K_a K_b$. Also note that ratio of concentrations is the same as the ratio of moles, since the volume is the same for both species.**

Solution: $100\ \text{mL NH}_2\text{OH} \times \frac{1\ \text{L}}{1000\ \text{mL}} = 0.1\ \text{L NH}_2\text{OH}$ then

$\frac{0.36\ \text{mol NH}_2\text{OH}}{1\ \text{L}} \times 0.1000\ \text{L} = 0.036\ \text{mol NH}_2\text{OH}$. $50.0\ \text{mL HCl} \times \frac{1\ \text{L}}{1000\ \text{mL}} = 0.0500\ \text{L HCl}$ then

$\frac{0.26\ \text{mol HCl}}{1\ \text{L}} \times 0.0500\ \text{L} = 0.013\ \text{mol HCl}$. Set up table to track changes:

	HCl (*aq*)	+ NH_2OH (*aq*)	→ NH_3OHCl (*aq*)
Before addition	0.00 mol	0.036 mol	0.00 mol
Addition	0.013 mol	–	–
After addition	≈ 0.00 mol	0.023 mol	0.013 mol

then $K_w = K_a K_b$ so $K_a = \frac{K_w}{K_b} = \frac{1.0 \times 10^{-14}}{1.10 \times 10^{-8}} = 9.\underline{0}909 \times 10^{-7}$ M then use Henderson–Hasselbalch Equation, since the solution is a buffer. Note that the ratio of moles is the same as the ratio of concentrations, since the volume for both terms is the same.

$$pH = pK_a + \log\frac{[base]}{[acid]} = -\log(9.\underline{0}909 \times 10^{-7}) + \log\frac{0.023\ \text{mol}}{0.013\ \text{mol}} = 6.2\underline{8}918 = 6.29.$$

Check: The units (none) are correct. The magnitude of the answer makes physical sense because pH should be more than the pK_a of the acid because there is more base than acid.

97. **Given:** 10.0 L of 75 ppm $CaCO_3$ and 55 ppm $MgCO_3$ (by mass) **Find:** mass Na_2CO_3 to precipitate 90.0 % of ions **Other:** $K_{sp}(CaCO_3) = 4.96 \times 10^{-9}$ and $K_{sp}(MgCO_3) = 6.82 \times 10^{-6}$

Conceptual Plan:

Assume that the density of water is 1.00 g/mL. L water → mL water → g water then

$\frac{1000\ \text{mL}}{1\ \text{L}}$ $\frac{1.00\ \text{g water}}{1\ \text{mL}}$

g water → g $CaCO_3$ → mol $CaCO_3$ → mol Ca^{2+} and g water → g $MgCO_3$ → mol $MgCO_3$ then

$\frac{75\ \text{g CaCO}_3}{10^6\ \text{g water}}$ $\frac{1\ \text{mol CaCO}_3}{100.09\ \text{g CaCO}_3}$ $\frac{1\ \text{mol Ca}^{2+}}{1\ \text{mol CaCO}_3}$ $\frac{55\ \text{g MgCO}_3}{10^6\ \text{g water}}$ $\frac{1\ \text{mol MgCO}_3}{84.32\ \text{g MgCO}_3}$

mol $MgCO_3$ → mol Mg^{2+} then comparing the two K_{sp} values, essentially all of the Ca^{2+} will

$\frac{1\ \text{mol Mg}^{2+}}{1\ \text{mol MgCO}_3}$

precipitate before the Mg^{2+} will begin to precipitate. Since 90.0 % of the ions are to be precipitates, there will be 10.0 % of the ions left in solution (all will be Mg^{2+}).

$(0.100)(\text{mol Ca}^{2+} + \text{mol Mg}^{2+})$

Calculate the moles of ions remaining in solution. Then mol Mg^{2+}, L → M Mg^{2+} then

$M = \frac{mol}{L}$

The solubility product constant (K_{sp}) is the equilibrium expression for a chemical equation representing the dissolution of an ionic compound. The expression of the solubility product constant of A_mX_n is: $K_{sp} = [A^{n+}]^m [X^{m-}]^n$. Use this equation to M Mg^{2+}, K_{sp} → M CO_3^{2-} then M CO_3^{2-}, L → mol CO_3^{2-}

for ionic compound, A_mX_n, $K_{sp} = [A^{n+}]^m [X^{m-}]^n$. $M = \frac{mol}{L}$

then mol CO_3^{2-} → mol Na_2CO_3 → g Na_2CO_3

$\frac{1\ \text{mol CO}_3^{2-}}{1\ \text{mol Na}_2\text{CO}_3}$ $\frac{105.99\ \text{g Na}_2\text{CO}_3}{1\ \text{mol Na}_2\text{CO}_3}$

Solution: $10.0\ \text{L} \times \frac{1000\ \text{mL}}{1\ \text{L}} \times \frac{1.00\ \text{g water}}{1\ \text{mL}} = 1.00 \times 10^4\ \text{g water}$ then

$1.00 \times 10^4\ \text{g water} \times \frac{75\ \text{g } CaCO_3}{10^6\ \text{g water}} \times \frac{1\ \text{mol } CaCO_3}{100.09\ \text{g } CaCO_3} \times \frac{1\ \text{mol } Ca^{2+}}{1\ \text{mol } CaCO_3} = 0.0074\underline{9}33\ \text{mol } Ca^{2+}$ and

$1.00 \times 10^4\ \text{g water} \times \frac{55\ \text{g } MgCO_3}{10^6\ \text{g water}} \times \frac{1\ \text{mol } MgCO_3}{84.32\ \text{g } MgCO_3} \times \frac{1\ \text{mol } Mg^{2+}}{1\ \text{mol } MgCO_3} = 0.0065\underline{2}28\ \text{mol } Mg^{2+}$ so the ions remaining in solution after 90.0 % precipitate out =

$(0.100)(\text{mol } Ca^{2+} + \text{mol } Mg^{2+}) = (0.100)(0.0074\underline{9}33\ \text{mol } Ca^{2+} + 0.0065\underline{2}28\ \text{mol } Mg^{2+}) = 0.0014\underline{0}161\ \text{mol ions}$

so $\frac{0.0014\underline{0}607\ \text{mol } Mg^{2+}}{10.0\ \text{L}} = 0.00014\underline{0}161\ \text{M } Mg^{2+}$. Then $K_{sp} = 6.82 \times 10^{-6}$, $A = Mg^{2+}$, $m = 1$, $X = CO_3^{2-}$, and $n = 1$, so $K_{sp} = 6.82 \times 10^{-6} = [Mg^{2+}][CO_3^{2-}] = (0.0014\underline{0}161)[CO_3^{2-}]$. Rearrange to solve for $[CO_3^{2-}]$.

So $[CO_3^{2-}] = 0.048\underline{6}583$ M. Then

$\frac{0.048\underline{6}583\ \text{mol } CO_3^{2-}}{1\ \text{L}} \times 10.0\ \text{L} \times \frac{1\ \text{mol } Na_2CO_3}{1\ \text{mol } CO_3^{2-}} \times \frac{105.99\ \text{g } Na_2CO_3}{1\ \text{mol } Na_2CO_3} = 51.6\ \text{g } Na_2CO_3$.

Check: The units (g) are correct. The mass is reasonable to put in a washing machine load.

99. If the concentration of the acid is greater than the concentration of the base, then the pH will be less than the pK_a. If the concentration of the acid is equal to the concentration of the base, then the pH will be equal to the pK_a. If the concentration of the acid is less than the concentration of the base, then the pH will be greater than the pK_a.
   a) $pH < pK_a$

   b) $pH > pK_a$

   c) $pH = pK_a$, the OH^- will convert half of the acid to base

   c) $pH > pK_a$, the OH^- will convert more than half of the acid to base

101. Only (a) is correct. The volume to the equivalence point will be the same since the number of moles of acid is the same. The pH profiles of the two titrations will be different.

103. a) The solubility will be unchanged since the pH is constant and there are no common ions added.

   b) The solubility will be less because extra fluoride ions are added, suppressing the solubility of the fluoride ionic compound.

   c) The solubility will increase because some of the fluoride ion will be converted to HF, and so more of the ionic compound can be dissolved.

Chapter 17
Free Energy and Thermodynamics

1. a and c are spontaneous processes.

3. Yes, the particles may also be distributed so that one is in the 0 J level and the other is in the 20 J level. The total energy is the sum of the energies of the two particles = 0 J + 20 J = 20 J. There are two such arrangements of the particles in this fashion, so this state will have the greatest entropy and be more likely.

5. a) $\Delta S > 0$ because a gas is being generated

   b) $\Delta S < 0$ because 2 moles of gas are being converted to 1 mole of gas

   c) $\Delta S < 0$ because a gas is being converted to a solid

   d) $\Delta S < 0$ because 4 moles of gas are being converted to 2 moles of gas

7. a) $\Delta S_{sys} > 0$, because 6 moles of gas are being converted to 7 moles of gas. Since $\Delta H < 0$ $\Delta S_{surr} > 0$ and the reaction is spontaneous at all temperatures.

   b) $\Delta S_{sys} < 0$ because 2 moles of different gases are being converted to 2 moles of one gas. Since $\Delta H > 0$ $\Delta S_{surr} < 0$ and the reaction is nonspontaneous at all temperatures.

   c) $\Delta S_{sys} < 0$, because 3 moles of gas are being converted to 2 moles of gas. Since $\Delta H > 0$ $\Delta S_{surr} < 0$ the reaction is nonspontaneous at all temperatures.

   d) $\Delta S_{sys} > 0$, because 9 moles of gas are being converted to 10 moles of gas. Since $\Delta H < 0$ $\Delta S_{surr} > 0$ the reaction is spontaneous at all temperatures.

9. a) **Given:** $\Delta H^\circ_{rxn} = -287$ kJ, $T = 298$ K **Find:** ΔS_{surr}

   Conceptual Plan: **kJ → J then ΔH°_{rxn}, T → ΔS_{surr}**

   $$\frac{1000\text{ J}}{1\text{ kJ}} \qquad \Delta S_{surr} = \frac{-\Delta H_{sys}}{T}$$

   Solution: $$-287\text{ kJ} \times \frac{1000\text{ J}}{1\text{ kJ}} = -287{,}000\text{ J then } \Delta S_{surr} = \frac{-\Delta H_{sys}}{T} = \frac{-(-287{,}000\text{ J})}{298\text{ K}} = 963\ \frac{\text{J}}{\text{K}}$$

   Check: The units (J/K) are correct. The magnitude of the answer (10^3 J/K) makes sense because the kJ and the temperature started with very similar values and then a factor of 10^3 was applied.

   b) **Given:** $\Delta H^\circ_{rxn} = -287$ kJ, $T = 77$ K **Find:** ΔS_{surr}

   Conceptual Plan: **kJ → J then ΔH°_{rxn}, T → ΔS_{surr}**

   $$\frac{1000\text{ J}}{1\text{ kJ}} \qquad \Delta S_{surr} = \frac{-\Delta H_{sys}}{T}$$

   Solution: $$-287\text{ kJ} \times \frac{1000\text{ J}}{1\text{ kJ}} = -287{,}000\text{ J then } \Delta S_{surr} = \frac{-\Delta H_{sys}}{T} = \frac{-(-287{,}000\text{ J})}{77\text{ K}} = 3730\ \frac{\text{J}}{\text{K}} = 3.73 \times 10^3\ \frac{\text{J}}{\text{K}}$$

   Check: The units (J/K) are correct. The magnitude of the answer (4 x 10^3 J/K) makes sense because the temperature is much lower than in part a, so the answer should increase.

   c) **Given:** $\Delta H^\circ_{rxn} = +127$ kJ, $T = 298$ K **Find:** ΔS_{surr}

   Conceptual Plan: **kJ → J then ΔH°_{rxn}, T → ΔS_{surr}**

   $$\frac{1000\text{ J}}{1\text{ kJ}} \qquad \Delta S_{surr} = \frac{-\Delta H_{sys}}{T}$$

   Solution: $$+127\text{ kJ} \times \frac{1000\text{ J}}{1\text{ kJ}} = +127{,}000\text{ J then } \Delta S_{surr} = \frac{-\Delta H_{sys}}{T} = \frac{-127{,}000\text{ J}}{298\text{ K}} = -426\ \frac{\text{J}}{\text{K}}$$

   Check: The units (J/K) are correct. The magnitude of the answer (– 400 J/K) makes sense because the kJ are less and of the opposite sign than part a and so the answer should decrease.

d) **Given:** $\Delta H^\circ_{rxn} = +\,127$ kJ, $T = 77$ K **Find:** ΔS_{surr}

Conceptual Plan: **kJ → J then $\Delta H^\circ_{rxn}, T$ → ΔS_{surr}**

$$\frac{1000\ \text{J}}{1\ \text{kJ}} \qquad \Delta S_{surr} = \frac{-\Delta H_{sys}}{T}$$

Solution: $+127\ \text{kJ} \times \frac{1000\ \text{J}}{1\ \text{kJ}} = +\,127{,}000\ \text{J}$ then $\Delta S_{surr} = \frac{-\Delta H_{sys}}{T} = \frac{-\,127{,}000\ \text{J}}{77\ \text{K}} = -\,1650\ \frac{\text{J}}{\text{K}} = -1.65 \times 10^3\ \frac{\text{J}}{\text{K}}$

Check: The units (J/K) are correct. The magnitude of the answer ($-\,2 \times 10^3$ J/K) makes sense because the temperature is much lower than in part c, so the answer should increase.

11. a) **Given:** $\Delta H^\circ_{rxn} = -\,125$ kJ, $\Delta S_{rxn} = +\,253$ J/K, $T = 298$ K **Find:** ΔS_{univ} and spontaneity

Conceptual Plan: **kJ → J then $\Delta H^\circ_{rxn}, T$ → ΔS_{surr} then $\Delta S_{rxn}, \Delta S_{surr}$ → ΔS_{univ}**

$$\frac{1000\ \text{J}}{1\ \text{kJ}} \qquad \Delta S_{surr} = \frac{-\Delta H_{sys}}{T} \qquad \Delta S_{univ} = \Delta S_{sys} + \Delta S_{surr}$$

Solution: $-125\ \text{kJ} \times \frac{1000\ \text{J}}{1\ \text{kJ}} = -\,125{,}000\ \text{J}$ then $\Delta S_{surr} = \frac{-\Delta H_{sys}}{T} = \frac{-(-\,125{,}000\ \text{J})}{298\ \text{K}} = 41\underline{9}.463\ \frac{\text{J}}{\text{K}}$ then

$\Delta S_{univ} = \Delta S_{sys} + \Delta S_{surr} = +\,253\ \frac{\text{J}}{\text{K}} + 41\underline{9}.463\ \frac{\text{J}}{\text{K}} = +\,672\ \frac{\text{J}}{\text{K}}$ so the reaction is spontaneous.

Check: The units (J/K) are correct. The magnitude of the answer (700 J/K) makes sense because both terms were positive and so the reaction is spontaneous.

b) **Given:** $\Delta H^\circ_{rxn} = +\,125$ kJ, $\Delta S_{rxn} = -\,253$ J/K, $T = 298$ K **Find:** ΔS_{univ} and spontaneity

Conceptual Plan: **kJ → J then $\Delta H^\circ_{rxn}, T$ → ΔS_{surr} then $\Delta S_{rxn}, \Delta S_{surr}$ → ΔS_{univ}**

$$\frac{1000\ \text{J}}{1\ \text{kJ}} \qquad \Delta S_{surr} = \frac{-\Delta H_{sys}}{T} \qquad \Delta S_{univ} = \Delta S_{sys} + \Delta S_{surr}$$

Solution: $+125\ \text{kJ} \times \frac{1000\ \text{J}}{1\ \text{kJ}} = +\,125{,}000\ \text{J}$ then $\Delta S_{surr} = \frac{-\Delta H_{sys}}{T} = \frac{-\,125{,}000\ \text{J}}{298\ \text{K}} = -\,41\underline{9}.463\ \frac{\text{J}}{\text{K}}$ then

$\Delta S_{univ} = \Delta S_{sys} + \Delta S_{surr} = -\,253\ \frac{\text{J}}{\text{K}} - 41\underline{9}.463\ \frac{\text{J}}{\text{K}} = -\,672\ \frac{\text{J}}{\text{K}}$ so the reaction is nonspontaneous.

Check: The units (J/K) are correct. The magnitude of the answer ($-$ 700 J/K) makes sense because both terms were negative and so the reaction is nonspontaneous.

c) **Given:** $\Delta H^\circ_{rxn} = -\,125$ kJ, $\Delta S_{rxn} = -\,253$ J/K, $T = 298$ K **Find:** ΔS_{univ} and spontaneity

Conceptual Plan: **kJ → J then $\Delta H^\circ_{rxn}, T$ → ΔS_{surr} then $\Delta S_{rxn}, \Delta S_{surr}$ → ΔS_{univ}**

$$\frac{1000\ \text{J}}{1\ \text{kJ}} \qquad \Delta S_{surr} = \frac{-\Delta H_{sys}}{T} \qquad \Delta S_{univ} = \Delta S_{sys} + \Delta S_{surr}$$

Solution: $-125\ \text{kJ} \times \frac{1000\ \text{J}}{1\ \text{kJ}} = -\,125{,}000\ \text{J}$ then $\Delta S_{surr} = \frac{-\Delta H_{sys}}{T} = \frac{-(-\,125{,}000\ \text{J})}{298\ \text{K}} = +\,41\underline{9}.463\ \frac{\text{J}}{\text{K}}$ then

$\Delta S_{univ} = \Delta S_{sys} + \Delta S_{surr} = -\,253\ \frac{\text{J}}{\text{K}} + 41\underline{9}.463\ \frac{\text{J}}{\text{K}} = +\,166\ \frac{\text{J}}{\text{K}}$ so the reaction is spontaneous.

Check: The units (J/K) are correct. The magnitude of the answer (200 J/K) makes sense because the larger term was positive and so the reaction is spontaneous.

d) **Given:** $\Delta H^\circ_{rxn} = -\,125$ kJ, $\Delta S_{rxn} = -\,253$ J/K, $T = 555$ K **Find:** ΔS_{univ} and spontaneity

Conceptual Plan: **kJ → J then $\Delta H^\circ_{rxn}, T$ → ΔS_{surr} then $\Delta S_{rxn}, \Delta S_{surr}$ → ΔS_{univ}**

$$\frac{1000\ \text{J}}{1\ \text{kJ}} \qquad \Delta S_{surr} = \frac{-\Delta H_{sys}}{T} \qquad \Delta S_{univ} = \Delta S_{sys} + \Delta S_{surr}$$

Solution: $-125\ \text{kJ} \times \frac{1000\ \text{J}}{1\ \text{kJ}} = -\,125{,}000\ \text{J}$ then $\Delta S_{surr} = \frac{-\Delta H_{sys}}{T} = \frac{-(-\,125{,}000\ \text{J})}{555\ \text{K}} = +\,22\underline{5}.225\ \frac{\text{J}}{\text{K}}$ then

$\Delta S_{univ} = \Delta S_{sys} + \Delta S_{surr} = -\,253\ \frac{\text{J}}{\text{K}} + 22\underline{5}.225\ \frac{\text{J}}{\text{K}} = -\,28\ \frac{\text{J}}{\text{K}}$ so the reaction is nonspontaneous.

Check: The units (J/K) are correct. The magnitude of the answer (– 30 J/K) makes sense because the larger term was negative and so the reaction is nonspontaneous.

13. a) **Given:** $\Delta H^\circ_{rxn} = -125$ kJ, $\Delta S_{rxn} = +253$ J/K, $T = 298$ K **Find:** ΔG and spontaneity

Conceptual Plan: J/K → kJ/K then $\Delta H^\circ_{rxn}, \Delta S_{rxn}, T$ → ΔG

$$\frac{1000\ \text{J}}{1\ \text{kJ}} \qquad \Delta G = \Delta H_{rxn} - T\Delta S_{rxn}$$

Solution: $+253\frac{\text{J}}{\text{K}} \times \frac{1\ \text{kJ}}{1000\ \text{J}} = +0.253\frac{\text{kJ}}{\text{K}}$ then

$\Delta G = \Delta H_{rxn} - T\Delta S_{rxn} = -125\ \text{kJ} - (298\ \text{K})\left(0.253\frac{\text{kJ}}{\text{K}}\right) = -200.\ \text{kJ} = -2.00 \times 10^2\ \text{kJ}$ so the reaction is spontaneous.

Check: The units (kJ) are correct. The magnitude of the answer (– 200 kJ) makes sense because both terms were negative and so the reaction is spontaneous.

b) **Given:** $\Delta H^\circ_{rxn} = +125$ kJ, $\Delta S_{rxn} = -253$ J/K, $T = 298$ K **Find:** ΔG and spontaneity

Conceptual Plan: J/K → kJ/K then $\Delta H^\circ_{rxn}, \Delta S_{rxn}, T$ → ΔG

$$\frac{1000\ \text{J}}{1\ \text{kJ}} \qquad \Delta G = \Delta H_{rxn} - T\Delta S_{rxn}$$

Solution: $-253\frac{\text{J}}{\text{K}} \times \frac{1\ \text{kJ}}{1000\ \text{J}} = -0.253\frac{\text{kJ}}{\text{K}}$ then

$\Delta G = \Delta H_{rxn} - T\Delta S_{rxn} = +125\ \text{kJ} - (298\ \text{K})\left(-0.253\frac{\text{kJ}}{\text{K}}\right) = +200.\ \text{kJ} = +2.00 \times 10^2\ \text{kJ}$ so the reaction is nonspontaneous.

Check: The units (kJ) are correct. The magnitude of the answer (200 kJ) makes sense because both terms were positive and so the reaction is nonspontaneous.

c) **Given:** $\Delta H^\circ_{rxn} = -125$ kJ, $\Delta S_{rxn} = -253$ J/K, $T = 298$ K **Find:** ΔG and spontaneity

Conceptual Plan: J/K → kJ/K then $\Delta H^\circ_{rxn}, \Delta S_{rxn}, T$ → ΔG

$$\frac{1000\ \text{J}}{1\ \text{kJ}} \qquad \Delta G = \Delta H_{rxn} - T\Delta S_{rxn}$$

Solution: $-253\frac{\text{J}}{\text{K}} \times \frac{1\ \text{kJ}}{1000\ \text{J}} = -0.253\frac{\text{kJ}}{\text{K}}$ then

$\Delta G = \Delta H_{rxn} - T\Delta S_{rxn} = -125\ \text{kJ} - (298\ \text{K})\left(-0.253\frac{\text{kJ}}{\text{K}}\right) = -4\underline{9}.606\ \text{kJ} = -5.0 \times 10^1\ \text{kJ}$ so the reaction is spontaneous.

Check: The units (kJ) are correct. The magnitude of the answer (–50 kJ) makes sense because the larger term was negative and so the reaction is spontaneous.

d) **Given:** $\Delta H^\circ_{rxn} = -125$ kJ, $\Delta S_{rxn} = -253$ J/K, $T = 555$ K **Find:** ΔG and spontaneity

Conceptual Plan: J/K → kJ/K then $\Delta H^\circ_{rxn}, \Delta S_{rxn}, T$ → ΔG

$$\frac{1000\ \text{J}}{1\ \text{kJ}} \qquad \Delta G = \Delta H_{rxn} - T\Delta S_{rxn}$$

Solution: $-253\frac{\text{J}}{\text{K}} \times \frac{1\ \text{kJ}}{1000\ \text{J}} = -0.253\frac{\text{kJ}}{\text{K}}$ then

$\Delta G = \Delta H_{rxn} - T\Delta S_{rxn} = -125\ \text{kJ} - (555\ \text{K})\left(-0.253\frac{\text{kJ}}{\text{K}}\right) = +15\ \text{kJ}$ so the reaction is nonspontaneous.

Check: The units (J/K) are correct. The magnitude of the answer (+ 15 kJ) makes sense because the larger term was positive and so the reaction is nonspontaneous.

15. **Given:** $\Delta H^\circ_{rxn} = -2217$ kJ, $\Delta S_{rxn} = +101.1$ J/K, $T = 25$ °C **Find:** ΔG and spontaneity

Conceptual Plan: **°C → K then J/K → kJ/K then $\Delta H^\circ_{rxn}, \Delta S_{rxn}, T$ → ΔG**

$K = 273.15 + °C$ $\frac{1 \text{kJ}}{1000 \text{ J}}$ $\Delta G = \Delta H_{rxn} - T\Delta S_{rxn}$

Solution: $T = 273.15 + 25\text{ °C} = 298\text{ K}$ then $+101.1\frac{\text{J}}{\text{K}} \times \frac{1\text{ kJ}}{1000\text{ J}} = +0.1011\frac{\text{kJ}}{\text{K}}$ then

$\Delta G = \Delta H_{rxn} - T\Delta S_{rxn} = -2217\text{ kJ} - (298\text{ K})\left(0.1011\frac{\text{kJ}}{\text{K}}\right) = -2247\text{ kJ} = -2.247 \times 10^6\text{ J}$ so the reaction is spontaneous.

Check: The units (kJ) are correct. The magnitude of the answer (– 2250 kJ) makes sense because both terms are negative, so the reaction is spontaneous.

17.

ΔH	ΔS	ΔG	Low Temp.	High Temp.
–	+	–	Spontaneous	Spontaneous
–	–	Temp dependent	Spontaneous	Nonspontaneous
+	+	Temp dependent	Nonspontaneous	Spontaneous
+	–	+	Nonspontaneous	Nonspontaneous

19. The molar entropy of a substance increases with increasing temperatures. The kinetic energy and the molecular motion increases. The substance will have access to an increased number of energy levels.

21. a) CO_2 (*g*) because it has greater molar mass/complexity.

b) CH_3OH (*g*) because it is in the gas phase.

c) CO_2 (*g*) because it has greater molar mass/complexity.

d) SiH_4 (*g*) because it has greater molar mass.

e) $CH_3CH_2CH_3$ (*g*) because it has greater molar mass/complexity.

f) NaBr (*aq*) because a solution has more entropy than a solid crystal.

23. a) He (*g*) < Ne (*g*) < SO_2 (*g*) < NH_3 (*g*) < CH_3CH_2OH (*g*). All are in the gas phase. From He to Ne there is an increase in molar mass, beyond that, the molecules increase in complexity.

b) H_2O (*s*) < H_2O (*l*) < H_2O (*g*). Entropy increases as we go from a solid to a liquid to a gas.

c) CH_4 (*g*) < CF_4 (*g*) < CCl_4 (*g*). Entropy increases as the molar mass increases.

25. a) **Given:** C_2H_4 (*g*) + H_2 (*g*) → C_2H_6 (*g*) **Find:** ΔS°_{rxn}

Conceptual Plan: $\Delta S^0_{rxn} = \sum n_p S^0(products) - \sum n_r S^0(reactants)$

Solution:

Reactant/Product	S^0 (J/mol K from Appendix IIB)
C_2H_4 (g)	219.3
H_2 (*g*)	130.7
C_2H_6 (*g*)	229.2

Be sure to pull data for the correct formula and phase.

$\Delta S^0_{rxn} = \sum n_p S^0(products) - \sum n_r S^0(reactants)$

$= [1(S^0(C_2H_6\ (g))] - [1(S^0(C_2H_4\ (g)) + 1(S^0(H_2\ (g))]$

$= [1(229.2\text{ J/K})] - [1(219.3\text{ J/K}) + 1(130.7\text{ J/K})]$ The moles of gas are decreasing.

$= [229.2\text{ J/K}] - [350.0\text{ J/K}]$

$= -120.8\text{ J/K}$

Check: The units (J/K) are correct. The answer is negative, which is consistent with 2 moles of gas going to 1 mole of gas.

b) **Given:** $C(s) + H_2O(g) \rightarrow CO(g) + H_2(g)$ **Find:** ΔS°_{rxn}

Conceptual Plan: $\Delta S^\circ_{rxn} = \sum n_p S^\circ(products) - \sum n_r S^\circ(reactants)$

Solution:

Reactant/Product	S° (J/molK from Appendix IIB)
$C(s)$	5.7
$H_2O(g)$	188.8
$CO(g)$	197.7
$H_2(g)$	130.7

Be sure to pull data for the correct formula and phase.

$$\Delta S^\circ_{rxn} = \sum n_p S^\circ(products) - \sum n_r S^\circ(reactants)$$
$$= [1(S^\circ(CO(g)) + 1(S^\circ(H_2(g))] - [1(S^\circ(C(s)) + 1(S^\circ(H_2O(g))]$$
$$= [1(197.7\ J/K) + 1(130.7\ J/K)] - [1(5.7\ J/K) + 1(188.8\ J/K)]$$ The moles of gas are increasing.
$$= [328.4\ J/K] - [194.5\ J/K]$$
$$= +133.9\ J/K$$

Check: The units (J/K) are correct. The answer is positive, which is consistent with 1 mole of gas going to 1 mole of gas.

c) **Given:** $CO(g) + H_2O(g) \rightarrow H_2(g) + CO_2(g)$ **Find:** ΔS°_{rxn}

Conceptual Plan: $\Delta S^\circ_{rxn} = \sum n_p S^\circ(products) - \sum n_r S^\circ(reactants)$

Solution:

Reactant/Product	S° (J/molK from Appendix IIB)
$CO(g)$	197.7
$H_2O(g)$	188.8
$H_2(g)$	130.7
$CO_2(g)$	213.8

Be sure to pull data for the correct formula and phase.

$$\Delta S^\circ_{rxn} = \sum n_p S^\circ(products) - \sum n_r S^\circ(reactants)$$
$$= [1(S^\circ(H_2(g)) + 1(S^\circ(CO_2(g))] - [1(S^\circ(CO(g)) + 1(S^\circ(H_2O(g))]$$
$$= [1(130.7\ J/K) + 1(213.8\ J/K)] - [1(197.7\ J/K) + 1(188.8\ J/K)]$$
$$= [344.5\ J/K] - [386.5\ J/K]$$
$$= -42.0\ J/K$$

The change is small because the number of moles of gas is constant.
Check: The units (J/K) are correct. The answer is small and negative, which is consistent with a constant number of moles of gas. Water molecules are bent and carbon dioxide molecules are linear, so the water has more complexity. Also, carbon monoxide is more complex than hydrogen gas.

d) **Given:** $2\ H_2S(g) + 3\ O_2(g) \rightarrow 2\ H_2O(l) + 2\ SO_2(g)$ **Find:** ΔS°_{rxn}

Conceptual Plan: $\Delta S^\circ_{rxn} = \sum n_p S^\circ(products) - \sum n_r S^\circ(reactants)$

Solution:

Reactant/Product	S° (J/molK from Appendix IIB)
$H_2S(g)$	205.8
$O_2(g)$	205.2
$H_2O(l)$	70.0
$SO_2(g)$	248.2

Be sure to pull data for the correct formula and phase.

$$\Delta S^0_{rxn} = \sum n_p S^0(products) - \sum n_r S^0(reactants)$$
$$= [2(S^0(H_2O\ (l)) + 2(S^0(SO_2\ (g))] - [2(S^0(H_2S\ (g)) + 3(S^0(O_2\ (g))]$$
$$= [2(70.0\ J/K) + 2(248.2\ J/K)] - [2(205.8\ J/K) + 3(205.2\ J/K)]$$
$$= [636.4\ J/K] - [1027.2\ J/K]$$
$$= -390.8\ J/K$$

The number of moles of gas is decreasing.
Check: The units (J/K) are correct. The answer is negative, which is consistent with a decrease in the number of moles of gas.

27. **Given:** CH_2Cl_2 (*g*) formed from elements in standard states **Find:** ΔS° and rationalize sign
Conceptual Plan: Write balanced reaction, then $\Delta S^0_{rxn} = \sum n_p S^0(products) - \sum n_r S^0(reactants)$
Solution: $C\ (s) + H_2\ (g) + Cl_2\ (g) \rightarrow CH_2Cl_2\ (g)$

Reactant/Product	S^0 (J/molK from Appendix IIB)
$C\ (s)$	5.7
$H_2\ (g)$	130.7
$Cl_2\ (g)$	223.1
$CH_2Cl_2\ (g)$	270.2

Be sure to pull data for the correct formula and phase.
$$\Delta S^0_{rxn} = \sum n_p S^0(products) - \sum n_r S^0(reactants)$$
$$= [1(S^0(CH_2Cl_2\ (g))] - [1(S^0(C\ (s)) + 1(S^0(H_2\ (g)) + 1(S^0(Cl_2\ (g))]$$
$$= [1(270.2\ J/K)] - [1(5.7\ J/K) + 1(130.7\ J/K) + 1(223.1\ J/K)]$$
$$= [270.2\ J/K] - [359.5\ J/K]$$
$$= -89.3\ J/K$$

The moles of gas are decreasing.
Check: The units (J/K) are correct. The answer is negative, which is consistent with 2 moles of gas going to 1 mole of gas.

29. **Given:** methanol (CH_3OH) combustion at 25 °C **Find:** ΔH°_{rxn}, ΔS°_{rxn}, ΔG°_{rxn}, and sponteneity
Conceptual Plan: write balanced reaction then $\Delta H^0_{rxn} = \sum n_p H^0_f(products) - \sum n_r H^0_f(reactants)$ **then**
$\Delta S^0_{rxn} = \sum n_p S^0(products) - \sum n_r S^0(reactants)$ **then °C → K then J/K → kJ/K then**

$K = 273.15 + °C$ $\qquad \dfrac{1\ kJ}{1000\ J}$

$\Delta H^\circ_{rxn}, \Delta S^\circ_{rxn}, T \rightarrow \Delta G^\circ$

$\Delta G = \Delta H_{rxn} - T\Delta S_{rxn}$

Solution: combustion is combination with oxygen to form carbon dioxide and water
$2\ CH_3OH\ (l) + 3\ O_2\ (g) \rightarrow 2\ CO_2\ (g) + 4\ H_2O\ (g)$

Reactant/Product	ΔH°_f (kJ/mol from Appendix IIB)
$CH_3OH\ (l)$	− 238.6
$O_2\ (g)$	0.0
$CO_2\ (g)$	− 393.5
$H_2O\ (g)$	− 241.8

Be sure to pull data for the correct formula and phase.
$$\Delta H^\circ_{rxn} = \Sigma n_p \Delta H^\circ_f(products) - \Sigma n_r \Delta H^\circ_f(reactants)$$
$$= [2(\Delta H^\circ_f(CO_2(g))) + 4(\Delta H^\circ_f(H_2O(g)))] - [2(\Delta H^\circ_f(CH_3OH(l))) + 3(\Delta H^\circ_f(O_2(g)))]$$
$$= [2(-393.5\ kJ) + 4(-241.8\ kJ)] - [2(-238.6\ kJ) + 3(0.0\ kJ)]$$
$$= [-1754.2\ kJ] - [-477.2\ kJ]$$
$$= -1277\ kJ$$

then

Reactant/Product	S^0 (J/molK from Appendix IIB)
CH_3OH (*l*)	126.8
O_2 (*g*)	205.2
CO_2 (*g*)	213.8
H_2O (*g*)	188.8

Be sure to pull data for the correct formula and phase.

$$\Delta S^\circ_{rxn} = \Sigma n_p S^\circ(products) - \Sigma n_r S^\circ(reactants)$$
$$= [2(S^\circ(CO_2(g))) + 4(S^\circ(H_2O(g)))] - [2(S^\circ(CH_3OH(l))) + 3(S^\circ(O_2(g)))]$$
$$= [2(213.8\ J/K) + 4(188.8\ J/K)] - [2(126.8\ J/K) + 3(205.2\ J/K)]$$ then
$$= [1182.8\ J/K] - [869.2\ J/K]$$
$$= 313.6\ J/K$$

$T = 273.15 + 25\ °C = 298\ K$ then $+313.6 \frac{J}{K} \times \frac{1\ kJ}{1000\ J} = +0.3136 \frac{kJ}{K}$ then

$\Delta G = \Delta H_{rxn} - T\Delta S_{rxn} = -1277\ kJ - (298\ K)\left(+0.3136 \frac{kJ}{K}\right) = -1370.\ kJ = -1.370 \times 10^6\ J$ so the reaction is spontaneous.

Check: The units (kJ, J/K, and kJ) are correct. Combustion reactions are exothermic and we see a large negative enthalpy. We expect a large positive entropy because we have an increase in the number of moles of gas. The free energy is the sum of two negative terms so we expect a large negative free energy and the reaction is spontaneous.

31. a) **Given:** $N_2O_4\ (g) \rightarrow 2\ NO_2\ (g)$ at 25 °C

Find: ΔH°_{rxn}, ΔS°_{rxn}, ΔG°_{rxn}, spontaneity and can temperature be changed to make it spontaneous ?

Conceptual Plan: $\Delta H^\circ_{rxn} = \sum n_p H^\circ_f(products) - \sum n_r H^\circ_f(reactants)$ **then**

$\Delta S^\circ_{rxn} = \sum n_p S^\circ(products) - \sum n_r S^\circ(reactants)$ **then °C → K then J/K → kJ/K then**

$K = 273.15 + °C$ $\qquad \frac{1\ kJ}{1000\ J}$

ΔH°_{rxn}, ΔS_{rxn}, T → ΔG

$\Delta G = \Delta H_{rxn} - T\Delta S_{rxn}$

Solution:

Reactant/Product	ΔH^0_f (kJ/mol from Appendix IIB)
N_2O_4 (*g*)	11.1
NO_2 (*g*)	33.2

Be sure to pull data for the correct formula and phase.

$$\Delta H^0_{rxn} = \sum n_p \Delta H^0_f(products) - \sum n_r \Delta H^0_f(reactants)$$
$$= [2(\Delta H^0_f(NO_2\ (g)))] - [1(\Delta H^0_f(N_2O_4\ (g)))]$$
$$= [2(33.2\ kJ)] - [1(11.1\ kJ)]$$ then
$$= [66.4\ kJ] - [11.1\ kJ]$$
$$= +55.3\ kJ$$

Reactant/Product	S^0 (J/molK from Appendix IIB)
N_2O_4 (*g*)	304.4
NO_2 (*g*)	240.1

Be sure to pull data for the correct formula and phase.

$$\Delta S^0_{rxn} = \sum n_p S^\circ(products) - \sum n_r S^\circ(reactants)$$
$$= [2(S^\circ(NO_2\ (g)))] - [1(S^\circ(N_2O_4\ (g)))]$$
$$= [2(240.1\ J/K)] - [1(304.4\ J/K)]$$ then $T = 273.15 + 25\ °C = 298\ K$ then
$$= [480.2\ J/K] - [304.4\ J/K]$$
$$= +175.8\ J/K$$

$$+175.8\frac{J}{K} \times \frac{1\ kJ}{1000\ J} = +0.1758\frac{kJ}{K}$$ then

$\Delta G^0 = \Delta H^0_{rxn} - T\Delta S^0_{rxn} = +55.3\ kJ - (298\ K)\left(+0.1758\frac{kJ}{K}\right) = +2.9\ kJ = +2.9 \times 10^3\ J$ so the reaction is nonspontaneous. It can be made spontaneous by raising the temperature.

Check: The units (kJ, J/K, and kJ) are correct. The reaction requires the breaking of a bond, so we expect that this will be an endothermic reaction. We expect a positive entropy change because we are increasing the number of moles of gas. Since the positive enthalpy term dominates at room temperature, the reaction is nonspontaneous. The second term can dominate if we raise the temperature high enough.

b) **Given:** $NH_4Cl\ (s) \rightarrow HCl\ (g) + NH_3\ (g)$ at 25 °C
Find: ΔH°_{rxn}, ΔS°_{rxn}, ΔG°_{rxn}, spontaneity and can temperature be changed to make it spontaneous?
Conceptual Plan: $\Delta H^0_{rxn} = \sum n_p H^0_f(products) - \sum n_r H^0_f(reactants)$ **then**

$\Delta S^0_{rxn} = \sum n_p S^0(products) - \sum n_r S^0(reactants)$ **then °C→K then J/K→ kJ/K then ΔH°_{rxn}, ΔS_{rxn}, T→ ΔG**

$K = 273.15 + °C$ $\quad \frac{1 kJ}{1000\ J}$ $\quad \Delta G = \Delta H_{rxn} - T\Delta S_{rxn}$

Solution:

Reactant/Product	ΔH^0_f(kJ/mol from Appendix IIB)
$NH_4Cl\ (s)$	− 314.4
$HCl\ (g)$	− 92.3
$NH_3\ (g)$	− 45.9

Be sure to pull data for the correct formula and phase.

$$\begin{aligned}\Delta H^0_{rxn} &= \sum n_p \Delta H^0_f(products) - \sum n_r \Delta H^0_f(reactants)\\ &= [1(\Delta H^0_f(HCl\ (g))) + 1(\Delta H^0_f(NH_3\ (g)))] - [1(\Delta H^0_f(NH_4Cl\ (g)))]\\ &= [1(-92.3\ kJ) + 1(-45.9\ kJ)] - [1(-314.4\ kJ)]\\ &= [-138.2\ kJ] - [-314.4\ kJ]\\ &= +176.2\ kJ\end{aligned}$$

then

Reactant/Product	S^0(J/molK from Appendix IIB)
$NH_4Cl\ (s)$	94.6
$HCl\ (g)$	186.9
$NH_3\ (g)$	192.8

Be sure to pull data for the correct formula and phase.

$$\begin{aligned}\Delta S^0_{rxn} &= \sum n_p S^0(products) - \sum n_r S^0(reactants)\\ &= [1(S^0(HCl\ (g))) + 1(S^0(NH_3\ (g)))] - [1(S^0(NH_4Cl\ (g)))]\\ &= [1(186.9\ J/K) + 1(192.8\ J/K)] - [1(94.6\ J/K)]\\ &= [379.7\ J/K] - [94.6\ J/K]\\ &= +285.1\ J/K\end{aligned}$$

then $T = 273.15 + 25\ °C = 298\ K$ then

$$+285.1\frac{J}{K} \times \frac{1\ kJ}{1000\ J} = +0.2851\frac{kJ}{K}$$ then

$\Delta G^0 = \Delta H^0_{rxn} - T\Delta S^0_{rxn} = +176.2\ kJ - (298\ K)\left(+0.2851\frac{kJ}{K}\right) = +91.2\ kJ = +9.12 \times 10^4\ J$ so the reaction is nonspontaneous. It can be made spontaneous by raising the temperature.

Check: The units (kJ, J/K, and kJ) are correct. The reaction requires the breaking of a bond, so we expect that this will be an endothermic reaction. We expect a positive entropy change because we are increasing the number of moles of gas. Since the positive enthalpy term dominates at room temperature, the reaction is nonspontaneous. The second term can dominate if we raise the temperature high enough.

c) **Given:** $3\ H_2\ (g) + Fe_2O_3\ (s) \rightarrow 2\ Fe\ (s) + 3\ H_2O\ (g)$ at 25 °C
Find: ΔH°_{rxn}, ΔS°_{rxn}, ΔG°_{rxn}, spontaneity and can temperature be changed to make it spontaneous ?
Conceptual Plan: $\Delta H^0_{rxn} = \sum n_p H^0_f(products) - \sum n_r H^0_f(reactants)$ **then**

$\Delta S^0_{rxn} = \sum n_p S^0(products) - \sum n_r S^0(reactants)$ **then °C → K then J/K → kJ/K then**

$K = 273.15 + °C$ $\qquad \frac{1\,kJ}{1000\,J}$

$\Delta H^\circ_{rxn}, \Delta S_{rxn}, T \rightarrow \Delta G$

$\Delta G = \Delta H_{rxn} - T\Delta S_{rxn}$

Solution:

Reactant/Product	ΔH^0_f(kJ/mol from Appendix IIB)
$H_2\ (g)$	0.0
$Fe_2O_3\ (s)$	− 824.2
$Fe\ (s)$	0.0
$H_2O\ (g)$	− 241.8

Be sure to pull data for the correct formula and phase.

$$\Delta H^0_{rxn} = \sum n_p \Delta H^0_f(products) - \sum n_r \Delta H^0_f(reactants)$$
$$= [2(\Delta H^0_f(Fe\ (s))) + 3(\Delta H^0_f(H_2O\ (g)))] - [3(\Delta H^0_f(H_2\ (g))) + 1(\Delta H^0_f(Fe_2O_3\ (s)))]$$
$$= [2(0.0\ kJ) + 3(-241.8\ kJ)] - [3(0.0\ kJ) + 1(-824.2\ kJ)]$$
$$= [-725.4\ kJ] - [-824.2\ kJ]$$
$$= +98.8\ kJ$$

then

Reactant/Product	S^0(J/molK from Appendix IIB)
$H_2\ (g)$	130.7
$Fe_2O_3\ (s)$	87.4
$Fe\ (s)$	27.3
$H_2O\ (g)$	188.8

Be sure to pull data for the correct formula and phase.

$$\Delta S^0_{rxn} = \sum n_p S^0(products) - \sum n_r S^0(reactants)$$
$$= [2(S^0(Fe\ (s))) + 3(S^0(H_2O\ (g)))] - [3(S^0(H_2\ (g))) + 1(S^0(Fe_2O_3\ (s)))]$$
$$= [2(27.3\ J/K) + 3(188.8\ J/K)] - [3(130.7\ J/K) + 1(87.4\ J/K)]$$
$$= [621.0\ J/K] - [479.5\ J/K]$$
$$= +141.5\ J/K$$

then

$T = 273.15 + 25\ °C = 298\ K$ then $+141.5 \frac{J}{K} \times \frac{1\ kJ}{1000\ J} = +0.1415 \frac{kJ}{K}$ then

$\Delta G^0 = \Delta H^0_{rxn} - T\Delta S^0_{rxn} = +98.8\ kJ - (298\ K)\left(+0.1415 \frac{kJ}{K}\right) = +56.6\ kJ = +5.66 \times 10^4\ J$ so the reaction is nonspontaneous. It can be made spontaneous by raising the temperature.

Check: The units (kJ, J/K, and kJ) are correct. The reaction requires the breaking of a bond, so we expect that this will be an endothermic reaction. We expect a positive entropy change because there is no change in the number of moles of gas, but the product gas is more complex. Since the positive enthalpy term dominates at room temperature, the reaction is nonspontaneous. The second term can dominate if we raise the temperature high enough. This process is the opposite of rusting, so we are not surprised that it is nonspontaneous.

d) **Given:** $N_2\ (g) + 3\ H_2\ (g) \rightarrow 2\ NH_3\ (g)$ at 25 °C
Find: ΔH°_{rxn}, ΔS°_{rxn}, ΔG°_{rxn}, spontaneity and can temperature be changed to make it spontaneous
Conceptual Plan: $\Delta H^0_{rxn} = \sum n_p H^0_f(products) - \sum n_r H^0_f(reactants)$ **then**

$\Delta S^\circ_{rxn} = \sum n_p S^\circ(products) - \sum n_r S^\circ(reactants)$ **then °C → K then J/K → kJ/K then**

$K = 273.15 + °C$ $\frac{1 kJ}{1000 J}$

$\Delta H°_{rxn}$, ΔS_{rxn}, T → ΔG

$\Delta G = \Delta H_{rxn} - T\Delta S_{rxn}$

Solution:

Reactant/Product	ΔH^0_f(kJ/mol from Appendix IIB)
N_2 (*g*)	0.0
H_2 (*g*)	0.0
NH_3 (*g*)	– 45.9

Be sure to pull data for the correct formula and phase.

$$\Delta H^0_{rxn} = \sum n_p \Delta H^0_f(products) - \sum n_r \Delta H^0_f(reactants)$$
$$= [2(\Delta H^0_f(NH_3\ (g)))] - [1(\Delta H^0_f(N_2\ (g))) + 3(\Delta H^0_f(H_2\ (g)))]$$
$$= [2(-45.9\ kJ)] - [1(0.0\ kJ) + 3(0.0\ kJ)]$$
$$= [-91.8\ kJ] - [0.0\ kJ]$$
$$= -91.8\ kJ$$

then

Reactant/Product	S^0(J/molK from Appendix IIB)
N_2 (*g*)	191.6
H_2 (*g*)	130.7
NH_3 (*g*)	192.8

Be sure to pull data for the correct formula and phase.

$$\Delta S^0_{rxn} = \sum n_p S^0(products) - \sum n_r S^0(reactants)$$
$$= [2(S^0(NH_3\ (g)))] - [1(S^0(N_2\ (g))) + 3(S^0(H_2\ (g)))]$$
$$= [2(192.8\ J/K)] - [1(191.6\ J/K) + 3(130.7\ J/K)]$$
$$= [385.6\ J/K] - [583.7\ J/K]$$
$$= -198.1\ J/K$$

then

$T = 273.15 + 25\ °C = 298\ K$ then $-198.1 \frac{J}{K} \times \frac{1\ kJ}{1000\ J} = -0.1981 \frac{kJ}{K}$ then

$\Delta G^0 = \Delta H^0_{rxn} - T\Delta S^0_{rxn} = -91.8\ kJ - (298\ K)\left(-0.1981 \frac{kJ}{K}\right) = -32.8\ kJ = -3.28 \times 10^4\ J$ so the reaction is spontaneous.

Check: The units (kJ, J/K, and kJ) are correct. The reaction requires the breaking of a bond, so we expect that this will be an endothermic reaction. We expect a positive entropy change because are increasing the number of moles of gas. Since the negative enthalpy term dominates at room temperature, the reaction is spontaneous. The second term can dominate if we raise the temperature high enough.

33. a) **Given:** N_2O_4 (*g*) → 2 NO_2 (*g*) at 25 °C **Find:** $\Delta G°_{rxn}$, spontaneity and compare to #31. Determine which method would show how free energy changes with temperature.

Conceptual Plan: $\Delta G^\circ_{rxn} = \sum n_p \Delta G^\circ_f(products) - \sum n_r \Delta G^\circ_f(reactants)$ **then compare to #31**

Solution:

Reactant/Product	ΔG^0_f(kJ/mol from Appendix IIB)
N_2O_4 (*g*)	99.8
NO_2 (*g*)	51.3

Be sure to pull data for the correct formula and phase.

$$\Delta G^0_{rxn} = \sum n_P \Delta G^0_f(products) - \sum n_R \Delta G^0_f(reactants)$$
$$= [2(\Delta G^0_f(NO_2\ (g)))] - [1(\Delta G^0_f(N_2O_4\ (g)))]$$
$$= [2(51.3\text{ kJ})] - [1(99.8\text{ kJ})]$$
$$= [102.6\text{ kJ}] - [99.8\text{ kJ}]$$
$$= +\ 2.8\text{ kJ}$$

so the reaction is nonspontaneous.

The value is similar to #31.

Check: The units (kJ) are correct. The free energy of the products is greater than the reactants, so the answer is positive and the reaction is nonspontaneous. The answer is the same as in #31 within the error of the calculation.

b) **Given:** $NH_4Cl\ (s) \rightarrow HCl\ (g) + NH_3\ (g)$ at 25 °C
Find: $\Delta G°_{rxn}$, spontaneity and compare to #31

Conceptual Plan: $\Delta G^0_{rxn} = \sum n_p \Delta G^0_f(products) - \sum n_r \Delta G^0_f(reactants)$ **then compare to #31**

Solution:

Reactant/Product	ΔG^0_f(kJ/mol from Appendix IIB)
$NH_4Cl\ (s)$	– 202.9
$HCl\ (g)$	– 95.3
$NH_3\ (g)$	– 16.4

Be sure to pull data for the correct formula and phase.

$$\Delta G^0_{rxn} = \sum n_P \Delta G^0_f(products) - \sum n_R \Delta G^0_f(reactants)$$
$$= [1(\Delta G^0_f(HCl\ (g))) + 1(\Delta G^0_f(NH_3\ (g)))] - [1(\Delta G^0_f(NH_4Cl\ (g)))]$$
$$= [1(-95.3\text{ kJ}) + 1(-16.4\text{ kJ})] - [1(-202.9\text{ kJ})]$$
$$= [-111.7\text{ kJ}] - [-202.9\text{ kJ}]$$
$$= +\ 91.2\text{ kJ}$$

so the reaction is nonspontaneous.

The result is the same as in #31.

Check: The units (kJ) are correct. The answer matches #31.

c) **Given:** $3\ H_2\ (g) + Fe_2O_3\ (s) \rightarrow 2\ Fe\ (s) + 3\ H_2O\ (g)$ at 25 °C
Find: $\Delta G°_{rxn}$, spontaneity and compare to #31

Conceptual Plan: $\Delta G^0_{rxn} = \sum n_p \Delta G^0_f(products) - \sum n_r \Delta G^0_f(reactants)$ **then compare to #31**

Solution:

Reactant/Product	ΔG^0_f(kJ/mol from Appendix IIB)
$H_2\ (g)$	0.0
$Fe_2O_3\ (s)$	– 742.2
$Fe\ (s)$	0.0
$H_2O\ (g)$	– 228.6

Be sure to pull data for the correct formula and phase.

$$\Delta G^0_{rxn} = \sum n_P \Delta G^0_f(products) - \sum n_R \Delta G^0_f(reactants)$$
$$= [2(\Delta G^0_f(Fe\ (s))) + 3(\Delta G^0_f(H_2O\ (g)))] - [3(\Delta G^0_f(H_2\ (g))) + 1(\Delta G^0_f(Fe_2O_3\ (s)))]$$
$$= [2(0.0\text{ kJ}) + 3(-228.6\text{ kJ})] - [3(0.0\text{ kJ}) + 1(-742.2\text{ kJ})]$$
$$= [-685.8\text{ kJ}] - [-742.2\text{ kJ}]$$
$$= +56.4\text{ kJ}$$

so the reaction is nonspontaneous. The value is similar to that in #31.

Check: The units (kJ) are correct. The answer is the same as in #31 within the error of the calculation.

d) **Given:** $N_2\ (g) + 3\ H_2\ (g) \rightarrow 2\ NH_3\ (g)$ at 25 °C
Find: $\Delta G°_{rxn}$, spontaneity and compare to #31

Conceptual Plan: $\Delta G^0_{rxn} = \sum n_p \Delta G^0_f(products) - \sum n_r \Delta G^0_f(reactants)$ **then compare to #31**

Solution:

Reactant/Product	ΔG_f^0(kJ/mol from Appendix IIB)
N_2 (*g*)	0.0
H_2 (*g*)	0.0
NH_3 (*g*)	− 16.4

Be sure to pull data for the correct formula and phase.

$$\Delta G^0_{rxn} = \sum n_P \Delta G^0_f(products) - \sum n_R \Delta G^0_f(reactants)$$
$$= [2(\Delta G^0_f(NH_3\ (g)))] - [1(\Delta G^0_f(N_2\ (g))) + 3(\Delta G^0_f(H_2\ (g)))]$$
$$= [2(-16.4\ kJ)] - [1(0.0\ kJ) + 3(0.0\ kJ)]$$
$$= [-32.8\ kJ] - [0.0\ kJ]$$
$$= -32.8\ kJ$$

so the reaction is spontaneous. The result is the same as in #31.
Check: The units (kJ) are correct. The answer matches #31.

Values calculated by the two methods are comparable. The method using $\Delta H°$ and $\Delta S°$ is longer, but it can be used to determine how $\Delta G°$ changes with temperature.

35. **Given:** 2 NO (*g*) + O_2 (*g*) → 2 NO_2 (*g*) **Find:** $\Delta G°_{rxn}$ and spontaneity at a) 298 K, b) 715 K, and 855 K

Conceptual Plan: $\Delta H^\circ_{rxn} = \sum n_p H^\circ_f(products) - \sum n_r H^\circ_f(reactants)$ **then**

$\Delta S^\circ_{rxn} = \sum n_p S^\circ(products) - \sum n_r S^\circ(reactants)$ **then J/K → kJ/K then** $\Delta H°_{rxn}, \Delta S_{rxn}, T$ **→** ΔG

$\frac{1 kJ}{1000 J}$ $\qquad$ $\Delta G = \Delta H_{rxn} - T\Delta S_{rxn}$

Solution:

Reactant/Product	ΔH_f^0(kJ/mol from Appendix IIB)
NO (*g*)	91.3
O_2 (*g*)	0.0
NO_2 (*g*)	33.2

Be sure to pull data for the correct formula and phase.

$$\Delta H^0_{rxn} = \sum n_p \Delta H^0_f(products) - \sum n_r \Delta H^0_f(reactants)$$
$$= [2(\Delta H^0_f(NO_2\ (g)))] - [2(\Delta H^0_f(NO\ (g))) + 1(\Delta H^0_f(O_2\ (g)))]$$
$$= [2(33.2\ kJ)] - [2(91.3\ kJ) + 1(0.0\ kJ)] \qquad \text{then}$$
$$= [66.4\ kJ] - [182.6\ kJ]$$
$$= -116.2\ kJ$$

Reactant/Product	S^0(J/molK from Appendix IIB)
NO (*g*)	210.8
O_2 (*g*)	205.2
NO_2 (*g*)	240.1

Be sure to pull data for the correct formula and phase.

$$\Delta S^0_{rxn} = \sum n_p S^0(products) - \sum n_r S^0(reactants)$$
$$= [2(S^0(NO_2\ (g)))] - [2(S^0(NO\ (g))) + 1(S^0(O_2\ (g)))]$$
$$= [2(240.1\ J/K)] - [2(210.8\ J/K) + 1(205.2\ J/K)] \qquad \text{then } -146.6\frac{J}{K} \times \frac{1\ kJ}{1000\ J} = -0.1466\frac{kJ}{K} \text{ then}$$
$$= [480.2\ J/K] - [626.8\ J/K]$$
$$= -146.6\ J/K$$

a) $\Delta G^0 = \Delta H^0_{rxn} - T\Delta S^0_{rxn} = -116.2\ kJ - (298\ K)\left(-0.1466\frac{kJ}{K}\right) = -72.5\ kJ = -7.25 \times 10^4\ J$ so the reaction is spontaneous.

b) $\Delta G^0 = \Delta H^0_{rxn} - T\Delta S^0_{rxn} = -116.2 \text{ kJ} - (715 \text{ K})\left(-0.1466 \frac{\text{kJ}}{\text{K}}\right) = -11.4 \text{ kJ} = -1.14 \times 10^4 \text{ J}$ so the reaction is spontaneous.

c) $\Delta G^0 = \Delta H^0_{rxn} - T\Delta S^0_{rxn} = -116.2 \text{ kJ} - (855 \text{ K})\left(-0.1466 \frac{\text{kJ}}{\text{K}}\right) = +9.1 \text{ kJ} = +9.1 \times 10^3 \text{ J}$ so the reaction is nonspontaneous.

Check: The units (kJ) are correct. The enthalpy term dominates at low temperatures, making the reaction spontaneous. As the temperature increases, the decrease in entropy starts to dominate and in the last case the reaction is nonspontaneous.

37. Since the first reaction has Fe_2O_3 as a product and the reaction of interest has it as a reactant, we need to reverse the first reaction. When the reaction direction is reversed, ΔG changes.

$Fe_2O_3\ (s) \rightarrow 2\ Fe\ (s) + 3/2\ O_2\ (g)$ $\qquad \Delta G^0 = +742.2 \text{ kJ}$

Since the second reaction has 1 mole CO as a reactant and the reaction of interest has 3 moles of CO as a reactant, we need to multiply the second reaction and the ΔG by 3.

$3\,[CO\ (g) + 1/2\ O_2\ (g) \rightarrow CO_2\ (g)]$ $\qquad \Delta G^0 = 3(-257.2 \text{ kJ}) = -771.6 \text{ kJ}$

Hess' Law states the ΔG of the net reaction is the sum of the ΔG of the steps.
The rewritten reactions are:

$Fe_2O_3\ (s) \rightarrow 2\ Fe\ (s) + \cancel{3/2\ O_2\ (g)}$ $\qquad \Delta G^0 = +742.2 \text{ kJ}$

$3\ CO\ (g) + \cancel{3/2\ O_2\ (g)} \rightarrow 3\ CO_2\ (g)$ $\qquad \Delta G^0 = -771.6 \text{ kJ}$

$Fe_2O_3\ (s) + 3\ CO\ (g) \rightarrow 2\ Fe\ (s) + 3\ CO_2\ (g)$ $\qquad \Delta G^0_{rxn} = -29.4 \text{ kJ}$

39. a) **Given:** $I_2\ (s) \rightarrow I_2\ (g)$ at 25.0 °C **Find:** ΔG°_{rxn}

Conceptual Plan: $\Delta G^0_{rxn} = \sum n_P \Delta G^0_f(products) - \sum n_R \Delta G^0_f(reactants)$

Solution:

Reactant/Product	ΔG^0_f (kJ/mol from Appendix IIB)
$I_2\ (s)$	0.0
$I_2\ (g)$	19.3

Be sure to pull data for the correct formula and phase.

$$\begin{aligned}\Delta G^0_{rxn} &= \sum n_p \Delta G^0_f(products) - \sum n_r \Delta G^0_f(reactants) \\ &= [1(\Delta G^0_f(I_2\ (g)))] - [1(\Delta G^0_f(I_2\ (s)))] \\ &= [1(19.3 \text{ kJ})] - [1(0.0 \text{ kJ})] \\ &= +19.3 \text{ kJ}\end{aligned}$$

so the reaction is nonspontaneous.

Check: The units (kJ) are correct. The answer is positive because gases have higher free energy than solids and the free energy change of the reaction is the same as free energy of formation of gaseous iodine.

b) **Given:** $I_2\ (s) \rightarrow I_2\ (g)$ at 25.0 °C (i) P_{I2} = 1.00 mmHg; (ii) P_{I2} = 0.100 mmHg **Find:** ΔG_{rxn}

Conceptual Plan: **°C → K and mmHg → atm then $\Delta G^\circ_{rxn}, P_{I2}, T$ → ΔG_{rxn}**

$K = 273.15 + °C \qquad \frac{1 \text{ atm}}{760 \text{ mm Hg}} \qquad \Delta G_{rxn} = \Delta G^0_{rxn} + RT \ln Q$ where $Q = P_{I2}$

Solution: $T = 273.15 + 25.0\ °C = 298.2$ K and (i) $1.00 \text{ mmHg} \times \frac{1 \text{ atm}}{760 \text{ mmHg}} = 0.00131579$ atm

then $\Delta G_{rxn} = \Delta G^0_{rxn} + RT \ln Q = \Delta G^0_{rxn} + RT \ln P_{I2} = +19.3 \text{ kJ} +$

$+\left(8.314 \frac{\text{J}}{\text{K} \cdot \text{mol}}\right)\left(\frac{1 \text{ kJ}}{1000 \text{ J}}\right)(298.2 \text{ K}) \ln(0.00131579) = +2.9 \text{ kJ}$ so the reaction is nonspontaneous. Then

(ii) $0.100 \text{ mmHg} \times \frac{1 \text{ atm}}{760 \text{ mmHg}} = 0.000131579$ atm then

$$\Delta G_{rxn} = \Delta G^0_{rxn} + RT\ln Q = \Delta G^0_{rxn} + RT\ln P_{I2} = +19.3 \text{ kJ} +$$

$$+\left(8.314 \frac{\text{J}}{\text{K}\cdot\text{mol}}\right)\left(\frac{1\text{ kJ}}{1000\text{ J}}\right)(298.2\text{ K})\ln(0.000131579) = -2.9\text{ kJ}$$ so the reaction is spontaneous.

Check: The units (kJ) are correct. The answer is positive at high pressures because the pressure is higher than the vapor pressure of iodine. Once the desired pressure is below the vapor pressure (0.31 mmHg at 25.0 °C), the reaction becomes spontaneous.

c) Iodine sublimes at room temperature because there is an equilibrium between the solid and the gas phases. The vapor pressure is low (0.31 mmHg at 25.0 °C), so a small amount of iodine can remain in the gas phase, which is consistent with the free energy values.

41. **Given:** CH_3OH (*g*) ⇌ CO (*g*) + 2 H_2 (*g*) at 25 °C, $P_{CH3OH} = 0.855$ atm, $P_{CO} = 0.125$ atm, $P_{H2} = 0.183$ atm
Find: ΔG
Conceptual Plan:
$\Delta G^\circ_{rxn} = \sum n_p \Delta G^\circ_f(products) - \sum n_r \Delta G^\circ_f(reactants)$ **then °C→ K then ΔG°_{rxn}, P_{CH3OH}, P_{CO}, P_{H2}, T → ΔG**

$K = 273.15 + °C \quad \Delta G_{rxn} = \Delta G^0_{rxn} + RT\ln Q$ where $Q = \frac{P_{CO}P^2_{H2}}{P_{CH3OH}}$

Solution:

Reactant/Product	ΔG^0_f(kJ/mol from Appendix IIB)
CH_3OH (*g*)	– 162.3
CO (*g*)	– 137.2
H_2 (*g*)	0.0

Be sure to pull data for the correct formula and phase.

$$\Delta G^0_{rxn} = \sum n_p \Delta G^0_f(products) - \sum n_r \Delta G^0_f(reactants)$$
$$= [1(\Delta G^0_f(CO\ (g))) + 2(\Delta G^0_f(H_2\ (g)))] - [1(\Delta G^0_f(CH_3OH\ (g)))]$$
$$= [1(-137.2\text{ kJ}) + 2(0.0\text{ kJ})] - [1(-162.3\text{ kJ})]$$
$$= [-137.2\text{ kJ}] - [-162.3\text{ kJ}]$$
$$= +25.1\text{ kJ}$$

$T = 273.15 + 25\text{ °C} = 298\text{ K}$ then

$$Q = \frac{P_{CO}P^2_{H2}}{P_{CH3OH}} = \frac{(0.125)(0.183)^2}{0.855} = 0.00489605$$ then

$$\Delta G_{rxn} = \Delta G^0_{rxn} + RT\ln Q = +25.1\text{ kJ} + \left(8.314 \frac{\text{J}}{\text{K}\cdot\text{mol}}\right)\left(\frac{1\text{ kJ}}{1000\text{ J}}\right)(298\text{ K})\ln(0.00489605) = +11.9\text{ kJ}$$

so the reaction is nonspontaneous.
Check: The units (kJ) are correct. The standard free energy for the reaction was positive and the fact that Q was less than one made the free energy smaller, but the reaction at these conditions is still not spontaneous.

43. a) **Given:** 2 CO (*g*) + O_2 (*g*) ⇌ 2 CO_2 (*g*) at 25 °C **Find:** K

Conceptual Plan: $\Delta G^\circ_{rxn} = \sum n_p \Delta G^\circ_f(products) - \sum n_r \Delta G^\circ_f(reactants)$ **then °C → K then ΔG°_{rxn}, T → K**

$K = 273.15 + °C \quad \Delta G^0_{rxn} = -RT\ln K$

Solution:

Reactant/Product	ΔG^0_f(kJ/mol from Appendix IIB)
CO (*g*)	– 137.2
O_2 (*g*)	0.0
CO_2 (*g*)	– 394.4

Be sure to pull data for the correct formula and phase.

$$\Delta G^0_{rxn} = \sum n_p \Delta G^0_f(products) - \sum n_r \Delta G^0_f(reactants)$$
$$= [2(\Delta G^0_f(CO_2\ (g)))] - [2(\Delta G^0_f(CO\ (g))) + 1(\Delta G^0_f(O_2\ (g)))]$$
$$= [2(-394.4\ kJ)] - [2(-137.2\ kJ) + 1(0.0\ kJ)]$$
$$= [-788.8\ kJ] - [-274.4\ kJ]$$
$$= -514.4\ kJ$$

$T = 273.15 + 25\ °C = 298\ K$ then

$\Delta G^0_{rxn} = -RT\ln K$ Rearrange to solve for K.

$$K = e^{\frac{-\Delta G^0_{rxn}}{RT}} = e^{\frac{-(-514.4\ kJ) \times \frac{1000\ J}{1\ kJ}}{\left(8.314 \frac{J}{K \cdot mol}\right)(298\ K)}} = e^{207.623} = 1.48 \times 10^{90}.$$

Check: The units (none) are correct. The standard free energy for the reaction was very negative and so we expect a very large K. The reaction is spontaneous and so mostly products are present at equilibrium.

b) **Given:** $2\ H_2S\ (g) \rightleftharpoons 2\ H_2\ (g) + S_2\ (g)$ at 25 °C **Find:** K

Conceptual Plan: $\Delta G^0_{rxn} = \sum n_p \Delta G^0_f(products) - \sum n_r \Delta G^0_f(reactants)$ **then °C → K then $\Delta G°_{rxn}$, T → K**

$K = 273.15 + °C$ $\quad \Delta G^0_{rxn} = -RT\ln K$

Solution:

Reactant/Product	ΔG^0_f (kJ/mol from Appendix IIB)
$H_2S\ (g)$	− 33.4
$H_2\ (g)$	0.0
$S_2\ (g)$	79.7

Be sure to pull data for the correct formula and phase.

$$\Delta G^0_{rxn} = \sum n_p \Delta G^0_f(products) - \sum n_r \Delta G^0_f(reactants)$$
$$= [2(\Delta G^0_f(H_2\ (g))) + 1(\Delta G^0_f(S_2\ (g)))] - [2(\Delta G^0_f(H_2S\ (g)))]$$
$$= [2(0.0\ kJ)] + 1(79.7\ kJ)] - [2(-33.4\ kJ)]$$
$$= [79.7\ kJ] - [-66.8\ kJ]$$
$$= +146.5\ kJ$$

$T = 273.15 + 25\ °C = 298\ K$ then

$\Delta G^0_{rxn} = -RT\ln K$ Rearrange to solve for K. $K = e^{\frac{-\Delta G^0_{rxn}}{RT}} = e^{\frac{-146.5\ kJ \times \frac{1000\ J}{1\ kJ}}{\left(8.314 \frac{J}{K \cdot mol}\right)(298\ K)}} = e^{-59.1305} = 2.09 \times 10^{-26}$.

Check: The units (none) are correct. The standard free energy for the reaction was positive and so we expect a small K. The reaction is nonspontaneous and so mostly reactants are present at equilibrium.

45. **Given:** $CO\ (g) + 2\ H_2\ (g) \rightleftharpoons CH_3OH\ (g)$ $K_p = 2.26 \times 10^4$ at 25 °C
Find: $\Delta G°_{rxn}$ at a) standard conditions, b) at equilibrium, and c) $P_{CH3OH} = 1.0$ atm, $P_{CO} = P_{H2} = 0.010$ atm
Conceptual Plan: °C → K then a) K, T → $\Delta G°_{rxn}$ then b) at equilibrium $\Delta G_{rxn} = 0$ then

$K = 273.15 + °C$ $\quad \Delta G^0_{rxn} = -RT\ln K$

c) **$\Delta G°_{rxn}$, P_{CH3OH}, P_{CO}, P_{H2}, T → ΔG**

$$\Delta G_{rxn} = \Delta G^0_{rxn} + RT\ln Q \quad \text{where} \quad Q = \frac{P_{CH3OH}}{P_{CO} P^2_{H2}}$$

Solution: $T = 273.15 + 25\ °C = 298\ K$ then

a) $\Delta G^0_{rxn} = -RT\ln K = -\left(8.314 \frac{J}{K \cdot mol}\right)\left(\frac{1\ kJ}{1000\ J}\right)(298\ K)\ln(2.26 \times 10^4) = -24.8\ kJ$;

b) at equilibrium $\Delta G_{rxn} = 0$

c) $Q = \frac{P_{CH3OH}}{P_{CO}P_{H2}^2} = \frac{1.0}{(0.010)(0.010)^2} = 1.0 \times 10^6$ then

$$\Delta G_{rxn} = \Delta G^0_{rxn} + RT\ln Q = -24.8 \text{ kJ} + \left(8.314\frac{\text{J}}{\text{K}\cdot\text{mol}}\right)\left(\frac{1 \text{ kJ}}{1000 \text{ J}}\right)(298 \text{ K})\ln(1.0\text{x } 10^6) = +9.4 \text{ kJ}$$

Check: The units (kJ) are correct. The *K* was greater than one so we expect a negative standard free energy for the reaction. At equilibrium, by definition, the free energy change is zero. Since the conditions give a Q > *K* then the reaction needs to proceed in the reverse direction, which means that the reaction is spontaneous in the reverse direction.

47. a) **Given:** $2\ CO\ (g) + O_2\ (g) \rightleftharpoons 2\ CO_2\ (g)$ at 25 °C **Find:** *K* at 525 K

Conceptual Plan: $\Delta H^\circ_{rxn} = \sum n_p H^\circ_f(products) - \sum n_r H^\circ_f(reactants)$ **then**

$\Delta S^\circ_{rxn} = \sum n_p S^\circ(products) - \sum n_r S^\circ(reactants)$ **then J/K → kJ/K then ΔH°_{rxn}, ΔS_{rxn}, T → ΔG**

$\frac{1 \text{kJ}}{1000 \text{ J}}$ $\Delta G = \Delta H_{rxn} - T\Delta S_{rxn}$

then ΔG°_{rxn}, T → K

$\Delta G^0_{rxn} = -RT\ln K$

Solution:

Reactant/Product	ΔH^0_f(kJ/mol from Appendix IIB)
CO (*g*)	−110.5
O_2 (*g*)	0.0
CO_2 (*g*)	−393.5

Be sure to pull data for the correct formula and phase.

$$\begin{aligned}\Delta H^0_{rxn} &= \sum n_p \Delta H^0_f(products) - \sum n_r \Delta H^0_f(reactants)\\ &= [2(\Delta H^0_f(CO_2\ (g)))] - [2(\Delta H^0_f(CO\ (g))) + 1(\Delta H^0_f(O_2\ (g)))]\\ &= [2(-393.5 \text{ kJ})] - [2(-110.5 \text{ kJ}) + 1(0.0 \text{ kJ})]\\ &= [-787.0 \text{ kJ}] - [-221.0 \text{ kJ}]\\ &= -566.0 \text{ kJ}\end{aligned}$$

then

Reactant/Product	S^0(J/molK from Appendix IIB)
CO (*g*)	197.7
O_2 (*g*)	205.2
CO_2 (*g*)	213.8

Be sure to pull data for the correct formula and phase.

$$\begin{aligned}\Delta S^0_{rxn} &= \sum n_p S^0(products) - \sum n_r S^0(reactants)\\ &= [2(S^0(CO_2\ (g)))] - [2(S^0(CO\ (g))) + 1(S^0(O_2\ (g)))]\\ &= [2(213.8 \text{ J/K})] - [2(197.7 \text{ J/K}) + 1(205.2 \text{ J/K})]\\ &= [427.6 \text{ J/K}] - [600.6 \text{ J/K}]\\ &= -173.0 \text{ J/K}\end{aligned}$$

then

$-173.0\frac{\text{J}}{\text{K}} \times \frac{1 \text{ kJ}}{1000 \text{ J}} = -0.1730\frac{\text{kJ}}{\text{K}}$ then

$\Delta G^0 = \Delta H^0_{rxn} - T\Delta S^0_{rxn} = -566.0 \text{ kJ} - (525 \text{ K})\left(-0.1730\frac{\text{kJ}}{\text{K}}\right) = -475.2 \text{ kJ} = -4.752 \times 10^5 \text{ J}$ then

$\Delta G^0_{rxn} = -RT\ln K$ Rearrange to solve for *K*.

$$K = e^{\frac{-\Delta G^0_{rxn}}{RT}} = e^{\frac{-(-4.752 \times 10^5 \text{ J})}{\left(8.314\frac{\text{J}}{\text{K}\cdot\text{mol}}\right)(525 \text{ K})}} = e^{108.864} = 1.90 \times 10^{47}.$$

Check: The units (none) are correct. The free energy change is very negative, indicating a spontaneous reaction. This results in a very large K.

b) **Given:** $2\ H_2S\ (g) \rightleftharpoons 2\ H_2\ (g) + S_2\ (g)$ at 25 °C **Find:** K at 525 K

Conceptual Plan: $\Delta H^\circ_{rxn} = \sum n_p H^\circ_f(products) - \sum n_r H^\circ_f(reactants)$ **then**

$\Delta S^\circ_{rxn} = \sum n_p S^\circ(products) - \sum n_r S^\circ(reactants)$ **then J/K → kJ/K then $\Delta H^\circ_{rxn}, \Delta S_{rxn}, T$ → ΔG**

$\frac{1\ kJ}{1000\ J}$ $\Delta G = \Delta H_{rxn} - T\Delta S_{rxn}$

then $\Delta G^\circ_{rxn}, T$ → K

$\Delta G^0_{rxn} = -RT \ln K$

Solution:

Reactant/Product	ΔH^0_f(kJ/mol from Appendix IIB)
$H_2S\ (g)$	– 20.6
$H_2\ (g)$	0.0
$S_2\ (g)$	128.6

Be sure to pull data for the correct formula and phase.

$$\Delta H^0_{rxn} = \sum n_p \Delta H^0_f(products) - \sum n_r \Delta H^0_f(reactants)$$
$$= [2(\Delta H^0_f(H_2\ (g))) + 1(\Delta H^0_f(S_2\ (g)))] - [2(\Delta H^0_f(H_2S\ (g)))]$$
$$= [2(0.0\ kJ) + 1(128.6\ kJ)] - [2(-20.6\ kJ)]$$
$$= [128.6\ kJ] - [-41.2\ kJ]$$
$$= +169.8\ kJ$$

then

Reactant/Product	S^0(J/molK from Appendix IIB)
$H_2S\ (g)$	205.8
$H_2\ (g)$	130.7
$S_2\ (g)$	228.2

Be sure to pull data for the correct formula and phase.

$$\Delta S^0_{rxn} = \sum n_p S^0(products) - \sum n_r S^0(reactants)$$
$$= [2(S^0(H_2\ (g))) + 1(S^0(S_2\ (g)))] - [2(S^0(H_2S\ (g)))]$$
$$= [2(130.7\ J/K) + 1(228.2\ J/K)] - [2(205.8\ J/K)]$$
$$= [489.6\ J/K] - [411.6\ J/K]$$
$$= +78.0\ J/K$$

then

$$+78.0\frac{J}{K} \times \frac{1\ kJ}{1000\ J} = +0.0780\frac{kJ}{K}$$ then

$$\Delta G^0 = \Delta H^0_{rxn} - T\Delta S^0_{rxn} = +169.8\ kJ - (525\ K)\left(+0.0780\frac{kJ}{K}\right) = +128.\underline{8}5\ kJ = +1.28\underline{8}5 \times 10^5\ J$$ then

$\Delta G^0_{rxn} = -RT \ln K$ Rearrange to solve for K.

$$K = e^{\frac{-\Delta G^0_{rxn}}{RT}} = e^{\frac{-1.28\underline{8}5 \times 10^5\ J}{\left(8.314\frac{J}{K \cdot mol}\right)(525\ K)}} = e^{-29.\underline{5}199} = 1.51 \times 10^{-13}.$$

Check: The units (none) are correct. The free energy change is positive, indicating a nonspontaneous reaction. This results in a very small K.

49. a) +, since vapors have higher entropy than liquids.

b) –, since solids have less entropy than liquids.

c) –, since there is only one microstate for the final macrostate and there are six microstates for the initial macrostate.

51. a) **Given:** $N_2(g) + O_2(g) \rightarrow 2\,NO(g)$ **Find:** ΔG°_{rxn}, and K_p at 25 °C

Conceptual Plan: $\Delta H^\circ_{rxn} = \sum n_p H^\circ_f(products) - \sum n_r H^\circ_f(reactants)$ **then**

$\Delta S^\circ_{rxn} = \sum n_p S^\circ(products) - \sum n_r S^\circ(reactants)$ **then °C → K then J/K → kJ/K then $\Delta H^\circ_{rxn}, \Delta S_{rxn}, T \rightarrow \Delta G$**

$K = 273.15 + °C$ $\quad \frac{1\,kJ}{1000\,J}$ $\quad \Delta G = \Delta H_{rxn} - T\Delta S_{rxn}$

then $\Delta G^\circ_{rxn}, T \rightarrow K$

$\Delta G^0_{rxn} = -RT\ln K$

Solution:

Reactant/Product	ΔH^0_f(kJ/mol from Appendix IIB)
$N_2(g)$	0.0
$O_2(g)$	0.0
$NO(g)$	91.3

Be sure to pull data for the correct formula and phase.

$$\begin{aligned}\Delta H^0_{rxn} &= \sum n_p\Delta H^0_f(products) - \sum n_r\Delta H^0_f(reactants)\\ &= [2(\Delta H^0_f(NO(g)))] - [1(\Delta H^0_f(N_2(g))) + 1(\Delta H^0_f(O_2(g)))]\\ &= [2(91.3\text{ kJ})] - [1(0.0\text{ kJ}) + 1(0.0\text{ kJ})]\\ &= [182.6\text{ kJ}] - [0.0\text{ kJ}]\\ &= +182.6\text{ kJ}\end{aligned}$$

then

Reactant/Product	S^0(J/molK from Appendix IIB)
$N_2(g)$	191.6
$O_2(g)$	205.2
$NO(g)$	210.8

Be sure to pull data for the correct formula and phase.

$$\begin{aligned}\Delta S^0_{rxn} &= \sum n_p S^0(products) - \sum n_r S^0(reactants)\\ &= [2(S^0(NO(g)))] - [1(S^0(N_2(g))) + 1(S^0(O_2(g)))]\\ &= [2(210.8\text{ J/K})] - [1(191.6\text{ J/K}) + 1(205.2\text{ J/K})]\\ &= [421.6\text{ J/K}] - [396.8\text{ J/K}]\\ &= +24.8\text{ J/K}\end{aligned}$$

then

$T = 273.15 + 25\text{ °C} = 298\text{ K}$ then $+24.8\frac{\text{J}}{\text{K}} \times \frac{1\text{ kJ}}{1000\text{ J}} = +0.0248\frac{\text{kJ}}{\text{K}}$ then

$\Delta G^0 = \Delta H^0_{rxn} - T\Delta S^0_{rxn} = +182.6\text{ kJ} - (298\text{ K})\left(0.0248\frac{\text{kJ}}{\text{K}}\right) = +175.2\text{ kJ} = +1.752 \times 10^5\text{ J}$ then

$\Delta G^0_{rxn} = -RT\ln K$ Rearrange to solve for K.

$$K = e^{\frac{-\Delta G^0_{rxn}}{RT}} = e^{\frac{-1.752\times10^5\text{ J}}{\left(8.314\frac{\text{J}}{\text{K}\cdot\text{mol}}\right)(298\text{ K})}} = e^{-70.7144} = 1.95 \times 10^{-31}$$

so the reaction is nonspontaneous and at equilibrium mostly reactants are present.

Check: The units (kJ and none) are correct. The enthalpy is twice the enthalpy of formation of NO. We expect a very small entropy change because the number of moles of gas is unchanged. Since the positive enthalpy term dominates at room temperature, the free energy change is very positive and the reaction in the forward direction is nonspontaneous. This results in a very small K.

b) **Given:** $N_2(g) + O_2(g) \rightarrow 2\,NO(g)$ **Find:** ΔG°_{rxn} at 2000 K

Conceptual Plan: use results from part a) $\Delta H^\circ_{rxn}, \Delta S_{rxn}, T \rightarrow \Delta G$ then $\Delta G^\circ_{rxn}, T \rightarrow K$

$\Delta G = \Delta H_{rxn} - T\Delta S_{rxn}$ $\quad \Delta G^0_{rxn} = -RT\ln K$

Solution: $\Delta G = \Delta H_{rxn} - T\Delta S_{rxn} = +182.6\text{ kJ} - (2000\text{ K})\left(0.0248\frac{\text{kJ}}{\text{K}}\right) = +133.0\text{ kJ} = +1.330 \times 10^5\text{ J}$ then

$\Delta G^0_{rxn} = -RT\ln K$ Rearrange to solve for *K*.

$$K = e^{\frac{-\Delta G^0_{rxn}}{RT}} = e^{\frac{-1.330 \times 10^5 \text{ J}}{\left(8.314 \frac{\text{J}}{\text{K}\cdot\text{mol}}\right)(2000 \text{ K})}} = e^{-7.998557} = 3.36 \times 10^{-4}$$ so the forward reaction is becoming more spontaneous.

Check: The units (kJ and none) are correct. As the temperature rises, the entropy term becomes more significant. The free energy change is reduced and the *K* increases. The reaction is still nonspontaneous.

53. **Given:** $C_2H_4\ (g) + X_2\ (g) \rightarrow C_2H_4X_2\ (g)$ where X = Cl, Br, and I
Find: ΔH°_{rxn}, ΔS°_{rxn}, ΔG°_{rxn} and *K* at 25 °C and spontaneity trends with X and temperature

Conceptual Plan: $\Delta H^\circ_{rxn} = \sum n_p H^\circ_f(products) - \sum n_r H^\circ_f(reactants)$ **then**

$\Delta S^\circ_{rxn} = \sum n_p S^\circ(products) - \sum n_r S^\circ(reactants)$ **then °C → K then J/K → kJ/K then**

$K = 273.15 + °C$ $\qquad \frac{1 \text{ kJ}}{1000 \text{ J}}$

$\Delta H^\circ_{rxn}, \Delta S_{rxn}, T \rightarrow \Delta G$

$\Delta G = \Delta H_{rxn} - T\Delta S_{rxn}$

Solution:

Reactant/Product	ΔH^0_f(kJ/mol from Appendix IIB)
$C_2H_4\ (g)$	52.4
$Cl_2\ (g)$	0.0
$C_2H_4Cl_2\ (g)$	– 129.7

Be sure to pull data for the correct formula and phase.

$$\begin{aligned}\Delta H^0_{rxn} &= \sum n_p \Delta H^0_f(products) - \sum n_r \Delta H^0_f(reactants)\\ &= [1(\Delta H^0_f(C_2H_4Cl_2\ (g)))] - [1(\Delta H^0_f(C_2H_4\ (g))) + 1(\Delta H^0_f(Cl_2\ (g)))]\\ &= [1(-129.7 \text{ kJ})] - [1(52.4 \text{ kJ}) + 1(0.0 \text{ kJ})]\\ &= [-129.7 \text{ kJ}] - [52.4 \text{ kJ}]\\ &= -182.1 \text{ kJ}\end{aligned}$$ then

Reactant/Product	S^0(J/molK from Appendix IIB)
$C_2H_4\ (g)$	219.3
$Cl_2\ (g)$	223.1
$C_2H_4Cl_2\ (g)$	308.0

Be sure to pull data for the correct formula and phase.

$$\begin{aligned}\Delta S^0_{rxn} &= \sum n_p S^0(products) - \sum n_r S^0(reactants)\\ &= [1(S^0(C_2H_4Cl_2\ (g)))] - [1(S^0(C_2H_4\ (g))) + 1(S^0(Cl_2\ (g)))]\\ &= [1(308.0 \text{ J/K})] - [1(219.3 \text{ J/K}) + 1(223.1 \text{ J/K})]\\ &= [308.0 \text{ J/K}] - [442.4 \text{ J/K}]\\ &= -134.4 \text{ J/K}\end{aligned}$$ then $T = 273.15 + 25$ °C $= 298$ K then

$$-134.4 \frac{\text{J}}{\text{K}} \times \frac{1 \text{ kJ}}{1000 \text{ J}} = -0.1344 \frac{\text{kJ}}{\text{K}}$$ then

$$\Delta G^0 = \Delta H^0_{rxn} - T\Delta S^0_{rxn} = -182.1 \text{ kJ} - (298 \text{ K})\left(-0.1344 \frac{\text{kJ}}{\text{K}}\right) = -142.0 \text{ kJ} = -1.420 \times 10^5 \text{ J}$$ then

$\Delta G^0_{rxn} = -RT\ln K$ Rearrange to solve for *K*. $K = e^{\frac{-\Delta G^0_{rxn}}{RT}} = e^{\frac{-(-1.420 \times 10^5 \text{ J})}{\left(8.314 \frac{\text{J}}{\text{K}\cdot\text{mol}}\right)(298 \text{ K})}} = e^{57.334} = 7.94 \times 10^{24}$ so the reaction is spontaneous.

Reactant/Product	ΔH_f^0(kJ/mol from Appendix IIB)
C_2H_4 (*g*)	52.4
Br_2 (*g*)	30.9
$C_2H_4Br_2$ (*g*)	– 38.3

Be sure to pull data for the correct formula and phase.

$$\Delta H^0_{rxn} = \sum n_p \Delta H_f^0(products) - \sum n_r \Delta H_f^0(reactants)$$
$$= [1(\Delta H_f^0(C_2H_4Br_2\ (g)))] - [1(\Delta H_f^0(C_2H_4\ (g))) + 1(\Delta H_f^0(Br_2\ (g)))]$$
$$= [1(-38.3\text{ kJ})] - [1(52.4\text{ kJ}) + 1(30.9\text{ kJ})]$$
$$= [-38.3\text{ kJ}] - [83.3\text{ kJ}]$$
$$= -121.6\text{ kJ}$$
then

Reactant/Product	S^0(J/molK from Appendix IIB)
C_2H_4 (*g*)	219.3
Br_2 (*g*)	245.5
$C_2H_4Br_2$ (*g*)	330.6

Be sure to pull data for the correct formula and phase.

$$\Delta S^0_{rxn} = \sum n_p S^0(products) - \sum n_r S^0(reactants)$$
$$= [1(S^0(C_2H_4Br_2\ (g)))] - [1(S^0(C_2H_4\ (g))) + 1(S^0(Br_2\ (g)))]$$
$$= [1(330.6\text{ J/K})] - [1(219.3\text{ J/K}) + 1(245.5\text{ J/K})]$$
$$= [330.6\text{ J/K}] - [464.8\text{ J/K}]$$
$$= -134.2\text{ J/K}$$

then $-134.2 \frac{\text{J}}{\text{K}} \times \frac{1\text{ kJ}}{1000\text{ J}} = -0.1342 \frac{\text{kJ}}{\text{K}}$

then $\Delta G^0 = \Delta H^0{}_{rxn} - T\Delta S^0{}_{rxn} = -121.6\text{ kJ} - (298\text{ K})\left(-0.1342 \frac{\text{kJ}}{\text{K}}\right) = -81.6\text{ kJ} = -8.16 \times 10^4\text{ J}$ then

$\Delta G^0_{rxn} = -RT\ln K$ Rearrange to solve for *K*. $K = e^{\frac{-\Delta G^0_{rxn}}{RT}} = e^{\frac{-(-8.16 \times 10^4\text{ J})}{\left(8.314 \frac{\text{J}}{\text{K}\cdot\text{mol}}\right)(298\text{ K})}} = e^{32.9389} = 2.02 \times 10^{14}$ so the reaction is spontaneous.

Reactant/Product	ΔH_f^0(kJ/mol from Appendix IIB)
C_2H_4 (*g*)	52.4
I_2 (*g*)	62.42
$C_2H_4I_2$ (*g*)	66.5

Be sure to pull data for the correct formula and phase.

$$\Delta H^0_{rxn} = \sum n_p \Delta H_f^0(products) - \sum n_r \Delta H_f^0(reactants)$$
$$= [1(\Delta H_f^0(C_2H_4I_2\ (g)))] - [1(\Delta H_f^0(C_2H_4\ (g))) + 1(\Delta H_f^0(I_2\ (g)))]$$
$$= [1(66.5\text{ kJ})] - [1(52.4\text{ kJ}) + 1(62.42\text{ kJ})]$$
$$= [66.5\text{ kJ}] - [114.\underline{8}2\text{ kJ}]$$
$$= -48.\underline{3}2\text{ kJ}$$
then

Reactant/Product	S^0(J/molK from Appendix IIB)
C_2H_4 (*g*)	219.3
I_2 (*g*)	260.69
$C_2H_4I_2$ (*g*)	347.8

Be sure to pull data for the correct formula and phase.

$$\Delta S^0_{rxn} = \sum n_p S^0(products) - \sum n_r S^0(reactants)$$
$$= [1(S^0(C_2H_4I_2\ (g)))] - [1(S^0(C_2H_4\ (g))) + 1(S^0(I_2\ (g)))]$$
$$= [1(347.8\text{ J/K})] - [1(219.3\text{ J/K}) + 1(260.69\text{ J/K})]$$
$$= [347.8\text{ J/K}] - [479.\underline{9}9\text{ J/K}]$$
$$= -132.2\text{ J/K}$$

then $-132.2 \frac{\text{J}}{\text{K}} \times \frac{1\text{ kJ}}{1000\text{ J}} = -0.1322 \frac{\text{kJ}}{\text{K}}$

then $\Delta G^0 = \Delta H^0_{rxn} - T\Delta S^0_{rxn} = -48.32 \text{ kJ} - (298 \text{ K})\left(-0.1322 \frac{\text{kJ}}{\text{K}}\right) = -8.9244 \text{ kJ} = -8.9244 \times 10^3 \text{ J}$ then

$\Delta G^0_{rxn} = -RT \ln K$ Rearrange to solve for K. $K = e^{\frac{-\Delta G^0_{rxn}}{RT}} = e^{\frac{-(-8.9244 \times 10^3 \text{ J})}{\left(8.314 \frac{\text{J}}{\text{K} \cdot \text{mol}}\right)(298 \text{ K})}} = e^{3.6021} = 37$ and the reaction is spontaneous.

Cl_2 is the most spontaneous in the forward direction, I_2 is the least. The entropy change in the reactions is very constant. The spontaneity is determined by the standard enthalpy of formation of the dihalogenated ethane. Higher temperatures make the forward reactions less spontaneous.
Check: The units (kJ and none) are correct. The enthalpy change becomes less negative as we move to larger halogens. The enthalpy term dominates at room temperature, the free energy change is the same sign as the enthalpy change. The more negative the free energy change, the larger the *K*.

55. a) **Given:** $N_2O\,(g) + NO_2\,(g) \rightleftharpoons 3\,NO\,(g)$ at 298 K **Find:** $\Delta G°_{rxn}$

Conceptual Plan: $\Delta G^0_{rxn} = \sum n_p \Delta G^0_f(products) - \sum n_r \Delta G^0_f(reactants)$

Solution:

Reactant/Product	ΔG^0_f(kJ/mol from Appendix IIB)
$N_2O\,(g)$	103.7
$NO_2\,(g)$	51.3
$NO\,(g)$	87.6

Be sure to pull data for the correct formula and phase.

$$\begin{aligned}\Delta G^0_{rxn} &= \sum n_p \Delta G^0_f(products) - \sum n_r \Delta G^0_f(reactants)\\ &= [3(\Delta G^0_f(\text{NO}\,(g)))] - [1(\Delta G^0_f(\text{N}_2\text{O}\,(g))) + 1(\Delta G^0_f(\text{NO}_2\,(g)))]\\ &= [3(87.6\text{kJ})] - [1(103.7\text{ kJ}) + 1(51.3\text{ kJ})]\\ &= [262.8\text{ kJ}] - [155.0\text{ kJ}]\\ &= +107.8\text{ kJ}\end{aligned}$$

The reaction is nonspontaneous.
Check: The units (kJ) are correct. The standard free energy for the reaction was positive and so the reaction is nonspontaneous.

b) **Given:** $P_{N2O} = P_{NO2} = 1.0$ atm initially **Find:** P_{N2O} when reaction ceases to be spontaneous
Conceptual Plan: **Reaction will no longer be spontaneous when Q = *K* so then $\Delta G°_{rxn}, T \rightarrow K$**

$\Delta G^0_{rxn} = -RT \ln K$

then solve equilibrium problem to get gas pressures, since *K* << 1 the amount of NO generated will be very, very small compared to 1.0 atm, so, within experimental error, $P_{N2O} = P_{NO2} = 1.0$ atm. Simply solve for P_{NO}.

$$K = \frac{P^3_{NO}}{P_{N2O}P_{NO2}}$$

Solution: $\Delta G^0_{rxn} = -RT \ln K$ Rearrange to solve for *K*.

$K = e^{\frac{-\Delta G^0_{rxn}}{RT}} = e^{\frac{-107.8 \text{ kJ} \times \frac{1000 \text{ J}}{1 \text{ kJ}}}{\left(8.314 \frac{\text{J}}{\text{K} \cdot \text{mol}}\right)(298 \text{ K})}} = e^{-43.5103} = 1.27 \times 10^{-19}$. Since $K = \frac{P^3_{NO}}{P_{N2O}P_{NO2}}$, rearrange to solve for P_{NO}. $P^3_{NO} = \sqrt[3]{K P_{N2O} P_{NO2}} = \sqrt[3]{(1.27 \times 10^{-19})(1.0)(1.0)} = 5.0 \times 10^{-7}$ atm.

Note that the assumption that P_{N2O} was very, very small was valid.
Check: The units (atm) are correct. Since the free energy change was positive, the *K* was very small. This leads us to expect that very little NO will be formed.

c) **Given:** $N_2O\ (g) + NO_2\ (g) \rightleftharpoons 3\ NO\ (g)$ **Find:** temperature for spontaneity

Conceptual Plan: $\Delta H^\circ_{rxn} = \sum n_p H^\circ_f(products) - \sum n_r H^\circ_f(reactants)$ **then**

$\Delta S^\circ_{rxn} = \sum n_p S^\circ(products) - \sum n_r S^\circ(reactants)$ **then J/K → kJ/K then ΔH°_{rxn}, ΔS_{rxn} → T**

$$\frac{1\,\text{kJ}}{1000\,\text{J}} \qquad \Delta G = \Delta H_{rxn} - T\Delta S_{rxn}$$

Solution:

Reactant/Product	ΔH^0_f(kJ/mol from Appendix IIB)
$N_2O\ (g)$	81.6
$NO_2\ (g)$	33.2
$NO\ (g)$	91.3

Be sure to pull data for the correct formula and phase.

$$\begin{aligned}\Delta H^0_{rxn} &= \sum n_p \Delta H^0_f(products) - \sum n_r \Delta H^0_f(reactants)\\ &= [3(\Delta H^0_f(\text{NO}\ (g)))] - [1(\Delta H^0_f(\text{N}_2\text{O}\ (g))) + 1(\Delta H^0_f(\text{NO}_2\ (g)))]\\ &= [3(91.3\text{ kJ})] - [1(81.6\text{ kJ}) + 1(33.2\text{ kJ})]\\ &= [273.9\text{ kJ}] - [114.8\text{ kJ}]\\ &= +159.1\text{ kJ}\end{aligned}$$

then

Reactant/Product	S^0(J/molK from Appendix IIB)
$N_2O\ (g)$	220.0
$NO_2\ (g)$	240.1
$NO\ (g)$	210.8

Be sure to pull data for the correct formula and phase.

$$\begin{aligned}\Delta S^0_{rxn} &= \sum n_p S^0(products) - \sum n_r S^0(reactants)\\ &= [3(S^0(\text{NO}\ (g)))] - [1(S^0(\text{N}_2\text{O}\ (g))) + 1(S^0(\text{NO}_2\ (g)))]\\ &= [3(210.8\text{ J/K})] - [1(220.0\text{ J/K}) + 1(240.1\text{ J/K})]\\ &= [632.4\text{ J/K}] - [460.1\text{ J/K}]\\ &= +172.3\text{ J/K}\end{aligned}$$

then

$+172.3\frac{\text{J}}{\text{K}} \times \frac{1\text{ kJ}}{1000\text{ J}} = +0.1723\frac{\text{kJ}}{\text{K}}$. Since $\Delta G = \Delta H_{rxn} - T\Delta S_{rxn}$, set $\Delta G = 0$ and rearrange to solve for T. $T = \frac{\Delta H_{rxn}}{\Delta S_{rxn}} = \frac{+159.1\text{ kJ}}{0.1723\frac{\text{kJ}}{\text{K}}} = +923.4\text{ K}$.

Check: The units (K) are correct. The reaction can be made more spontaneous by raising the temperature, because the entropy change is positive (increase in the number of moles of gas).

57. a) **Given:** ATP (aq) + $H_2O\ (l)$ → ADP (aq) + $P_i\ (aq)$ $\Delta G^\circ_{rxn} = -30.5$ kJ at 298 K **Find:** K

Conceptual Plan: **ΔG°_{rxn}, T → K**

$$\Delta G^0_{rxn} = -RT\ln K$$

Solution: $\Delta G^0_{rxn} = -RT\ln K$ Rearrange to solve for K.

$$K = e^{\frac{-\Delta G^0_{rxn}}{RT}} = e^{\frac{-(-30.5\text{ kJ})\times\frac{1000\text{ J}}{1\text{ kJ}}}{\left(8.314\frac{\text{J}}{\text{K}\cdot\text{mol}}\right)(298\text{ K})}} = e^{12.3104} = 2.22 \times 10^5.$$

Check: The units (none) are correct. The free energy change is negative and the reaction is spontaneous. This results in a large K.

b) **Given:** oxidation of glucose drives reforming of ATP

Find: ΔG°_{rxn} of oxidation of glucose and moles ATP formed per mole of glucose

Conceptual Plan: **write balanced reaction for glucose oxidation then**

$\Delta G^0_{rxn} = \sum n_p \Delta G^0_f(products) - \sum n_r \Delta G^0_f(reactants)$ **then $\Delta G°_{rxn}$s → moles ATP/ mole glucose**

$$\frac{\Delta G^0_{rxn}\ glucose\ oxidation}{\Delta G^0_{rxn}\ ATP\ hydrolysis}$$

Solution: $C_6H_{12}O_6\ (s) + 6\ O_2\ (g) \rightarrow 6\ CO_2\ (g) + 6\ H_2O\ (l)$

Reactant/Product	ΔG^0_f(kJ/mol from Appendix IIB)
$C_6H_{12}O_6\ (s)$	− 910.4
$O_2\ (g)$	0.0
$CO_2\ (g)$	− 394.4
$H_2O\ (l)$	− 237.1

Be sure to pull data for the correct formula and phase.

$$\Delta G^0_{rxn} = \sum n_p \Delta G^0_f(products) - \sum n_r \Delta G^0_f(reactants)$$
$$= [6(\Delta G^0_f(CO_2\ (g))) + 6(\Delta G^0_f(H_2O\ (l)))] - [1(\Delta G^0_f(C_6H_{12}O_6\ (s))) + 6(\Delta G^0_f(O_2\ (g)))]$$
$$= [6(-394.4\ kJ) + 6(-237.1\ kJ)] - [1(-910.4\ kJ) + 6(0.0\ kJ)]$$
$$= [-3789.0\ kJ] - [-910.4\ kJ]$$
$$= -2878.6\ kJ$$

So the reaction is very spontaneous.
$$\frac{2878.6\ \frac{\cancel{kJ\ generated}}{mole\ glucose\ oxidized}}{30.5\ \frac{\cancel{kJ\ needed}}{mole\ ATP\ reformed}} = 94.4\ \frac{mole\ ATP\ reformed}{mole\ glucose\ oxidized}$$

Check: The units (mol) are correct. The free energy change for the glucose oxidation is large compared to the ATP hydrolysis, so we expect to reform many moles of ATP.

59. a) **Given:** $2\ CO\ (g) + 2\ NO\ (g) \rightarrow N_2\ (g) + 2\ CO_2\ (g)$ **Find:** $\Delta G°_{rxn}$ and effect of increasing T on ΔG

Conceptual Plan: $\Delta G^0_{rxn} = \sum n_p \Delta G^0_f(products) - \sum n_r \Delta G^0_f(reactants)$

Solution:

Reactant/Product	ΔG^0_f(kJ/mol from Appendix IIB)
CO (g)	− 137.2
NO (g)	87.6
$N_2\ (g)$	0.0
$CO_2\ (g)$	− 394.4

Be sure to pull data for the correct formula and phase.

$$\Delta G^0_{rxn} = \sum n_p \Delta G^0_f(products) - \sum n_r \Delta G^0_f(reactants)$$
$$= [1(\Delta G^0_f(N_2\ (g))) + 2(\Delta G^0_f(CO_2\ (g)))] - [2(\Delta G^0_f(CO\ (g))) + 2(\Delta G^0_f(NO\ (g)))]$$
$$= [1(0.0\ kJ) + 2(-394.4\ kJ)] - [2(-137.2\ kJ) + 2(87.6\ kJ)]$$
$$= [-788.8\ kJ] - [-99.2\ kJ]$$
$$= -689.6\ kJ$$

Since the number of moles of gas is decreasing, the entropy change is negative and so ΔG will become more positive with increasing temperature.

Check: The units (kJ) are correct. The free energy change is negative since the carbon dioxide has such a low free energy of formation.

b) **Given:** $5\ H_2\ (g) + 2\ NO\ (g) \rightarrow 2\ NH_3\ (g) + 2\ H_2O\ (g)$ **Find:** $\Delta G°_{rxn}$ and effect of increasing T on ΔG

Conceptual Plan: $\Delta G^0_{rxn} = \sum n_p \Delta G^0_f(products) - \sum n_r \Delta G^0_f(reactants)$

Solution:

Reactant/Product	ΔG^0_f(kJ/mol from Appendix IIB)
$H_2\ (g)$	0.0
NO (g)	87.6
$NH_3\ (g)$	− 16.4
$H_2O\ (g)$	− 228.6

Be sure to pull data for the correct formula and phase.

$$\Delta G^0_{rxn} = \sum n_p \Delta G^0_f(products) - \sum n_r \Delta G^0_f(reactants)$$
$$= [2(\Delta G^0_f(NH_3\ (g))) + 2(\Delta G^0_f(H_2O\ (g)))] - [5(\Delta G^0_f(H_2\ (g))) + 2(\Delta G^0_f(NO\ (g)))]$$
$$= [2(-16.4\text{ kJ}) + 2(-228.6\text{ kJ})] - [5(0.0\text{ kJ}) + 2(87.6\text{ kJ})]$$
$$= [-490.0\text{ kJ}] - [175.2\text{ kJ}]$$
$$= -665.2\text{ kJ}$$

Since the number of moles of gas is decreasing, the entropy change is negative and so ΔG will become more positive with increasing temperature.
Check: The units (kJ) are correct. The free energy change is negative since ammonia and water have such a low free energy of formation.

c) **Given:** $2\ H_2\ (g) + 2\ NO\ (g) \rightarrow N_2\ (g) + 2\ H_2O\ (g)$ **Find:** ΔG°_{rxn} and effect of increasing T on ΔG

Conceptual Plan: $\Delta G^\circ_{rxn} = \sum n_p \Delta G^\circ_f(products) - \sum n_r \Delta G^\circ_f(reactants)$

Solution:

Reactant/Product	ΔG^0_f(kJ/mol from Appendix IIB)
$H_2\ (g)$	0.0
$NO\ (g)$	87.6
$N_2\ (g)$	0.0
$H_2O\ (g)$	−228.6

Be sure to pull data for the correct formula and phase.

$$\Delta G^0_{rxn} = \sum n_p \Delta G^0_f(products) - \sum n_r \Delta G^0_f(reactants)$$
$$= [1(\Delta G^0_f(N_2\ (g))) + 2(\Delta G^0_f(H_2O\ (g)))] - [2(\Delta G^0_f(H_2\ (g))) + 2(\Delta G^0_f(NO\ (g)))]$$
$$= [1(0.0\text{ kJ}) + 2(-228.6\text{ kJ})] - [2(0.0\text{ kJ}) + 2(87.6\text{ kJ})]$$
$$= [-457.2\text{ kJ}] - [175.2\text{ kJ}]$$
$$= -632.4\text{ kJ}$$

Since the number of moles of gas is decreasing, the entropy change is negative and so ΔG will become more positive with increasing temperature.
Check: The units (kJ) are correct. The free energy change is negative since water has such a low free energy of formation.

d) **Given:** $2\ NH_3\ (g) + 2\ O_2\ (g) \rightarrow N_2O\ (g) + 3\ H_2O\ (g)$**Find:** ΔG°_{rxn} and effect of increasing T on ΔG

Conceptual Plan: $\Delta G^\circ_{rxn} = \sum n_p \Delta G^\circ_f(products) - \sum n_r \Delta G^\circ_f(reactants)$

Solution:

Reactant/Product	ΔG^0_f(kJ/mol from Appendix IIB)
$NH_3\ (g)$	−16.4
$O_2\ (g)$	0.0
$N_2O\ (g)$	103.7
$H_2O\ (g)$	−228.6

Be sure to pull data for the correct formula and phase.

$$\Delta G^0_{rxn} = \sum n_p \Delta G^0_f(products) - \sum n_r \Delta G^0_f(reactants)$$
$$= [1(\Delta G^0_f(N_2O\ (g))) + 3(\Delta G^0_f(H_2O\ (g)))] - [2(\Delta G^0_f(NH_3\ (g))) + 2(\Delta G^0_f(O_2\ (g)))]$$
$$= [1(103.7\text{ kJ}) + 3(-228.6\text{ kJ})] - [2(-16.4\text{ kJ}) + 2(0.0\text{ kJ})]$$
$$= [-582.1\text{ kJ}] - [-32.8\text{ kJ}]$$
$$= -549.3\text{ kJ}$$

Since the number of moles of gas is constant the entropy change will be small and slightly negative and so the magnitude of ΔG will decrease with increasing temperature.
Check: The units (kJ) are correct. The free energy change is negative since water has such a low free energy of formation. The entropy change is negative once the S° values are reviewed ($\Delta S_{rxn} = -9.6$ J/K).

61. With one exception, the formation of any oxide of nitrogen at 298 K requires more moles of gas as reactants than are formed as products. For example 1 mole of N_2O requires 0.5 moles of O_2 and 1 mole of N_2; 1 mole of

N_2O_3 requires 1 mole of N_2 and 1.5 moles of O_2, and so on. The exception is NO, where 1 mole of NO requires 0.5 moles of O_2 and 0.5 moles of N_2: ½ $N_2(g)$ + ½ $O_2(g) \rightarrow NO(g)$ This reaction has a positive ΔS because what is essentially mixing of the N and O has taken place in the product.

63. a) **Given:** glutamate (*aq*) + NH_3 (*aq*) → glutamine (*aq*) + H_2O (*l*) $\Delta G°_{rxn}$ = + 14.2 kJ at 298 K **Find:** K
Conceptual Plan: $\Delta G°_{rxn}, T \rightarrow K$

$$\Delta G^0_{rxn} = -RT \ln K$$

Solution: $\Delta G^0_{rxn} = -RT \ln K$ Rearrange to solve for K.

$$K = e^{\frac{-\Delta G^0_{rxn}}{RT}} = e^{\frac{-14.2 \text{ kJ} \times \frac{1000 \text{ J}}{1 \text{ kJ}}}{\left(8.314 \frac{\text{J}}{\text{K}\cdot\text{mol}}\right)(298 \text{ K})}} = e^{-5.73142} = 3.24 \times 10^{-3}.$$

Check: The units (none) are correct. The free energy change is positive and the reaction is nonspontaneous. This results in a small K.

b) **Given:** pair ATP hydrolysis with glutamate/NH_3 reaction **Find:** show coupled reactions, $\Delta G°_{rxn}$ and K
Conceptual Plan: use reaction mechanism shown, where A = NH_3 and B = glutamate ($C_5H_8O_4N^-$), then calculate $\Delta G°_{rxn}$ by adding free energies of reactions then $\Delta G°_{rxn}, T \rightarrow K$

$$\Delta G^0_{rxn} = -RT \ln K$$

Solution:

$$NH_3\ (aq) + ATP\ (aq) + \cancel{H_2O\ (l)} \rightarrow \cancel{NH_3-P_i(aq)} + ADP\ (aq) \qquad \Delta G^0_{rxn} = -\ 30.5 \text{ kJ}$$

$$\cancel{NH_3-P_i(aq)} + C_5H_8O_4N^-\ (aq) \rightarrow C_5H_9O_3N_2^-\ (aq) + \cancel{H_2O\ (l)} + P_i(aq) \qquad \Delta G^0_{rxn} = +\ 14.2 \text{ kJ}$$

$$NH_3\ (aq) + C_5H_8O_4N^-\ (aq) + ATP\ (aq) \rightarrow C_5H_9O_3N_2^-\ (aq) + ADP\ (aq) + P_i(aq) \quad \Delta G^0_{rxn} = -16.3 \text{ kJ}$$

then $\Delta G^0_{rxn} = -RT \ln K$ Rearrange to solve for K.

$$K = e^{\frac{-\Delta G^0_{rxn}}{RT}} = e^{\frac{-(-16.3 \text{ kJ}) \times \frac{1000 \text{ J}}{1 \text{ kJ}}}{\left(8.314 \frac{\text{J}}{\text{K}\cdot\text{mol}}\right)(298 \text{ K})}} = e^{6.57902} = 7.20 \times 10^2.$$

Check: The units (none) are correct. The free energy change is negative and the reaction is spontaneous. This results in a large K.

65. a) **Given:** ½ H_2 (*g*) + ½ Cl_2 (*g*) → HCl (*g*), define standard state as 2 atm **Find:** $\Delta G°_f$
Conceptual Plan: $\Delta G°_f, P_{H2}, P_{Cl2}, P_{HCl}, T \rightarrow$ new $\Delta G°_f$

$$\Delta G_{rxn} = \Delta G^0_{rxn} + RT \ln Q \quad \text{where } Q = \frac{P_{HCl}}{P_{H2}^{1/2}\, P_{Cl2}^{1/2}}$$

Solution: $\Delta G°_f = -\ 95.3$ kJ/mol and $Q = \frac{P_{HCl}}{P_{H2}^{1/2}\, P_{Cl2}^{1/2}} = \frac{2}{2^{1/2}\, 2^{1/2}} = 1$ then

$$\Delta G_{rxn} = \Delta G^0_{rxn} + RT \ln Q = -\ 95.3 \frac{\text{kJ}}{\text{mol}} + \left(8.314 \frac{\text{J}}{\text{K}\cdot\text{mol}}\right)\left(\frac{1 \text{ kJ}}{1000 \text{ J}}\right)(298 \text{ K}) \ln(1) = -\ 95.3 \frac{\text{kJ}}{\text{mol}} =$$

$$= -\ 95{,}300 \frac{\text{J}}{\text{mol}}$$

Since the number of moles of reactants and products are the same, the decrease in volume affects the entropy of both equally, so there is no change in $\Delta G°_f$.
Check: The units (kJ) are correct. The Q is one so $\Delta G°_f$ is unchanged under the new standard conditions.

b) **Given:** N_2 (*g*) + ½ O_2 (*g*) → N_2O (*g*), define standard state as 2 atm **Find:** $\Delta G°_f$
Conceptual Plan: $\Delta G°_f, P_{N2}, P_{O2}, P_{N2O}, T \rightarrow$ new $\Delta G°_f$

$$\Delta G_{rxn} = \Delta G^0_{rxn} + RT \ln Q \quad \text{where } Q = \frac{P_{N2O}}{P_{N2}\, P_{O2}^{1/2}}$$

Solution: $\Delta G°_f = +\,103.7$ kJ/mol and $Q = \dfrac{P_{N2O}}{P_{N2}\,P_{O2}^{1/2}} = \dfrac{2}{2\;2^{1/2}} = \dfrac{1}{\sqrt{2}}$ then

$$\Delta G_{rxn} = \Delta G^0_{rxn} + RT\ln Q = +\,103.7\,\frac{\text{kJ}}{\text{mol}} + \left(8.314\,\frac{\text{J}}{\text{K}\cdot\text{mol}}\right)\left(\frac{1\text{ kJ}}{1000\text{ J}}\right)(298\text{ K})\ln\left(\frac{1}{\sqrt{2}}\right) = +\,102.8\,\frac{\text{kJ}}{\text{mol}} =$$

$$= +\,102{,}800\,\frac{\text{J}}{\text{mol}}$$

The entropy of the reactants (1.5 mol) is decreased more than the entropy of the product (1 mol). Since the product is relatively more favored at lower volume, $\Delta G°_f$ is less positive.

Check: The units (kJ) are correct. The Q is less than one so $\Delta G°_f$ is reduced under the new standard conditions.

c) **Given:** ½ H_2 (*g*) → H (*g*), define standard state as 2 atm **Find:** $\Delta G°_f$

Conceptual Plan: $\Delta G°_f, P_{H2}, P_H, T$ → **new** $\Delta G°_f$

$$\Delta G_{rxn} = \Delta G^0_{rxn} + RT\ln Q \quad \text{where } Q = \frac{P_H}{P_{H2}^{1/2}}$$

Solution: $\Delta G°_f = +\,203.3$ kJ/mol and $Q = \dfrac{P_H}{P_{H2}^{1/2}} = \dfrac{2}{2^{1/2}} = \sqrt{2}$ then

$$\Delta G_{rxn} = \Delta G^0_{rxn} + RT\ln Q = +\,203.3\,\frac{\text{kJ}}{\text{mol}} + \left(8.314\,\frac{\text{J}}{\text{K}\cdot\text{mol}}\right)\left(\frac{1\text{ kJ}}{1000\text{ J}}\right)(298\text{ K})\ln\left(\sqrt{2}\right) = +\,204.2\,\frac{\text{kJ}}{\text{mol}} =$$

$$= +\,204{,}200\,\frac{\text{J}}{\text{mol}}$$

The entropy of the product (1 mol) is decreased more than the entropy of the reactant (1/2 mol). Since the product is relatively less favored, $\Delta G°_f$ is more positive.

Check: The units (kJ) are correct. The Q is greater than one so $\Delta G°_f$ is increased under the new standard conditions.

67. a) **Given:** NH_4NO_3 (*s*) → HNO_3 (*g*) + NH_3 (*g*) **Find:** $\Delta G°_{rxn}$

Conceptual Plan: $\Delta G^0_{rxn} = \sum n_p \Delta G^0_f(products) - \sum n_r \Delta G^0_f(reactants)$

Solution:

Reactant/Product	ΔG^0_f(kJ/mol from Appendix IIB)
NH_4NO_3 (*s*)	– 183.9
HNO_3 (*g*)	– 73.5
NH_3 (*g*)	– 16.4

Be sure to pull data for the correct formula and phase.

$$\Delta G^0_{rxn} = \sum n_p \Delta G^0_f(products) - \sum n_r \Delta G^0_f(reactants)$$
$$= [1(\Delta G^0_f(HNO_3\,(g))) + 1(\Delta G^0_f(NH_3\,(g)))] - [1(\Delta G^0_f(NH_4NO_3\,(s)))]$$
$$= [1(-\,73.5\text{kJ}) + 1(-\,16.4\text{ kJ})] - [1(-\,183.9\text{ kJ})]$$
$$= [-\,89.9\text{ kJ}] - [-\,183.9\text{ kJ}]$$
$$= +\,94.0\text{ kJ}$$

Check: The units (kJ) are correct. The free energy change is positive since ammonium nitrate has such a low free energy of formation.

b) **Given:** NH_4NO_3 (*s*) → N_2O (*g*) + 2 H_2O (*g*) **Find:** $\Delta G°_{rxn}$

Conceptual Plan: $\Delta G^0_{rxn} = \sum n_p \Delta G^0_f(products) - \sum n_r \Delta G^0_f(reactants)$

Solution:

Reactant/Product	ΔG^0_f(kJ/mol from Appendix IIB)
NH_4NO_3 (*s*)	– 183.9
N_2O (*g*)	103.7
H_2O (*g*)	– 228.6

Be sure to pull data for the correct formula and phase.

$$\Delta G^0_{rxn} = \sum n_p \Delta G^0_f(products) - \sum n_r \Delta G^0_f(reactants)$$
$$= [1(\Delta G^0_f(N_2O\ (g))) + 2(\Delta G^0_f(H_2O\ (g)))] - [1(\Delta G^0_f(NH_4NO_3\ (s)))]$$
$$= [1(103.7\ kJ) + 2(-228.6\ kJ)] - [1(-183.9\ kJ)]$$
$$= [-353.5\ kJ] - [-183.9\ kJ]$$
$$= -169.6\ kJ$$

Check: The units (kJ) are correct. The free energy change is negative since water has such a low free energy of formation.

c) **Given:** $NH_4NO_3\ (s) \rightarrow N_2\ (g) + ½\ O_2\ (g) + 2\ H_2O\ (g)$ **Find:** $\Delta G°_{rxn}$

Conceptual Plan: $\Delta G^0_{rxn} = \sum n_p \Delta G^0_f(products) - \sum n_r \Delta G^0_f(reactants)$

Solution:

Reactant/Product	ΔG^0_f (kJ/mol from Appendix IIB)
$NH_4NO_3\ (s)$	− 183.9
$N_2\ (g)$	0.0
$O_2\ (g)$	0.0
$H_2O\ (g)$	− 228.6

Be sure to pull data for the correct formula and phase.

$$\Delta G^0_{rxn} = \sum n_p \Delta G^0_f(products) - \sum n_r \Delta G^0_f(reactants)$$
$$= [1(\Delta G^0_f(N_2\ (g))) + 1/2(\Delta G^0_f(O_2\ (g))) + 2(\Delta G^0_f(H_2O\ (g)))] - [1(\Delta G^0_f(NH_4NO_3\ (s)))]$$
$$= [1(0.0\ kJ) + 1/2(0.0\ kJ) + 2(-228.6\ kJ)] - [1(-183.9\ kJ)]$$
$$= [-457.2\ kJ] - [-183.9\ kJ]$$
$$= -273.3\ kJ$$

Check: The units (kJ) are correct. The free energy change is negative since water has such a low free energy of formation.

The second and third reactions are spontaneous and so we would expect decomposition products of N_2O, N_2, O_2, and H_2O in the gas phase. It is still possible for ammonium nitrate to remain as a solid because the thermodynamics of the reaction say nothing of the kinetics of the reaction (reaction can be extremely slow). Since all of the products are gases, the decomposition of ammonium nitrate will result in a large increase in volume (explosion). Some of the products aid in combustion, which could facilitate the combustion of materials near the ammonium nitrate. Also, N_2O is known as laughing gas, which has anesthetic and toxic effects on humans. The solid should not be kept in tightly sealed containers.

69. c) The spontaneity of a reaction says nothing about the speed of a reaction. It only states which direction the reaction will go as it approaches equilibrium.

71. b) has the largest decrease in the number of microstates from the initial to the final state. In a) there are initially $\frac{9!}{4!\,4!\,1!} = 90$ microstates and $\frac{9!}{3!\,3!\,3!} = 1680$ microstates at the end, so $\Delta S > 0$. In b) there are initially $\frac{9!}{4!\,2!\,3!} = 1260$ microstates and $\frac{9!}{6!\,3!\,0!} = 84$ microstates at the end, so $\Delta S < 0$. In c) there are initially $\frac{9!}{3!\,4!\,2!} = 1260$ microstates and $\frac{9!}{3!\,4!\,2!} = 1260$ microstates at the end, so $\Delta S = 0$. Also the final state in b) has the least entropy.

73. c) Since the vapor pressure of water at 298 K is 23.78 mmHg or 0.03129 atm. As long as the desired pressure (0.010 atm) is less than the equilibrium vapor pressure of water, the reaction will be spontaneous.

Chapter 18
Electrochemistry

1. **Conceptual Plan: Separate the overall reaction into two half-reactions: one for oxidation and one for reduction. → Balance each half-reaction with respect to mass in the following order: 1) balance all elements other than H and O; 2) balance O by adding H_2O; and 3) balance H by adding H^+. → Balance each half-reaction with respect to charge by adding electrons. (The sum of the charges on both sides of the equation should be made equal by adding electrons as necessary.) → Make the number of electrons in both half-reactions equal by multiplying one or both half-reactions by a small whole number. → Add the two half-reactions together, canceling electrons and other species as necessary. → Verify that the reaction is balanced both with respect to mass and with respect to charge.**

Solution:

a) Separate: $K(s) \rightarrow K^+(aq)$ and $Cr^{3+}(aq) \rightarrow Cr(s)$

Balance elements: $K(s) \rightarrow K^+(aq)$ and $Cr^{3+}(aq) \rightarrow Cr(s)$

Add electrons: $K(s) \rightarrow K^+(aq) + e^-$ and $Cr^{3+}(aq) + 3\,e^- \rightarrow Cr(s)$

Equalize electrons: $3\,K(s) \rightarrow 3\,K^+(aq) + 3\,e^-$ and $Cr^{3+}(aq) + 3\,e^- \rightarrow Cr(s)$

Add half-reactions: $3\,K(s) + Cr^{3+}(aq) + \cancel{3\,e^-} \rightarrow 3\,K^+(aq) + \cancel{3\,e^-} + Cr(s)$

Cancel electrons: $3\,K(s) + Cr^{3+}(aq) \rightarrow 3\,K^+(aq) + Cr(s)$

Check:

Reactants	Products
3 K atoms	3 K atoms
1 Cr atom	1 Cr atom
+3 charge	+3 charge

b) Separate: $Al(s) \rightarrow Al^{3+}(aq)$ and $Fe^{2+}(aq) \rightarrow Fe(s)$

Balance elements: $Al(s) \rightarrow Al^{3+}(aq)$ and $Fe^{2+}(aq) \rightarrow Fe(s)$

Add electrons: $Al(s) \rightarrow Al^{3+}(aq) + 3\,e^-$ and $Fe^{2+}(aq) + 2\,e^- \rightarrow Fe(s)$

Equalize electrons: $2\,Al(s) \rightarrow 2\,Al^{3+}(aq) + 6\,e^-$ and $3\,Fe^{2+}(aq) + 6\,e^- \rightarrow 3\,Fe(s)$

Add half-reactions: $2\,Al(s) + 3\,Fe^{2+}(aq) + \cancel{6\,e^-} \rightarrow 2\,Al^{3+}(aq) + \cancel{6\,e^-} + 3\,Fe(s)$

Cancel electrons: $2\,Al(s) + 3\,Fe^{2+}(aq) \rightarrow 2\,Al^{3+}(aq) + 3\,Fe(s)$

Check:

Reactants	Products
2 Al atoms	2 Al atoms
3 Fe atom	3 Fe atom
+6 charge	+6 charge

c) Separate: $BrO_3^-(aq) \rightarrow Br^-(aq)$ and $N_2H_4(g) \rightarrow N_2(g)$

Balance non H & O elements: $BrO_3^-(aq) \rightarrow Br^-(aq)$ and $N_2H_4(g) \rightarrow N_2(g)$

Balance O with H_2O: $BrO_3^-(aq) \rightarrow Br^-(aq) + 3\,H_2O(l)$ and $N_2H_4(g) \rightarrow N_2(g)$

Balance H with H^+: $BrO_3^-(aq) + 6\,H^+(aq) \rightarrow Br^-(aq) + 3\,H_2O(l)$ and $N_2H_4(g) \rightarrow N_2(g) + 4\,H^+(aq)$

Add electrons:

$BrO_3^-(aq) + 6\,H^+(aq) + 6\,e^- \rightarrow Br^-(aq) + 3\,H_2O(l)$ and $N_2H_4(g) \rightarrow N_2(g) + 4\,H^+(aq) + 4\,e^-$

Equalize electrons:

$2\,BrO_3^-(aq) + 12\,H^+(aq) + 12\,e^- \rightarrow 2\,Br^-(aq) + 6\,H_2O(l)$ and $3\,N_2H_4(g) \rightarrow 3\,N_2(g) + 12\,H^+(aq) + 12\,e^-$

Add half-reactions: $2\,BrO_3^-(aq) + \cancel{12\,H^+(aq)} + 3\,N_2H_4(g) + \cancel{12\,e^-} \rightarrow 2\,Br^-(aq) + 6\,H_2O(l) + 3\,N_2(g) + \cancel{12\,H^+(aq)} + \cancel{12\,e^-}$

Cancel electrons & others: $2\,BrO_3^-(aq) + 3\,N_2H_4(g) \rightarrow 2\,Br^-(aq) + 6\,H_2O(l) + 3\,N_2(g)$

Check:

Reactants	Products
3 Br atoms	3 Br atoms
6 O atoms	6 O atoms
12 H atoms	12 H atoms
6 N atoms	6 N atoms
–2 charge	–2 charge

3. **Conceptual Plan: Separate the overall reaction into two half-reactions: one for oxidation and one for reduction. → Balance each half-reaction with respect to mass in the following order: 1) balance all elements other than H and O; 2) balance O by adding H_2O; and 3) balance H by adding H^+. → Balance each half-reaction with respect to charge by adding electrons. (The sum of the charges on both sides of the equation should be made equal by adding electrons as necessary.) → Make the number of electrons**

in both half-reactions equal by multiplying one or both half-reactions by a small whole number. → Add the two half-reactions together, canceling electrons and other species as necessary. → Verify that the reaction is balanced both with respect to mass and with respect to charge.

Solution:

a) Separate: $PbO_2\,(s) \rightarrow Pb^{2+}\,(aq)$ and $I^-\,(aq) \rightarrow I_2\,(s)$

Balance non H & O elements: $PbO_2\,(s) \rightarrow Pb^{2+}\,(aq)$ and $2\,I^-\,(aq) \rightarrow I_2\,(s)$

Balance O with H_2O: $PbO_2\,(s) \rightarrow Pb^{2+}\,(aq) + 2\,H_2O\,(l)$ and $2\,I^-\,(aq) \rightarrow I_2\,(s)$

Balance H with H^+: $PbO_2\,(s) + 4\,H^+\,(aq) \rightarrow Pb^{2+}\,(aq) + 2\,H_2O\,(l)$ and $2\,I^-\,(aq) \rightarrow I_2\,(s)$

Add electrons: $PbO_2\,(s) + 4\,H^+\,(aq) + 2\,e^- \rightarrow Pb^{2+}\,(aq) + 2\,H_2O\,(l)$ and $2\,I^-\,(aq) \rightarrow I_2\,(s) + 2\,e^-$

Equalize electrons: $PbO_2\,(s) + 4\,H^+\,(aq) + 2\,e^- \rightarrow Pb^{2+}\,(aq) + 2\,H_2O\,(l)$ and $2\,I^-\,(aq) \rightarrow I_2\,(s) + 2\,e^-$

Add half-reactions: $PbO_2\,(s) + 4\,H^+\,(aq) + \cancel{2\,e^-} + 2\,I^-\,(aq) \rightarrow Pb^{2+}\,(aq) + 2\,H_2O\,(l) + I_2\,(s) + \cancel{2\,e^-}$

Cancel electrons & others: $PbO_2\,(s) + 4\,H^+\,(aq) + 2\,I^-\,(aq) \rightarrow Pb^{2+}\,(aq) + 2\,H_2O\,(l) + I_2\,(s)$

Check:

Reactants	Products
1 Pb atom	1 Pb atom
2 O atoms	2 O atoms
4 H atoms	4 H atoms
2 I atoms	2 I atoms
+2 charge	+2 charge

b) Separate: $MnO_4^-\,(aq) \rightarrow Mn^{2+}\,(aq)$ and $SO_3^{2-}\,(aq) \rightarrow SO_4^{2-}\,(aq)$

Balance non H & O elements: $MnO_4^-\,(aq) \rightarrow Mn^{2+}\,(aq)$ and $SO_3^{2-}\,(aq) \rightarrow SO_4^{2-}\,(aq)$

Balance O with H_2O: $MnO_4^-\,(aq) \rightarrow Mn^{2+}\,(aq) + 4\,H_2O\,(l)$ and $SO_3^{2-}\,(aq) + H_2O\,(l) \rightarrow SO_4^{2-}\,(aq)$

Balance H with H^+:

$MnO_4^-\,(aq) + 8\,H^+\,(aq) \rightarrow Mn^{2+}\,(aq) + 4\,H_2O\,(l)$ and $SO_3^{2-}\,(aq) + H_2O\,(l) \rightarrow SO_4^{2-}\,(aq) + 2\,H^+\,(aq)$

Add electrons: $MnO_4^-\,(aq) + 8\,H^+\,(aq) + 5\,e^- \rightarrow Mn^{2+}\,(aq) + 4\,H_2O\,(l)$ and

$SO_3^{2-}\,(aq) + H_2O\,(l) \rightarrow SO_4^{2-}\,(aq) + 2\,H^+\,(aq) + 2\,e^-$

Equalize electrons: $2\,MnO_4^-\,(aq) + 16\,H^+\,(aq) + 10\,e^- \rightarrow 2\,Mn^{2+}\,(aq) + 8\,H_2O\,(l)$ and

$5\,SO_3^{2-}\,(aq) + 5\,H_2O\,(l) \rightarrow 5\,SO_4^{2-}\,(aq) + 10\,H^+\,(aq) + 10\,e^-$

Add half-reactions: $2\,MnO_4^-\,(aq) + 6\,\cancel{16}\,H^+\,(aq) + \cancel{10\,e^-} + 5\,SO_3^{2-}\,(aq) + \cancel{5\,H_2O(l)} \rightarrow$

$2\,Mn^{2+}\,(aq) + 3\,\cancel{8}\,H_2O(l) + 5\,SO_4^{2-}\,(aq) + \cancel{10\,H^+}\,(aq) + \cancel{10\,e^-}$

Cancel electrons: $2\,MnO_4^-\,(aq) + 6\,H^+\,(aq) + 5\,SO_3^{2-}\,(aq) \rightarrow 2\,Mn^{2+}\,(aq) + 3\,H_2O\,(l) + 5\,SO_4^{2-}\,(aq)$

Check:

Reactants	Products
2 Mn atoms	2 Mn atoms
23 O atoms	23 O atoms
6 H atoms	6 H atoms
5 S atoms	5 S atoms
–6 charge	–6 charge

c) Separate: $S_2O_3^{2-}\,(aq) \rightarrow SO_4^{2-}\,(aq)$ and $Cl_2\,(g) \rightarrow Cl^-\,(aq)$

Balance non H & O elements: $S_2O_3^{2-}\,(aq) \rightarrow 2\,SO_4^{2-}\,(aq)$ and $Cl_2\,(g) \rightarrow 2\,Cl^-\,(aq)$

Balance O with H_2O: $S_2O_3^{2-}\,(aq) + 5\,H_2O\,(l) \rightarrow 2\,SO_4^{2-}\,(aq)$ and $Cl_2\,(g) \rightarrow 2\,Cl^-\,(aq)$

Balance H with H^+: $S_2O_3^{2-}\,(aq) + 5\,H_2O\,(l) \rightarrow 2\,SO_4^{2-}\,(aq) + 10\,H^+\,(aq)$ and $Cl_2\,(g) \rightarrow 2\,Cl^-\,(aq)$

Add electrons:

$S_2O_3^{2-}\,(aq) + 5\,H_2O\,(l) \rightarrow 2\,SO_4^{2-}\,(aq) + 10\,H^+\,(aq) + 8\,e^-$ and $Cl_2\,(g) + 2\,e^- \rightarrow 2\,Cl^-\,(aq)$

Equalize electrons:

$S_2O_3^{2-}\,(aq) + 5\,H_2O\,(l) \rightarrow 2\,SO_4^{2-}\,(aq) + 10\,H^+\,(aq) + 8\,e^-$ and $4\,Cl_2\,(g) + 8\,e^- \rightarrow 8\,Cl^-\,(aq)$

Add half-reactions:

$S_2O_3^{2-}\,(aq) + 5\,H_2O\,(l) + 4\,Cl_2\,(g) + \cancel{8\,e^-} \rightarrow 2\,SO_4^{2-}\,(aq) + 10\,H^+\,(aq) + \cancel{8\,e^-} + 8\,Cl^-\,(aq)$

Cancel electrons: $S_2O_3^{2-}\,(aq) + 5\,H_2O\,(l) + 4\,Cl_2\,(g) \rightarrow 2\,SO_4^{2-}\,(aq) + 10\,H^+\,(aq) + 8\,Cl^-\,(aq)$

Check:

Reactants	Products
2 S atoms	2 S atoms
8 O atoms	8 O atoms
10 H atoms	10 H atoms
8 Cl atoms	8 Cl atoms
–2 charge	–2 charge

5. **Conceptual Plan: Separate the overall reaction into two half-reactions: one for oxidation and one for reduction. → Balance each half-reaction with respect to mass in the following order: 1) balance all**

elements other than H and O; 2) balance O by adding H_2O; 3) balance H by adding H^+; and 4) Neutralize H^+ by adding enough OH^- to neutralize each H^+. Add the same number of OH^- ions to each side of the equation. → Balance each half-reaction with respect to charge by adding electrons. (The sum of the charges on both sides of the equation should be made equal by adding electrons as necessary.) → Make the number of electrons in both half-reactions equal by multiplying one or both half-reactions by a small whole number. → Add the two half-reactions together, canceling electrons and other species as necessary. → Verify that the reaction is balanced both with respect to mass and with respect to charge.

Solution:

a) Separate: $ClO_2\,(aq) \rightarrow ClO_2^-\,(aq)$ and $H_2O_2\,(aq) \rightarrow O_2\,(g)$

Balance non H & O elements: $ClO_2\,(aq) \rightarrow ClO_2^-\,(aq)$ and $H_2O_2\,(aq) \rightarrow O_2\,(g)$

Balance O with H_2O: $ClO_2\,(aq) \rightarrow ClO_2^-\,(aq)$ and $H_2O_2\,(aq) \rightarrow O_2\,(g)$

Balance H with H^+: $ClO_2\,(aq) \rightarrow ClO_2^-\,(aq)$ and $H_2O_2\,(aq) \rightarrow O_2\,(g) + 2\,H^+\,(aq)$

Neutralize H^+ with OH^-:

$ClO_2\,(aq) \rightarrow ClO_2^-\,(aq)$ and $H_2O_2\,(aq) + 2\,OH^-\,(aq) \rightarrow O_2\,(g) + \underbrace{2\,H^+\,(aq) + 2\,OH^-\,(aq)}_{2\,H_2O\,(l)}$

Add electrons: $ClO_2\,(aq) + e^- \rightarrow ClO_2^-\,(aq)$ and $H_2O_2\,(aq) + 2\,OH^-\,(aq) \rightarrow O_2\,(g) + 2\,H_2O\,(l) + 2\,e^-$

Equalize electrons:

$2\,ClO_2\,(aq) + 2\,e^- \rightarrow 2\,ClO_2^-\,(aq)$ and $H_2O_2\,(aq) + 2\,OH^-\,(aq) \rightarrow O_2\,(g) + 2\,H_2O\,(l) + 2\,e^-$

Add half-reactions:

$2\,ClO_2\,(aq) + \cancel{2\,e^-} + H_2O_2\,(aq) + 2\,OH^-\,(aq) \rightarrow 2\,ClO_2^-\,(aq) + O_2\,(g) + 2\,H_2O\,(l) + \cancel{2\,e^-}$

Cancel electrons: $2\,ClO_2\,(aq) + H_2O_2\,(aq) + 2\,OH^-\,(aq) \rightarrow 2\,ClO_2^-\,(aq) + O_2\,(g) + 2\,H_2O\,(l)$

Check:

Reactants	Products
2 Cl atoms	2 Cl atoms
8 O atoms	8 O atoms
4 H atoms	4 H atoms
–2 charge	–2 charge

b) Separate: $MnO_4^-\,(aq) \rightarrow MnO_2\,(s)$ and $Al\,(s) \rightarrow Al(OH)_4^-\,(aq)$

Balance non H & O elements: $MnO_4^-\,(aq) \rightarrow MnO_2\,(s)$ and $Al\,(s) \rightarrow Al(OH)_4^-\,(aq)$

Balance O with H_2O: $MnO_4^-\,(aq) \rightarrow MnO_2\,(s) + 2\,H_2O\,(l)$ and $Al\,(s) + 4\,H_2O\,(l) \rightarrow Al(OH)_4^-\,(aq)$

Balance H with H^+:

$MnO_4^-\,(aq) + 4\,H^+\,(aq) \rightarrow MnO_2\,(s) + 2\,H_2O\,(l)$ and $Al\,(s) + 4\,H_2O\,(l) \rightarrow Al(OH)_4^-\,(aq) + 4\,H^+\,(aq)$

Neutralize H^+ with OH^-: $MnO_4^-\,(aq) + \underbrace{4\,H^+\,(aq) + 4\,OH^-\,(aq)}_{\cancel{4}2\,H_2O\,(l)} \rightarrow MnO_2\,(s) + \cancel{2\,H_2O\,(l)} + 4\,OH^-\,(aq)$

and $Al\,(s) + \cancel{4\,H_2O\,(l)} + 4\,OH^-\,(aq) \rightarrow Al(OH)_4^-\,(aq) + \underbrace{4\,H^+\,(aq) + 4\,OH^-\,(aq)}_{\cancel{4\,H_2O\,(l)}}$

Add electrons: $MnO_4^-\,(aq) + 2\,H_2O\,(l) + 3\,e^- \rightarrow MnO_2\,(s) + 4\,OH^-\,(aq)$ and

$Al\,(s) + 4\,OH^-\,(aq) \rightarrow Al(OH)_4^-\,(aq) + 3\,e^-$

Equalize electrons: $MnO_4^-\,(aq) + 2\,H_2O\,(l) + 3\,e^- \rightarrow MnO_2\,(s) + 4\,OH^-\,(aq)$ and

$Al\,(s) + 4\,OH^-\,(aq) \rightarrow Al(OH)_4^-\,(aq) + 3\,e^-$

Add half-reactions:

$MnO_4^-\,(aq) + 2\,H_2O\,(l) + \cancel{3\,e^-} + Al\,(s) + \cancel{4\,OH^-\,(aq)} \rightarrow MnO_2\,(s) + \cancel{4\,OH^-\,(aq)} + Al(OH)_4^-\,(aq) + \cancel{3\,e^-}$

Cancel electrons: $MnO_4^-\,(aq) + 2\,H_2O\,(l) + Al\,(s) \rightarrow MnO_2\,(s) + Al(OH)_4^-\,(aq)$

Check:

Reactants	Products
1 Mn atom	1 Mn atom
6 O atoms	6 O atoms
4 H atoms	4 H atoms
1 Al atom	1 Al atom
–1 charge	–1 charge

c) Separate: $Cl_2\,(g) \rightarrow Cl^-\,(aq)$ and $Cl_2\,(g) \rightarrow ClO^-\,(aq)$

Balance non H & O elements: $Cl_2\,(g) \rightarrow 2\,Cl^-\,(aq)$ and $Cl_2\,(g) \rightarrow 2\,ClO^-\,(aq)$

Balance O with H_2O: $Cl_2\,(g) \rightarrow 2\,Cl^-\,(aq)$ and $Cl_2\,(g) + 2\,H_2O\,(l) \rightarrow 2\,ClO^-\,(aq)$

Balance H with H^+: $Cl_2\,(g) \rightarrow 2\,Cl^-\,(aq)$ and $Cl_2\,(g) + 2\,H_2O\,(l) \rightarrow 2\,ClO^-\,(aq) + 4\,H^+\,(aq)$

Neutralize H^+ with OH^-:

Cl_2 (*g*) → 2 Cl^- (*aq*) and Cl_2 (*g*) + ~~2 H_2O(*l*)~~ + 4 OH^- (*aq*) → 2 ClO^- (*aq*)+ 4 H^+ (*aq*) + 4 OH^- (*aq*)

$\underbrace{\quad\quad\quad\quad}_{2\!\!\!/\,4\, H_2O\,(l)}$

Add electrons: Cl_2 (*g*) + 2 e^- → 2 Cl^- (*aq*) and Cl_2 (*g*) + 4 OH^- (*aq*) → 2 ClO^- (*aq*) + 2 H_2O (*l*) + 2 e^-

Equalize electrons: Cl_2 (*g*) + 2 e^- → 2 Cl^- (*aq*) and Cl_2 (*g*) + 4 OH^- (*aq*) → 2 ClO^- (*aq*) + 2 H_2O (*l*) + 2 e^-

Add half-reactions: Cl_2 (*g*) + ~~2 e^-~~ + Cl_2 (*g*) + 4 OH^- (*aq*) → 2 Cl^- (*aq*) + 2 ClO^- (*aq*) + 2 H_2O (*l*) + ~~2 e^-~~

Cancel electrons: 2 Cl_2 (*g*) + 4 OH^- (*aq*) → 2 Cl^- (*aq*) + 2 ClO^- (*aq*) + 2 H_2O (*l*)

Simplify: Cl_2 (*g*) + 2 OH^- (*aq*) → Cl^- (*aq*) + ClO^- (*aq*) + H_2O (*l*)

Check:

Reactants	Products
2 Cl atoms	2 Cl atoms
2 O atoms	2 O atoms
2 H atoms	2 H atoms
–2 charge	–2 charge

7. **Given:** voltaic cell overall redox reaction
Find: Sketch voltaic cell, labeling anode, cathode, all species, and direction of electron flow
Conceptual Plan: Separate overall reaction into 2 half-cell reactions and add electrons as needed to balance reactions. Put anode reaction on the left (oxidation = electrons as product) and cathode reaction on the right (reduction = electrons as reactant). Electrons flow from anode to cathode.
Solution:

a) 2 Ag^+ (*aq*)+ Pb (*s*) → 2 Ag (*s*) + Pb^{2+} (*aq*) separates to 2 Ag^+ (*aq*) → 2 Ag (*s*) and Pb (*s*) → Pb^{2+} (*aq*) then add electrons to balance to get the cathode reaction: 2 Ag^+ (*aq*) + 2 e^- → 2 Ag (*s*) and the anode reaction: Pb (*s*) → Pb^{2+} (*aq*) + 2 e^-.

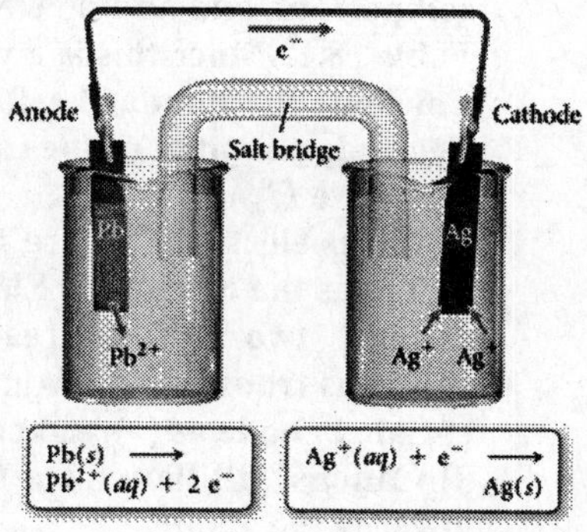

Since we have Pb (s) as the reactant for the oxidation, it will be our anode. Since we have Ag (s) as the product for the reduction, it will be our cathode. Simplify the cathode reaction, dividing all terms by 2.

b) 2 ClO_2 (*g*)+ 2 I^- (*aq*) → 2 ClO_2^- (*aq*) + I_2 (*s*) separates to 2 ClO_2 (*g*) → 2 ClO_2^- (*aq*) and 2 I^- (*aq*) → I_2 (*s*) then add electrons to balance to get the cathode reaction: 2 ClO_2 (*g*) + 2 e^- → 2 ClO_2^- (*aq*) and the anode reaction: 2 I^- (*aq*) → I_2 (*s*) + 2 e^-.

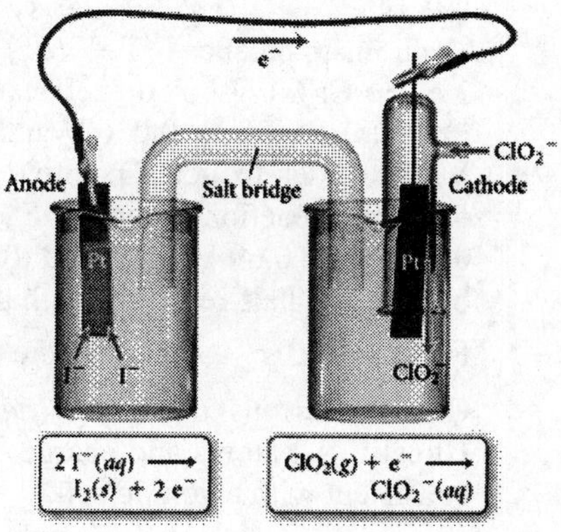

Since we have I^- (*aq*) as the reactant for the oxidation, we will need to use Pt as our anode. Since we have ClO_2^- (*aq*) as the product for the reduction, we will need to use Pt as our cathode. Since ClO_2 (*g*) is our reactant for the reduction, we need to use an electrode assembly like that is used for a SHE. Simplify the cathode reaction, dividing all terms by 2.

c) O_2 (*g*)+ 4 H^+ (*aq*) + 2 Zn (*s*) → 2 H_2O (*l*) + 2 Zn^{2+} (*aq*) separates to O_2 (*g*)+ 4 H^+ (*aq*) → 2 H_2O (*l*) and 2 Zn (*s*) → 2 Zn^{2+} (*aq*) then add electrons to balance to get the cathode reaction: O_2 (*g*)+ 4 H^+ (*aq*) + 4 e^- → 2 H_2O (*l*) and the anode reaction: 2 Zn (*s*) → 2 Zn^{2+} (*aq*) + 4 e^-.

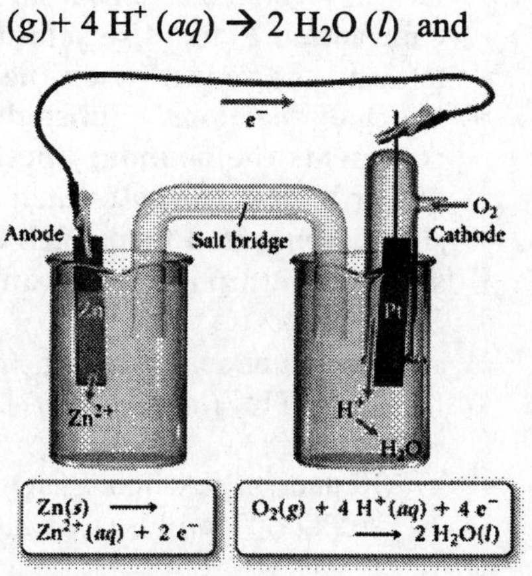

Since we have Zn (*s*) as the reactant for the oxidation, it will be our anode. Since we have H_2O (*l*) as the product for the reduction, we will need to use Pt as our cathode. Since O_2 (*g*) is our reactant for the reduction, we need to use an electrode assembly like what is used for a SHE. Simplify the anode reaction, dividing all terms by 2.

9. **Given:** overall reactions from #7 **Find:** E°_{cell}
Conceptual Plan: Look up half-reactions from solution of problem #7 in Table 18.1. Calculate the standard cell potential by subtracting the electrode potential of the anode from the electrode potential of the cathode: $E^\circ_{cell} = E^\circ_{cathode} - E^\circ_{anode}$.
Solution:
a) $Ag^+ (aq) + e^- \rightarrow Ag (s)$ $E^\circ_{red} = 0.80$ V $= E^\circ_{cathode}$ and $Pb (s) \rightarrow Pb^{2+} (aq) + 2 e^-$ $E^\circ_{red} = -0.13$ V $= E^\circ_{anode}$. Then $E^\circ_{cell} = E^\circ_{cathode} - E^\circ_{anode} = 0.80$ V $- (-0.13$ V$) = 0.93$ V.

b) $ClO_2 (g) + e^- \rightarrow ClO_2^- (aq)$ $E^\circ_{red} = 0.95$ V $= E^\circ_{cathode}$ and $2 I^- (aq) \rightarrow I_2 (s) + 2 e^-$ $E^\circ_{red} = -0.54$ V $= E^\circ_{anode}$. Then $E^\circ_{cell} = E^\circ_{cathode} - E^\circ_{anode} = 0.95$ V $- 0.54$ V $= 0.41$ V.

c) $O_2 (g) + 4 H^+ (aq) + 4 e^- \rightarrow 2 H_2O (l)$ $E^\circ_{red} = 1.23$ V and $Zn (s) \rightarrow Zn^{2+} (aq) + 2 e^-$ $E^\circ_{red} = -0.76$ V $= E^\circ_{anode}$. Then $E^\circ_{cell} = E^\circ_{cathode} - E^\circ_{anode} = 1.23$ V $- (-0.76$ V$) = 1.99$ V.
Check: The units (V) are correct. All of the voltages are positive, which is consistent with a voltaic cell.

11. **Given:** voltaic cell drawing **Find:** a) determine electron flow direction, anode, and cathode; b) write balanced overall reaction and calculate E°_{cell}; c) label electrodes as + and –; and d) directions of anions and cations from salt bridge
Conceptual Plan: Look at each half-cell and write a reduction reaction, by using electrode and solution composition and adding electrons to balance. Look up half-reactions standard reduction potentials in Table 18.1. Since this is a voltaic cell, the cell potentials must be assigned to give a positive E°_{cell}. Calculate the standard cell potential by subtracting the electrode potential of the anode from the electrode potential of the cathode: $E^\circ_{cell} = E^\circ_{cathode} - E^\circ_{anode}$, choosing the electrode assignments to give a positive E°_{cell}.
a) Label electrode where the oxidation occurs as the anode. Label the electrode where the reduction occurs as the cathode. Electrons flow from anode to cathode.
b) Take two half-cell reactions and multiply the reactions as necessary to equalize the number of electrons transferred. Add the two half-cell reactions and cancel electrons and any other species.
c) Label anode as (–) and cathode as (+).
d) Cations will flow from the salt bridge towards the cathode and the anions will flow from salt bridge towards the anode.
Solution:
left side: $Fe^{3+} (aq) \rightarrow Fe (s)$ and right side: $Cr^{3+} (aq) \rightarrow Cr (s)$ add electrons to balance $Fe^{3+} (aq) + 3 e^- \rightarrow Fe (s)$ and right side: $Cr^{3+} (aq) + 3 e^- \rightarrow Cr (s)$. Look up cell standard reduction potentials: $Fe^{3+} (aq) + 3 e^- \rightarrow Fe (s)$ $E^\circ_{red} = -0.036$ V and $Cr^{3+} (aq) + 3 e^- \rightarrow Cr (s)$ $E^\circ_{red} = -0.73$ V. In order to get a positive cell potential, the second reaction is the oxidation reaction (anode). $E^\circ_{cell} = E^\circ_{cathode} - E^\circ_{anode} = -0.036$ V $- (-0.73$ V$) = +0.69$ V. (a, c, and d)

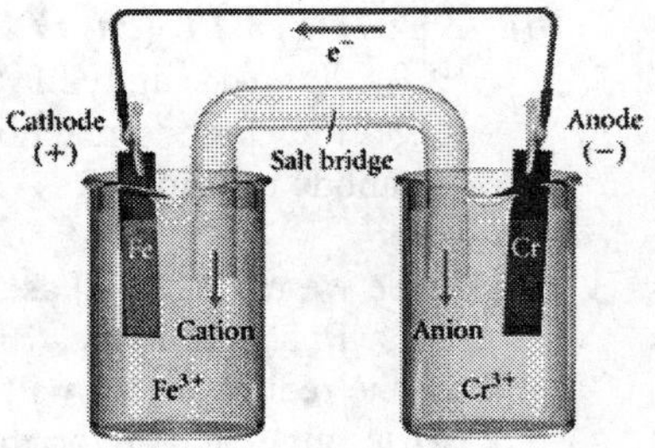

b) Add two half-reactions with the second reaction reversed.
$Fe^{3+} (aq) + \cancel{3e^-} + Cr (s) \rightarrow Fe (s) + Cr^{3+} (aq) + \cancel{3e^-}$.
Cancel electrons to get: $Fe^{3+} (aq) + Cr (s) \rightarrow Fe (s) + Cr^{3+} (aq)$.
Check: All atoms and charge are balanced. The units (V) are correct. The cell potential is positive which is consistent with a voltaic cell.

13. **Given:** overall reactions from #7 **Find:** line notation
Conceptual Plan: Use solution from problem #7. Write the oxidation half-reaction components on the left and the reduction on the right. A double vertical line (||), indicating the salt bridge, separates the two half-reactions. Substances in different phases are separated by a single vertical line (|), which represents the boundary between the phases. For some redox reactions, the reactants and products of one or both of the half-reactions may be in the same phase. In these cases, the reactants and products are simply separated from each other with a comma in the line diagram. Such cells use an inert electrode, such as platinum (Pt) or graphite, as the anode or cathode (or both).
Solution:
a) Reduction reaction: $Ag^+ (aq) + e^- \rightarrow Ag (s)$ and the oxidation reaction: $Pb (s) \rightarrow Pb^{2+} (aq) + 2 e^-$ so $Pb (s)|Pb^{2+} (aq)||Ag^+ (aq)|Ag (s)$

b) Reduction reaction: $ClO_2 (g) + e^- \rightarrow ClO_2^- (aq)$ and the oxidation reaction: $2 I^- (aq) \rightarrow I_2 (s) + 2 e^-$ so $Pt (s)|I^- (aq)| I_2 (s)|| ClO_2 (g)|ClO_2^- (aq)|Pt (s)$

c) Reduction reaction: O_2 (*g*) + 4 H^+ (*aq*) + 4 e^- → 2 H_2O (*l*) and the oxidation reaction: Zn (*s*) → Zn^{2+} (*aq*) + 2 e^- so Zn (*s*)|Zn^{2+} (*aq*) ||O_2 (*g*)|H^+ (*aq*), H_2O (*l*)|Pt (*s*)

15. **Given:** Sn (*s*)|Sn^{2+} (*aq*) || NO_3^- (*aq*), H^+ (*aq*)), H_2O (*l*)|NO (*g*)|Pt (*s*)
Find: Sketch voltaic cell, labeling anode, cathode, all species, direction of electron flow, and $E°_{cell}$
Conceptual Plan: Separate overall reaction into 2 half-cell reactions knowing that the oxidation half-reaction components are on the left and the reduction half-reaction components are on the right. Add electrons as needed to balance reactions. Multiply the half-reactions by the appropriate factors to have an equal number of electrons transferred. Add the half-cell reactions and cancel electrons. Put anode reaction on the left (oxidation = electrons as product) and cathode reaction on the right (reduction = electrons as reactant). Electrons flow from anode to cathode. Look up half-reactions from solution of problem #8 in Table 18.1. Calculate the standard cell potential by subtracting the electrode potential of the anode from the electrode potential of the cathode: $E°_{cell} = E°_{cathode} - E°_{anode}$.
Solution: Oxidation reaction (anode): Sn (*s*) → Sn^{2+} (*aq*) + 2 e^- $E°_{red}$ = – 0.14 V and Reduction reaction (cathode): NO_3^- (*aq*) + 4 H^+ (*aq*) + 3 e^- → NO (*g*) + 2 H_2O (*l*) $E°_{red}$ = 0.96 V. $E°_{cell} = E°_{cathode} - E°_{anode}$ = 0.96 V – (– 0.14 V) = 1.10 V. Multiply first reaction by 3 and the second reaction by 2 so that 6 electrons are transferred. 3 Sn (*s*) → 3 Sn^{2+} (*aq*) + 6 e^- and 2 NO_3^- (*aq*) + 8 H^+ (*aq*) + 6 e^- → 2 NO (*g*) + 4 H_2O (*l*). Add the two half-reactions and cancel electrons 3 Sn (*s*)+ 2 NO_3^- (*aq*) + 8 H^+ (*aq*) + ~~6 e^-~~ → 3 Sn^{2+} (*aq*) + ~~6 e^-~~ + 2 NO (*g*) + 4 H_2O (*l*). So balanced reaction is: 3 Sn (*s*) + 2 NO_3^- (*aq*) + 8 H^+ (*aq*) → 3 Sn^{2+} (*aq*) + 2 NO (*g*) + 4 H_2O (*l*).

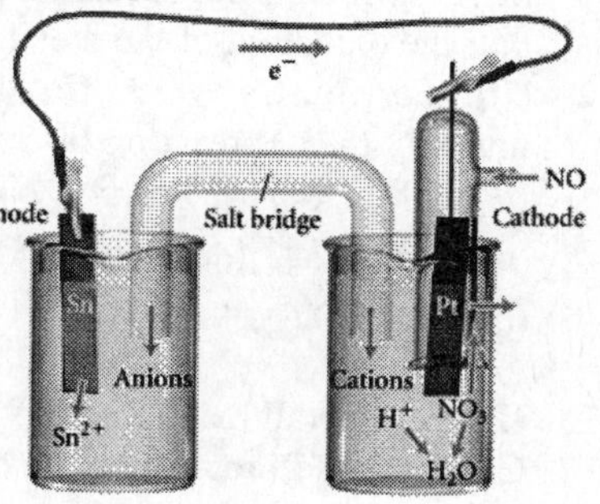

Check: All atoms and charge are balanced. The units (V) are correct. The cell potential is positive which is consistent with a voltaic cell.

17. **Given:** overall reactions **Find:** spontaneity in forward direction
Conceptual Plan: Separate overall reaction into 2 half-cell reactions and add electrons as needed to balance reactions. Look up half-reactions in Table 18.1. Calculate the standard cell potential by subtracting the electrode potential of the anode from the electrode potential of the cathode: $E°_{cell} = E°_{cathode} - E°_{anode}$. If $E°_{cell} > 0$ the reaction is spontaneous in the forward direction.
Solution:

a) Ni (*s*) + Zn^{2+} (*aq*) → Ni^{2+} (*aq*) + Zn (*s*) separates to Ni (*s*) → Ni^{2+} (*aq*) and Zn^{2+} (*aq*) → Zn (*s*) add electrons Ni (*s*) → Ni^{2+} (*aq*) + 2 e^- and Zn^{2+} (*aq*) + 2 e^- → Zn (*s*). Look up cell potentials. Ni is oxidized so $E°_{red}$ = – 0.23 V = $E°_{anode}$. Zn^{2+} is reduced so $E°_{cathode}$ = – 0.76 V. Then $E°_{cell} = E°_{cathode} - E°_{anode}$ = – 0.76 V – (– 0.23 V) = – 0.53 V and so the reaction is nonspontaneous.

b) Ni (*s*) + Pb^{2+} (*aq*) → Ni^{2+} (*aq*) + Pb (*s*) separates to Ni (*s*) → Ni^{2+} (*aq*) and Pb^{2+} (*aq*) → Pb (*s*) add electrons Ni (*s*) → Ni^{2+} (*aq*) + 2 e^- and Pb^{2+} (*aq*) + 2 e^- → Pb (*s*). Look up cell potentials. Ni is oxidized so $E°_{red}$ = – 0.23 V = $E°_{anode}$. Pb^{2+} is reduced so $E°_{red}$ = – 0.13 V = $E°_{cathode}$. Then $E°_{cell} = E°_{cathode} - E°_{anode}$ = – 0.13 V – (– 0.23 V) = + 0.10 V and so the reaction is spontaneous.

c) Al (*s*) + 3 Ag^+ (*aq*) → Al^{3+} (*aq*) + 3 Ag (*s*) separates to Al (*s*) → Al^{3+} (*aq*) and 3 Ag^+ (*aq*) → 3 Ag (*s*) add electrons Al (*s*) → Al^{3+} (*aq*) + 3 e^- and 3 Ag^+ (*aq*) + 3 e^- → 3 Ag (*s*). Simplify the Ag reaction to: Ag^+ (*aq*) + e^- → Ag (*s*). Look up cell potentials. Al is oxidized so $E°_{red}$ = – 1.66 V = $E°_{anode}$. Ag^+ is reduced so $E°_{red}$ = 0.80 V = $E°_{cathode}$. Then $E°_{cell} = E°_{cathode} - E°_{anode}$ = 0.80 V – (– 1.66 V) = + 2.46 V and so the reaction is spontaneous.

d) Pb (*s*) + Mn^{2+} (*aq*) → Pb^{2+} (*aq*) + Mn (*s*) separates to Pb (*s*) → Pb^{2+} (*aq*) and Mn^{2+} (*aq*) → Mn (*s*) add electrons Pb (*s*) → Pb^{2+} (*aq*) + 2 e^- and Mn^{2+} (*aq*) + 2 e^- → Mn (*s*). Look up cell potentials. Pb is oxidized so $E°_{red}$ = – 0.13 V = $E°_{anode}$. Mn^{2+} is reduced so $E°_{red}$ = – 1.18 V= $E°_{cathode}$. Then $E°_{cell} = E°_{cathode} - E°_{anode}$ = – 1.18 V – (– 0.13 V) = – 1.05 V and so the reaction is nonspontaneous.

Check: The units (V) are correct. If the voltage is positive, the reaction is spontaneous.

19. In order for a metal to be able to reduce an ion, it must be above it in Table 18.1 (need positive $E°_{cell} = E°_{cathode} - E°_{anode}$). So we need a metal that is above Mn^{2+}, but below Mg^{2+}. Aluminum is the only one in the table that meets these criteria.

21. In general, metals whose reduction half-reactions lie below the reduction of H^+ to H_2 in Table 18.1 will dissolve in acids, while metals above it will not. a) Al and c) Pb meet this criterion. To write the balanced redox

reactions, pair the oxidation of the metal with the reduction of H^+ to H_2 ($2\ H^+ (aq) + 2\ e^- \rightarrow H_2 (g)$). For Al, $Al\ (s) \rightarrow Al^{3+} (aq) + 3\ e^-$. In order to balance the number of electrons transferred we need to multiply the Al reaction by 2 and the H^+ reaction by 3. So, $2\ Al\ (s) \rightarrow 2\ Al^{3+} (aq) + 6\ e^-$ and $6\ H^+ (aq) + 6\ e^- \rightarrow 3\ H_2 (g)$. Adding the two reactions: $2\ Al\ (s) + 6\ H^+ (aq) + \cancel{6e^-} \rightarrow 2\ Al^{3+} (aq) + \cancel{6e^-} + 3\ H_2 (g)$. Simplify to $2\ Al\ (s) + 6\ H^+(aq) \rightarrow 2\ Al^{3+} (aq) + 3\ H_2 (g)$. For Pb, $Pb\ (s) \rightarrow Pb^{2+} (aq) + 2\ e^-$. Since each reaction involves 2 electrons we can add the two reactions. $Pb\ (s) + 2\ H^+ (aq) + \cancel{2e^-} \rightarrow Pb^{2+} (aq) + \cancel{2e^-} + H_2 (g)$. Simplify to $Pb\ (s) + 2\ H^+ (aq) \rightarrow Pb^{2+} (aq) + H_2 (g)$.

23. Nitric acid (HNO_3) oxidizes metals through the following reduction half-reaction: $NO_3^- (aq) + 4\ H^+ (aq) + 3\ e^- \rightarrow NO\ (g) + 2\ H_2O\ (l)$ $E^\circ_{red} = 0.96$ V. Since this half-reaction is above the reduction of H^+ in Table 18.1, HNO_3 can oxidize metals (such as copper, for example) that cannot be oxidized by HCl. a) Cu will be oxidized, but b) Au (which has a reduction potential of 1.50 V) will not be oxidized. To write the balanced redox reactions, pair the oxidation of the metal with the reduction of nitric acid ($NO_3^- (aq) + 4\ H^+ (aq) + 3\ e^- \rightarrow NO\ (g) + 2\ H_2O\ (l)$). For Cu, $Cu\ (s) \rightarrow Cu^{2+} (aq) + 2\ e^-$. In order to balance the number of electrons transferred we need to multiply the Cu reaction by 3 and the nitric acid reaction by 2. So, $3\ Cu\ (s) \rightarrow 3\ Cu^{2+} (aq) + 6\ e^-$ and $2\ NO_3^- (aq) + 8\ H^+ (aq) + 6\ e^- \rightarrow 2\ NO\ (g) + 4\ H_2O\ (l)$. Adding the two reactions: $3\ Cu\ (s) + 2\ NO_3^- (aq) + 8\ H^+ (aq) + \cancel{6e^-} \rightarrow 3\ Cu^{2+} (aq) + \cancel{6e^-} + 2\ NO\ (g) + 4\ H_2O\ (l)$. Simplify to $3\ Cu\ (s) + 2\ NO_3^- (aq) + 8\ H^+ (aq) \rightarrow 3\ Cu^{2+} (aq) + 2\ NO\ (g) + 4\ H_2O\ (l)$.

25. **Given:** overall reactions **Find:** E°_{cell} and spontaneity in forward direction
Conceptual Plan: Separate overall reaction into 2 half-cell reactions and add electrons as needed to balance reactions. Look up half-reactions in Table 18.1. Calculate the standard cell potential by subtracting the electrode potential of the anode from the electrode potential of the cathode: $E^\circ_{cell} = E^\circ_{cathode} - E^\circ_{anode}$. If $E^\circ_{cell} > 0$ the reaction is spontaneous in the forward direction.
Solution:

a) $2\ Cu\ (s) + Mn^{2+} (aq) \rightarrow 2\ Cu^+ (aq) + Mn\ (s)$ separates to $2\ Cu\ (s) \rightarrow 2\ Cu^+ (aq)$ and $Mn^{2+} (aq) \rightarrow Mn\ (s)$ add electrons $2\ Cu\ (s) \rightarrow 2\ Cu^+ (aq) + 2\ e^-$ and $Mn^{2+} (aq) + 2\ e^- \rightarrow Mn\ (s)$. Simplify the Cu reaction to: $Cu\ (s) \rightarrow Cu^+ (aq) + e^-$. Look up cell potentials. Cu is oxidized so $E^\circ_{red} = -0.52\ V = E^\circ_{anode}$. Mn^{2+} is reduced so $E^\circ_{red} = -1.18\ V = E^\circ_{cathode}$. Then $E^\circ_{cell} = E^\circ_{cathode} - E^\circ_{anode} = -0.52\ V - 1.18\ V = -1.70$ V and so the reaction is nonspontaneous.

b) $MnO_2\ (s) + 4\ H^+(aq) + Zn\ (s) \rightarrow Mn^{2+} (aq) + 2\ H_2O\ (l) + Zn^{2+} (aq)$ separates to $MnO_2\ (s) + 4\ H^+(aq) \rightarrow Mn^{2+} (aq) + 2\ H_2O\ (l)$ and $Zn\ (s) \rightarrow Zn^{2+} (aq)$ add electrons $MnO_2\ (s) + 4\ H^+(aq) + 2\ e^- \rightarrow Mn^{2+} (aq) + 2\ H_2O\ (l)$ and $Zn\ (s) \rightarrow Zn^{2+} (aq) + 2\ e^-$. Look up cell potentials. Zn is oxidized so $E^\circ_{red} = -0.76\ V = E^\circ_{anode}$. Mn is reduced so $E^\circ_{red} = 1.21\ V = E^\circ_{cathode}$. Then $E^\circ_{cell} = E^\circ_{cathode} - E^\circ_{anode} = 1.21\ V - (-0.76\ V) = +1.97$ V and so the reaction is spontaneous.

c) $Cl_2\ (g) + 2\ F^- (aq) \rightarrow 2\ Cl^- (aq) + F_2\ (g)$ separates to $Cl_2\ (g) \rightarrow 2\ Cl^- (aq)$ and $2\ F^- (aq) \rightarrow F_2\ (g)$ add electrons $Cl_2\ (g) + 2\ e^- \rightarrow 2\ Cl^- (aq)$ and $2\ F^- (aq) \rightarrow F_2\ (g) + 2\ e^-$. Look up cell potentials. F is oxidized so $E^\circ_{red} = 2.87\ V = E^\circ_{anode}$. Cl is reduced so $E^\circ_{red} = 1.36\ V = E^\circ_{cathode}$. Then $E^\circ_{cell} = E^\circ_{cathode} - E^\circ_{anode} = 1.36\ V - 2.87\ V + = -1.51$ V and so the reaction is nonspontaneous.

Check: The units (V) are correct. If the voltage is positive, the reaction is spontaneous.

27. a) Pb^{2+}. The strongest oxidizing agent is the one with the reduction reaction that is closest to the top of Table 18.1.

29. **Given:** overall reactions **Find:** ΔG°_{rxn} and spontaneity in forward direction
Conceptual Plan: Separate overall reaction into 2 half-cell reactions and add electrons as needed to balance reactions. Look up half-reactions in Table 18.1. Calculate the standard cell potential by subtracting the electrode potential of the anode from the electrode potential of the cathode: $E^\circ_{cell} = E^\circ_{cathode} - E^\circ_{anode}$, then calculate ΔG°_{rxn} using $\Delta G^0_{rxn} = -nFE^0_{cell}$.
Solution:

a) $Pb^{2+} (aq) + Mg\ (s) \rightarrow Pb\ (s) + Mg^{2+} (aq)$ separates to $Pb^{2+} (aq) \rightarrow Pb\ (s)$ and $Mg\ (s) \rightarrow Mg^{2+} (aq)$ add electrons $Pb^{2+} (aq) + 2\ e^- \rightarrow Pb\ (s)$ and $Mg\ (s) \rightarrow Mg^{2+} (aq) + 2\ e^-$. Look up cell potentials. Mg is oxidized so $E^\circ_{ox} = -E^\circ_{red} = -2.37\ V = E^\circ_{anode}$. Pb^{2+} is reduced so $E^\circ_{red} = -0.13\ V = E^\circ_{cathode}$. Then $E^\circ_{cell} = E^\circ_{cathode} - E^\circ_{anode} = -0.13\ V - (-2.37\ V) = +2.24$ V. $n = 2$ so $\Delta G^0_{rxn} = -nFE^0_{cell}$

$$= -2\ \cancel{mol\ e^-} \times \frac{96{,}485\ C}{\cancel{mol\ e^-}} \times 2.24\ V = -2 \times 96{,}485\ \cancel{C} \times 2.24 \frac{J}{\cancel{C}} = -4.32 \times 10^5\ J = -432\ kJ$$

b) $Br_2\ (l) + 2\ Cl^-\ (aq) \rightarrow 2\ Br^-\ (aq) + Cl_2\ (g)$ separates to $Br_2\ (g) \rightarrow 2\ Br^-\ (aq)$ and $2\ Cl^-\ (aq) \rightarrow Cl_2\ (g)$ add electrons $Br_2\ (g) + 2\ e^- \rightarrow 2\ Br^-\ (aq)$ and $2\ Cl^-\ (aq) \rightarrow Cl_2\ (g) + 2\ e^-$. Look up cell potentials. Cl is oxidized so $E^\circ_{red} = 1.36\ V = E^\circ_{anode}$. Br is reduced so $E^\circ_{red} = 1.09\ V = E^\circ_{cathode}$. Then $E^\circ_{cell} = E^\circ_{cathode} - E^\circ_{anode} = 1.09\ V - 1.36\ V = -0.27\ V$. $n = 2$ so $\Delta G^0_{rxn} = -nFE^0_{cell}$

$$= -2\ \cancel{mol\ e^-} \times \frac{96{,}485\ C}{\cancel{mol\ e^-}} \times -0.27\ V = -2 \times 96{,}485\ \cancel{C} \times -0.27 \frac{J}{\cancel{C}} = 5.2 \times 10^4\ J = 52\ kJ.$$

c) $MnO_2\ (s) + 4\ H^+(aq) + Cu\ (s) \rightarrow Mn^{2+}\ (aq) + 2\ H_2O\ (l) + Cu^{2+}\ (aq)$ separates to $MnO_2\ (s) + 4\ H^+(aq) \rightarrow Mn^{2+}\ (aq) + 2\ H_2O\ (l)$ and $Cu\ (s) \rightarrow Cu^{2+}\ (aq)$ add electrons $MnO_2\ (s) + 4\ H^+(aq) + 2\ e^- \rightarrow Mn^{2+}\ (aq) + 2\ H_2O\ (l)$ and $Cu\ (s) \rightarrow Cu^{2+}\ (aq) + 2\ e^-$. Look up cell potentials. Cu is oxidized so $E^\circ_{red} = 0.34\ V = E^\circ_{anode}$. Mn is reduced so $E^\circ_{red} = 1.21\ V = E^\circ_{cathode}$. Then $E^\circ_{cell} = E^\circ_{cathode} - E^\circ_{anode} = 1.21\ V - 0.34\ V = +0.87\ V$. $n = 2$ so $\Delta G^0_{rxn} = -nFE^0_{cell} = -2\ \cancel{mol\ e^-} \times \frac{96{,}485\ C}{\cancel{mol\ e^-}} \times 0.87\ V = -2 \times 96{,}485\ \cancel{C} \times 0.87 \frac{J}{\cancel{C}} = -1.7 \times 10^5\ J$

$= -1.7 \times 10^2\ kJ$.

Check: The units (kJ) are correct. If the voltage is positive, the reaction is spontaneous and the free energy change is negative.

31. **Given:** overall reactions from #29 **Find:** K

Conceptual Plan: °C → K then $\Delta G^\circ_{rxn}, T \rightarrow K$

$K = 273.15 + °C$ $\quad \Delta G^0_{rxn} = -RT\ln K$

Solution: $T = 273.15 + 25\ °C = 298\ K$ then

a) $\Delta G^0_{rxn} = -RT\ln K$ Rearrange to solve for K. $K = e^{\frac{-\Delta G^0_{rxn}}{RT}} = e^{\frac{-(-432)\ \cancel{kJ} \times \frac{1000\ \cancel{J}}{1\ \cancel{kJ}}}{\left(8.314 \frac{\cancel{J}}{\cancel{K} \cdot mol}\right)(298\ \cancel{K})}} = e^{174.364} = 5.31 \times 10^{75}$.

b) $\Delta G^0_{rxn} = -RT\ln K$ Rearrange to solve for K. $K = e^{\frac{-\Delta G^0_{rxn}}{RT}} = e^{\frac{-52\ \cancel{kJ} \times \frac{1000\ \cancel{J}}{1\ \cancel{kJ}}}{\left(8.314 \frac{\cancel{J}}{\cancel{K} \cdot mol}\right)(298\ \cancel{K})}} = e^{-20.988} = 7.7 \times 10^{-10}$.

c) $\Delta G^0_{rxn} = -RT\ln K$ Rearrange to solve for K. $K = e^{\frac{-\Delta G^0_{rxn}}{RT}} = e^{\frac{-(-170\ \cancel{kJ}) \times \frac{1000\ \cancel{J}}{1\ \cancel{kJ}}}{\left(8.314 \frac{\cancel{J}}{\cancel{K} \cdot mol}\right)(298\ \cancel{K})}} = e^{68.616} = 6.3 \times 10^{29}$.

Check: The units (none) are correct. If the voltage is positive, the reaction is spontaneous and the free energy change is negative and the equilibrium constant is large.

33. **Given:** $Ni^{2+}\ (aq) + Cd\ (s) \rightarrow$ **Find:** K

Conceptual Plan: Write 2 half-cell reactions and add electrons as needed to balance reactions. Look up half-reactions in Table 18.1. Calculate the standard cell potential by subtracting the electrode potential of the anode from the electrode potential of the cathode: $E^\circ_{cell} = E^\circ_{cathode} - E^\circ_{anode}$, then °C → K then $E^\circ_{cell}, n, T \rightarrow K$

$K = 273.15 + °C$ $\quad \Delta G^0_{rxn} = -RT\ln K = -nFE^0_{cell}$

Solution: $Ni^{2+}\ (aq) + 2\ e^- \rightarrow Ni\ (s)$ and $Cd\ (s) \rightarrow Cd^{2+}\ (aq) + 2\ e^-$. Look up cell potentials. Cd is oxidized so $E^\circ_{ox} = -E^\circ_{red} = -0.40\ V = E^\circ_{anode}$. Ni^{2+} is reduced so $E^\circ_{red} = -0.23\ V = E^\circ_{cathode}$. Then $E^\circ_{cell} = E^\circ_{cathode} - E^\circ_{anode} = -0.23\ V - (-0.40\ V) = +0.17\ V$. The overall reaction is $Ni^{2+}\ (aq) + Cd\ (s) \rightarrow Ni\ (s) + Cd^{2+}\ (aq)$. $n = 2$ and $T = 273.15 + 25\ °C = 298\ K$ then $\Delta G^0_{rxn} = -RT\ln K = -nFE^0_{cell}$. Rearrange to solve for K.

$$K = e^{\frac{nFE^0_{cell}}{RT}} = e^{\frac{2\ \cancel{mol\ e^-} \times \frac{96{,}485\ \cancel{C}}{\cancel{mol\ e^-}} \times 0.17 \frac{\cancel{J}}{\cancel{C}}}{\left(8.314 \frac{\cancel{J}}{\cancel{K} \cdot mol}\right)(298\ \cancel{K})}} = e^{13.241} = 5.6 \times 10^5.$$

Check: The units (none) are correct. If the voltage is positive, the reaction is spontaneous and the equilibrium constant is large.

35. **Given:** $n = 2$ and $K = 25$ **Find:** ΔG°_{rxn} and E°_{cell}
Conceptual Plan: $K, T \rightarrow \Delta G^\circ_{rxn}$ **and** $\Delta G^\circ_{rxn}, n \rightarrow E^\circ_{cell}$

$$\Delta G^0_{rxn} = -RT \ln K \qquad \Delta G^0_{rxn} = -nFE^0_{cell}$$

Solution: $\Delta G^0_{rxn} = -RT \ln K = -\left(8.314 \frac{\text{J}}{\cancel{\text{K}} \cdot \text{mol}}\right)(298\ \cancel{\text{K}}) \ln 25 = -7.9\underline{7}500 \times 10^3\ \text{J} = -8.0\ \text{kJ}$ and

$\Delta G^0_{rxn} = -nFE^0_{cell}$. Rearrange to solve for E°_{cell}.

$$E^0_{cell} = \frac{\Delta G^0_{rxn}}{-nF} = \frac{-7.9\underline{7}500 \times 10^3\ \text{J}}{-2\ \cancel{\text{mol e}^-} \times \frac{96{,}485\ \text{C}}{\cancel{\text{mol e}^-}}} = 0.041 \frac{\text{V}\cancel{\text{C}}}{\cancel{\text{C}}} = 0.041\ \text{V}.$$

Check: The units (kJ and V) are correct. If $K > 1$ then the voltage is positive, the free energy change is negative.

37. **Given:** Sn^{2+} (*aq*) + Mn (*s*) → Sn (*s*) + Mn^{2+} (*aq*) **Find:** a) E°_{cell}; b) E_{cell} when $[Sn^{2+}] = 0.0100$ M; $[Mn^{2+}] = 2.00$ M; and c) E_{cell} when $[Sn^{2+}] = 2.00$ M; $[Mn^{2+}] = 0.0100$ M
Conceptual Plan: a) Separate overall reaction into 2 half-cell reactions and add electrons as needed to balance reactions. Look up half-reactions in Table 18.1. Calculate the standard cell potential by subtracting the electrode potential of the anode from the electrode potential of the cathode: $E^\circ_{cell} = E^\circ_{cathode} - E^\circ_{anode}$. b) and c) E°_{cell}, $[Sn^{2+}]$, $[Mn^{2+}]$, $n \rightarrow E_{cell}$

$$E_{cell} = E^\circ_{cell} - \frac{0.0592\ V}{n} \log Q \text{ where } Q = \frac{[Mn^{2+}]}{[Sn^{2+}]}$$

Solution: a) separate overall reaction to: Sn^{2+} (*aq*) → Sn (*s*) and Mn (*s*) → Mn^{2+} (*aq*) add electrons Sn^{2+} (*aq*) + 2 e^- → Sn (*s*) and Mn (*s*) → Mn^{2+} (*aq*) + 2 e^-. Look up cell potentials. Mn is oxidized so $E^\circ_{red} = -1.18\ \text{V} = E^\circ_{anode}$. Sn^{2+} is reduced so $E^\circ_{red} = -0.14\ \text{V} = E^\circ_{cathode}$. Then $E^\circ_{cell} = E^\circ_{cathode} - E^\circ_{anode} = -0.14\ \text{V} - (-1.18\ \text{V}) = +1.04$ V.

b) $$Q = \frac{[Mn^{2+}]}{[Sn^{2+}]} = \frac{2.00\ \cancel{M}}{0.0100\ \cancel{M}} = 200. \text{ and } n = 2 \text{ then}$$

$$E_{cell} = E^0_{cell} - \frac{0.0592\ \text{V}}{n} \log Q = 1.04\ \text{V} - \frac{0.0592\ \text{V}}{2} \log 200. = +0.97\ \text{V}$$

c) $$Q = \frac{[Mn^{2+}]}{[Sn^{2+}]} = \frac{0.0100\ \cancel{M}}{2.00\ \cancel{M}} = 0.00500 \text{ and } n = 2 \text{ then}$$

$$E_{cell} = E^0_{cell} - \frac{0.0592\ \text{V}}{n} \log Q = 1.04\ \text{V} - \frac{0.0592\ \text{V}}{2} \log 0.00500 = +1.11\ \text{V}$$

Check: The units (V, V, and V) are correct. The Sn^{2+} reduction reaction is above the Mn^{2+} reduction reaction, so the standard cell potential will be positive. Having more products than reactants reduces the cell potential. Having more reactants than products raises the cell potential.

39. **Given:** Pb (*s*) → Pb^{2+} (*aq*, 0.10 M) + 2 e^- and MnO_4^- (*aq*, 1.50 M) + 4 H^+ (*aq*, 2.0 M) + 3 e^- → MnO_2 (*s*) + 2 H_2O (*l*) **Find:** E_{cell}
Conceptual Plan: Look up half-reactions in Table 18.1. Calculate the standard cell potential by subtracting the electrode potential of the anode from the electrode potential of the cathode: $E^\circ_{cell} = E^\circ_{cathode} - E^\circ_{anode}$. Equalize the number of electrons transferred by multiplying the first reaction by 3 and the second reaction by 2. Add the two half-cell reactions and cancel the electrons.
Then E°_{cell}, $[Pb^{2+}]$, $[MnO_4^-]$, $[H^+]$, $n \rightarrow E_{cell}$

$$E_{cell} = E^\circ_{cell} - \frac{0.0592\ V}{n} \log Q \text{ where } Q = \frac{[Pb^{2+}]^3}{[MnO_4^-]^2[H^+]^8}$$

Solution: Pb is oxidized so $E^\circ_{red} = -0.13\ \text{V} = E^\circ_{anode}$. Mn is reduced so $E^\circ_{red} = 1.68\ \text{V} = E^\circ_{cathode}$. Then $E^\circ_{cell} = E^\circ_{cathode} - E^\circ_{anode} = 1.68\ \text{V} - (-0.13\ \text{V}) = +1.81$ V. Equalizing the electrons: 3 Pb (*s*) → 3 Pb^{2+} (*aq*) + 6 e^- and 2 MnO_4^- (*aq*) + 8 H^+ (*aq*) + 6 e^- → 2 MnO_2 (*s*) + 4 H_2O (*l*). Adding the two reactions: 3 Pb (*s*)+ 2 MnO_4^- (*aq*) + 8 H^+ (*aq*) + $\cancel{6\ e^-}$ → 3 Pb^{2+} (*aq*) + $\cancel{6\ e^-}$ + 2 MnO_2 (*s*) + 4 H_2O (*l*). Cancel the electrons: 3 Pb (*s*)+ 2 MnO_4^-

$(aq) + 8\ H^+\ (aq) \rightarrow 3\ Pb^{2+}\ (aq) + 2\ MnO_2\ (s) + 4\ H_2O\ (l)$. So $n = 6$ and $Q = \dfrac{[Pb^{2+}]^3}{[MnO_4^-]^2[H^+]^8} = \dfrac{(0.10)^3}{(1.50)^2(2.0)^8}$

$= 1.7361 \times 10^{-6}$ then $E_{cell} = E^0_{cell} - \dfrac{0.0592\ V}{n} \log Q = 1.81\ V - \dfrac{0.0592\ V}{6} \log 1.7361 \times 10^{-6} = +1.87\ V$.

Check: The units (V) are correct. The MnO_4^- reduction reaction is above the Pb^{2+} reduction reaction, so the standard cell potential will be positive. Having more reactants than products raises the cell potential.

41. **Given:** Zn/Zn^{2+} and Ni/Ni^{2+} half-cells in voltaic cell; initially $[Ni^{2+}] = 1.50$ M, and $[Zn^{2+}] = 0.100$ M
Find: a) initial E_{cell}; b) E_{cell} when $[Ni^{2+}] = 0.500$ M; and c) $[Ni^{2+}]$ and $[Zn^{2+}]$ when $E_{cell} = 0.45$ V
Conceptual Plan: a) Write 2 half-cell reactions and add electrons as needed to balance reactions. Look up half-reactions in Table 18.1. Calculate the standard cell potential by subtracting the electrode potential of the anode from the electrode potential of the cathode: $E^\circ_{cell} = E^\circ_{cathode} - E^\circ_{anode}$. Choose the direction of the half-cell reactions so that $E^\circ_{cell} > 0$. Add two half-cell reactions and cancel electrons to generate overall reaction. Define Q based on overall reaction. Then E°_{cell}, $[Ni^{2+}]$, $[Zn^{2+}]$, $n \rightarrow E_{cell}$

$$E_{cell} = E^\circ_{cell} - \frac{0.0592\ V}{n} \log Q$$

b) When $[Ni^{2+}] = 0.500$ M, then $[Zn^{2+}] = 1.100$ M (since the stoichiometric coefficients for Ni^{2+}: Zn^{2+} are 1:1, and the $[Ni^{2+}]$ drops by 1.00 M, the other concentration must rise by 1.00 M). Then E°_{cell}, $[Ni^{2+}]$, $[Zn^{2+}]$, $n \rightarrow E_{cell}$

$$E_{cell} = E^\circ_{cell} - \frac{0.0592\ V}{n} \log Q$$

c) E°_{cell}, E_{cell}, $n \rightarrow [Zn^{2+}] / [Ni^{2+}] \rightarrow [Ni^{2+}]$, $[Zn^{2+}]$

$$E_{cell} = E^\circ_{cell} - \frac{0.0592\ V}{n} \log Q \qquad [Ni^{2+}] + [Zn^{2+}] = 1.50\ M + 0.100\ M = 1.60\ M$$

Solution:
a) $Zn^{2+}\ (aq) + 2\ e^- \rightarrow Zn\ (s)$ and $Ni^{2+}\ (aq) + 2\ e^- \rightarrow Ni\ (s)$. Look up cell potentials. For Zn, $E^\circ_{red} = -0.76$ V. For Ni, $E^\circ_{red} = -0.23$ V. In order to get a positive E°_{cell} Zn is oxidized so $E^\circ_{red} = -0.76\ V = E^\circ_{anode}$. Ni^{2+} is reduced so $E^\circ_{red} = -0.23\ V = E^\circ_{cathode}$. Then $E^\circ_{cell} = E^\circ_{cathode} - E^\circ_{anode} = -0.23\ V - (-0.76\ V) = +0.53$ V. Adding the two half-cell reactions: $Zn\ (s) + Ni^{2+}\ (aq) + \cancel{2e^-} \rightarrow Zn^{2+}\ (aq) + \cancel{2e^-} + Ni\ (s)$. The overall reaction is: $Zn\ (s) + Ni^{2+}\ (aq) \rightarrow Zn^{2+}\ (aq) + Ni\ (s)$. Then $Q = \dfrac{[Zn^{2+}]}{[Ni^{2+}]} = \dfrac{0.100}{1.50} = 0.0666667$ and $n = 2$ then $E_{cell} = E^0_{cell} - \dfrac{0.0592\ V}{n} \log Q = 0.53\ V - \dfrac{0.0592\ V}{2} \log 0.0666667 = +0.56\ V$.

b) $Q = \dfrac{[Zn^{2+}]}{[Ni^{2+}]} = \dfrac{1.100}{0.500} = 2.20$ then $E_{cell} = E^0_{cell} - \dfrac{0.0592\ V}{n} \log Q = 0.53\ V - \dfrac{0.0592\ V}{2} \log 2.20 = +0.52\ V$.

c) $E_{cell} = E^0_{cell} - \dfrac{0.0592\ V}{n} \log Q$ so $0.45\ V = 0.53\ V - \dfrac{0.0592\ V}{2} \log Q \rightarrow 0.08\ \cancel{V} = \dfrac{0.0592\ \cancel{V}}{2} \log Q \rightarrow$

$\log Q = 2.70270 \rightarrow Q = 10^{2.70270} = 504.32$ then $Q = 504.32 = \dfrac{[Zn^{2+}]}{1.60M - [Zn^{2+}]}$ solving for $[Zn^{2+}]$

$(504.32)(1.60M - [Zn^{2+}]) = [Zn^{2+}] \rightarrow [Zn^{2+}] = \dfrac{806.906\ M}{505.32} = 1.59628\ M = 1.60\ M$ then

$[Ni^{2+}] = 1.60\ M - 1.59628\ M = 0.003\ M$.

Check: The units (V, V, and V) are correct. The standard cell potential is positive and since there are more reactants than products this raises the cell potential. As the reaction proceeds, reactants are converted to products so the cell potential drops for parts b) and c).

43. **Given:** Zn/Zn^{2+} concentration cell, with $[Zn^{2+}] = 2.0$ M in one half-cell and $[Zn^{2+}] = 1.0 \times 10^{-3}$ M in other half-cell **Find:** Sketch voltaic cell, labeling anode, cathode, reactions at electrodes, all species, and direction of electron flow
Conceptual Plan: In a concentration cell, the half-cell with the higher concentration is always the half-cell where the reduction takes place (contains the cathode). The 2 half-cell reactions are the same, only reversed. Put anode reaction on the left (oxidation = electrons as product) and cathode reaction on the right (reduction = electrons as reactant). Electrons flow from anode to cathode.

Solution:

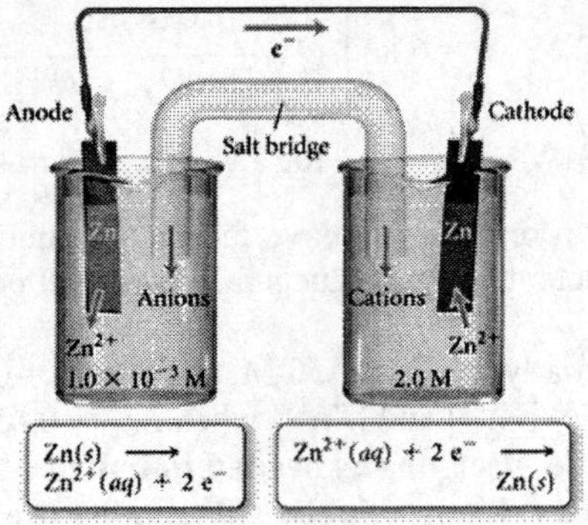

Check: The figure looks similar to the right side of Figure 18.10.

45. **Given:** Sn/Sn^{2+} concentration cell with $E_{cell} = 0.10$ V **Find:** ratio of $[Sn^{2+}]$ in two half-cells
Conceptual Plan: Determine n, then E^o_{cell}, E_{cell}, n → Q = ratio of $[Sn^{2+}]$ in two half-cells

$$E_{cell} = E^o_{cell} - \frac{0.0592\ V}{n}\log Q$$

Solution: Since Sn^{2+} (aq) + 2 e^- → Sn (s), n = 2. In a concentration cell, E^o_{cell} = 0 V. So $E_{cell} = E^0_{cell} - \frac{0.0592\ V}{n}\log Q$ so $0.10\ V = 0.00\ V - \frac{0.0592\ V}{2}\log Q$ → $0.10\ \cancel{V} = -\frac{0.0592\ \cancel{V}}{2}\log Q$ →

$\log Q = -3.\underline{3}784$ → $Q = 10^{-3.3784} = 4.2 \times 10^{-4} = \frac{[Sn^{2+}](ox)}{[Sn^{2+}](red)}$.

Check: The units (none) are correct. Since the concentration in the reduction reaction half-cell is always greater than the concentration in the oxidation half-cell in a voltaic concentration cell, the Q or ratio of two cells is less than 1.

47. **Given:** alkaline battery **Find:** optimum mass ratio of Zn to MnO_2
Conceptual Plan: Look up alkaline battery reactions. Use stoichiometry to get mole ratio. Then

$$\frac{1\ \text{mol Zn}}{2\ \text{mol MnO}_2}$$

$Zn\ (s) + 2\ OH^-\ (aq) \rightarrow Zn(OH)_2\ (s) + 2\ e^-$

$2\ MnO_2\ (s) + 2\ H_2O\ (l) + 2\ e^- \rightarrow 2\ MnO(OH)\ (s) + 2\ OH^-\ (aq).$

mol Zn → g Zn then mol MnO_2 → g MnO_2

$$\frac{65.41\ \text{g Zn}}{1\ \text{mol Zn}} \qquad \frac{1\ \text{mol MnO}_2}{86.94\ \text{g MnO}_2}$$

Solution: $\frac{1\ \cancel{\text{mol Zn}}}{2\ \cancel{\text{mol MnO}_2}} \times \frac{65.41\ \text{g Zn}}{1\ \cancel{\text{mol Zn}}} \times \frac{1\ \cancel{\text{mol MnO}_2}}{86.94\ \text{g MnO}_2} = 0.3762 \frac{\text{g Zn}}{\text{g MnO}_2}$.

Check: The units (mass ratio) are correct. Since more moles of MnO_2 are needed and the molar mass is larger, the ratio is less than 1.

49. **Given:** $CH_4\ (g) + 2\ O_2\ (g)$ → $CO_2\ (g) + 2\ H_2O\ (g)$ **Find:** E^o_{cell}
Conceptual Plan: $\Delta G^0_{rxn} = \sum n_p \Delta G^0_f(products) - \sum n_r \Delta G^0_f(reactants)$ **and determine n then ΔG^o_{rxn}, n → E^o_{cell}**

$$\Delta G^0_{rxn} = -nFE^0_{cell}$$

Solution:

Reactant/Product	ΔG^0_f(kJ/mol from Appendix IIB)
$CH_4\ (g)$	− 50.5
$O_2\ (g)$	0.0
$CO_2\ (g)$	− 394.4
$H_2O\ (g)$	− 228.6

Be sure to pull data for the correct formula and phase.

$$\Delta G^0_{rxn} = \sum n_p \Delta G^0_f(products) - \sum n_r \Delta G^0_f(reactants)$$
$$= [1(\Delta G^0_f(CO_2\ (g))) + 2(\Delta G^0_f(H_2O\ (g)))] - [1(\Delta G^0_f(CH_4\ (g))) + 2(\Delta G^0_f(O_2\ (g)))]$$
$$= [1(-394.4\text{ kJ}) + 2(-228.6\text{ kJ})] - [1(-50.5\text{ kJ}) + 2(0.0\text{ kJ})]$$
$$= [-851.6\text{ kJ}] - [-50.5\text{ kJ}]$$
$$= -801.1\text{ kJ} = -8.011 \times 10^5\text{ J}$$

and since one C atom goes from an oxidation state of −4 to +4 and 4 O atoms are going from 0 to −2, then $n = 8$ and $\Delta G^0_{rxn} = -nFE^0_{cell}$. Rearrange to solve for E°_{cell}.

$$E^0_{cell} = \frac{\Delta G^0_{rxn}}{-nF} = \frac{-8.011 \times 10^5\text{ J}}{-8\ \cancel{\text{mole}^-} \times \frac{96{,}485\text{ C}}{\cancel{\text{mole}^-}}} = 1.038 \frac{\text{V}\cancel{\text{C}}}{\cancel{\text{C}}} = 1.038\text{ V}.$$

Check: The units (V) are correct. The cell voltage is positive which is consistent with a spontaneous reaction.

51. In order for a metal to be able to protect iron, it must, more easily oxidized than iron or be below it in Table 18.1. a) Zn and c) Mn meet this criterion.

53. **Given:** electrolytic cell sketch **Find:** a) Label anode and cathode and indicate half-reactions; b) Indicate direction of electron flow; and c) Label battery terminals and calculate minimum voltage to drive reaction
Conceptual Plan: a) Write 2 half-cell reactions and add electrons as needed to balance reactions. Look up half-reactions in Table 18.1. Calculate the standard cell potential by subtracting the electrode potential of the anode from the electrode potential of the cathode: $E^\circ_{cell} = E^\circ_{cathode} - E^\circ_{anode}$. Choose the direction of the half-cell reactions so that $E^\circ_{cell} < 0$. b) Electrons flow from anode to cathode. c) Each half-cell reaction moves forward, so concentration change direction can be determined.
Solution:
a) Ni^{2+} (*aq*) + 2 e^- → Ni (*s*) and Cd^{2+} (*aq*) + 2 e^- → Cd (*s*). Look up cell potentials. For Ni, $E^\circ_{red} = -0.23$ V. For Cd, $E^\circ_{red} = -0.40$ V. In order to get a negative cell potential, Ni is oxidized so $E^\circ_{red} = -0.23\text{ V} = E^\circ_{anode}$. Cd^{2+} is reduced so $E^\circ_{red} = -0.23\text{ V} = E^\circ_{cathode}$. Then $E^\circ_{cell} = E^\circ_{cathode} - E^\circ_{anode} = -0.40\text{ V} - (-0.23\text{ V}) = -0.17$ V. Since oxidation occurs at the anode, the Ni is the anode and the reaction is Ni (*s*) → Ni^{2+} (*aq*) + 2 e^-. Since reduction takes place at the cathode, Cd is the cathode and the reaction is Cd^{2+} (*aq*) + 2 e^- → Cd (*s*).

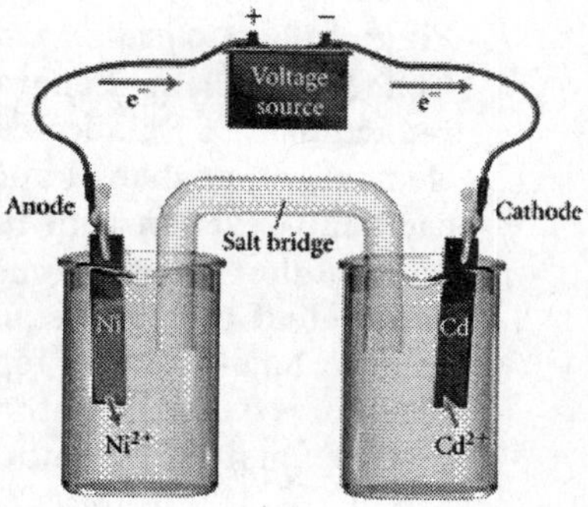

c) Since reduction is occurring at the cathode, the battery terminal closest to the cathode is the negative terminal. Since the cell potential from part a) is = − 0.17 V, a minimum of 0.17 V must be applied by the battery.
Check: The reaction is nonspontaneous, since the reduction of Ni^{2+} is above Cd^{2+}. Electrons still flow from the anode to the cathode. The reaction can be made spontaneous with the application of electrical energy.

55. **Given:** electrolysis cell to electroplate Cu onto a metal surface **Find:** Draw cell, labeling anode and cathode; and write half-reactions
Conceptual Plan: Write 2 half-cell reactions and add electrons as needed to balance reactions. The cathode reaction will be the reduction of Cu^{2+} to the metal. The anode will be the reverse reaction.
Solution:

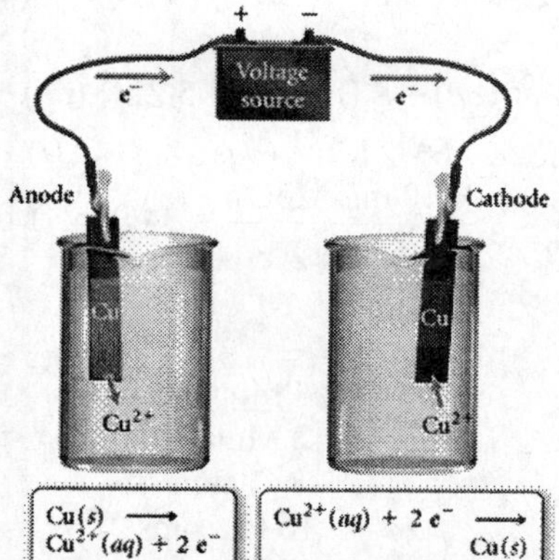

Check: The metal to be plated is the cathode, since metal ions are converted to Cu (*s*) on the surface of the metal.

57. **Given:** Cu electroplating of 225 mg Cu at a current of 7.8 A; Cu^{2+} (*aq*) + 2 e^- → Cu (*s*) **Find:** time
Conceptual Plan: mg Cu → g Cu → mol Cu → mol e⁻ → C → s

$$\frac{1\text{ g}}{1000\text{ mg}} \quad \frac{1\text{ mol Cu}}{63.55\text{ g Cu}} \quad \frac{2\text{ mol e}^-}{1\text{ mol Cu}} \quad \frac{96{,}485\text{ C}}{1\text{ mol e}^-} \quad \frac{1\text{ s}}{7.8\text{ C}}$$

Solution: $$225\text{ mg Cu} \times \frac{1\text{ g Cu}}{1000\text{ mg Cu}} \times \frac{1\text{ mol Cu}}{63.55\text{ g Cu}} \times \frac{2\text{ mol e}^-}{1\text{ mol Cu}} \times \frac{96{,}485\text{ C}}{1\text{ mol e}^-} \times \frac{1\text{ s}}{7.8\text{ C}} = 88\text{ s}.$$

Check: The units (s) are correct. Since far less than a mole of Cu is electroplated, the time is short.

59. **Given:** Na electrolysis, 1.0 kg in one hour **Find:** current
Conceptual Plan: Na^+ (*l*) + e^- → Na (*l*) $\frac{\text{kg Na}}{\text{hr}} \rightarrow \frac{\text{g Na}}{\text{hr}} \rightarrow \frac{\text{mol Na}}{\text{hr}} \rightarrow \frac{\text{mol e}^-}{\text{hr}} \rightarrow \frac{\text{C}}{\text{hr}} \rightarrow \frac{\text{C}}{\text{min}} \rightarrow \frac{\text{C}}{\text{s}}$

$$\frac{1000\text{ g}}{1\text{ kg}} \quad \frac{1\text{ mol Na}}{22.99\text{ g Na}} \quad \frac{1\text{ mol e}^-}{1\text{ mol Na}} \quad \frac{96{,}485\text{ C}}{1\text{ mol e}^-} \quad \frac{1\text{ hr}}{60\text{ min}} \quad \frac{1\text{ min}}{60\text{ s}}$$

Solution:
$$\frac{1.0\text{ kg Na}}{1\text{ hr}} \times \frac{1000\text{ g Na}}{1\text{ kg Na}} \times \frac{1\text{ mol Na}}{22.99\text{ g Na}} \times \frac{1\text{ mol e}^-}{1\text{ mol Na}} \times \frac{96{,}485\text{ C}}{1\text{ mol e}^-} \times \frac{1\text{ hr}}{60\text{ min}} \times \frac{1\text{ min}}{60\text{ s}} = 1.2 \times 10^3\ \frac{\text{C}}{\text{s}} = 1.2 \times 10^3\text{ A}$$

Check: The units (A) are correct. Since the amount per hour is so large we expect a very large current.

61. **Given:** MnO_4^- (*aq*) + Zn (*s*) → Mn^{2+} (*aq*) + Zn^{2+} (*aq*) 0.500 M $KMnO_4$ and 2.85 g Zn
Find: balance equation and volume $KMnO_4$ solution
Conceptual Plan: Separate the overall reaction into two half-reactions: one for oxidation and one for reduction. → Balance each half-reaction with respect to mass in the following order: 1) balance all elements other than H and O; 2) balance O by adding H_2O; and 3) balance H by adding H^+. → Balance each half-reaction with respect to charge by adding electrons. (The sum of the charges on both sides of the equation should be made equal by adding electrons as necessary.) → Make the number of electrons in both half-reactions equal by multiplying one or both half-reactions by a small whole number. → Add the two half-reactions together, canceling electrons and other species as necessary. → Verify that the reaction is balanced both with respect to mass and with respect to charge. Then
g Zn → mol Zn → mol MnO_4^- → L MnO_4^- → mL MnO_4^-

$$\frac{1\text{ mol Zn}}{65.41\text{ g Zn}} \quad \frac{2\text{ mol MnO}_4^-}{5\text{ mol Zn}} \quad \frac{1\text{ L MnO}_4^-}{0.500\text{ mol MnO}_4^-} \quad \frac{1000\text{ mL MnO}_4^-}{1\text{ L MnO}_4^-}$$

Solution:

Separate: MnO_4^- (*aq*) → Mn^{2+} (*aq*) and Zn (*s*) → Zn^{2+} (*aq*)
Balance non H & O elements: MnO_4^- (*aq*) → Mn^{2+} (*aq*) and Zn (*s*) → Zn^{2+} (*aq*)
Balance O with H_2O: MnO_4^- (*aq*) → Mn^{2+} (*aq*) + 4 H_2O (*l*) and Zn (*s*) → Zn^{2+} (*aq*)
Balance H with H^+: MnO_4^- (*aq*) + 8 H^+ (*aq*) → Mn^{2+} (*aq*) + 4 H_2O (*l*) and Zn (*s*) → Zn^{2+} (*aq*)
Add electrons: MnO_4^- (*aq*) + 8 H^+ (*aq*) + 5 e^- → Mn^{2+} (*aq*) + 4 H_2O (*l*) and Zn (*s*) → Zn^{2+} (*aq*) + 2 e^-
Equalize electrons:
2 MnO_4^- (*aq*) + 16 H^+ (*aq*) + 10 e^- → 2 Mn^{2+} (*aq*) + 8 H_2O (*l*) and 5 Zn (*s*) → 5 Zn^{2+} (*aq*) + 10 e^-
Add half-reactions:
2 MnO_4^- (*aq*) + 16 H^+ (*aq*) + 10 e^- + 5 Zn (*s*) → 2 Mn^{2+} (*aq*) + 8 H_2O (*l*) + 5 Zn^{2+} (*aq*) + 10 e^-
Cancel electrons: 2 MnO_4^- (*aq*) + 16 H^+ (*aq*) +5 Zn (*s*) → 2 Mn^{2+} (*aq*) + 8 H_2O (*l*) + 5 Zn^{2+} (*aq*)

$$2.85\text{ g Zn} \times \frac{1\text{ mol Zn}}{65.41\text{ g Zn}} \times \frac{2\text{ mol MnO}_4^-}{5\text{ mol Zn}} \times \frac{1\text{ L MnO}_4^-}{0.500\text{ mol MnO}_4^-} \times \frac{1000\text{ mL MnO}_4^-}{1\text{ L MnO}_4^-} = 34.9\text{ mL MnO}_4^- =$$

= 34.9 mL $KMnO_4$.

Check:

Reactants	Products
2 Mn atoms	2 Mn atoms
8 O atoms	8 O atoms
16 H atoms	16 H atoms
5 Zn atoms	5 Zn atoms
+14 charge	+14 charge

The units (mL) are correct. Since far less than a mole of zinc is used, less than a mole of permanganate is consumed, so the volume is less than a liter.

63. **Given:** beaker with Al strip and Zn^{2+} ions **Find:** draw sketch after Al is submerged for a few minutes
Conceptual Plan: Write 2 half-cell reactions and add electrons as needed to balance reactions. Look up half-reactions in Table 18.1. Calculate the standard cell potential by subtracting the electrode potential of the anode from the electrode potential of the cathode: $E^\circ_{cell} = E^\circ_{cathode} - E^\circ_{anode}$. If $E^\circ_{cell} > 0$ the reaction is spontaneous in the forward direction and Al will dissolve and Cu will deposit.
Solution: Al (s) → Al^{3+} (aq) and Cu^{2+} (aq) → Cu (s) add electrons Al (s) → Al^{3+} (aq) + 3 e^- and Cu^{2+} (aq) + 2 e^- → Cu (s). Look up cell potentials. Al is oxidized so $E^\circ_{ox} = -E^\circ_{red} = -(-1.66\text{ V}) = 1.66$ V. Cu^{2+} is reduced so $E^\circ_{red} = 0.34$ V. Then $E^\circ_{cell} = E^\circ_{ox} + E^\circ_{red} = 1.66\text{ V} + 0.34\text{ V} = +2.00$ V and so the reaction is spontaneous. Al will dissolve to generate Al^{3+} (aq) and Cu (s) will deposit.

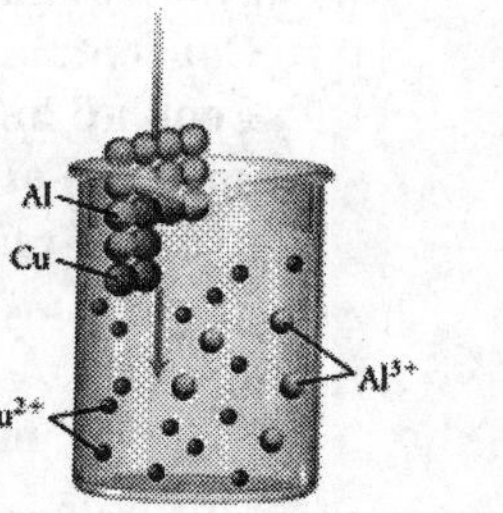

Check: The units (V) are correct. If the voltage is positive, the reaction is spontaneous so Al will dissolve and Cu will deposit.

65. **Given:** a) 2.15 g Al; b) 4.85 g Cu; and c) 2.42 g Ag in 3.5 M HI
Find: if metal dissolves, write balanced reaction and minimum amount of HI needed to dissolve metal
Conceptual Plan: In general, metals whose reduction half-reactions lie below the reduction of H^+ to H_2 in Table 18.1 will dissolve in acids, while metals above it will not. Stop here if metal does not dissolve. To write the balanced redox reactions, pair the oxidation of the metal with the reduction of H^+ to H_2 (2 H^+(aq) + 2 e^- → H_2(g)). Balance the number of electrons transferred. Add the two reactions. Cancel electrons. Then g metal → mol metal → mol H^+ → L HI → mL HI

$$\mathfrak{M} \qquad \frac{x\text{ mol H}^+}{y\text{ mol metal}} \qquad \frac{1\text{ L HI}}{3.5\text{ mol HI}} \qquad \frac{1000\text{ mL HI}}{1\text{ L HI}}$$

Solution: a) Al meets this criterion. For Al, Al (s) → Al^{3+} (aq) + 3 e^-. We need to multiply the Al reaction by 2 and the H^+ reaction by 3. So, 2 Al (s) → 2 Al^{3+} (aq) + 6 e^- and 6 H^+ (aq) + 6 e^- → 3 H_2 (g). Adding the half-reactions together: 2 Al (s) + 6 H^+ (aq) + $\cancel{6\,e^-}$ → 2 Al^{3+} (aq) + $\cancel{6\,e^-}$ + 3 H_2 (g). Simplify to 2 Al (s) + 6 H^+ (aq) → 2 Al^{3+} (aq) + 3 H_2 (g). Then

$$2.15\,\cancel{\text{g Al}} \times \frac{1\,\cancel{\text{mol Al}}}{26.98\,\cancel{\text{g Al}}} \times \frac{6\,\cancel{\text{mol H}^+}}{2\,\cancel{\text{mol Al}}} \times \frac{1\,\cancel{\text{L HI}}}{3.5\,\cancel{\text{mol HI}}} \times \frac{1000\text{ mL HI}}{1\,\cancel{\text{L HI}}} = 68.3\text{ mL HI}$$

b) Cu does not meet this criterion, so it will not dissolve in HI.

c) Ag does not meet this criterion, so it will not dissolve in HI.
Check: Only metals with negative reduction potentials will dissolve. The volume of acid needed is fairly small since the amount of metal is much less than 1 mole and the concentration of acid is high.

67. **Given:** Pt (s)|H_2 (g, 1 atm)|H^+ (aq, ? M)||Cu^{2+} (aq, 1.0 M)|Cu (s), $E_{cell} = 355$ mV **Find:** pH
Conceptual Plan: Write half-reactions from line notation. Look up half-reactions in Table 18.1. Calculate the standard cell potential by subtracting the electrode potential of the anode from the electrode potential of the cathode: $E^\circ_{cell} = E^\circ_{cathode} - E^\circ_{anode}$. Add the two half-cell reactions and cancel the electrons. Then mV → V E°_{cell}, E_{cell}, P_{H2}, $[Cu^{2+}]$, n → $[H^+]$ → pH

$$\frac{1\text{ V}}{1000\text{ mV}} \qquad E_{cell} = E^\circ_{cell} - \frac{0.0592\text{ V}}{n}\log Q \qquad \text{pH} = -\log[H^+]$$

Solution: The half-reactions are: H_2 (g) → 2 H^+(aq) + 2 e^- and Cu^{2+} (aq) + 2 e^- → Cu (s). H is oxidized so $E^\circ_{red} = -0.00\text{ V} = E^\circ_{anode}$. Cu is reduced so $E^\circ_{red} = 0.34\text{ V} = E^\circ_{cathode}$. Then $E^\circ_{cell} = E^\circ_{cathode} - E^\circ_{anode} = 0.34\text{ V} - (-0.00\text{ V}) = +0.34$ V. Adding the two reactions: H_2 (g) + Cu^{2+} (aq) + $\cancel{2\,e^-}$ → 2 H^+ (aq) + $\cancel{2\,e^-}$ + Cu (s).

Cancel the electrons: H_2 (g) + Cu^{2+} (aq) → 2 H^+ (aq) + Cu (s). Then $355\,\cancel{\text{mV}} \times \frac{1\text{ V}}{1000\,\cancel{\text{mV}}} = 0.355\text{ V}$. So n = 2 and $Q = \frac{[H^+]^2}{P_{H2}[Cu^{2+}]} = \frac{(x)^2}{(1)(1.0)} = x^2$ then $E_{cell} = E^0_{cell} - \frac{0.0592\text{ V}}{n}\log Q$ substitute in values and solve for

x. $0.355\text{ V} = 0.34\text{ V} - \frac{0.0592\text{ V}}{2}\log x^2$ → $0.015\text{ V} = -\frac{0.0592\text{ V}}{2}\log x^2$ → $-0.50676 = \log x^2$ →

$x^2 = 10^{-0.50676} = 0.31135$ → $x = 0.55798$ then $\text{pH} = -\log[\text{H}^+] = -\log[0.55798] = 0.25338 = 0.3$.

Check: The units (none) are correct. The pH is acidic, which is consistent with dissolving a metal in acid.

69. **Given:** Mg oxidation and Cu^{2+} reduction; initially $[Mg^{2+}] = 1.0 \times 10^{-4}$ M and $[Cu^{2+}] = 1.5$ M in 1.0 L half-cells
Find: a) initial E_{cell}; b) E_{cell} after 5.0 A for 8.0 hr; and c) how long can battery deliver 5.0A
Conceptual Plan: a) Write the 2 half-cell reactions and add electrons as needed to balance reactions. Look up half-reactions in Table 18.1. Calculate the standard cell potential by subtracting the electrode potential of the anode from the electrode potential of the cathode: $E^\circ_{cell} = E^\circ_{cathode} - E^\circ_{anode}$. Add the two half-cell reactions and cancel electrons and determine n. Then E°_{cell}, $[Mg^{2+}]$, $[Cu^{2+}]$, n → E_{cell}

$$E_{cell} = E^\circ_{cell} - \frac{0.0592\ V}{n}\log Q$$

b) hr → min → s → C → mol e⁻ → mol Cu reduced → $[Cu^{2+}]$ and

$$\frac{60\text{ min}}{1\text{ hr}} \quad \frac{60\text{ s}}{1\text{ min}} \quad \frac{5.0\text{ C}}{1\text{ s}} \quad \frac{1\text{ mol e}^-}{96{,}485\text{ C}} \quad \frac{1\text{ mol Cu}^{2+}}{2\text{ mol e}^-} \quad \text{since V} = 1.0\text{L} \quad [Cu^{2+}] = [Cu^{2+}] - \frac{\text{mol Cu}^{2+}\text{ reduced}}{1.0\text{ L}}$$

mol Cu reduced → mol Mg oxidized → $[Mg^{2+}]$

$$\frac{1\text{ mol Mg oxidized}}{1\text{ mol Cu}^{2+}\text{ reduced}} \quad \text{since V} = 1.0\text{L} \quad [Mg^{2+}] = [Mg^{2+}] + \frac{\text{mol Mg oxidized}}{1.0\text{ L}}$$

c) $[Cu^{2+}]$ → mol e⁻ → C → s → min → hr

$$\frac{1\text{ mol e}^-}{2\text{ mol Cu}^{2+}} \quad \frac{96{,}485\text{ C}}{1\text{ mol e}^-} \quad \frac{1\text{ s}}{5.0\text{ C}} \quad \frac{1\text{ min}}{60\text{ s}} \quad \frac{1\text{ hr}}{60\text{ min}}$$

Solution: a) write half-reactions and add electrons Cu^{2+} (*aq*) + 2 e⁻ → Cu (*s*) and Mg (*s*) → Mg^{2+} (*aq*) + 2 e⁻. Look up cell potentials. Mg is oxidized so $E^\circ_{red} = -2.37\text{ V} = E^\circ_{anode}$. Cu^{2+} is reduced so $E^\circ_{red} = 0.34\text{ V} = E^\circ_{cathode}$. Then $E^\circ_{cell} = E^\circ_{cathode} - E^\circ_{anode} = 0.34\text{ V} - (-2.37\text{ V}) = +2.71\text{ V}$. Add the two half-cell reactions: Cu^{2+} (*aq*) + ~~2e⁻~~ + Mg (*s*) → Cu (*s*) + Mg^{2+} (*aq*) + ~~2e⁻~~. Simplify to Cu^{2+} (*aq*) + Mg (*s*) → Cu (*s*) + Mg^{2+} (*aq*).

So $Q = \frac{[Mg^{2+}]}{[Cu^{2+}]} = \frac{1.0 \times 10^{-4}}{1.5} = 6.6667 \times 10^{-5}$ and $n = 2$ then

$$E_{cell} = E^0_{cell} - \frac{0.0592\text{ V}}{n}\log Q = 2.71\text{ V} - \frac{0.0592\text{ V}}{2}\log 6.6667 \times 10^{-5} = +2.83361\text{ V} = +2.83\text{ V}.$$

b) $8.0\text{ hr} \times \frac{60\text{ min}}{1\text{ hr}} \times \frac{60\text{ s}}{1\text{ min}} \times \frac{5.0\text{ C}}{1\text{ s}} \times \frac{1\text{ mol e}^-}{96{,}485\text{ C}} \times \frac{1\text{ mol Cu}^{2+}}{2\text{ mol e}^-} = 0.74623\text{ mol Cu}^{2+}$ and

$$[Cu^{2+}] = [Cu^{2+}] - \frac{\text{mol Cu}^{2+}\text{ reduced}}{1.0\text{ L}} = 1.5\text{ M} - \frac{0.74623\text{ mol Cu}^{2+}}{1.0\text{ L}} = 0.75377\text{ M Cu}^{2+}\text{ and}$$

$$0.74623\text{ mol Cu}^{2+} \times \frac{1\text{ mol Mg oxidized}}{1\text{ mol Cu}^{2+}\text{ reduced}} = 0.74623\text{ mol Mg oxidized and}$$

$$[Mg^{2+}] = [Mg^{2+}] + \frac{\text{mol Mg oxidized}}{1.0\text{ L}} = 1.0 \times 10^{-4}\ M + \frac{0.74623\text{ mol Mg oxidized}}{1.0\text{ L}} = 0.74633\text{ M Mg}^{2+}$$

$$Q = \frac{[Mg^{2+}]}{[Cu^{2+}]} = \frac{0.74633}{0.75377} = 0.99013\text{ and } n = 2\text{ then}$$

$$E_{cell} = E^0_{cell} - \frac{0.0592\text{ V}}{n}\log Q = 2.71\text{ V} - \frac{0.0592\text{ V}}{2}\log 0.99013 = +2.71013\text{ V} = +2.71\text{ V}.$$

c) In 1.0 L there are initially 1.5 moles of Cu^{2+}. So

$$1.5\text{ mol Cu}^{2+} \times \frac{2\text{ mol e}^-}{1\text{ mol Cu}^{2+}} \times \frac{96{,}485\text{ C}}{1\text{ mol e}^-} \times \frac{1\text{ s}}{5.0\text{ C}} \times \frac{1\text{ min}}{60\text{ s}} \times \frac{1\text{ hr}}{60\text{ min}} = 16\text{ hr}$$

Check: The units (V, V, and hr) are correct. The Cu^{2+} reduction reaction is above the Mg^{2+} reduction reaction, so the standard cell potential will be positive. Having more reactants than products increases the cell potential.

As the reaction proceeds the potential drops. The concentrations drop by ½ in 8 hours (part b), so it is all consumed in 16 hours.

71. **Given:** Cu (*s*)|CuI (*s*)|I^- (*aq*, 1.0 M)||Cu^+ (*aq*, 1.0 M)|Cu (*s*), K_{sp} (CuI) = 1.1 x 10^{-12} **Find:** E_{cell}
Conceptual Plan: Write half-reactions from line notation. Since this is a concentration cell E°_{cell} = 0.00 V. Then K_{sp}, [I^-] → [Cu^+](ox) then E°_{cell}, [Cu^+](ox), [Cu^+](red), *n* → E_{cell}

$$K_{sp} = [Cu^+][I^-] \qquad E_{cell} = E^\circ_{cell} - \frac{0.0592\ V}{n}\log Q$$

Solution: The half-reactions are: Cu (*s*) → Cu^+(*aq*) + e^- and Cu^+ (*aq*) + e^- → Cu (*s*). Since this is a concentration cell E°_{cell} = 0.00 V and *n* = 1. Since K_{sp} = [Cu^+][I^-], rearrange to solve for [Cu^+](ox).

$$[Cu^+](ox) = \frac{K_{sp}}{[I^-]} = \frac{1.1 \times 10^{-12}}{1.0} = 1.1 \times 10^{-12}\ M \text{ then } Q = \frac{[Cu^+](ox)}{[Cu^+](red)} = \frac{1.1 \times 10^{-12}}{1.0} = 1.1 \times 10^{-12} \text{ then}$$

$$E_{cell} = E^0_{cell} - \frac{0.0592\ V}{n}\log Q = 0.00\ V - \frac{0.0592\ V}{1}\log(1.1 \times 10^{-12}) = 0.71\ V.$$

Check: The units (V) are correct. Since [Cu^+](ox) is so low and [Cu^+](red) is high, the Q is very small so the voltage increase is significant.

73. **Given:** a) disproportionation of Mn^{2+} (*aq*) to Mn (*s*) and MnO_2 (*s*); and b) disproportionation of MnO_2 (*s*) to Mn^{2+} (*aq*) and MnO_4^- (*s*) in acidic solution **Find:** ΔG°_{rxn} and *K*
Conceptual Plan: Separate the overall reaction into two half-reactions: one for oxidation and one for reduction. → Balance each half-reaction with respect to mass in the following order: 1) balance all elements other than H and O; 2) balance O by adding H_2O; and 3) balance H by adding H^+. → Balance each half-reaction with respect to charge by adding electrons. (The sum of the charges on both sides of the equation should be made equal by adding electrons as necessary.) → Make the number of electrons in both half-reactions equal by multiplying one or both half-reactions by a small whole number. → Add the two half-reactions together, canceling electrons and other species as necessary. → Verify that the reaction is balanced both with respect to mass and with respect to charge. Look up half-reactions in Table 18.1. Calculate the standard cell potential by subtracting the electrode potential of the anode from the electrode potential of the cathode: $E^\circ_{cell} = E^\circ_{cathode} - E^\circ_{anode}$.
Then calculate ΔG°_{rxn} using $\Delta G^0_{rxn} = -nFE^0_{cell}$. Finally °C → K then ΔG°_{rxn}, *T* → *K*

$$K = 273.15 + °C \qquad \Delta G^0_{rxn} = -RT\ln K$$

Solution:

a) Separate: Mn^{2+} (*aq*) → MnO_2 (*s*) and Mn^{2+} (*aq*) → Mn (*s*)
Balance non H & O elements: Mn^{2+} (*aq*) → MnO_2 (*s*) and Mn^{2+} (*aq*) → Mn (*s*)
Balance O with H_2O: Mn^{2+} (*aq*) + 2 H_2O (*l*) → MnO_2 (*s*) and Mn^{2+} (*aq*) → Mn (*s*)
Balance H with H^+: Mn^{2+} (*aq*) + 2 H_2O (*l*) → MnO_2 (*s*) + 4 H^+ (*aq*) and Mn^{2+} (*aq*) → Mn (*s*)
Add electrons: Mn^{2+} (*aq*) + 2 H_2O (*l*) → MnO_2 (*s*) + 4 H^+ (*aq*) + 2 e^- and Mn^{2+} (*aq*) + 2 e^- → Mn (*s*)
Equalize electrons: Mn^{2+} (*aq*) + 2 H_2O (*l*) → MnO_2 (*s*) + 4 H^+ (*aq*) + 2 e^- and Mn^{2+} (*aq*) + 2 e^- → Mn (*s*)
Add half-reactions: Mn^{2+} (*aq*) + 2 H_2O (*l*) + Mn^{2+} (*aq*) + ~~2 e^-~~ → MnO_2 (*s*) + 4 H^+ (*aq*) + ~~2 e^-~~ + Mn (*s*)
Cancel electrons: 2 Mn^{2+} (*aq*) + 2 H_2O (*l*) → MnO_2 (*s*) + 4 H^+ (*aq*) + Mn (*s*)
Look up cell potentials. Mn is oxidized in the first half-cell reaction so $E^\circ_{ox} = -E^\circ_{red} = -$ 1.21 V. Mn is reduced in the second half-cell reaction so $E^\circ_{red} = -$ 1.18 V. Then $E^\circ_{cell} = E^\circ_{ox} + E^\circ_{red} = -$ 1.21 V − 1.18 V

$$= -\ 2.39\ V.\ n = 2 \text{ so} \qquad \Delta G^0_{rxn} = -nFE^0_{cell} = -2\ \cancel{mol\ e^-} \times \frac{96{,}485\ C}{\cancel{mol\ e^-}} \times -2.39\ V$$

$$= -2 \times 96{,}485\ \cancel{C} \times -2.39\frac{J}{\cancel{C}} = 4.6\underline{1}198 \times 10^5\ J = 461\ kJ \quad \text{and } T = 273.15 + 25\ °C = 298\ K \text{ then}$$

$\Delta G^0_{rxn} = -RT\ln K$ Rearrange to solve for *K*.

$$K = e^{\frac{-\Delta G^0_{rxn}}{RT}} = e^{\frac{-4.6\underline{1}198 \times 10^5\ \cancel{J}}{\left(8.314\frac{\cancel{J}}{\cancel{K}\ mol}\right)(298\ \cancel{K})}} = e^{-18\underline{6}.149} = 1.43 \times 10^{-81}$$

Check:

Reactants	Products
2 Mn atoms	2 Mn atoms
2 O atoms	2 O atoms
4 H atoms	4 H atoms
+4 charge	+4 charge

The units (kJ and none) are correct. If the voltage is negative, the reaction is nonspontaneous and the free energy change is very positive and the equilibrium constant is extremely small.

b) Separate: $MnO_2\ (s) \rightarrow Mn^{2+}\ (aq)$ and $MnO_2\ (s) \rightarrow MnO_4^-\ (aq)$
Balance non H & O elements: $MnO_2\ (s) \rightarrow Mn^{2+}\ (aq)$ and $MnO_2\ (s) \rightarrow MnO_4^-\ (aq)$
Balance O with H_2O: $MnO_2\ (s) \rightarrow Mn^{2+}\ (aq) + 2\ H_2O\ (l)$ and $MnO_2\ (s) + 2\ H_2O\ (l) \rightarrow MnO_4^-\ (aq)$
Balance H with H^+:
$MnO_2\ (s) + 4\ H^+\ (aq) \rightarrow Mn^{2+}\ (aq) + 2\ H_2O\ (l)$ and $MnO_2\ (s) + 2\ H_2O\ (l) \rightarrow MnO_4^-\ (aq) + 4\ H^+\ (aq)$
Add electrons: $MnO_2\ (s) + 4\ H^+(aq) + 2\ e^- \rightarrow Mn^{2+}\ (aq) + 2\ H_2O\ (l)$ and $MnO_2\ (s) + 2\ H_2O\ (l) \rightarrow MnO_4^-\ (aq) + 4\ H^+\ (aq) + 3e^-$
Equalize electrons: $3\ MnO_2\ (s) + 12\ H^+\ (aq) + 6\ e^- \rightarrow 3\ Mn^{2+}\ (aq) + 6\ H_2O\ (l)$ and
$2\ MnO_2\ (s) + 4\ H_2O\ (l) \rightarrow 2\ MnO_4^-\ (aq) + 8\ H^+\ (aq) + 6\ e^-$
Add half-reactions: $3\ MnO_2\ (s) + 4\ \cancel{12\ H^+(aq)} + \cancel{6e^-} + 2\ MnO_2\ (s) + \cancel{4\ H_2O(l)} \rightarrow$
$3\ Mn^{2+}\ (aq) + 2\ \cancel{4}\ H_2O(l) + 2\ MnO_4^-\ (aq) + \cancel{8\ H^+(aq)} + \cancel{6e^-}$
Cancel electrons & species: $5\ MnO_2\ (s) + 4\ H^+\ (aq) \rightarrow 3\ Mn^{2+}\ (aq) + 2\ H_2O\ (l) + 2\ MnO_4^-\ (aq)$
Look up cell potentials. Mn is reduced in the first half-cell reaction so $E^\circ_{red} = 1.21$ V. Mn is oxidized in the second half-cell reaction so $E^\circ_{ox} = -E^\circ_{red} = -1.68$ V. Then $E^\circ_{cell} = E^\circ_{ox} + E^\circ_{red} = 1.21\ V - 1.68\ V = -0.47$ V. $n = 6$ so $\Delta G^0_{rxn} = -nFE^0_{cell} = -6\ \cancel{mol\,e^-} \times \frac{96{,}485\ C}{\cancel{mol\,e^-}} \times -0.47\ V = -6 \times 96{,}485\ \cancel{C} \times -0.47 \frac{J}{\cancel{C}}$
$= 2.7209 \times 10^5\ J = 270\ kJ = 2.7 \times 10^2\ kJ$. $T = 273.15 + 25\ °C = 298\ K$ then $\Delta G^0_{rxn} = -RT \ln K$.

Rearrange to solve for K. $K = e^{\frac{-\Delta G^0_{rxn}}{RT}} = e^{\frac{-2.7209 \times 10^5\ \cancel{J}}{\left(8.314 \frac{\cancel{J}}{\cancel{K}\cdot mol}\right)(298\ \cancel{K})}} = e^{-10.982} = 2.0 \times 10^{-48}$

Check:

Reactants	Products
5 Mn atoms	5 Mn atoms
10 O atoms	10 O atoms
4 H atoms	4 H atoms
+4 charge	+4 charge

The units (kJ and none) are correct. If the voltage is negative, the reaction is nonspontaneous and the free energy change is very positive and the equilibrium constant is extremely small. The voltage is less than in part a) so the free energy change is not as large and the equilibrium constant is not as small.

75. **Given:** Metal, M, 50.9 g/mol, 1.20 g of metal reduced in 23.6 minutes at 6.42 A from molten chloride
Find: empirical formula of chloride
Conceptual Plan: min → s → C → mol e⁻ and g M → mol M then mol e⁻, mol M → charge → MCl_x

$\frac{60\ s}{1\ min}$ $\frac{6.42\ C}{1\ s}$ $\frac{1\ mol\ e^-}{96{,}485\ C}$ $\frac{1\ mol\ M}{50.9\ g\ M}$ $\frac{1\ mol\ e^-}{1\ mol\ M}$

Solution: $23.6\ \cancel{min} \times \frac{60\ \cancel{s}}{1\ \cancel{min}} \times \frac{6.42\ \cancel{C}}{1\ \cancel{s}} \times \frac{1\ mol\ e^-}{96{,}485\ \cancel{C}} = 0.0942190\ mol\ e^-$ and

$1.20\ \cancel{g\,M} \times \frac{1\ mol\ M}{50.9\ \cancel{g\,M}} = 0.0235756\ mol\ M$ then $\frac{0.0942190\ mol\ e^-}{0.02357561\ mol\ M} = 3.99646 \frac{e^-}{M}$ so the empirical formula is MCl_4.
Check: The units (none) are correct. The result was an integer within the error of the measurements. The formula is typical for a metal salt. It could be vanadium, which has a +4 oxidation state.

77. **Given:** hydrogen–oxygen fuel cell; 1.2×10^3 kWh of electricity/month **Find:** V of $H_2\ (g)$ at STP/month
Conceptual Plan: Write half-reactions. Look up half-reactions in Table 18.1. The reaction on the left is the oxidation. Calculate the standard cell potential by subtracting the electrode potential of the anode from the electrode potential of the cathode: $E^\circ_{cell} = E^\circ_{cathode} - E^\circ_{anode}$. Add the two half-cell reactions and cancel the electrons. Then kWh → J → C → mol e⁻ → mol H_2 → V

$\frac{3.60 \times 10^6\ J}{1\ kWh}$ $\frac{1\ C}{0.41\ J}$ $\frac{1\ mol\ e^-}{96{,}485\ C}$ $\frac{2\ mol\ H_2}{4\ mol\ e^-}$ at STP $\frac{22.414\ L}{1\ mol\ H_2}$

Solution: $2\ H_2\ (g) + 4\ OH^-\ (aq) \rightarrow 4\ H_2O\ (l) + 4\ e^-$ where $E^\circ_{red} = -0.83\ V = E^\circ_{anode}$; and $O_2\ (g) + 2\ H_2O\ (l) + 4\ e^- \rightarrow 4\ OH^-\ (aq)$ where $E^\circ_{red} = 0.40\ V = E^\circ_{cathode}$. $E^\circ_{cell} = E^\circ_{cathode} - E^\circ_{anode} = 0.40\ V - (-0.83\ V) = 1.23\ V =$

1.23 J/C and $n = 4$. Net reaction is: $2\ H_2\ (g) + O_2\ (g) \rightarrow 2\ H_2O\ (l)$. Then

$$1.2 \times 10^3\ \text{kWh} \times \frac{3.60 \times 10^6\ \text{J}}{1\ \text{kWh}} \times \frac{1\ \text{C}}{1.23\ \text{J}} \times \frac{1\ \text{mole}^-}{96{,}485\ \text{C}} \times \frac{2\ \text{mol}H_2}{4\ \text{mole}^-} \times \frac{22.414\ \text{L}}{1\ \text{mol}H_2} = 4.1 \times 10^6\ \text{L}.$$

Check: The units (L) are correct. A large volume is expected since we are trying to generate a large amount of electricity.

79. **Given:** Au^{3+}/Au electroplating; surface area = 49.8 cm^2, Au thickness = 1.00×10^{-3} cm, density = 19.3 g/cm^3; at 3.25 A **Find:** time

Conceptual Plan: Write the half-cell reaction and add electrons as needed to balance reactions. Then surface area, thickness → V → g Au → mol Au → mol e$^-$ → C → s

$$V = surface\ area \times thickness \qquad \frac{19.3\ \text{g Au}}{1\ \text{cm}^3\ \text{Au}} \qquad \frac{1\ \text{mol Au}}{196.97\ \text{g Au}} \qquad \frac{3\ \text{mol e}^-}{1\ \text{mol Au}} \qquad \frac{96{,}485\ \text{C}}{1\ \text{mol e}^-} \qquad \frac{1\ \text{s}}{3.25\ \text{C}}$$

Solution: write half-reaction and add electrons $Au^{3+}\ (aq) + 3\ e^- \rightarrow Au\ (s)$.

$V = surface\,area \times thickness = (49.8\ \text{cm}^2)(1.00 \times 10^{-3}\ \text{cm}) = 0.0498\ \text{cm}^3$ then

$$0.0498\ \text{cm}^3\ \text{Au} \times \frac{19.3\ \text{g Au}}{1\ \text{cm}^3\ \text{Au}} \times \frac{1\ \text{mol Au}}{196.97\ \text{g Au}} \times \frac{3\ \text{mole}^-}{1\ \text{mol Au}} \times \frac{96{,}485\ \text{C}}{1\ \text{mole}^-} \times \frac{1\ \text{s}}{3.25\ \text{C}} = 435\ \text{s}.$$

Check: The units (s) are correct. Since the layer is so thin there is far less than a mole of gold, so the time is not very long. In order to be an economical process, it must be fairly quick.

81. **Given:** $C_2O_4^{2-} \rightarrow CO_2$ and $MnO_4^-\ (aq) \rightarrow Mn^{2+}\ (aq)$; 50.1 mL of MnO_4^- to titrate 0.339 g $Na_2C_2O_4$; and 4.62 g U sample titrated by 32.3 mL MnO_4^-; and $UO^{2+} \rightarrow UO_2^{2+}$ **Find:** percent U in sample

Conceptual Plan: Separate the overall reaction into two half-reactions: one for oxidation and one for reduction. → Balance each half-reaction with respect to mass in the following order: 1) balance all elements other than H and O; 2) balance O by adding H_2O; and 3) balance H by adding H^+. → Balance each half-reaction with respect to charge by adding electrons. (The sum of the charges on both sides of the equation should be made equal by adding electrons as necessary.) → Make the number of electrons in both half-reactions equal by multiplying one or both half-reactions by a small whole number. → Add the two half-reactions together, canceling electrons and other species as necessary. → Verify that the reaction is balanced both with respect to mass and with respect to charge. Then mL MnO_4^- → LMnO_4^- and g $Na_2C_2O_4$ → mol $Na_2C_2O_4$ → mol MnO_4^- then

$$\frac{1\ \text{L}\ MnO_4^-}{1000\ \text{mL}\ MnO_4^-} \qquad \frac{1\ \text{mol}\ Na_2C_2O_4}{134.00\ \text{g}\ Na_2C_2O_4} \qquad \frac{2\ \text{mol}\ MnO_4^-}{5\ \text{mol Zn}} \qquad \frac{1\ \text{L}\ MnO_4^-}{0.500\ \text{mol}\ MnO_4^-}$$

L MnO_4^-, mol MnO_4^- → M MnO_4^- then write U half-reactions and balance as above. →

$$M = \frac{\text{mol}\ MnO_4^-}{\text{L}}$$

Make the number of electrons in both half-reactions equal by multiplying one or both half-reactions by a small whole number. → Add the two half-reactions together, canceling electrons and other species as necessary. → Verify that the reaction is balanced both with respect to mass and with respect to charge. Then mL MnO_4^-, M MnO_4^- → mol MnO_4^- → mol U → g U then g U, g sample → % U

$$M = \frac{\text{mol}\ MnO_4^-}{\text{L}} \qquad \frac{5\ \text{mol U}}{2\ \text{mol}\ MnO_4^-} \qquad \frac{238.03\ \text{g U}}{1\ \text{mol U}} \qquad \text{percent U} = \frac{\text{g U}}{\text{g sample}} \times 100\%$$

Solution:

Separate: $MnO_4^-\ (aq) \rightarrow Mn^{2+}\ (aq)$ and $C_2O_4^{2-}\ (aq) \rightarrow CO_2\ (g)$

Balance non H & O elements: $MnO_4^-\ (aq) \rightarrow Mn^{2+}\ (aq)$ and $C_2O_4^{2-}\ (aq) \rightarrow 2\ CO_2\ (g)$

Balance O with H_2O: $MnO_4^-\ (aq) \rightarrow Mn^{2+}\ (aq) + 4\ H_2O\ (l)$ and $C_2O_4^{2-}\ (aq) \rightarrow 2\ CO_2\ (g)$

Balance H with H^+: $MnO_4^-\ (aq) + 8\ H^+\ (aq) \rightarrow Mn^{2+}\ (aq) + 4\ H_2O\ (l)$ and $C_2O_4^{2-}\ (aq) \rightarrow 2\ CO_2\ (g)$

Add electrons: $MnO_4^-\ (aq) + 8\ H^+\ (aq) + 5\ e^- \rightarrow Mn^{2+}\ (aq) + 4\ H_2O\ (l)$ and $C_2O_4^{2-}\ (aq) \rightarrow 2\ CO_2\ (g) + 2\ e^-$

Equalize electrons:

$2\ MnO_4^-\ (aq) + 16\ H^+\ (aq) + 10\ e^- \rightarrow 2\ Mn^{2+}\ (aq) + 8\ H_2O\ (l)$ and $5\ C_2O_4^{2-}\ (aq) \rightarrow 10\ CO_2\ (g) + 10\ e^-$

Add half-reactions:

$2\ MnO_4^-\ (aq) + 16\ H^+\ (aq) + 10\ e^- + 5\ C_2O_4^{2-}\ (aq) \rightarrow 2\ Mn^{2+}\ (aq) + 8\ H_2O\ (l) + 10\ CO_2\ (g) + 10\ e^-$

Cancel electrons: $2\ MnO_4^-\ (aq) + 16\ H^+\ (aq) + 5\ C_2O_4^{2-}\ (aq) \rightarrow 2\ Mn^{2+}\ (aq) + 8\ H_2O\ (l) + 10\ CO_2\ (g)$

then

$$50.1\ \cancel{\text{mL MnO}_4^-} \times \frac{1\ \text{L MnO}_4^-}{1000\ \cancel{\text{mL MnO}_4^-}} = 0.0501\ \text{L MnO}_4^-$$

$$0.339\ \cancel{\text{g Na}_2\text{C}_2\text{O}_4} \times \frac{1\ \cancel{\text{mol Na}_2\text{C}_2\text{O}_4}}{134.00\ \cancel{\text{g Na}_2\text{C}_2\text{O}_4}} \times \frac{2\ \text{mol MnO}_4^-}{5\ \cancel{\text{mol Na}_2\text{C}_2\text{O}_4}} = 0.0010\underline{1}194\ \text{mol MnO}_4^-$$

$$M = \frac{0.0010\underline{1}194\ \text{mol MnO}_4^-}{0.0501\ \text{L}} = 0.020\underline{1}984\ \text{M MnO}_4^-$$

Separate: $MnO_4^-\ (aq) \rightarrow Mn^{2+}\ (aq)$ and $UO^{2+}\ (aq) \rightarrow UO_2^{2+}\ (aq)$

Balance non H & O elements: $MnO_4^-\ (aq) \rightarrow Mn^{2+}\ (aq)$ and $UO^{2+}\ (aq) \rightarrow UO_2^{2+}\ (aq)$

Balance O with H_2O: $MnO_4^-\ (aq) \rightarrow Mn^{2+}\ (aq) + 4\ H_2O\ (l)$ and $UO^{2+}\ (aq) + H_2O\ (l) \rightarrow UO_2^{2+}\ (aq)$

Balance H with H^+:

$MnO_4^-\ (aq) + 8\ H^+\ (aq) \rightarrow Mn^{2+}\ (aq) + 4\ H_2O\ (l)$ and $UO^{2+}\ (aq) + H_2O\ (l) \rightarrow UO_2^{2+}\ (aq) + 2\ H^+\ (aq)$

Add electrons: $MnO_4^-\ (aq) + 8\ H^+\ (aq) + 5\ e^- \rightarrow Mn^{2+}\ (aq) + 4\ H_2O\ (l)$ and

$UO^{2+}\ (aq) + H_2O\ (l) \rightarrow UO_2^{2+}\ (aq) + 2\ H^+\ (aq) + 2\ e^-$

Equalize electrons: $2\ MnO_4^-\ (aq) + 16\ H^+\ (aq) + 10\ e^- \rightarrow 2\ Mn^{2+}\ (aq) + 8\ H_2O\ (l)$ and

$5\ UO^{2+}\ (aq) + 5\ H_2O\ (l) \rightarrow 5\ UO_2^{2+}\ (aq) + 10\ H^+\ (aq) + 10\ e^-$

Add half-reactions: $2\ MnO_4^-\ (aq) + 6\ \cancel{16}\ H^+(aq) + \cancel{10\ e^-} + 5\ UO^{2+}\ (aq) + \cancel{5\ H_2O(l)} \rightarrow$

$2\ Mn^{2+}\ (aq) + 3\ \cancel{5\ H_2O(l)} + 5\ UO_2^{2+}\ (aq) + \cancel{10\ H^+(aq)} + \cancel{10\ e^-}$

Cancel electrons & species:

$2\ MnO_4^-\ (aq) + 6\ H^+\ (aq) + 5\ UO^{2+}\ (aq) \rightarrow 2\ Mn^{2+}\ (aq) + 3\ H_2O\ (l) + 5\ UO_2^{2+}\ (aq)$

$$32.3\ \cancel{\text{mL MnO}_4^-} \times \frac{0.020\underline{1}984\ \cancel{\text{mol MnO}_4^-}}{1000\ \cancel{\text{mL MnO}_4^-}} \times \frac{5\ \cancel{\text{mol U}}}{2\ \cancel{\text{mol MnO}_4^-}} \times \frac{238.03\ \text{g U}}{1\ \cancel{\text{mol U}}} = 0.38\underline{8}232\ \text{g U then}$$

$$\text{percent U} = \frac{\text{g U}}{\text{g sample}} \times 100\% = \frac{0.38\underline{8}232\ \text{g U}}{4.63\ \text{g sample}} \times 100\% = 8.39\ \%.$$

Check:

	Reactants	Products
first reaction	2 Mn atoms	2 Mn atoms
	28 O atoms	28 O atoms
	16 H atoms	16 H atoms
	10 C atoms	10 C atoms
	+4 charge	+4 charge
second reaction	**Reactants**	**Products**
	2 Mn atoms	2 Mn atoms
	13 O atoms	13 O atoms
	6 H atoms	6 H atoms
	5 U atoms	5 U atoms
	+14 charge	+14 charge

The reactions are balanced. The units (%) are correct. The percentage is between 0 and 100 %.

83. a) Looking for anion reductions that are in between the reduction potentials of Cl_2 and Br_2. The only one that meets this criterion is the dichromate ion.

Chapter 19
Radioactivity and Nuclear Chemistry

1. **Conceptual Plan: Begin with the symbol for parent nuclide on the left side of the equation and the symbol for a particle on the right side (except for electron capture). → Equalize the sum of the mass numbers and the sum of the atomic numbers on both sides of the equation by writing the appropriate mass number and atomic number for the unknown daughter nuclide. → Using the periodic table, deduce the identity of the unknown daughter nuclide from the atomic number and write its symbol.**
Solution:
a) U-234 (alpha decay) ${}^{234}_{92}\text{U} \rightarrow {}^{?}_{?}? + {}^{4}_{2}\text{He}$ then ${}^{234}_{92}\text{U} \rightarrow {}^{230}_{90}? + {}^{4}_{2}\text{He}$ then ${}^{234}_{92}\text{U} \rightarrow {}^{230}_{90}\text{Th} + {}^{4}_{2}\text{He}$

b) Th-230 (alpha decay) ${}^{230}_{90}\text{Th} \rightarrow {}^{?}_{?}? + {}^{4}_{2}\text{He}$ then ${}^{230}_{90}\text{Th} \rightarrow {}^{226}_{88}? + {}^{4}_{2}\text{He}$ then ${}^{230}_{90}\text{Th} \rightarrow {}^{226}_{88}\text{Ra} + {}^{4}_{2}\text{He}$

c) Pb-214 (beta decay) ${}^{214}_{82}\text{Pb} \rightarrow {}^{?}_{?}? + {}^{0}_{-1}\text{e}$ then ${}^{214}_{82}\text{Pb} \rightarrow {}^{214}_{83}? + {}^{0}_{-1}\text{e}$ then ${}^{214}_{82}\text{Pb} \rightarrow {}^{214}_{83}\text{Bi} + {}^{0}_{-1}\text{e}$

d) N-13 (positron emission) ${}^{13}_{7}\text{N} \rightarrow {}^{?}_{?}? + {}^{0}_{+1}\text{e}$ then ${}^{13}_{7}\text{N} \rightarrow {}^{13}_{6}? + {}^{0}_{+1}\text{e}$ then ${}^{13}_{7}\text{N} \rightarrow {}^{13}_{6}\text{C} + {}^{0}_{+1}\text{e}$

e) Cr-51 (electron capture) ${}^{51}_{24}\text{Cr} + {}^{0}_{-1}\text{e} \rightarrow {}^{?}_{?}?$ then ${}^{51}_{24}\text{Cr} + {}^{0}_{-1}\text{e} \rightarrow {}^{51}_{23}?$ then ${}^{51}_{24}\text{Cr} + {}^{0}_{-1}\text{e} \rightarrow {}^{51}_{23}\text{V}$
Check: a) 234 = 230 + 4, 92 = 90 + 2, and Thorium is atomic number 90. b) 230 = 226 + 4, 90 = 88 + 2, and Radium is atomic number 88. c) 214 = 214 + 0, 82 = 83 – 1, and Bismuth is atomic number 83. d) 13 = 13 + 0, 7 = 6 + 1, and Carbon is atomic number 6. e) 51 + 0 = 51, 24 – 1 = 23, and Vanadium is atomic number 23.

3. **Given:** Th-232 decay series: α, β, β, α **Find:** balanced decay reactions
Conceptual Plan: Begin with the symbol for parent nuclide on the left side of the equation and the symbol for a particle on the right side (except for electron capture). → Equalize the sum of the mass numbers and the sum of the atomic numbers on both sides of the equation by writing the appropriate mass number and atomic number for the unknown daughter nuclide. → Using the periodic table, deduce the identity of the unknown daughter nuclide from the atomic number and write its symbol. → Use the product of this reaction to write the next reaction.
Solution:
Th-232 (alpha decay) ${}^{232}_{90}\text{Th} \rightarrow {}^{?}_{?}? + {}^{4}_{2}\text{He}$ then ${}^{232}_{90}\text{Th} \rightarrow {}^{228}_{88}? + {}^{4}_{2}\text{He}$ then ${}^{232}_{90}\text{Th} \rightarrow {}^{228}_{88}\text{Ra} + {}^{4}_{2}\text{He}$
Ra-228 (beta decay) ${}^{228}_{88}\text{Ra} \rightarrow {}^{?}_{?}? + {}^{0}_{-1}\text{e}$ then ${}^{228}_{88}\text{Ra} \rightarrow {}^{228}_{89}? + {}^{0}_{-1}\text{e}$ then ${}^{228}_{88}\text{Ra} \rightarrow {}^{228}_{89}\text{Ac} + {}^{0}_{-1}\text{e}$
Ac-228 (beta decay) ${}^{228}_{89}\text{Ac} \rightarrow {}^{?}_{?}? + {}^{0}_{-1}\text{e}$ then ${}^{228}_{89}\text{Ac} \rightarrow {}^{228}_{90}? + {}^{0}_{-1}\text{e}$ then ${}^{228}_{89}\text{Ac} \rightarrow {}^{228}_{90}\text{Th} + {}^{0}_{-1}\text{e}$
Th-228 (alpha decay) ${}^{228}_{90}\text{Th} \rightarrow {}^{?}_{?}? + {}^{4}_{2}\text{He}$ then ${}^{228}_{90}\text{Th} \rightarrow {}^{224}_{88}? + {}^{4}_{2}\text{He}$ then ${}^{228}_{90}\text{Th} \rightarrow {}^{224}_{88}\text{Ra} + {}^{4}_{2}\text{He}$
Thus the decay series is: ${}^{232}_{90}\text{Th} \rightarrow {}^{228}_{88}\text{Ra} + {}^{4}_{2}\text{He}$, ${}^{228}_{88}\text{Ra} \rightarrow {}^{228}_{89}\text{Ac} + {}^{0}_{-1}\text{e}$, ${}^{228}_{89}\text{Ac} \rightarrow {}^{228}_{90}\text{Th} + {}^{0}_{-1}\text{e}$, ${}^{228}_{90}\text{Th} \rightarrow {}^{224}_{88}\text{Ra} + {}^{4}_{2}\text{He}$.
Check: 232 = 228 + 4, 90 = 88 + 2, and Radium is atomic number 88. 228 = 228 + 0, 88 = 89 – 1, and Actinium is atomic number 89. 228 = 228 + 0, 89 = 90 – 1, and Thorium is atomic number 90. 228 = 224 + 4, 90 = 88 + 2, and Radium is atomic number 88.

5. **Conceptual Plan: Equalize the sum of the mass numbers and the sum of the atomic numbers on both sides of the equation by writing the appropriate mass number and atomic number for the unknown species. → Using the periodic table and the list of particles, deduce the identity of the unknown species from the atomic number and write its symbol.**
Solution:
a) ${}^{?}_{?}? \rightarrow {}^{217}_{85}\text{At} + {}^{4}_{2}\text{He}$ becomes ${}^{221}_{87}? \rightarrow {}^{217}_{85}\text{At} + {}^{4}_{2}\text{He}$ then ${}^{221}_{87}\text{Fr} \rightarrow {}^{217}_{85}\text{At} + {}^{4}_{2}\text{He}$

b) ${}^{241}_{94}\text{Pu} \rightarrow {}^{241}_{95}\text{Am} + {}^{?}_{?}?$ becomes ${}^{241}_{94}\text{Pu} \rightarrow {}^{241}_{95}\text{Am} + {}^{0}_{-1}?$ then ${}^{241}_{94}\text{Pu} \rightarrow {}^{241}_{95}\text{Am} + {}^{0}_{-1}\text{e}$

c) ${}^{19}_{11}\text{Na} \rightarrow {}^{19}_{10}\text{Ne} + {}^{?}_{?}?$ becomes ${}^{19}_{11}\text{Na} \rightarrow {}^{19}_{10}\text{Ne} + {}^{0}_{1}?$ then ${}^{19}_{11}\text{Na} \rightarrow {}^{19}_{10}\text{Ne} + {}^{0}_{+1}\text{e}$

d) ${}^{75}_{34}\text{Se} + {}^{?}_{?}? \rightarrow {}^{75}_{33}\text{As}$ becomes ${}^{75}_{34}\text{Se} + {}^{0}_{-1}? \rightarrow {}^{75}_{33}\text{As}$ then ${}^{75}_{34}\text{Se} + {}^{0}_{-1}\text{e} \rightarrow {}^{75}_{33}\text{As}$

Check: a) 221 = 217 + 4, 87 = 85 + 2, and Francium is atomic number 87. b) 241 = 241 + 0, 94 = 95 – 1, and the particle is a beta particle. c) 19 = 19 + 0, 11 = 10 + 1, and the particle is a positron. d) 75 = 75 + 0, 34 – 1 = 33, and the particle is an electron.

7. a) stable, N/Z ratio is close to 1, acceptable for low Z atoms

   b) not stable, N/Z ratio much too high for low Z atom

   c) not stable, N/Z ratio is less than 1, much too low

   d) stable, N/Z ratio is acceptable for this Z

9. Sc, V, and Mn, each have odd numbers of protons. Atoms with an odd number of protons typically have fewer stable isotopes than those with an even number of protons.

11. a) beta decay, since N/Z is too high

    b) positron emission, since N/Z is too low

    c) positron emission, since N/Z is too low

    d) positron emission, since N/Z is too low

13. a) Cs-125, since it is closer to the proper N/Z

    b) Fe-62, since it is closer to the proper N/Z

15. Since the half-life of U-235 is 703 million years, 1/2 will be present after 703 million years, 1/4 will be present after 2 half-lives and 1/8 will be present after 3 half-lives or 2110 million years or 2.11×10^9 years.

17. **Given:** $t_{1/2}$ for isotope decay = 3.8 days; 1.55 g isotope initially **Find:** mass of isotope after 5.5 days
Conceptual plan: radioactive decay implies first order kinetics, $t_{1/2} \rightarrow k$ then

$$t_{1/2} = \frac{0.693}{k}$$

$m_{isotope\ 0}$, t, k → $m_{isotope\ t}$

$$\ln N_t = -kt + \ln N_0$$

Solution: $t_{1/2} = \dfrac{0.693}{k}$ rearrange to solve for k. $k = \dfrac{0.693}{t_{1/2}} = \dfrac{0.693}{3.8 \text{ days}} = 0.18237 \text{ day}^{-1}$. Since

$\ln N_t = -kt + \ln N_0 = -(0.18237 \text{ day}^{-1})(5.5 \text{ day}) + \ln(1.55 \text{ g}) = -0.56478 \rightarrow N_t = e^{-0.56478} = 0.57 \text{ g}$.
Check: The units (g) are correct. The amount is consistent with a time between one and two half-lives.

19. **Given:** F-18 initial decay rate = 1.5×10^5 /s, $t_{1/2}$ for F-18 = 1.83 h **Find:** t to decay rate of 1.0×10^2 /s
Conceptual plan: radioactive decay implies first order kinetics, $t_{1/2} \rightarrow k$ then $Rate_0$, $Rate_t$, $k \rightarrow t$

$$t_{1/2} = \frac{0.693}{k} \qquad \ln\frac{Rate_t}{Rate_0} = -kt$$

Solution: $t_{1/2} = \dfrac{0.693}{k}$ rearrange to solve for k. $k = \dfrac{0.693}{t_{1/2}} = \dfrac{0.693}{1.83 \text{ h}} = 0.378689 \text{ h}^{-1}$. Since $\ln\dfrac{Rate_t}{Rate_0} = -kt$

rearrange to solve for t $t = -\dfrac{1}{k}\ln\dfrac{Rate_t}{Rate_0} = -\dfrac{1}{0.378689 \text{ h}^{-1}}\ln\dfrac{1.0 \times 10^2 \text{ /s}}{1.5 \times 10^5 \text{ /s}} = 19.3 \text{ h}$.
Check: The units (h) are correct. The time is between 10 and 11 half-lives and the rate is just under $1/2^{10}$ of the original amount.

21. **Given:** boat analysis, C-14/C-12 = 72.5 % of living organism **Find:** t **Other:** $t_{1/2}$ for decay of C-14 = 5730 years

Conceptual plan: radioactive decay implies first order kinetics, $t_{1/2} \rightarrow k$ then 72.5 % of $m_{C\text{-}14\ 0}$, $k \rightarrow t$

$$t_{1/2} = \frac{0.693}{k} \qquad \ln N_t = -kt + \ln N_0$$

Solution: $t_{1/2} = \frac{0.693}{k}$ rearrange to solve for k. $k = \frac{0.693}{t_{1/2}} = \frac{0.693}{5730\ \text{yr}} = 1.2\underline{0}942 \times 10^{-4}\ \text{yr}^{-1}$ then

$[\text{C-14}]_t = 0.725\ [\text{C-14}]_0$. Since $\ln m_{C\text{-}14\ t} = -kt + \ln m_{C\text{-}14\ 0}$ rearrange to solve for t.

$$t = -\frac{1}{k}\ln\frac{m_{C\text{-}14\ t}}{m_{C\text{-}14\ 0}} = -\frac{1}{1.2\underline{0}942 \times 10^{-4}\ \text{yr}^{-1}}\ln\frac{0.725\ \cancel{m_{C\text{-}14\ 0}}}{\cancel{m_{C\text{-}14\ 0}}} = 2.66 \times 10^3\ \text{yr}.$$

Check: The units (yr) are correct. The time to 72.5 % decay is consistent a time less than one half-life.

23. **Given:** skull analysis, C-14 decay rate = 15.3 dis/min·gC in living organisms and 0.85 dis/min·gC in skull
Find: t **Other:** $t_{1/2}$ for decay of C-14 = 5730 years
Conceptual plan: radioactive decay implies first order kinetics, $t_{1/2} \rightarrow k$ then Rate_0, Rate_t, $k \rightarrow t$

$$t_{1/2} = \frac{0.693}{k} \qquad \ln\frac{\text{Rate}_t}{\text{Rate}_0} = -kt$$

Solution: $t_{1/2} = \frac{0.693}{k}$ rearrange to solve for k. $k = \frac{0.693}{t_{1/2}} = \frac{0.693}{5730\ \text{yr}} = 1.2\underline{0}942 \times 10^{-4}\ \text{yr}^{-1}$

Since $\ln\frac{\text{Rate}_t}{\text{Rate}_0} = -kt$, rearrange to solve for t.

$$t = -\frac{1}{k}\ln\frac{\text{Rate}_t}{\text{Rate}_0} = -\frac{1}{1.2\underline{0}942 \times 10^{-4}\ \text{yr}^{-1}}\ln\frac{0.85\ \cancel{\text{dis/min}\cdot\text{gC}}}{15.3\ \cancel{\text{dis/min}\cdot\text{gC}}} = 2.39 \times 10^4\ \text{yr}.$$

Check: The units (yr) are correct. The rate is 6 % of initial value and the time is consistent a time just more than four half-lives.

25. **Given:** rock analysis, 0.438 g Pb-206 to every 1.00 g U-238, no Pb-206 initially **Find:** age of rock
Other: $t_{1/2}$ for decay of U-238 to Pb-206 = 4.5 x 10^9 years
Conceptual plan: radioactive decay implies first order kinetics, $t_{1/2} \rightarrow k$ then

$$t_{1/2} = \frac{0.693}{k}$$

g Pb-206 → mol Pb-206 → mol U-238 → g U-238 then $m_{U\text{-}238\ 0}$, $m_{U\text{-}238\ t}$, $k \rightarrow t$

$$\frac{1\ \text{mol Pb-206}}{206\ \text{g Pb-206}} \qquad \frac{1\ \text{mol U-238}}{1\ \text{mol Pb-206}} \qquad \frac{238\ \text{g U-238}}{1\ \text{mol U-238}} \qquad \ln N_t = -kt + \ln N_0$$

Solution: $t_{1/2} = \frac{0.693}{k}$ rearrange to solve for k. $k = \frac{0.693}{t_{1/2}} = \frac{0.693}{4.5 \times 10^9\ \text{yr}} = 1.\underline{5}4 \times 10^{-10}\ \text{yr}^{-1}$ then

$$0.438\ \cancel{\text{g Pb-206}} \times \frac{1\ \cancel{\text{mol Pb-206}}}{206\ \cancel{\text{g Pb-206}}} \times \frac{1\ \cancel{\text{mol U-238}}}{1\ \cancel{\text{mol Pb-206}}} \times \frac{238\ \text{g U-238}}{1\ \cancel{\text{mol U-238}}} = 0.50\underline{6}039\ \text{g U-238}.$$ Since

$ln\frac{m_{U\text{-}238\ t}}{m_{U\text{-}238\ 0}} = -kt$, rearrange to solve for t.

$$t = -\frac{1}{k}\ln\frac{m_{U\text{-}238\ t}}{m_{U\text{-}238\ 0}} = -\frac{1}{1.\underline{5}4 \times 10^{-10}\ \text{yr}^{-1}}\ln\frac{1.00\ \cancel{\text{g U-238}}}{(1.00 + 0.50\underline{6}039)\ \cancel{\text{g U-238}}} = 2.7 \times 10^9\ \text{yr}.$$

Check: The units (yr) are correct. The amount of Pb-206 is less than half of the initial U-238 amount and time is less than one half-life.

27. **Given:** U-235 fission induced by neutrons to Xe-144 and Sr-90 **Find:** number of neutrons produced
Conceptual Plan: Write species given on the appropriate side of the equation. → Equalize the sum of the mass numbers and the sum of the atomic numbers on both sides of the equation by writing the stoichiometric coefficient in front of the desired species.
Solution: ${}^{235}_{92}\text{U} + {}^{1}_{0}\text{n} \rightarrow {}^{144}_{54}\text{Xe} + {}^{90}_{38}\text{Sr} + ?\,{}^{1}_{0}\text{n}$ becomes ${}^{235}_{92}\text{U} + {}^{1}_{0}\text{n} \rightarrow {}^{144}_{54}\text{Xe} + {}^{90}_{38}\text{Sr} + 2\,{}^{1}_{0}\text{n}$ so 2 neutrons are produced.
Check: 235 + 1 = 144 + 90 + 2, 92 + 0 = 54 + 38 + 0, and no other particle is necessary to balance the equation.

29. **Given:** fusion of 2 H-2 atoms to form He-3 and 1 neutron **Find:** balanced equation
Conceptual Plan: Write species given on the appropriate side of the equation. →Equalize the sum of the mass numbers and the sum of the atomic numbers on both sides of the equation by writing the stoichiometric coefficient in front of the desired species.
Solution: $2\,{}^{2}_{1}\text{H} \rightarrow {}^{3}_{2}\text{He} + {}^{1}_{0}\text{n}$.
Check: 2(2) = 3 + 1, 2(1) = 2 + 0, and no other particle is necessary to balance the equation.

31. **Given:** U-238 bombarded by neutrons to form U-239 which undergoes 2 beta decays to form Pu-239
Find: balanced equations
Conceptual Plan: Write species given on the appropriate side of the equation. →Equalize the sum of the mass numbers and the sum of the atomic numbers on both sides of the equation by writing the stoichiometric coefficient in front of the desired species. → Use the product of this reaction to write the next reaction until process is complete.
Solution: ${}^{238}_{92}\text{U} + ?\,{}^{1}_{0}\text{n} \rightarrow {}^{239}_{92}\text{U}$ becomes ${}^{238}_{92}\text{U} + {}^{1}_{0}\text{n} \rightarrow {}^{239}_{92}\text{U}$ then
beta decay ${}^{239}_{92}\text{U} \rightarrow {}^{?}_{?}? + {}^{0}_{-1}\text{e}$ becomes ${}^{239}_{92}\text{U} \rightarrow {}^{239}_{93}? + {}^{0}_{-1}\text{e}$ then ${}^{239}_{92}\text{U} \rightarrow {}^{239}_{93}\text{Np} + {}^{0}_{-1}\text{e}$ then
beta decay ${}^{239}_{93}\text{Np} \rightarrow {}^{?}_{?}? + {}^{0}_{-1}\text{e}$ becomes ${}^{239}_{93}\text{Np} \rightarrow {}^{239}_{94}? + {}^{0}_{-1}\text{e}$ then ${}^{239}_{93}\text{Np} \rightarrow {}^{239}_{94}\text{Pu} + {}^{0}_{-1}\text{e}$.
The entire process is ${}^{238}_{92}\text{U} + {}^{1}_{0}\text{n} \rightarrow {}^{239}_{92}\text{U}$, ${}^{239}_{92}\text{U} \rightarrow {}^{239}_{93}\text{Np} + {}^{0}_{-1}\text{e}$, ${}^{239}_{93}\text{Np} \rightarrow {}^{239}_{94}\text{Pu} + {}^{0}_{-1}\text{e}$.
Check: 238 + 1 = 239, 92 + 0 = 92, and no other particle is necessary to balance the equation. 239 = 239 + 0, 92 = 93 – 1, and Neptunium is atomic number 93. 239 = 239 + 0, 93 = 94 – 1, and Plutonium is atomic number 94.

33. **Given:** 1.0 g of matter converted to energy **Find:** energy
Conceptual plan: g → kg → E

$$\frac{1\,\text{kg}}{1000\,\text{g}} \qquad E = mc^2$$

Solution: $1.0\,\text{g} \times \frac{1\,\text{kg}}{1000\,\text{g}} = 0.0010\,\text{kg}$ then $E = mc^2 = (0.0010\,\text{kg})\left(2.9979 \times 10^{8}\,\frac{\text{m}}{\text{s}}\right)^2 = 9.0 \times 10^{13}\,\text{J}$

Check: The units (J) are correct. The magnitude of the answer makes physical sense because we are converting a large quantity of amus to energy.

35. **Given:** a) O-16 = 15.9949145 amu; b) Ni-58 = 57.935346 amu; and c) Xe-129 = 128.904780 amu
Find: mass defect and nuclear binding energy per nucleon
Conceptual plan: ${}^{A}_{Z}\mathbf{X}$**, isotope mass → mass defect → nuclear binding energy per nucleon**

$$\text{mass defect} = Z(\text{mass } {}^{1}_{1}\text{H}) + (A - Z)(\text{mass } {}^{1}_{0}\text{n}) - \text{mass of isotope} \qquad \frac{931.5\,\text{MeV}}{(1\,\text{amu})(A\,\text{nucleons})}$$

Solution: mass defect = $Z(\text{mass } {}^{1}_{1}\text{H}) + (A - Z)(\text{mass } {}^{1}_{0}\text{n}) - \text{mass of isotope}$.

a) O-16 mass defect = 8(1.00783 amu) + (16 – 8)(1.00866 amu) – 15.9949145 amu = 0.1370055 amu
= 0.13701 amu and $0.1370055\,\text{amu} \times \frac{931.5\,\text{MeV}}{(1\,\text{amu})(16\,\text{nucleons})} = 7.976\,\frac{\text{MeV}}{\text{nucleons}}$.

b) Ni-58 mass defect = 28(1.00783 amu) + (58 – 28)(1.00866 amu) – 57.935346 amu = 0.543694 amu
= 0.54369 amu and $0.543694\,\text{amu} \times \frac{931.5\,\text{MeV}}{(1\,\text{amu})(58\,\text{nucleons})} = 8.732\,\frac{\text{MeV}}{\text{nucleons}}$.

c) Xe-129 mass defect = 54(1.00783 amu) + (129 – 54)(1.00866 amu) – 128.904780 amu
= 1.16754 amu and $1.16754\,\text{amu} \times \frac{931.5\,\text{MeV}}{(1\,\text{amu})(129\,\text{nucleons})} = 8.431\,\frac{\text{MeV}}{\text{nucleons}}$.

Check: The units (amu and MeV/nucleon) are correct. The mass defect increases with an increasing number of nucleons, but the MeV/nucleon does not change by as much (on a relative basis).

37. **Given:** ${}^{235}_{92}\text{U} + {}^{1}_{0}\text{n} \rightarrow {}^{144}_{54}\text{Xe} + {}^{90}_{38}\text{Sr} + 2\,{}^{1}_{0}\text{n}$, U-235 = 235.043922 amu, Xe-144 = 143.9385 amu, and Sr-90 = 89.907738 amu **Find:** energy per g of U-235

Conceptual plan: mass of products & reactants → mass defect → mass defect / g of U-235 then

$$\text{mass defect} = \sum \text{mass of reactants} - \sum \text{mass of products} \qquad \frac{\text{mass defect}}{235.043922 \text{ g U-235}}$$

g → kg → E

$$\frac{1\text{kg}}{1000 \text{ g}} \qquad E = m\,c^2$$

Solution: $\text{mass defect} = \sum \text{mass of reactants} - \sum \text{mass of products}$ notice that we can cancel a neutron from each side to get: ${}^{235}_{92}\text{U} \rightarrow {}^{144}_{54}\text{Xe} + {}^{90}_{38}\text{Sr} + {}^{1}_{0}\text{n}$ and

$\text{mass defect} = 235.043922 \text{ g} - (143.9385 \text{ g} + 89.907738 \text{ g} + 1.00866 \text{ g}) = 0.189024 \text{ g}$

then $\dfrac{0.189024 \text{ g}}{235.043922 \text{ g U-235}} \times \dfrac{1\text{kg}}{1000 \text{ g}} = 8.04207 \times 10^{-7} \dfrac{\text{kg}}{\text{g U-235}}$ then

$$E = m\,c^2 = \left(8.04207 \times 10^{-7} \frac{\text{kg}}{\text{g U-235}}\right)\left(2.9979 \times 10^8 \frac{\text{m}}{\text{s}}\right)^2 = 7.228 \times 10^{10} \frac{\text{J}}{\text{g U-235}}.$$

Check: The units (J) are correct. A large amount of energy is expected per gram of fuel in a nuclear reactor.

39. **Given:** $2\,{}^{2}_{1}\text{H} \rightarrow {}^{3}_{2}\text{He} + {}^{1}_{0}\text{n}$, H-2 = 2.014102 amu, and He-3 = 3.016029 amu **Find:** energy per g reactant
Conceptual plan: mass of products & reactants → mass defect → mass defect / g of H-2 then

$$\text{mass defect} = \sum \text{mass of reactants} - \sum \text{mass of products} \qquad \frac{\text{mass defect}}{2(2.014102 \text{ g H-2})}$$

g → kg → E

$$\frac{1\text{kg}}{1000 \text{ g}} \qquad E = m\,c^2$$

Solution: $\text{mass defect} = \sum \text{mass of reactants} - \sum \text{mass of products}$ and

$\text{mass defect} = 2(2.014102 \text{ g}) - (3.016029 \text{ g} + 1.00866 \text{ g}) = 0.003515 \text{ g}$

then $\dfrac{0.003515 \text{ g}}{2(2.014102 \text{ g H-2})} \times \dfrac{1\text{kg}}{1000 \text{ g}} = 8.72597 \times 10^{-7} \dfrac{\text{kg}}{\text{g H-2}}$ then

$$E = m\,c^2 = \left(8.72597 \times 10^{-7} \frac{\text{kg}}{\text{g H-2}}\right)\left(2.9979 \times 10^8 \frac{\text{m}}{\text{s}}\right)^2 = 7.84 \times 10^{10} \frac{\text{J}}{\text{g H-2}}.$$

Check: The units (J) are correct. A large amount of energy is expected per gram of fuel in a nuclear reactor.

41. **Given:** 75 kg human exposed to 32.8 rad and falling from chair **Find:** energy absorbed in each case
Conceptual plan: rad, kg → J and assume d = 0.50 m chair height then mass, d → J

$$\frac{0.01 \text{ J}}{1 \text{ kg body tissue}} \qquad E = F \cdot d = m\,g\,d$$

Solution: $32.8 \text{ rad} = 32.8 \dfrac{0.01\text{J}}{1 \text{ kg body tissue}} \times 75 \text{ kg} = 25 \text{ J}$ and

$$E = F \cdot d = m\,g\,d = 75 \text{ kg} \times 9.8 \frac{\text{m}}{\text{s}^2} \times 0.50 \text{ m} = 370 \text{ kg} \frac{\text{m}^2}{\text{s}^2} = 370 \text{ J}.$$

Check: The units (J and J) are correct. Allowable radiation exposures are low, since the radiation is very ionizing and, thus, damaging to tissue. Falling may have more energy, but it is not ionizing.

43. **Given:** $t_{1/2}$ for F-18 = 1.83 h, 65 % of F-18 makes it to the hospital traveling at 60.0 miles/hour
Find: distance between hospital and cyclotron

Conceptual plan: $t_{1/2} \rightarrow k$ **then** $m_{\text{F-18 }0}, m_{\text{F-18 }t}, k \rightarrow t$ **then h → mi**

$$t_{1/2} = \frac{0.693}{k} \qquad \ln \frac{m_{\text{F-18 }t}}{m_{\text{F-18 }0}} = -k\,t \qquad \frac{60.0 \text{ mi}}{1 \text{ h}}$$

Solution: $t_{1/2} = \dfrac{0.693}{k}$ rearrange to solve for k. $k = \dfrac{0.693}{t_{1/2}} = \dfrac{0.693}{1.83 \text{ h}} = 0.378689 \text{ h}^{-1}$. Since $\ln \dfrac{m_{\text{F-18 }t}}{m_{\text{F-18 }0}} = -k\,t$

rearrange to solve for t $t = -\frac{1}{k}\ln\frac{m_{F\text{-}18\ t}}{m_{F\text{-}18\ 0}} = -\frac{1}{0.378689\ h^{-1}}\ln\frac{0.65\ \cancel{m_{F\text{-}18\ 0}}}{\cancel{m_{F\text{-}18\ 0}}} = 1.1376\ h$. Then

$1.1376\ \cancel{h}\ x\ \frac{60.0\ mi}{1\ \cancel{h}} = 68\ mi$.

Check: The units (mi) are correct. The time less than one half-life, so the distance is less than 1.83 times the speed of travel.

45. **Given:** a) Ru-114, b) Ra-216, c) Zn-58, and d) Ne-31 **Find:** write nuclear equation for most likely decay
Conceptual Plan: Decide on most likely decay mode depending on N/Z (too large = beta decay, too low = positron emission) → Write the symbol for parent nuclide on the left side of the equation and the symbol for a particle on the right side. → Equalize the sum of the mass numbers and the sum of the atomic numbers on both sides of the equation by writing the appropriate mass number and atomic number for the unknown daughter nuclide. → Using the periodic table, deduce the identity of the unknown daughter nuclide from the atomic number and write its symbol.
Solution:

a) Ru-114 will undergo beta decay ${}^{114}_{44}Ru \rightarrow {}^{?}_{?}? + {}^{0}_{-1}e$ then ${}^{114}_{44}Ru \rightarrow {}^{114}_{45}? + {}^{0}_{-1}e$ then ${}^{114}_{44}Ru \rightarrow {}^{114}_{45}Rh + {}^{0}_{-1}e$

b) Ra-216 will undergo positron emission ${}^{216}_{88}Ra \rightarrow {}^{?}_{?}? + {}^{0}_{+1}e$ then ${}^{216}_{88}Ra \rightarrow {}^{216}_{87}? + {}^{0}_{+1}e$ then ${}^{216}_{88}Ra \rightarrow {}^{216}_{87}Fr + {}^{0}_{+1}e$

c) Zn-58 will undergo positron emission ${}^{58}_{30}Zn \rightarrow {}^{?}_{?}? + {}^{0}_{+1}e$ then ${}^{58}_{30}Zn \rightarrow {}^{58}_{29}? + {}^{0}_{+1}e$ then ${}^{58}_{30}Zn \rightarrow {}^{58}_{29}Cu + {}^{0}_{+1}e$

d) Ne-31 will undergo beta decay ${}^{31}_{10}Ne \rightarrow {}^{?}_{?}? + {}^{0}_{-1}e$ then ${}^{31}_{10}Ne \rightarrow {}^{31}_{11}? + {}^{0}_{-1}e$ then ${}^{31}_{10}Ne \rightarrow {}^{31}_{11}Na + {}^{0}_{-1}e$

Check: a) 114 = 114 + 0, 44 = 45 – 1, and Rhodium is atomic number 45. b) 216 = 216 + 0, 88 = 87 + 1, and Francium is atomic number 87. c) 58 = 58 + 0, 30 = 29 + 1, and Copper is atomic number 29. d) 31 = 31 + 0, 10 = 11 – 1, and Sodium is atomic number 11.

47. **Given:** Bi-210, $t_{1/2}$ = 5.0 days, 1.2 g Bi-210, 209.984105 amu, 5.5 % absorbed
Find: beta emissions in 15.5 days and dose (in Ci)
Conceptual plan: $\boldsymbol{t_{1/2} \rightarrow k}$ **then** $\boldsymbol{m_{Bi\text{-}210\ 0}, t, k \rightarrow m_{Bi\text{-}210\ t}}$ **then**

$t_{1/2} = \frac{0.693}{k}$ $\ln N_t = -kt + \ln N_0$

$\boldsymbol{g_0, g_t}$ **→ g decayed → mol decayed → beta decays then day → h → min → s then**

$g_0 - g_t = g\ decayed$ $\frac{1\ mol\ Bi\text{-}210}{209.984105\ g\ Bi\text{-}210}$ $\frac{6.022\ x\ 10^{23}\ beta\ decays}{1\ mol\ Bi\text{-}210}$ $\frac{1\ day}{24\ h}$ $\frac{60\ min}{1\ h}$ $\frac{60\ s}{1\ min}$

beta decays, s → beta decays / s → Ci available → Ci absorbed

take ratio $\frac{1\ Ci}{3.7\ x\ 10^{10}\ \frac{decays}{s}}$ $\frac{5.5\ Ci\ absorbed}{100\ Ci\ emitted}$

Solution: $t_{1/2} = \frac{0.693}{k}$ rearrange to solve for k. $k = \frac{0.693}{t_{1/2}} = \frac{0.693}{5.0\ days} = 0.1386\ day^{-1}$. Since

$\ln m_{Bi\text{-}210\ t} = -kt + \ln m_{Bi\text{-}210\ 0} = -(0.1386\ \cancel{day^{-1}})(13.5\ \cancel{day}) + \ln(1.2\ g) = -1.6888$ →

$m_{Bi\text{-}210\ t} = e^{-1.6888} = 0.18475\ g$. then $g_0 - g_t = g\ decayed = 1.2\ g - 0.18475\ g = 1.0153\ g\ Bi\text{-}210$ then

$1.0153\ \cancel{g\ Bi\text{-}210}\ x\ \frac{1\ \cancel{mol\ Bi\text{-}210}}{209.984105\ \cancel{g\ Bi\text{-}210}}\ x\ \frac{6.022\ x\ 10^{23}\ beta\ decays}{1\ \cancel{mol\ Bi\text{-}210}} = 2.9116\ x\ 10^{21}\ beta\ decays$

$= 2.9\ x\ 10^{21}$ beta decays then $13.5\ \cancel{day}\ x\ \frac{24\ \cancel{h}}{1\ \cancel{day}}\ x\ \frac{60\ \cancel{min}}{1\ \cancel{h}}\ x\ \frac{60\ s}{1\ \cancel{min}} = 1.1664\ x\ 10^{6}\ s$ then

$\frac{2.9116\ x\ 10^{21}\ \cancel{beta\ decays}}{1.1664\ x\ 10^{6}\ \cancel{s}}\ x\ \frac{1\ Ci}{3.7\ x\ 10^{10}\ \frac{\cancel{decays}}{\cancel{s}}} = 6.7466\ x\ 10^{4}\ \cancel{Ci\ emitted}\ x\ \frac{5.5\ Ci\ absorbed}{100\ \cancel{Ci\ emitted}} = 3700\ Ci$.

Check: The units (decays and Ci) are correct. The amount that decays is large since the time is over 3 half-lives and we have a relatively large amount of the isotope. Since the decay is large, the dosage is large.

49. **Given:** Ra-226 (226.05402 amu) decays to Rn-224, $t_{1/2} = 1.6 \times 10^3$ yr, 25.0 g Ra-226, $T = 25.0$ °C, $P = 1.0$ atm
Find: V of Rn-224 gas produced in 5.0 day
Conceptual plan: day → yr then $t_{1/2}$ → k then $m_{Ra\text{-}226\ 0}$, t, k → $m_{Ra\text{-}226\ t}$

$$\frac{1\ \text{yr}}{365.24\ \text{day}} \qquad t_{1/2} = \frac{0.693}{k} \qquad \ln N_t = -kt + \ln N_0$$

then g_0, g_t → g decayed → mol decayed → mol Rn-224 formed then °C → K then

$$g_0 - g_t = \text{g decayed} \qquad \frac{1\ \text{mol Ra-226}}{226.05402\ \text{g Ra-226}} \qquad \frac{1\ \text{mol Rn-224}}{1\ \text{mol Ra-226}} \qquad K = °C + 273.15$$

then P, n, T → V

$$PV = nRT$$

Solution: $5.0\ \text{day} \times \frac{1\ \text{yr}}{365.24\ \text{day}} = 0.01\underline{3}690\ \text{yr}$ then $t_{1/2} = \frac{0.693}{k}$ rearrange to solve for k.

$$k = \frac{0.693}{t_{1/2}} = \frac{0.693}{1.6 \times 10^3\ \text{yr}} = 4.\underline{3}3125 \times 10^{-4}\ \text{yr}^{-1}.\ \text{Since}$$

$$\ln m_{Ra\text{-}226\ t} = -kt + \ln m_{Ra\text{-}226\ 0} = -(4.\underline{3}3125 \times 10^{-4}\ \text{yr}^{-1})(0.01\underline{3}690\ \text{yr}) + \ln(25.0\ \text{g}) = 3.2\underline{1}887 \rightarrow$$

$m_{Ra\text{-}226\ t} = e^{3.2\underline{1}887} = 24.\underline{9}999$ g. Then $mg_0 - mg_t$ = mg decayed = 25.0 g − 24.$\underline{9}$999 g = 0.$\underline{0}$00148 g Ra-226 then

$$0.\underline{0}00148\ \text{g Ra-226} \times \frac{1\ \text{mol Ra-226}}{226.05402\ \text{g Ra-226}} \times \frac{1\ \text{mol Rn-224}}{1\ \text{mol Ra-226}} = \underline{6}.5576 \times 10^{-7}\ \text{mol Rn-224 then}$$

and $T = 25.0\ °C + 273.15 = 298.2$ K, then $PV = nRT$ Rearrange to solve for V.

$$V = \frac{nRT}{P} = \frac{6.5576 \times 10^{-7}\ \text{mol} \times 0.08206\ \frac{\text{L} \cdot \text{atm}}{\text{mol} \cdot \text{K}} \times 298.2\ \text{K}}{1.0\ \text{atm}} = \underline{1}.6047 \times 10^{-5}\ \text{L} = 1.6 \times 10^{-5}\ \text{L}.$$

Two significant figures are reported as requested in problem.
Check: The units (L) are correct. The amount of gas is small since the time is so small compared to the half-life.

51. **Given:** ${}^{0}_{+1}\text{e} + {}^{0}_{-1}\text{e} \rightarrow 2\ {}^{0}_{0}\gamma$ **Find:** a) energy (in kJ/mol) and b) wavelength of gamma ray photons
Conceptual plan:
a) mass of products & reactants → mass defect (g) → kg → kg/mol → E (J/mol) → E (kJ/mol)

$$\text{mass defect} = \sum \text{mass of reactants} - \sum \text{mass of products} \qquad \frac{1\ \text{kg}}{1000\ \text{g}} \qquad \div 2\ \text{mol} \qquad E = mc^2 \qquad \frac{1\ \text{kJ}}{1000\ \text{J}}$$

b) Answer in part a) is for 2 moles of γ, so E (J/2 mol γ) → E (J/ γ photon) → λ

$$\frac{1\ \text{mol photons}}{6.022 \times 10^{23}\ \text{photons}} \qquad E = \frac{hc}{\lambda}$$

Solution: a) mass defect $= \sum \text{mass of reactants} - \sum \text{mass of products} = (0.00055\ \text{g} + 0.00055\ \text{g}) - 0\ \text{g}$

$$= 0.00110\ \text{g then } \frac{0.00110\ \text{g}}{2\ \text{mol}} \times \frac{1\ \text{kg}}{1000\ \text{g}} = 5.50 \times 10^{-7}\ \frac{\text{kg}}{\text{mol}}\ \text{then}$$

$$E = mc^2 = \left(5.50 \times 10^{-7}\ \frac{\text{kg}}{\text{mol}}\right)\left(2.9979 \times 10^8\ \frac{\text{m}}{\text{s}}\right)^2 = 4.9\underline{4}307 \times 10^{10}\ \frac{\text{J}}{\text{mol}} \times \frac{1\ \text{kJ}}{1000\ \text{J}} = 4.94 \times 10^7\ \frac{\text{kJ}}{\text{mol}}.$$

b) $E = 4.9\underline{4}307 \times 10^{10}\ \frac{\text{J}}{\text{mol}\ \gamma} \times \frac{1\ \text{mol}\ \gamma}{6.022 \times 10^{23}\ \gamma\text{photons}} = 8.2\underline{0}835 \times 10^{-14}\ \frac{\text{J}}{\gamma\text{photons}}$. Then $E = \frac{hc}{\lambda}$. Rearrange

to solve for λ. $$\lambda = \frac{hc}{E} = \frac{(6.626 \times 10^{-34}\ \text{J} \cdot \text{s})\left(2.9979 \times 10^8\ \frac{\text{m}}{\text{s}}\right)}{8.2\underline{0}835 \times 10^{-14}\ \frac{\text{J}}{\gamma\text{photons}}} = 2.42 \times 10^{-12}\ \text{m} = 2.42\ \text{pm}.$$

Check: The units (kJ/mol and pm) are correct. A large amount of energy is expected per mole of mass lost. The photon is in the gamma ray region of the electromagnetic spectrum.

53. **Given:** $^{3}He = 3.016030$ amu **Find:** nuclear binding energy per nucleon
Conceptual plan: $^{A}_{Z}X$**, isotope mass → mass defect → nuclear binding energy per nucleon**

$$\text{mass defect} = Z(\text{mass } ^{1}_{1}H) + (A - Z)(\text{mass } ^{1}_{0}n) - \text{mass of isotope} \qquad \frac{931.5 \text{ MeV}}{(1 \text{ amu})(A \text{ nucleons})}$$

Solution: mass defect = $Z(\text{mass } ^{1}_{1}H) + (A - Z)(\text{mass } ^{1}_{0}n) -$ mass of isotope.
He-3 mass defect = 2(1.00783 amu) + (3 – 2)(1.00866 amu) – 3.016030 amu = 0.00829 amu

and $0.00829 \text{ amu} \times \frac{931.5 \text{ MeV}}{1 \text{ amu}} = 7.72 \text{ MeV}$.

Check: The units (MeV) are correct. The number of nucleons is small, so the MeV is not that large.

55. **Given:** $t_{1/2}$ for decay of $^{238}U = 4.5 \times 10^{9}$ years, 1.6 g rock, 29 dis/s all radioactivity from U-238
Find: percent by mass ^{238}U in rock
Conceptual plan:
$t_{1/2}$ **→** k **and s → min → h → day → yr then Rate,** k **→ N → mol** ^{238}U **→ g** ^{238}U

$$t_{1/2} = \frac{0.693}{k} \quad \frac{1 \text{ min}}{60 \text{ s}} \quad \frac{1 \text{ h}}{60 \text{ min}} \quad \frac{1 \text{ day}}{24 \text{ h}} \quad \frac{1 \text{ yr}}{365.24 \text{ day}} \qquad \text{Rate} = k\,N \quad \frac{1 \text{ mol dis}}{6.022 \times 10^{23} \text{ dis}} \quad \frac{238 \text{ g } ^{238}U}{1 \text{ mol } ^{238}U}$$

then g ^{238}U**, g rock → percent by mass** ^{238}U

$$\text{percent by mass } ^{238}U = \frac{\text{g } ^{238}U}{\text{g rock}} \times 100\ \%$$

Solution: $t_{1/2} = \frac{0.693}{k}$ rearrange to solve for k. $k = \frac{0.693}{t_{1/2}} = \frac{0.693}{4.5 \times 10^{9} \text{ yr}} = 1.54 \times 10^{-10} \text{ yr}^{-1}$ and

$$1 \text{ s} \times \frac{1 \text{ min}}{60 \text{ s}} \times \frac{1 \text{ h}}{60 \text{ min}} \times \frac{1 \text{ day}}{24 \text{ h}} \times \frac{1 \text{ yr}}{365.24 \text{ day}} = 3.16889554 \times 10^{-8} \text{ yr}.$$ Rate = $k\,N$ Rearrange to solve for N.

$$N = \frac{\text{Rate}}{k} = \frac{29 \dfrac{\text{dis}}{3.16889554 \times 10^{-8} \text{ yr}}}{1.54 \times 10^{-10} \text{ yr}^{-1}} = 5.9425 \times 10^{18} \text{ dis then}$$

$$5.9425 \times 10^{18} \text{ dis} \times \frac{1 \text{ mol dis}}{6.022 \times 10^{23} \text{ dis}} \times \frac{238 \text{ g } ^{238}U}{1 \text{ mol } ^{238}U} = 2.3486 \times 10^{-3} \text{ g } ^{238}U \text{ then}$$

$$\text{percent by mass } ^{238}U = \frac{\text{g } ^{238}U}{\text{g rock}} \times 100\ \% = \frac{2.3486 \times 10^{-3} \text{ g } ^{238}U}{1.6 \text{ g rock}} \times 100\ \% = 0.15\ \%.$$

Check: The units (%) are correct. The mass percent is low because the dis/s is low.

57. a) **Given:** 72,500 kg Al *(s)* and 10 Al *(s)* + 6 NH_4ClO_4 *(s)* → 4 Al_2O_3 *(s)* + 2 $AlCl_3$ *(s)* + 12 H_2O *(g)* + 3 N_2 *(g)* and 608,000 kg O_2 *(g)* that reacts with hydrogen to form gaseous water
Find: energy generated (ΔH°_{rxn})
Conceptual plan: write balanced reaction for O_2 ***(g)*** **then**

$$\Delta H^{0}_{rxn} = \sum n_p \Delta H^{0}_{f}(products) - \sum n_r \Delta H^{0}_{f}(reactants) \quad \textbf{then}$$

kg → g → mol → energy then add the results from the two reactions

$$\frac{1000 \text{ g}}{1 \text{ kg}} \quad \mathfrak{M} \quad \Delta H^{\circ}_{rxn}$$

Solution:

Reactant/Product	ΔH^{0}_{f} (kJ/mol from Appendix IIB)
Al *(s)*	0.0
NH_4ClO_4 *(s)*	−295
Al_2O_3 *(s)*	−1675.7
$AlCl_3$ *(s)*	−704.2
H_2O *(g)*	− 241.8
N_2 *(g)*	0.0

Be sure to pull data for the correct formula and phase.

$$\Delta H^0_{rxn} = \sum n_p \Delta H^0_f(products) - \sum n_r \Delta H^0_f(reactants)$$

$$= [4(\Delta H^0_f(Al_2O_3\ (s))) + 2(\Delta H^0_f(AlCl_3\ (s))) + 12(\Delta H^0_f(H_2O\ (g))) + 3(\Delta H^0_f(N_2\ (g)))] + $$
$$- [10(\Delta H^0_f(Al\ (s))) + 6(\Delta H^0_f(NH_4ClO_4\ (s)))]$$
$$= [4(-1675.7\ kJ) + 2(-704.2\ kJ) + 12(-241.8\ kJ) + 3(0.0\ kJ)] - [10(0.0\ kJ) + 6(-295\ kJ)]$$
$$= [-11012.8\ kJ] - [-1770.\ kJ]$$
$$= -924\underline{2}.8\ kJ$$

then $72{,}500\ \cancel{kg\ Al} \times \frac{1000\ \cancel{g\ Al}}{1\ \cancel{kg\ Al}} \times \frac{1\ \cancel{mol\ Al}}{26.98\ \cancel{g\ Al}} \times \frac{924\underline{2}.8\ kJ}{10\ \cancel{mol\ Al}} = 2.4\underline{8}3703 \times 10^9\ kJ$.

balanced reaction: $H_2\ (g) + \frac{1}{2}\ O_2\ (g) \rightarrow H_2O\ (g)$ $\Delta H^\circ_{rxn} = \Delta H^\circ_f(H_2O\ (g)) = -241.8$ kJ/mol then

$608{,}000\ \cancel{kg\ O_2} \times \frac{1000\ \cancel{g\ O_2}}{1\ \cancel{kg\ O_2}} \times \frac{1\ \cancel{mol\ O_2}}{32.00\ \cancel{g\ O_2}} \times \frac{241.8\ kJ}{0.5\ \cancel{mol\ O_2}} = 9.1\underline{8}84 \times 10^9\ kJ$. So the total is

$2.4\underline{8}3703 \times 10^9\ kJ + 9.1\underline{8}84 \times 10^9\ kJ = 1.16\underline{7}2103 \times 10^{10}\ kJ = 1.167 \times 10^{10}\ kJ$.

Check: The units (kJ) are correct. The answer is very large because the reactions are very exothermic and the weight of reactants is so large.

b) **Given:** ${}^{1}_{1}H + {}^{-1}_{-1}p + {}^{0}_{+1}e \rightarrow {}^{0}_{0}\gamma$ **Find:** mass of antimatter to give same energy as part a)

Conceptual plan: since the reaction is an annihilation reaction, no matter will be left, so the mass of antimatter is the same as the mass of the hydrogen. so kJ → J → kg → g

$$\frac{1000\ J}{1\ kJ} \qquad E = m\,c^2 \qquad \frac{1000\ g}{1\ kg}$$

Solution: $1.16\underline{7}2103 \times 10^{10}\ \cancel{kJ} \times \frac{1000\ J}{1\ \cancel{kJ}} = 1.16\underline{7}2103 \times 10^{13}\ J$. Since $E = m\,c^2$, rearrange to solve for m.

$$m = \frac{E}{c^2} = \frac{1.16\underline{7}2103 \times 10^{13}\ kg \frac{\cancel{m^2}}{\cancel{s^2}}}{\left(2.9979 \times 10^8\ \frac{\cancel{m}}{\cancel{s}}\right)^2} = 1.299 \times 10^{-4}\ \cancel{kg} \times \frac{1000\ g}{1\ \cancel{kg}} = 0.1299\ g.$$

Check: The units (g) are correct. A small mass is expected since nuclear reactions to generate a large amount of energy.

59. **Given:** ${}^{235}_{92}U \rightarrow {}^{206}_{82}Pb$ and ${}^{232}_{90}Th \rightarrow {}^{206}_{82}Pb$ **Find:** decay series

Conceptual Plan: Write species given on the appropriate side of the equation. →Equalize the sum of the mass numbers and the sum of the atomic numbers on both sides of the equation by writing the stoichiometric coefficient in front of the desired species.

Solution: ${}^{235}_{92}U \rightarrow {}^{?}_{82}Pb + ?{}^{4}_{2}He + ?{}^{0}_{-1}e$ becomes ${}^{235}_{92}U \rightarrow {}^{207}_{82}Pb + 7{}^{4}_{2}He + 4{}^{0}_{-1}e$.

${}^{232}_{90}Th \rightarrow {}^{?}_{82}Pb + ?{}^{4}_{2}He + ?{}^{0}_{-1}e$ becomes ${}^{232}_{90}Th \rightarrow {}^{208}_{82}Pb + 6{}^{4}_{2}He + 4{}^{0}_{-1}e$.

U-235 forms Pb-207 in 7 α-decays and 4 β-decays and Th-232 forms Pb-208 in 6 α-decays and 4 β-decays.

Check: 235 = 207 + 7(4) + 4(0), and 92 = 82 + 7(2) + 4(–1). 232 = 208 + 6(4) + 4(0), and 90 = 82 + 6(2) + 4(–1). The mass of the Pb can be determined because alpha particles are large and need to be included as integer values.

61. 7. Since $1/2^6 = 1.6\%$ and $1/2^7 = 0.8\ \%$.

63. The gamma emitter is a greater threat while you sleep because it can penetrate more tissue. The alpha particles will not penetrate the wall to enter your bedroom. The alpha emitter is a greater threat if you ingest it since it is more ionizing.

Chapter 20
Organic Chemistry

1. a) C_5H_{12} is an alkane, since it follows the general formula C_nH_{2n+2}, where n = 5.

    b) C_3H_6 is an alkene, since it follows the general formula C_nH_{2n}, where n = 3.

    c) C_7H_{12} is an alkyne, since it follows the general formula C_nH_{2n-2}, where n = 7.

    d) $C_{11}H_{22}$ is an alkene, since it follows the general formula C_nH_{2n}, where n = 11.

3. $CH_3-CH_2-CH_2-CH_2-CH_2-CH_2-CH_3$,

$$CH_3-CH-CH_2-CH_2-CH_2-CH_3$$
$$\underset{CH_3}{|}$$
,

$$CH_3-CH_2-CH-CH_2-CH_2-CH_3$$
$$\underset{CH_3}{|}$$
,
$$CH_3-\underset{\underset{CH_3}{|}}{\overset{\overset{CH_3}{|}}{C}}-CH_2-CH_2-CH_3$$
,
$$CH_3-CH-CH-CH_2-CH_3$$
$$\underset{CH_3}{|}\ \underset{CH_3}{|}$$
,

$$CH_3-CH_2-\overset{\overset{CH_3}{|}}{\underset{\underset{CH_3}{|}}{C}}-CH_2-CH_3$$
,
$$CH_3-CH-CH_2-CH-CH_3$$
$$\underset{CH_3}{|}\qquad\underset{CH_3}{|}$$
$$CH_3-CH_2-CH-CH_2-CH_3$$
$$\underset{CH_2-CH_3}{|}$$
, and

$$CH_3-\overset{\overset{CH_3}{|}}{\underset{\underset{CH_3}{|}CH_3}{C}}-CH-CH_3 .$$

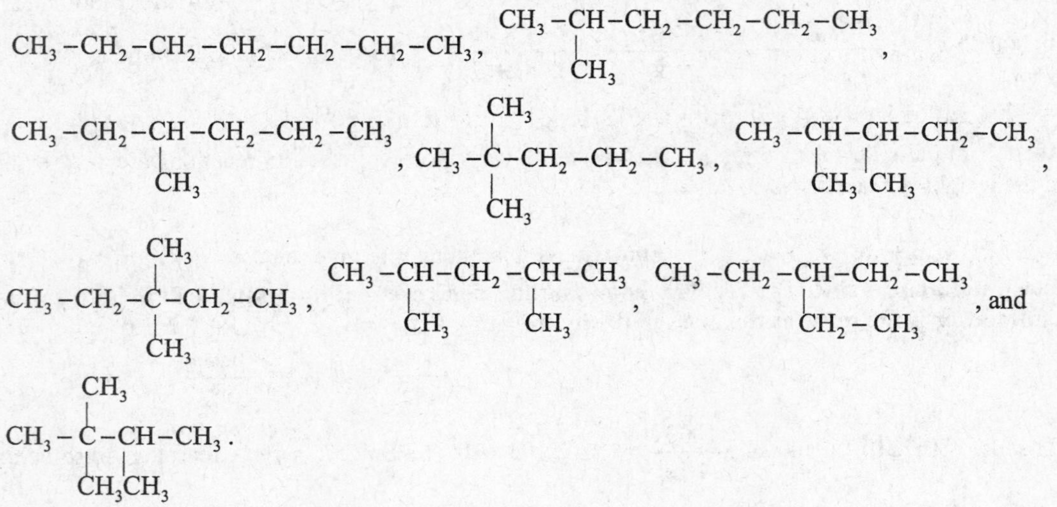

5. a) No, this molecule will not because all four of the substituents are Cl atoms.

    b) Yes, this molecule will because the third carbon has four different substituents groups.

    c) Yes, this molecule will because the second carbon has four different substituents groups.

    d) No, each carbon has at most three different substituents groups.

7. a) They are enantiomers, because they are mirror images of each other.

    b) They are the same, because you can get the second molecule by rotating the first molecule counterclockwise about the C–H bond.

    c) They are enantiomers, because they are mirror images of each other.

9. **Given:** alkane structures   **Find:** name
    Conceptual plan: Count the number of carbon atoms in the longest continuous carbon chain to determine the base name of the compound. Find the prefix corresponding to this number of atoms in Table 20. 5 and add the ending -ane to form the base name. → Consider every branch from the base chain to be a substituent. Name each substituent according to Table 20.6. →Beginning with the end closest to the branching, number the base chain and assign a number to each substituent. (If two substituents occur at equal distances from each end, go to the next substituent to determine from which end to start numbering.) →Write the name of the compound in the following format: (subst. #)-(subst. name)(base name). → If there are two or more substituents, give each one a number and list them alphabetically with hyphens between words and numbers. →If a compound has two or more identical substituents, designate the number of identical substituents with the prefix di- (2), tri- (3), or tetra- (4) before the substituent's name. Separate the numbers indicating the positions of the substituents relative to each other with a comma. The prefixes are not taken into account when alphabetizing.
    Solution:
    a) $CH_3-CH_2-CH_2-CH_2-CH_3$ has 5 carbons as the longest continuous chain. The prefix for 5 is penta-.

There are no substituent groups on any of the carbons, so the name is pentane.

b) $\boxed{CH_3-CH_2-CH-CH_3}$ has 4 carbons as the longest continuous chain. The prefix for 4 is but- and the
$\qquad\qquad\qquad\qquad CH_3$

base name is butane. The only substituent group is a methyl group. $\boxed{C^4H_3-C^3H_2-C^2H-C^1H_3}$ If we start
$\qquad\qquad\qquad\qquad\qquad\qquad\qquad\qquad\qquad\qquad\boxed{CH_3}$

numbering the chain at the end closest to the methyl group, the methyl substituent is assigned the number 2. The name of the compound is 2-methylbutane.

c)
$$CH_3$$
$$|$$
$$CH_3 \qquad CH-CH_3$$
$$| \qquad\qquad |$$
$$\boxed{CH_3-CH-CH_2-CH-CH_2-CH_2-CH_3}$$
has 7 carbons as the longest continuous chain. The prefix for 7

is hept- and the base name is heptane. The substituent groups are methyl and isopropyl groups.
$$\boxed{CH_3}$$
$$|$$
$$\boxed{CH_3} \qquad CH-CH_3$$
$$| \qquad\qquad |$$
$$\boxed{C^1H_3-C^2H-C^3H_2-C^4H-C^5H_2-C^6H_2-C^7H_3}$$
If we start numbering the chain at the end closest to the methyl group, the methyl substituent is assigned the number 2 and the isopropyl group is assigned the number 4. Since i comes before m, the name of the compound is 4-isopropyl-2-methylheptane.

d) $\boxed{CH_3-CH-CH_2-CH-CH_2-CH_3}$ has 6 carbons as the longest continuous chain. The prefix for 6 is
$\qquad\qquad\quad CH_3 \qquad CH_2-CH_3$

hex- and the base name is hexane. The only substituent groups are methyl and ethyl groups.
$\boxed{C^1H_3-C^2H-C^3H_2-C^4H-C^5H_2-C^6H_3}$
$\qquad\qquad\quad\boxed{CH_3}\qquad\quad\boxed{CH_2-CH_3}$ If we start numbering the chain at the end closest to the methyl

group, the methyl substituent is assigned the number 2 and the ethyl group is assigned the number 4. Since e comes before m, the name of the compound is 4-ethyl-2-methylhexane.

11. **Given:** alkane names **Find:** structure
Conceptual plan: Find the number of carbon atoms corresponding to prefix of the base name in Table 20.5. → Draw the base chain and number the carbons from left to right. → Using Table 20.6 and the prefix di- (2), tri- (3), or tetra- (4) before the substituent's name, determine each substituent. → Add the substituent to the proper carbon position in the chain. → Add hydrogen atoms to the base chain so that each carbon has 4 bonds.
Solution:
a) 3-ethylhexane. The base name hexane designates that there are 6 carbon atoms in the base chain. $C^1-C^2-C^3-C^4-C^5-C^6$. 3-ethyl designates that a $-CH_2CH_3$ group in the 3rd position.

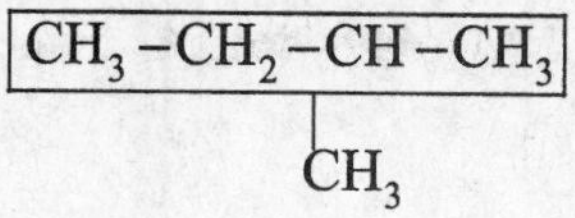

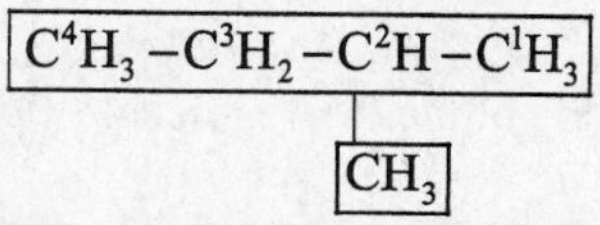

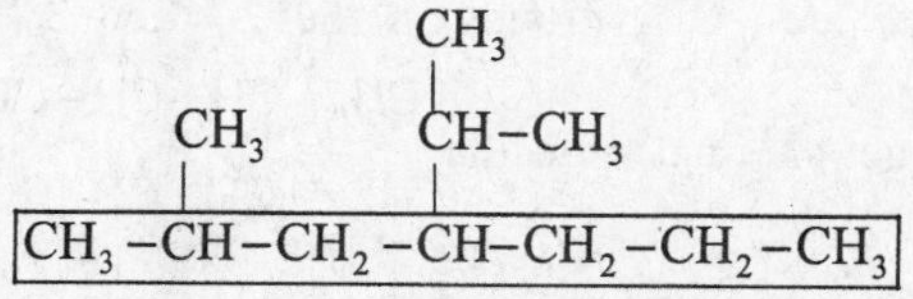

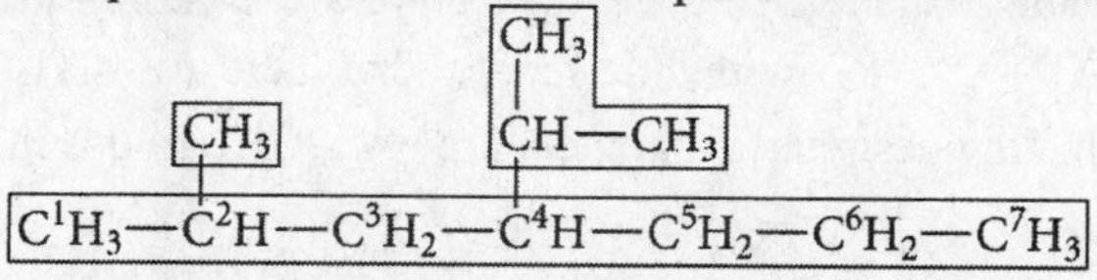

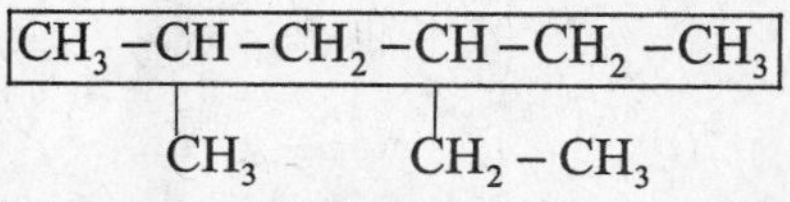

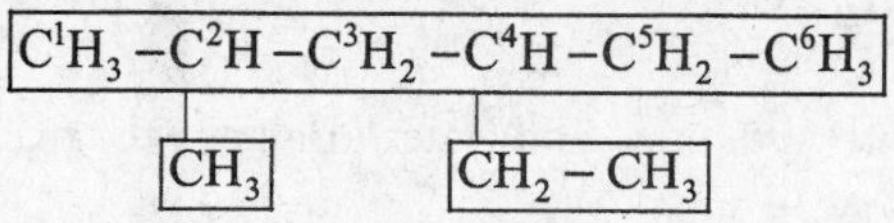

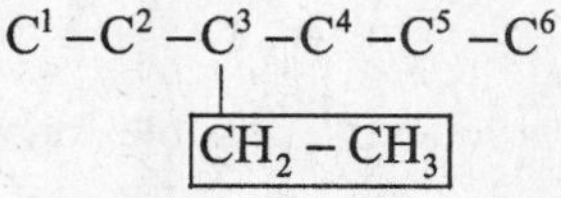

$C^1-C^2-C^3-C^4-C^5-C^6$
$\qquad\qquad |$
$\qquad\quad \boxed{CH_2-CH_3}$. Add hydrogens to the base chain to get the final molecule

$CH_3-CH_2-CH-CH_2-CH_2-CH_3$
$\qquad\qquad\quad |$
$\qquad\qquad CH_2-CH_3$

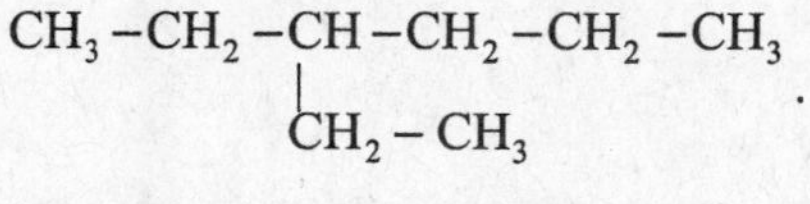

b) 3-ethyl-3-methylpentane. The base name pentane designates that there are 5 carbon atoms in the base chain. $C^1-C^2-C^3-C^4-C^5$. 3-ethyl designates that a $-CH_2CH_3$ group in the 3rd position; and 3-methyl designates that a $-CH_3$ group in the 3rd position .

$$\begin{array}{c} \boxed{CH_3} \\ | \\ C^1-C^2-C^3-C^4-C^5 \\ | \\ \boxed{CH_2-CH_3} \end{array}$$

. Add hydrogens to the base chain

to get the final molecule

$$\begin{array}{c} CH_3 \\ | \\ CH_3-CH_2-CH-CH_2-CH_3 \\ | \\ CH_2-CH_3 \end{array}$$

.

c) 2,3-dimethylbutane. The base name butane designates that there are 4 carbon atoms in the base chain. $C^1-C^2-C^3-C^4$. 2,3-dimethyl designates that there are $-CH_3$ groups in the 2nd and 3rd positions.

$$\begin{array}{c} C^1-C^2-C^3-C^4 \\ \quad | \quad\;\; | \\ \boxed{CH_3}\boxed{CH_3} \end{array}$$

. Add hydrogens to the base chain to get the final molecule

$$\begin{array}{c} CH_3-CH-CH-CH_3 \\ \quad | \quad\;\; | \\ CH_3\;CH_3 \end{array}$$

.

d) 4,7-diethyl-2,2-dimethylnonane. The base name nonane designates that there are 9 carbon atoms in the base chain. $C^1-C^2-C^3-C^4-C^5-C^6-C^7-C^8-C^9$. 4,7-diethyl designates that there are $-CH_2CH_3$ groups in the 4th and 7th positions; and 2,2-dimethyl designates that there are two $-CH_3$ groups in the 2nd position.

$$\begin{array}{c} \boxed{CH_3} \\ | \\ C^1-C^2-C^3-C^4-C^5-C^6-C^7-C^8-C^9 \\ | \qquad\quad | \qquad\qquad | \\ \boxed{CH_3}\quad \boxed{CH_2-CH_3}\;\boxed{CH_2-CH_3} \end{array}$$

. Add hydrogens to the base chain to get the final molecule

$$\begin{array}{c} CH_3 \\ | \\ CH_3-C-CH_2-CH-CH_2-CH_2-CH-CH_2-CH_3 \\ | \qquad\quad | \qquad\qquad\quad | \\ CH_3 \quad CH_2-CH_3 \quad\; CH_2-CH_3 \end{array}$$

.

13. Hydrocarbon combustion in the presence of oxygen forms carbon dioxide and water. Balance the reaction.
a) $CH_3CH_2CH_3\ (g) + 5\ O_2\ (g) \rightarrow 3\ CO_2\ (g) + 4\ H_2O\ (g)$.

b) $CH_3CH_2CH{=}CH_2\ (g) + 6\ O_2\ (g) \rightarrow 4\ CO_2\ (g) + 4\ H_2O\ (g)$.

c) $2\ CH{\equiv}CH\ (g) + 5\ O_2\ (g) \rightarrow 4\ CO_2\ (g) + 2\ H_2O\ (g)$.

15. Halogen substitution reactions remove a hydrogen atom from the alkane and replace it with a halogen atom and generate a hydrohalic acid. Assume one substitution on the hydrocarbon.

a) $CH_3CH_3 + Br_2 \rightarrow CH_3CH_2Br + HBr$. Only one carbon-containing product is possible since the C–C bond freely rotates.

b) $CH_3CH_2CH_3 + Cl_2 \rightarrow [CH_3CH_2CH_2Cl \text{ and } CH_3CHClCH_3] + HCl$. Two carbon-containing products are possible, either on the end carbon or the middle carbon, since the C–C bond freely rotates and the end carbons are equivalent before reaction.

c) $CH_2Cl_2 + Br_2 \rightarrow CHBrCl_2 + HBr$. Only one carbon-containing product is possible since halogen substitution reactions only remove hydrogen atoms.

d)

$$\begin{array}{c} CH_3-CH-CH_3 \\ | \\ CH_3 \end{array} + Cl_2 \rightarrow \left[\begin{array}{c} H \\ | \\ CH_3-C-CH_2Cl \\ | \\ CH_3 \end{array} \text{ and } \begin{array}{c} Cl \\ | \\ CH_3-C-CH_3 \\ | \\ CH_3 \end{array}\right] + HCl$$

Two carbon-containing products are possible, either on the end carbon or the middle carbon, since the C–C bond freely rotates and the end carbons are all equivalent before reaction.

17. $CH_2{=}CH-CH_2-CH_2-CH_2-CH_3$, $CH_3-CH{=}CH-CH_2-CH_2-CH_3$ and

$CH_3-CH_2-CH=CH-CH_2-CH_3$ are the only structural isomers. Remember that cis–trans isomerism generates geometric isomers, not structural isomers.

19. **Given:** alkene structures **Find:** name

Conceptual plan: Count the number of carbon atoms in the longest continuous carbon chain that contains the multiple bond to determine the base name of the compound. Find the prefix corresponding to this number of atoms in Table 20.5 and add the ending -ene to form the base name. → Consider every branch from the base chain to be a substituent. Name each substituent according to Table 20.6. →Beginning with the end closest to the multiple bond, number the base chain and assign a number to each substituent. →Write the name of the compound in the following format: (subst. #)-(subst. name)(base name). → If there are two or more substituents, give each one a number and list them alphabetically with hyphens between words and numbers. →If a compound has two or more identical substituents, designate the number of identical substituents with the prefix di- (2), tri- (3), or tetra- (4) before the substituent's name. Separate the numbers indicating the positions of the substituents relative to each other with a comma. The prefixes are not taken into account when alphabetizing.

Solution:

a) $\boxed{CH_2=CH-CH_2-CH_3}$ has 4 carbons as the longest continuous chain. The prefix for 4 is but- and the base name is butene. There are no substituent groups. $\boxed{C^1H_2=C^2H-C^3H_2-C^4H_3}$ Start numbering on the left since it is closer to the double bond. Since the double bond is between position number 1 and 2, the name of the compound is 1-butene.

b)
$$\begin{array}{cc} CH_3 & CH_3 \\ | & | \end{array}$$
$\boxed{CH_3-CH-C=CH-CH_3}$ has 5 carbons as the longest continuous chain. The prefix for 5 is pent- and the base name is pentene. The substituent groups are two methyl groups.

$\boxed{CH_3}$ $\boxed{CH_3}$

$\boxed{C^5H_3-C^4H-C^3=C^2H-C^1H_3}$ If we start numbering the chain at the end closest to the double bond, the double bond is between position number 2 and 3, and the methyl substituents are assigned the numbers 3 and 4. The name of the compound is 3, 4-dimethyl-2-pentene.

c)
$\boxed{CH_2=CH-CH-CH_2-CH_2-CH_3}$
CH_3-CH has 6 carbons as the longest continuous chain. The prefix for 6 is
$\quad\quad CH_3$

hex- and the base name is hexene. The only substituent group is an isopropyl group. $\boxed{C^1H_2=C^2H-C^3H-C^4H_2-C^5H_2-C^6H_3}$

$\boxed{\begin{array}{c} CH_3-CH \\ | \\ CH_3 \end{array}}$ Start numbering on the left since it is closer to the double bond.

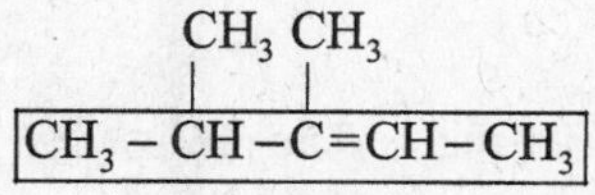

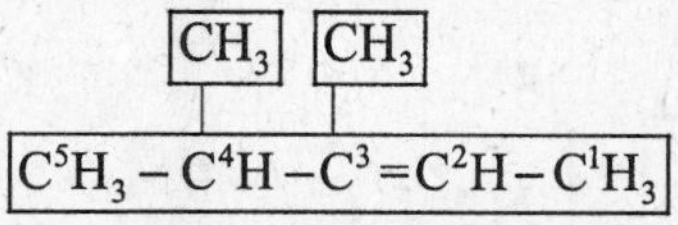

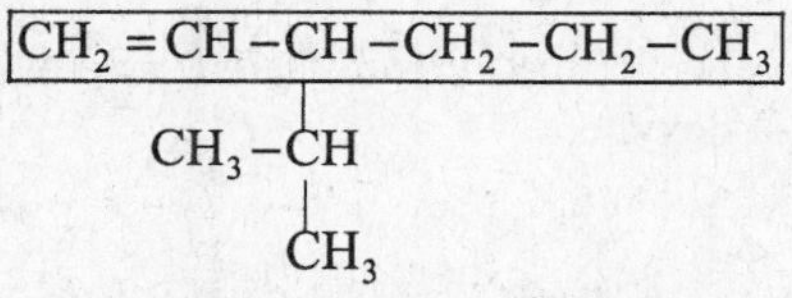

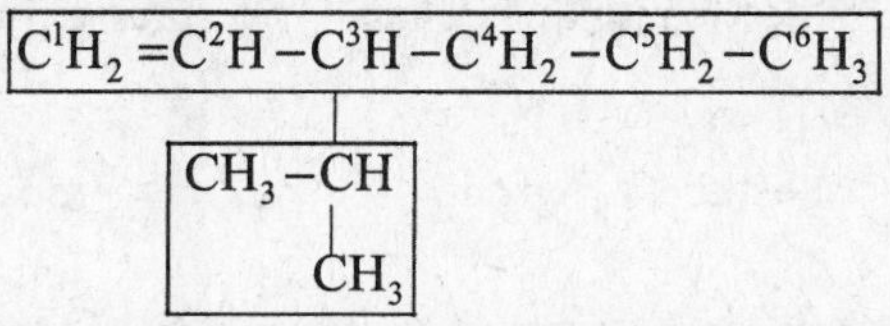

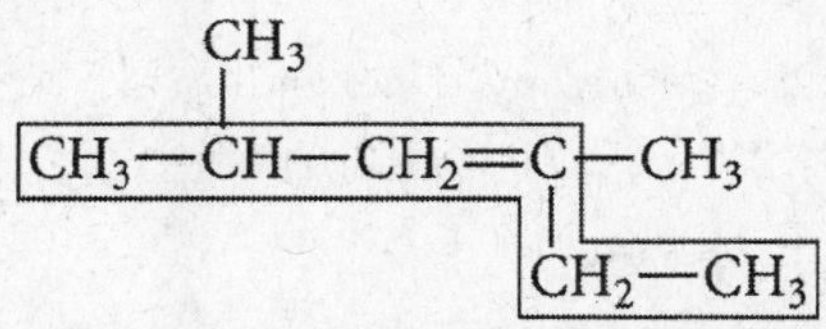

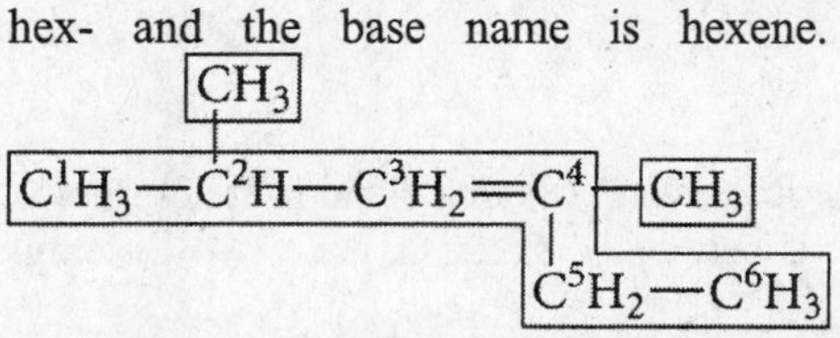

Since the double bond is between position number 1 and 2, and the isopropyl group is at position 3, the name of the compound is 3-isopropyl-1-hexene.

d)
$$\begin{array}{c} CH_3 \\ | \end{array}$$
$\boxed{CH_3-CH-CH_2=C-CH_3}$
$\quad\quad\quad\quad\quad | $
$\boxed{CH_2-CH_3}$ has 6 carbons as the longest continuous chain. The prefix for 6 is hex- and the base name is hexene. The only substituent groups are two methyl groups.

$\boxed{CH_3}$

$\boxed{C^1H_3-C^2H-C^3H_2=C^4-CH_3}$
$\quad\quad\quad\quad\quad | $
$\boxed{C^5H_2-C^6H_3}$ Since the double bond is in the middle of the chain, it will be at position 3 in both numbering schemes. If we start numbering the chain at the end closest to the left methyl

group (closest to an end), the methyl substituents are assigned the numbers 2 and 4. The name of the compound is 2,4-dimethyl-3-hexene.

21. **Given:** alkyne structures **Find:** name
Conceptual plan: **Count the number of carbon atoms in the longest continuous carbon chain that contains the multiple bond to determine the base name of the compound. Find the prefix corresponding to this number of atoms in Table 20.5 and add the ending -yne to form the base name. → Consider every branch from the base chain to be a substituent. Name each substituent according to Table 20.6. →Beginning with the end closest to the multiple bond, number the base chain and assign a number to each substituent. →Write the name of the compound in the following format: (subst. #)-(subst. name)(base name). → If there are two or more substituents, give each one a number and list them alphabetically with hyphens between words and numbers. →If a compound has two or more identical substituents, designate the number of identical substituents with the prefix di- (2), tri- (3), or tetra- (4) before the substituent's name. Separate the numbers indicating the positions of the substituents relative to each other with a comma. The prefixes are not taken into account when alphabetizing.**
Solution:

a) $\boxed{CH_3 - C \equiv C - CH_3}$ has 4 carbons as the longest continuous chain. The prefix for 4 is but- and the base

name is butyne. There are no substituent groups. $\boxed{C^1H_3 - C^2 \equiv C^3 - C^4H_3}$ Since the triple bond is in the

middle of the molecule it does not matter which end the numbering is started. Since the triple bond is between position number 2 and 3, the name of the compound is 2-butyne.

b) $\boxed{CH_3 - C \equiv C - \overset{\displaystyle CH_3}{\underset{\displaystyle CH_3}{C}} - CH_2 - CH_3}$ has 6 carbons as the longest continuous chain. The prefix for 6 is hex- and

the base name is hexyne. The only substituent groups are two methyl groups.

$\boxed{C^1H_3 - C^2 \equiv C^3 - \overset{\displaystyle CH_3}{\underset{\displaystyle CH_3}{C^4}} - C^5H_2 - C^6H_3}$ If we start numbering the chain at the end closest to the triple

bond, the triple bond is between position number 2 and 3, and the methyl substituents are assigned the numbers 4 and 4. The name of the compound is 4,4-dimethyl-2-hexyne.

c) $\boxed{CH \equiv C - \overset{\displaystyle CH - CH_3}{\underset{\displaystyle CH_3}{CH}} - CH_2 - CH_2 - CH_3}$ has 6 carbons as the longest continuous chain. The prefix for 6 is hex-

and the base name is hexyne. The only substituent group is an isopropyl group.

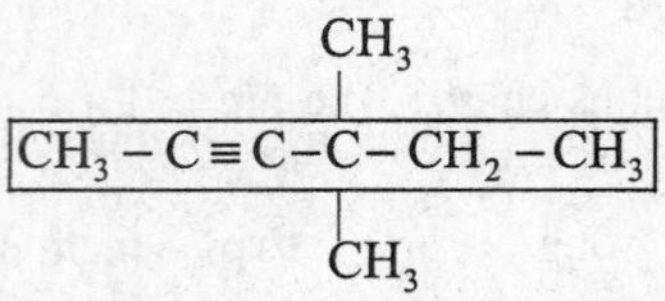

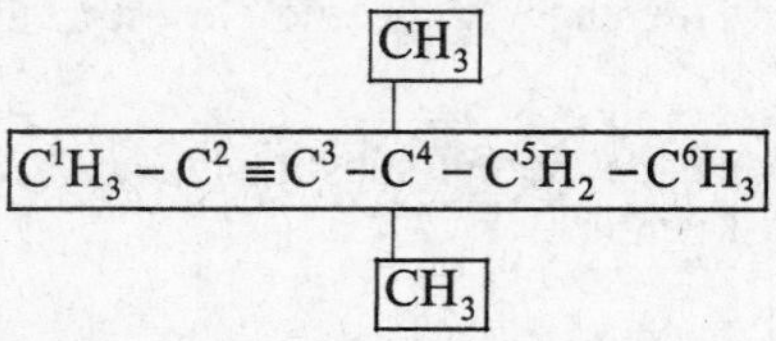

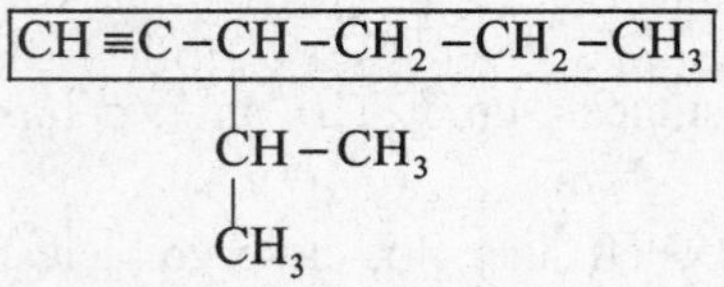

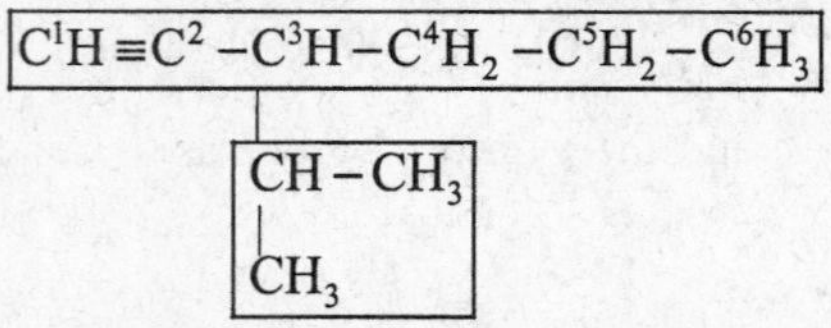

$\boxed{C^1H \equiv C^2 - \overset{\displaystyle CH - CH_3}{\underset{\displaystyle CH_3}{C^3H}} - C^4H_2 - C^5H_2 - C^6H_3}$ Start numbering at the end closest to the triple bond. The isopropyl

substituent is assigned number 3. The name of the compound is 3-isopropyl-1-hexyne.

d) $\boxed{CH_3 - \overset{\displaystyle}{\underset{\displaystyle CH_2 - CH_3}{CH}} - C \equiv C - \overset{\displaystyle CH_3}{\underset{\displaystyle CH_2 - CH_3}{CH}} - CH_2}$ has 9 carbons as the longest continuous chain. The prefix for 9 is

non- and the base name is nonyne. The substituent groups are two methyl groups.

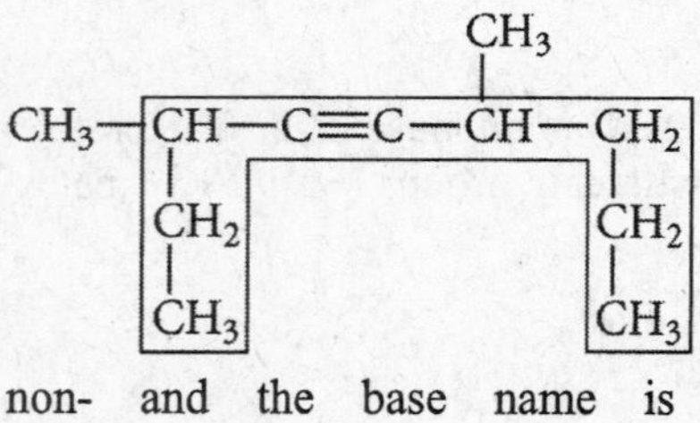

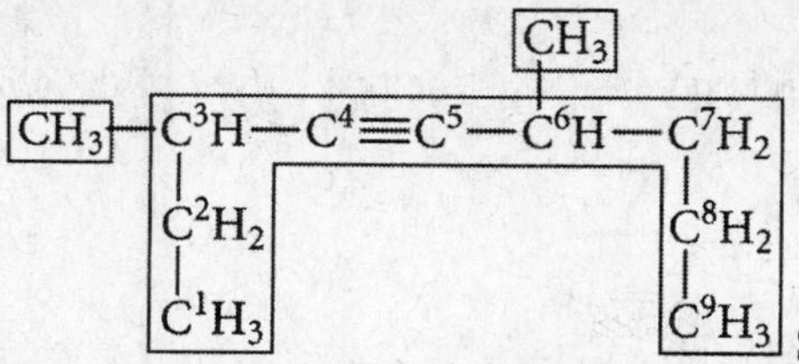

Start numbering at the bottom left and count clockwise along the chain. The triple bond is between positions 4 and 5 and the methyl substituents are assigned the numbers 3 and 6. The name of the compound is 3,6-dimethyl-4-nonyne.

23. **Given:** hydrocarbon names **Find:** structure
Conceptual plan: Find the number of carbon atoms corresponding to prefix of the base name in Table 20.5. → Draw the base chain and number the carbons from left to right. → Determine the multiple bond type from the ending of the base name (-ene = double bond, and -yne = triple bond) and place it in the appropriate position in the chain. → Using Table 20.6 and the prefix di- (2), tri- (3), or tetra- (4) before the substituent's name, to determine each substituent. → Add the substituent to the proper carbon position in the chain. → Add hydrogen atoms to the base chain so that each carbon has 4 bonds.
Solution:

a) 4-octyne. The base name octyne designates that there are 8 carbon atoms in the base chain. $C^1-C^2-C^3-C^4-C^5-C^6-C^7-C^8$. The -yne ending designates that there is a triple bond and the 4 prefix designates that it is between the 4th and 5th positions. $C^1-C^2-C^3-C^4\equiv C^5-C^6-C^7-C^8$. Add hydrogens to the base chain to get the final molecule $CH_3-CH_2-CH_2-C\equiv C-CH_2-CH_2-CH_3$.

b) 3-nonene. The base name nonene designates that there are 9 carbon atoms in the base chain. $C^1-C^2-C^3-C^4-C^5-C^6-C^7-C^8-C^9$. The -ene ending designates that there is a double bond and the 3 prefix designates that it is between the 3rd and 4th positions. $C^1-C^2-C^3=C^4-C^5-C^6-C^7-C^8-C^9$. Add hydrogens to the base chain to get the final molecule
$CH_3-CH_2-CH=CH-CH_2-CH_2-CH_2-CH_2-CH_3$.

c) 3,3-dimethyl-1-pentyne. The base name pentyne designates that there are 5 carbon atoms in the base chain. $C^1-C^2-C^3-C^4-C^5$. The -yne ending designates that there is a triple bond and the 1 prefix designates that it is between the 1st and 2nd positions. $C^1\equiv C^2-C^3-C^4-C^5$ 3,3-dimethyl designates that there are two –CH_3 groups in the 3rd position. $C^1\equiv C^2-\underset{\boxed{CH_3}}{\overset{\boxed{CH_3}}{C^3}}-C^4-C^5$. Add hydrogens to the base chain to get the final molecule $CH\equiv C-\underset{CH_3}{\overset{CH_3}{C}}-CH_2-CH_3$.

d) 5-ethyl-3,6-dimethyl-2-heptene. The base name heptene designates that there are 7 carbon atoms in the base chain. $C^1-C^2-C^3-C^4-C^5-C^6-C^7$. The -ene ending designates that there is a double bond and the 2 prefix designates that it is between the 2nd and 3rd positions. $C^1-C^2=C^3-C^4-C^5-C^6-C^7$. 5-ethyl designates that there is a –CH_2CH_3 group in the 5th position; and 3,6-dimethyl designates that are –CH_3 groups in the 3rd and 6th positions. $C^1-C^2=\underset{\boxed{CH_3}}{C^3}-C^4-\underset{\boxed{CH_2-CH_3}}{C^5}-\overset{\boxed{CH_3}}{C^6}-C^7$. Add hydrogens to the base chain to get the final molecule $CH_3-CH=\underset{CH_3}{C}-CH_2-\underset{CH_2-CH_3}{CH}-\overset{CH_3}{CH}-CH_3$.

25. Alkene addition reactions convert a double bond to a single bond and place the two halves of the other reactant on the two carbons that were in the double bond.

a) $$CH_3-CH{=}CH-CH_3 + Cl_2 \rightarrow \begin{array}{l} CH_3-CH-CH-CH_3 \\ \qquad\;\; | \qquad\; | \\ \qquad\; Cl \quad\;\; Cl \end{array}.$$

b) $$CH_3-CH_2-CH{=}CH-CH_3 + Br_2 \rightarrow \begin{array}{l} CH_3-CH_2-CH-CH-CH_3 \\ \qquad\qquad\quad\;\; | \qquad\; | \\ \qquad\qquad\quad\; Br \quad\; Br \end{array}.$$

27. Hydrogenation reactions convert a double bond to a single bond and place the hydrogen atoms on each of the two carbons that were in the double bond.

a) $CH_2{=}CH-CH_3 + H_2 \rightarrow CH_3-CH_2-CH_3$.

b) $$\begin{array}{l} CH_3-CH-CH{=}CH_2 \\ \qquad\;\; | \\ \qquad CH_3 \end{array} + H_2 \rightarrow \begin{array}{l} CH_3-CH-CH_2-CH_3 \\ \qquad\;\; | \\ \qquad CH_3 \end{array}.$$

c) $$\begin{array}{l} CH_3-CH-C{=}CH_2 \\ \qquad\;\; | \qquad | \\ \qquad CH_3\;\, CH_3 \end{array} + H_2 \rightarrow \begin{array}{l} CH_3-CH-CH-CH_3 \\ \qquad\;\; | \qquad\; | \\ \qquad CH_3\;\; CH_3 \end{array}.$$

29. **Given:** monosubstituted benzene structures **Find:** name

Conceptual plan: Consider the branch from the benzene ring to be a substituent. Name the substituent according to Table 20.6 or with the base name of a halogen with an "o" added at the end. →Write the name of the compound in the following format: (name of substituent)benzene.

Solution:

a) $-CH_3$ or methyl is the substituent group, so the name is methylbenzene.

b) –Br or bromo is the substituent, so the name is bromobenzene.

c) –Cl or chloro is the substituent, so the name is chlorobenzene.

31. **Given:** disubstituted benzene structures **Find:** name

Conceptual plan: Consider the branches from the benzene ring to be substituents. Name the substituent according to Table 20.6 or with the base name of a halogen with an "o" added at the end. → Number the benzene ring starting with the substituent that is first alphabetically and count in the direction that gets to the second substituent with the lower position number. → Write the name of the compound in the following format: (name of substituent)benzene, listing them alphabetically with hyphens between words and numbers. →If a compound has two identical substituents, designate the number of identical substituents with the prefix di- (2) before the substituent's name. Separate the numbers indicating the positions of the substituents relative to each other with a comma. → Alternate names are 1,2 = *ortho* or *o*; 1,3 = *meta* or *m*; and 1,4 = *para* or *p* replacing the numbers.

Solution:

a) Both of the substituent groups are –Br or bromo groups. Start by giving one bromo an assignment of 1 and count in either direction to give the second bromo group an assignment of 4. The name is 1,4-dibromobenzene or *p*-dibromobenzene.

b) Both of the substituent groups are $-CH_2CH_3$ or ethyl groups. Start by giving the top ethyl an assignment of 1 and count in the clockwise direction to give the second ethyl group an assignment of 3. The name is 1,3-diethylbenzene or *m*-diethylbenzene.

c) One substituent group is –Cl or chloro and the other is –F or fluoro. Start by giving the chloro group an assignment of 1 and count in the counterclockwise direction to give the fluoro group an assignment of 2. The name is 1-chloro-2-fluorobenzene or *o*-chlorofluorobenzene.

33. **Given:** substituted benzene names **Find:** structures

Conceptual plan: Start with a benzene ring. Identify the structure of the substituent(s) according to Table 20.6 or with the base name of a halogen with an "o" added at the end. → If only one substituent is present, simply attach it to the benzene ring. →If a compound has two identical substituents, they are

designated with the prefix di- (2) before the substituent's name. → If there are two substituents, place the first one on the benzene ring and count clockwise around the ring to determine the location to attach the second substituent. Note, 1,2 = *ortho* or *o*; 1,3 = *meta* or *m*; and 1,4 = *para* or *p* replacing the numbers.

Solution:

a) isopropylbenzene. The isopropyl group is $-\underset{\displaystyle CH_3}{\underset{|}{CH}}-CH_3$ 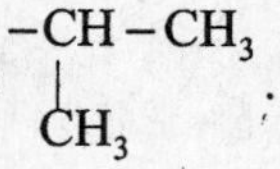. Attach this to a benzene ring to make

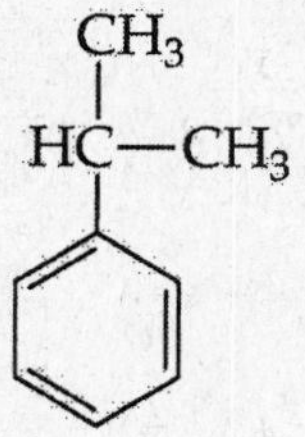

b) *meta*-dibromobenzene. There are two bromo or –Br groups. The positions are 1 and 3, since the designation is *meta.* Attach the –Br's at the 1st and 3rd positions to make

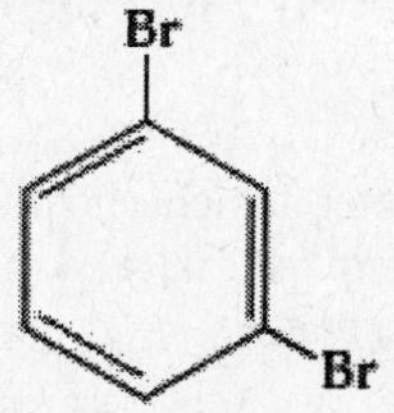

c) 1-chloro-4-methylbenzene. One substituent group is –Cl or chloro and the other is $-CH_3$ or methyl. Start by giving the chloro group an assignment of 1 and count in the clockwise direction to give the methyl group an assignment of 4. Attach the two groups to make

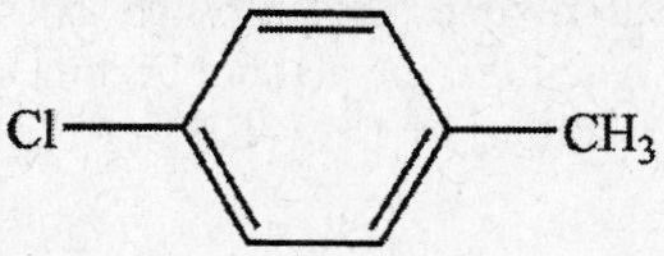

35. **Given:** alcohol structures **Find:** name

Conceptual plan: Alcohols are named like alkanes with the following differences: (1) The base chain is the longest continuous carbon chain that contains the –OH functional group; (2) The base name has the ending -ol; (3) The base chain is numbered to give the –OH group the lowest possible number; and (3) A number indicating the position of the –OH group is inserted just before the base name.

Solution:

a) $\boxed{CH_3-CH_2-CH_2}-OH$ The longest carbon chain has 3 carbons. The prefix for 3 is prop-, so the base name is propanol. The alcohol group is on the end, or 1st carbon, so the name is 1-propanol.

b) 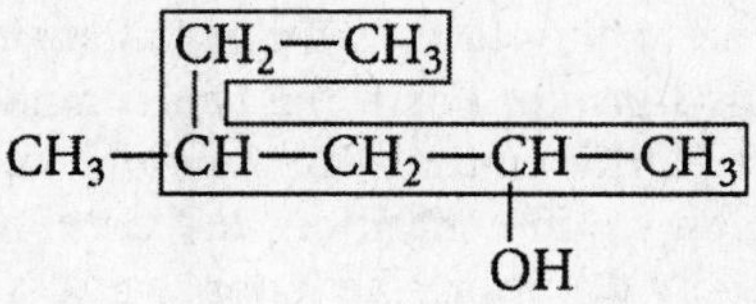

The longest carbon chain has 6 carbons. The prefix for 6 is hex-, so the base name is hexanol. There is a methyl substituent group.

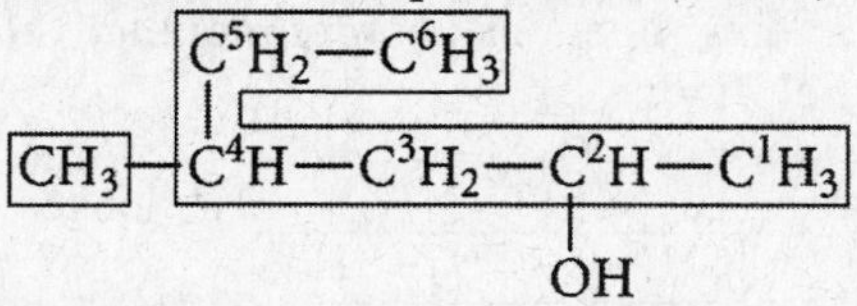

Start numbering at the end closest to the alcohol group. The alcohol group is assigned number 2 and the methyl group is assigned number 4. The name is 4-methyl-2-hexanol.

c) 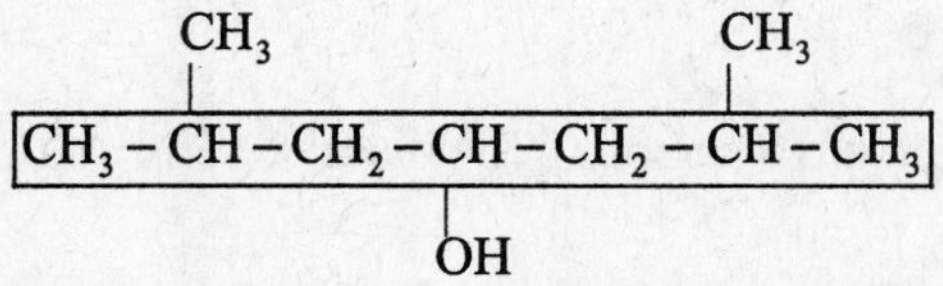

The longest carbon chain has 7 carbons. The prefix for 7 is hept-, so the base name is heptanol. There are two methyl substituent groups.

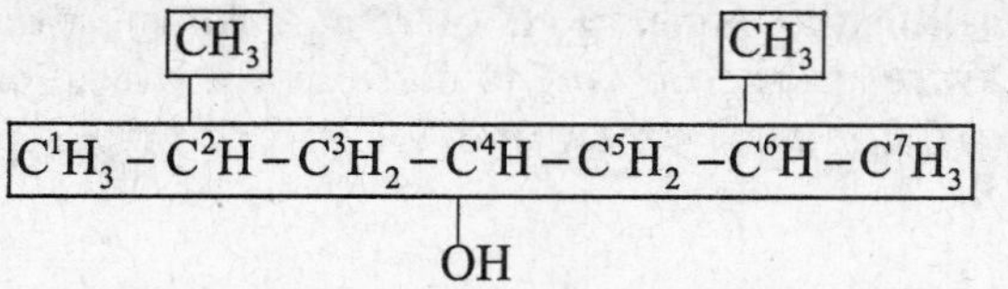

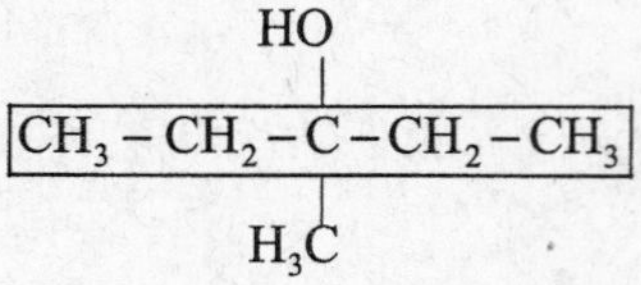

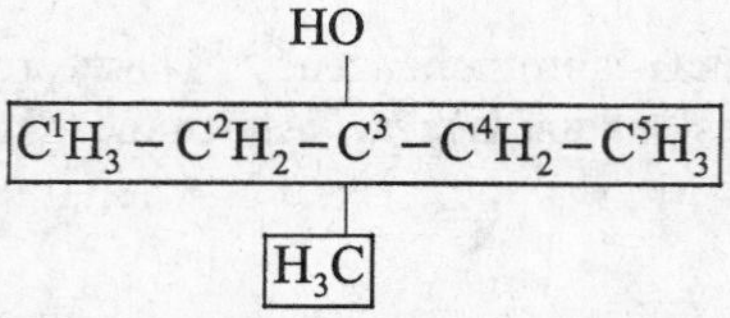

It does not matter which end the numbering is started,

since substitutions are symmetrically attached. The alcohol group is assigned number 4 and the methyl groups are assigned numbers 2 and 6. The name is 2,6-dimethyl-4-heptanol.

d) $CH_3 - CH_2 - \overset{HO}{\underset{H_3C}{C}} - CH_2 - CH_3$ The longest carbon chain has 5 carbons. The prefix for 5 is pent-, so the

base name is pentanol. There is a methyl substituent group. $C^1H_3 - C^2H_2 - \overset{HO}{\underset{H_3C}{C^3}} - C^4H_2 - C^5H_3$ It does not

matter which end the numbering is started, since substitutions are symmetrically attached. The alcohol group is assigned number 3 and the methyl group is assigned number 3. The name is 3-methyl-3-pentanol.

37. In a substitution reaction, an alcohol reacts with an acid, such as HBr, to form halogenated hydrocarbons and water. In an elimination (or dehydration) reaction, concentrated acids, such as H_2SO_4, react with alcohols to eliminate water, forming an alkene. In an oxidation reaction, carbon atoms gain oxygen atoms and/or lose hydrogen atoms. In these reactions, the alcohol becomes a carboxylic acid.

a) This is a substitution reaction, so $CH_3 - CH_2 - CH_2 - OH + HBr \rightarrow CH_3 - CH_2 - CH_2 - Br + H_2O$.

b) This is an elimination reaction, so $CH_3 - \underset{CH_3}{\overset{|}{CH}} - CH_2 - OH \xrightarrow{H_2SO_4} CH_3 - \underset{CH_3}{\overset{|}{C}} = CH_2 + H_2O$.

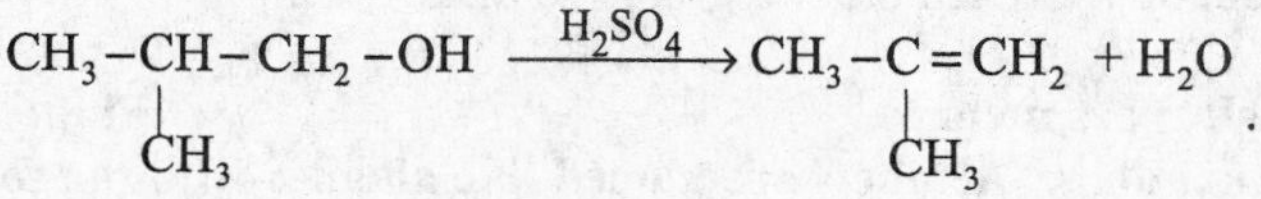

c) This is an oxidation reaction, so $CH_3 - \overset{CH_3}{\underset{CH_3}{\overset{|}{\underset{|}{C}}}} - CH_2 - CH_2 - OH \xrightarrow[H_2SO_4]{Na_2Cr_2O_7} CH_3 - \overset{CH_3}{\underset{CH_3}{\overset{|}{\underset{|}{C}}}} - CH_2 - \overset{O}{\overset{||}{C}} - OH$.

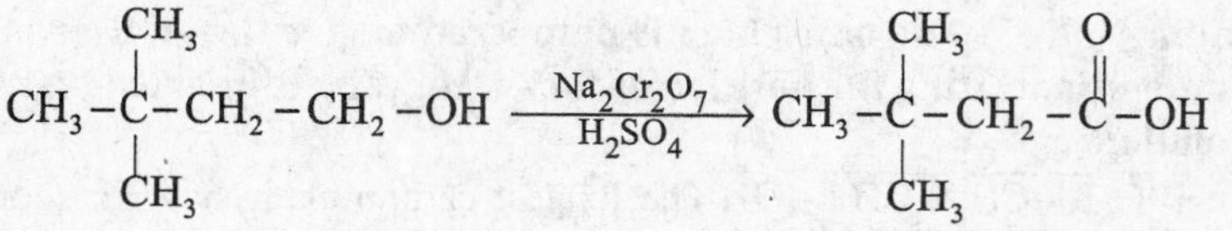

39. **Given:** aldehyde or ketone structures   **Find:** name
Conceptual Plan: Simple aldehydes are systematically named according to the number of carbon atoms in the longest continuous carbon chain that contains the carbonyl group. Form the base name from the name of the corresponding alkane by dropping the -e and add the ending -al. Simple ketones are systematically named according to the longest continuous carbon chain containing the carbonyl group. Form the base name from the name of the corresponding alkane by dropping the letter -e and adding the ending -one. For ketones, number the chain to give the carbonyl group the lowest possible number.
Solution:

a) $CH_3 - \overset{O}{\overset{||}{C}} - CH_2 - CH_3$ The longest carbon chain has 4 carbons. The prefix for 4 is but-, and since this is a

ketone the base name is butanone. The position of the carbonyl carbon does not need to be specified, since there is only one place for it to be in the molecule (if it were on the end carbon it would be an aldehyde, not a ketone). Since there are no other substituent groups, the name is butanone.

b) $CH_3 - CH_2 - CH_2 - CH_2 - \overset{O}{\overset{||}{CH}}$ The longest carbon chain has 5 carbons. The prefix for 5 is pent-, and since

this is an aldehyde the base name is pentanal. Since there are no other substituent groups, the name is pentanal.

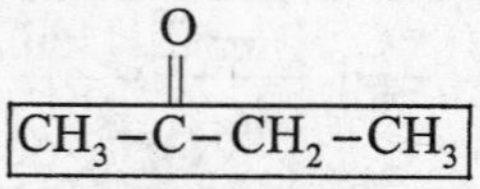

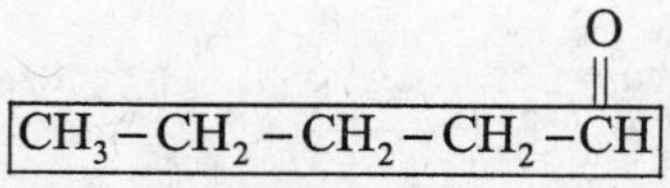

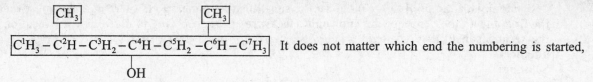

c) 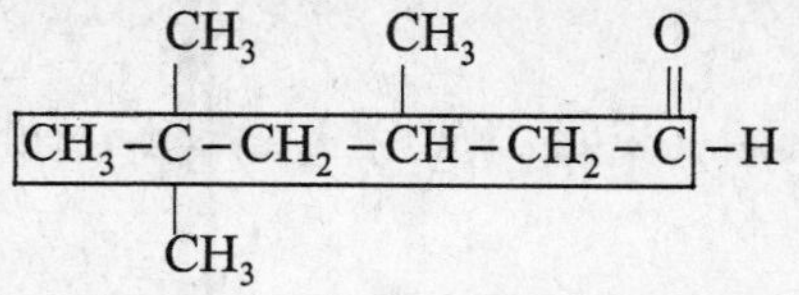 The longest carbon chain has 6 carbons. The prefix for 6 is hex-, and since this is an aldehyde the base name is hexanal. There are 3 methyl substituent groups.

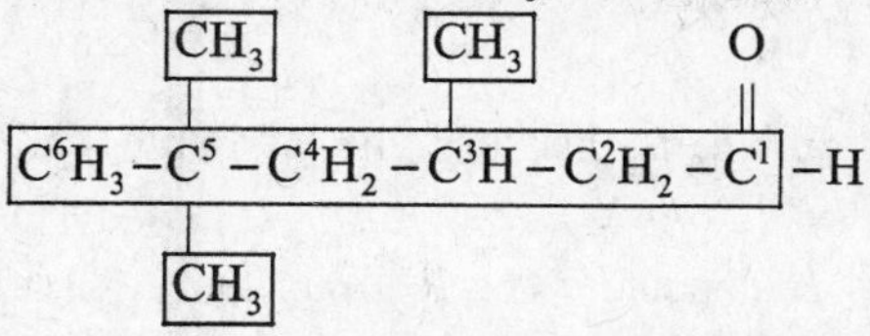 Start numbering at the carbonyl carbon. The methyl groups are at positions 3, 5, and 5. The name of the molecule is 3,5,5-trimethylhexanal.

d) 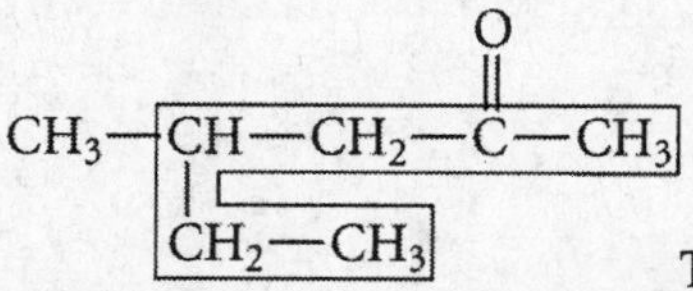 The longest carbon chain has 6 carbons. The prefix for 6 is hex-, and since this is a ketone the base name is hexanone. There is one methyl substituent group.

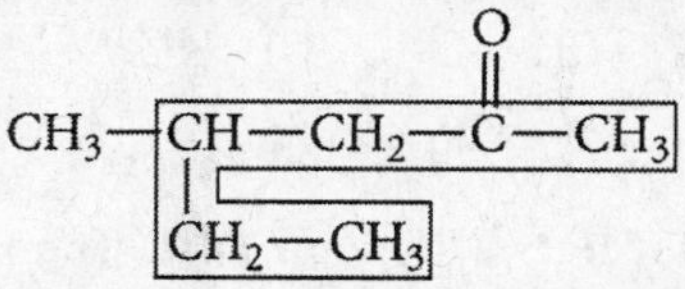 Start numbering at the end closest to the carbonyl carbon. The carbonyl carbon is in position 2 and the methyl group is at position 4. The name of the molecule is 4-methyl-2-hexanone.

41. This is an addition reaction, $CH_3-CH_2-CH_2-\overset{O}{\overset{\|}{C}}-H + H-C\equiv N \xrightarrow{NaCN} CH_3-CH_2-CH_2-\underset{H}{\overset{OH}{C}}-C\equiv N$.

43. **Given:** carboxylic acid or ester structures **Find:** name
Conceptual Plan: Carboxylic acids are systematically named according to the number of carbon atoms in the longest chain containing the –COOH functional group. Form the base name by dropping the -e from the name of the corresponding alkane, and adding the ending -oic acid. Esters are systematically named as if they were derived from a carboxylic acid by replacing the H on the OH with an alkyl group. The R group from the parent acid forms the base name of the compound. Change the -ic on the name of the corresponding carboxylic acid to -ate, and drop "acid." The R group that replaced the H on the carboxylic acid is named as an alkyl group with the ending -yl.
Solution:

a) 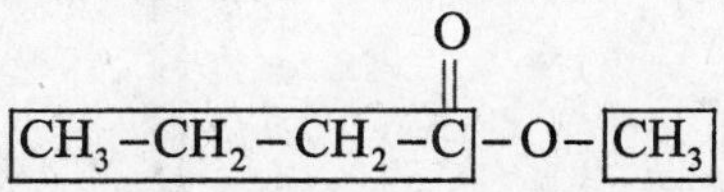 The carbon chain that has the carbonyl carbon has 4 carbons. The prefix for 4 is but-, so the base name is butanoate. The R group that replaces the H of the carboxylic acid is a methyl group. The name is methylbutanoate.

b) 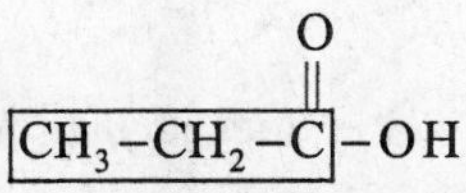 The carbon chain has 3 carbons. The prefix for 3 is prop-, since this is a carboxylic acid, and there are no other substituents, the name is propanoic acid.

c) 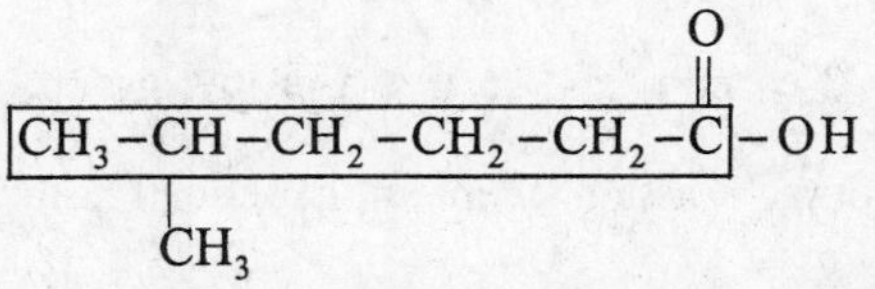 The carbon chain has 6 carbons. The prefix for 6 is hex-, since this is a carboxylic acid, the base name is hexanoic acid. There is one methyl substituent group.

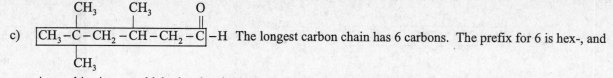

$$\boxed{C^6H_3-C^5H-C^4H_2-C^3H_2-C^2H_2-C^1}-OH \quad (C^1 = O;\ CH_3 \text{ on } C^5)$$

Number the chain starting with the carbonyl carbon. The methyl is in the 5th position. The name of the molecule is 5-methylhexanoic acid.

d) $\boxed{CH_3-CH_2-CH_2-CH_2-\overset{O}{\overset{\|}{C}}}-O-\boxed{CH_2-CH_3}$ The carbon chain that has the carbonyl carbon has 5 carbons. The prefix for 5 is pent-, so the base name is pentanoate. The R group that replaces the H of the carboxylic acid is an ethyl group. The name is ethylpentanoate.

45. The reactions are condensation reactions. The two non-carbonyl oxygen groups react linking the two molecules (or parts of a molecule) with concomitant generation of a water molecule.

$$CH_3-CH_2-CH_2-CH_2-\overset{O}{\overset{\|}{C}}-OH + CH_3-CH_2-OH \xrightarrow{H_2SO_4}$$

$$CH_3-CH_2-CH_2-CH_2-\overset{O}{\overset{\|}{C}}-O-CH_2-CH_3 + H_2O.$$

47. **Given:** ether structures **Find:** name

Conceptual Plan: Common names for ethers have the following format: (R group 1)(R group 2) ether, where the alkyl groups are those found in Table 20.5. List the names alphabetically. If the two R groups are different, use each of their names. If the two R groups are the same, use the prefix di-.

Solution:

a) $\boxed{CH_3-CH_2-CH_2}-O-\boxed{CH_2-CH_3}$ The left carbon chain has 3 carbons and so it is a propyl group. The right carbon chain has 2 carbons and so it is an ethyl group. Listing them alphabetically, the name is ethyl propyl ether.

b) $\boxed{CH_3-CH_2-CH_2-CH_2-CH_2}-O-\boxed{CH_2-CH_3}$ The left carbon chain has 5 carbons and so it is a pentyl group. The right carbon chain has 2 carbons and so it is an ethyl group. Listing them alphabetically, the name is ethyl pentyl ether.

c) $\boxed{CH_3-CH_2-CH_2}-O-\boxed{CH_2-CH_2-CH_3}$ Both carbon chains have 3 carbons and so they are propyl groups. The name is dipropyl ether.

d) $\boxed{CH_3-CH_2}-O-\boxed{CH_2-CH_2-CH_2-CH_3}$ The left carbon chain has 2 carbons and so it is an ethyl group. The right carbon chain has 4 carbons and so it is an butyl group. Listing them alphabetically, the name is butyl ethyl ether.

49. **Given:** amine structures **Find:** name

Conceptual Plan: Amines are systematically named by alphabetically listing the alkyl groups that are attached to the nitrogen atom and adding an -amine ending. The alkyl groups are those found in Table 20.5. Use a prefix of di- or tri- if the alkyl groups are the same.

Solution:

a) $\boxed{CH_3-CH_2}-\underset{H}{N}-\boxed{CH_2-CH_3}$ Both carbon chains have 2 carbons and so they are ethyl groups. The name is diethylamine.

b) $\boxed{CH_3-CH_2-CH_2}-\underset{H}{N}-\boxed{CH_3}$ The left carbon chain has 3 carbons and so it is a propyl group. The right carbon chain has 1 carbon and so it is a methyl group. Listing them alphabetically, the name is methylpropylamine.

c) $CH_3-CH_2-CH_2-N(CH_3)-CH_2-CH_2-CH_2-CH_3$ The left carbon chain has 3 carbons and so it is a propyl group. The top carbon chain has 1 carbon and so it is a methyl group. The right carbon chain has 4 carbons and so it is a butyl group. Listing them alphabetically, the name is butyl methylpropylamine.

51. Since the acid is a carboxylic acid, this reaction is a condensation reaction,
$CH_3CH_2NH_2\ (aq) + CH_3CH_2COOH\ (aq) \rightarrow CH_3CH_2CONHCH_2CH_3\ (aq) + H_2O\ (l)$.

53. Since the monomer is 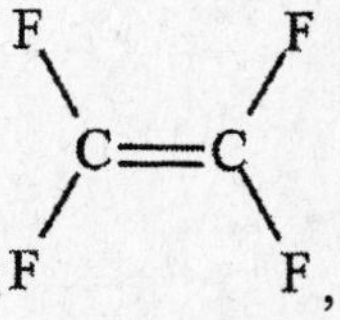, as the polymer forms the double bond breaks to link with another monomer. The structure is 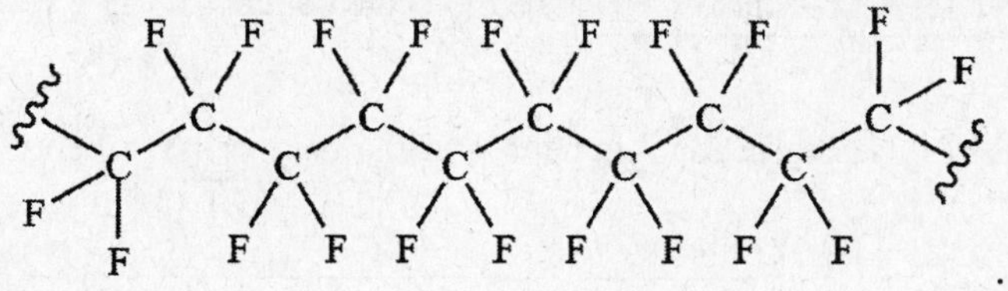 .

55. In a condensation polymer atoms are eliminated. Here water is eliminated. The structure is

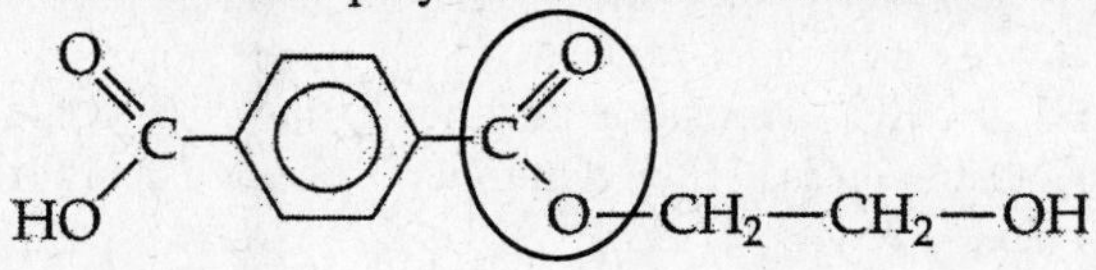

57. **Given:** structures **Find:** identify molecule type and name
Conceptual Plan: Look for functional groups to identify molecule type. Review appropriate set of naming rules to name molecule.
Solution:

a) $CH_3-CH(CH_3)-CH_2-C(=O)-O-CH_3$ The compound has the formula of RCOOR', so it is an ester.

$\boxed{CH_3-CH(CH_3)-CH_2-C}(=O)-O-\boxed{CH_3}$ The left carbon chain has 4 carbons, which translates to but- and a base name of butanoate. There is a methyl substituent on this chain. The right carbon chain has 2 carbons, so it is a methyl group. $\boxed{C^4H_3-C^3H(\boxed{CH_3})-C^2H_2-C^1}(=O)-O-\boxed{CH_3}$ Number the left chain starting with the carbonyl carbon. The methyl group is at position 3. The name of the compound is methyl-3-methylbutanoate.

b) $\boxed{CH_3-CH_2-CH(CH_3)-CH_2}-O-\boxed{CH_2-CH_3}$ The compound has the formula of ROR', so it is an ether. The left carbon chain has 4 carbons and so it is a butyl group. There is a methyl substituent on this chain. The right carbon chain has 2 carbons and so it is an ethyl group. $\boxed{C^4H_3-C^3H_2-C^2H(\boxed{CH_3})-C^1H_2}-O-\boxed{CH_2-CH_3}$ Number the left chain starting with the carbon next to the oxygen. The methyl substituent is in position 2, so the alkyl group is 2-methylbutyl. Listing the alkyl group alphabetically the name is ethyl 2-methylbutyl ether.

c) 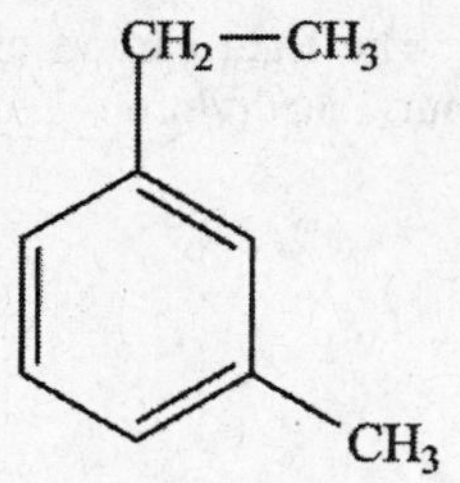

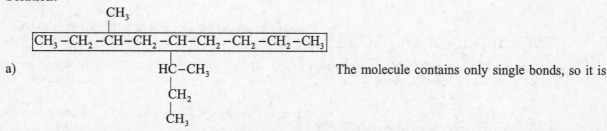

The molecule is a disubstituted benzene, so it is an aromatic hydrocarbon. The substituents are an ethyl group and a methyl group. If the ethyl position is assigned the 1st position, then the methyl group is in the 3rd position. Listing the functional groups alphabetically, the name is 1-ethyl-3-methylbenzene or *m*-ethylmethylbenzene.

$$\begin{array}{c} CH_2-CH_3 \\ | \\ \boxed{CH_3-C\equiv C-CH-CH-CH_2-CH_3} \\ | \\ CH_3 \end{array}$$

d) The molecule contains a triple bond, so it is an alkyne. The longest carbon chain is 7. The prefix for 7 is hept- and the base name is heptyne. The substituent groups are a methyl and an ethyl group.

$$\boxed{C^1H_3-C^2\equiv C^3-C^4H-C^5H-C^6H_2-C^7H_3} \quad (\boxed{CH_2-CH_3} \text{ on } C^5,\ \boxed{CH_3} \text{ on } C^4)$$

If we start numbering the chain at the end closest to the triple bond, the triple bond is between position number 2 and 3, and the methyl substituent is assigned the number 4 and the ethyl substituent is assigned number 5. Since e is before m, the name of the compound is 5-ethyl-4-methyl-2-heptyne.

e)

$$\boxed{CH_3-CH_2-CH_2-\overset{\overset{\displaystyle O}{\|}}{C}H}$$

The molecule has the formula RCHO, so is it is an aldehyde. The longest carbon chain has 4 carbons. The prefix for 4 is but-, and since this is an aldehyde the base name is butanal. Since there are no other substituent groups, the name is butanal.

f)

$$\begin{array}{c} \quad OH \\ \quad | \\ \boxed{CH_3-CH-CH_2} \\ | \\ H_3C \end{array}$$

The molecule has the formula ROH, so it is an alcohol. The longest carbon chain has 3 carbons. The prefix for 3 is prop-, so the base name is propanol. There is a methyl substituent group.

$$\begin{array}{c} \qquad\quad OH \\ \qquad\quad | \\ \boxed{C^3H_3-C^2H-C^1H_2} \\ | \\ \boxed{H_3C} \end{array}$$

Start numbering with the carbon with the –OH group. The alcohol group is assigned number 1 and the methyl group is assigned number 2. The name is 2-methyl-1-propanol.

59. **Given:** structures **Find:** name
Conceptual Plan: Look for functional groups to identify molecule type. Review appropriate set of naming rules to name molecule.
Solution:

a)

$$\begin{array}{c} \quad CH_3 \\ \quad | \\ \boxed{CH_3-CH_2-CH-CH_2-CH-CH_2-CH_2-CH_2-CH_3} \\ | \\ HC-CH_3 \\ | \\ CH_2 \\ | \\ CH_3 \end{array}$$

The molecule contains only single bonds, so it is an alkane. The longest continuous chain has 9 carbons. The prefix for 9 is non- and the base name is nonane. The substituent groups are a methyl group and an isobutyl group.

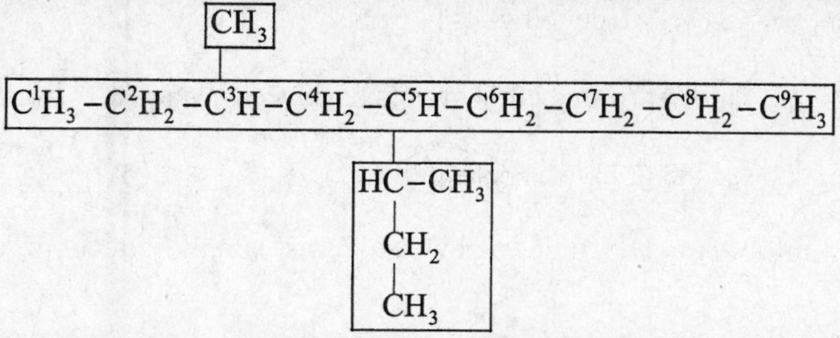

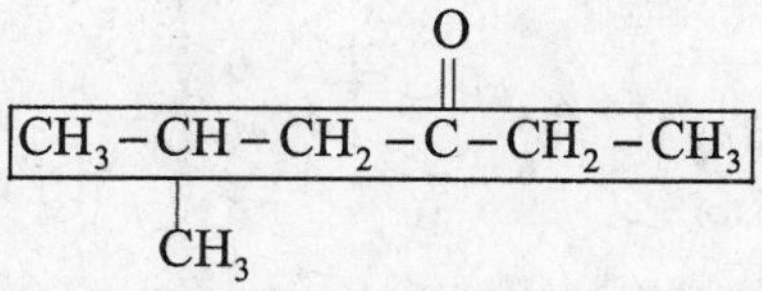

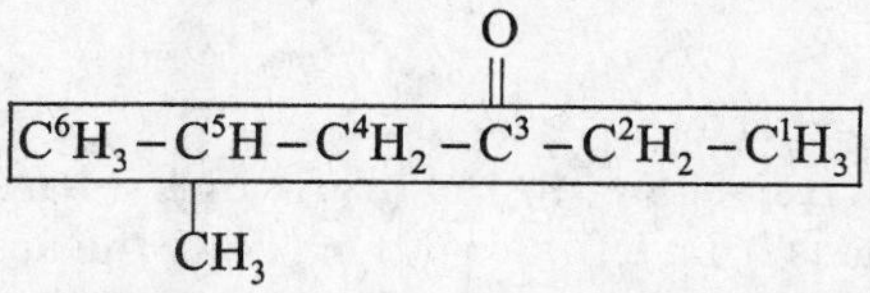

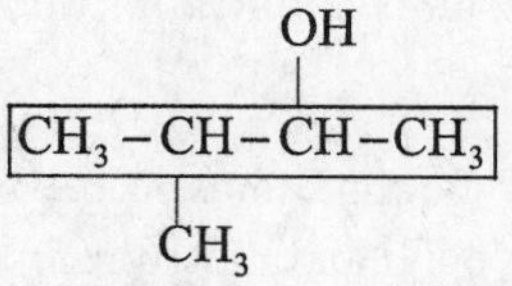

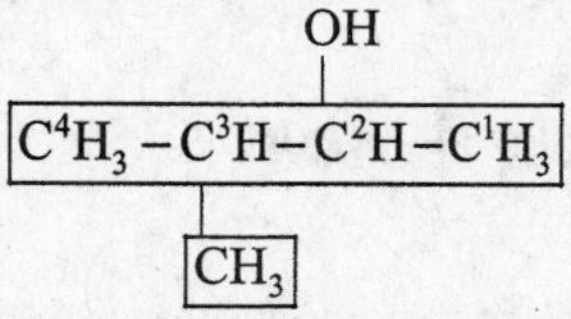

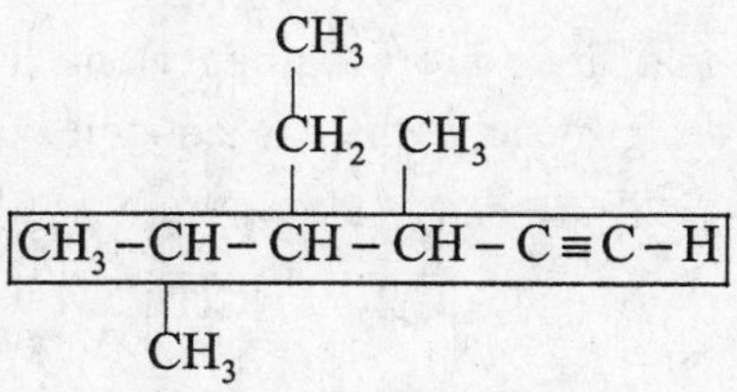

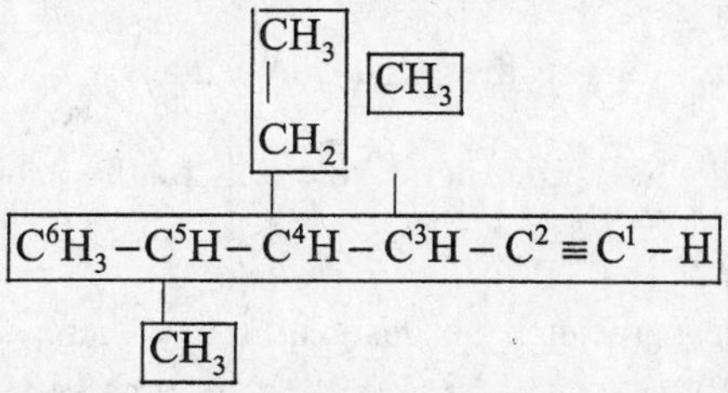

If we start numbering the chain at the end

closest to the methyl group, the methyl substituent is assigned the number 3 and the isobutyl group is assigned the number 5. List the groups alphabetically to get the name of the compound is 5-isobutyl-3-methylnonane.

b) $CH_3-CH-CH_2-\overset{\overset{O}{\|}}{C}-CH_2-CH_3$ The molecule contains a carbonyl carbon in the interior of the
        |
      CH_3

molecule and so it is a ketone. The longest carbon chain has 6 carbons. The prefix for 6 is hex-, since this is a ketone the base name is hexanone. There is one methyl substituent group.

$C^6H_3-C^5H-C^4H_2-\overset{\overset{O}{\|}}{C^3}-C^2H_2-C^1H_3$ Start numbering at the right side of the molecule. The carbonyl
         |
       CH_3

carbon is at position number 3. The methyl group is at position 5. The name of the molecule is 5- methyl-3-hexanone.

c) $CH_3-\overset{\overset{OH}{|}}{CH}-CH-CH_3$ The molecule has the formula ROH, so is it is an alcohol. The longest carbon
          |
        CH_3

chain has 4 carbons. The prefix for 4 is but-, so the base name is butanol. The only substituent group is

a methyl group. $C^4H_3-C^3H-\overset{\overset{OH}{|}}{C^2H}-C^1H_3$ Start numbering with the end with the –OH group. The
                |
            CH_3

alcohol group is assigned number 2 and the methyl group is assigned number 3. The name is 3-methyl-2-butanol.

d)
        CH_3
        |
        CH_2 CH_3
        | |
$CH_3-CH-CH-CH-C\equiv C-H$ The molecule contains a triple bond, so it is an alkyne. The longest
     |
   CH_3

carbon chain is 6. The prefix for 6 is hex- and the base name is hexyne. The substituent groups are two

               CH_3
               | CH_3
               CH_2
               |
$C^6H_3-C^5H-C^4H-C^3H-C^2\equiv C^1-H$
        |
methyl and an ethyl group. CH_3 If we start numbering the chain at the
end closest to the triple bond, the triple bond is between position number 1 and 2, and the methyl

substituents are assigned the numbers 3 and 5 and the ethyl substituent is assigned number 4. Since e is before m, the name of the compound is 4-ethyl-3,5-dimethyl-1-hexyne.

61. a) These structures are structural isomers. The methyl and ethyl groups have been swapped.

b) These structures are isomers. The iodo second group has been moved.

c) These structures are the same molecule. The first methyl group is simply drawn in a different orientation.

63. **Given:** 15.5 kg 2-butene hydrogenation **Find:** minimum g of H_2 gas

Conceptual Plan: Write a balanced equation. kg C_4H_8 → g C_4H_8 → mol C_4H_8 → mol H_2 → g H_2

$$\frac{1000\text{ g}}{1\text{ kg}} \qquad \frac{1\text{ mol } C_4H_8}{56.10\text{ g } C_4H_8} \qquad \frac{1\text{ mol } H_2}{1\text{ mol } C_4H_8} \qquad \frac{2.02\text{ g } H_2}{1\text{ mol } H_2}$$

Solution: $C_4H_8\,(g) + H_2\,(g) \rightarrow C_4H_{10}\,(g)$ then

$$15.5\,\cancel{\text{kg } C_4H_8} \times \frac{1000\,\cancel{\text{g } C_4H_8}}{1\,\cancel{\text{kg } C_4H_8}} \times \frac{1\,\cancel{\text{mol } C_4H_8}}{56.10\,\cancel{\text{g } C_4H_8}} \times \frac{1\,\cancel{\text{mol } H_2}}{1\,\cancel{\text{mol } C_4H_8}} \times \frac{2.02\text{ g } H_2}{1\,\cancel{\text{mol } H_2}} = 558\text{ g } H_2$$

Check: The units (g) are correct. The magnitude of the answer (600) makes physical sense because we are reacting 15,500 g of butane. Hydrogen is lighter and the molar ratio is 1:1 so the amount of hydrogen will be significant, but less than the mass of butene.

65. **Given:** alkene names **Find:** structure

Conceptual plan: Find the number of carbon atoms corresponding to the prefix of the base name in Table 20.5. → Draw the base chain and number the carbons from left to right. → Place the double bond in the appropriate position in the chain. → Using Table 20.6 and the prefix di- (2), tri- (3), or tetra- (4) before the substituent's name, determine each substituent. → Add the substituent to the proper carbon position in the chain. → Add hydrogen atoms to the base chain so that each carbon has 4 bonds. → Look for substitution around double bond to determine if cis-trans isomerism is applicable and look for carbons with four different alkyl groups attached.

Solution:

a) 3-methyl-1-pentene. The base name pentene designates that there are 5 carbon atoms in the base chain. $C^1-C^2-C^3-C^4-C^5$. The -ene ending designates that there is a double bond and the 1 prefix designates that it is between the 1st and 2nd positions. $C^1=C^2-C^3-C^4-C^5$ 3-methyl designates that there is a $-CH_3$ group in the 3rd position. $\begin{array}{c} \quad\quad\quad \boxed{CH_3} \\ \quad\quad\quad | \\ C^1=C^2-C^3-C^4-C^5 \end{array}$. Add hydrogens to the base chain to get the final molecule $\begin{array}{c} \quad\quad CH_3 \\ \quad\quad | \\ CH_2=CH-CH-CH_2-CH_3 \end{array}$. There is stereoisomerism since the 3rd carbon has four different groups attached.

a) 3,5-dimethyl-2-hexene. The base name hexene designates that there are 6 carbon atoms in the base chain. $C^1-C^2-C^3-C^4-C^5-C^6$. The -ene ending designates that there is a double bond and the 2 prefix designates that it is between the 2nd and 3rd positions. $C^1-C^2=C^3-C^4-C^5-C^6$. 3,5-dimethyl designates that there are $-CH_3$ groups in the 3rd and 5th positions. $\begin{array}{c} \quad\quad \boxed{CH_3} \quad\; \boxed{CH_3} \\ \quad\quad | \quad\quad\quad | \\ C^1-C^2=C^3-C^4-C^5-C^6 \end{array}$. Add hydrogens to the base chain to get the final molecule $\begin{array}{c} \quad\quad\; CH_3 \quad\quad CH_3 \\ \quad\quad\; | \quad\quad\quad | \\ CH_3-CH=C-CH_2-CH-CH_3 \end{array}$. Cis–trans isomerism is possible around the double bond since all of the groups surrounding the double bond are different.

b) 3-propyl-2-hexene. The base name hexene designates that there are 6 carbon atoms in the base chain. $C^1-C^2-C^3-C^4-C^5-C^6$. The -ene ending designates that there is a double bond and the 2 prefix designates that it is between the 2nd and 3rd positions. $C^1-C^2=C^3-C^4-C^5-C^6$. 3-propyl designates that there is a $-CH_2CH_2CH_3$ group in the 3rd position. $\boxed{CH_2-CH_2-CH_3}$. Add

$$C^1-C^2=\overset{|}{C^3}-C^4-C^5-C^6$$

hydrogens to the base chain to get the final molecule $CH_3-CH=\overset{\overset{\displaystyle CH_2-CH_2-CH_3}{|}}{C}-CH_2-CH-CH_3$. Cis–trans

isomerism is not possible around the double bond since the two groups on the 3rd carbon are the same.

67. The 11 isomers are $\overset{\overset{\displaystyle O}{\|}}{HC}-CH_2-CH_2-CH_3$ = aldehyde, $CH_3-\overset{\overset{\displaystyle O}{\|}}{C}-CH_2-CH_3$ = ketone,

$CH_2=C-CH_2-CH_2-OH$ = alcohol & alkene, $CH_3-CH=CH-CH_2-OH$ = alcohol & alkene,

$CH_3-CH_2-CH=CH-OH$ = alcohol & alkene, $CH_3-CH_2-O-CH=CH_2$ = ether & alkene,

$CH_3-O-CH_2-CH=CH_2$ = ether & alkene, $CH_3-O-CH=CH-CH_3$ = ether & alkene,

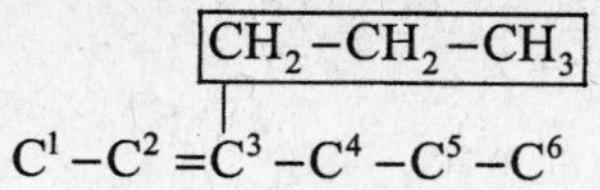

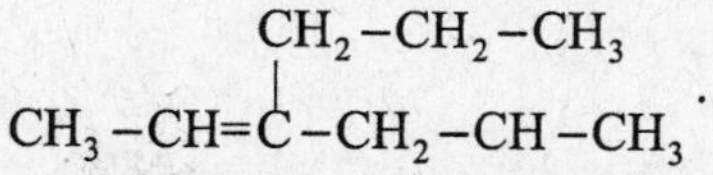

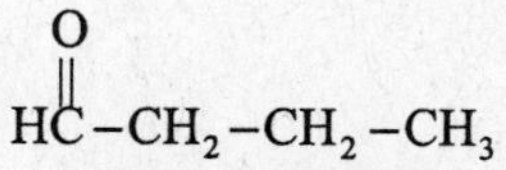

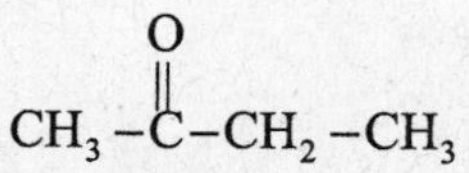

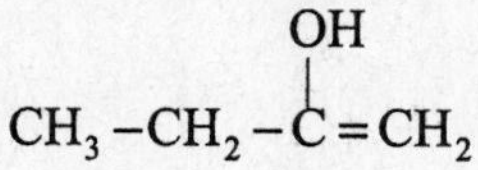

$CH_3-CH_2-\overset{\overset{\displaystyle OH}{|}}{C}=CH_2$ = alcohol & alkene, $CH_3-\overset{\overset{\displaystyle OH}{|}}{CH}-CH=CH_2$ = alcohol & alkene, and

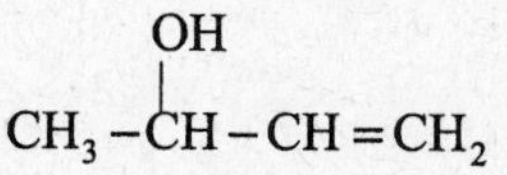

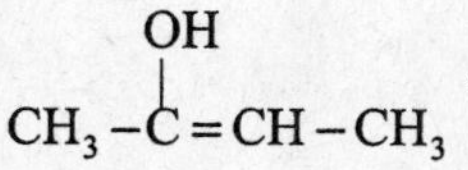

$CH_3-\overset{\overset{\displaystyle OH}{|}}{C}=CH-CH_3$ = alcohol & alkene. There is also a 12th isomer that is cyclic: $\begin{matrix} H_2C-CH_2 \\ | \quad\quad | \\ H_2C \quad CH_2 \\ \backslash \quad / \\ O \end{matrix}$ = ether.

69. This is an internal condensation reaction. It only requires heat to cause it to happen.

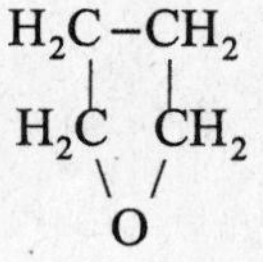

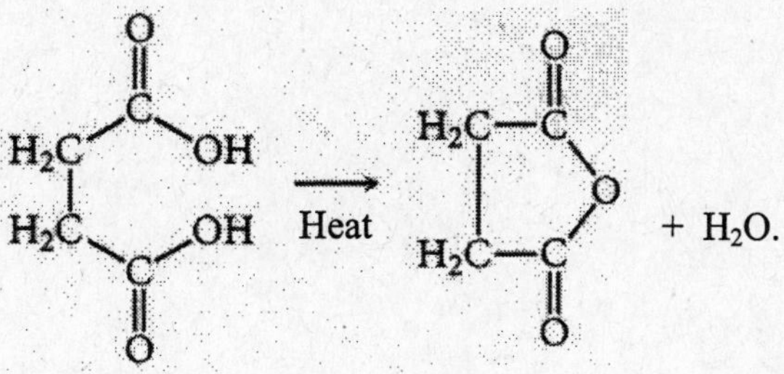

b) This is a dehydration reaction, $CH_3-CH-CH_2-\overset{\overset{\displaystyle CH_2-OH}{|}}{\underset{\underset{\displaystyle CH_2-CH_3}{|}}{CH}}-CH_3 \xrightarrow{H_2SO_4} CH_3-CH-CH_2-\overset{\overset{\displaystyle CH_2}{\|}}{\underset{\underset{\displaystyle CH_2-CH_3}{|}}{C}}-CH_3 + H_2O.$

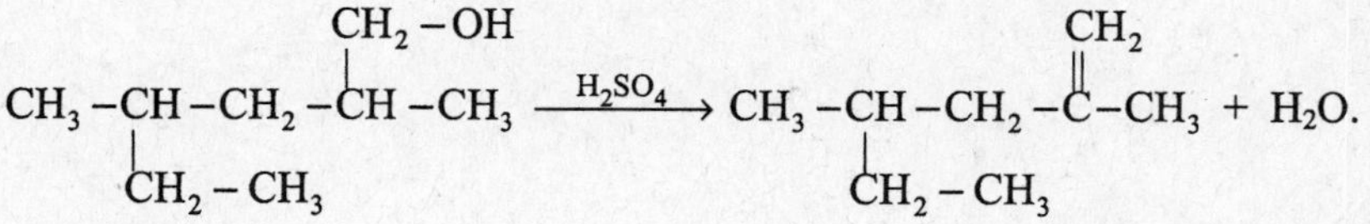

71. a) Since there are 6 primary hydrogen atoms and 2 secondary hydrogen atoms, if they are equally reactive we expect a ratio of 6:2 or 3:1.

b) Assume that we generate 100 product molecules. The yield = (# hydrogen atoms)(reactivity). So $1° = 45 =$ (3)(reactivity 1°) and $2° = 55 =$ (1)(reactivity 2°). Taking the ratio, 2°: 1° = 55: (45/3) = 55 : 15 = 11: 3. The 2° hydrogens are much more reactive.

73. The chiral products (that have four different groups on a carbon) are: $CH_2Cl-CHCl-CH_2-CH_3$ (2nd carbon), $CH_2Cl-CH_2-CHCl-CH_3$ (3rd carbon), and $CH_3-CHCl-CHCl-CH_3$ (2nd and 3rd carbons).

75. b) and d) are chiral since they have carbons with four different groups attached.